APPLIED STATISTICS

Regression and Analysis of Variance

Bayo Lawal and Felix Famoye

University Press of America,® Inc.
Lanham · Boulder · New York · Toronto · Plymouth, UK

Copyright © 2013 by
University Press of America,® Inc.
4501 Forbes Boulevard
Suite 200
Lanham, Maryland 20706
UPA Acquisitions Department (301) 459-3366

10 Thornbury Road
Plymouth PL6 7PP
United Kingdom

Library of Congress Control Number: 2013942359
ISBN: 978-0-7618-6171-3 (paperback : alk. paper)

To

Nike and all our children

and

Busola, Folake, Fisayo and Folarin

Contents

List of Tables   ix

List of Figures   xiii

Preface   xv

1 Introduction   **1**
1.1 Some Basic Distributions and Concepts   2
1.2 Tests of Hypothesis and p-values   7
1.3 Dependent and Independent Variables   8
1.4 Classical Regression Models   8
1.5 Uses of Regression   9
1.6 Exercises   10

2 Simple Linear Regression   **13**
2.1 Introduction   13
2.2 Model Assumptions   13
2.3 Estimating Simple Linear Regression Parameters   14
2.4 Properties of the Estimators   19
2.5 Exercises   21

3 Inferences on Parameter Estimates   **23**
3.1 Introduction   23
3.2 Analysis of Variance   23
3.3 Hypotheses Involving Model Parameters   25
3.4 Confidence Intervals for Regression Parameters   26
3.5 Predictions from Estimated Regression Equation   27
3.6 The Coefficient of Determination R^2 30
3.7 The Coefficient of Correlation ρ 31
3.8 Exercises   34

4 Multiple Linear Regression   **39**
4.1 Introduction   39
4.2 Parameter Estimation in MR Models   39
4.3 Hypotheses Involving Model Parameters   44
4.4 Other Hypotheses Testing in Multiple Regression   44
4.5 Partial F Tests   47
4.6 Multiple Partial F-Tests   50
4.7 Multiple Linear Regression in Matrix Form   52
4.8 Constrained Least Squares   52
4.9 Exercises   56

5 Regression Diagnostics and Remedial Methods   **61**
5.1 Introduction   61

5.2	Residuals and Diagnostics	62
5.3	Detecting Normality of Error Terms	67
5.4	Detecting Outliers	69
5.5	Detecting Influential Observations	70
5.6	Detecting Multicollinearity	79
5.7	Weighted Least Squares	84
5.8	Transformation Approach	87
5.9	Robust Regression	95
5.10	Quantile Regression	104
5.11	Autocorrelated Error Terms	113
5.12	Methods for Solving Multicollinearity Problems	124
5.13	Exercises	135

6 Multiple and Partial Correlations — **141**

6.1	Introduction	141
6.2	Partial Correlation Coefficient	142
6.3	Testing Hypothesis Involving Partial Correlations	145
6.4	Multiple Partial Correlation	147
6.5	Semi-partial Correlation Coefficients	148
6.6	Exercises	149

7 Model Selection Strategies — **153**

7.1	Introduction	153
7.2	Approaches	153
7.3	Guidelines When Selecting Candidate Models	154
7.4	All Possible Regression	156
7.5	The Forward Selection Strategy	158
7.6	The Backward Selection Strategy	161
7.7	The Stepwise Regression Procedure	162
7.8	Model Validation	168
7.9	Exercises	178

8 Use of Dummy Variables in Regression Analysis — **183**

8.1	Introduction	183
8.2	Cell Reference Scheme	184
8.3	Effect Coding Scheme	187
8.4	Categorical Variable with more than Two Levels: Cell Reference Scheme	190
8.5	Categorical Variable with more than Two Levels: Effect Coding Scheme	195
8.6	Models with Two or More Categorical Independent Variables	197
8.7	Use of Dummy Variables in Piece-wise Regression	199
8.8	Exercises	203

9 Polynomial Regression — **209**

9.1	Introduction	209
9.2	Lack of Fit Tests	212
9.3	Model with Two Explanatory Variables	216
9.4	Orthogonal Polynomials Approach	222
9.5	Non-parametric Regression	224
9.6	Exercises	226

10 Logistic Regression — **229**

10.1	Introduction	229
10.2	Binary Response Model	230
10.3	Linear Logistic Regression	231
10.4	Multiple Logistic Regression	242

10.5 A More Complex Example 256
10.6 Multi Category Response Variable 261
10.7 Exercises 268

11 Count Data Regression Models   **273**
11.1 Introduction 273
11.2 Poisson Regression Model 273
11.3 Over-dispersion in Poisson Regression 284
11.4 Negative Binomial Regression Model 287
11.5 The Generalized Poisson Regression Model 291
11.6 Zero-inflated Count Models 292
11.7 Zero-truncated Count Models 300
11.8 Case Study: The HMO Data 305
11.9 Exercises 312

12 Regression with Censored or Truncated Data   **319**
12.1 Survival Analysis 319
12.2 Start and End Times 319
12.3 Describing Event Times 321
12.4 Estimating the Survival Function $S(t)$ 321
12.5 Hazard Function 332
12.6 Proportional Hazards Model 337
12.7 Truncated Regression Model 339
12.8 Exercises 346

13 Nonlinear Regression   **349**
13.1 Introduction 349
13.2 Estimating Nonlinear Models 352
13.3 Problems with the Newton Procedure 362
13.4 Exercises 365

14 One-Way Analysis of Variance   **367**
14.1 Introduction 367
14.2 Regression Approach Methods 370
14.3 ANOVA Model Approach 375
14.4 Multiple Comparisons Procedures 381
14.5 Partitioning the Treatments SS 386
14.6 One-Way ANOVA with a Quantitative Treatment 388
14.7 Correspondence Between Regression and ANOVA 392
14.8 Exercises 396

15 Two-Factor Analysis of Variance   **401**
15.1 The Mean Model 401
15.2 Additive Factor Effects 402
15.3 Two-Level Factor Factorial Experiments: The 2^n Series . . . 403
15.4 Structure of Observations in a Two-Way ANOVA 408
15.5 Analysis with Quantitative Factor Levels 411
15.6 Two-Way ANOVA in Randomized Block Designs 416
15.7 Constructing Possible Contrasts for Interaction 420
15.8 Random Effects Models in Two-Factor Studies 423
15.9 Exercises 432

16 Analysis of Covariance   **437**
16.1 Introduction 437
16.2 One-Way ANOVA Review 438

16.3 Model and Assumptions 438
16.4 Hypothesis of Interest 439
16.5 Using ANOVA Model 445
16.6 Factorial Case: Two Factors 447
16.7 Other Covariance Models 454
16.8 Exercises 454

17 Randomized Complete Block Design **457**
17.1 Introduction 457
17.2 Relative Efficiency of the RCBD 466
17.3 Tukey's Test for Additivity 467
17.4 Random Effects in RCBD 469
17.5 Missing Values in RCBD 473
17.6 Exercises 476

18 Non Orthogonal Classification **479**
18.1 Introduction 479
18.2 Models 480
18.3 Types of Sum of Squares and Estimable Functions . 481
18.4 Least Squares Estimates 487
18.5 Regression Implementation 489
18.6 Case of No Significant Interaction Effects 492
18.7 Case of Empty Cell Entries 501
18.8 Exercises 504

Appendix **507**

Bibliography **519**

Index **523**

About the Authors **527**

List of Tables

1.1	Two types of errors	8
2.1	Age and SBP values for a sample of 30 individuals	17
3.1	Analysis of variance table	23
3.2	Analysis of variance table for Example 2.1	24
4.1	Data on certain species of mammals	42
4.2	Data for example with three explanatory variables	44
4.3	Extra SS for each of the variables as they enter the model	46
4.4	Types I and II SS implicit that β_0 is also in the models	46
4.5	ANOVA table based on the order in (i)	48
4.6	ANOVA table based on the order in (i)	48
4.7	ANOVA table based on the order in (i)	49
4.8	ANOVA table for the variable-added-last tests	50
5.1	Data relating performance to training	64
5.2	Diastolic blood pressure and age among a sample of 54 healthy women	66
5.3	Residuals and expected values under normality	68
5.4	SBP and ages of thirty individuals	70
5.5	Influence of removing the flagged observations	76
5.6	Heat evolved in calories during hardening of cement	79
5.7	Sample means of x compared with residual variances	85
5.8	Parameter estimates and standard errors for both OLS and WLS models	87
5.9	Mileage against weight of cars in pounds	91
5.10	USA population from 1790 to 1970 in steps of 10 years	105
5.11	Acorn data from Schroeder & Vangilder	108
5.12	Import of crude oil to the USA from 1974-1993	113
5.13	Regression results for various forms of estimation	120
5.14	Fit of possible models to our data	123
5.15	Data for Problem 5.11	137
5.16	Simulated data for Problem 5.13	138
6.1	All possible partials for our data example	142
7.1	Heat evolved in calories during hardening of cement	156
7.2	Best subsets based on four optimal selection criteria	166
7.3	Job proficiency estimation data set for 25 applicants	170
7.4	Job proficiency prediction data set for 25 applicants	170
8.1	Data on a random sample of 32 birth records	183
8.2	Number of days as a function of amount of money and state	191
8.3	Advertising and sales data	199

9.1	Skin response (y) as a function of concentration (x) in ml/l	209
9.2	Replicated observations and their corresponding response values	213
9.3	Revised analysis of variance table	214
9.4	Revised analysis of variance table	215
9.5	Revised analysis of variance table	216
9.6	Data on product quality	217
9.7	Set-up for the calculations	223
10.1	Effect of different concentrations of nicotine sulphate on DM, Hubert (1992)	230
10.2	Died versus alive for the data in Table 10.1	238
10.3	Course grade and ethics examination performance for 27 subjects	240
10.4	Occurrence of vasoconstriction in the skin of fingers	249
10.5	50 years survival for men and women after graduation	252
11.1	Data on frequency of falls	275
11.2	Occurrence of esophageal cancer	277
11.3	Parameter estimates under Poisson and logit models	281
11.4	Number of inedible popcorn kernels	282
11.5	Number of C. caretta hatchlings dying from sun exposure (*: structural zeros)	293
11.6	Number of running shoes purchased from online running logs	302
11.7	Hmo data on length of stay	306
11.8	Number of doctor and non-doctor consultations, and admissions	313
11.9	Number of prescribed and non-prescribed medications	313
11.10	Total number of prescribed and non-prescribed medications	313
11.11	Total number of doctor and non-doctor consultations	314
11.12	Data originally presented in Dieckman (1981)	314
11.13	Distribution of occupants in houses	314
12.1	Line plots of time to event for ten subjects	320
12.2	Survival times for ten subjects	321
12.3	Time to death(event) for all the ten subjects, starting from 0	322
12.4	Patient data from Dunn and Clark (2009)	323
12.5	Times to death for 45 breast cancer patients	327
12.6	Data for the HMO HIV+ study	334
12.7	Results of fitting the AFT models to our data	336
13.1	Reaction velocity and substrate concentration data	353
13.2	Drug responsiveness data	359
13.3	Length and age of 20 female mussels from Virginia	362
14.1	A layout with 4 treatments and 5 replications	368
14.2	Tensile strength as functions of percentage of cotton	389
14.3	ANOVA table for the data in Table 14.2	389
14.4	Calculation of components SS	389
14.5	Revised analysis of variance table	390
15.1	Observed mean responses for a 2×3 table	401
15.2	A two-factor example	403
15.3	Structure of ANOVA table	404
15.4	Population means and simple effects in a 2^2 factorial	404
15.5	Simple and interaction effects for four different tables of means	405
15.6	Coded data for this example	411
15.7	Treatment sums formed from data in Table 15.6	411
15.8	Initial analysis of variance	412
15.9	Factor A contrasts	414

15.10 Full analysis of variance table   415
15.11 Table of parameter estimates   415
15.12 Data for the 4 × 3 factorial experiment in this example   417
15.13 Two-way interaction table for times and varieties   418
15.14 Full analysis of variance table for the data in Table 15.12   418
15.15 Table of treatment means   418
15.16 Table of population means   421
15.17 Table of expected mean squares under model IIa   424
15.18 Table of expected mean squares under model IIb   427
15.19 Table of expected mean squares under model IIc   430

16.1   Data for the experiment on steel coupons   438
16.2   Data on methods of instruction   448
16.3   Example of unequal cell counts data   452

17.1   A CRD layout with 4 treatments and 5 replications   457
17.2   Typical table of observations   459
17.3   Analysis of variance for a randomized block design   460
17.4   The yields in (kg/plot) for the experiment in Example 17.1   461
17.5   Analysis of variance table   462
17.6   Muscle tissue data   463
17.7   Expected mean squares under the mixed model (blocks random)   470
17.8   Expected mean squares when blocks and treatments are random   473
17.9   ANOVA table for a missing value analysis   474

18.1   Table of means   480
18.2   Times when subjects began to suffer from hay fever (* = missing)   481
18.3   Male and female volunteers (* = missing)   492
18.4   The otherwise balanced data with missing cell   501
18.5   Table of population means   501

List of Figures

1.1	A normal density plot	2
2.1	Plot of regression line overlayed with the observed values	19
3.1	Plot of 95% fiducial confidence intervals	31
5.1	Patterns of residuals plots	63
5.2	Graph of Residuals versus predicted	65
5.3	Plot of residuals against predicted	67
5.4	Plot of studentized residuals against the predicted values	71
5.5	Bivariate scatter plot matrix for the explanatory variables	81
5.6	Plot of weighted residuals against predicted	86
5.7	Plot of LL and RMSE values against values of λ	90
5.8	Residual plots against x under the reciprocal transformation	90
5.9	Histogram plot of the residuals	91
5.10	Scatter plot of y versus x	92
5.11	Graph of residuals versus predicted	92
5.12	Box-Cox result plots for the mileage-weight of cars data	94
5.13	Graph of residuals versus predicted with transformed $y(\lambda = 0)$	95
5.14	Q-Q plot of robust residuals for y	99
5.15	Distribution of robust residuals for y	100
5.16	Residuals plots for the OLS and median regression	107
5.17	Three estimated quantiles for the data in Table 5.10	108
5.18	Residuals plot for the median regression on the acorn data	111
5.19	Residuals plot for the median regression on the acorn data	112
5.20	Estimated three quantiles for the Acorn data	112
5.21	Residual time plot (residual versus time)	115
5.22	Residuals and predicted plots	126
5.23	Ridge trace of estimated regression coefficients	132
7.1	Bivariate scatter plot matrix for the five variables	171
7.2	Plot of standardized residuals against predicted values	173
7.3	Bivariate scatter plot matrix for the validation data set	174
7.4	Plot of standardized residuals against predicted values	176
8.1	Plot of estimated smoking status profiles	190
8.2	Plot of residuals from the separate regression lines	191
8.3	Plot of estimated state profiles	195
8.4	Scatter plot of sales versus advertisement budget	199
8.5	Plot of residuals against x_1 over the entire range of x_1	200
8.6	Graph of predicted versus x_1 under model (8.44)	202
8.7	Graph of predicted versus x_1 under the quadratic model	202
9.1	Plots of cubic and quadratic polynomials	210

9.2   Scatter plot of the data overlayed with a possible plot   210
9.3   Plot of predicted versus x under the estimated model   213
9.4   Response surface plot for the estimated model   219
9.5   The **loess** fit to the data in Table 9.2   226

10.1   Plot of logistic function   232
10.2   Fitted logistic model of logistic function   236
10.3   Estimated quadratic logistic model   244
10.4   Estimated reciprocal logistic model plot   245
10.5   Overlay plot of both linear and quadratic estimated functions   247
10.6   Plot of quadratic and CLL models   248
10.7   Predicted probabilities against age for both sexes when trt $= 0$ 261
10.8   Predicted probabilities against age for both sexes when trt $= 1$ 262
10.9   Predicted probabilities against age for both treatments when sex $= 0$ 262
10.10 Predicted probabilities against age for both treatments when sex $= 1$ 263

11.1   Plot of estimated mean of cancer occurrence against age   281
11.2   Plot of estimated probabilities of cancer occurrence against age   282

12.1   Plot of estimated survival function   324
12.2   Plots of LS and LLS   325
12.3   Plot of estimated survival function under the LT method   326
12.4   Plot of estimated LS and LLS functions under the LT method   327
12.5   Plot of estimated survival curves under both K-M and LT methods   331
12.6   Plot of estimated survival function against time (age=35)   340
12.7   Plot of estimated survival function against time (age=44)   340

13.1   Exponential growth and decay curves   350
13.2   Negative exponential growth curves   351
13.3   Two-term exponential growth curves   351
13.4   Predicted M-M model superimposed over observed values   359
13.5   Estimated drug responsiveness curve   361

14.1   Plot of estimated regression function   392

15.1   Identical simple effects- Table 15.5(a)   406
15.2   Unequal simple effects with the same signs- Table 15.5(b)   406
15.3   Unequal simple effects with opposite signs- Table 15.5(c)   407
15.4   Unequal simple effects with same signs- Table 15.5(d)   407
15.5   Plot of the significant AB interaction term   417
15.6   Time and variety interaction plot   419

18.1   Plot of the interaction effects   490

Preface

Regression and analysis of variance are widely used in many fields, including but not limited to engineering, business, humanities, social sciences, natural sciences, health and medical sciences. Regression analysis can be used to model, examine, and explore relationships leading to prediction or forecasting. It can be used to identify factors affecting a response variable through statistical tests. Analysis of variance, which is closely related to regression analysis, provides a statistical test of whether or not the means of several groups are all equal.

The purpose of the book is to write a text in regression and analysis of variance suitable for junior and senior undergraduate students in statistics, graduate students in other disciplines other than statistics, and researchers and practitioners who have good knowledge of introductory statistical techniques. The book presents a treatment of the methods of regression and analysis of variance. The book focuses on conceptual understanding of statistical methods in regression and analysis of variance as well the use of SAS® statistical software to perform data analysis.

The book offers a treatment of regression analysis and analysis of variance. Chapter 1 covers some basic distributions and the concepts of hypothesis testing, dependent and independent variables. Simple linear regression and its inferences are covered in chapters 2 and 3. Chapter 4 discusses multiple linear regressions while regression diagnostics and their remedial measures are discussed in chapter 5. Multiple and partial correlations are the topics of chapter 6. The different model selection strategies are discussed in chapter 7 while the use of dummy variables in regression analysis is covered in chapter 8. Special types of regression are discussed in chapters 9 through 13. The logistic regression and count data regression models are discussed in chapters 10 and 11, respectively. Regression with censored or truncated data and nonlinear regression are discussed, respectively, in chapters 12 and 13 to round up the topics in regression analysis. Chapter 14 is on one-way analysis of variance, followed by two-factor analysis of variance in chapter 15 and analysis of covariance in chapter 16. In chapter 17, the topics are on randomized complete block design and the final chapter, chapter 18 covers non-orthogonal classification.

Many different examples are presented in the book to illustrate the diversity of applications of regression and analysis of variance. The various applications of regression and ANOVA often require some computations. Hence, almost all examples in the book are accompanied with their corresponding SAS® codes while the R programs are made available on the website (http://people.cst.cmich.edu/famoy1kf/appliedstat/) dedicated to the book. The SAS® programs are accompanied with partial output. To facilitate data entry, many of the data sets for examples and exercises are provided on the book's website (http://people.cst.cmich.edu/famoy1kf/appliedstat/). The readme.txt file on the web site provides information about the data sets. The book's website (http://people.cst.cmich.edu/famoy1kf/appliedstat/) will be used to convey any changes to the book and as errata become known to us, these will be added to the errata section of the book's website.

The book is intended for use in undergraduate and graduate courses in regression and analysis of variance. It can also be used as a second course in Applied Statistics. The amount of material presented in the book that is needed for a particular course depends on the course objectives and the amount of available time. Some possible course topics are: A two-semester undergraduate course sequence in (i) Applied Statistics I with topics from chapters 2-9 and (ii) Applied Statistics II with topics from chapters 10, 11, 14-18; one-semester undergraduate course in Regression Analysis with topics from chapters 2-9; one-semester undergraduate course in Applied Statistics with topics from chapters 2-8, 10-12; one-semester graduate course (in programs other than statistics) in Regression Analysis with topics from chapters 2-8, 10-12; and one-semester graduate course (in programs other than statistics) in Applied Statistics with topics from chapters 2-8, 14-17.

Acknowledgments

We gratefully acknowledge the support and guidance of Ms. Laura Espinoza, Ms. Piper Owens, Laura Grzybowski, Lindsey Frederick and Megan Barnett of UPA throughout the publication of this work.

Special thanks go to Dr. Sunghoon Chung at Central Michigan University, Mt. Pleasant, Michigan who graciously provided us with a list of errors (and corrections) that were encountered in the use of the draft for his Applied Statistics course.

We are grateful to Dr. M.R. Watnik, Dr. G. Heinz, Dr. R. Johnson, Dr. S. Chu, Dr. D.A. Dickey, Dr. M.C. Meyer, D.L. Guber, Dr. N. Binnie, Dr. M. Friendly and Dr. S. Long for their kind permission to use their data in developing problems in the book. We are indebted to John Wiley and Sons, Brooks/Cole, Oxford University Press, Sage Publisher, Elsevier Limited, McGraw-Hill Companies, Dr. M. Friendly, and Dr. S. Long for permission to use copyrighted materials.

We thank the SAS® Institute, for the use of their software for the analyses of all the data examples in this text, the results of which are presented in the text.

Bayo Lawal   *Felix Famoye*
Department of Statistics & Mathematical Sciences   Department of Mathematics
Kwara State University   Central Michigan University
Malete, Nigeria   Mount Pleasant, Michigan, USA

June 5, 2013

Chapter 1

Introduction

Regression analysis involves methods for examining the relationships between two or more variables. A good understanding of these relationships among variables can be used for either prediction, where values of one or more variables are used to predict that of another variable. For instance, we may want to predict the GPA of freshmen students in a small college from the knowledge of their scores in say the mathematics part of the SAT and the verbal part of the SAT. In this case, the variable to be predicted, y, is called the *response* or *dependent* variable and it is the GPA of all freshmen students in the college, while the *independent* or *explanatory* variables are scores in the mathematics part of the SAT and the verbal part of the SAT, denoted by x_1 and x_2 respectively. In this example, we have two explanatory variables. Knowledge of scores in the mathematics part of the SAT and the verbal part of the SAT will therefore allow us to predict a student's GPA.

Regression methods have also been used to adjust for the effect of one variable, while controlling the value(s) of other variable(s). This often gives us a better understanding of the relationship between the dependent variable and the other explanatory variables. However, the classifications of the dependent and explanatory variables give rise to various methods of analysis of ensuing data in a regression analysis. We present below these characterizations of the variables and their equivalent statistical analysis.

1. If the dependent variable is continuous and all the independent variables are continuous, this leads to the classical regression analysis. A variant of these occurs if some of the independent variables are continuous and some are categorical. The resulting analysis is still the general multiple regression analysis. These are discussed in Chapters 2 to 9.

2. If the dependent variable is continuous and all the independent variables are categorical. This leads to the analysis of variance (ANOVA) models in Chapters 14-15, 17

3. If the dependent variable is continuous and all the independent variables are a mixture of categorical variables and a concomitant (controlling) continuous variable, the ensuing analysis is the analysis of covariance (ANCOVA), discussed in Chapter 16.

4. If the dependent variable is a binary categorical variable, and the independent variables are either continuous, categorical or both, then the resulting analysis will be the **linear logistic regression analysis**, discussed in Chapter 10.

5. If the dependent variable is a multi-level categorical variable, and the independent variables are either continuous, categorical or both, then the resulting analysis will be either baseline category model if the dependent variable is nominal or cumulative logit model if the dependent variable is ordinal.

6. If the dependent variable consists of counts, and the independent variables are either continuous, categorical or both, then the resulting analysis will be count data regression, discussed in Chapter 11. One example is the **Poisson Regression** analysis.

We present in the remainder of this chapter, some of the basic concepts required for a proper understanding of the material in the rest of this book. We discuss below, some of the distributions that we will encounter in this text and how to employ them with the statistical software SAS.

1.1   Some Basic Distributions and Concepts

In this section, we briefly review some of the probability distributions we will encounter in the later chapters. The probability density function (pdf) for a continuous random variable Y is denoted by $f(y)$ while the probability mass function of a discrete random variable Y will be written as $P(Y = y)$. The cumulative distribution function (cdf) is $F(y) = P(Y \leq y)$. If $P(Y \leq y) = F(b) = p$, where p is a given probability, the value b is called the quantile from the pdf of Y. This is the inverse of the cdf and $b = F^{-1}(p)$.

In SAS, there are functions to compute the pdf, cdf, and the quantile from many distributions. The following are some of the functions:

- **pdf**('dist-name', y, para1, para2, ...) for pdf

- **cdf**('dist-name', y, para1, para2, ...) for cdf

- **quantile**('dist-name', y, para1, para2, ...) for quantile

In the above commands, 'dist-name' is the distribution name with quotes, para1, para2, ... are the distribution parameters. For example, to compute the cdf of normal random variable Y with mean μ and standard deviation σ at the point y, we use **cdf**('normal', y, μ, σ). When a parameter is optional in SAS, such a parameter is placed between $<$ and $>$. For example, if para3 and para4 are optional, we use the syntax **cdf**('dist-name', y, para1, para2 $<$, para3, para4$>$).

1.1.1   The Normal or Gaussian Distribution

The classical regression analysis utilizes the normal or Gaussian distribution with parameter μ and σ^2, where μ is the mean and σ^2 is the variance. We often write this distribution as $Y \sim N(\mu, \sigma^2)$, and the distribution is defined as

$$f(y) = \frac{1}{\sigma\sqrt{2\pi}} e^{-\frac{(y-\mu)^2}{2\sigma^2}}, \quad -\infty < y < \infty, \tag{1.1}$$

for $-\infty < \mu < \infty$ and $\sigma > 0$. The plot of a normal distribution with parameters $\mu = 10$ and variance $\sigma^2 = 4$ is presented in Figure 1.1.

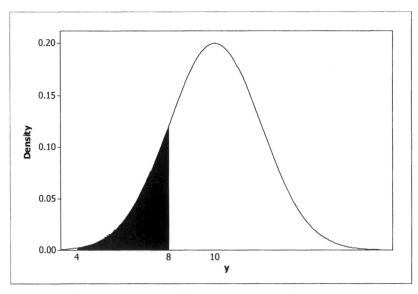

Figure 1.1: A normal density plot

The shaded area for instance, gives the probability that $4 < Y < 8$. In general, to compute $P(a \leq Y \leq b)$ by hand, we must first transform Y to Z with the transformation $Z = (Y - \mu)/\sigma$, which is distributed as $N(0, 1)$. That is, Z is a standard normal with $\mu = 0$ and variance $\sigma^2 = 1$. Hence,

$$P(a \leq Y \leq b) = P\left(\frac{a-\mu}{\sigma} \leq Z \leq \frac{b-\mu}{\sigma}\right).$$

In our example above for instance, $\sigma^2 = 4$, implies that $\sigma = 2$. Hence,

$$P(4 < Y < 8) = P\left(\frac{4-10}{2} < Z < \frac{8-10}{2}\right) = P(-3.00 < Z < -1.00) = 0.1574.$$

The above answer is obtained from a normal probability table (see Table 1 in the Appendix). In addition to the **cdf** function, SAS has the function **probnorm**(y) to compute the cdf of a $N(0,1)$ distribution. The **probit**(p) function in SAS computes the quantile from the standard normal distribution. The **cdf** function **cdf**('normal', y, μ, σ) is employed in the following to solve the above problem.

```
data norm;
y1=cdf('normal',8,10,2)-cdf('normal',4,10,2);
output;
proc print;
var y1;
run;
```

```
          Obs        y1
           1      0.15731
```

1.1.2 Distributions Arising from Sampling Distributions

If we let $r_1, r_2, \ldots, x_n$ be realized values from a simple random sample of size n drawn from a normal population with mean μ and variance σ^2 which is unknown, then the sample mean and variance are computed as

$$x = \frac{1}{n}\sum_{i=1}^{n} x_i \quad \text{and} \quad S^2 = \frac{1}{n-1}\sum_{i=1}^{n}(x_i - \bar{x})^2.$$

Student's t Distribution

A random variable t^* with the Student's t distribution is given by

$$t^* = \frac{\bar{x} - \mu}{S/\sqrt{n}}.$$

The distribution is based on $n - 1$ degrees of freedom and the random variable can be written as t_{n-1}. Other SAS functions for the Student's t distribution are **probt** for the cdf and **tinv** for the quantile. The syntax for these functions are **probt**$(y, df <, nc >)$ and **tinv**$(p, df <, nc >)$, where $0 \le p \le 1$, df is the degrees of freedom and nc is the non-centrality parameter, which is zero in many applications. The syntax for the **cdf** function is **cdf**('t', $y, df <, nc >)$.

Example 1.1

Compute the following:
　　1. $P(t_{14} > 2.05) = ?$　　2. $P(t_8 \le ?) = 0.253$.
We can use SAS functions **probt** or **cdf** to solve (1) and **tinv** or **quantile** to solve (2). The following SAS program with the output accomplishes these. Both functions **probt** and **cdf** return the same result. Also, the functions **tinv** and **quantile** return the same value.

```
data tt;
pb1=1-probt(2.05, 14);
pb2=1-cdf('t', 2.05, 14);
av1=tinv(0.253, 8);
av2=quantile('t', 0.253, 8);
proc print;
var pb1 pb2 av1 av2;
run;
```

```
       Obs       pb1         pb2          av1          av2
        1      0.029790    0.029790    -0.69624     -0.69624
```

The Chi-squared (χ^2) Distribution

The random variable

$$\frac{(n-1)S^2}{\sigma^2} \sim \chi^2_{n-1}, \tag{1.2}$$

has a chi-square distribution. That is, the expression on the left hand side of equation (1.2) is said to be distributed χ^2 with $n-1$ degrees of freedom. The distribution is not symmetric and thus upper and lower quantiles of the distributions need to be tabulated. The other SAS functions to compute the cdf and quantile are **probchi** and **cinv** respectively. The syntax for the functions are **probchi**(y, df $<$, nc $>$) and **cinv**(p, df $<$, nc $>$), where df is the degrees of freedom and nc is the non-centrality parameter and $0 \leq p \leq 1$. The syntax for the **cdf** function is **cdf**('chisq', y, df $<$, nc $>$).

Example 1.2

Compute the following:
    1. $P(\chi^2_9 \leq 11.5) = ?$ 2. $P(\chi^2_{13} \leq ?) = 0.085$.
The above problems are again solved with the chi-squared distribution SAS functions described in this section. The resulting program and output are displayed below.

```
data chi;
pb1=probchi(11.5, 9);
pb2=cdf('chisq', 11.5, 9);
av1=cinv(0.085, 13);
av2=quantile('chisq', 0.085, 13);
proc print;
var pb1 pb2 av1 av2;
run;
            Obs       pb1        pb2       av1        av2
             1      0.75701    0.75701   6.74102    6.74102
```

The F Distribution

Consider random samples n_1 and n_2 from two independent normal populations $N(\mu_1, \sigma_1^2)$ and $N(\mu_2, \sigma_2^2)$, where S_1^2 and S_2^2 are the respective sample variances from each of the samples. The ratio

$$F^* = \frac{S_1^2/\sigma_1^2}{S_2^2/\sigma_2^2}, \tag{1.3}$$

has an F distribution with $n_1 - 1$ and $n_2 - 1$ degrees of freedom, and we can write this more succinctly as

$$\frac{S_1^2 \sigma_2^2}{S_2^2 \sigma_1^2} \sim F_{n_1-1, n_2-1}.$$

The other SAS functions to implement the F distribution are **probf** for the cdf and **finv** for the quantile. The syntax for the functions are **probf**(y, ndf, ddf $<$, nc $>$) and **finv**(p, ndf, ddf $<$, nc $>$), where ndf and ddf are the numerator and denominator degrees of freedom respectively, nc is the non-centrality parameter and $0 \leq p \leq 1$. The syntax for the **cdf** function is **cdf**('F', y, ndf, ddf $<$, nc $>$).

Example 1.3

Compute the following:
    1. $P(F_{5,20} \geq 5.87) = ?$ 2. $P(F_{4,10} \geq ?) = 0.05$.
The above problems are solved with the F distribution SAS functions described in this section. The resulting program and output are displayed below. One can use the command **format** pb1 pb2 av1 av2 **f10.4** to write the values to four decimal places.

```
data ff;
pb1=1-probf(5.87, 5, 20);
pb2=1-cdf('f', 5.87, 5, 20);
av1=finv(0.95, 4, 10);
av2=quantile('f', 0.95, 4, 10);
proc print;
```

```
var pb1 pb2 av1 av2;
run;
            Obs          pb1          pb2       av1       av2
             1      .001703512   .001703512   3.47805   3.47805
```

Notice that for problem 2, we have changed this to its complement by using $1.00 - 0.05 = 0.95$.

1.1.3 The Binomial Distribution

Consider an experiment consisting of a sequence of n independent trials in which the outcome at each trial is binary or dichotomous. At each trial, the probability that a particular event A (success) occurs is π and the probability that the event does not occur (failure) is therefore $(1 - \pi)$. We shall also assume that this probability of success is constant from one trial to another. If the number of times the event A occurs in the sequence of n trials is represented by the random variable X, then, X is said to follow the binomial distribution with parameters n and π (fixed). The probability mass function for X is

$$
P(X = x) = \begin{cases} \binom{n}{x} \pi^x (1 - \pi)^{n-x}, & x = 0, 1, \dots, n \\ \\ 0, & \text{otherwise}, \end{cases} \tag{1.4}
$$

where $0 < \pi < 1$. The mean and variance of the binomial distribution are respectively given by $n\pi$ and $n\pi(1 - \pi)$.

Another SAS function to compute the cdf of binomial distribution is **probbnml**. The syntax for the function is **probbnml**(p, n, x), where p is the success probability and n is the number of trials. To find $P(X = x)$, one can find $P(X \leq x)$ and $P(X \leq x - 1)$ and take the difference. Alternatively, one can use the SAS function **pdf**, which has a similar syntax as the function **cdf**.

Example 1.4

Suppose the probability that a person suffering from migraine headache will obtain relief with a certain analgesic drug is 0.8. Five randomly selected sufferers from migraine headache are given this drug. We wish to find the following probabilities that the number obtaining relief will be:

    (a) Exactly zero   (b) Exactly two   (c) More than 3
    (d) Four or fewer (e) Three or four (f) At least 2.

Solution:

Let X be the binomial random variable with parameters $n = 5$ and $\pi = 0.8$. We can solve the problems in (a) to (f) by the use of the cumulative distribution function of the binomial random variable. The **cdf** $F(x) = P(X \leq x)$ for this problem is generated below with the accompanying SAS program.

```
data binomial;
do x=0 to 5;
pb1=probbnml(0.8, 5, x);
cd1=round(pb1, .0001);
pb2=cdf('binom', x, 0.8, 5);
cd2=round(pb2, .0001);
output;
end;
proc print noobs;
var x cd1 cd2;
run;
          x       cd1        cd2
          0      0.0003     0.0003
          1      0.0067     0.0067
          2      0.0579     0.0579
          3      0.2627     0.2627
          4      0.6723     0.6723
          5      1.0000     1.0000
```

In the program, we have employed the SAS function **cdf**('binom', x, p, n) for computing the cumulative probabilities and the function **pdf**('binom', x, p, n) for computing the probabilities of a binomial random variable with the specified parameters. The **ROUND** statement asked SAS to round the probabilities to four decimal places.

(a) The number obtaining relief will be exactly zero is $P(X = 0) = 0.0003$.

(b) The number obtaining relief will be exactly two is $P(X = 2) = 0.0512$.

(c) The number obtaining relief will be more than 3 is $P(X > 3) = 1 - P(X \le 3) = 0.7373$.

(d) The number obtaining relief will be four or fewer is $P(X \le 4) = 0.6723$.

(e) The number obtaining relief will be three or four is $P(3 \le X \le 4) = P(X \le 4) - P(X \le 2) = 0.6144$.

(f) The number obtaining relief will be at least 2 is $P(X \ge 2) = 1 - P(X \le 1) = 0.9933$.

The SAS program for implementing the above solutions are presented below.

```
data bb;
a=pdf('binom', 0, 0.8, 5);
b=pdf('binom', 2, 0.8, 5);
c=1-cdf('binom', 3, 0.8, 5);
d=cdf('binom', 4, 0.8, 5);
e=cdf('binom', 4, 0.8, 5)-cdf('binom', 2, 0.8, 5);
f=1-cdf('binom', 1, 0.8, 5);
proc print;
var a b c d e f;
format a b c d e f f6.4;
run;
          Obs       a        b        c        d        e        f
           1     0.0003   0.0512   0.7373   0.6723   0.6144   0.9933
```

Normal Approximation to Binomial

When n is large and π is close to 0.5, the normal approximation to the binomial is very satisfactory. A rule of thumb is to use normal approximation when $n\pi$ and $n(1 - \pi)$ are both greater than 5. To use normal approximation, one needs to apply continuity correction. Continuity correction is the process where 0.5 is added or subtracted from the given value. Thus, the probability $P(X \le b)$ is approximated by $P(X \le b + 0.5)$, $P(X \ge a)$ is approximated by $P(X \ge a - 0.5)$, and $P(X = c)$ is approximated by $P(c - 0.5 \le X \le c + 0.5)$. For any strict inequality, first convert to weak inequality and apply the correction as done here.

1.1.4   The Poisson Distribution

The Poisson distribution is very important in the study of categorical data and is often a realization of a rare event. If $Y_i \sim P(\mu)$, $i = 1, 2, \ldots$, then the probability mass function for this random variable is given by

$$P(Y = y) = \frac{e^{-\mu}\mu^y}{y!}, \quad y = 0, 1, 2, \ldots \tag{1.5}$$

The mean of the Poisson distribution is μ and this is also the variance. It can be easily shown that

$$\frac{P(Y = y + 1)}{P(Y = y)} = \frac{\mu}{y + 1}. \tag{1.6}$$

Some applications of the Poisson distribution include the number of typing errors per page; number of telephone calls per hour; number of home injuries over time; number of bugs per leaf; and number of passengers, excluding driver, per car. The Poisson distribution provides a good approximation to the binomial when $n \ge 20$ and $\mu \le 0.05$. Another SAS function for the cdf of Poisson distribution is **poisson**(μ, x), where μ is the mean of the Poisson distribution. The syntax for the **cdf** function is **cdf**('poisson', x, μ).

Example 1.5

Suppose the number of bugs per leaf is a random variable having a Poisson distribution with mean 3.5. Compute the probability that
    (a) a leaf will have exactly 4 bugs   (b) a leaf will have at most 1 bug
    (c) a leaf will have at least 3 bugs.

Solution:

We can solve the problems in (a) to (c) by using the cumulative distribution function of the Poisson random variable. The **cdf** $F(x) = P(X \leq x)$ for this problem is generated with the accompanying SAS program. We generated probabilities for x values less than or equal to 5 since the largest x value in the example is 4.

```
data poisson;
do x=0 to 5;
cd1=poisson(3.5, x);
cd2=cdf('poisson', x, 3.5);
output;
end;
proc print noobs;
var x cd1 cd2;
run;
```

x	cd1	cd2
0	0.03020	0.03020
1	0.13589	0.13589
2	0.32085	0.32085
3	0.53663	0.53663
4	0.72544	0.72544
5	0.85761	0.85761

In the program, we used the SAS function **poisson**(μ, x) for generating the cumulative distribution function of a Poisson random variable with the specified parameter.

(a) A leaf will have exactly 4 bugs is $P(X = 4) = 0.18881$.

(b) A leaf will have at most 1 bug is $P(X \leq 1) = 0.13589$.

(c) A leaf will have at least 3 bugs is $P(X \geq 3) = 1 - P(X \leq 2) = 1 - 0.32085 = 0.67915$.

The functions **cdf** and **pdf** are used in the following SAS program.

```
data po;
a=pdf('poisson', 4, 3.5);
b=cdf('poisson', 1, 3.5);
c=1 - cdf('poisson', 2, 3.5);
proc print;
var a b c;
run;
```

Obs	a	b	c
1	0.18881	0.13589	0.67915

1.2 Tests of Hypothesis and p-values

A *statistical hypothesis* is a statement about the population. We have two types of hypotheses- H_0, the null hypothesis and H_a, the alternative (research) hypothesis. Using information from the sample data, we make one of the following decisions- reject H_0 or fail to reject (FTR) H_0. The process of making one of these two decisions is called *testing statistical hypothesis*. In some introductory books, the following five-step procedure is suggested to test hypothesis:

1. State the null and the alternative hypotheses. For compound hypotheses about the population mean, the alternative hypothesis takes one of the three forms $H_a : \mu < \mu_0$, $H_a : \mu > \mu_0$ or $H_a : \mu \neq \mu_0$.

2. Choose a test statistic.

3. State the rejection (or critical) region.

4. Calculate the observed value of the test statistic and make your decision.

5. Give an appropriate conclusion in terms of the original problem.

When a decision is made, one is liable to either Type I or Type II error. The decision table is presented in Table 1.1.
The probabilities of Type I and Type II errors are $\alpha = P(\text{Type I error}) = P(\text{Rejecting } H_0 \text{ when } H_0 \text{ is true})$ and $\beta = P(\text{Type II error}) = P(\text{FTR } H_0 \text{ when } H_a \text{ is true})$. Note that α is also called the significance level of the test.

Decision	True state of nature	
	H_0 is true	H_a is true
Reject H_0	Type I error	Correct Decision
FTR H_0	Correct Decision	Type II error

Table 1.1: Two types of errors

If H_0 is rejected, conclude that the claim in H_a is true. Remember to use the language of the original problem. If the decision is fail to reject H_0, conclude that there is no sufficient evidence to support the statement in H_a. If we fail to reject H_0, we do not conclude that H_0 is true since the probability of Type II error, which is the one we are liable to is generally unknown.

The p-value or observed significance level for a test is the probability of seeing a result from a random sample that is as extreme as or more extreme than the one obtained from your random sample if the null hypothesis H_0 is true. The decision rule in all cases would be to reject H_0 if p-value $< \alpha$, and fail to reject H_0 if p-value $\geq \alpha$, where α is the nominal level or the level of significance chosen a priori. Values of α are usually 0.01, 0.05, or 0.10, with 0.05 the most commonly used value. In most of the analysis in this text, the p-value from the statistical hypothesis would be computed automatically by SAS statistical software.

1.3 Dependent and Independent Variables

In a statistics class having 30 students, suppose we record the ages, gender, weights and heights of the students. Suppose further that our goal is to be able to predict the weight y_i of this group of students from our knowledge of their ages (x_1), heights (x_2) and gender $(x_3$, where $x_3=1$ if male and 0 if female). Then we have the following measurements and observations on each student:

$$y = \text{weight of student}, \quad x_1 = \text{age}, \quad x_2 = \text{height}, \quad x_3 = \text{gender}.$$

If for instance, we believe that the variables x_1, x_2 and x_3 predict y, then we would describe x_1, x_2 and x_3 as the *explanatory, independent* or *regressor* variables, while y would similarly be described as the *dependent* or *response* variable. As we can see in this case, the explanatory variables can be of the continuous type (measurements, like age and height in this case) or categorical, like gender in the example. However, the response or dependent variable for the classical regression analysis must be continuous. We shall examine other cases where the dependent variable is categorical (binary), which leads to the logistic regression in Chapter 10; or where the dependent variable are counts or frequencies, which leads to Poisson regression in Chapter 11. These two types of regression analysis are discussed in the latter chapters of this text. However, most of our discussion in this text focuses on the case where the dependent variable is continuous.

1.4 Classical Regression Models

Our goal is to build a model that relates the dependent variable y, with the explanatory variables. In our case, we wish to build a model of the form

$$y = \beta_0 + \beta_1 x_1 + \beta_2 x_2 + \beta_3 x_3 + \varepsilon. \tag{1.7}$$

Here, $\beta_0, \beta_1, \beta_2$ and β_3 are the parameters of the model in (1.7) and would need to be estimated from available data. To achieve this, we would collect observations (either by measurement or by mere observation or classification) on a random sample of n subjects or objects, such that each explanatory and dependent variable is indexed by $i = 1, 2, \ldots, n$. The ε or error term is included in model (1.7) to distinguish our regression model from a mere deterministic or mathematical model. For example, consider building a deterministic model with only one of the explanatory variables, say, height, so that we have

$$y = \beta_0 + \beta_1 x_1. \tag{1.8}$$

Let us suppose that $\beta_0 = 15$ and $\beta_1 = 20$, so that we have the following deterministic model

$$y = 15 + 10x_1. \tag{1.9}$$

Then the model in (1.9) implies that every subject with a height of $x_1 = 10$ (coded of course) units will have a corresponding weight of 115 pounds. Let us examine this for a moment. Does every body who is 5ft 6inches tall weigh the same? Certainly not. What we are saying is that every subject that has a coded height of 10 units will have a corresponding weight of 115 pounds, a prediction that does not make much sense since every body with that height does not have a weight of 115 pounds. So, how could we account for individual weight differences for all those with a height of 10 coded units? The introduction of the error term to modify the model in (1.8) would take care of this as we would now have

$$y = \beta_0 + \beta_1 x_1 + \varepsilon. \tag{1.10}$$

Thus for an i-th subject with a coded height of 10 units, the expected weight would be $115 + \varepsilon_i$, while for the j-th subject with the same coded height, the expected weight would also be $115 + \varepsilon_j$. Clearly the two expected weights are not the same unless $\varepsilon_i = \varepsilon_j$. The models in (1.7) and (1.10) are therefore referred to as *regression* or *stochastic* models because they both incorporate the random error terms ε.

Linear and Nonlinear Models

A linear regression model means that the regression function is linear in its parameters, and it can be written in the form

$$y = c_0 \beta_0 + c_1 \beta_1 + c_2 \beta_2 + \cdots + c_k \beta_k + \varepsilon, \tag{1.11}$$

where $c_0, c_1, \ldots, c_k$ are coefficients involving the explanatory variables. The model in (1.7) for example is a linear model and has, $c_0 = 1, c_1 = x_1, c_2 = x_2$ and $c_3 = x_3$. Other examples of linear models are the following:

$$\left.\begin{aligned}
y &= \beta_0 + \beta_1 x_1 \\[2mm]
y &= \beta_0 + \beta_1 x + \beta_2 x^2 \\[2mm]
y &= \beta_0 + \beta_1 x_1 + \beta_2 x_2 + \beta_3 x_2^2 + \beta_4 x_1 x_3 \\[2mm]
y &= \beta_0 + \beta_1 x_1^{5/2} + \beta_2 x_3 + \beta_3 e^{x1x3}
\end{aligned}\right\} \tag{1.12}$$

Each of the models in (1.12) is a linear model, because each is linear in its parameters. However, a model of the form

$$y = \beta_0 e^{\beta_1 x} + \varepsilon, \tag{1.13}$$

is nonlinear because it cannot be written in the form of (1.11). Other nonlinear models include the following:

$$\left.\begin{aligned}
y &= \beta_0 + \beta_2 e^{\beta_1 x} + \varepsilon \\[2mm]
y &= \frac{\alpha}{1 + \beta e^{-kx}} + \varepsilon \\[2mm]
y &= \alpha \exp(-\beta e^{-kx}) + \varepsilon \\[2mm]
y &= \alpha (1 + \beta e^{-kx})^{-1/\delta} + \varepsilon
\end{aligned}\right\} \tag{1.14}$$

The last three models in (1.14) are respectively, the logistic, the Gompertz and the Richards growth models. Most of the discussion in this text will be devoted to linear models. However, we shall briefly discuss nonlinear models in Chapter 13.

1.5 Uses of Regression

Regression analysis is used to examine possible relationships between two or more variables. For example, it can be used to examine the relationship between sales and advertising expenditures. When we develop a regression equation, this equation can be used to (a) describe the relationship between the dependent variable and the independent variable(s), (b) predict (or forecast) the dependent variable from the independent variable(s), and (c) determine the value of the independent variable(s) needed to produce a certain level of the response (dependent) variable. The regression equation can be used to estimate the mean value of the dependent variable for given values of the independent variables. Also, regression analysis enables us to place a bound on the error of prediction.

1.6   Exercises

1.1 The grade point averages of a large population of college students are approximately normally distributed with a mean of 2.65 and a standard deviation of 0.8.

   (a) What fraction of the students will possess a grade point average in excess of 3.0?

   (b) If students possessing a grade point average equal to or less than 1.96 are dropped from the college, what percentage of the students will be dropped?

   (c) Find the 90th percentile. Interpret the value.

   (d) Suppose that three students are randomly selected from the student body. What is the probability that exactly two of them will possess a grade point average in excess of 3.0?

1.2 Students' scores in a Calculus test are normally distributed with a mean $\mu = 50$ and a standard deviation $\sigma = 8$.

   (a) Suppose a student receives a score of 46. What fraction of the students taking the test score better?

   (b) Suppose the top 15% of the students are to be assigned a grade of A. Find a numerical limit for the grade.

   (c) Suppose the bottom 10% of the students are to be assigned a grade of E. Find a numerical limit for the grade.

1.3 Suppose X has a normal distribution and it is known that $P(X < 2.0) = 0.3085$ and $P(X > 3.0) = 0.2266$. Find the mean μ and the standard deviation σ for X.

1.4 The probability that a patient recovers from a certain disease is 0.8. Suppose twelve people are known to have contracted this disease and let X, a binomial random variable, denote the number that survives.

   (a) What is the probability that exactly 8 survive?

   (b) What is the probability that at least 9 survive?

   (c) What is the probability that between 4 and 8 (both inclusive) survive?

   (d) What is the probability that at most 5 survive?

   (e) Among eighty patients who contracted the disease, how many would you expect to survive?

1.5 According to a poll, 36% of adults abstain from drinking. A random sample of 49 adults is selected and let X denote the number of adults that abstain from drinking.

   (a) Find the probability that less than 25 adults abstain from drinking. Do not evaluate your result.

   (b) Find the mean, the variance, and the standard deviation of X.

   (c) By using an appropriate continuous distribution, approximate the probability in (a).

   (d) Is the approximation in (c) justified? Explain.

1.6 The number of typing errors, X, made by a particular typist has a Poisson distribution with a mean of four errors per page.

   (a) Find the probability that a certain page has exactly two typing errors.

   (b) Find the proportion of the distribution of X that lies within 0.25 standard deviation of the mean.

   (c) If more than two errors show on a given page, the typist must retype the whole page. What is the probability that a certain page does not have to be retyped?

1.7 In a particular department store, customers arrive at a checkout counter according to a Poisson distribution at an average of 4 per half hour.

   (a) During a given half hour, what is the probability that exactly 2 customers arrive?

   (b) During a given half hour, what is the probability that at least 2 customers arrive?

   (c) Is it likely that X will exceed 10? Explain.

1.8 A manufacturer of car tires finds that, on the average, 5% of the tires are defective. Suppose a random sample of 475 tires is selected.

(a) What is the exact probability that more than 17 tires will be defective? Do not evaluate your result.

(b) Using an appropriate discrete distribution approximation, write down the probability in (a). Do not evaluate your result.

(c) Using an appropriate continuous distribution approximation, compute the probability in (a).

(d) Is the approximation in (c) justified? Explain.

1.9 Differentiating with respect to θ the expressions on both sides of the equation

$$1 = \sum_{x=0}^{n} \binom{n}{x} \theta^x (1 - \theta)^{n-x},$$

show that the mean of the binomial distribution is given by $n\theta$. Differentiating again with respect to θ, show that the variance is given by $n\theta(1 - \theta)$.

1.10 Consider the binomial probability distribution

$$P(X = x) = \binom{n}{x} \theta^x (1 - \theta)^{n-x}.$$

Suppose the parameter $n \to \infty$ in such a way that $n\theta$ can be approximated by a fixed quantity λ, show that the limit of the binomial probability distribution is given by

$$P(X = x) = \lambda^x e^{-\lambda}/x!$$

1.11 Using a statistical software, compute the probabilities in problems 1.8(a), 1.8(b), and 1.8(c).

1.12 Using a statistical software, compute the following probabilities:

(a) $P(12 < X < 16)$, where X is $N(\mu = 15, \sigma^2 = 7)$.

(b) $P(-1.3333 < Z < 2.11456)$, where Z is standard normal.

(c) $P(Z \geq -0.33333)$, where Z is standard normal.

(d) $P(-1 \leq t_{25} < 3.45)$

(e) $P(\chi^2_{10} \geq 15.556)$

(f) $P(2.56 \leq F_{5,10} \leq 4.86)$.

1.13 Using a statistical software, find the value of constant a in the following problems:

(a) $P(X > a) = 0.88886$, where X is $N(\mu = 14, \sigma^2 = 7)$.

(b) $P(Z \leq a) = 0.46994$, where Z is standard normal.

(c) $P(t_{20} \geq a) = 0.111111$

(d) $P(\chi^2_{15} \leq a) = 0.77778$

(e) $P(F_{10,20} \geq a) = 0.25$.

1.14 In a Mid-western university, 12% of the student population voted during the last student government association election. Suppose a random sample of 50 students are selected from the university and let X denote the number (from 50) that voted in the last election. Using a statistical software, answer the following questions:

(a) Find the probability that at most 8 students voted.

(b) Find the probability that between 10 and 20 (both inclusive) students voted.

(c) Find the probability that exactly 15 voted.

(d) Using an appropriate discrete distribution, approximate the probability in (a).

(e) Using an appropriate continuous distribution, approximate the probability in (a).

(f) Compute the errors of approximations in (d) and (e). Are the approximations justified? Comment on your results.

Chapter 2

Simple Linear Regression

2.1 Introduction

The simple linear regression model is the simplest model with only one explanatory variable. If y_i and x_i are pairs of observations from a simple random sample of n subjects, then the simple linear regression model can be written as

$$y_i = \beta_0 + \beta_1 x_i + \varepsilon_i, \text{ for } i = 1, 2, \ldots, n. \tag{2.1}$$

Here,

- The model represents the equation of a straight line and is sometimes referred to as a first-order linear regression model.

- β_0 and β_1 are the unknown parameters of the model and would have to be estimated from available data. β_0 represents the intercept of the straight line while β_1 represents the slope of the straight line.

- y_i is the response for observation i with x_i its corresponding explanatory variable value.

- ε_i is the random error term attributable to observation i.

- $(y_1, x_1), (y_2, x_2), \ldots, (y_n, x_n)$ therefore constitute pairs of n observations on the subjects.

2.2 Model Assumptions

The model assumptions must be verified in the light of available data. Any model violations must be examined and correction sought. We shall discuss the verification of these assumptions and further discuss measures for addressing any violations. For the simple linear model in (2.1), the following are the underlying assumptions:

1. The random error term ε has a mean of 0. That is, $E(\varepsilon_i) = 0$, for all i and hence, $E(Y_i) = \beta_0 + \beta_1 x_i$. This is sometimes denoted as

$$\mu_{Y|x} = \beta_0 + \beta_1 x. \tag{2.2}$$

2. The random error term ε has a constant variance σ^2. That is, $\text{Var}(\varepsilon_i) = \sigma^2$, for all i.

   This assumption implies that the variances do not depend on the values of x. In other words, the variance is constant from one observation to another. This homogeneity of variances is often referred to as **homoscedasticity** and therefore $\sigma^2_{Y|x} \equiv \sigma^2$, for all i. Here $\sigma^2_{Y|x}$ is read as 'the conditional variance of Y given x'. Further,

$$\text{Var}(\beta_0 + \beta_1 x_1 + \varepsilon_i) = \text{Var}(\varepsilon_i) = \sigma^2. \tag{2.3}$$

3. The random error term ε is distributed normal. That is, combining assumptions (1) and (2), we have $\varepsilon \sim N(0, \sigma^2)$.

4. The error terms are assumed to be uncorrelated. That is,

$$\text{Cov}(\varepsilon_i, \varepsilon_j) = 0, \quad \text{for all } i \neq j. \tag{2.4}$$

A consequence of this is that any two dependent variables y_i and y_j are uncorrelated, or simply stated are independently distributed.

2.3   Estimating Simple Linear Regression Parameters

Several methods are used to estimate the parameters of the linear regression model, of which the model in (2.1) is a simple linear regression model. Our focus here is to obtain point estimates for the parameters of our model using available data from our random sample. By far the most used method is the *ordinary least squares* (OLS) method developed by Gauss. We list below two of the methods that have been used to estimate the parameters of a regression model.

(a) The method of ordinary least squares (OLS)

(b) The method of maximum likelihood estimation (MLE)

We now discuss these methods in the following subsections.

2.3.1   The Ordinary Least Squares (OLS) Method

For the simple regression model,

$$y_i = \beta_0 + \beta_1 x_i + \varepsilon_i, \quad i = 1, 2, \ldots, n, \tag{2.5}$$

the ordinary least squares (least squares for short) method minimizes the sum of squared deviations with respect to the parameters β_0 and β_1. That is, the method sought to minimize

$$Q = \sum_{i=1}^{n} \varepsilon_i^2 = \sum_{i=1}^{n} (y_i - \beta_0 - \beta_1 x_i)^2, \tag{2.6}$$

with respect to β_0 and β_1. This implies that we need to obtain $\dfrac{\partial Q}{\partial \beta_0}$ and $\dfrac{\partial Q}{\partial \beta_1}$ and set them to zero. Hence,

$$\frac{\partial Q}{\partial \beta_0} = 2 \sum_{i=1}^{n} (y_i - \beta_0 - \beta_1 x_i)(-1), \tag{2.7a}$$

$$\frac{\partial Q}{\partial \beta_1} = 2 \sum_{i=1}^{n} (y_i - \beta_0 - \beta_1 x_i)(-x_i). \tag{2.7b}$$

Setting the equations in (2.7a) and (2.7b) to zero, and replacing the parameters by their estimates, we have

$$\sum_{i=1}^{n} y_i - n\hat{\beta}_0 - \hat{\beta}_1 \sum_{i=1}^{n} x_i = 0, \tag{2.8a}$$

$$\sum_{i=1}^{n} x_i y_i - \hat{\beta}_0 \sum_{i=1}^{n} x_i - \hat{\beta}_1 \sum_{i=1}^{n} x_i^2 = 0. \tag{2.8b}$$

Multiplying equation (2.8a) by $\sum x_i$ and equation (2.8b) by n, and dropping the range of i on the summation sign for brevity, we have

$$\sum x_i \sum y_i - n\hat{\beta}_0 \sum x_i - \hat{\beta}_1 \left(\sum x_i \right)^2 = 0, \tag{2.9a}$$

$$n \sum x_i y_i - n\hat{\beta}_0 \sum x_i - n\hat{\beta}_1 \sum x_i^2 = 0. \tag{2.9b}$$

Subtracting equation (2.9b) from equation (2.9a), we have

$$\sum x_i \sum y_i - \hat{\beta}_1 \left(\sum x_i \right)^2 - n \sum x_i y_i + n \hat{\beta}_1 \sum x_i^2 = 0. \tag{2.10}$$

That is,

$$n \hat{\beta}_1 \left[\sum x_i^2 - \frac{1}{n} \left(\sum x_i \right)^2 \right] = n \sum x_i y_i - \sum x_i \sum y_i.$$

Dividing through by n, we have,

$$\hat{\beta}_1 \left[\sum x_i^2 - \frac{1}{n} \left(\sum x_i \right)^2 \right] = \sum x_i y_i - \frac{1}{n} \sum x_i \sum y_i,$$

and hence,

$$\hat{\beta}_1 = \frac{\sum x_i y_i - \frac{1}{n} \sum x_i \sum y_i}{\sum x_i^2 - \frac{1}{n} (\sum x_i)^2} = \frac{\sum (x_i - \bar{x})(y_i - \bar{y})}{\sum (x_i - \bar{x})^2} = \frac{\sum x_i y_i - n \bar{x} \bar{y}}{\sum x_i^2 - n \bar{x}^2}. \tag{2.11}$$

That is,

$$\hat{\beta}_1 = \begin{cases} \dfrac{\sum x_i y_i - \frac{1}{n} \sum x_i \sum y_i}{\sum x_i^2 - \frac{1}{n} \left(\sum x_i \right)^2} \\[4mm] \dfrac{\sum (x_i - \bar{x})(y_i - \bar{y})}{\sum (x_i - \bar{x})^2} \\[4mm] \dfrac{\sum x_i y_i - n \bar{x} \bar{y}}{\sum x_i^2 - n \bar{x}^2} \end{cases} \tag{2.12}$$

Writing

$$S_{xy} = \sum x_i y_i - \frac{1}{n} \sum x_i \sum y_i = \sum (x_i - \bar{x})(y_i - \bar{y}) = \sum x_i y_i - n \bar{x} \bar{y}$$

and

$$S_{xx} = \sum x_i^2 - \frac{(\sum x_i)^2}{n} = \sum (x_i - \bar{x})^2 = \sum x_i^2 - n \bar{x}^2,$$

we can therefore write the parameter estimate of β_1 more succinctly as

$$\hat{\beta}_1 = \frac{S_{xy}}{S_{xx}}. \tag{2.13}$$

From equation (2.8a), we have $n \bar{y} - n \hat{\beta}_0 - n \hat{\beta}_1 \bar{x} = 0$. Hence, the parameter estimate of β_0 can be obtained as

$$\hat{\beta}_0 = \bar{y} - \hat{\beta}_1 \bar{x}. \tag{2.14}$$

The estimated regression equation is therefore given by

$$\hat{y}_i = \hat{\beta}_0 + \hat{\beta}_1 x_i = \bar{y} - \hat{\beta}_1 \bar{x} + \hat{\beta}_1 x_i = \bar{y} + \hat{\beta}_1 (x_i - \bar{x}). \tag{2.15}$$

2.3.2   Maximum Likelihood Estimation

If we consider the simple linear regression model in (2.1), we stated that one of the assumptions of the linear model is that the error term ε_i has a normal distribution with mean μ and constant variance σ^2. That is, we assume that, $\varepsilon_i \sim N(0, \sigma^2)$. That is, the error term has the distribution

$$\frac{1}{\sqrt{2\pi\sigma^2}} \exp \left(-\frac{\varepsilon_i^2}{2\sigma^2} \right). \tag{2.16}$$

From (2.1) therefore, we can write $\varepsilon_i = (y_i - \beta_0 - \beta_1 x_i)$, hence substituting this in (2.16), we have

$$f(\varepsilon_i) = \frac{1}{\sqrt{2\pi\sigma^2}} \exp\left(-\frac{(y_i - \beta_0 - \beta_1 x_i)^2}{2\sigma^2}\right). \tag{2.17}$$

Since the parameters will be estimated from n pairs of observations (x_i, y_i), the likelihood function of β_0, β_1 and σ^2, is the joint probability density function of $\varepsilon_1, \varepsilon_2, \ldots, \varepsilon_n$ from (2.17) and it is given as

$$\begin{aligned} L(\beta_0, \beta_1, \sigma^2) &= \prod_{i=1}^{n} \left[\frac{1}{\sqrt{2\pi\sigma^2}} \exp\left\{-\frac{1}{2\sigma^2}(y_i - \beta_0 - \beta_1 x_i)^2\right\} \right] \\ &= \left(\frac{1}{\sqrt{2\pi\sigma^2}}\right)^n \exp\left\{-\frac{1}{2\sigma^2} \sum_{i=1}^{n}(y_i - \beta_0 - \beta_1 x_i)^2\right\}. \end{aligned} \tag{2.18}$$

Hence, the log-likelihood is given by

$$\ln L = \ln L(\beta_0, \beta_1, \sigma^2) = -\frac{n}{2}\log 2\pi - \frac{n}{2}\log \sigma^2 - \frac{1}{2\sigma^2}\sum_{i=1}^{n}(y_i - \beta_0 - \beta_1 x_i)^2.$$

Taking partial derivatives with respect to the parameters β_0, β_1 and σ^2, we have

$$\frac{\partial L}{\partial \beta_0} = \frac{1}{\sigma^2}\sum_{i=1}^{n}(y_i - \beta_0 - \beta_1 x_i), \tag{2.19a}$$

$$\frac{\partial L}{\partial \beta_1} = \frac{1}{\sigma^2}\sum_{i=1}^{n}(y_i - \beta_0 - \beta_1 x_i)x_i, \tag{2.19b}$$

$$\frac{\partial L}{\partial \sigma^2} = -\frac{n}{2\sigma^2} + \frac{1}{(\sigma^2)^2}\sum_{i=1}^{n}(y_i - \beta_0 - \beta_1 x_i)^2. \tag{2.19c}$$

Setting the partial derivatives in (2.19a) to (2.19c) to zero and solving, we have

$$\hat{\beta}_0 = \bar{y} - \hat{\beta}_1 \bar{x}, \tag{2.20a}$$

$$\hat{\beta}_1 = \frac{S_{xy}}{S_{xx}}, \tag{2.20b}$$

$$\hat{\sigma}^2 = \frac{1}{n}\sum_{i=1}^{n}(y_i - \hat{\beta}_0 - \hat{\beta}_1 x_i)^2. \tag{2.20c}$$

We see that both equations (2.19a) and (2.19b) when solved lead respectively to the same normal equations as the OLS approach and thus give the same estimates as the OLS approach as indicated in expressions (2.20a) and (2.20b) respectively for parameters β_0 and β_1. However, as we shall see in the next section, the parameter estimate $\hat{\sigma}^2$ for the variance as given in (2.20c) is a biased estimator for σ^2. That is, $\mathrm{E}(\hat{\sigma}^2) \neq \sigma^2$.

Example 2.1

The following data is on the age (x) and the systolic blood pressure (y) of 30 randomly selected individuals.
For the data, we can obtain the summary statistics as follows, where age is denoted by x and the dependent variable SBP is similarly denoted as y.

Subj.	x	y	xy	x^2	y^2
1	42	130	5460	1764	16900
2	46	115	5290	2116	13225
$\vdots$	$\vdots$	$\vdots$	$\vdots$	$\vdots$	$\vdots$
29	65	140	9100	4225	19600
30	48	130	6240	2304	16900
Total	1948	4405	290154	132250	653723

Subject	Age	SBP	Subject	Age	SBP	Subject	Age	SBP
1	42	130	11	64	155	21	71	158
2	46	115	12	81	160	22	76	158
3	42	148	13	41	125	23	44	130
4	71	100	14	61	150	24	55	144
5	80	156	15	75	165	25	80	162
6	74	162	16	53	135	26	63	150
7	70	151	17	77	153	27	82	160
8	80	156	18	60	146	28	53	140
9	85	162	19	82	156	29	65	140
10	72	158	20	55	150	30	48	130

Table 2.1: Age and SBP values for a sample of 30 individuals

That is, we can present these summary statistics as,

$$\sum x = 1948.0, \quad \sum y = 4405.0, \quad \sum x^2 = 132250.0, \quad \sum y^2 = 653723.0, \quad \sum xy = 290154.0.$$

Hence, the corrected sum of squares and cross products are computed as

$$S_{xy} = \sum xy - \frac{\sum x \sum y}{n} \quad = 290154.0 - \frac{1948 \times 4405}{30} \quad = 4122.6667$$

$$S_{xx} = \sum x^2 - \frac{(\sum x)^2}{n} \quad = 132250.0 - \frac{1948^2}{30} \quad = 5759.8667$$

$$S_{yy} = \sum y^2 - \frac{(\sum y)^2}{n} \quad = 653723.0 - \frac{4405^2}{30} \quad = 6922.1667.$$

In SAS, the **proc corr** with options **sscp** and **csscp** will produce the following equivalent results:

```
data table1;
input subj age sbp;
datalines;
1 42 130
2 46 115
.........
29 65 140
30 48 130
;
proc print;
run;
proc corr sscp csscp nocorr;
var age sbp;
run;
                        The CORR Procedure

            2 Variables:    age     sbp

                    SSCP Matrix

                        age             sbp
            age    132250.0000     290154.0000
            sbp    290154.0000     653723.0000

                    CSSCP Matrix

                        age             sbp
            age    5759.866667     4122.666667
            sbp    4122.666667     6922.166667

                    Simple Statistics

Variable       N       Mean     Std Dev      Sum     Minimum     Maximum

age           30    64.93333    14.09312     1948    41.00000    85.00000
sbp           30   146.83333    15.44977     4405   100.00000   165.00000
```

The sum of squares and cross products (SSCP) matrix and the corrected sum of squares and cross products (CSSCP) matrix in the above output are often referred to as the summary statistics for the data. They are produced by the

proc corr in SAS. Similar summary statistics (uncorrected sums of squares and cross products) can also be obtained from within **proc reg** by the use of options **xpx i**, giving the following results:

```
set table1;
proc reg;
model sbp = age / xpx i;
run;
                              The REG Procedure
                      Model Crossproducts X'X X'Y Y'Y

          Variable        Intercept              age              sbp

          Intercept              30             1948             4405
          age                  1948           132250           290154
          sbp                  4405           290154           653723

The REG Procedure
                          Model: MODEL1
                      Dependent Variable: sbp

                  Number of Observations Read        30
                  Number of Observations Used        30

              X'X Inverse, Parameter Estimates, and SSE

          Variable        Intercept              age              sbp

          Intercept     0.7653533647     -0.011273409     100.35682539
          age          -0.011273409      0.0001736151      0.7157573092
          sbp         100.35682539       0.7157573092   3971.3378666

                        Analysis of Variance

                                  Sum of          Mean
          Source         DF      Squares        Square    F Value    Pr > F

          Model           1   2950.82880    2950.82880      20.80    <.0001
          Error          28   3971.33787     141.83350
          Corrected Total 29   6922.16667

                  Root MSE              11.90939    R-Square     0.4263
                  Dependent Mean       146.83333    Adj R-Sq     0.4058
                  Coeff Var              8.11082

                        Parameter Estimates

                         Parameter      Standard
          Variable   DF    Estimate         Error    t Value    Pr > |t|

          Intercept   1   100.35683      10.41886       9.63     <.0001
          age         1     0.71576       0.15692       4.56     <.0001
```

Obtaining the Parameter Estimates

The parameter estimates for our regression from (2.13) and (2.14) can now be computed as follows:

$$\hat{\beta}_1 = \frac{S_{xy}}{S_{xx}} = \frac{4122.6667}{5759.8667} = 0.7158,$$

and

$$\hat{\beta}_0 = \bar{y} - \hat{\beta}_1\bar{x} = 146.8333 - 0.7158(64.9333) = 100.3540.$$

The estimated regression equation therefore is

$$\hat{y}_i = 100.3540 + 0.7158x_i. \tag{2.21}$$

Since $\bar{y} = 146.8333$ and $\bar{x} = 64.9333$, we can therefore write an equivalent estimated regression model as

$$\hat{y}_i = 146.8333 + 0.7158(x_i - 64.9333). \tag{2.22}$$

The regression equation in (2.21) can be used to predict the SPB for a subject who is 50 years old or subjects who are 50 years old. This is done by replacing x_i in (2.21) by 50 to obtain $\hat{y} = 100.3540 + 0.7158(5) = 136.144$. We present in Figure 2.1 the plot of the estimated regression line overlayed with the scatter plot of observed values.

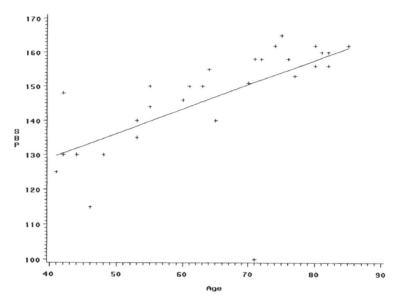

Figure 2.1: Plot of regression line overlayed with the observed values

Estimated Parameter Interpretation

For a unit increase in x (age), a subject's SBP (y) increases by 0.7158 units. Thus a two unit increase in age, would increase the SBP by $(2 \times 0.7158) = 1.4316$ units. The intercept $\hat{\beta}_0 = 100.354$ has no practical meaning in this example since 100.354 is the SBP of a subject whose age is zero. Observe that x cannot be zero in the data.

2.4 Properties of the Estimators

Unbiasedness

The OLS estimators $\hat{\beta}_0$ and $\hat{\beta}_1$ are unbiased estimators respectively for the population parameters. To show this, the expected value of $\hat{\beta}_1$ can be calculated by writing $\hat{\beta}_1$ as

$$\hat{\beta}_1 = \frac{S_{xy}}{S_{xx}} = \frac{\sum(x_i - \bar{x})(y_i - \bar{y})}{S_{xx}} = \frac{\sum(x_i - \bar{x})y_i}{S_{xx}} - \frac{\sum(x_i - \bar{x})\bar{y}}{S_{xx}} = \frac{\sum(x_i - \bar{x})y_i}{S_{xx}},$$

since $\sum(x_i - \bar{x}) = 0$. Hence,

$$
\begin{aligned}
E\left(\hat{\beta}_1\right) &= E\left(\frac{S_{xy}}{S_{xx}}\right) = \sum_{i=1}^{n}\frac{(x_i - \bar{x})E(Y_i)}{S_{xx}} = \frac{1}{S_{xx}}\sum_{i=1}^{n}(x_i - \bar{x})(\beta_0 + \beta_1 x_i) \\
&= \frac{\beta_0}{S_{xx}}\sum_{i=1}^{n}(x_i - \bar{x}) + \frac{\beta_1}{S_{xx}}\sum_{i=1}^{n}x_i(x_i - \bar{x}) = 0 + \frac{\beta_1}{S_{xx}}\sum_{i=1}^{n}(x_i^2 - x_i\bar{x}) \\
&= \frac{\beta_1}{S_{xx}}\left(\sum_{i=1}^{n}x_i^2 - n\bar{x}^2\right) = \frac{\beta_1}{S_{xx}}S_{xx} = \beta_1.
\end{aligned}
$$

Since $E(\hat{\beta}_1) = \beta_1$, therefore, the estimator $\hat{\beta}_1$ is an unbiased estimator for the unknown parameter β_1 of the population model. Similarly, for the estimator $\hat{\beta}_0$, we have,

$$
\begin{aligned}
E(\hat{\beta}_0) &= E(\bar{Y} - \hat{\beta}_1\bar{x}) = \frac{1}{n}E\left(\sum_{i=1}^{n}Y_i\right) - \beta_1\bar{x} = \frac{1}{n}\sum_{i=1}^{n}(\beta_0 + \beta_1 x_i) - \beta_1\bar{x} \\
&= \frac{1}{n}(n\beta_0) + \frac{1}{n}(\beta_1 n\bar{x}) - \beta_1\bar{x} = \beta_0.
\end{aligned}
$$

Again, since $E(\hat{\beta}_0) = \beta_0$, therefore, the estimator $\hat{\beta}_0$ is unbiased for β_0. Note that $E(\hat{\beta}_1\bar{x}) = \beta_1\bar{x}$ because of our previous result for $\hat{\beta}_1$ being unbiased. Observe that x is not a random variable and so $E(\bar{x}) = \bar{x}$.

Variances of the Parameter Estimators

For the variance of $\hat{\beta}_1$ and writing $\sum$ for $\sum_{i=1}^{n}$, we have

$$\text{Var}(\hat{\beta}_1) \;=\; \frac{\text{Var}[\sum(x_i - \bar{x})Y_i]}{(S_{xx})^2} = \frac{\sum(x_i - \bar{x})^2\text{Var}(Y_i)}{(S_{xx})^2} = \frac{\sigma^2\sum(x_i - \bar{x})^2}{(S_{xx})^2} = \frac{\sigma^2 S_{xx}}{(S_{xx})^2} = \frac{\sigma^2}{S_{xx}}.$$

Similarly, to find the variance of $\hat{\beta}_0$, we can write

$$\hat{\beta}_0 = \bar{y} - \hat{\beta}_1\bar{x},$$

and hence,

$$\text{Var}(\hat{\beta}_0) \;=\; \text{Var}(\bar{Y} - \hat{\beta}_1\bar{x}) = \text{Var}(\bar{Y}) + \bar{x}^2\text{Var}(\hat{\beta}_1) - 2\bar{x}\text{Cov}(\bar{Y}, \hat{\beta}_1) = \frac{\sigma^2}{n} + \bar{x}^2\frac{\sigma^2}{S_{xx}} + 0,$$

since $\text{Cov}(\bar{Y}, \hat{\beta}_1) = 0$. Hence,

$$\text{Var}(\hat{\beta}_0) = \sigma^2\left[\frac{1}{n} + \frac{\bar{x}^2}{S_{xx}}\right] = \sigma^2\left[\frac{S_{xx} + n\bar{x}^2}{nS_{xx}}\right] = \frac{\sum x_i^2}{nS_{xx}}\sigma^2.$$

Since $S_{xx} = \sum x_i^2 - n\bar{x}^2$, hence $\sum x_i^2 = n\bar{x}^2 + S_{xx}$. Therefore, the result follows. The covariance of $\hat{\beta}_1$ and $\hat{\beta}_0$ is given as

$$\begin{aligned}\text{Cov}(\hat{\beta}_0, \hat{\beta}_1) \;&=\; \text{Cov}(\bar{Y} - \hat{\beta}_1\bar{x}, \hat{\beta}_1) = \text{Cov}(-\hat{\beta}_1\bar{x}, \hat{\beta}_1), \quad \text{since } \text{Cov}(\bar{Y}, \hat{\beta}_1) = 0 \\ &=\; -\bar{x}\text{Cov}(\hat{\beta}_1, \hat{\beta}_1) = -\bar{x}\text{Var}(\hat{\beta}_1) = \frac{-\bar{x}\sigma^2}{S_{xx}}.\end{aligned}$$

We can summarize the results on variances and covariance of the estimators as follows:

$$\text{Var}(\hat{\beta}_1) = \frac{\sigma^2}{S_{xx}}, \tag{2.23a}$$

$$\text{Var}(\hat{\beta}_0) = \frac{\sum x_i^2}{nS_{xx}}\sigma^2, \tag{2.23b}$$

$$\text{Cov}(\hat{\beta}_0, \hat{\beta}_1) = \frac{-\bar{x}\sigma^2}{S_{xx}}. \tag{2.23c}$$

Sampling Distribution of $\hat{\beta}_1$ and $\hat{\beta}_0$

From $\hat{\beta}_1 = \sum(x_i - \bar{x})y_i/S_{xx}$, we notice that $\hat{\beta}_1$ is a linear combination of the response variable y_i. That is, $\hat{\beta}_1 = \sum c_i y_i$ where $c_i = (x_i - \bar{x})/S_{xx}$ are a function of x_i. Since the x_i are fixed quantities, the c_i are also fixed. Since the y_i are assumed to be normal, then the sampling distribution of $\hat{\beta}_1$ is normal with mean β_1 and variance σ^2/S_{xx} in (2.23a). The sampling distribution of $\hat{\beta}_0 = \bar{y} - \hat{\beta}_1\bar{x}$ is normal with mean β_0 and the variance is given by (2.23b).

Estimated Variances

The variances of $\hat{\beta}_1$ and $\hat{\beta}_0$ are functions of the population variance σ^2. We shall estimate this population variance by S^2 in the next chapter, and the estimated variances and standard errors for $\hat{\beta}_1$ and $\hat{\beta}_0$ would be

$$\widehat{\text{Var}}(\hat{\beta}_1) = \frac{S^2}{S_{xx}}, \tag{2.24a}$$

$$\widehat{\text{Var}}(\hat{\beta}_0) = \frac{\sum x_i^2}{nS_{xx}}S^2, \tag{2.24b}$$

with corresponding standard errors $S/\sqrt{S_{xx}}$ and $S\sqrt{\frac{\sum x_i^2}{nS_{xx}}}$, respectively.

Recall that the maximum likelihood estimate of σ^2 is given by (2.20c) as

$$\hat{\sigma}^2 = \frac{1}{n}\sum_{i=1}^{n}(y_i - \hat{y}_i)^2 = \frac{1}{n}\sum_{i=1}^{n}(y_i - \hat{\beta}_0 - \hat{\beta}_1 x_i)^2.$$

By taking the expectation of $\hat{\sigma}^2$, it is not difficult to show that this expectation is $(n-2)\sigma^2/n$ which implies that $\hat{\sigma}^2$ is a biased estimator of σ^2. [See exercise 2.11]

2.5   Exercises

2.1 A data set that relates the amount of fire damage in major residential fires to the distance between the residence and the nearest fire station is analyzed. The variable x is the distance from the fire station in miles and y is the fire damage in thousands of dollars. The result of fitting a model of the form $y_i = \beta_0 + \beta_1 x_i + \varepsilon_i$ to the data is in the following SAS output:

The REG Procedure
Model: MODEL1
Dependent Variable: y

Analysis of Variance

Source	DF	Sum of Squares	Mean Square	F Value	Pr > F
Model	1	841.76636	841.76636	156.89	<.0001
Error	13	69.75098	5.36546		
Corrected Total	14	911.51733			

Root MSE	2.31635	R-Square	0.9235
Dependent Mean	26.41333	Adj R-Sq	0.9176
Coeff Var	8.76961		

Parameter Estimates

Variable	DF	Parameter Estimate	Standard Error	t Value	Pr > |t|
Intercept	1	10.27793	1.42028	7.24	<.0001
x	1	4.91933	0.39275	12.53	<.0001

(a) Write down the estimated least squares regression equation for the data. Interpret the regression coefficients.

(b) How many residential areas were involved in the study?

2.2 A simple linear regression analysis for $n = 20$ data points produced the following results.

$$\bar{x} = 2.50, \quad \bar{y} = 10.60, \quad S_{xx} = 4.77, \quad S_{yy} = 59.21, \quad S_{xy} = 16.22.$$

Determine the equation of the best fitting straight line for these data.

2.3 The following summary statistics come from data relating the weight (x) to the systolic blood pressure (y) of $n = 26$ randomly selected males in the age group 25-30 years.

$$\sum x = 4743, \quad \sum x^2 = 880545, \quad \sum y = 3786, \quad \sum y^2 = 555802, \quad \sum xy = 697076.$$

Fit a regression of the form $y_i = \beta_0 + \beta_1 x_i + \varepsilon_i$ to the data. Give estimates of your parameters.

2.4 Show that for the linear model $y_i = \beta_0 + \beta_1 x_i + \varepsilon_i$, $\sum_i e_i = 0$, where $e_i = y_i - \hat{y}_i$ are the fitted residuals. Explain the difference between the e_i's and the ε_i's.

2.5 Find the least squares estimate of the parameter β in the regression equation $y_i = \beta x_i$.

2.6 Twenty employees were selected randomly from the production-and-research department of a company and the following results were obtained from their satisfaction (y) and productivity (x).

$$\sum x_i = 122, \quad \sum x_i^2 = 834, \quad \sum y_i = 125, \quad \sum y_i^2 = 911, \quad \sum x_i y_i = 863.$$

(a) Find the least squares line. What are your assumptions, if any, about satisfaction (y) and productivity (x)?

(b) Interpret the regression coefficients.

2.7 The following summary statistics are on the number of wins (y) and the earned run averages (x) of fourteen American League teams after a regular season.

$$\sum x_i = 54.5, \quad \sum x_i^2 = 214.91, \quad \sum y_i = 1133.0, \quad \sum y_i^2 = 92815.0, \quad \sum x_i y_i = 4395.0.$$

(a) Obtain the least squares line of y on x.

(b) Interpret the regression coefficients in (a).

2.8 Using the formulas for $\hat{\beta}_1$ in (2.13) and $\hat{\beta}_0$ in (2.14), show that

$$\sum_{i=1}^{n} \left[y_i - \left(\hat{\beta}_0 + \hat{\beta}_1 x_i \right) \right]^2 = S_{yy} - \hat{\beta}_1 S_{xy}.$$

2.9 The data in the following table is on the final standings of fifteen Eastern Conference National Basketball Association teams during a regular season. The variables are the winning percentage for each team (PCT), average points for each team (PF), and the average points against each team (PA).

PCT	PF	PA	PCT	PF	PA	PCT	PF	PA
80.5	100.5	90.3	50.0	100.2	97.3	40.2	97.3	100.4
72.0	97.5	90.1	48.8	96.6	96.2	39.0	97.1	101.4
63.4	104.5	99.0	45.1	98.2	100.0	31.7	97.0	103.9
54.9	96.4	96.7	43.9	104.0	105.4	28.0	96.9	103.5
52.4	98.8	99.2	41.5	95.8	100.9	18.3	91.4	100.0

(a) Obtain the least squares line of PCT on PF.

(b) Interpret the regression coefficients in (a).

(c) Obtain the least squares of PCT on PA.

(d) Interpret the regression coefficients in (c).

2.10 For a simple linear regression model $y_i = \beta_0 + \beta_1 x_i + \varepsilon_i$, let $\bar{y}$ be the sample mean of the y_i observations and $\hat{\beta}_1$ be the least squares estimate of β_1. Show that $\text{Cov}(\bar{Y}, \hat{\beta}_1) = 0$.

2.11 In a simple linear regression model $y_i = \beta_0 + \beta_1 x_i + \varepsilon_i$, show that the $E(\hat{\sigma}^2)$ for the maximum likelihood estimator of σ^2 in (2.20c) is given by $(n-2)\sigma^2/n$.

Chapter 3

Inferences on Parameter Estimates

3.1 Introduction

Because of the assumption that the error terms are normally distributed, this enables us to make inferences on the regression parameters. In the previous chapter, we found the estimated variances of the parameter estimators to be

$$\widehat{\text{Var}}(\hat{\beta}_1) = \frac{S^2}{S_{xx}},$$

and

$$\widehat{\text{Var}}(\hat{\beta}_0) = \frac{\sum x_i^2}{nS_{xx}}S^2.$$

In the above expressions, the variance estimator S^2 is unknown. To find this from available data, we would need to carry out the analysis of variance and subsequently produce the analysis of variance table. We carry this out in the next section.

3.2 Analysis of Variance

For the simple linear regression model in (2.5), the total sum of squares (total variation in the response variable) is designated as S_{yy} and is defined as

$$S_{yy} = \sum (y_i - \bar{y})^2 = \sum y^2 - n\bar{y}^2 = \sum y^2 - \frac{1}{n}\left(\sum y\right)^2.$$

The total sum of squares (SS) is based on $n-1$ degrees of freedom. The regression SS is defined as

$$\text{Reg } SS = \hat{\beta}_1 S_{xy} = \frac{S_{xy}^2}{S_{xx}}.$$

The regression SS is based on **the number of parameters in the model** -1. In this case, we have only two parameters β_0 and β_1 and hence, the degree of freedom would be $2 - 1 = 1$. The analysis of variance is therefore as displayed in Table 3.1.

Source	df	SS	MS	F
Reg	1	$\hat{\beta}_1 S_{xy}$	$\hat{\beta}_1 S_{xy}$	F^*
Error	$n-2$	SSE	MSE	
Total	$n-1$	S_{yy}		

Table 3.1: Analysis of variance table

In Table 3.1,

1. *MS* refers to Mean Square, which is the ratio of the *SS* and the corresponding degrees of freedom for the specific source line.

2. *SSE* is the error *SS* and is obtained as the difference between the Total *SS* and the Regression *SS* and is based on $(n-1) - 1 = (n-2)$ degrees of freedom. That is,

$$SSE = S_{yy} - \hat{\beta}_1 S_{xy} \text{ on } (n-2) \text{ degrees of freedom.}$$

3. *MSE* is the mean square error (or error mean square), which is the ratio of error *SS* and the corresponding error *df*. That is,

$$MSE = \frac{SSE}{n-2}.$$

4. F^* is the ratio of the regression mean square and error mean square. That is,

$$F^* = \frac{\text{Reg } MS}{\text{Error } MS} = \frac{\hat{\beta}_1 S_{xy}}{MSE}.$$

5. F^* will be distributed as an *F* distribution with 1 and $n-2$ degrees of freedom and this can be written as $F_{1,\,n-2}$.

Example 2.1 continued:

For the data in Table 2.1 from Chapter 2, we have computed that $S_{xy} = 4122.6667$, $S_{xy} = 4122.6667$ and $S_{yy} = 6922.1667$.

The regression *SS* is computed from $\beta_1 S_{xy} = S_{xy}^2 / S_{xx}$ as

$$\text{Reg } SS = \frac{S_{xy}^2}{S_{xx}} = \frac{4122.6667^2}{5759.8667} = 2950.8288,$$

and is based on 1 *df*. The Total *SS* is $S_{yy} = 6922.16667$ on $(30 - 1) = 29$ *df*. Hence, the error *SS* is obtained by subtraction, viz:

$$\text{Error } SS = \text{Total } SS - \text{Reg } SS = 6922.1667 - 2950.8288 = 3977.3379,$$

and is based on $n - 2 = 30 - 2 = 28$ degrees of freedom. Thus, we have the analysis of variance table presented in Table 3.2.

Source	*df*	*SS*	*MS*	*F*
Reg	1	2950.8288	2950.8288	20.805
Error	28	3971.3379	141.8335	
Total	29	6922.1667		

Table 3.2: Analysis of variance table for Example 2.1

As explained before, we have

1. $MSE = 3971.3379/28 = 141.8335 = S^2$

2. $F^* = 2950.8288/141.8335 = 20.0849$

3. $S^2 = 141.8335$ is an estimate of the variance of y_i, that is, estimate of $\text{Var}(\varepsilon_i) = \sigma^2$. Thus, $\hat{\sigma}^2 = S^2 = 141.8335$. We see here that the population variance σ^2 is estimated by

$$S^2 = \frac{\text{Error } SS}{n-2} = \frac{1}{n-2} \sum_{i=1}^{n} (y_i - \hat{y}_i)^2. \tag{3.1}$$

4. The total sum of squares can be partitioned into the following components:

$$\text{Total } SS = \text{Reg } SS + \text{Error } SS.$$

That is,

$$\sum_{i=1}^{n}(y_i - \bar{y})^2 = \sum_{i=1}^{n}(\hat{y}_i - \bar{y})^2 + \sum_{i=1}^{n}(y_i - \hat{y}_i)^2$$

$$(n - 1) = 1 + (n - 2). \tag{3.2}$$

The results in (3.2) give the actual expressions for the partitioning as well as the corresponding partitioning of the degrees of freedom. We present the results from **proc reg** in SAS for this example. We note that the results are very close to the results from the hand calculation.

```
set Table 1;
proc reg;
model sbp=age;
run;
```

```
                              Analysis of Variance

                                  Sum of          Mean
Source                 DF        Squares        Square    F Value    Pr > F

Model                   1     2950.82880     2950.82880     20.80    <.0001
Error                  28     3971.33787      141.83350
Corrected Total        29     6922.16667

                Root MSE            11.90939    R-Square    0.4263
                Dependent Mean     146.83333    Adj R-Sq    0.4058
                Coeff Var            8.11082


                            Parameter Estimates

                      Parameter     Standard
Variable      DF      Estimate        Error    t Value    Pr > |t|

Intercept      1     100.35683     10.41886       9.63    <.0001
age            1       0.71576      0.15692       4.56    <.0001
```

3.3 Hypotheses Involving Model Parameters

3.3.1 Hypothesis Concerning β_1

We would like to test the hypotheses that

$$H_0 : \beta_1 = 0$$
$$H_a : \beta_1 \neq 0. \tag{3.3}$$

If the null hypothesis were true, then our model becomes $y_i = \beta_0$, that is, a straight line parallel to the x-axis. To test the null hypothesis, we saw from our earlier discussion that $\text{Var}(\hat{\beta}_1) = \dfrac{\sigma^2}{S_{xx}}$. Hence the estimated standard error for $\hat{\beta}_1$ is computed as

$$\text{s.e.}(\hat{\beta}_1) = \sqrt{\frac{S^2}{S_{xx}}} = \sqrt{\frac{141.8335}{5759.8667}} = 0.1569. \tag{3.4}$$

The test statistic for testing the hypotheses in (3.3) is

$$t^* = \frac{\hat{\beta}_1 - \beta_{10}}{\text{s.e.}(\hat{\beta}_1)}, \tag{3.5}$$

which follows a Student's t distribution with $n - 2$ degrees of freedom. In this case, the hypothesized value of $\beta_1 = \beta_{10} = 0$. Hence, the computed statistic for this case becomes

$$t^* = \frac{0.7158 - 0}{0.1569} = 4.5621.$$

The value 4.5621 of the test statistic will be compared to the tabulated Student's t distribution with $n - 2$ degrees of freedom. For $\alpha = 0.05$, we have $\alpha/2 = 0.025$. Hence, we will compare the value with the tabulated $t_{0.025, \, 28}$, 28 being the error d.f. From Table 2 in the Appendix, $t_{0.025, \, 28} = 2.048$. Since our computed value of 4.5621 is much greater than 2.048, we would therefore conclude that $\beta_1 \neq 0$. That is, the explanatory variable x (Age) is important in the model. For the simple linear model (one explanatory variable), the hypotheses in (3.3) can also be conducted from the F^* value in the analysis of variance table. In this case the decision rule would be to reject H_0 if $F^* > f_{0.025, \, 1, \, 28}$, where F is the F distribution with 1 and 28 degrees of freedom. Again, from the Table 4 in the Appendix, $f_{0.025, \, 1, \, 28} = 5.61$. Again since the computed F value $F^* = 20.805 \gg 5.61$, we would reject H_0 and arrive at a similar conclusion. We may note here that $(t^*)^2 = 4.5621^2 = 20.81 = F^*$. That is, $t_{n-2}^2 = F_{1, \, n-2}$. Alternatively, we may simply use the SAS output to test the hypotheses. Here, we would reject H_0 if the computed p-value < 0.05.

3.3.2 Hypothesis Concerning β_0

Similarly, the hypotheses concerning β_0, the intercept of the regression model can be formulated as

$$H_0 : \beta_0 = 0$$
$$H_a : \beta_0 \neq 0. \tag{3.6}$$

If the null hypothesis were true, then our model becomes $y_i = \beta_1 x_i$, that is, a straight line passing through the origin. To test the null hypothesis, we again note from our earlier discussion that $\mathrm{Var}(\hat{\beta}_0) = \dfrac{\sum x_i^2}{n S_{xx}} \sigma^2$. Hence the estimated standard error for $\hat{\beta}_0$ is computed as

$$\mathrm{s.e.}(\hat{\beta}_0) = \sqrt{\frac{\sum x_i^2 \times S^2}{n S_{xx}}} = \sqrt{\frac{132250.0 \times 141.8335}{30 \times 5759.8667}} = 10.4189. \tag{3.7}$$

The test statistic for testing the hypotheses in (3.6) is

$$t^* = \frac{\hat{\beta}_0 - \beta_{00}}{\mathrm{s.e.}(\hat{\beta}_0)}. \tag{3.8}$$

In this example, the hypothesized value of $\beta_0 = \beta_{00} = 0$. Hence, the computed test statistic for this case becomes

$$t^* = \frac{100.3540 - 0}{10.4189} = 9.6320.$$

The decision rule rejects H_0 if computed $t^* > t_{0.025, \, 28} = 2.048$. Again, since $9.6320 \gg 2.048$, we would reject H_0 and conclude that $\beta_0 \neq 0$. That is, the regression line does not pass through the origin. We observe that SAS provides these test statistics for the two parameter estimates as well as the corresponding p-values for making the decisions. We might add here that to make decisions with p-values, we have the following decision rule:

Reject H_0 if p-value $< \alpha$ (with $\alpha = 0.05$) and fail to reject H_0 otherwise.

For both tests, the computed p-values from SAS are < 0.0001 respectively, which both strongly indicate that the null hypotheses H_0 in both cases are untenable. The estimated regression equation therefore is

$$\hat{y}_i = 100.3540 + 0.7158 \, x_i. \tag{3.9}$$

3.4 Confidence Intervals for Regression Parameters

For our random sample of 30 individuals, we obtain, $\hat{\beta}_1 = 0.7158$ and $\hat{\beta}_0 = 100.3540$. It is obvious from an introductory statistics class on sampling distributions that, if we were to take another sample of 30 individuals from the same population, it is almost certain that we would obtain different values for the parameter estimates. Hence, the above parameter estimates are statistics and we must therefore build a level of confidence around them. Usually, we would settle for a 95% confidence interval, but we shall see how other levels of confidence can similarly be obtained.

3.4.1 Confidence Intervals for β_0

A $100(1-\alpha)\%$ confidence interval for β_0 is given by

$$\hat{\beta}_0 \pm t_{\alpha/2,\, n-2} \text{ s.e.}(\hat{\beta}_0), \tag{3.10}$$

where the tabulated $t_{\alpha/2,\, n-2}$ in Table 2 is the $(1-\alpha/2)$ percentile of Student's t distribution with $n-2$ degrees of freedom. In our case therefore, a 95% confidence interval would have $t_{0.025,28} = 2.048$. Hence, a 95% confidence interval for β_0 is computed as

$$100.3540 \pm 2.048\,(10.4189) = 100.3540 \pm 21.3379 = (79.0161, 121.6919).$$

3.4.2 Confidence Intervals for β_1

Similarly, a $100(1-\alpha)\%$ confidence interval for β_1 is given by

$$\hat{\beta}_1 \pm t_{\alpha/2,\, n-2} \text{ s.e.}(\hat{\beta}_1). \tag{3.11}$$

A 95% confidence interval is therefore similarly computed as

$$0.7158 \pm 2.048\,(0.1569) = 0.7158 \pm 0.3213 = (0.3945, 1.0371).$$

A 99% confidence interval ($\alpha = 0.01$) for either of the parameters will be computed similarly, but with $t_{0.005,28} = 2.763$ substituted in the above equations. Both the 95% and 99% confidence intervals can be obtained in SAS by specifying the options **clb alpha = 0.05** or **clb alpha = 0.01**, and **alpha = .05** is the default in SAS. The two outputs below together with the accompanying options in the model statements generate both the 95% and the 99% confidence intervals for the parameters.

```
proc reg;
model sbp=age/xpx i clb;
run;
                          Parameter Estimates

                  Parameter    Standard
Variable   DF     Estimate       Error   t Value  Pr > |t|    95% Confidence Limits

Intercept   1    100.35683     10.41886    9.63    <.0001    79.01475    121.69890
age         1      0.71576      0.15692    4.56    <.0001     0.39432      1.03720

proc reg;
model sbp=age/xpx i clb alpha=.01;
run;
                          Parameter Estimates

                  Parameter    Standard
Variable   DF     Estimate       Error   t Value  Pr > |t|    99% Confidence Limits

Intercept   1    100.35683     10.41886    9.63    <.0001    71.56677    129.14688
age         1      0.71576      0.15692    4.56    <.0001     0.28214      1.14937
```

Observe that none of the confidence intervals enclose zero and hence both parameters β_0 and β_1 differ from zero at the respective confidence levels. From the computer output, we are 95% confident that β_1 lies between 0.3943 and 1.0372.

3.5 Predictions from Estimated Regression Equation

The estimated regression equation is

$$\hat{y}_i = 100.3540 + 0.7158\, x_i.$$

Thus, when $x = 42$, the age of the first subject in our data, we have the predicted or estimated value of $\hat{y}_1 = 100.3540 + (0.7158)(42) = 130.4176$. The observed y_1 value was 130, hence, the difference between the observed and the predicted value, denoted by e_1 is called the residual for the first observation. In this case, $e_1 = 130 - 130.4176 = -0.4176$. In general, the residuals are computed as

$$e_i = y_i - \hat{y}_i. \tag{3.12}$$

3.5.1 Properties of e_i

The following are the properties of the residuals:

1.

$$\sum_{i=1}^{n} e_i = 0$$

Proof:

$$e_i = y_i - \hat{y}_i = y_i - (\hat{\beta}_0 + \hat{\beta}_1 x_i) = y_i - (\bar{y} - \hat{\beta}_1 \bar{x} + \hat{\beta}_1 x_i) = (y_i - \bar{y}) + \hat{\beta}_1 (x_i - \bar{x}).$$

Thus, summing over all the observations, we have

$$\sum_{i=1}^{n} e_i = \sum_{i=1}^{n} (y_i - \bar{y}) + \hat{\beta}_1 \sum_{i=1}^{n} (x_i - \bar{x}) = 0.$$

Hence, $\sum e_i = 0$.

2.

$$S^2 = MSE = \frac{\sum e_i^2}{n - 2}.$$

3.

$$\widehat{Var}(e_i) = \hat{\sigma}^2 = S^2 = MSE.$$

The predicted and residuals can be generated in SAS by specifying the option **P** and **R** in the model option statements. We shall illustrate these later.

3.5.2 Predicting the Mean of y at x_0

In using an estimated regression equation to predict at a given value of x_0, with $a \leq x_0 \leq b$, where a and b are the minimum and maximum values of the explanatory variable x, care must be taken as to whether we wish to predict for an individual value of y (for instance, for an individual having age of x_0) or for all individuals in the population having age x_0.

At $x = x_0$, we have,

$$\hat{y}_0 = \hat{\beta}_0 + \hat{\beta}_1 x_0,$$

with variance given by

$$\text{Var}(\hat{y}_0) = \sigma^2 \left[\frac{1}{n} + \frac{(x_0 - \bar{x})^2}{S_{xx}} \right]. \tag{3.13}$$

Proof:

$$\hat{y}_0 = \bar{y} - \hat{\beta}_1 (x_0 - \bar{x})$$

Hence, variance of $\hat{y}_0$, that is, $\text{Var}(\hat{y}_0)$ equals

$$\text{Var}(\hat{y}_0) = \text{Var}[\bar{Y} - \hat{\beta}_1 (x_0 - \bar{x})] = \text{Var}(\bar{Y}) + (x_0 - \bar{x})^2 \text{Var}(\hat{\beta}_1) - 2\text{Cov}[\bar{Y}, (x_0 - \bar{x})\hat{\beta}_1]$$

$$= \frac{\sigma^2}{n} + (x_0 - \bar{x})^2 \frac{\sigma^2}{S_{xx}} - 0 = \sigma^2 \left[\frac{1}{n} + \frac{(x_0 - \bar{x})^2}{S_{xx}} \right],$$

since $\text{Cov}[\bar{Y}, (x_0 - \bar{x})\hat{\beta}_1] = 0$. An estimated variance of the predicted mean value of y is therefore given by

$$\widehat{\text{Var}}(\hat{y}_0) = S^2 \left[\frac{1}{n} + \frac{(x_0 - \bar{x})^2}{S_{xx}} \right].$$

Hence, the estimated standard error is computed as

$$\text{s.e.}(\hat{y}_0) = \sqrt{S^2 \left[\frac{1}{n} + \frac{(x_0 - \bar{x})^2}{S_{xx}}\right]}. \tag{3.14}$$

A $100(1 - \alpha)\%$ confidence interval for the predicted mean of y at $x = x_0$ is therefore computed as

$$\hat{y}_0 \pm t_{\alpha/2,\, n-2} \,\text{s.e.}(\hat{y}_0) = \hat{y}_0 \pm t_{\alpha/2,\, n-2} \times \sqrt{S^2 \left[\frac{1}{n} + \frac{(x_0 - \bar{x})^2}{S_{xx}}\right]}, \tag{3.15}$$

where $S^2 = MSE$.

3.5.3 Predicting an Individual Value of y at x_0

In this case, the predicted value at $x = x_0$ is still given by,

$$\hat{y}_0 = \hat{\beta}_0 + \hat{\beta}_1 x_0,$$

but with estimated standard error given by

$$\text{s.e.}(\hat{y}_0) = \sqrt{S^2 \left[1 + \frac{1}{n} + \frac{(x_0 - \bar{x})^2}{nS_{xx}}\right]}. \tag{3.16}$$

Proof:

$$\hat{y}_0 = \bar{y} - \hat{\beta}_1(x_0 - \bar{x}) + e_0.$$

Hence, variance of $\hat{y}_0$, that is, $\text{Var}(\hat{y}_0)$ equals

$$
\begin{aligned}
\text{Var}(\hat{y}_0) &= \text{Var}[\bar{Y} - \hat{\beta}_1(x_0 - \bar{x}) + e_0] \\
&= \text{Var}(\bar{Y}) + (x_0 - \bar{x})^2 \text{Var}(\hat{\beta}_1) + \text{Var}(e_0) - 2\text{Cov}[\bar{Y}, (x_0 - \bar{x})\hat{\beta}_1] \\
&\quad + 2\text{Cov}[\bar{Y}, e_0] - 2\text{Cov}[e_0, (x_0 - \bar{x})\hat{\beta}_1] \\
&= \frac{\sigma^2}{n} + (x_0 - \bar{x})^2 \frac{\sigma^2}{S_{xx}} + \sigma^2 - 0 + 0 - 0 = \sigma^2 \left[1 + \frac{1}{n} + \frac{(x_0 - \bar{x})^2}{S_{xx}}\right],
\end{aligned}
$$

since all the covariances are zeros in this case. Hence, an estimated variance of the predicted individual value of y is therefore given by

$$\widehat{\text{Var}}(\hat{y}_0) = S^2 \left[1 + \frac{1}{n} + \frac{(x_0 - \bar{x})^2}{S_{xx}}\right].$$

Hence, the estimated standard error is computed as

$$\text{s.e.}(\hat{y}_0) = \sqrt{S^2 \left[1 + \frac{1}{n} + \frac{(x_0 - \bar{x})^2}{S_{xx}}\right]}.$$

Similarly, a $100(1 - \alpha)\%$ prediction interval for the predicted individual value of y at $x = x_0$ is therefore computed as

$$\hat{y}_0 \pm t_{\alpha/2,\, n-2} \,\text{s.e.}(\hat{y}_0) = \hat{y}_0 \pm t_{\alpha/2,\, n-2} \times \sqrt{S^2 \left[1 + \frac{1}{n} + \frac{(x_0 - \bar{x})^2}{S_{xx}}\right]}. \tag{3.17}$$

This prediction interval is often referred to as the *fiducial confidence interval*.

Example 2.1 continued:

Consider $x = x_0 = 42$ for the data in Table 2.1. We have $\hat{y}_1 = 100.3540 + (0.7158)(42) = 130.4176$. For the mean prediction, the standard error is computed as

$$\text{s.e.}(\hat{y}_0) = \sqrt{S^2\left[\frac{1}{n} + \frac{(x_0 - \bar{x})^2}{S_{xx}}\right]} = \sqrt{141.8335\left[\frac{1}{30} + \frac{(42 - 64.9333)^2}{5759.8667}\right]} = 4.2046,$$

and a 95% confidence interval for the mean predicted value of y at $x_0 = 42$ equals

$$130.4176 \pm 2.048\,(4.2046) = 130.4176 \pm 8.6110 = (121.8066, 139.0286).$$

Similarly, for the data in Table 2.1, the standard error for the individual prediction at $x_0 = 42$ is computed as

$$\text{s.e.}(\hat{y}_0) = \sqrt{S^2\left[1 + \frac{1}{n} + \frac{(x_0 - \bar{x})^2}{S_{xx}}\right]} = \sqrt{141.8335\left[1 + \frac{1}{30} + \frac{(42 - 64.9333)^2}{5759.8667}\right]} = 12.6298,$$

and a 95% confidence interval for the predicted individual value of y at $x_0 = 42$ equals

$$130.4176 \pm 2.048\,(12.6298) = 130.4176 \pm 25.8658 = (104.5518, 156.2834).$$

We present the SAS options to generate **p**=predicted, **r**=residuals, standard error of mean y prediction (**stdp**=stm), the 95% confidence interval for the mean predictions (**l95m**=ml, **u95m**=mu), the standard error for the individual prediction (**stdi**=sti) and the corresponding 95% fiducial confidence intervals with (**l95**=l1, **u95**=l2). A partial SAS output is included. The results for the case when age=42, that is, $x_0 = 42$, the first observation, agree with the calculations we carried out earlier in this section.

```
set tab1;
proc reg data=tab1;
model sbp=age/noprint;
output out=aa p=pred r=resid l95m=ml u95m=mu stdp=stm l95=l1 u95=l2 stdi=sti;
run;
proc print data=aa noobs;
var age sbp pred resid stm ml mu sti l1 l2;
format pred resid stm ml mu sti l1 l2 8.4;
run;
```

age	sbp	pred	resid	stm	ml	mu	sti	l1	l2
42	130	130.4186	-0.4186	4.2046	121.8059	139.0314	12.6298	104.5476	156.2896
46	115	133.2817	-18.2817	3.6817	125.7400	140.8233	12.4655	107.7473	158.8161
42	148	130.4186	17.5814	4.2046	121.8059	139.0314	12.6298	104.5476	156.2896
71	100	151.1756	-51.1756	2.3736	146.3135	156.0377	12.1436	126.3005	176.0507
80	156	157.6174	-1.6174	3.2121	151.0377	164.1971	12.3350	132.3504	182.8844
..	...	....	....	......	....	....	....	....	....
63	150	145.4495	4.5505	2.1954	140.9524	149.9466	12.1101	120.6432	170.2558
82	160	159.0489	0.9511	3.4497	151.9826	166.1152	12.3989	133.6509	184.4470
53	140	138.2920	1.7080	2.8696	132.4139	144.1700	12.2502	113.1985	163.3854
65	140	146.8811	-6.8811	2.1744	142.4271	151.3350	12.1063	122.0825	171.6796
48	130	134.7132	-4.7132	3.4334	127.6801	141.7463	12.3944	109.3243	160.1020

We also present in Figure 3.1 the plots of the 95% confidence intervals for both the means and individual predictions.

3.6   The Coefficient of Determination R^2

The coefficient of determination, R^2, measures the proportion of the total variation in y, the dependent variable, which is accounted for by the regression model. This is computed as

$$R^2 = \frac{\text{Reg } SS}{\text{Total } SS} = \frac{S_{xy}^2}{S_{xx}S_{yy}} = \frac{\sum(\hat{y}_i - \bar{y})^2}{\sum(y_i - \bar{y})^2}.$$

We can re-write R^2 as

$$R^2 = \frac{S_{yy} - SSE}{S_{yy}} = 1 - \frac{SSE}{S_{yy}},$$

where SSE represents the sum of squares for error or simply, the error SS.

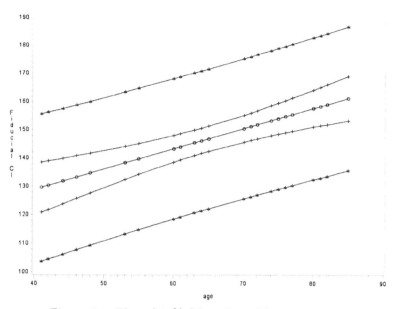

Figure 3.1: Plot of 95% fiducial confidence intervals

Properties of R^2

1. $0 \leq R^2 \leq 1$. A value of 1 is obtained when we have a perfect fit. That is, when all the residuals are zero, which would happen if and only if $\hat{y}_i = y_i$ for all i.

2. R^2 gives the percentage of the total variation in y that is accounted for by our regression model. What would be an acceptable value of R^2 then?

In our data example,

$$R^2 = \frac{2950.8288}{6922.1667} = 0.4263.$$

That is, our regression model accounted for about $0.4263 \times 100 = 42.63\%$ of the total variation in y. In other words, age accounted for about 42.63% of the total variation in systolic blood pressure (SBP). At this point, this value look very small, indicating that our model might not be good enough. We shall examine this again at a later chapter. SAS routinely gives the value of R^2 as simply **R-square** in its output.

3.7 The Coefficient of Correlation ρ

Consider two random variables X and Y having the joint bivariate normal distribution

$$f(x,y) = \frac{1}{2\pi\sigma_1\sigma_2\sqrt{1-\rho^2}}\exp\left\{-\frac{1}{2(1-\rho^2)}\left[\frac{y-\mu_1}{\sigma_1}\right]^2 + \left[\frac{x-\mu_2}{\mu_2}\right]^2 - 2\rho\left[\frac{y-\mu_1}{\mu_1}\right]\left[\frac{x-\mu_2}{\mu_2}\right]\right\}.$$

Here, $\mu_1, \mu_2, \sigma_1, \sigma_2$ relate to the means and standard deviations of x and y respectively. The parameter *rho* (ρ) is the coefficient of correlation between x and y, where

$$\rho = \frac{\sigma_{xy}}{\sigma_y\sigma_x}. \tag{3.18}$$

The parameter ρ can be estimated by the sample correlation coefficient, r, which is defined as

$$r = \frac{S_{xy}}{\sqrt{S_{yy}S_{xx}}}. \tag{3.19}$$

The correlation coefficient r measures the strength of the *linear* association between x and y. Thus, from (3.19), we have

$$r^2 = \frac{S_{xy}^2}{S_{xx}S_{yy}} = \hat{\beta}_1 \frac{S_{xy}}{S_{yy}} = \frac{\text{Reg } SS}{\text{Total } SS} = R^2. \tag{3.20}$$

Thus r^2 equals the coefficient of determination discussed earlier. Since $0 \le r^2 \le 1$, this implies that

$$-1 \le r \le 1.$$

Further, the sign of r is that carried by $\hat{\beta}_1$ in the regression analysis. The correlation coefficient r is dimensionless and if r is very close to zero, then the linear association between x and y can be assumed to be very weak. If the interest centers on the association between x and y, then one might wish to conduct an hypothesis test involving the correlation coefficient ρ. The hypotheses

$$\begin{aligned} H_0 &: \rho = 0 \\ H_a &: \rho \ne 0, \end{aligned} \tag{3.21}$$

can be tested by computing the test statistic

$$t^* = \frac{r\sqrt{n-2}}{\sqrt{1-r^2}}. \tag{3.22}$$

Under the null hypothesis, $H_0 : \rho = 0$, the statistic t^* can be shown to be distributed as Student's t distribution with $(n-2)$ degrees of freedom.

Example 2.1 continued:

For the data in Example 2.1,

$$r = \frac{4122.6667}{\sqrt{5759.8667 \times 6922.1667}} = 0.6529.$$

Alternatively, we could use $r = \pm\sqrt{R^2} = \pm\sqrt{0.4263} = 0.6529$, since $\hat{\beta}_1$ is positive. Hence a test of the hypotheses in (3.21) has the test statistic

$$t^* = \frac{0.6529\sqrt{28}}{\sqrt{1-0.4263}} = 4.5612.$$

The decision rule rejects H_0 when $| t^* | \ge t_{0.025,\ 28} = 2.048$. Since $4.5612 > 2.048$, we would therefore reject the null hypothesis and conclude that $\rho \ne 0$, implying that a significant linear association exists between the variables x and y at the $\alpha = .05$ level of significance. We may observe that the t^* computed above equals the t-value obtained for testing the hypothesis that $\beta_1 = 0$ versus the alternative in (3.3). The hypotheses in (3.21) can similarly be formulated as a one-tailed alternative, with the appropriate change in the decision rule.

For a specified value of ρ under the null hypothesis, we can set up the hypotheses

$$\begin{aligned} H_0 &: \rho = \rho_0 \\ H_a &: \rho \ne \rho_0. \end{aligned} \tag{3.23}$$

We would need to take advantage of Fisher's Z transformation (an approximate normality transformation)

$$z' = \frac{1}{2}\ln\left(\frac{1+r}{1-r}\right), \tag{3.24}$$

The mean and variance of z' are respectively given by

$$E(Z') = \frac{1}{2}\ln\left(\frac{1+\rho_0}{1-\rho_0}\right) \text{ and } V(Z') = \frac{1}{n-3}.$$

Therefore, the corresponding test statistic for the hypothesis in (3.23) is

$$z^* = \frac{\sqrt{n-3}}{2}\left[\ln\left(\frac{1+r}{1-r}\right) - \ln\left(\frac{1+\rho_0}{1-\rho_0}\right)\right]. \tag{3.25}$$

The test statistic is distributed as $N(0, 1)$, a standard normal distribution and the decision rules would be to reject H_0 when

$$\begin{cases} z^* \leq -z_\alpha & \text{for alternative hypothesis } H_a : \rho < \rho_0 \\ z^* \geq z_\alpha & \text{for alternative hypothesis } H_a : \rho > \rho_0 \\ |z^*| \geq z_{\alpha/2} & \text{for alternative hypothesis } H_a : \rho \neq \rho_0, \end{cases} \tag{3.26}$$

where z_α is such that $P(Z > z_\alpha) = \alpha$. Here, for instance, if $\alpha = .05$, then, $z_{0.05} = 1.645$, $-z_{0.05} = -1.645$ and $z_{.025} = 1.96$.

Example 2.1 continued:

For the data in Example 2.1, suppose we wish to test the left-tailed hypothesis at $\alpha = .05$ level,

$$\begin{aligned} H_0 &: \rho \leq 0.60 \\ H_a &: \rho > 0.60. \end{aligned} \tag{3.27}$$

Since $r = 0.6529$ and $\rho_0 = 0.60$, we have

$$z^* = \frac{\sqrt{27}}{2} \left[\ln\left(\frac{1 + 0.6529}{1 - 0.6529}\right) - \ln\left(\frac{1 + 0.60}{1 - 0.60}\right) \right] = 0.4531.$$

From the decision rule in (3.26), we would reject H_0 if $z^* > 1.645$. Since, z^* is not greater than 1.645, we would therefore fail to reject H_0. We do not have sufficient evidence to say that the population correlation coefficient is above 0.60.

Confidence Interval for ρ

A $100(1 - \alpha)\%$ confidence interval for the population correlation coefficient ρ is obtained by first computing

$$z' \pm z_{\alpha/2}\text{se}(z') = \frac{1}{2} \ln\left(\frac{1 + r}{1 - r}\right) \pm \frac{z_{\alpha/2}}{\sqrt{n - 3}}. \tag{3.28}$$

If we let the lower and upper limits in (3.28) be denoted by ω_1 and ω_2 respectively, then, the corresponding lower and upper $100(1 - \alpha)\%$ confidence interval for ρ is obtained as

$$L_1 = \frac{A - 1}{A + 1}, \quad \text{and} \quad L_2 = \frac{B - 1}{B + 1}, \tag{3.29}$$

where $A = e^{\omega_1}$ and $B = e^{\omega_2}$.

Example 2.1 continued:

For the data in Example 2.1, suppose we wish to obtain a 95% confidence interval for ρ, then using (3.28), we have,

$$z' \pm z_{\alpha/2}\text{se}(z') = \frac{1}{2} \ln\left(\frac{1 + 0.6529}{1 - 0.6529}\right) \pm \frac{1.96}{\sqrt{27}} = 0.7803 \pm 0.3772 = (0.4031, 1.1575).$$

Thus,

$$\omega_1 = 0.4031 \Rightarrow A = e^{0.4031} = 1.4965 \quad \text{and} \quad \omega_2 = 1.1575 \Rightarrow B = e^{1.1575} = 3.1820.$$

Hence, the lower and upper 95% confidence intervals are L_1 and L_2, which from (3.29) are computed as

$$L_1 = \frac{1.4965 - 1}{1.4965 + 1} = \frac{0.4965}{2.4965} = 0.1989 \quad \text{and} \quad L_2 = \frac{3.1820 - 1}{3.1820 + 1} = \frac{2.1820}{4.1820} = 0.5218.$$

A 95% confidence interval for ρ is therefore given by $(0.1989, 0.5218)$. That is, $0.1989 \leq \rho \leq 0.5218$.

3.8   Exercises

3.1 For the data in Exercise 2.1, answer the following:

(a) What is the purpose of testing whether or not $\beta_1 = 0$? If we reject $\beta_1 = 0$, does this imply a good fit?

(b) Test the hypothesis that $H_0 : \beta_1 = 0$ versus $H_a : \beta_1 \neq 0$. Interpret $\hat{\beta}_1$.

(c) Obtain r, the sample correlation coefficient and interpret the value.

3.2 For the data in Exercise 2.1, Obtain both the regression SS and the error SS. Give an estimate of σ^2.

3.3 For the data in Exercise 2.1, answer the following questions:

(a) Obtain a 95% confidence interval for β_1, and interpret the interval.

(b) Conduct a test at $\alpha = 0.01$ to determine whether or not the regression line passes through the origin. State the alternative hypothesis, the decision rule and your conclusion.

(c) Obtain a 90% interval estimate of the mean amount of fire damage to individual residences that are 4 miles from the fire station.

(d) Obtain a 90% prediction interval for the amount of fire damage to an individual residence that is 4 miles from the fire station.

(e) What assumption(s) are necessary in other to carry out the above analyses in (a)-(d).

3.4 Twenty employees were selected randomly from the human resource department of a company and the following results were obtained from their satisfaction (y) and productivity (x): $\sum x_i = 130$, $\sum x_i^2 = 870$, $\sum y_i = 135$, $\sum y_i^2 = 951$, $\sum x_i y_i = 906$.

(a) Find the least squares line. What are your assumptions, if any, about satisfaction (y) and productivity (x)?

(b) Test the hypothesis, at a 5% significance level, that productivity does not positively influence a worker's satisfaction.

(c) Find a 95% confidence interval for the slope of the regression equation.

3.5 The following data relate the amount of fire damage in major residential fires to the distance between the residence and the nearest fire station.

x	y	x	y	x	y	x	y	x	y
3.4	26.2	2.3	23.1	0.7	14.1	4.3	31.3	6.1	43.2
1.8	17.8	3.1	27.5	3.0	22.3	2.1	24.0	4.8	36.4
4.6	31.3	5.5	36.0	2.6	19.6	1.1	17.3	3.8	26.1

We wish to fit a model of the form $y_i = \beta_0 + \beta_1 x_i + \varepsilon_i$ to the data. A SAS output for fitting the above model follows.

```
                          The REG Procedure
                          Model: MODEL1
                       Dependent Variable: y

                Number of Observations Read        15
                Number of Observations Used        15

                       Analysis of Variance

                            Sum of         Mean
        Source        DF    Squares       Square     F Value    Pr > F

        Model          1   841.76636    841.76636     156.89    <.0001
        Error         13    69.75098      5.36546
        Corrected Total 14  911.51733

              Root MSE            2.31635    R-Square    0.9235
              Dependent Mean     26.41333    Adj R-Sq    0.9176
```

```
                    Coeff Var               8.76961

                            Parameter Estimates

                    Parameter    Standard
Variable    DF      Estimate      Error    t Value   Pr > |t|    90% Confidence Limits

Intercept    1      10.27793     1.42028     7.24     <.0001     7.76271     12.79315
x            1       4.91933     0.39275    12.53     <.0001     4.22380      5.61486

                            The CORR Procedure

                    2  Variables:    x      y

                            Simple Statistics

Variable        N        Mean       Std Dev        Sum       Minimum       Maximum

x              15      3.28000      1.57625     49.20000     0.70000       6.10000
y              15     26.41333      8.06898    396.20000    14.10000      43.20000

                Pearson Correlation Coefficients, N = 15
                    Prob > |r|  under H0: Rho=0

                                    x            y
                    x            1.00000      0.96098
                                              <.0001

                    y            0.96098      1.00000
                                 <.0001
```

Study this output carefully and answer the following questions:

(a) Does the estimated regression equation provide a good fit to the data? Use the estimated model to predict the damage of two residential areas that are 3.5 miles and 0.5 miles from the nearest fire stations. Do these make sense?

(b) Interpret R^2 and explain how it was calculated?

(c) Test the hypothesis (at $\alpha = 0.05$) that $H_0 : \beta_1 = 0$ against $H_a : \beta_1 \neq 0$.

(d) What is the purpose of testing whether or not $\beta_1 = 0$? If we reject $\beta_1 = 0$, does this imply a good fit?

(e) Obtain r, the sample correlation coefficient and interpret it.

(f) Obtain a 90% confidence interval for β_1 and interpret it.

(g) If $e_i = y_i - \hat{y}_i$, $i = 1, 2, \ldots, 15$, find e_8.

(h) Compute $\sum_{i=1}^{15} e_i$.

3.6 The following summary statistics come from data relating the weight (x) and systolic blood pressure (y) of 26 randomly selected males in the age group 25-30 years.

$$\sum x = 4743, \quad \sum x^2 = 880545, \quad \sum y = 3786, \quad \sum y^2 = 555802, \quad \sum xy = 697076.$$

(a) For a regression of the form $y_i = \beta_0 + \beta_1 x_i + \varepsilon_i$, set up the analysis of variance table.

(b) Estimate the mean systolic blood pressure for a unit gain in weight. Obtain a 95% confidence interval, and interpret this.

(c) Conduct a test to determine whether or not there is linear association between weight and systolic blood pressure at $\alpha = 0.01$. State the alternative, the decision rule and your conclusion.

(d) Obtain a 90% interval estimate of the mean systolic blood pressure for individuals whose weight is 166.

(e) Obtain a 90% prediction intervals for the systolic blood pressure for an individual whose weight is 166.

(f) What assumption(s) are necessary in order to carry out the above analyses in (a)-(e).

3.7 A simple linear regression analysis for $n = 20$ data points produced the following results:

$$S_{xx} = 4.77, \quad S_{xy} = 16.22, \quad S_{yy} = 59.21, \quad \bar{x} = 2.50, \quad \bar{y} = 10.60.$$

(a) Set up the analysis of variance table.

(b) From (a), give both the regression sum of squares and the error sum of squares. Give an estimate of σ^2.

(c) Obtain R^2 for the above regression.

3.8 The following data relate to placement scores (x) and statistics grades (y) for fifteen freshmen in a college.

Student	x	y	Student	x	y	Student	x	y
1	21	69	6	19	80	11	16	80
2	17	72	7	15	65	12	25	93
3	21	94	8	23	88	13	8	55
4	11	61	9	13	54	14	14	60
5	15	62	10	19	75	15	17	64

We wish to fit a model of the form $y_i = \beta_0 + \beta_1 x_i + \varepsilon_i$ to the data. The result of fitting this model to the above data is produced in the following SAS output.

```
                    The REG Procedure
                     Model: MODEL1
                 Dependent Variable: y

         Number of Observations Read        15
         Number of Observations Used        15

                  Analysis of Variance

                            Sum of          Mean
   Source           DF      Squares        Square    F Value   Pr > F

   Model             1    1754.56903    1754.56903     34.60   <.0001
   Error            13     659.16430      50.70495
   Corrected Total  14    2413.73333

            Root MSE              7.12074    R-Square    0.7269
            Dependent Mean       71.46667    Adj R-Sq    0.7059
            Coeff Var             9.96372

                  Parameter Estimates

                    Parameter     Standard
   Variable    DF    Estimate        Error    t Value    Pr > |t|

   Intercept    1    29.88222      7.30439       4.09      0.0013
   x            1     2.45577      0.41747       5.88      <.0001

                  The CORR Procedure

             2  Variables:    x      y

                  Simple Statistics

Variable     N       Mean      Std Dev        Sum     Minimum     Maximum

x           15   16.93333      4.55861  254.00000     8.00000    25.00000
y           15   71.46667     13.13048       1072    54.00000    94.00000

            Pearson Correlation Coefficients, N = 15
                Prob > |r| under H0: Rho=0

                         x             y
             x     1.00000       0.85259
                                 <.0001

             y     0.85259       1.00000
                   <.0001
```

(a) Write down the estimated least squares regression equation for the data. Use this to predict the statistics grades of two students who have scores of 18 and 26 respectively.

(b) Test the hypothesis that $H_0 : \beta_1 = 0$ versus $H_a : \beta_1 \neq 0$. Interpret $\hat{\beta}_1$.

(c) Obtain r, the sample correlation coefficient and interpret the value.

3.9 Refer to the SAS output in Exercise 3.8, show (*no calculations*) how the error sum of squares of 659.1643 on 13 degrees of freedom was obtained.

3.10 The marketing department of a company wanted to estimate the relation between x, the number of showings of a TV commercial in a sales territory and y, the territory sales of one of their products. Data were obtained from ten territories and the following information were given: $\sum x_i = 20$, $\sum x_i^2 = 60$, $\sum y_i = 21.82$, $\sum y_i^2 = 56.617$, and $r^2 = 0.905381$,
where r is the correlation coefficient between x and y. Assume a regression model $y_i = \beta_0 + \beta_1 x_i + \varepsilon_i$.

(a) Set up the analysis of variance table for this problem.

(b) Obtain the estimated regression function by taking r to be positive.

3.11 A supermarket chain conducted an experiment to investigate the effect of price (x) on weekly demand y (in pounds) for a house brand of coffee. Eight supermarket stores that had nearly equal past records of demand for the product were used in the experiment. Eight prices in dollars (x) were randomly assigned to the stores and were advertised using the same procedures. The number of pounds of coffee sold or demanded (y) during the following week was recorded for each of the stores and is shown in the following table:

y	1120	999	932	884	807	760	701	688
x	3.00	3.10	3.20	3.30	3.40	3.50	3.60	3.70

Assume the regression model $y_i = \beta_0 + \beta_1 x_i + \varepsilon_i$.

(a) Obtain the estimated regression function.

(b) Obtain a point estimate of the change in the mean sales when the price increases by (i) $1.00 and (ii) $2.50.

(c) Obtain a 95% confidence interval for the change in the mean response when the price increases by $1.00.

(d) For which price does the sales deviate the least from the least squares line in absolute value?

(e) Which parameter of the regression model is estimated by *MSE*? In what units is *MSE* expressed?

(f) Write down R^2. Interpret this measure in the context of this regression application.

(g) Calculate r from (f). What sign did you attach to r and why? What does this sign signify?

(h) Suppose that a supermarket that had been selling coffee for $3.70/pound is considering a raise in price to $3.80/pound. The supermarket decided that the price increase will be desirable as long as average sales will not drop below 550 pounds/week. Compute a point estimate for the estimated average sales in subsequent weeks if the price is raised to $3.80/pound. Can the supermarket be 95% certain that sales will be at least 550 pounds? (Assume that a linear regression model is correct)

(i) What is the interpretation of the intercept for this analysis?

(j) Test the overall regression via an F-test.

3.12 A chemist is interested in determining the weight loss y of a particular compound as a function of the amount of time (x) the compound is exposed to the air. The simple linear regression model is considered for a sample of size $n = 12$. The summary data are:
$\sum x_i = 66.00$, $\sum x_i^2 = 378.00$, $\sum y_i = 66.10$, $\sum y_i^2 = 396.57$, $\sum x_i y_i = 383.30$.

(a) Obtain a point estimate of the change in the mean response per unit increase in x.

(b) Obtain a point estimate of the mean response when $x = 5.0$.

(c) Set up the analysis of variance (ANOVA) table.

(d) Obtain a point estimate of σ, the standard deviation of the error terms.

(e) Calculate R^2, the coefficient of simple determination. Interpret your result.

(f) Obtain r, the coefficient of simple correlation from (e). What sign did you attach to r and why? In what units is r expressed?

3.13 A researcher wished to estimate the relation between y, the cost of the most recent reconditioning (in $'000) of a large incinerator and x, the number of operating hours (in thousands) since the preceding reconditioning. Data were obtained from 10 large incinerators and the following summary statistics were obtained:
$\sum x_i = 24$, $\sum x_i^2 = 59.98$, $\sum y_i = 52$, $\sum y_i^2 = 278.18$, and $r^2 = 0.88147$,
where r is the correlation coefficient between x and y. Assume a regression model $y_i = \beta_0 + \beta_1 x_i + \varepsilon_i$.

(a) Set up the analysis of variance table for this problem.

(b) Obtain the estimated regression function. (Take r to be positive).

3.14 Using the fact that
$$Z = \frac{\sqrt{n-3}}{2} \left[\ln\left(\frac{1+r}{1-r}\right) - \ln\left(\frac{1+\rho}{1-\rho}\right) \right]$$

has a standard normal distribution, show that the confidence limits for the correlation coefficient ρ is given by $\left(\frac{A-1}{A+1}, \frac{B-1}{B+1}\right)$, where $A = e^{\omega_1}$, $B = e^{\omega_2}$, and ω_1 and ω_2 are respectively the lower and upper limits in equation (3.28).

3.15 Use the data in Exercise 3.5.

(a) Test at $\alpha = 0.05$ if there is a significant relationship between the amount of fire damage and distance from the fire station.

(b) Obtain the 90% confidence interval for the population correlation coefficient ρ. Interpret the interval.

Chapter 4

Multiple Linear Regression

4.1 Introduction

A multiple regression (MR) analysis has a regression equation with at least two explanatory variables. The simplest multiple regression equation has two explanatory variables viz,

$$y_i = \beta_0 + \beta_1 x_{i1} + \beta_2 x_{i2} + \varepsilon_i, \quad i = 1, 2, \ldots, n. \tag{4.1}$$

If we decide to drop the i subscript on the explanatory variables, then we have

$$y_i = \beta_0 + \beta_1 x_1 + \beta_2 x_2 + \varepsilon_i, \quad i = 1, 2, \ldots, n. \tag{4.2}$$

A multiple regression model with $p - 1$ explanatory variables has the form

$$
\begin{aligned}
y_i &= \beta_0 + \beta_1 x_{i1} + \beta_2 x_{i2} + \cdots + \beta_{p-1} x_{i,p-1} + \varepsilon_i, \quad i = 1, 2, \ldots, n \\
&= \beta_0 + \sum_{j=1}^{p-1} \beta_j x_{ij} + \varepsilon_i, \quad i = 1, 2, \ldots, n,
\end{aligned}
\tag{4.3}
$$

where as in the previous chapter, the following assumptions about the model in (4.3) are also applicable.

1. The ε_i are random error terms.

2. The ε_i are identically and independently distributed as normal with mean 0 and constant variance σ^2. Thus we assume that the Y_i, the dependent variables, are statistically independent of one another and

$$Y \sim N(\mu_Y, \sigma^2) \quad \text{and} \quad \varepsilon \sim N(0, \sigma^2).$$

3. The explanatory variables $x_1, x_2, \ldots, x_{p-1}$ are assumed fixed.

4. The mean of Y is a linear function of the explanatory variables (either for all $p - 1$ variables or a subset of the $p - 1$ variables) $x_1, x_2, \ldots, x_{p-1}$.

5. The variance of Y as indicated in (2) above is the same for any fixed combination of the explanatory variables $x_1, x_2, \cdots, x_{p-1}$. That is,

$$\sigma_Y^2 = \text{Var}(Y | x_1, x_2, \cdots, x_{p-1}) = \sigma^2. \tag{4.4}$$

4.2 Parameter Estimation in MR Models

The model in (4.3) is a general multiple regression equation with $p - 1$ explanatory variables. The $\beta_0, \beta_1, \beta_2, \ldots, \beta_{p-1}$ are the *regression coefficients* or *parameters* that would have to be estimated from available data. The explanatory or independent variables $x_1, x_2, \ldots, x_{p-1}$, may well be basic variables or some may be functions of the basic variables such as $x_3 = x_1^2$ or $x_6 = x_1 * x_4$, etc. To estimate the parameters of the model, we would again use the principle of Ordinary Least Squares, by minimizing Q, with respect to $\beta_0, \beta_1, \ldots, \beta_{p-1}$. That is, we minimize

$$Q = \sum_{i=1}^{n} \varepsilon_i^2 = \sum_{i=1}^{n} (y_i - \beta_0 - \beta_1 x_{i1} - \beta_2 x_{i2} - \cdots - \beta_{p-1} x_{i,p-1})^2, \tag{4.5}$$

with respect to $\beta_0, \beta_1, \ldots, \beta_{p-1}$. This will be accomplished by taking the partial derivatives of Q with respect to $\beta_0, \beta_1, \ldots, \beta_{p-1}$ and setting them to zero. That is, set the following partial derivatives $\dfrac{\partial Q}{\partial \beta_0}, \dfrac{\partial Q}{\partial \beta_1}, \dfrac{\partial Q}{\partial \beta_2}, \cdots, \dfrac{\partial Q}{\partial \beta_{p-1}}$ to zero. For the case when $p - 1 = 2$, these reduce to the following normal equations (we have assumed here that n set of observations are available on y, x_1 and x_2).

$$\left. \begin{array}{l} \sum y = n\hat{\beta}_0 + \hat{\beta}_1 \sum x_1 + \hat{\beta}_2 \sum x_2 \\[2mm] \sum x_1 y = \hat{\beta}_0 \sum x_1 + \hat{\beta}_1 \sum x_1^2 + \hat{\beta}_2 \sum x_1 x_2 \\[2mm] \sum x_2 y = \hat{\beta}_0 \sum x_2 + \hat{\beta}_1 \sum x_1 x_2 + \hat{\beta}_2 \sum x_2^2. \end{array} \right\} \tag{4.6}$$

Solving the set of equations in (4.6), we have

$$\hat{\beta}_1 = \frac{S_{x_1 y} S_{x_2} - S_{x_2 y} S_{x_1 x_2}}{D}$$

$$\hat{\beta}_2 = \frac{S_{x_2 y} S_{x_1} - S_{x_1 y} S_{x_1 x_2}}{D}$$

$$\hat{\beta}_0 = \bar{y} - \hat{\beta}_1 \bar{x}_1 - \hat{\beta}_2 \bar{x}_2,$$

where

$$S_{x_1} = \sum x_1^2 - \frac{(\sum x_1)^2}{n}, \qquad S_{x_2} = \sum x_2^2 - \frac{(\sum x_2)^2}{n}$$

$$S_y = \sum y^2 - \frac{(\sum y)^2}{n}, \qquad S_{x_1 y} = \sum x_1 y - \frac{\sum x_1 \sum y}{n}$$

$$S_{x_2 y} = \sum x_2 y - \frac{\sum x_2 \sum y}{n}, \quad S_{x_1 x_2} = \sum x_1 x_2 - \frac{\sum x_1 \sum x_2}{n}$$

$$D = S_{x_1} S_{x_2} - (S_{x_1 x_2})^2.$$

The estimated Regression equation is given by

$$\hat{y}_i = \hat{\beta}_0 + \hat{\beta}_1 x_{1i} + \hat{\beta}_2 x_{2i}. \tag{4.7}$$

4.2.1   Properties of Least Squares Estimates

1. Each of the parameter estimates $\hat{\beta}_0, \hat{\beta}_1, \hat{\beta}_2, \ldots, \hat{\beta}_{p-1}$ are linear functions of the observed y values.

2. Gauss-Markov theorem states that the least squares estimators are the best minimum-variance unbiased estimators.

3. The multiple correlation coefficient can be computed as

$$r_{y,\hat{y}} = \frac{\sum (y_i - \bar{y})(\hat{y}_i - \bar{\hat{y}})}{\sqrt{\sum (y_i - \bar{y})^2 \sum (\hat{y}_i - \bar{\hat{y}})^2}}.$$

4.2.2 Partitioning the Total SS

As in the simple linear regression, the total sum of squares (SST), $\sum(y_i - \bar{y})^2$, can be partitioned into the regression sum of squares (SSR) and the error sum of squares (SSE) as follows:

$$\sum_{i=1}^{n}(y_i - \bar{y})^2 = \sum_{i=1}^{n}(\hat{y}_i - \bar{y})^2 + \sum_{i=1}^{n}(y_i - \hat{y}_i)^2, \text{ which is}$$
$$SST = SSR + SSE.$$

The degrees of freedom for the SST, SSR and SSE are respectively $n-1, p-1$ and $n-p$. We have $p-1$ explanatory variables in the model and hence p parameters, including the intercept β_0.

The analysis of variance table therefore takes the form

Source	df	SS	MS	F
Regression	$p-1$	SSR	$MSR = \frac{SSR}{p-1}$	F^*
Error	$n-p$	SSE	$MSE = \frac{SSE}{n-p}$	
Total	$n-1$	SST		

In the above table, F^* is computed as the ratio of the regression means square and the error mean square. We may also note here that $SSE = SST - SSR$ from above. The regression sum of squares is obtained for instance for the case when $k = 2$ as

$$SSR = \hat{\beta}_1 S_{x_1 y} + \hat{\beta}_2 S_{x_2 y}.$$

4.2.3 The Coefficient of Multiple Determination

The *coefficient of multiple determination*, R^2, again is computed as

$$R^2 = \frac{SSR}{SST} = 1 - \frac{SSE}{SST}, \quad 0 \le R^2 \le 1. \tag{4.8}$$

As for the simple linear regression model, the coefficient of determination tells us the percentage of the total variation in y, the dependent variable, that is explained or accounted for by the explanatory variable (Regression model). It is a sample statistic and gives an idea of how well our model fits the data and thus represent a measure of utility of the model.

The adjusted R^2, denoted here as R_a^2 is defined as:

$$R_a^2 = 1 - \frac{MSE}{MST} = 1 - \frac{n-1}{n-p}\left(\frac{SSE}{SST}\right) = 1 - \frac{n-1}{n-p}\left(1 - R^2\right),$$

where MST is the total mean square. R_a^2 takes into account the number of β parameters in the model and the sample size n on which the model is based. Note $R_a^2 < R^2$ and further, it penalizes the model if spurious variables are added to the model unlike the R^2 which in any case would increase as more and more variables are added to the model.

Example 4.1

The data in Table 4.1 was collected by a wildlife biologist who was interested in a certain species of mammals. The data is on 25 offspring of that species under the age of 12 months. The variables observed are y = weight of the offspring (in pounds), x_1 = age of the offspring (in months), and x_2 = length of the offspring (in inches). The data are displayed below.

As shown in Chapter 1, we can use SAS to obtain both the corrected and uncorrected sum of squares and cross-products to facilitate obtaining the summary statistics for the data. These are displayed in the following SAS output.

Obs	y	x_1	x_2	Obs	y	x_1	x_2	Obs	y	x_1	x_2
1	16.4	10	23.1	10	16.3	11	23.5	19	16.4	10	23.1
2	16.3	7	22.8	11	15.0	8	22.5	20	14.1	3	21.0
3	18.2	12	24.0	12	16.8	9	22.6	21	17.9	10	23.1
4	14.8	6	22.1	13	16.0	8	23.2	22	17.7	11	22.8
5	14.7	5	21.5	14	18.4	11	23.4	23	16.3	10	23.5
6	16.4	11	23.2	15	13.5	5	21.2	24	16.6	8	22.3
7	17.3	10	22.7	16	12.4	2	20.6	25	15.2	6	22.5
8	13.0	1	20.2	17	15.6	9	22.2				
9	18.8	12	24.2	18	16.6	4	23.4				

Table 4.1: Data on certain species of mammals

```
                          SSCP Matrix

                    y                  x1                  x2
      y       6487.89000          3297.80000          9086.77000
      x1      3297.80000          1827.00000          4561.20000
      x2      9086.77000          4561.20000         12780.43000

                          CSSCP Matrix

                    y                  x1                  x2
      y         65.4704000         108.2280000          35.7584000
      x1       108.2280000         242.9600000          66.1880000
      x2        35.7584000          66.1880000          24.9864000

                       Simple Statistics

Variable      N        Mean     Std Dev        Sum     Minimum     Maximum

y            25     16.02800     1.65165    400.70000    12.40000    18.80000
x1           25      7.96000     3.18172    199.00000     1.00000    12.00000
x2           25     22.58800     1.02034    564.70000    20.20000    24.20000
```

From the analysis we have

$$S_{yy} = 65.4704, \qquad S_{x_1 y} = 108.2280, \quad S_{x_2 y} = 35.7584,$$
$$S_{x_1} = 242.9600, \quad S_{x_1 x_2} = 66.1888, \qquad S_{x_2} = 24.9864,$$
$$\bar{y} = 16.0280, \qquad \bar{x}_1 = 7.9600, \qquad \bar{x}_2 = 22.5880.$$

Hence, $D = S_{x_1} S_{x_2} - (S_{x_1 x_2})^2 = 242.96(24.9864) - 66.1888^2 = 1689.7385$. Also,

$$\hat{\beta}_1 = \frac{S_{x_1 y} S_{x_2} - S_{x_2 y} S_{x_1 x_2}}{D} \quad = \frac{337.4225}{1689.7385} = 0.1997$$

$$\hat{\beta}_2 = \frac{S_{x_2 y} S_{x_1} - S_{x_1 y} S_{x_1 x_2}}{D} \quad = \frac{1524.3794}{1689.7385} = 0.9021.$$

Therefore,
$$\hat{\beta}_0 = \bar{y} - \hat{\beta}_1 \bar{x}_1 - \hat{\beta}_2 \bar{x}_2 = 16.0280 - 0.1997(7.9600) - 0.9021(22.5880) = -5.9382.$$

The estimated regression equation is therefore

$$\hat{y}_i = -5.9382 + 0.1997 x_1 + 0.9021 x_2.$$

The SST is $S_{yy} = 65.4704$ on $(25 - 1) = 24$ degrees of freedom. The SSR is computed as

$$SSR = \hat{\beta}_1 S_{x_1 y} + \hat{\beta}_2 S_{x_2 y} = (0.1997)(108.2280) + (0.9021)(35.7584) = 53.8708.$$

The SSR is based on $(3 - 1) = 2$ degrees of freedom (number of parameters being estimated minus one). The SSE is then obtained by subtraction. That is, SSE equals SST minus $SSR = (65.4704 - 53.8708) = 11.5996$, and is based on $(24 - 2) = 22$ degrees of freedom. Hence the analysis of variance table is displayed as

Source	df	SS	MS	F
Regression	2	53.8708	26.9354	51.082
Error	22	11,5996	0.5273	
Total	24	65.4704		

Are the explanatory variables x_1 and x_2 important in the model? The hypothesis to test this is often referred to as the *global* or *omnibus* test, viz:

$$H_0 : \beta_1 = \beta_2 = 0$$
$$H_a : \text{at least one of these } \beta \text{ parameters is not zero.}$$

The above hypotheses are tested with the computed F^* value, where decision rules are

$$\text{Reject } H_0 \text{ if } F^* \geq F_{.025, 2, 22} = 4.38.$$

Since $F^* = 51.082 \ggg 4.38$, we would therefore strongly reject H_0. That is, at least one of β_1 and β_2 is not zero. We may note here that this decision does not preclude the fact that one of the β's might be zero. What it says is that both the β parameters can not be zero simultaneously. Based on our result therefore, we do know that at least one of the explanatory variables is important for our model. The coefficient of determination R^2 is computed as

$$R^2 - \frac{\text{Reg } SS}{\text{Total } SS} = \frac{53.8708}{65.4708} = 0.8228.$$

Similarly, the adjusted R^2 is computed as

$$R_a^2 = 1 - \frac{MSE}{\text{Total } MS} = 1 - \frac{0.5273}{65.4704/24} = 1 - 0.1933 = 0.8067.$$

The following SAS program and the corresponding partial output give the implementation of the model for the data in Table 4.1.

```
data ex31;
input y x1 x2 @@;
datalines;
16.4 10 23.1 16.3 7 22.8 18.2 12 24.0
..................................
16.3 10 23.5 16.6 8 22.3 15.2 6 2.5
;
proc reg;
model y=x1 x2/ clb;
run;
```

```
                        Analysis of Variance

                               Sum of        Mean
    Source            DF       Squares      Square    F Value    Pr > F

    Model              2       53.87131    26.93565     51.09    <.0001
    Error             22       11.59909     0.52723
    Corrected Total   24       65.47040


            Root MSE            0.72611    R-Square     0.8228
            Dependent Mean     16.02800    Adj R-Sq     0.8067
            Coeff Var           4.53024


                        Parameter Estimates

               Parameter    Standard
Variable   DF    Estimate      Error   t Value   Pr > |t|   95% Confidence Limits

Intercept   1    -5.93896    5.63608     -1.05     0.3034    -17.62747    5.74954
x1          1     0.19969    0.08829      2.26     0.0339      0.01658    0.38280
x2          1     0.90213    0.27532      3.28     0.0034      0.33115    1.47312
```

The results preceding the SAS output are very close to those obtained from SAS which are computed to a better accuracy. The global hypotheses could be tested from the given *p*-value $< .0001$. That is, we would strongly reject H_0.

4.3   Hypotheses Involving Model Parameters

Hypothesis concerning β_0, β_1 and β_2

As in the previous chapter, we would also like to test the hypotheses that

$$H_0 : \beta_j = 0$$
$$H_a : \beta_j \neq 0, \tag{4.9}$$

for $j = 0, 1, 2$. From SAS, the estimated standard errors for $\hat{\beta}_0, \hat{\beta}_1$ and $\hat{\beta}_2$ are given as $5.6361, 0.0883$, and 0.2753 respectively. The test statistic for testing the hypotheses in (4.9) is

$$t = \frac{\hat{\beta}_j - 0}{\text{s.e.}(\hat{\beta}_j)}. \tag{4.10}$$

Thus, to test the hypotheses that $H_0 : \beta_2 = 0$ versus $H_a : \beta_2 \neq 0$, we compute

$$t = \frac{0.9021}{0.2753} = 3.28.$$

We would then compare this at $\alpha = 0.05$ level of significance with a Student's t value of $t_{.025, 22} = 2.074$. Since $3.28 > 2.074$, we would therefore reject H_0 and conclude that β_2 is not zero and therefore the explanatory variable x_2 is important in our model given that x_1 is already in the model. The SAS output gives us both the computed t statistics for all the parameters (under Parameter Estimates) as well as the corresponding p-values. For testing the hypotheses for $j = 0, 1, 2$ in SAS, the corresponding p-values are $0.3034, 0.0339$, and 0.0034 respectively. The p-values indicate that the null hypothesis would be rejected for $j = 1, 2$ but we would fail to reject H_0 for $j = 0$. But of course, we would like to keep the intercept in the model and the result for $j = 0$ for all intents and purposes is often ignored in multiple regression analysis.

4.4   Other Hypotheses Testing in Multiple Regression

Example 4.2

In this example, we have three explanatory variables x_1, x_2 and x_3. Let us proceed to fit the following three models in turn to the data in Table 4.2.

$$y_i = \beta_0 + \beta_1 x_1 + \varepsilon_i \tag{4.11a}$$
$$y_i = \beta_0 + \beta_1 x_1 + \beta_2 x_2 + \varepsilon_i \tag{4.11b}$$
$$y_i = \beta_0 + \beta_1 x_1 + \beta_2 x_2 + \beta_3 x_3 + \varepsilon_i. \tag{4.11c}$$

Obs	y	x_1	x_2	x_3	Obs	y	x_1	x_2	x_3	Obs	y	x_1	x_2	x_3
1	33.2	3.5	9	6.1	9	30.1	3.1	5	5.8	17	34.2	6.2	7	5.5
2	40.3	5.3	20	6.4	10	52.9	7.2	47	8.3	18	48.0	7.0	40	7.0
3	38.7	5.1	18	7.4	11	38.2	4.5	25	5.0	19	38.0	4.0	35	6.0
4	46.8	5.8	33	6.7	12	31.8	4.9	11	6.4	20	35.9	4.5	23	3.5
5	41.4	4.2	31	7.5	13	43.3	8.0	23	7.6	21	40.4	5.9	33	4.9
6	37.5	6.0	13	5.9	14	44.1	6.5	35	7.0	22	36.8	5.6	27	4.3
7	39.0	6.8	25	6.0	15	42.8	6.6	39	5.0	23	45.2	4.8	34	8.0
8	40.7	5.5	30	4.0	16	33.6	3.7	21	4.4	24	35.1	3.9	15	5.0

Table 4.2: Data for example with three explanatory variables

In the three partial SAS outputs presented below, for models (4.11a), (4.11b) and (4.11c), we see that the total sum of squares remains the same, being 689.2600 on 23 degrees of freedom.

Model (a)
```
                    Analysis of Variance

                        Sum of          Mean
Source          DF      Squares         Square    F Value    Pr > F
-------------------------------------------------------------------------
Model            1      306.73233       306.73233   17.64    0.0004
Error           22      382.52767        17.38762
Corrected Total 23      689.26000
-------------------------------------------------------------------------
```

Model (b)
```
                    Analysis of Variance

                        Sum of          Mean
Source          DF      Squares         Square    F Value    Pr > F
-------------------------------------------------------------------------
Model            2      570.52677       285.26339   50.45    <.0001
Error           21      118.73323         5.65396
Corrected Total 23      689.26000
-------------------------------------------------------------------------
```

Model (c)
```
                    Analysis of Variance

                        Sum of          Mean
Source          DF      Squares         Square    F Value    Pr > F
-------------------------------------------------------------------------
Model            3      627.81700       209.27233   68.12    <.0001
Error           20       61.44300         3.07215
Corrected Total  23     689.26000
-------------------------------------------------------------------------
```

However, the following can be summarized:

(a) For the model in (4.11a), we have

$$SS(x_1|\beta_0) = 306.7323, \quad SSE = 382.5277,$$

where $SS(x_1|\beta_0)$ is the regression SS when variable x_1 entered the model given that β_0 is already in the model. For brevity, we would simply write this as $SS(x_1)$, with the implicit assumption that β_0 is already in the model. In this model, both $SS(x_1)$ and SSE are based on 1 and 22 degrees of freedom respectively.

(b) For model (4.11b), we also have

$$SS(x_1, x_2) = 570.5268, \quad SSE = 118.7332.$$

Here, $SS(x_1, x_2)$ is the regression SS when both variables x_1 and x_2 are in the model. Here, both $SS(x_1, x_2)$ and SSE are based on 2 and 21 degrees of freedom respectively. The extra SS due to x_2 therefore, given that x_1 is already in the model, written as $SS(x_2|x_1)$ is

$$SS(x_2|x_1) = SS(x_1, x_2) - SS(x_1) = 570.5268 - 306.7323 = 263.7945. \tag{4.12}$$

The extra sum of squares $SS(x_2|x_1)$ is accordingly based on $(2-1) = 1$ degree of freedom.

(c) Similarly from the model in (4.11c), we have

$$SS(x_1, x_2, x_3) = 627.8170, \quad SSE = 61.44300,$$

where as before $SS(x_1, x_2, x_3)$ is the regression SS when all the three variables x_1, x_2 and x_3 are in the model. Both the regression sum of squares and error sum of squares are based, in this case, on 3 and 20 degrees of freedom respectively. The extra SS due to x_3, given that both x_1 and x_2 are already in the model, written as $SS(x_3|x_1, x_2)$, is

$$SS(x_3|x_1, x_2) = SS(x_1, x_2, x_3) - SS(x_1, x_2) = 627.8170 - 570.5268 = 57.2902. \tag{4.13}$$

This extra sum of squares, $SS(x_3|x_1, x_2)$ is based on $(3-2) = 1$ degree of freedom.

It is important to emphasize at this point that the extra sum of squares obtained above depends on the order in which the variables entered the regression equation. To see this, we list below the $3! = 6$ ways by which the three variables can be arranged or entered in our regression analysis. These are (i) x_1, x_2, x_3, (ii) x_1, x_3, x_2; (iii) x_2, x_1, x_3; (iv) x_2, x_3, x_1; (v) x_3, x_1, x_2; and (vi) x_3, x_2, x_1. The extra SS obtained for each of the above possible order of the variables are displayed in Table 4.3.

Order	Variables In	$SS(x_i)$	$SS(x_j\|x_i)$	$SS(x_k\|x_i, x_j)$
(i)	x_1, x_2, x_3	306.7323	263.7945	57.2902
(ii)	x_1, x_3, x_2	306.7323	90.4592	230.6255
(iii)	x_2, x_1, x_3	508.0688	62.4579	57.2902
(iv)	x_2, x_3, x_1	508.0688	85.3297	34.4185
(v)	x_3, x_1, x_2	214.7616	182.4300	230.6255
(vi)	x_3, x_2, x_1	214.7616	378.6369	34.4185

Table 4.3: Extra SS for each of the variables as they enter the model

In Table 4.3, order (i) represents $i = 1, j = 2$ and $k = 3$, while order (v) represents $i = 3, j = 1$ and $k = 2$. We immediately notice that the order of entry of the variables into the model considerably affects the extra sum of squares accrued or contributed by that variable. Take the variable x_1 for instance. When it entered the model first as in (i) and (ii), its contribution to the regression SS is 306.7323. However, when it entered as a second variable as in (iii) and (v), its contributions to the regression sum of squares are respectively, 62.4579 and 182.4300. These correspond respectively to $SS(x_1|x_2)$ and $SS(x_1|x_3)$. Both are not as high as in (i) and (ii) and secondly, its contributions depend on which of the other variables entered first (in this case x_2 or x_3).

On the other hand, when x_1 entered the model last as in (iv) and (vi), its contributions to the overall regression SS is 34.4185, corresponding to $SS(x_1|x_2, x_3)$ and $SS(x_1|x_3, x_2)$ respectively. Because the order of entry of a variable into a regression model is important, the extra SS obtained this way is characterized as a sequential SS or the Type I SS. Thus, Type I SS are sequential. In other words, the variable that entered first is most favored, followed by the next, and so on. In order to put the contributions of all the variables to the regression SS on a level playing field, we often obtain the Type II SS, which is the sum of squares obtained when the variable in question is assumed to have entered the regression model last. In our example, these would be $SS(x_1|x_2, x_3)$, $SS(x_2|x_1, x_3)$ and $SS(x_3|x_1, x_2)$, which are respectively 34.4185, 230.6255 and 57.2902 in this example. We present in Table 4.4 these sum of squares (SS).

Types I and II Sum of Squares (SS)

Consider the model

$$y_i = \beta_0 + \beta_1 x_1 + \beta_2 x_2 + \beta_3 x_3 + \varepsilon_i.$$

Then, the form of **Type I** and **Type II** sum of squares are presented in Table 4.4.

Effect	Type I SS (sequential)	Type II SS (partial)
x_1	$SS(x_1)$	$SS(x_1 \mid x_2, x_3)$
x_2	$SS(x_2 \mid x_1)$	$SS(x_2 \mid x_1, x_3)$
x_3	$SS(x_3 \mid x_1, x_2)$	$SS(x_3 \mid x_1, x_2)$

Table 4.4: Types I and II SS implicit that β_0 is also in the models

The Type I SS for the intercept β_0 is simply $(\sum y_i)^2/n$, or the correction factor. Note that in all cases, the sum of squares for regression (SSR) under the Type I SS partitions into the previous sum of squares. For instance, consider the case when the order of entry is (i) in Table 4.3, that is, x_1, x_2, x_3, then

$$SS(x_1) = 306.7323$$
$$SS(x_1, x_2) = 570.5268$$
$$SS(x_1, x_2, x_3) = 627.8170.$$

Thus,

$$SS(x_1, x_2) = SS(x_1) + SS(x_2|x_1) = 306.7323 + 263.7945 = 570.5268,$$

and,

$$SS(x_1, x_2, x_3) = SS(x_1) + SS(x_2|x_1) + SS(x_3|x_1, x_2)$$
$$= 306.7323 + 263.7945 + 57.2902$$
$$= 627.8170.$$

The Type I SS is order dependent and as we have seen, if the variables in the model are given in a different order, the Type I SS will change. The Type I SS can be described as the SS due to a reduction in the error SS by adding that variable to a model that already contains all the variables preceding it in the **model** statement. This is not the case for the Type II SS. The Type II SS for a variable corresponds to a reduction in error SS due to the addition of that variable in the model that already contained all the other variables in the model list. SAS can produce both the Type I and Type II SS by specifying **ss1** and **ss2** as model options. The result of the application of this to our example with these options are presented below.

```
data ex2;
infile 'h:\s321\exam2.dat';
input y x1 x2 x3 @@;
proc reg data=ex2;
model y=x1 x2 x3 / ss1 ss2;
run;
                        Parameter Estimates

                  Parameter    Standard
Variable   DF      Estimate      Error   t Value  Pr > |t|    Type I SS    Type II SS
-----------------------------------------------------------------------------------
Intercept   1      17.84693     2.00188    8.92   <.0001         37446      244.17168
x1          1       1.10313     0.32957    3.35    0.0032     306.73233      34.41851
x2          1       0.32152     0.03711    8.66   <.0001      263.79445     230.62548
x3          1       1.28894     0.29848    4.32    0.0003      57.29022      57.29022
```

4.5 Partial F Tests

Before we discuss the partial F-tests in details, let us first distinguish between *variables-added-in-order* and *variables-added-last* tests in the context of our example involving three explanatory variables in Table 4.2.

4.5.1 Tests Based on Variables-Added-in-Order

For the variables-added-in-order test, we would have the analysis of variance as in Table 4.5. The hypotheses to be tested are

$$H_0 : \beta_1 = 0$$
$$H_a : \beta_1 \neq 0. \tag{4.14}$$

The ANOVA table can be summarized as follows:

(a) To test the hypotheses in (4.14), it has been suggested that for this model, only the x_1 variable is in the model and the F statistic should be computed as

$$F^* = \frac{SS(x_1)/1}{SSE/22}, \tag{4.15}$$

Source	df	SS	MS	F
x_1	1	306.7323	306.7323	17.64
Error	22	382.5277	17.3876	
Total	23	689.2600		

Table 4.5: ANOVA table based on the order in (i)

since in this case the SSE would have been based on $n - 2 = 24 - 2 = 22$ degrees of freedom. That is,

$$F^* = \frac{306.7323/1}{(263.7945 + 57.2902 + 61.4430)/22} = \frac{306.7323}{382.5277/22} = 17.64.$$

It is often argued that the test be conducted with the full regression (all three variables in the model) mean square as the divisor. That is, use instead,

$$F^* = \frac{SS(x_1)/1}{MSE} = \frac{306.7323/1}{3.0722} = 99.84.$$

(b) To test the hypotheses,

$$H_0 : \beta_2|\beta_1 = 0$$
$$H_a : \beta_2|\beta_1 \neq 0,$$
(4.16)

we could use the result in Table 4.6 to compute the statistic

$$F^* = \frac{SS(x_2|x_1)/1}{SSE/21} = \frac{[SS(x_1, x_2) - SS(x_1)]/1}{SSE/21},$$
(4.17)

where $SSE = S_{yy} - SS(x_1, x_2)$, the error SS based on 21 degrees of freedom. This would lead to

$$F^* = \frac{263.7945/1}{118.7332/21} = 46.66.$$

Source	df	SS	MS	F
x_1	1	306.7323	306.7323	
$x_2|x_1$	1	263.7945	263.7945	46.66
Error	21	118.7332	5.6540	
Total	23	689.2600		

Table 4.6: ANOVA table based on the order in (i)

Alternative to using the result in Table 4.6, we could also use the error mean square under the full model as the denominator in the computation of the F statistic (Table 4.7). This would lead to

$$F^* = \frac{263.7945/1}{3.0722} = 85.87.$$

(c) For the hypotheses in (4.19), these can be tested by again computing (using Table 4.7)

$$F* = \frac{[SS(x_1, x_2, x_3) - SS(x_1, x_2)]/1}{MSE} = \frac{SS(x_3|x_1, x_2)/1}{MSE},$$
(4.18)

$$H_0 : \beta_3|(\beta_1, \beta_2) = 0$$
$$H_a : \beta_3|(\beta_1, \beta_2) \neq 0.$$
(4.19)

Source	df	SS	MS	F
x_1	1	306.7323	306.7323	99.84
$x_2\|x_1$	1	263.7945	263.7945	85.87
$x_3\|x_1, x_2$	1	57.2902	57.2902	18.65
Error	20	61.4430	3.0722	
Total	23	689.2600		

Table 4.7: ANOVA table based on the order in (i)

Thus we have the computed F in this case being

$$F^* = \frac{57.2902/1}{3.0722} = 18.65.$$

In SAS, **proc glm** can be used to obtain tests that are based on variables-added-in-order by specifying **SS1** (that is, use Type I SS) in the model option statement. We have presented the results in the following output.

```
data one;
infile ''C:\Research\Book\Data\table42.prn" truncover;
input id y x1 x2 x3;
proc glm data=one;
model y = x1 x2 x3 / ss1;
run;
                        The GLM Procedure

Source             DF      Type I SS      Mean Square    F Value    Pr > F
x1                  1     306.7323285    306.7323285      99.84    <.0001
x2                  1     263.7944455    263.7944455      85.87    <.0001
x3                  1      57.2902224     57.2902224      18.65    0.0003
```

The results obtained agree with those obtained in Table 4.7 and the following conclusions can thus be derived.

- The hypotheses in (4.14) would be rejected, indicating that variable x_1 is important in the model.

- The hypotheses in (4.16) would also be rejected, again indicating that x_2 is very important for inclusion in the model that already contained x_1.

- The hypotheses in (4.19) would also be rejected (p-value $= 0.0003$). This again indicates that variable x_3 is important in the model that already contained x_1 and x_2.

4.5.2   Tests Based on Variables-Added-Last

For the three explanatory variables we have in this example, the three hypotheses of interest are:

$$H_0 : \beta_1|(\beta_2, \beta_3) = 0$$
$$H_a : \beta_1|(\beta_2, \beta_3) \neq 0 \tag{4.20}$$

$$H_0 : \beta_2|(\beta_1, \beta_3) = 0$$
$$H_a : \beta_2|(\beta_1, \beta_3) \neq 0 \tag{4.21}$$

$$H_0 : \beta_3|(\beta_1, \beta_2) = 0$$
$$H_a : \beta_3|(\beta_1, \beta_2) \neq 0. \tag{4.22}$$

These three hypotheses can be tested from the analysis of variance table presented in Table 4.8.

The hypotheses in (4.20) to (4.22) are tested by computing

$$F^* = \frac{\text{Type II } SS \text{ for } x_j}{MSE}, \; j = 1, 2, 3.$$

Source	df	SS	MS	F
$x_1\|x_2, x_3$	1	34.4185	34.4185	11.203
$x_2\|x_1, x_3$	1	230.6255	230.6255	75.069
$x_3\|x_1, x_2$	1	57.2902	57.2902	18.648
Error	20	61.4430	3.0722	
Total	23	689.2600		

Table 4.8: ANOVA table for the variable-added-last tests

In particular, for the hypotheses in (4.20), that is for x_1, the computation becomes

$$F^* = \frac{34.4185}{3.0722} = 11.20.$$

The SAS **proc glm** can be used to obtain tests that are based on variables-added-last by specifying SS3 (that is, use type III SS) in the model option statement. We have presented the results from the output below.

```
proc glm data=one;
model y = x1 x2 x3 / ss3;
run;

Source                DF    Type III SS    Mean Square    F Value    Pr > F
x1                     1    34.4185079     34.4185079     11.20      0.0032
x2                     1    230.6254757    230.6254757    75.07      <.0001
x3                     1    57.2902224     57.2902224     18.65      0.0003
```

In all the cases for this example, the hypotheses are all strongly rejected, indicating that all the three variables are needed in the model.

4.6   Multiple Partial F-Tests

Consider a general multiple regression equation containing k explanatory variables, which is given as

$$y_i = \beta_0 + \beta_1 x_1 + \beta_2 x_2 + \cdots + \beta_k x_k + \varepsilon_i. \tag{4.23}$$

Now suppose we wish to test whether the addition of variables $x_{k+1}, x_{k+2}, \ldots, x_{p-1}$, such that $p - 1 > k$, significantly improves the model which already has $x_1, x_2, \ldots, x_k$. The full regression model in this case would be

$$y_i = \beta_0 + \beta_1 x_1 + \beta_2 x_2 + \cdots + \beta_k x_k + \beta_{k+1} x_{k+1} + \beta_{k+2} x_{k+2} + \cdots + \beta_{p-1} x_{p-1} + \varepsilon_i. \tag{4.24}$$

In this case, the hypotheses of interest would be

$$\begin{aligned} H_0 &: \beta_{k+1} = \beta_{k+2} = \beta_{k+3} = \cdots = \beta_{p-1} = 0 \\ H_a &: \text{at least one of these } \beta \text{ parameters is not zero.} \end{aligned} \tag{4.25}$$

The model in (4.24) will be referred to as the full model, while the model in (4.23) will be referred to as the reduced model under H_0, the null hypothesis. To conduct the hypotheses in (4.25), we do the following:

1. Fit the full model in (4.24) and obtain the regression sum of squares, $SS(\text{Full})$, and the corresponding error sum of squares (SSE), which is based on $(n - p)$ degrees of freedom, viz:

$$SS(x_1, x_2, \ldots, x_k, x_{k+1}, \ldots, x_{p-1}) = SS(\text{Full}), \quad \text{on } p \text{ degrees of freedom}$$
$$SSE = SST - SS(\text{Full}).$$

2. If H_0 were true, or under H_0, we will fit the reduced model in (4.23) to obtain the regression sum of squares, and the extra SS due to $(x_{k+1}, x_{k+2}, \ldots, x_{p-1} | x_1, x_2, \ldots, x_k)$, and it will be based on $(p - 1 - k)$ degrees of freedom.

$$SS(x_1, x_2, \ldots, x_k) = SS(\text{Reduced}), \quad \text{on } k \text{ degrees of freedom}$$
$$SS(x_{k+1}, x_{k+2}, \ldots, x_{p-1} | x_1, x_2, \ldots, x_k) = SS(\text{Full}) - SS(\text{Reduced}).$$

3. The test statistic for conducting the hypotheses in (4.25) is computed as

$$F^* = \frac{SS(x_{k+1}, x_{k+2}, \ldots, x_{p-1} | x_1, x_2, \ldots, x_k)/(p-1-k)}{MSE(\text{Full})}$$
$$= \frac{[SS(\text{Full}) - SS(\text{Reduced})]/(p-1-k)}{MSE(\text{Full})}.$$

We notice here that $SS(\text{Full}) - SS(\text{Reduced}) = SS(x_{k+1}, x_{k+2}, \ldots, x_{p-1} | x_1, x_2, \ldots, x_k)$ is the extra sum of squares for fitting $x_{k+1}, \ldots, x_{p-1}$ given that $x_1, x_2, \ldots, x_k$ are already in the model and is based on $(p-1-k)$ degrees of freedom and $MSE = SSE/(n-p)$, the full model error mean square.

4. The null hypothesis in (4.25) is rejected if $F^* \geq f_{\alpha/2, (p-1-k), (n-p)}$.

Example 4.2 continued

For the data in Example 4.2 involving three explanatory variables x_1, x_2, x_3, suppose we wish to test the hypothesis whether the addition of variables x_2, x_3 is significant given that we already had variable x_1 in the model. That is, we wish to test the hypotheses

$$H_0 : \beta_2 = \beta_3 = 0$$
$$H_a : \text{at least one of the } \beta \text{ is not equal to zero.}$$

The full and reduced model under the above null hypothesis are respectively,

$$y_i = \beta_0 + \beta_1 x_1 + \beta_2 x_2 + \beta_3 x_3 + \varepsilon_i,$$

and

$$y_i = \beta_0 + \beta_1 x_1 + \varepsilon_i.$$

Under the full model, $SS(x_1, x_2, x_3) = 627.8170$ on 3 degrees of freedom and $SSE = 61.4430$ on 20 degrees of freedom. Hence, $MSE(\text{Full}) = 3.0722$. Under the reduced model, $SS(x_1) = 306.7323$ on 1 degree of freedom, hence the extra sum of squares is $SS(x_1, x_2, x_3) - SS(x_1) = 627.8170 - 306.7323 = 321.0847$ on 2 degrees of freedom. Therefore, the F^* statistic is computed as

$$F^* = \frac{321.0847/2}{3.0722} = 52.26.$$

Alternatively, the test statistic F^* can be computed as

$$F^* = \frac{[SS(x_2|x_1) + SS(x_3|(x_1, x_2))]/2}{MSE} = \frac{(263.7945 + 57.2902)/2}{3.0722} = \frac{160.5424}{3.0722} = 52.26.$$

The decision rule calls for the rejection of H_0 if $F^* \geq f_{.025, 2, 20} = 4.46$. Since $52.26 \ggg 4.46$, we would therefore strongly reject H_0 and conclude that at least one of the β parameters (β_2, β_3) is not zero.

We can implement the above hypothesis in SAS with the **test** statement after the model statement in **proc reg**.

```
proc reg data=one;
model y = x1 x2 x3 / ss1 ss2;
test x2, x3;
run;
```

Test 1 Results for Dependent Variable y

Source	DF	Mean Square	F Value	Pr > F
Numerator	2	160.54233	52.26	<.0001
Denominator	20	3.07215		

4.7   Multiple Linear Regression in Matrix Form

The multiple linear regression model

$$y_i = \beta_0 + \beta_1 x_{i1} + \beta_2 x_{i2} + \cdots + \beta_{p-1} x_{i,p-1} + \varepsilon_i, \quad i = 1, 2, \ldots, n,$$

in (4.3) can be expressed in matrix form as

$$\mathbf{Y} = \mathbf{X}\boldsymbol{\beta} + \boldsymbol{\varepsilon}, \tag{4.26}$$

where $\mathbf{Y}$ is an $n \times 1$ column vector of responses with entries $y_1, y_2, \ldots, y_n$, $\boldsymbol{\beta}$ is a $p \times 1$ column vector of parameters with entries $\beta_0, \beta_1, \ldots, \beta_{p-1}$, $\boldsymbol{\varepsilon}$ is an $n \times 1$ column vector with entries $\varepsilon_1, \varepsilon_2, \ldots, \varepsilon_n$ and $\mathbf{X}$ is an $n \times p$ matrix given as

$$\mathbf{X} = \begin{bmatrix} 1 & X_{11} & X_{12} & X_{13} & \ldots & X_{1,p-1} \\ 1 & X_{21} & X_{22} & X_{23} & \ldots & X_{2,p-1} \\ 1 & X_{31} & X_{32} & X_{33} & \ldots & X_{3,p-1} \\ \vdots & \vdots & \vdots & \vdots & \ddots & \vdots \\ 1 & X_{n1} & X_{n2} & X_{n3} & \ldots & X_{n,p-1} \end{bmatrix}. \tag{4.27}$$

In the representation in (4.27), $\mathbf{X}$ is a matrix of constants and $\boldsymbol{\varepsilon}$ is a vector of independent normal random variables with $\mathrm{E}(\boldsymbol{\varepsilon}) = 0$ and the variance-covariance matrix $\sigma^2 \mathbf{I}$, where $\mathbf{I}$ is an $n \times n$ identity matrix.

In matrix terms, the normal equations for obtaining $\hat{\boldsymbol{\beta}}$, the estimates of $\boldsymbol{\beta}$ are

$$\mathbf{X}'\mathbf{X}\hat{\boldsymbol{\beta}} = \mathbf{X}'\mathbf{Y},$$

which lead to

$$\hat{\boldsymbol{\beta}} = (\mathbf{X}'\mathbf{X})^{-1}\mathbf{X}'\mathbf{Y}.$$

Therefore, the fitted values of $\mathbf{Y}$ are given by

$$\hat{\mathbf{Y}} = \mathbf{X}\hat{\boldsymbol{\beta}} = \mathbf{X}(\mathbf{X}'\mathbf{X})^{-1}\mathbf{X}'\mathbf{Y} = \mathbf{H}\mathbf{Y},$$

where

$$\mathbf{H} = \mathbf{X}(\mathbf{X}'\mathbf{X})^{-1}\mathbf{X}',$$

a square $n \times n$ matrix $\mathbf{H}$ which is called the *hat matrix*. This matrix is very useful in regression diagnostics which are considered in Chapter 5. The hat matrix $\mathbf{H}$ is symmetric and it is idempotent because $\mathbf{H}\mathbf{H} = \mathbf{H}$.

4.8   Constrained Least Squares

In many applications, particularly in economics, it is necessary to impose some linear restrictions or constraints on the parameters of the model, which in turn would lead to what is often described as *Constrained Least Squares* or *Restricted Least Squares*. The latter is more often used and is often designated as (RLS)-restricted least squares.

Consider again Example 4.2 in this chapter. The model of interest is

$$y_i = \beta_0 + \beta_1 x_1 + \beta_2 x_2 + \beta_3 x_3 + \varepsilon_i. \tag{4.28}$$

We have seen how we can use the concept of partial F tests to conduct a test of the following hypotheses in (4.29) for instance,

$$\begin{aligned} H_0 &: \beta_2 = \beta_3 \\ H_a &: \beta_2 \neq \beta_3. \end{aligned} \tag{4.29}$$

Then we can write the above in the form

$$\mathbf{C}\boldsymbol{\beta} = \mathbf{r},$$

where for instance,

$$\begin{bmatrix} 0 & 0 & 1 & -1 \end{bmatrix} \begin{bmatrix} \beta_0 \\ \beta_1 \\ \beta_2 \\ \beta_3 \end{bmatrix} = \begin{bmatrix} \mathbf{r} \end{bmatrix} \quad \text{and } \mathbf{r} = [0].$$

$\mathbf{r}$ may not necessarily be zero. Under (4.29) therefore, we have the reduced model

$$\begin{aligned} y_i &= \beta_0 + \beta_1 x_1 + \beta_2(x_2 + x_3) + \varepsilon_i \\ &= \gamma_0 + \gamma_1 z_1 + \gamma_2 z_2 + \varepsilon_i. \end{aligned} \tag{4.30}$$

For instance, if the hypotheses of interest had been the following:

$$\begin{aligned} H_0 &: \beta_2 - \beta_3 = 1; \quad \beta_1 + \beta_2 = 0 \\ H_a &: \beta_2 - \beta_3 \neq 1 \text{ or } \beta_1 + \beta_2 \neq 0. \end{aligned} \tag{4.31}$$

Thus, we wish to restrict the least squares parameter estimates such that

$$\begin{cases} \beta_2 - \beta_3 = 1 \\ \beta_1 + \beta_2 = 0. \end{cases}$$

Then we could write the above in matrix notation as

$$\begin{bmatrix} 0 & 0 & 1 & -1 \\ 0 & 1 & 1 & 0 \end{bmatrix} \begin{bmatrix} \beta_0 \\ \beta_1 \\ \beta_2 \\ \beta_3 \end{bmatrix} = \begin{bmatrix} 1 \\ 0 \end{bmatrix}.$$

Both hypotheses in (4.29) and (4.31) are implemented in SAS for the data in Table 4.2 as tests 1 and 2 respectively. Clearly, the null hypothesis in (4.29) is not tenable and we can conclude that $\beta_2 \neq \beta_3$.

```
proc reg data=one;
model y = x1 x2 x3;
test1: test x2 - x3;
test2: test x2 - x3=1, x1 + x2=0;
run;
```

```
                        The REG Procedure
                          Model: MODEL1
                      Dependent Variable: y

               Number of Observations Read        24
               Number of Observations Used        24


                        Analysis of Variance

                                  Sum of        Mean
     Source              DF       Squares      Square    F Value    Pr > F

     Model                3     627.81700    209.27233      68.12    <.0001
     Error               20      61.44300      3.07215
     Corrected Total     23     689.26000


              Root MSE              1.75276    R-Square     0.9109
              Dependent Mean       39.50000    Adj R-Sq     0.8975
              Coeff Var             4.43735


                        Parameter Estimates

                      Parameter     Standard
     Variable    DF    Estimate        Error    t Value    Pr > |t|

     Intercept    1    17.84693      2.00188       8.92      <.0001
     x1           1     1.10313      0.32957       3.35      0.0032
     x2           1     0.32152      0.03711       8.66      <.0001
     x3           1     1.28894      0.29848       4.32      0.0003


                        The REG Procedure
                          Model: MODEL1
```

```
        Test test1 Results for Dependent Variable y

                                Mean
        Source          DF      Square      F Value     Pr > F

        Numerator        1     30.85215      10.04      0.0048
        Denominator     20      3.07215

                    The REG Procedure
                      Model: MODEL1

        Test test2 Results for Dependent Variable y

                                Mean
        Source          DF      Square      F Value     Pr > F

        Numerator        2    120.37261      39.18      <.0001
        Denominator     20      3.07215
```

We may note here the numerator degrees of freedom for both tests. It is 1 for the first test, since there is only one constraint equation on the parameters, whereas it is 2 for the second test where there are now two restrictions on the parameters.

Suppose we assume that the null hypothesis in either (4.29) or (4.31) is true, then we would need to estimate the parameter β subject to the constraint(s) imposed in the equation(s). It is clear that the least-squares estimators would not satisfy the constraint(s) and we would therefore employ the method of **constrained least-squares estimators.** We give below in matrix notation the RLS estimator as

$$\beta_{RLS} = \hat{\beta}_{OLS} + (\mathbf{X}'\mathbf{X})^{-1}\mathbf{C}'\left[\mathbf{C}((\mathbf{X}'\mathbf{X})^{-1}\mathbf{C}'\right]^{-1}(\mathbf{r} - \mathbf{C}\hat{\beta}_{OLS}), \tag{4.32}$$

where $\mathbf{Y} = \mathbf{X}\beta + \varepsilon$, is the OLS model and $\mathbf{C}\beta = \mathbf{r}$ are the constraint(s) equations.

A general test for the linear restrictions is therefore given by the statistic

$$GT = \frac{(SSE_{RLS} - SSE_{OLS})/m}{SSE_{OLS}/(n-p)}, \tag{4.33}$$

and statistic GT has an F distribution with m and $n-p$ degrees of freedom. Here, m, is the number of restrictions, and p is the number of parameters in the OLS model. The linear restriction in (4.29) is implemented with the following SAS codes and partial output.

```
proc reg data=one;
model y = x1 x2 x3;
restrict x2-x3=0;
run;
                    The REG Procedure
                      Model: MODEL1
                   Dependent Variable: y

NOTE: Restrictions have been applied to parameter estimates.

            Number of Observations Read        24
            Number of Observations Used        24

                    Analysis of Variance

                        Sum of      Mean
        Source      DF  Squares     Square    F Value    Pr > F

        Model        2  596.96485  298.48242   67.91    <.0001
        Error       21   92.29515    4.39501
        Corrected Total 23  689.26000

            Root MSE            2.09643   R-Square   0.8661
            Dependent Mean     39.50000   Adj R-Sq   0.8533
            Coeff Var           5.30741

                    Parameter Estimates

                     Parameter    Standard
        Variable  DF  Estimate     Error    t Value   Pr > |t|

        Intercept  1  21.73833    1.89102    11.50    <.0001
```

x1	1	1.29361	0.38758	3.34	0.0031
x2	1	0.34997	0.04307	8.13	<.0001
x3	1	0.34997	0.04307	8.13	<.0001
RESTRICT	-1	-31.89112	12.03669	-2.65	0.0048*

```
                * Probability computed using beta distribution.
```

The estimated restricted least squares model therefore is

$$\hat{y} = 21.7383 + 1.2936x_1 + 0.3500\,(x_2 + x_3).$$

We observe here that the error sum of squares for the restricted model is 92.2952 which is larger than the 61.4430 error sum of squares under the least squares model. Generally, the $SSE_{RLS} > SSE_{OLS}$. The general test gives a test statistic value of $(92.2952 - 61.4430)/3.07215 = 10.043$. Thus comparing this with an F distribution with 1 and 20 degrees of freedom, we would find that this result agrees with the test conducted under the **restrict** line in the above output. The p-value is computed in SAS and the following are the codes and the generated output. The results are identical. Note here that 3.07215 is the error mean square under the full or unrestricted OLS model.

```
data new;
x=1-probf(10.0425,1,20,0);
proc print;
var x;
format x f8.4;
                        Obs           x
                         1        0.0048
```

The linear restrictions in (4.31) are also implemented in SAS and the following are the codes and partial output.

```
proc reg data=one;
model y = x1 x2 x3;
restrict x2-x3=1, x1+x2=0;
run;
                        The REG Procedure
                          Model: MODEL1
                      Dependent Variable: y

NOTE: Restrictions have been applied to parameter estimates.

                Number of Observations Read          24
                Number of Observations Used          24

                        Analysis of Variance

                              Sum of          Mean
        Source          DF    Squares        Square    F Value    Pr > F

        Model            1   387.07179     387.07179     28.18    <.0001
        Error           22   302.18821      13.73583
        Corrected Total 23   689.26000

                Root MSE              3.70619    R-Square    0.5616
                Dependent Mean       39.50000    Adj R-Sq    0.5416
                Coeff Var             9.38275

                        Parameter Estimates

                        Parameter    Standard
        Variable    DF   Estimate      Error    t Value   Pr > |t|

        Intercept    1   33.56924     1.94268     17.28    <.0001
        x1           1   -0.46578     0.06993     -6.66    <.0001
        x2           1    0.46578     0.06993      6.66    <.0001
        x3           1   -0.53422     0.06993     -7.64    <.0001
        RESTRICT    -1  -78.54945    21.80747     -3.60    <.0001*
        RESTRICT    -1   60.50984    21.08309      2.87     0.0019*

            * Probability computed using beta distribution.
```

The estimated restricted least squares model therefore is

$$\hat{y} = 33.5692 - 0.4658x_1 + 0.4658x_2 - 0.5342x_3.$$

Of course in both cases, the restricted estimated parameters satisfy the constraints imposed.

4.9   Exercises

4.1 The data in the file **cars2004.txt** is taken from 2004 new car and truck data submitted by Johnson, R.W. to the Journal of Statistics Education Data Archive. For this problem, use the first 25 cases of the whole data set and the following variables: The city miles per gallon (y), the horsepower (x_1), the weight in pounds (x_2), and the length in inches (x_3). A model of the form

$$y_i = \beta_0 + \beta_1 x_1 + \beta_2 x_2 + \beta_3 x_3 + \varepsilon_i,$$

is suggested. Use a software of your choice to answer the following questions.

(a) Find the following sum of squares:

   (i) $SS(x_1 \mid x_2, x_3)$.
   (ii) $SS(x_2 \mid x_1, x_3)$.
   (iii) $SS(x_2 \mid x_1)$.
   (iv) $SS(x_3 \mid x_1, x_2)$, where SS refers to the sum of squares.

(b) Based on your results in (a), which of the variables x_1, x_2, x_3 would you consider to be the most important predictor of y?

4.2 Consider a regression model of the form

$$y_i = \beta_0 + \beta_1 x_{i1} + \beta_2 x_{i2} + \beta_3 x_{i3}^2 + \varepsilon_i, \ i = 1, 2, \cdots, 24.$$

State the reduced models for testing whether or not

(a) $\beta_1 = \beta_3 = 0$.
(b) $\beta_3 = 5$.
(c) $\beta_1 = \beta_2$.

4.3 The data in the file **cars2004.txt** is taken from 2004 new car and truck data submitted by Johnson, R.W. to the Journal of Statistics Education Data Archive. For this problem, use the first 20 cases of the whole data set and the following variables: The highway miles per gallon (y), the horsepower (x_1), the weight in pounds (x_2), and the length in inches (x_3). A mean response model of the form

$$y_i = \beta_0 + \beta_1 x_1 + \beta_2 x_2 + \beta_3 x_3 + \varepsilon_i,$$

is suggested. Use a software of your choice to answer the following questions.

(a) At the 5% level, do the data provide sufficient evidence to conclude that taken together, horsepower, weight, and length are useful for predicting highway miles per gallon?

(b) Interpret the estimate of the parameter β_1.

(c) Complete the following ANOVA table.

Source	df	SS	MS	F
x_1	1	—	—	—
$x_2 \mid x_1$	1	—	—	—
$x_3 \mid x_1, x_2$	1	—	—	—
Error	16	89.1055	5.5691	
Total	—	—		

   (i) Test for the significance of each independent variable as it enters the model. State the null hypothesis for each test in terms of the regression parameters.
   (ii) Test for the significance of the addition of both x_2 and x_3 to a model already containing x_1. State the null hypothesis in terms of the parameters of the model.

(d) What model would you recommend for the data set in this problem? Justify your answer.

4.4 Shown below is the estimated regression equation for a model involving two explanatory variables and 10 observations.

$$\hat{y} = 29.1270 + 0.5906x_1 + 0.4980x_2.$$

Complete the following ANOVA table and use this to answer the questions that follow:

Source	df	SS	MS	F
Reg	–	6216.375	–	–
Error	–	–	–	
Total	–	6724.125		

(a) Find *SSE*.

(b) Compute R^2.

(c) Does the estimated regression equation explain a large amount of the variability in the data? Explain.

4.5 The data below represents the sales price (y, $'000), square footage x_1, number of rooms x_2, number of bedrooms x_3, age x_4, and number of bathrooms x_5 for each of 23 single-family residences sold during the past year in a Michigan city.

y	x_1	x_2	x_3	x_4	x_5	y	x_1	x_2	x_3	x_4	x_5
93.5	1008	5	2	35	1.0	89.5	1008	6	3	35	2.0
89.0	1290	6	3	36	1.0	145.0	1950	8	3	52	1.5
90.5	860	8	2	36	1.0	192.5	2086	7	3	12	2.0
89.9	912	5	3	41	1.0	125.0	2011	9	4	76	1.5
92.0	1204	6	3	40	1.0	100.0	1465	6	3	102	1.0
95.0	1204	5	3	10	1.5	98.5	1232	5	2	69	1.5
120.5	1764	8	4	64	1.5	141.0	1736	7	3	67	1.0
126.0	1600	7	3	19	2.0	119.4	1296	6	3	11	1.5
109.0	1255	5	3	16	2.0	165.0	1996	7	3	9	2.5
189.0	3600	10	5	17	2.5	127.9	1874	5	2	14	2.0
86.0	864	5	3	37	1.0	120.0	1580	5	3	11	1.0
78.0	720	4	2	41	1.0						

Assume the regression model $y_i = \beta_0 + \beta_1 x_1 + \beta_2 x_2 + \beta_3 x_3 + \beta_4 x_4 + \beta_5 x_5 + \varepsilon_i$.

(a) Which independent variable has the least correlation with y?

(b) State the estimated regression function.

(c) Test whether $\beta_2 = 0$ using $\alpha = 0.025$. State the null and the alternative hypotheses, the decision rule, the value of the test statistic, and your conclusion. Give the p-value.

(e) Is there evidence that the model is useful for predicting y? Test using $\alpha = 0.05$.

(f) If the test in (e) is significant, which of the model terms contribute to the overall significance?

4.6 A fisheries commission wants to estimate the number of bass caught in a given lake during a season in order to restock the lake with appropriate number of young fish. The commission samples a number of lakes and records y, the seasonal catch (thousands of bass per square mile of lake area); x_1, the number of lake shore residences per square mile of lake area; x_2, the size of lake in square miles; $x_3 = 1$, if the lake has public access, 0 if not; and x_4, a structure index. The data is given in the following table.

Assume the regression model $y_i = \beta_0 + \beta_1 x_1 + \beta_2 x_2 + \beta_3 x_3 + \beta_4 x_4 + \varepsilon_i$.

(a) Which independent variable has the least correlation with y?

Obs	y	x_1	x_2	x_3	x_4	Obs	y	x_1	x_2	x_3	x_4
1	3.6	92.2	0.21	0	81	11	2.4	64.6	0.91	1	40
2	0.8	86.7	0.30	0	26	12	1.9	50.0	1.10	1	22
3	2.5	80.2	0.31	0	52	13	2.0	50.0	1.24	1	50
4	2.9	87.2	0.40	0	64	14	1.9	51.2	1.47	1	37
5	1.4	64.9	0.44	0	40	15	3.1	40.1	2.21	1	61
6	0.9	90.1	0.56	0	22	16	2.6	45.0	2.46	1	39
7	3.2	60.7	0.78	0	80	17	3.4	50.0	2.80	1	53
8	2.7	50.9	1.21	0	60	18	3.6	70.0	0.78	1	61
9	2.2	86.1	0.34	1	30	19	2.9	75.0	0.66	1	50
10	5.9	90.0	0.40	1	90	20	3.3	80.4	0.52	1	74

(b) State the estimated regression function.

(c) Obtain a point estimate of the change in mean response when x_3 increases by two units while other variables are held constant.

(d) Obtain a 95% confidence interval in place of the point estimate in (c). Interpret your interval.

(e) Test whether or not $\beta_4 = 0$ controlling the α risk at 0.025 when $\beta_4 = 0$. State the null and the alternative hypotheses, the decision rule, the value of the test statistic, and your conclusion. What can you say concerning the variable whose elements are denoted by x_4?

4.7 The data in the file **baseball.txt** is taken from baseball players' salaries data submitted by Watnik, M.R. to the Journal of Statistics Education Data Archive (see also Watnik, 1998). For this problem, use the first 24 cases of the whole data set and the following variables: The 1992 salary in thousands of dollars (y), the batting average (x_1), the number of hits (x_2), and the number of home runs (x_3). Assume the regression model

$$y_i = \beta_0 + \beta_1 x_1 + \beta_2 x_2 + \beta_3 x_3 + \varepsilon_i.$$

(a) Which independent variable is most highly correlated with y?

(b) Does the data indicate that salary y tend to increase with number of hits? Explain.

(c) State the estimated regression function.

(d) Interpret β_2. In what units is β_2 expressed?

(e) Obtain a point estimate of the change in y when x_3 increases by one unit while other variables are held constant.

(f) Obtain a 95% confidence interval in place of the point estimate in (e). Interpret your interval.

(g) Test whether or not $\beta_3 = 0$ controlling the α risk at 0.01 when $\beta_3 = 0$. State the null and the alternative hypotheses, the decision rule, the value of the test statistic, and your conclusion. What is the implication of your conclusion?

4.8 The coefficient of determination R^2 always increase when new independent variable is added to the model. However, including many variables in a model reduces the degrees of freedom available for estimating σ^2. Suppose you want to use 18 independent variables to model a college student's GPA. You fit the model

$$E(Y) = \beta_0 + \beta_1 x_1 + \beta_2 x_2 + \cdots + \beta_{18} x_{18},$$

where $y =$ GPA and $x_1, x_2, \ldots, x_{18}$ are the 18 independent variables. Data for twenty students ($n = 20$) are used to fit the model, and you obtain $R^2 = 0.95$.

(a) Is there a sufficient evidence to indicate that the model is adequate for predicting GPA? State the observed value of the test statistic, the decision rule, and your conclusion. Use $\alpha = 0.05$.

(b) Calculate R_a^2, adjusted R-square. Interpret its value.

4.9 The data in the following table from 24 men is a subset of body dimension measurements data used in Heinz et al. (2003). The dependent variable is weight (y, in kg) and the three independent variables are chest diameter (x_1, in cm), age (x_2, in years) and height (x_3, in cm). Assume the regression model $y_i = \beta_0 + \beta_1 x_1 + \beta_2 x_2 + \beta_3 x_3 + \varepsilon_i$.

  (a) Which independent variable is most highly correlated with y?

  (b) Does the data indicate that weight y tend to increase with age? Explain.

  (c) State the estimated regression function.

  (d) Interpret β_2. In what units is β_2 expressed?

  (e) Obtain a point estimate of the change in y when x_3 increases by one unit while other variables are held constant.

  (f) Obtain a 95% confidence interval in place of the point estimate in (e). Interpret your interval.

  (g) Test whether or not $\beta_1 = 0$ controlling the α risk at 0.01 when $\beta_1 = 0$. State the null and the alternative hypotheses, the decision rule, the value of the test statistic, and your conclusion. What is the implication of your conclusion?

Obs	y	x_1	x_2	x_3	Obs	y	x_1	x_2	x_3
1	65.6	28.0	21	174.0	13	90.0	31.4	20	192.0
2	71.8	30.8	23	175.3	14	74.6	28.0	26	176.0
3	80.7	31.7	28	193.5	15	71.0	28.6	23	174.0
4	72.6	28.2	23	186.5	16	79.6	30.6	22	184.0
5	78.8	29.4	22	187.2	17	93.8	29.4	30	192.7
6	74.8	31.3	21	181.5	18	70.0	30.0	22	171.5
7	86.4	31.7	26	184.0	19	72.4	29.5	29	173.0
8	78.4	28.8	27	184.5	20	85.9	32.6	22	176.0
9	62.0	27.5	23	175.0	21	78.8	32.0	22	176.0
10	81.6	28.0	21	184.0	22	77.8	32.0	20	180.5
11	76.6	30.3	23	180.0	23	66.2	30.9	22	172.7
12	83.6	29.7	22	177.8	24	86.4	32.8	24	176.0

Chapter 5

Regression Diagnostics and Remedial Methods

5.1 Introduction

From our discussion in Chapter 2, the following are the assumptions that must hold true for any regression model

1. The error term ε has zero mean and is randomly distributed around the mean. $E(\varepsilon_i) = 0$ for all i.

2. The error term ε has constant variance σ^2 (*homoscedasticity*). That is, $\text{Var}(\varepsilon_i) = \sigma^2$ for all i.

3. The error terms ε are identically normally distributed.
   Combining (1), (2) and (3), we have $\varepsilon_i \sim N(0, \sigma^2)$.

4. The structural form of the model is linear.

5. The error terms ε_i and ε_j are uncorrelated for $i \neq j$. $\text{Cov}(\varepsilon_i, \varepsilon_j) = 0$ for all $i \neq j$.

   Thus, combining (2) and (5), we have that the variance-covariance matrix has the structure

$$\begin{bmatrix} \sigma^2 & 0 & 0 & \ldots & 0 \\ 0 & \sigma^2 & 0 & \ldots & 0 \\ 0 & 0 & \sigma^2 & \ldots & 0 \\ \vdots & \vdots & \vdots & \ddots & \vdots \\ 0 & 0 & 0 & \ldots & \sigma^2 \end{bmatrix} = \sigma^2 \mathbf{I},$$

where $\mathbf{I}$ is an $n \times n$ identity matrix.

In this chapter, we shall discuss in turn the effects of the violations of each of the above assumptions on a regression model and their impact on inferences on parameter estimates as well as on the stability of the regression models. Other diagnostics issues in this chapter also relate to the following:

6. Multicollinearity, and

7. Outliers and influential observations.

   One of the most important tools for investigating the violations of the above assumptions are *residuals*. We will discuss in the next section some of the residuals that we shall employ in this chapter. In sections 5.2 to 5.6, we present diagnostic measures for the various violations in a regression model. Some of the remedial measures to these violations are discussed in sections 5.7 to 5.12.

5.2   Residuals and Diagnostics

For the general regression model of the form

$$y_i = \beta_0 + \sum_{j=1}^{p-1} \beta_j x_{ij} + \varepsilon_i, \quad \text{for } i = 1, 2, \ldots, n, \tag{5.1}$$

the raw residuals, e_i, are defined as $e_i = y_i - \hat{y}_i, \quad i = 1, 2, \ldots, n$, where y_i is the observed value of y at the i-th observation and $\hat{y}_i$ is the corresponding fitted or estimated regression value from the fitted regression model. The residuals e_i have mean zero and approximate variance

$$\text{Var}(e) = \frac{\sum (e_i - \bar{e})^2}{n-p} = \frac{\sum e_i^2}{n-p} = \frac{\text{SSE}}{n-p} = \text{MSE}.$$

We observe here that the residuals are not themselves independent but for as long as $n \gg k$, the residuals can be readily used for model adequacy and all other diagnostics. We now define the standardized residuals and the studentized residuals. The studentized deleted residuals will be defined in section 5.5.

Standardized Residuals

The standardized residuals, denoted by d_i is obtained by standardizing the raw residuals by the square root of its variance. That is,

$$d_i = \frac{e_i}{\sqrt{MSE}}, \quad i = 1, 2, \ldots, n. \tag{5.2}$$

The standardized residuals have mean zero and an approximate variance of 1. They are also called semi-studentized residuals.

Studentized Residuals

The studentized residuals denoted by r_i is defined as

$$r_i = \frac{e_i}{\text{s.e.}(e_i)} = \frac{e_i}{S\sqrt{(1-h_i)}}, \tag{5.3}$$

where $S = \sqrt{MSE}$ and h_i are the leverages (which are discussed is section 5.5), the diagonal elements of the $n \times n$ hat matrix defined as

$$\mathbf{H} = \mathbf{X}(\mathbf{X'X})^{-1}\mathbf{X'}.$$

The hat matrix is defined in section 4.7.

For a simple linear regression, the model can be written in a matrix form as

$$\begin{bmatrix} y_1 \\ y_2 \\ \vdots \\ y_n \end{bmatrix} = \begin{bmatrix} 1 & x_1 \\ 1 & x_2 \\ \vdots & \vdots \\ 1 & x_n \end{bmatrix} \begin{bmatrix} \beta_0 \\ \beta_1 \end{bmatrix} + \begin{bmatrix} \varepsilon_1 \\ \varepsilon_2 \\ \vdots \\ \varepsilon_n \end{bmatrix},$$

with

$$(\mathbf{X'X})^{-1} = \frac{\begin{bmatrix} \sum x_i^2 & -\sum x_i \\ -\sum x_i & n \end{bmatrix}}{n \sum x_i^2 - (\sum x_i)^2} = \frac{\begin{bmatrix} \sum x_i^2 & -\sum x_i \\ -\sum x_i & n \end{bmatrix}}{n S_{xx}},$$

since $n S_{xx} = n \sum x_i^2 - (\sum x_i)^2$. For the simple linear regression model, the diagonals of $\mathbf{H}$ matrix reduce to

$$h_i = \frac{1}{n} + \frac{(x_i - \bar{x})^2}{S_{xx}}, \tag{5.4}$$

and $\sum h_i = p$, the number of parameters in the model (including β_0). We shall discuss other residuals in later sections of this chapter.

Diagnostics for the residuals

The true error in a regression model is $\varepsilon_i = y_i - \mathrm{E}(Y_i)$. The residual (or the observed error) is defined as $e_i = y_i - \hat{y}_i$. The true error ε_i are independent normal random variables, with mean 0 and constant variance σ^2. When the model is appropriate, the observed residuals, e_i, should satisfy the assumptions for ε_i. Residual analysis can be used to detect (1) whether the regression function is linear, (2) whether the error terms have constant variance, (3) whether the error terms are independent, (4) whether the error terms are normally distributed, (5) whether there is existence of outlying observations, and (6) whether one or more important predictor variables are excluded from the regression model.

Graphical Residual Analysis:

Some common residual plots include (a) the plot of e_i against x_i, (b) the plot of e_i against the predicted $\hat{y}_i$, (c) the plot of e_i against time, (d) the plot of e_i against omitted independent variables, (e) box plot, dot plot, or stem-and-leaf plot of e_i, and (f) normal probability plot of e_i.

We shall consider the following residual plots of e_i against $\hat{y}_i$, the fitted values. The plots allow us to detect some common types of model inadequacies.

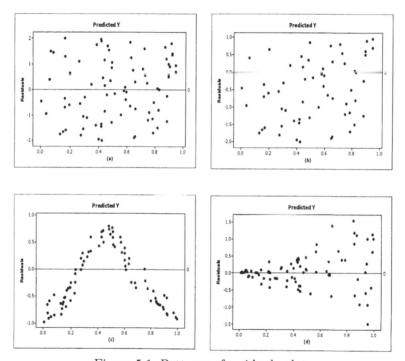

Figure 5.1: Patterns of residuals plots

In Figure 5.1,

- Pattern (a) is the ideal or satisfactory.

- Pattern (b) indicates that the assumption of normality of error terms is violated.

- Pattern (c) indicates that a curvature term like x^2 is missing in the model. That is, nonlinearity.

- Pattern (d) shows that the variance varies with the explanatory or predicted values. In other words, the assumption of constant variance is violated or we have what is often referred to as *heteroscedasticity*. The residual plot in (d) "fans out", which indicates that the variance is increasing. We can also have a reverse situation in which the residual plot "funnels in" indicating that the variance is decreasing.

In a simple linear regression, the plot of residuals (e_i) or standardized residuals (r_i) against the predicted values ($\hat{y}_i$) gives similar information as the residuals plot against the independent variable (x_i). A plot of e_i against $\hat{y}_i$ can be used to detect unequal error variances, presence of outlying y observations and correlated error terms. A plot of e_i against an x_{ij} that is not in the regression model can be used to detect the omission of an important independent variable. If the plot shows any relationship between the response and the new variable, term(s) of the new variable can be added to the model. When the data is measured over time, a plot of e_i against a time sequence can be used to detect correlated error terms.

Example 5.1

The data below represent experimental data for 10 sales trainees on number of days of training received (x) and performance score (y) in a battery of simulated sales situations. Consider a regression model relating performance score y to days of training, x. That is, a model of the form

$$y_i = \beta_0 + \beta_1 x_i + \varepsilon_i. \tag{5.5}$$

Trainee	1	2	3	4	5	6	7	8	9	10
x	0.5	0.5	1.0	1.0	1.5	1.5	2.0	2.0	2.5	2.5
y	46	51	71	75	92	99	105	112	121	125

Table 5.1: Data relating performance to training

```
*Program for Example 5.1 in the book;
data one;
input x y @@;
datalines;
0.5 46 0.5 51 1.0 71 1.0 75 1.5 92 1.5 99 2.0 105 2.0 112 2.5 121 2.5 125
;

proc plot data=one;
plot y*x='*';
run;
proc reg data=one;
model y = x;
output out=two p=pred r=resid;
run;
```

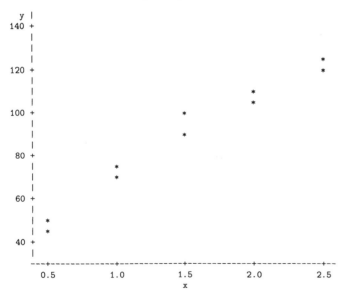

```
                    Plot of y*x.  Symbol used is '*'.

     y |
    140 +
        |
        |
        |                                                  *
    120 +                                                  *
        |
        |                                        *
        |                                        *
    100 +                             *
        |
        |                             *
        |
     80 +
        |                   *
        |                   *
        |
     60 +
        |
        |      *
        |      *
     40 +
        |
        ---+--------------+--------------+--------------+--------------+--
         0.5            1.0            1.5            2.0            2.5
                                        x
```

                        Model: MODEL1
                    Dependent Variable: y

                    Analysis of Variance

Source	DF	Sum of Squares	Mean Square	F Value	Pr > F
Model	1	6808.05000	6808.05000	254.45	<.0001
Error	8	214.05000	26.75625		
Corrected Total	9	7022.10000			

Root MSE	5.17264	R-Square	0.9695
Dependent Mean	89.70000	Adj R-Sq	0.9657
Coeff Var	5.76660		

Parameter Estimates

Variable	DF	Parameter Estimate	Standard Error	t Value	Pr > |t|
Intercept	1	34.35000	3.83614	8.95	<.0001
x	1	36.90000	2.31328	15.95	<.0001

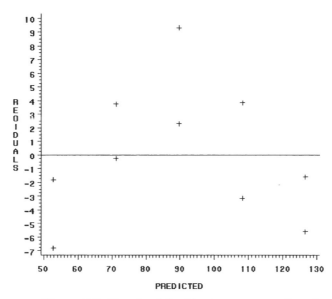

Figure 5.2: Graph of Residuals versus predicted

We observe in Figure 5.2 that the pattern of the residual plot against $\hat{y}_i$, the predicted values is like that of (c) in Figure 5.1, that is, parabolic, indicating that the model in (5.5) needs an x^2 or x^3 term.

Example 5.2

A health researcher interested in studying the relationship between diastolic blood pressure and age among healthy adult women 20 to 60 years old collected data on 54 subjects. The data are presented in Table 5.2. A model of the form $y_i = \beta_0 + \beta_1\, x_i + \varepsilon_i$ is suggested.

The suggested regression model is implemented in SAS with the following results and a plot of residuals versus predicted values is given in Figure 5.3.

```
/* Program to analyze data in Table 5.1 for Example 5.2 */
data one;
input x y @@;
y1=1/y;
datalines;
.......
;
run;
proc reg data=one;
model y = x;
output out=two p=pred r=resid;
run;
goptions vsize=5inches hsize=5inches;
symbol c=b i=spline value=none line=1 height=0.75;
```

Obs	x	y	Obs	x	y	Obs	x	y	Obs	x	y
1	27	73	15	32	76	29	40	70	43	54	71
2	21	66	16	33	69	30	42	72	44	57	99
3	22	63	17	31	66	31	43	80	45	52	86
4	26	79	18	34	73	32	46	83	46	53	79
5	25	68	19	37	78	33	43	75	47	56	92
6	28	67	20	38	87	34	49	80	48	52	85
7	24	75	21	33	76	35	40	90	49	57	109
8	25	71	22	35	79	36	48	70	50	50	71
9	23	70	23	30	73	37	42	85	51	59	90
10	20	65	24	37	68	38	44	71	52	50	91
11	29	79	25	31	80	39	46	80	53	52	100
12	24	72	26	39	75	40	47	96	54	58	80
13	20	70	27	46	89	41	45	92			
14	38	91	28	49	101	42	55	76			

Table 5.2: Diastolic blood pressure and age among a sample of 54 healthy women

```
axis1 label=(angle=-90 rotate=90 'RESIDUALS');
axis2 label=('PREDICTED');

proc gplot data=two;
plot resid*pred ='plus' / vaxis=axis1 haxis=axis2 noframe vref=0;
run;
```
```
                          The REG Procedure
                           Model: MODEL1
                        Dependent Variable: y

                  Number of Observations Read        54
                  Number of Observations Used        54

                        Analysis of Variance

                                 Sum of        Mean
        Source           DF      Squares      Square    F Value    Pr > F

        Model             1    2374.96833  2374.96833     35.79    <.0001
        Error            52    3450.36501    66.35317
        Corrected Total  53    5825.33333

                Root MSE              8.14575   R-Square    0.4077
                Dependent Mean       79.11111   Adj R-Sq    0.3963
                Coeff Var            10.29659

                        Parameter Estimates

                        Parameter      Standard
        Variable    DF    Estimate        Error    t Value    Pr > |t|

        Intercept    1    56.15693      3.99367      14.06     <.0001
        x            1     0.58003      0.09695       5.98     <.0001
```

The plot of the residuals versus predicted in Figure 5.3 is of type (d), indicating that the assumption of constancy of variance (homoscedasticity) is not valid for the model employed for these data.

Hypothesis Testing on Residuals:

Runs test for randomness on the residuals can be used to check if the error terms are independent. Goodness of fit tests like the Kolmogorov-Smirnov or Shapiro-Wilk can be used to check if the error terms are normally distributed. Tests like Goldfeld-Quandt (Goldfeld and Quandt, 1965), Harrison-McCabe (Harrison and McCabe, 1979) and the Breusch-Pagan (Breusch and Pagan, 1979) can be used to check if the error terms have constant variance. Tests like Rainbow (Utts, 1982) and Harvey-Collier (Harvey and Collier, 1977) for linearity can be used to test if the regression function is linear. A test for heteroscedasticity in SAS (White, 1980) tests whether the first and second moments of the regression model are correctly specified. We will not go into the details of any of these tests as they can be easily

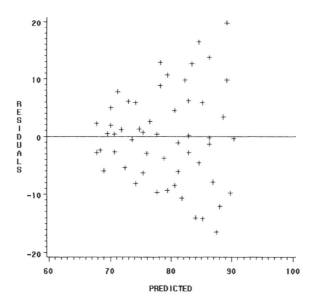

Figure 5.3: Plot of residuals against predicted

implemented in SAS or R.

5.3 Detecting Normality of Error Terms

Available procedures for detecting the normality of the random error terms in a regression model include graphical displays like the stem and leaf plot, the box plot, and the normal probability plot of the residuals. Other procedures are the various tests for normality.

For the data in Table 5.2 for instance, the stem and leaf display, and the box plot for the residuals $e_i = y_i - \hat{y}_i$ from the ordinary least squares regression are displayed below with the use of **proc univariate** in SAS. The program and a partial output are provided. The option **plot** will display the stem and leaf display, and a box plot, while the option **normal** will conduct a normality test as well as display a normal probability plot. The moments, basic statistics and tests for location which we do not need are also provided. The output gives four tests for normality: these are the Shapiro-Wilk, the Kolmogorov-Smirnov, the Cramer-von Mises and the Anderson-Darling tests. Of the four, the Anderson-Darling (AD) test will be adopted through out in this book. Basically, the hypotheses of interest are

$$H_0 : \text{The } \varepsilon_i \text{ are independent and normally distributed}$$
$$H_a : H_0 \text{ is not true.} \tag{5.6}$$

The decision rule rejects H_0 if the p-value under the Anderson-Darling test ≤ 0.05. The computed AD p-value for the residuals is > 0.25, hence we would therefore fail to reject H_0 and conclude that the residuals are normally distributed. That is the $\varepsilon_i \sim$ Normal.

```
options nodate nonumber;
data one;
input x y @@;
datalines;
........
;

proc reg data=one;
model y = x;
output out=two p=pred r=resid;
run;
proc univariate data=two normal plot;
var resid;
run;
                    The UNIVARIATE Procedure
                   Variable:  resid  (Residual)
```

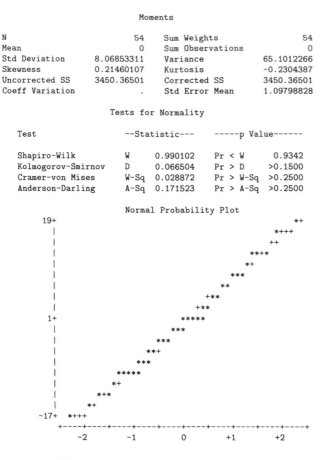

```
                      Moments

N                        54    Sum Weights              54
Mean                      0    Sum Observations          0
Std Deviation     8.06853311   Variance          65.1012266
Skewness          0.21460107   Kurtosis          -0.2304387
Uncorrected SS    3450.36501   Corrected SS      3450.36501
Coeff Variation           .    Std Error Mean    1.09798828

                  Tests for Normality

Test                  --Statistic---    -----p Value------

Shapiro-Wilk          W    0.990102    Pr < W        0.9342
Kolmogorov-Smirnov    D    0.066504    Pr > D       >0.1500
Cramer-von Mises      W-Sq 0.028872    Pr > W-Sq    >0.2500
Anderson-Darling      A-Sq 0.171523    Pr > A-Sq    >0.2500
```

The stem and leaf display as well as the box-plot display of the residuals also indicate to some extent, that the residuals follow a normal distribution.

The Normal Probability Plot

The normal probability plot derives its power from the order statistics theory. Here we would plot the residuals

$$(y_h - \hat{y}_h) \quad \text{against} \quad \sqrt{\text{MSE}} \left[z \left(\frac{h - 0.375}{n + 0.25} \right) \right],$$

where $e_h = (y_h - \hat{y}_h)$ is the h-th smallest residual and $z(B)$ denotes the $(B)100$ percentile of the standard normal variate.

Obs (i)	x	y	e_i	Rank (h)	Expected
1	27	73	1.1822	33	2.0929
2	21	66	−2.3376	22	−2.0929
3	22	63	−5.9176	14	-5.46477
4	26	79	7.7623	45	7.5368
⋮	⋮	⋮	⋮	⋮	⋮
51	59	90	−0.3787	26	−0.5650
52	50	91	5.8415	41	5.4647
53	52	100	13.6815	52	13.5276
54	58	80	−9.7987	6	−10.2703

Table 5.3: Residuals and expected values under normality

For the example, Table 5.3 displays the result of the expected values under normality for the first four and last four observations in the data. For example, for the first observation, the residual $e_1 = 1.1822$ and this is ranked 33rd ($h = 33$) among all the 54 residuals. Hence, we calculate

$$\frac{h - 0.375}{n + 0.25} = \frac{33 - 0.375}{54 + 0.25} = 0.6014,$$

and the expected value under normality is $\sqrt{\text{MSE}}\, z(0.6014) = 8.14575(0.25693) = 2.0929$.

Similar calculations give the results in the last column of Table 5.3. Ideally, the plot of the ranked residuals against the expected normal values should be a straight line. Deviations from a straight line reflect departures from normality. The normal probability plot presented in the previous SAS output indicates that the residuals are indeed normally distributed, as the plot clearly follows a straight line. The normal probability plot interpretation can sometimes be very subjective, and in all cases, the use of one of the suggested normality tests is recommended. The tests are all based on empirical results and not subject to how one views the normal probability plots.

5.4 Detecting Outliers

In section 5.2, we define the raw residuals e_i, the standardized residuals d_i and the studentized residuals r_i for observation i, where $i = 1, 2, \ldots, n$. We will examine what each of these residuals can tell us about the impact or effect of that particular observation on our regression model.

Each of the residuals defined in section 5.2 can be used to detect outliers. Outliers are observations with very large residuals, that is $\hat{y} \gg y_i$. An observation y is outlying when its corresponding residual is far greater than all other residuals in absolute value. In general, we consider a residual to be an outlier if the residual is three or four standard deviations away from the mean of the residuals. It is easier to work with the standardized or studentized residuals. An additional outlier diagnostics is provided by the **PRESS** (prediction error sum of squares) residuals, denoted here as $p_{(i)}$, and is defined as

$$p_{(i)} = \frac{e_i}{1 - h_i}, \quad i = 1, 2, \ldots, n, \tag{5.7}$$

where h_i is the leverage defined in section 5.2. The $p_{(i)}$ are obtained by deleting the ith observation, fit the regression model containing the remaining $(n - 1)$ observations and then obtain the predicted value of y_i corresponding to the deleted observation. In this case, the corresponding prediction error would be

$$p_{(i)} = y_i - \hat{y}_{(i)},$$

where $\hat{y}_{(i)}$ is the predicted value of the i-th observation based on all the observations except the i-th observation. The $p_{(i)}$ are sometimes referred to as *deleted residuals*.

The three residuals in section 5.2 and the leverages can be generated or obtained in SAS by specifying the following options in an **output** command:

$$\mathbf{R} = e_i, \quad \mathbf{STUDENT} = r_i, \quad \mathbf{PRESS} = p_{(i)}, \quad \text{and} \quad \mathbf{H} = h_i$$

Example 5.3

The following data in Table 5.4 is on age (x) in years and the systolic blood pressure (y) of thirty individuals. The following is the SAS program that fits a regression model of the form $y_i = \beta_0 + \beta_1 x_i + \varepsilon_i$. Some of the diagnostic statistics are specified in the **output** statement and the statistics are printed.

```
*Program for Example 5.3 with the data in Table 5.4;
data one;
input age sbp @@;
datalines;
.........
;
proc reg data=one;
model sbp=age;
output out=two p=pred r=resid student=ri press=pre h=lev;
run;
data new;
```

x	y	x	y	x	y	x	y	x	y
42	130	70	151	41	125	82	156	80	162
46	115	80	156	61	150	55	150	63	150
42	148	85	162	75	165	71	158	82	160
71	100	72	158	53	135	76	158	53	140
80	156	64	155	77	153	44	130	65	140
74	162	81	160	60	146	55	144	48	130

Table 5.4: SBP and ages of thirty individuals

```
set two;
di = resid/11.9094;
proc print data=new;
var age sbp pred resid di ri pre lev;
format pred resid di ri pre lev f8.4;
run;
```

```
Obs   age   sbp      pred      resid         di         ri        pre       lev
  1    42   130   130.4186    -0.4186    -0.0352    -0.0376    -0.4782    0.1246
  2    46   115   133.2817   -18.2817    -1.5351    -1.6141   -20.2134    0.0956
  3    42   148   130.4186    17.5814     1.4763     1.5779    20.0848    0.1246
  4    71   100   151.1756   -51.1756    -4.2971    -4.3851   -53.2925    0.0397
  5    80   156   157.6174    -1.6174    -0.1358    -0.1410    -1.7443    0.0727
.................................................................................
 26    63   150   145.4495     4.5505     0.3821     0.3888     4.7105    0.0340
 27    82   160   159.0489     0.9511     0.0799     0.0834     1.0382    0.0839
 28    53   140   138.2920     1.7080     0.1434     0.1478     1.8133    0.0581
 29    65   140   146.8811    -6.8811    -0.5778    -0.5877    -7.1183    0.0333
 30    48   130   134.7132    -4.7132    -0.3958    -0.4133    -5.1404    0.0831
```

For observation 1 for instance, we have

$$\text{RESID} = e_i = \text{SBP} - \text{PRED} = 130 - 130.4186 \qquad = -0.4186$$

$$\text{RI} = r_i = \frac{e_i}{S\sqrt{(1-h_i)}} = \frac{-0.4186}{11.9094\sqrt{(1-0.1246)}} = -0.0376$$

$$\text{PRE} = p_{(i)} = \frac{e_i}{(1-h_i)} = \frac{-0.4186}{(1-0.1246)} \qquad = -0.4782.$$

While SAS does not produce the standardized residuals d_i, denoted by 'di' in the SAS print out, this is generated by dividing the residuals e_i by $S = 11.9094$. Thus d_1 for instance equals -0.0352. From the above results we notice that observation numbered 4, has both its standardized residual $d_4 = -4.2971$ and studentized residual $r_4 = -4.3851$. In both cases, $|r_4|$ and $|d_4|$ are greater than 3. Hence, it is quite likely that observation 4 (Age = 71, sbp = 100) is a possible outlier. Similarly, a plot of the studentized residuals r_i against the predicted values in Figure 5.4 indicates that one observation is outside the range $-3 \le r_i \le 3$, indicating that the particular observation is a possible outlier.

5.5   Detecting Influential Observations

When you perform ordinary least squares regression, all observations in the data set are given equal weight. Unfortunately, not all observations have an equal impact on the regression estimates. Observations that have a strong influence on the regression estimates are called *influential* observations.

One approach for examining the influence of an observation is to delete that observation from the analysis and measure the effect on predicted values and the parameter estimates. Thus, an observation is said to be an influential observation if its exclusion from the proposed model results in inferences that are very different from the inferences reached when this particular observation was included in the given model. In practice, one does not need to repeat the analysis with each observation deleted in turn. The statistics required to detect influential observations can all be computed from the initial analysis using the complete set of observations. Measures used to detect influential observations are called *influence diagnostics*. Seven such measures are discussed in this section.

1. Studentized Deleted Residuals or the **R-student**,

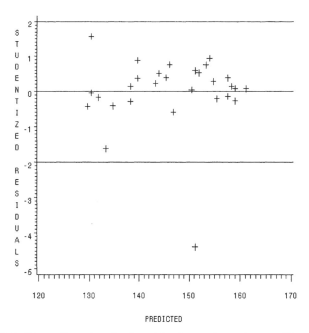

Figure 5.4: Plot of studentized residuals against the predicted values

2. Cook's D,

3. DFBETAS,

4. DFFITS,

5. Leverage,

6. Covariance ratios, and

7. PRESS statistic.

Influential observations have a strong impact on the parameter estimates. The following are three common effects of influential observations:

- a change in the slope

- a shift in the regression line (affecting the intercept)

- the slope being determined primarily by one point.

Studentized Deleted Residuals

In order to check if the observation y_i is an outlier, we compute the residual $e_{(i)} = y_i - \hat{y}_{i(i)}$ when the ith observation is deleted. We also compute the residual $e_i = y_i - \hat{y}_i$ when the ith observation is included. The observation y_i is outlying when $e_{(i)}$ tends to be larger than e_i. The method involves deleting the ith observation, fit a model based on $n-1$ observations and compute $\hat{y}_{i(i)}$, based on the levels of the independent variables for case i, as the estimate of y_i. The difference $e_{(i)} = y_i - \hat{y}_{i(i)}$ is called the deleted residual. In order to avoid refitting the regression models when the ith case is deleted, the deleted residual can be computed from a regression model based on all observations by $e_{(i)} = e_i/(1 - h_i)$. The deleted residual is the same as the PRESS residual in (5.7). The standard deviation of $e_{(i)}$ is $s_{e_{(i)}} = \sqrt{\text{MSE}_{(i)}/(1 - h_i)}$, where $\text{MSE}_{(i)}$ is the mean square error when the ith observation is deleted from the regression model.

The **studentized deleted residuals** or **R-student** is defined as

$$d_i^* = \frac{e_i}{\sqrt{\text{MSE}_{(i)}(1 - h_i)}} = e_i \left[\frac{n - p - 1}{\text{SSE}(1 - h_i) - e_i^2} \right]^{1/2}, \tag{5.8}$$

where SSE is the error sum of squares, e_i is the raw residual for observation i, and p is the number of parameters (including the β_0) in the regression model based on the n cases.

Cook's D Statistic

Cook's D statistic can be thought of as measuring the simultaneous change in parameter estimates when an observation is deleted from the general model, say,

$$y_i = \beta_0 + \beta_1 x_{i1} + \beta_2 x_{i2} + \cdots + \beta_{p-1} x_{i,p-1} + \varepsilon_i. \tag{5.9}$$

To obtain Cook's statistic, we do the following:

1. Compute parameter estimates with the entire data set. That is, obtain, $\hat{\beta}_0, \hat{\beta}_1, \hat{\beta}_2, \ldots, \hat{\beta}_{p-1}$ from the model in (5.9).

2. Remove, say, observation i, from the data and obtain new parameter estimates,

$$\hat{\beta}_0^*, \hat{\beta}_1^*, \hat{\beta}_2^*, \ldots, \hat{\beta}_{p-1}^*$$

   from the remaining $(n-1)$ observations.

3. Cook's Distance (Cook's D) statistic allows us to measure the change(s) in the parameter estimates for all such possible regressions.

4. However, we do not need to carry out n regressions that omit observation i in turn. Instead we calculate Cook's D statistic defined as

$$D_i = \frac{1}{p} \left(\frac{h_i}{1 - h_i} \right) r_i^2, \tag{5.10}$$

where r_i^2 is the studentized residuals and p is the number of parameters in the model. A large value of D_i indicates that the estimates of some or all of the parameters will change substantially if observation i is omitted from the model. Clearly, D_i will be large for large values of e_i, h_i or both.

If the change is "large" then that observation is considered overly influential. *Cook's D statistic is approximately distributed as an F with p numerator degrees of freedom and $n-p$ denominator degrees of freedom. Because relatively small changes in parameter estimates can be a sign of difficulty, the confidence level is usually set at a value much smaller than 95% (making it somewhat easier to detect influential observations).* At the 50-th percentile, Cook's D statistic is significant if

$$D_i > f_{.5,\,p,\,n-p} \approx 1.$$

It has been shown in the literature, (Fox, 1991), that another possible cut-off value for Cook's D statistic is

$$D_i > \frac{4}{n} \quad \text{where } n \text{ is the total sample size.}$$

DFBETAS and DFFITS

DFBETAS: The DFBETAS measure the change in parameter estimates when an observation is deleted from the analysis. The DFBETAS are defined, using the model in (5.9), by

$$\text{DFBETAS}_{j(i)} = \frac{\hat{\beta}_j - \hat{\beta}_j^*}{\text{s.e.}(\hat{\beta}_j)} = \frac{\hat{\beta}_j - \hat{\beta}_{j(i)}}{\text{s.e.}(\hat{\beta}_j)}, \quad j = 0, 1, \ldots, p,$$

where

- $\hat{\beta}_j$ is the parameter estimate for the jth predictor variable

- $\hat{\beta}_j^* = \hat{\beta}_{j(i)}$ is the parameter estimate for the jth predictor variable with the ith observation deleted from the model.

- s.e.$(\hat{\beta}_j)$ is the standard error of the jth parameter estimate.

DFFITS: The DFFITS measure for each ith observation, the change in predicted values when the ith observation is deleted from the analysis. That is, it measures the differences in fitted values of the models when the ith observation is deleted, for $i = 1, 2, \ldots, n$. Thus we need to,

(a) Compute $\hat{y}_i$ for the model in (5.9) when all the observations n are used.

(b) Delete observation i from the model, and again using the model in (5.9) obtain $\hat{y}_{(i)}$, the predicted value for observation i. Then, the DFFITS are defined by

$$\text{DFFITS}_i = \frac{\hat{y}_i - \hat{y}_{(i)}}{s(\hat{y}_i)} = d_i^* \sqrt{\frac{h_i}{1 - h_i}} = e_i \left[\frac{n - p - 1}{\text{SSE}(1 - h_i) - e_i^2} \right]^{1/2} \sqrt{\frac{h_i}{1 - h_i}}, \tag{5.11}$$

where

- $\hat{y}_i$ is the ith predicted value.
- $\hat{y}_{(i)}$ is the ith predicted value when the ith observation is deleted from the analysis.
- $s(\hat{y}_i)$ is the standard error of the ith predicted value.
- d_i^* is the studentized deleted residual (**R-student**) for the ith observation.

Suggested cutoff values for the DFBETAS and DFFITS are given by $\mid \text{DFBETAS}_{j(i)} \mid > 2/\sqrt{n}$ and $\mid \text{DFFITS}_i \mid > 2\sqrt{p/n}$.

Leverage Values

A simple diagnostic measure is the leverage which measures how influential an observation is with respect to the set of $(p - 1)$ predictors in the model. Leverage values, h_i, range from 0 to 1 and are closer to 1 when you are farther away from the center of the data. The leverage values are defined as the diagonal elements of the hat matrix $\mathbf{H} = \mathbf{X}(\mathbf{X}'\mathbf{X})^{-1}\mathbf{X}'$, and are referred to as "hat-values."

- Large leverage values may indicate that an observation is influential.

- The sum of all leverage values is equal to p, the number of parameters in the model. That is, $\sum_{i=0}^{n} h_i = p$. Thus, an average value for the leverage would be p/n.

- A suggested cutoff value for the leverage is

$$h_i > \frac{2p}{n} \qquad \text{or} \qquad h_i > \frac{3p}{n}. \tag{5.12}$$

An observation flagged by either of these cut-off points for the leverage indicates that at least a predictor has a leverage that is very unusual when compared to leverages obtained for other predictors in the model.

Covariance ratios

The covariance ratio for the ith observation is defined as the ratio of the generalized variance leaving out the ith observation to the generalized variance using all n cases. This is given by

$$\text{Covratio}_i = \frac{\text{MSE}_{(i)}}{\text{MSE}} \left(\frac{1}{1 - h_i} \right),$$

where $\text{MSE}_{(i)}$ is the mean square error when the ith observation is deleted, MSE is the mean square error when all n cases are used and h_i is the leverage for the ith observation. The covariance ratio provides an overall measure of the ith observation on the precision of the regression coefficient estimates. The ith observation has no effect on the precision when its covariance ratio is near 1. When the covariance ratio exceeds 1, it increases precision and when it is less than 1, it decreases precision. In a regression model with $(p-1)$ independent variables, it is suggested that an observation has effects on the precision when its covariance ratio lies outside the interval $1 \pm 3p/n$.

PRESS Statistic

The PRESS residual measures the deviation of the ith observation about the regression line formed when that observation is deleted from the analysis. In section 5.4, it is defined as $p(i) = y_i - \hat{y}_{(i)}$, where y_i is the ith value of the dependent variable and $\hat{y}_{(i)}$ is the ith predicted value from the regression equation when the ith observation is deleted from the analysis. The PRESS residual measures how well the regression model predicts the ith observation as though it were a new observation. The PRESS statistic is calculated as the sum of the PRESS residuals:

$$\text{PRESS} = \sum_{i=1}^{n} (y_i - \hat{y}_{(i)})^2 = \sum_{i=1}^{n} \left(\frac{e_i}{1 - h_i} \right)^2 .$$

Example 5.3 continued

Let us consider again the data in Table 5.4, relating the SBP (y) of thirty individuals to their ages. We have seen that observation 4 is a possible outlier in the previous section. Is it an influential observation? To answer this question, we again fit our regression model with the following output:

```
*Program for Example 5.3 with the data in Table 5.4;
data one;
input age sbp @@;
datalines;
.....
;
proc reg data=one;
model sbp=age / influence;
output out=two r=resid p=pred student=stud rstudent=rstud h=lev cookd=cooks dffits=ddfit press=pres;
run;
proc print data=two;
var age sbp pred resid stud rstud cooks ddfit pres lev;
format pred resid stud rstud cooks ddfit pres lev f8.4;
run;
data new;
hh=(2*2)/30; cookf=finv(0.5, 2, 28); dff = 2*sqrt(2/30); dfbeta=2/sqrt(30);
run;
proc print data=new;
run;
```

```
                          The REG Procedure
                           Model: MODEL1
                       Dependent Variable: sbp

                          Analysis of Variance

                                  Sum of        Mean
        Source         DF        Squares       Square     F Value    Pr > F

        Model           1     2950.82880    2950.82880      20.80    <.0001
        Error          28     3971.33787     141.83350
        Corrected Total 29    6922.16667
```

Output Statistics

Obs	Residual	RStudent	Hat Diag H	Cov Ratio	DFFITS	DFBETAS Intercept	age
1	-0.4186	-0.0369	0.1246	1.2285	-0.0139	-0.0132	0.0119
2	-18.2817	-1.6644	0.0956	0.9781	-0.5410	-0.4937	0.4366
3	17.5814	1.6233	0.1246	1.0198	0.6125	0.5788	-0.5243
4	-51.1756	-7.6935	0.0397	0.1099	-1.5648	0.3146	-0.6276
5	-1.6174	-0.1385	0.0727	1.1582	-0.0388	0.0225	-0.0286
6	8.6771	0.7405	0.0476	1.0847	0.1656	-0.0597	0.0907
7	0.5402	0.0454	0.0378	1.1175	0.0090	-0.0013	0.0031
8	-1.6174	-0.1385	0.0727	1.1582	-0.0388	0.0225	-0.0286
9	0.8038	0.0700	0.1032	1.1988	0.0237	-0.0163	0.0195
10	6.1086	0.5172	0.0420	1.1007	0.1083	-0.0280	0.0492
11	8.8347	0.7486	0.0335	1.0679	0.1393	0.0382	-0.0094
12	1.6668	0.1432	0.0781	1.1648	0.0417	-0.0252	0.0316
13	-4.7029	-0.4177	0.1328	1.2242	-0.1635	-0.1554	0.1415
14	5.9820	0.5047	0.0360	1.0949	0.0976	0.0456	-0.0266
15	10.9614	0.9429	0.0509	1.0621	0.2184	-0.0887	0.1284
16	-3.2920	-0.2801	0.0581	1.1351	-0.0695	-0.0554	0.0454
17	-2.4701	-0.2101	0.0586	1.1387	-0.0524	0.0254	-0.0344
18	2.6977	0.2270	0.0376	1.1132	0.0448	0.0235	-0.0150
19	-3.0489	-0.2630	0.0839	1.1680	-0.0796	0.0500	-0.0618
20	10.2765	0.8820	0.0505	1.0701	0.2033	0.1503	-0.1185

21	6.8244	0.5778	0.0397	1.0927	0.1175	-0.0236	0.0471
22	3.2456	0.2756	0.0546	1.1312	0.0662	-0.0296	0.0413
23	-1.8501	-0.1617	0.1094	1.2052	-0.0567	-0.0528	0.0473
24	4.2765	0.3627	0.0505	1.1216	0.0836	0.0618	-0.0487
25	4.3826	0.3763	0.0727	1.1477	0.1054	-0.0610	0.0776
26	4.5505	0.3828	0.0340	1.1013	0.0718	0.0245	-0.0099
27	0.9511	0.0819	0.0839	1.1734	0.0248	-0.0156	0.0193
28	1.7080	0.1452	0.0581	1.1400	0.0360	0.0287	-0.0235
29	-6.8811	-0.5807	0.0333	1.0853	-0.1078	-0.0220	-0.0005
30	-4.7132	-0.4071	0.0831	1.1587	-0.1226	-0.1090	0.0949

```
                Sum of Residuals                        0
                Sum of Squared Residuals        3971.33787
                Predicted Residual SS (PRESS)   4439.35322
```

Obs	age	sbp	pred	resid	stud	rstud	cooks	ddfit	pres	lev
1	42	130	130.4186	-0.4186	-0.0376	-0.0369	0.0001	-0.0139	-0.4782	0.1246
2	46	115	133.2817	-18.2817	-1.6141	-1.6644	0.1377	-0.5410	-20.2134	0.0956
3	42	148	130.4186	17.5814	1.5779	1.6233	0.1773	0.6125	20.0848	0.1246
4	71	100	151.1756	-51.1756	-4.3851	-7.6935	0.3977	-1.5648	-53.2925	0.0397
5	80	156	157.6174	-1.6174	-0.1410	-0.1385	0.0008	-0.0388	-1.7443	0.0727
6	74	162	153.3229	8.6771	0.7466	0.7405	0.0139	0.1656	9.1109	0.0476
7	70	151	150.4598	0.5402	0.0462	0.0454	0.0000	0.0090	0.5614	0.0378
8	80	156	157.6174	-1.6174	-0.1410	-0.1385	0.0008	-0.0388	-1.7443	0.0727
9	85	162	161.1962	0.8038	0.0713	0.0700	0.0003	0.0237	0.8963	0.1032
10	72	158	151.8914	6.1086	0.5241	0.5172	0.0060	0.1083	6.3765	0.0420
11	64	155	146.1653	8.8347	0.7546	0.7486	0.0099	0.1393	9.1408	0.0335
12	81	160	158.3332	1.6668	0.1458	0.1432	0.0009	0.0417	1.8081	0.0781
13	41	125	129.7029	-4.7029	-0.4240	-0.4177	0.0138	-0.1635	-5.4229	0.1328
14	61	150	144.0180	5.9820	0.5116	0.5047	0.0049	0.0976	6.2055	0.0360
15	75	165	154.0386	10.9614	0.9448	0.9429	0.0239	0.2184	11.5496	0.0509
16	53	135	138.2920	-3.2920	-0.2848	-0.2801	0.0025	-0.0605	-3.4010	0.0581
17	77	153	155.4701	-2.4701	-0.2138	-0.2101	0.0014	-0.0524	-2.6239	0.0586
18	60	146	143.3023	2.6977	0.2309	0.2270	0.0010	0.0448	2.8030	0.0376
19	82	156	159.0489	-3.0489	-0.2675	-0.2630	0.0033	-0.0796	-3.3282	0.0839
20	55	150	139.7235	10.2765	0.8855	0.8820	0.0208	0.2033	10.8227	0.0505
21	71	158	151.1756	6.8244	0.5848	0.5778	0.0071	0.1175	7.1067	0.0397
22	76	158	154.7544	3.2456	0.2803	0.2756	0.0023	0.0662	3.4331	0.0546
23	44	130	131.8501	-1.8501	-0.1646	-0.1617	0.0017	-0.0567	-2.0774	0.1094
24	55	144	139.7235	4.2765	0.3685	0.3627	0.0036	0.0836	4.5038	0.0505
25	80	162	157.6174	4.3826	0.3822	0.3763	0.0057	0.1054	4.7264	0.0727
26	63	150	145.4495	4.5505	0.3888	0.3828	0.0027	0.0718	4.7105	0.0340
27	82	160	159.0489	0.9511	0.0834	0.0819	0.0003	0.0248	1.0382	0.0839
28	53	140	138.2920	1.7080	0.1478	0.1452	0.0007	0.0360	1.8133	0.0581
29	65	140	146.8811	-6.8811	-0.5877	-0.5807	0.0060	-0.1078	-7.1183	0.0333
30	48	130	134.7132	-4.7132	-0.4133	-0.4071	0.0077	-0.1226	-5.1404	0.0831

Obs	hh	cookf	dff	dfbeta
1	0.13333	0.71059	0.51640	0.36515

In the SAS program, the options **influence** generate the influential statistics on the top half page of the output. The statistics generated are, the raw residuals e_i, the studentized deleted residuals d_i^*, the leverages h_i, the covariance ratios, the DFFITS and the DFBETAS for both the intercept and the explanatory variable (Age) in this case.

The PRESS statistic appears at the end of the output from the **influence** option in SAS. The PRESS statistic measures the effect of each observation on the regression. When the PRESS statistic is large compared to the error (residuals) sum of squares, it indicates the presence of influential observations. The residuals sum of squares is given right above the PRESS statistic in the SAS output.

We may note here that the influence option does not generate Cook's D_i as well as the PRESS statistic. These are generated into an output file 'two' and printed in the second half of the output. Further, we have asked SAS to compute cut-off points for the leverages $2p/n = 2 \times 2/30 = 0.1333$. Similar cut-off points are computed for Cook's statistic, DFFITS and DFBETAS. From this output we have the following conclusions.

1. Based on the leverages, $h_i > 0.1333$, none of the observations can be considered influential.

2. Based on the Cook's statistic, $D_i > 0.7106$, the 50-th percentile of the F distribution with 2 and 28 degrees of freedom, none of the observations can be considered influential.

3. Based on the DFFITS statistic, $|\text{DFFITS}_i| > 0.5164$, observations 2, 3 and 4 are possible influential observations.

4. Based on the studentized deleted residuals, $|d_i^*| > 3$, observation 4 has $|d_i^*| = 7.6935 \gg 2$. Hence observation 4 is a possible influential observation.

5. Based on the DFBETAS statistics, observations 2 and 3 have values that are greater in absolute values than 0.36515, the cutoff point. Hence these observations are possibly influential.

In the above output, for observation $i = 1$ for instance, we have

$$d_1^* = e_1 \left[\frac{n - p - 1}{\text{SSE}(1 - h_1) - e_1^2} \right]^{1/2} = -0.4186 \left[\frac{30 - 2 - 1}{3971.3379(1 - 0.1246) - (-0.4186)^2} \right]^{1/2} = -0.0369$$

$$D_1 = \frac{1}{p} \left(\frac{h_1}{1 - h_1} \right) r_1^2 = \frac{1}{2} \left(\frac{0.1246}{1 - 0.1246} \right) (-0.0376)^2 = 0.0001$$

$$\text{DFFITS}_1 = d_1^* \sqrt{\frac{h_1}{1 - h_1}} = (-0.0369) \sqrt{\frac{0.1246}{1 - 0.1246}} = -0.0139.$$

Similar computations lead to values for the other observations. We have presented in Table 5.5 the results of removing each of the flagged observations on the model in terms of various statistics considered in this chapter.

Observations	$\hat{\beta}_0$	$S_{\hat{\beta}_0}$	$\hat{\beta}_1$	$S_{\hat{\beta}_1}$	MSE	R^2
2,3 & 4 in	100.3568	10.4189	0.7158	0.1569	141.8335	0.4263
4 out	98.4885	5.9434	0.7719	0.0897	46.0763	0.7326
2 out	105.3450	10.5395	0.6493	0.1573	133.4001	0.3868
3 out	94.4947	10.7520	0.7957	0.1603	134.0081	0.4772
2 & 4 out	103.6622	5.2054	0.7032	0.0779	32.5011	0.7582

Table 5.5: Influence of removing the flagged observations

Example 5.4 (Multiple Regression):

Are a person's brain size and body size predictive of his/her intelligence? Data on y based on the performance IQ (PIQ) scores from Wechsler Adult Intelligence Scale (revised), brain size (x_1) based on the count from MRI scans (given as count/10000), and body size measured in height (x_2) in inches and weight (x_3) in pounds on 38 college students are displayed below.

y	x_1	x_2	x_3	y	x_1	x_2	x_3	y	x_1	x_2	x_3
124	81.69	64.5	118	90	87.89	66.0	146	72	79.35	63.0	106
150	103.84	73.3	143	96	86.54	68.0	135	124	86.67	66.5	159
128	96.54	68.8	172	120	85.22	68.5	127	132	85.78	62.5	127
134	95.15	65.0	147	102	94.51	73.5	178	137	94.96	67.0	191
110	92.88	69.0	146	84	80.80	66.3	136	110	99.79	75.5	192
131	99.13	64.5	138	86	88.91	70.0	180	86	88.00	69.0	181
98	85.43	66.0	175	84	90.59	76.5	186	81	83.43	66.5	143
84	90.49	66.3	134	134	79.06	62.0	122	128	94.81	66.5	153
147	95.55	68.8	172	128	95.50	68.0	132	124	94.94	70.5	144
124	83.39	64.5	118	102	83.18	63.0	114	94	89.40	64.5	139
128	107.95	70.0	151	131	93.55	72.0	171	74	93.00	74.0	148
124	92.41	69.0	155	84	79.86	68.0	140	89	93.59	75.5	179
147	85.65	70.5	155	110	106.25	77.0	187				

A model of the form $y_i = \beta_0 + \beta_1 x_1 + \beta_2 x_2 + \beta_3 x_3 + \varepsilon_i$ is fitted to the data.

```
/* Program to analyze the data for Example 5.4 */
data one;
infile "C:\Research\Book\Data\example54.txt";
input y x1 x2 x3;
run;
```

```
proc reg data=one;
model y=x1-x3;
test1: test x3;
run;
```

The REG Procedure
Model: MODEL1
Dependent Variable: y

Analysis of Variance

Source	DF	Sum of Squares	Mean Square	F Value	Pr > F
Model	3	5572.74444	1857.58148	4.74	0.0072
Error	34	13322	391.81789		
Corrected Total	37	18895			

Root MSE	19.79439	R-Square	0.2949	
Dependent Mean	111.34211	Adj R-Sq	0.2327	
Coeff Var	17.77799			

Parameter Estimates

Variable	DF	Parameter Estimate	Standard Error	t Value	Pr > |t|
Intercept	1	111.35361	62.97110	1.77	0.0860
x1	1	2.06037	0.56345	3.66	0.0009
x2	1	-2.73193	1.22943	-2.22	0.0330
x3	1	0.00055994	0.19707	0.00	0.9977

The REG Procedure
Model: MODEL1

Test test1 Results for Dependent Variable y

Source	DF	Mean Square	F Value	Pr > F
Numerator	1	0.00316	0.00	0.9977
Denominator	34	391.81789		

It is very important before carrying out diagnostics tests to first ascertain the most parsimonious model. In this example, the partial F-test for the hypotheses

$$H_0 : \beta_3 = 0$$
$$H_a : \beta_3 \neq 0,$$

$$(5.13)$$

gives a p-value of 0.9977 indicating that given that X_1 and X_2 are already in the model, X_3 can be dropped. Hence our revised or reduced model would be

$$y_i = \beta_0 + \beta_1 x_1 + \beta_2 x_2 + \varepsilon_i.$$

$$(5.14)$$

```
/* Program to analyze the data for Example 5.4 */
data one;
infile "C:\Research\Book\Data\example54.txt";
input y x1 x2 x3;
run;
proc reg data=one;
model y=x1-x2/influence;
output out=two r=resid p=pred student=stud rstudent=rstud h=lev cookd=cooks dffits=dffit;
run;
proc print data=two;
var y pred resid stud rstud lev cooks dffit;
format pred resid stud rstud lev cooks dffit 8.4;
run;
data new;
hh=(2*3)/38;
cookf=finv(0.5, 3, 35);
dff = 2*sqrt(3/38);
proc print data=new;
run;
```

The REG Procedure
Model: MODEL1
Dependent Variable: y

```
                              Analysis of Variance

                                    Sum of          Mean
        Source              DF      Squares         Square    F Value   Pr > F

        Model                2    5572.74127     2786.37064     7.32    0.0022
        Error               35       13322        380.62318
        Corrected Total     37       18895

                  Root MSE            19.50957    R-Square     0.2949
                  Dependent Mean     111.34211    Adj R-Sq     0.2546
                  Coeff Var           17.52218

                            Parameter Estimates

                          Parameter      Standard
        Variable    DF     Estimate        Error    t Value   Pr > |t|

        Intercept    1     111.27567      55.86732     1.99    0.0542
        x1           1       2.06065       0.54664     3.77    0.0006
        x2           1      -2.72993       0.99319    -2.75    0.0094

  Obs   y      pred     resid       stud     rstud      lev     cooks    dffit
    1  124   103.5297   20.4703    1.0883    1.0912   0.0704   0.0299   0.3003
    2  150   125.1497   24.8503    1.3549    1.3719   0.1163   0.0805   0.4976
    3  128   122.3917    5.6083    0.2949    0.2910   0.0500   0.0015   0.0667
    4  134   129.9011    4.0989    0.2212    0.2182   0.0981   0.0018   0.0719
    5  110   114.3037   -4.3037   -0.2238   -0.2208   0.0289   0.0005  -0.0381
    6  131   139.4675   -8.4675   -0.4787   -0.4733   0.1779   0.0165  -0.2202
    7   98   107.1417   -9.1417   -0.4787   -0.4733   0.0418   0.0033  -0.0989
    8   84   116.7496  -32.7496   -1.7109   -1.7615   0.0373   0.0378  -0.3469
    9  147   120.3516   26.6484    1.3957    1.4156   0.0422   0.0286   0.2973
   10  124   107.0328   16.9672    0.8970    0.8944   0.0599   0.0171   0.2257
   11  128   142.6278  -14.6278   -0.8496   -0.8462   0.2213   0.0684  -0.4510
   12  124   113.3352   10.6648    0.5544    0.5489   0.0279   0.0029   0.0929
   13  147    95.3103   51.6897    2.7546    3.0678   0.0749   0.2047   0.8728
   14   90   112.2109  -22.2109   -1.1597   -1.1656   0.0363   0.0169  -0.2262
   15   96   103.9691   -7.9691   -0.4163   -0.4113   0.0373   0.0022  -0.0809
   16  120    99.8841   20.1159    1.0581    1.0600   0.0504   0.0198   0.2443
   17  102   105.3779   -3.3779   -0.1797   -0.1772   0.0720   0.0008  -0.0494
   18   84    96.7819  -12.7819   -0.6828   -0.6775   0.0794   0.0134  -0.1990
   19   86   103.3930  -17.3930   -0.9099   -0.9076   0.0399   0.0115  -0.1850
   20   84    89.1104   -5.1104   -0.2922   -0.2884   0.1966   0.0070  -0.1427
   21  134   104.9351   29.0649    1.5826    1.6189   0.1139   0.1073   0.5804
   22  128   122.4326    5.5674    0.2925    0.2887   0.0485   0.0015   0.0651
   23  102   110.6950   -8.6950   -0.4642   -0.4590   0.0784   0.0061  -0.1339
   24  131   107.4946   23.5054    1.2353    1.2450   0.0487   0.0261   0.2818
   25   84    90.2040   -6.2040   -0.3373   -0.3329   0.1110   0.0047  -0.1176
   26  110   120.0152  -10.0152   -0.5680   -0.5624   0.1832   0.0241  -0.2664
   27   72   102.8027  -30.8027   -1.6644   -1.7095   0.1001   0.1027  -0.5702
   28  124   108.3319   15.6681    0.8178    0.8138   0.0356   0.0082   0.1563
   29  132   117.4176   14.5824    0.7824    0.7780   0.0873   0.0195   0.2407
   30  137   124.0497   12.9503    0.6833    0.6780   0.0562   0.0093   0.1654
   31  110   110.7983   -0.7983   -0.0434   -0.0428   0.1131   0.0001  -0.0153
   32   86   104.2478  -18.2478   -0.9523   -0.9510   0.0354   0.0111  -0.1822
   33   81   101.6554  -20.6554   -1.0884   -1.0913   0.0537   0.0224  -0.2601
   34  128   125.1056    2.8944    0.1532    0.1511   0.0626   0.0005   0.0391
   35  124   114.4538    9.5462    0.4986    0.4932   0.0369   0.0032   0.0966
   36   94   119.4173  -25.4173   -1.3431   -1.3592   0.0590   0.0377  -0.3405
   37   74   100.9014  -26.9014   -1.4450   -1.4687   0.0895   0.0684  -0.4604
   38   89    98.0223   -9.0223   -0.4953   -0.4899   0.1282   0.0120  -0.1879

                    Obs      hh       cookf      dff
                     1    0.15789   0.80425   0.56195
```

1. Based on the leverages, h_i, observations 6, 11 and 20 are suspects.

2. Based on the Cook's statistic, D_i, none of the observations is suspect.

3. Based on the DFFITS statistic, $|\text{DFFITS}_i| > 0.562$, observations 13, 21 and 27 are possible influential observations.

4. Based on the studentized deleted residuals, $|d_i^*| > 2$, observations 13 is suspect.

After reaching this stage where some observations have been identified as possible influential observations, one should check to see if the observations are in error. If not, one can use robust regression (discussed in section 5.9) to reduce the influence of the observations.

5.6   Detecting Multicollinearity

Suppose we have a dependent variable y with $x_1, x_2, \ldots, x_{p-1}$ explanatory variables. One of the problems with building a linear model involving y and the $p - 1$ explanatory variables exists when the explanatory variables are truly not independent or rather when there are strong inter-correlations among some or all of the explanatory variables. This problem is referred to as *multicollinearity*. Simply put, multicollinearity exists when two or more of the explanatory variables in the regression model are correlated. Thus the inclusion of a large number of explanatory or independent variables in a regression model often leads to multicollinearity. Some possible signs of multicollinearity in a regression analysis output that should alert us are the following:

- There are significant correlations between pairs of independent variables in the model.

- The omnibus hypothesis $H_0 : \beta_1 = \beta_2 = \cdots = \beta_{p-1} = 0$ being tested by the overall F test is significant, indicating that at least one of the β parameters is not zero but all or nearly all the t individual tests are non-significant.

- Parameter estimates from the model have signs that are not consistent with our expectations. If for instance, we expect the estimated $\hat{\beta}_i$ to be positive but the estimate turns out to be negative, then we need to be concerned.

Other Effects of Multicollinearity are

- The estimates $\hat{\beta}_j$ are subject to numerical error and are unreliable. This is sometimes reflected in large changes in their magnitudes with small changes in data. Sometimes the signs of the $\hat{\beta}_j$ are reversed.

- Most of the coefficients have large standard errors and as a result are statistically non-significant even if the overall F-statistic is significant

Example 5.5

The data in Table 5.6, taken from an anonymous source, relate to the heat evolved in calories during hardening of cement on a per gram basis (y) along with the percentages of four ingredients: tricalcium aluminate (x_1), tricalcium silicate (x_2), tetracalcium alumino ferrite (x_3), and decalcium silicate (x_4). A regression model of the form $y_i = \beta_0 + \beta_1 x_1 + \beta_2 x_2 + \beta_3 x_3 + \beta_4 x_4 + \varepsilon_i$ is to be fitted.

No	x_1	x_2	x_3	x_4	y
1	7	26	6	60	78.5
2	1	29	15	52	74.3
3	11	56	8	20	104.3
4	11	31	8	47	87.6
5	7	52	6	33	95.9
6	11	55	9	22	109.2
7	3	71	17	6	102.7
8	1	31	22	44	72.5
9	2	54	18	22	93.1
10	21	47	4	26	115.9
11	1	40	23	34	83.8
12	11	66	9	12	113.3
13	10	68	8	12	109.4

Table 5.6: Heat evolved in calories during hardening of cement

```
/* Program to analyze the data for Example 5.5 */
data one;
input x1-x4 y @@;
datalines;
.....
;
```

```
proc corr data=one;
var x1-x4;
run;
proc reg data=one;
model y=x1-x4;
run;
```

The CORR Procedure

4 Variables: x1   x2   x3   x4

Simple Statistics

Variable	N	Mean	Std Dev	Sum	Minimum	Maximum
x1	13	7.46154	5.88239	97.00000	1.00000	21.00000
x2	13	48.15385	15.56088	626.00000	26.00000	71.00000
x3	13	11.76923	6.40513	153.00000	4.00000	23.00000
x4	13	30.00000	16.73818	390.00000	6.00000	60.00000

Pearson Correlation Coefficients, N = 13
Prob > |r| under H0: Rho=0

	x1	x2	x3	x4
x1	1.00000	0.22858	-0.82413	-0.24545
		0.4526	0.0005	0.4189
x2	0.22858	1.00000	-0.13924	-0.97295
	0.4526		0.6501	<.0001
x3	-0.82413	-0.13924	1.00000	0.02954
	0.0005	0.6501		0.9237
x4	-0.24545	-0.97295	0.02954	1.00000
	0.4189	<.0001	0.9237	

The REG Procedure
Model: MODEL1
Dependent Variable: y

Analysis of Variance

Source	DF	Sum of Squares	Mean Square	F Value	Pr > F
Model	4	2667.89944	666.97486	111.48	<.0001
Error	8	47.86364	5.98295		
Corrected Total	12	2715.76308			

Root MSE	2.44601	R-Square	0.9824
Dependent Mean	95.42308	Adj R-Sq	0.9736
Coeff Var	2.56333		

Parameter Estimates

Variable	DF	Parameter Estimate	Standard Error	t Value	Pr > |t|
Intercept	1	62.40537	70.07096	0.89	0.3991
x1	1	1.55110	0.74477	2.08	0.0708
x2	1	0.51017	0.72379	0.70	0.5009
x3	1	0.10191	0.75471	0.14	0.8959
x4	1	-0.14406	0.70905	-0.20	0.8441

In the above example involving four explanatory variables, the correlation analysis obtained from the use of **proc corr** in SAS indicates the following:

1. Variables x_1 and x_3 are significantly correlated, $r_{x_1 x_3}$ being -0.8241 (p-value $= 0.0005$). Similarly, variables x_2 and x_4 are significantly correlated, $r_{x_2 x_4}$ being -0.9730 (p-value $= < 0.0001$). Since some of these inter-correlations are significant, we should therefore expect multicollinearity to manifest itself in this example.

2. The analysis of variance table indicates an overall computed F-value of 111.48 with a corresponding p-value of < 0.0001, clearly indicating that at least one of $\beta_1, \beta_2, \beta_3$ or β_4 is not zero. However, the individual computed t values under parameter estimates indicate that none of the individual parameters is significantly different from zero because all the p-values are greater than $\alpha = 0.05$ level of significance. This contradicts the omnibus test and we should therefore expect to encounter multicollinearity in this example.

We present in Figure 5.5, the bivariate scatter plot matrix for the four explanatory variables x_1 to x_4. The plot suggests that variables x_2 and x_4 are strongly linearly correlated as well as variables x_1 and x_3. We will discuss in this section two methods of diagnosing multicollinearity in a regression model involving k explanatory variables.

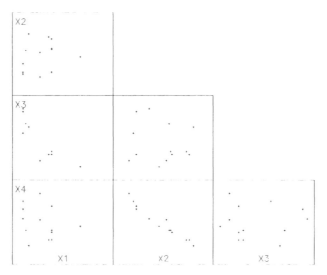

Figure 5.5: Bivariate scatter plot matrix for the explanatory variables

From Example 5.5, consider constructing $\binom{4}{1}$ multiple regressions only on the explanatory variables, namely,

$$x_1 = \beta_{01} + \beta_{21}x_2 + \beta_{31}x_3 + \beta_{41}x_4 + \varepsilon_1 \tag{5.15a}$$

$$x_2 = \beta_{02} + \beta_{12}x_1 + \beta_{32}x_3 + \beta_{42}x_4 + \varepsilon_2 \tag{5.15b}$$

$$x_3 = \beta_{03} + \beta_{13}x_1 + \beta_{23}x_2 + \beta_{43}x_4 + \varepsilon_3 \tag{5.15c}$$

$$x_4 = \beta_{04} + \beta_{14}x_1 + \beta_{24}x_2 + \beta_{34}x_3 + \varepsilon_4. \tag{5.15d}$$

From (5.15a) for instance, we can obtain $R_1^2 = R^2(x_1|x_2, x_3, x_4)$, the multiple coefficient of determination. Similarly, we can obtain corresponding multiple coefficients of determination $R_2^2 = R^2(x_2|x_1, x_3, x_4)$, $R_3^2 = R^2(x_3|x_1, x_2, x_4)$ and $R_4^2 = R^2(x_4|x_1, x_2, x_3)$ corresponding respectively, to the regression equations in (5.15b), (5.15c) and (5.15d). For $p-1$ explanatory variables, we would therefore obtain $R_j^2, j = 1, 2, \ldots, p-1$. In this example, we have $R_j^2, j = 1, 2, 3, 4$.

5.6.1 Variance Inflation Factor (VIF)

The variance inflation factor (VIF) for the j-th regression coefficient which can be used to measure collinearity is defined as

$$\mathrm{VIF}_j = \frac{1}{1 - R_j^2}, \qquad j = 1, 2, \ldots, p-1, \tag{5.16}$$

where R_j^2 is the multiple coefficient of determination obtained from the model

$$x_j = \beta_0 + \beta_1 x_1 + \beta_2 x_2 + \cdots + \beta_{j-1}x_{j-1} + \beta_{j+1}x_{j+1} + \cdots + \beta_{p-1}x_{p-1} + \varepsilon_j.$$

```
proc reg data=one;
model x1=x2-x4;
model x2=x1 x3 x4;
model x3=x1 x2 x4;
model x4=x1-x3;
run;
```

	Root MSE	1.09475	R-Square	0.9740
	Root MSE	1.12648	R-Square	0.9961
	Root MSE	1.08033	R-Square	0.9787
	Root MSE	1.14990	R-Square	0.9965

When the regressions in (5.15a) to (5.15d) are performed on the data set in this example with the above SAS program, the multiple coefficients of determination obtained for each regression are

$$R_1^2 = 0.9740, \quad R_2^2 = 0.9961, \quad R_3^2 = 0.9787, \quad \text{and} \quad R_4^2 = 0.9965.$$

Consequently, the variance inflation factors employing (5.16) are

$$\text{VIF}_1 = 38.462, \quad \text{VIF}_2 = 256.410, \quad \text{VIF}_3 = 46.948, \text{ and VIF}_4 = 285.714.$$

The above can also be accomplished in SAS with the following program by specifying the options **VIF, TOL** in the model statement. We present a partial output below:

```
proc reg data=one;
model y=x1-x4 / vif tol;
run;
```

```
                        The REG Procedure
                           Model: MODEL1
                       Dependent Variable: y

                       Analysis of Variance

                                Sum of       Mean
       Source            DF     Squares      Square    F Value   Pr > F

       Model              4    2667.89944   666.97486   111.48   <.0001
       Error              8      47.86364     5.98295
       Corrected Total   12    2715.76308

              Root MSE              2.44601   R-Square   0.9824
              Dependent Mean       95.42308   Adj R-Sq   0.9736
              Coeff Var             2.56333

                       Parameter Estimates

              Parameter   Standard                              Variance
Variable   DF  Estimate    Error   t Value  Pr > |t|  Tolerance  Inflation

Intercept  1   62.40537   70.07096   0.89    0.3991       .          0
x1         1    1.55110    0.74477   2.08    0.0708    0.02598    38.49621
x2         1    0.51017    0.72379   0.70    0.5009    0.00393   254.42317
x3         1    0.10191    0.75471   0.14    0.8959    0.02134    46.86839
x4         1   -0.14406    0.70905  -0.20    0.8441    0.00354   282.51286
```

The VIFs computed earlier agree closely with those obtained from SAS. The option **TOL** also computes *tolerance*, which is defined as

$$\text{TOL}_j = \frac{1}{\text{VIF}_j} = 1 - R_j^2.$$

Multicollinearity is adjudged to be present for values of $R_j^2 > 0.90$. That is, for $R_j > 0.95$. If $R_j > 0.95$, this implies that, we would have extreme multicollinearity if either

$$\text{VIF}_j > 10 \quad \text{or if} \quad \text{TOL}_j < 0.1. \tag{5.17}$$

Multicollinearity is therefore assumed present whenever the $\text{VIF}_j > 10$. We can also use either the TOL_j or R_j^2. In our case, we would use VIF and for the data in this example, the computed VIFs all exceed 10, hence multicollinearity is present.

5.6.2   Condition Index Diagnostics

Let the inter-correlations between the four explanatory variables in Example 5.5 be represented by the correlation matrix in (5.18).

$$\begin{array}{c} \\ x_1 \\ x_2 \\ x_3 \\ x_4 \end{array} \begin{array}{cccc} x_1 & x_2 & x_3 & x_4 \\ \begin{pmatrix} 1 & r_{12} & r_{13} & r_{14} \\ r_{21} & 1 & r_{23} & r_{24} \\ r_{31} & r_{32} & 1 & r_{34} \\ r_{41} & r_{42} & r_{43} & 1 \end{pmatrix}. \end{array} \tag{5.18}$$

That is, let

$$R_{ij} = \begin{bmatrix} 1 & r_{12} & r_{13} & r_{14} \\ r_{21} & 1 & r_{23} & r_{24} \\ r_{31} & r_{32} & 1 & r_{34} \\ r_{41} & r_{42} & r_{43} & 1 \end{bmatrix}, \qquad i = 1, 2, \ldots, 4; j = 1, 2, \ldots, 4.$$

Our diagnostics involve obtaining the *eigenvalues* of the explanatory variables' correlation matrix R_{ij}. That is, we need to solve

$$|R_{ij} - \lambda_j \mathbf{I}| = 0.$$

An eigenvalue that is closer to zero is associated with *near collinearity* among the explanatory variables. Thus, when an eigenvalue is exactly zero, this implies that there is perfect collinearity among the explanatory variables. With $p - 1$ explanatory variables, the number of zero (or near zero) eigenvalues gives the number of collinearities or near collinearities present in the data. However, multicollinearity is often measured by computing the *condition number* of the correlation matrix, which is defined as the quantity

$$\phi = \frac{\lambda_{\max}}{\lambda_{\min}}. \tag{5.19}$$

If the eigenvalues are ordered from largest (λ_1) to smallest, then the condition index for each eigenvalue is therefore computed as

$$\phi = \sqrt{\frac{\lambda_1}{\lambda_j}}, \quad j = 1, 2, \ldots, p - 1. \tag{5.20}$$

We may note here that

$$\sum_{i=1}^{p-1} \lambda_j = p - 1.$$

We can implement this procedure in SAS by the use of the option **collin** in the model statement. The use of the option **collinoint** performs the same analysis with the intercept adjusted out. The **collinoint** option adjusted out the intercept variable before the eigenvalues and eigenvectors are extracted. Both the **collin** and **collinoint** options give the analysis of the structure of $\mathbf{X'X}$. The **collin** option does not center the variables but will include the intercept variable and $\mathbf{X'X}$ is scaled to have 1's on the diagonal. On the other hand, the **collinoint** option scales and centers the variables but will first adjust out the intercept variable. The latter option is often preferred. We have presented a partial output from the regression analysis below.

```
proc reg data=one;
model y=x1-x4 / collin collinoint;
run;
```

Collinearity Diagnostics

Number	Eigenvalue	Condition Index
1	4.11970	1.00000
2	0.55389	2.72721
3	0.28870	3.77753
4	0.03764	10.46207
5	0.00006614	249.57825

Collinearity Diagnostics

Number	Intercept	x1	x2	x3	x4
		----Proportion of Variation----			
1	0.00000551	0.00036889	0.00001833	0.00021022	0.00003641
2	8.812348E-8	0.01004	0.00001265	0.00266	0.00010070
3	3.060952E-7	0.00057551	0.00031981	0.00159	0.00168
4	0.00012679	0.05745	0.00278	0.04569	0.00088373
5	0.99987	0.93157	0.99687	0.94985	0.99730

Collinearity Diagnostics (intercept adjusted)

		Condition
Number	Eigenvalue	Index
1	2.23570	1.00000
2	1.57607	1.19102
3	0.18661	3.46134
4	0.00162	37.10634

Collinearity Diagnostics (intercept adjusted)

	-----------------Proportion of Variation---------------			
Number	x1	x2	x3	x4
1	0.00263	0.00055897	0.00148	0.00047533
2	0.00427	0.00042729	0.00495	0.00045729
3	0.06352	0.00208	0.04650	0.00072440
4	0.92958	0.99693	0.94707	0.99834

Others have defined the *condition number* to be $\sqrt{\lambda_1/\lambda_k}$ rather than as it is defined in (5.19). In either case, serious multicollinearity is observed if the latter is greater than 30 or if the former is greater than 1000. In this example, the condition number by the second definition is $37.1063 > 30$. Hence, there is collinearity present in these data and there is a need to correct or further investigate this circumstance. Furthermore on the output, under the collinearity diagnostics, the 0.92958 under variable x_1 in the last row indicates that 92.958% of the variance of the x_1 parameter is associated with eigenvalue 4 (# 4). Since there are four variables, hence there are 4 eigenvalues of the correlation matrix arranged from largest to smallest. The magnitude of multicollinearity can be detected from the relative magnitude of these eigenvalues. Eigenvalues close to zero reveals that there is linear dependence or exact collinearity. There is only one in our example, thus we can assume that we have only one set of very strong relationship among the four variables.

5.7   Weighted Least Squares

The weighted least squares method is often used mainly to stabilize the variance of the error term ε, so that the constant variance assumption can be satisfied. As the name implies, we need to find appropriate weights ω_i for each observation so that the variance can be stabilized.

Example 5.2 continued

In section 5.2 (see Figure 5.3), we observe that the data in this example have unequal error variances. To correct this, we could do one of two things, namely,

1. Conduct a weighted least squares and obtain the appropriate weighted least squares parameter estimates (it is important to note here that the ordinary least squares assigns weights of 1 to each observation).

2. Find a suitable variance stabilizing transformation.

In this section, we apply weighted least squares and defer transformation method to the next section.

   To employ the weighted least squares method, we must first obtain the weights ω_i. For the data in the example, it is important to first partition the data into various classes based on the explanatory variable x. If each value of x in the data is replicated, then we would ideally use each value of x to represent each class. However, in our example, each value of x is not replicated, and we therefore subset the data into four (near equal number of observations) classes, $20 \leq x_j < 30$ consisting of $n_1 = 13$ observations, $30 \leq x_j < 40$, consisting of $n_2 = 13$ observations, $40 \leq x_j < 50$, consisting of $n_3 = 15$ observations, and $50 \leq x_j < 60$ also consisting of $n_4 = 13$ observations.

   For each class, we obtain the sample variances s_j^2 of the residuals from the regression of y on x from each class. These variances are then compared with different functions of $\bar{x}$, namely, $\bar{x}$, $\bar{x}^2$ and $\sqrt{\bar{x}}$. For our data, the estimated residual variances (error mean squares) s_j^2, for each of the groups $j = 1, 2, 3, 4$ are obtained respectively as 16.0990292, 39.9956409, 85.4293436, and 121.222301.

Using the midpoint of each range of ages as the value of $\bar{x}$, we obtain Table 5.7.

```
/* Program to analyze data in Table 5.1 for Example 5.2 */
data one;
input x y @@;
datalines;
.....
;
```

| Group | | Sample | | Weights | | |
j	Age	Size	$\bar{x}_j$	$s_j^2/\bar{x}_j$	$s_j^2/\bar{x}_j^2$	$s_j^2/\sqrt{\bar{x}_j}$
1	$20 \leq x_j < 30$	13	25	0.6440	0.0258	3.2198
2	$30 \leq x_j < 40$	13	35	1.1427	0.0326	6.7605
3	$40 \leq x_j < 50$	15	45	1.8984	0.0422	12.7351
4	$50 \leq x_j < 60$	13	55	2.2040	0.0400	16.3456

Table 5.7: Sample means of x compared with residual variances

```
run;
proc means data=one;
var x y;
run;
data new;
set one;
if 20<=x<30 then do; agegp=1; wt=1/(25*25); end;
if 30<=x<40 then do; agegp=2; wt=1/(35*35); end;
if 40<=x<50 then do; agegp=3; wt=1/(45*45); end;
if 50<=x<60 then do; agegp=4; wt=1/(55*55); end;
run;

proc reg data=new;
by agegp;
model y=x;
output out=two r=resid;
run;
proc means data=two n mean var;
class agegp;
var x resid;
run;
proc reg data=new;
weight wt;
model y=x;
output out=three p=pred r=resid;
run;
data new2;
set three;
resida=sqrt(wt)*resid;
run;
goptions vsize=5in hsize=5in;
symbol c=b;
axis1 label=(angle=-90 rotate=90 'RESIDUALS');
axis2 label=('PREDICTED');
proc gplot data=new2;
plot resida*pred='plus' / vaxis=axis1 haxis=axis2 noframe vref=0;
run;
```

```
 Variable    N          Mean        Std Dev       Minimum        Maximum
 x          54     39.5740741    11.5409065    20.0000000     59.0000000
 y          54     79.1111111    10.4838900    63.0000000    109.0000000

                       The MEANS Procedure

            N
 agegp     Obs   Variable   Label       N          Mean        Variance
     1      13   x                      13    24.1538462       8.4743590
                 resid      Residual    13    4.372571E-15    16.0990292

     2      13   x                      13    34.4615385       9.4358974
                 resid      Residual    13   -5.46571E-15     39.9956409

     3      15   x                      15    44.6666667       8.8095238
                 resid      Residual    15           0        85.4293436

     4      13   x                      13    54.2307692       9.0256410
                 resid      Residual    13   -2.18629E-15    121.2223011
```

We can adopt here the procedure that every y observation in a group receives a weight which is the reciprocal of the estimated variance for that group. That is, $\omega_j = 1/s_j^2$, or we can assume that a reasonable approximation to the weight of each group is

$$\omega_j = \frac{1}{\bar{x}_j^2},$$

since the weight $s_i^2/\bar{x}^2$ seems to be stabilized from Table 5.7. For our example with one explanatory variable, x, the normal equations for the weighted least squares regression are

$$\sum \omega_i y_i = \hat{\beta}_{w0} \sum \omega_i + \hat{\beta}_{w1} \sum \omega_i x_i$$
$$\sum \omega_i x_i y_i = \hat{\beta}_{w0} \sum \omega_i x_i + \hat{\beta}_{w1} \sum \omega_i x_i^2. \tag{5.21}$$

Hence, the weighted least squares estimates from (5.21) are obtained as

$$\hat{\beta}_{w1} = \frac{\sum \omega_i x_i y_i - \dfrac{\sum \omega_i x_i \sum \omega_i y_i}{\sum \omega_i}}{\sum \omega_i x_i^2 - \dfrac{(\sum \omega_i x_i)^2}{\sum \omega_i}} \tag{5.22a}$$

$$\hat{\beta}_{w0} = \frac{\sum \omega_i y_i - \hat{\beta}_{w1} \sum \omega_i x_i}{\sum \omega_i}. \tag{5.22b}$$

We next obtain the weighted least squares residuals, which are computed with the following expression:

$$\sqrt{\omega}_j (y_i - \hat{y}_i) = \sqrt{\omega}_j \times \text{new residuals}.$$

A plot of the modified residuals against x or predicted will indicate whether the weighted least squares approach has stabilized the variances. This plot is presented in Figure 5.6. Clearly, the pattern of the distribution of the residuals around the mean of zero is now random, and indicates that the variance have now being stabilized across the x values.

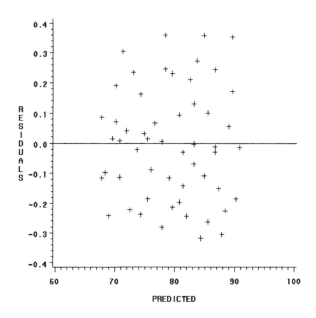

Figure 5.6: Plot of weighted residuals against predicted

The parameter estimates and their corresponding standard errors under both the ordinary least squares (OLS) regression and the weighted least squares (WLS) regression are presented in Table 5.8. While the parameter estimates are very similar, we notice that the standard errors of the parameter estimates are much smaller than their OLS counterparts. Further, the error mean squares (estimates of population variances) under both models are respectively, 66.35317 and 0.03575. Clearly, the WLS gives a much smaller error variance and it is therefore a better model. The estimated WLS regression equation is

$$\hat{y}_i = 56.1304 + 0.5867\, x_i. \tag{5.23}$$

Unweighted LS (OLS)				Weighted LS			
$\hat{\beta}_0$	$\mathrm{se}_{\hat{\beta}_0}$	$\hat{\beta}_1$	$\mathrm{se}_{\hat{\beta}_1}$	$\hat{\beta}_0$	$\mathrm{se}_{\hat{\beta}_0}$	$\hat{\beta}_1$	$\mathrm{se}_{\hat{\beta}_1}$
56.1569	3.9937	0.5800	0.0970	56.1304	2.9670	0.5867	0.0850

Table 5.8: Parameter estimates and standard errors for both OLS and WLS models

Alternatively, some authors may advocate using instead weights

$$\omega_i = \frac{1}{x_i^2}, \quad i = 1, 2, \ldots, n.$$

If we employ this, we would have the following parameter estimates and corresponding estimated standard errors in parentheses:

$$\hat{\beta}_0 = 55.83104(2.78093), \quad \hat{\beta}_1 = 0.58883(0.08158).$$

Clearly, the estimated standard errors here are slightly smaller than those obtained for the earlier weights. Either case will be fine however. The corresponding weighted least squares regression equation is therefore,

$$\hat{y}_i = 55.8310 + 0.5888\, x_i. \tag{5.24}$$

The groupings used above can also be obtained in SAS by utilizing PROC RANK. This is accomplished with the following SAS statements. It also generates the residuals for each group together with their standard deviations and variances.

```
set one;
*-- Examine residual variance in relation to age;
proc sort data=one;
by x;
run;
*-- Break up into 4 age groups;
proc rank data=one groups=4 out=wttss;
var x;
ranks agegp;
run;
proc print data=wttss;
run;
proc reg data=wttss;
model y=x/noprint;
by agegp;
output out=aa r=resid;
run;
proc univariate noprint data=aa;
by agegp;
var resid;
output out=sumry n=n mean=mean std=std var=var qrange=iqr;
run;
proc print data=sumry;
run;
      Obs    agegp    n    mean     std      var       iqr
       1       0     13      0    4.0124   16.099    4.8366
       2       1     13      0    6.3242   39.996    8.0564
       3       2     15      0    9.2428   85.429   16.4541
       4       3     13      0   11.0101  121.222   19.9560
```

5.8   Transformation Approach

In this section, we consider the transformation method as a remedial measure to address non-normality of error terms and unequal error variances. A transformation of y may remedy both problems. In some situations, we may need to transform the independent variables especially to obtain a linear regression function. For the data in Example 5.2, the WLS regression seems to work very well. In this section, we explore an alternative to weighted least squares approach by finding a suitable transformation of y.

The Box-Cox Transformations

The Box-Cox (1964) approach identifies a suitable transformation from the family of power transformations of the form

$$y' = y^\lambda, \tag{5.25}$$

where λ is to be estimated from the data. Possible values of λ could include the following:

$$
\begin{aligned}
\lambda &= 2 & y' &= y^2 \\
\lambda &= 1 & y' &= y \\
\lambda &= 0.5 & y' &= \sqrt{y} \\
\lambda &= 0 & y' &= \ln y \\
\lambda &= -0.5 & y' &= \frac{1}{\sqrt{y}} \\
\lambda &= -1 & y' &= \frac{1}{y}.
\end{aligned}
\tag{5.26}
$$

To determine λ, a regression of the following form is carried out.

$$y_i^\lambda = \beta_0 + \beta_1 x_i + \varepsilon_i, \tag{5.27}$$

where $\beta_0, \beta_1, \lambda$ and σ^2, the variance of ε are to be estimated from available data. One procedure to accomplish this is to use a numerical search for values of λ ranging from -3 to 3 (the default in SAS). For instance, we can use values of $\lambda = -3, -1.8. -1.6, \ldots, 1.6, 1.8, 3.0$. For each value of λ, Kutner et al. (2005, page 135) have suggested that we first standardize the y^λ, such that the magnitude of the error sum of squares (SSE) or root mean square error (RMSE) does not depend on the value of λ. They suggested that we use the power *transformation* for non-negative dependent observations,

$$
y_i^{(\lambda)} = \begin{cases} K_1(y_i^\lambda - 1), & \lambda \neq 0 \\ K_2(\ln y_i), & \lambda = 0, \end{cases}
\tag{5.28}
$$

where

$$K_2 = \left(\prod_{i=1}^{n} y_i \right)^{1/n} = \exp \left[\frac{1}{n} \sum_{i=1}^{n} \ln(y_i) \right], \tag{5.29a}$$

$$K_1 = \frac{1}{\lambda K_2^{\lambda-1}}. \tag{5.29b}$$

K_2 here is the geometric mean of the y_i observations. We assume that the transformed values $y_i^{(\lambda)}$ follow a normal linear model with parameters β and σ^2 for some values of λ. The log-likelihood which SAS uses can be shown to be

$$\ln L(\lambda) = u - \frac{n}{2} \ln [\text{SSE}(y_i^{(\lambda)})], \tag{5.30}$$

where $u = \frac{n}{2} \ln(2\pi/n) - \frac{n}{2}$ is a constant not involving λ and SSE is the error sum of squares from a regression of y' on the explanatory variable x. For any given model, the root mean square error is defined by

$$\text{RMSE} = \sqrt{\frac{\text{SSE}}{\text{error d.f.}}}.$$

Since the degrees of freedom is constant for a given model, plotting $\ln L(\lambda)$ versus values of λ will indicate the value of λ that optimizes the log likelihood profile. We can also plot the RMSE against values of λ to obtain similar results. This is the approach employed in SAS, although others have suggested instead, plotting SSE against values of λ. The results would be the same because of the relationship between SSE and RMSE. The **proc transreg** in SAS can be employed to implement the Box-Cox transformation. This is done in the following program with a partial output.

```
*set one;
ods output boxcox=b;
ods exclude boxcox;
proc transreg data=one ss2 details;
model boxcox(y/lambda=-3 to 3 by 0.05 alpha=0.05)=identity(x);
output out=two residuals predicted;
run;
proc print data=b; run;
```

Dependent	Lambda	Convenient	RSquare	RMSE	LogLike	CI
BoxCox(y)	-3.00		0.43	7.447013	-108.422	*
BoxCox(y)	-2.95		0.43	7.441208	-108.380	*
BoxCox(y)	-2.90		0.43	7.435738	-108.340	*
BoxCox(y)	-2.85		0.43	7.430603	-108.303	*
. . . .						
BoxCox(y)	-2.25		0.43	7.395413	-108.046	*
BoxCox(y)	-2.20		0.43	7.394706	-108.041	*
BoxCox(y)	-2.15		0.43	7.394346	-108.039	*
BoxCox(y)	-2.10		0.43	7.394334	-108.039	<
BoxCox(y)	-2.05		0.43	7.394670	-108.041	*
BoxCox(y)	-2.00		0.43	7.395356	-108.046	*
BoxCox(y)	-1.95		0.43	7.396394	-108.054	*
. . . .						
BoxCox(y)	-1.15		0.43	7.461868	-108.530	*
BoxCox(y)	-1.10		0.43	7.469094	-108.582	*
BoxCox(y)	-1.05		0.43	7.476700	-108.637	*
BoxCox(y)	-1.00	+	0.43	7.484687	-108.694	*
BoxCox(y)	-0.95		0.43	7.493058	-108.755	*
BoxCox(y)	-0.90		0.43	7.501814	-108.818	*
BoxCox(y)	-0.85		0.43	7.510957	-108.884	*
. . . .						
BoxCox(y)	-0.35		0.42	7.624209	-109.692	*
BoxCox(y)	-0.30		0.42	7.637770	-109.788	*
BoxCox(y)	-0.25		0.42	7.651747	-109.886	*
BoxCox(y)	-0.20		0.42	7.666143	-109.988	
BoxCox(y)	-0.15		0.42	7.680963	-110.092	
BoxCox(y)	-0.10		0.42	7.696207	-110.199	
BoxCox(y)	-0.05		0.42	7.711880	-110.309	
BoxCox(y)	0.00		0.42	7.727984	-110.422	
BoxCox(y)	0.05		0.42	7.744522	-110.537	
. . . .						

The setup generates 121 values of λ and corresponding values of RMSE, log-likelihood profiles as well as R^2. The plot of the likelihood profiles (LL) against the λ values is presented in the first half of Figure 5.7. The maximum value of the log likelihood is attained at a value of $\lambda = -2.10$. The confidence intervals however include -2 and -1. Hence transformations of the form $y' = 1/y$, and $y' = 1/y^2$ will be explored further in our analysis. Both choices of λ are within the confidence band as indicated by the asterisks in the SAS output. In the second half of Figure 5.7 is the plot of root mean square error values against the values of lambda. Best result is obtained for value(s) of λ that minimizes the RMSE.

A transformation $y' = 1/y$, that is, the reciprocal transformation, has the model in the form

$$y' = \frac{1}{y_i} = \beta_0 + \beta_1 x_i + \varepsilon_i. \tag{5.31}$$

Implementation of the model in (5.31) gives the following results.

```
                        Analysis of Variance

                                Sum of          Mean
Source              DF          Squares        Square    F Value   Pr > F

Model                1       0.00005716     0.00005716     38.68   <.0001
Error               52       0.00007686     0.00000148
Corrected Total     53       0.00013402


            Root MSE              0.00122    R-Square    0.4265
            Dependent Mean        0.01285    Adj R-Sq    0.4155
            Coeff Var             9.46443


                        Parameter Estimates

                     Parameter       Standard
        Variable  DF   Estimate        Error      t Value   Pr > |t|
```

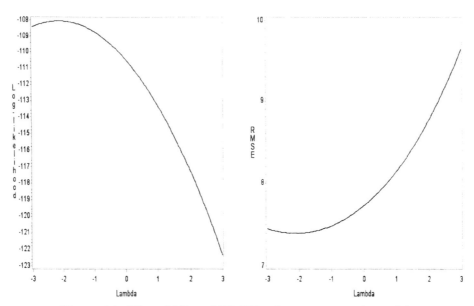

Figure 5.7: Plot of LL and RMSE values against values of λ

Intercept	1	0.01641	0.00059604	27.53	<.0001
x	1	-0.00008999	0.00001447	-6.22	<.0001

That is, the model has the estimated regression equation

$$\frac{1}{\hat{y}_i} = 0.01641 - 0.00009\,x_i.$$

Examination of the residuals from this model suggest that observation 49, which has an observed value of 109 has expected value of 88.6753, a residual of 20.3247 and a standardized residual of 2.1584, which we may consider to be somewhat large (see section 5.4). The residual plot under this model against the predictor variable x is presented in Figure 5.8. The pattern of the distribution of these residuals are clearly different from the earlier one in Figure 5.3. In other words, the transformation has been effective even for this model with $R^2 = 0.4265$. But how effective?

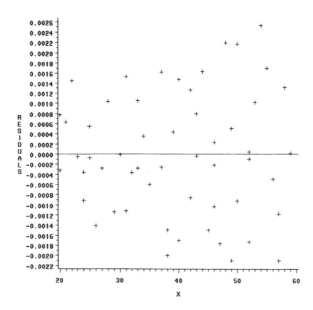

Figure 5.8: Residual plots against x under the reciprocal transformation

Figure 5.9 is the histogram plot of the residuals generated from the regression analysis where the Box-Cox transformation is applied. The normal curve is transposed over this and as we can see again, the histogram looks symmetric. Further, **proc capability**, also amongst other things computes the goodness of fit tests for the normal distribution. These results confirm that the assumption of normality of error terms can reasonably be made for this model.

```
proc capability data=bb graphics noprint;
spec llsl=2 lusl=2;
histogram resid/ normal;
run;
```

                 Goodness-of-Fit Tests for Normal Distribution

        Test   ----Statistic-----   DF   ------p Value------
        Kolmogorov-Smirnov   D   0.10214568   Pr > D   >0.150
        Cramer-von Mises   W-Sq 0.04579473   Pr > W-Sq   >0.250
        Anderson-Darling   A-Sq 0.25734422   Pr > A-Sq   >0.250
        Chi-Square   Chi-Sq 0.78269866   4   Pr > Chi-Sq   0.941

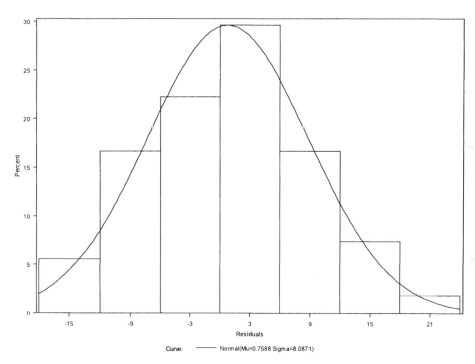

Figure 5.9: Histogram plot of the residuals

Example 5.6 (Gas Consumption):

We consider the following example which relates gas consumptions in miles per gallon (MPG) and the weight of 12 cars in pounds. The data are presented in Table 5.9.

Car	y	x	Car	y	x
1	28.7	2289	7	23.9	2657
2	29.2	2113	8	30.5	2106
3	34.2	2180	9	18.1	3226
4	27.9	2448	10	19.5	3213
5	33.3	2026	11	14.3	3607
6	26.4	2702	12	20.9	2888

Table 5.9: Mileage against weight of cars in pounds

In Figure 5.10 is displayed the scatter plot of y against x. The plot indicates negative association. That is, as the weight of the car, x, increases, y, the miles per gallon decreases.

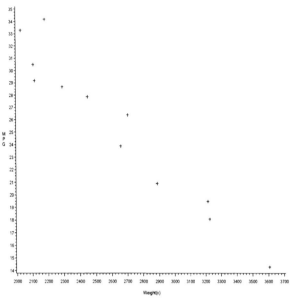

Figure 5.10: Scatter plot of y versus x

A model of the form

$$y_i = \beta_0 + \beta_1 x_i + \varepsilon_i, \tag{5.32}$$

is proposed. The implementation of this model and the plot of the residuals from the model against the predicted values are displayed in Figure 5.11.

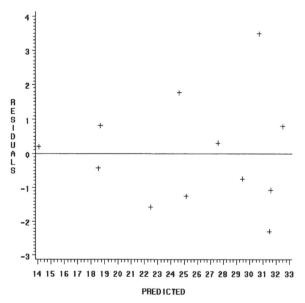

Figure 5.11: Graph of residuals versus predicted

The plot indicates the megaphone shape ("fans out") indicating that the variances are not constant. To determine λ, one carries out a regression of the following form (similar to equation 5.27)

$$y_i^\lambda = \beta_0 + \beta_1 x_i + \varepsilon_i, \tag{5.33}$$

where $\beta_0, \beta_1, \lambda$ and σ^2, the variance of ε are to be estimated from available data. One procedure to accomplish this is to use a numerical search for values of λ ranging from -2 to 2. For instance we can use values of $\lambda = -2, -1.9, -1.8, \ldots, 1.8, 1.9, 2.0$. The implementation of **proc transreg** on the data for the specified values of λ produces the following results.

```
*Program for the data in Table 5.8 for Example 5.6;
data one;
input y x @@;
label x='weight' y='mpg';
datalines;
......
;
ods output boxcox=outb details=d;
ods exclude boxcox;
proc transreg data=one details ss2;
model boxcox(y/lambda=-2 to 2 by 0.1) = identity(x);
run;
proc print data=outb; run;

goptions vsize=5inches hsize=5inches;
symbol c=b i=spline value=none line=1 height=0.75;
axis1 label=(angle=-90 rotate=90 'rmse');
axis2 label=('Lambda');
run;
proc gplot data=outb;
plot rmse*lambda / vaxis=axis1 haxis=axis2 noframe;
run;
proc transreg data=one details ss2;
model boxcox(y/lambda 0) identity(x);
run;
```

Obs	Dependent	Lambda	Convenient	RSquare	RMSE	LogLike	CI
1	BoxCox(y)	-2.0		0.87	3.342351	-14.4801	
2	BoxCox(y)	-1.9		0.88	3.187795	-13.9120	
3	BoxCox(y)	-1.8		0.89	3.040934	-13.3460	
4	BoxCox(y)	-1.7		0.89	2.901574	-12.7830	
5	BoxCox(y)	-1.6		0.90	2.769545	-12.2242	
6	BoxCox(y)	-1.5		0.90	2.644701	-11.6707	
7	BoxCox(y)	-1.4		0.91	2.526920	-11.1240	
8	BoxCox(y)	-1.3		0.91	2.416103	-10.5859	
9	BoxCox(y)	-1.2		0.92	2.312174	-10.0583	
10	BoxCox(y)	-1.1		0.92	2.215077	-9.5434	
11	BoxCox(y)	-1.0		0.93	2.124776	-9.0440	
12	BoxCox(y)	-0.9		0.93	2.041251	-8.5628	
13	BoxCox(y)	-0.8		0.93	1.964495	-8.1028	
14	BoxCox(y)	-0.7		0.94	1.894510	-7.6675	
15	BoxCox(y)	-0.6		0.94	1.831300	-7.2603	*
16	BoxCox(y)	-0.5		0.94	1.774867	-6.8847	*
17	BoxCox(y)	-0.4		0.94	1.725204	-6.5441	*
18	BoxCox(y)	-0.3		0.95	1.682284	-6.2418	*
19	BoxCox(y)	-0.2		0.95	1.646059	-5.9806	*
20	BoxCox(y)	-0.1		0.95	1.616448	-5.7628	*
21	BoxCox(y)	0.0		0.95	1.593338	-5.5900	*
22	BoxCox(y)	0.1		0.95	1.576576	-5.4631	*
23	BoxCox(y)	0.2		0.95	1.565971	-5.3821	*
24	BoxCox(y)	0.3		0.95	1.561296	-5.3462	<
25	BoxCox(y)	0.4		0.95	1.562291	-5.3538	*
26	BoxCox(y)	0.5		0.95	1.568672	-5.4028	*
27	BoxCox(y)	0.6		0.94	1.580135	-5.4901	*
28	BoxCox(y)	0.7		0.94	1.596365	-5.6128	*
29	BoxCox(y)	0.8		0.94	1.617047	-5.7672	*
30	BoxCox(y)	0.9		0.94	1.641868	-5.9500	*
31	BoxCox(y)	1.0	+	0.94	1.670526	-6.1577	*
32	BoxCox(y)	1.1		0.93	1.702736	-6.3868	*
33	BoxCox(y)	1.2		0.93	1.738230	-6.6344	*
34	BoxCox(y)	1.3		0.93	1.776763	-6.8975	*
35	BoxCox(y)	1.4		0.92	1.818111	-7.1736	*
36	BoxCox(y)	1.5		0.92	1.862074	-7.4603	
37	BoxCox(y)	1.6		0.91	1.908476	-7.7557	
38	BoxCox(y)	1.7		0.91	1.957160	-8.0579	
39	BoxCox(y)	1.8		0.91	2.007991	-8.3656	
40	BoxCox(y)	1.9		0.90	2.060855	-8.6775	
41	BoxCox(y)	2.0		0.90	2.115654	-8.9924	

Results from the SAS printout indicate that RMSE as a function of λ is fairly stable in the range $-0.2 \le \lambda \le 1.0$. The convenient $\lambda = 1.0$ is indicated by the '+' sign. Of course, this is not good as this is our original value of y. The confidence region indicates values of λ in $-0.6 \le \lambda \le 1.4$. Clearly, λ values of $-0.5, 0, 0.5, 1.0$ are all included in this range and are all acceptable. Of course, SAS gives as the best value $\lambda = 0.30$, this is indicated with the $<$ sign in the output.

The plots of the log-likelihood and the RMSE as functions of λ are presented in the first and second halves respectively of Figure 5.12.

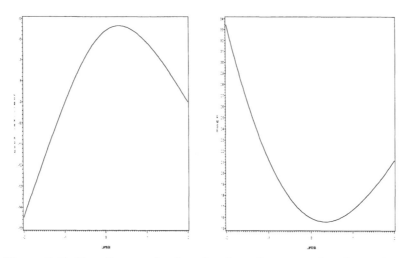

Figure 5.12: Box-Cox result plots for the mileage-weight of cars data

Hence, we would like to see how the transformations with $\lambda = -0.5$, $\lambda = 0$, $\lambda = 0.3$ and $\lambda = 0.5$ behave. Of course, $\lambda = -0.5$ implies we use $y' = 1/\sqrt{y}$, while $\lambda = 0$ by definition implies we use $y' = \log(y)$ and $\lambda = 0.5$ implies the transformation $y' = \sqrt{y}$. When these four transformations are applied to the data in this example, the following mean square error values (estimates of population variance) and **spec** are obtained. Here, **spec** is an option which tests whether the first and second moments of the model are correctly specified.

λ	y'	MSE	**spec**
-0.5	$1/\sqrt{y}$	0.000052	0.7674
0.0	$\log y$	0.00413	0.6687
0.3	$y^{0.3}$	0.02402	0.4047
0.5	$\sqrt{y}$	0.02481	0.3737

Both transformations utilizing $\lambda = -0.5$ and 0.0 give very high **spec** p-values. Hence, either would be desirable. Both gives R^2 values of 0.9415 and 0.9477 respectively. Of the two, $\lambda = 0$, that is, $y' = \log y$ is much easier to interpret than the inverse square root transformation.

With our selected value of $\lambda = 0$, we present the estimated parameters under this model together with the residual plot in Figure 5.13. The parameter estimates are

$$\hat{\beta}_0 = 4.52423, \qquad \hat{\beta}_1 = -0.00050,$$

and the estimated regression equation is therefore given by

$$\log \hat{y}_i = 4.52423 - 0.0005011 x_i. \tag{5.34}$$

For instance, when $x = 2550$, we have $\log \hat{y} = 3.2464$. Hence, $\hat{y} = \exp(3.2464) = 25.6977$. The 95% mean prediction and fiducial intervals for $\log \hat{y}$ are $(3.2047, 3.2882)$ and $(3.0973, 3.3956)$ respectively. Hence the corresponding mean prediction interval is

$$(\exp(3.2047), \exp(3.2882)) = (24.6481, 26.7946).$$

The corresponding 95% fiducial confidence interval for $\hat{y}$ is,

$$(\exp(3.0973), \exp(3.3956)) = (22.1381, 29.8325).$$

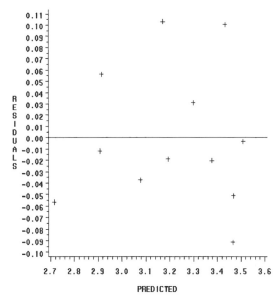

Figure 5.13: Graph of residuals versus predicted with transformed $y(\lambda = 0)$

5.9   Robust Regression

We have seen from our discussions in this chapter that we can combat the problem of heteroscedasticity (non-constant variance) by employing weighted least squares or by the use of an appropriate transformation, such as the Box-Cox class of transformations. However for data exhibiting outliers or influential observations from our analysis or non-normality of error terms, robust regression provides an alternative to the ordinary least squares. Once we have established that an individual observation(s) is not erroneous with regards to recording or model adequacy, robust regression provides a means of dampening the influence of these observations. The robust regression applies 'weights' to lower the influence of those observations that produce large residuals. It is therefore a form of weighted least squares regression. Thus it deals with observations with very high leverages and those that are outliers. Robust regression therefore is a compromise between having to delete the outlying cases, and allowing them to remain in the model with a plan to pay some penalty in terms of violating the OLS assumptions.

The approach is to use a *weight* or *influence* function ϕ, where, the $\hat{\beta}$ are the solutions of

$$\sum_{i=1}^{n} \phi\left(\frac{e_i}{\sigma}\right) \mathbf{x}_i = 0. \tag{5.35}$$

The expression in (5.35) is analogous to that of the ordinary least squares, where we have, $\sigma = 1$ and $\sum_{i=1}^{n} e_i \mathbf{x}_i = 0$, that is, $\phi(e_i) = e_i$ in this case. The robust regression approach chooses the weight function ϕ which dampens the influence of large residuals. Thus outliers get lower weights, which dampens their influence. Before we discuss the implementation of robust regression, we illustrate a situation in which an outlying observation is deleted, that is a weight of zero is applied to the observation.

Example 5.3 continued

In section 5.4, we diagnosed observation 4 to be a possible outlier. What then do we do with this observation? We now need to carefully examine this observation to see if it were due to recording error, wrong analysis or measurements or due to some other explainable reasons. Suggestions to correcting outliers include outright removal of observation from analysis regardless of whether the cause can be assigned to some attributable event, or just the correction of only those outliers that can be traced to specific causes. The latter correction can be very dangerous, because if indeed the outlier is an unusual but legitimate observation, removing it might prove inappropriate. Outliers have been known in many situations to be indeed more important indicators of the properties of the model under examination, as it may be the model and not the outlying observation that is suspect. Omission of predictor variables or higher-order terms might sometimes influence the model to not adequately capture the outlying observations. If we wish to fit a new regression model without the outlying observation in the model, we can accomplish this in SAS as follows:

- First, a model of the form $y_i = \beta_0 + \beta_1 x_i + \varepsilon_i$ is fitted to the data containing all the 30 observations. The model gives an $R^2 = 0.4263$ and an MSE $= S^2 = 141.8335$.

- Next, we fit the same model with the 4th observation deleted. This is accomplished in SAS with the command **reweight**. We have stated here that observation with AGE $= 71$ and SBP $= 100$ should be deleted from the model. Thus the model is being applied to 29 observations in this case. Usually, only either the x or y would be sufficient in the reweight command. In our case, there are more than one observation with age $= 71$ and we would therefore need to separate our desired observation with the two variable values being specified. The model gives an $R^2 = 0.7326$ and a new error mean square MSE $= 46.0763$. The MSE here is much smaller than that from the entire observations. Similarly, we note the dramatic increase in R^2 from 0.4263 to 0.7326.

```
data one;
input age sbp @@;
no=_n_;
datalines;
......
;
proc reg data=one;
model sbp=age;
run;
proc reg data=one;
title 'Table 5.4 with 1 observation deleted';
model sbp=age;
reweight age=71 and sbp=100;
run;
```

```
                        The REG Procedure
                         Model: MODEL1
                      Dependent Variable: sbp


                      Analysis of Variance

                                Sum of         Mean
   Source              DF       Squares        Square     F Value    Pr > F

   Model                1     2950.82880     2950.82880     20.80    <.0001
   Error               28     3971.33787      141.83350
   Corrected Total     29     6922.16667


            Root MSE              11.90939    R-Square     0.4263
            Dependent Mean       146.83333    Adj R-Sq     0.4058
            Coeff Var              8.11082


                      Parameter Estimates

                      Parameter     Standard
      Variable   DF    Estimate       Error     t Value    Pr > |t|

      Intercept   1    100.35683     10.41886      9.63     <.0001
      age         1      0.71576      0.15692      4.56     <.0001

                Table 5.4 with 1 observation deleted


                        The REG Procedure
                        Model: MODEL1.1
                      Dependent Variable: sbp


            Number of Observations Read          30
            Number of Observations Used          29


                      Weight: REWEIGHT

                      Analysis of Variance

                                Sum of         Mean
   Source              DF       Squares        Square     F Value    Pr > F

   Model                1     3409.11203     3409.11203     73.99    <.0001
   Error               27     1244.06038       46.07631
   Corrected Total     28     4653.17241


            Root MSE               6.78795    R-Square     0.7326
```

```
        Dependent Mean      148.44828    Adj R-Sq      0.7227
        Coeff Var             4.57261
```

```
                    Parameter Estimates

                      Parameter      Standard
        Variable   DF   Estimate        Error    t Value   Pr > |t|

        Intercept   1    98.48846      5.94337     16.57    <.0001
        age         1     0.77189      0.08974      8.60    <.0001
```

5.9.1   The M Estimator

The solutions arising from the expression in (5.35) are called the M estimators. The M estimator was introduced by Huber (1973) and is the simplest procedure of all known robust regression methods. It detects outliers (vertical extreme values due to the dependent variable y), but usually fails to detect leverages which are horizontal extreme values due to the independent variables. Other estimation methods employed in SAS are the *Least Trimmed Squares* (LTS) (Hetmansperger, 1984) which is considered as one of the *high breakdown value estimation* procedures. The least trimmed squares minimizes the sum of the h smallest squared residual, viz:

$$\sqrt{\left(\frac{1}{h}\sum_{i-1}^{h} r_{i:n}^2\right)},$$

where h is such that

$$\frac{n}{2}+1 \le h \le \frac{3n}{4}+\frac{p+1}{4}, \tag{5.36}$$

with $n =$ sample size and $p =$ number of parameters.

Other robust estimation methods are the S estimation and the MM estimation, the latter being more an efficient method than the others. We shall employ both the M and MM methods of estimation in our discussion. It is always better to start with the M estimation approach because of its ease and speed of implementation. The **proc robustreg** in SAS handles all the four procedures and we will now see how **proc robustreg** can be applied to our data in Table 5.4 for Example 5.3.

```
proc robustreg data=one;
model sbp=age/diagnostics leverage;
output out=aa r=resid p=pred sr=stdr weight=wt leverage=lev;
run;
proc print data=aa;
var age sbp pred resid stdr wt;
format pred resid stdr wt 8.4;
run;
```

```
                The ROBUSTREG Procedure

                  Model Information

        Data Set                          WORK.ONE
        Dependent Variable                     sbp
        Number of Independent Variables          1
        Number of Observations                  30
        Method                       M Estimation

            Number of Observations Read        30
            Number of Observations Used        30

                Summary Statistics

                                                 Standard
Variable       Q1      Median       Q3      Mean  Deviation       MAD
age       53.0000     67.5000  77.0000   64.9333    14.0931   18.5325
sbp         140.0       150.5    158.0     146.8    15.4498   12.6021

                Parameter Estimates

                      Standard   95% Confidence    Chi-
    Parameter DF Estimate  Error       Limits      Square Pr > ChiSq

    Intercept  1  98.9827  5.5116  88.1801 109.7853  322.52    <.0001
    age        1   0.7654  0.0830   0.6027   0.9281   85.01    <.0001
```

```
Scale     1    6.3326
```

```
                    Goodness-of-Fit

             Statistic        Value

             R-Square         0.5855
             AICR            34.8473
             BICR            39.0311
             Deviance      1292.423
```

In SAS, **proc robustreg** uses the M method as its default. We have requested for **diagnostics** option which gives us a diagnostic table for outliers. The first part of the result gives us Q_1, Q_2 and Q_3, the quartiles for both the explanatory and response variables, as well as the mean absolute deviation, defined by $\text{MAD} = \dfrac{1}{n}\sum_{i=1}^{n}|x_i - \bar{x}|$. Often large differences between the standard deviation and MAD for any of the variables indicate big jumps. In this case there does not seem to be too many jumps between the standard deviations and their respective MAD values.

The table of estimated robust parameters are next presented in the SAS output together with their accompanying standard errors and 95% confidence intervals. Based on these, the estimated regression equation is

$$\hat{y}_i = 98.9827 + 0.7654 x_i. \tag{5.37}$$

These parameter estimates compares with those from OLS, where we have $\hat{\beta}_0 = 100.3568$ and $\hat{\beta}_1 = 0.7158$ with $R^2 = 0.4263$.

The diagnostics produced by **proc robustreg** indicate that observations 2 and 4 are outliers with standardized robust residuals of -3.0305 and -8.4209 respectively.

```
                        Diagnostics

                   Robust                   Standardized
        Mahalanobis  MCD                        Robust
  Obs    Distance   Distance   Leverage       Residual     Outlier

   2      1.3434     1.6416                     -3.0305        *
   4      0.4305     0.0029                     -8.4209        *

                   Diagnostics Summary

            Observation
            Type          Proportion    Cutoff

            Outlier         0.0667       3.0000
            Leverage        0.0000       2.2414
```

We present below the predicted, residuals, and standardized residuals, as well as the weights applied to each observation based on their influential values.

```
     Obs   age   sbp     pred      resid      stdr       wt
     -------------------------------------------------------
      1     42   130   131.1293   -1.1293    -0.1783    0.9971
      2     46   115   134.1909  -19.1909    -3.0305    0.3382
      3     42   148   131.1293   16.8707     2.6641    0.4578
      4     71   100   153.3258  -53.3258    -8.4209    0.0000
      5     80   156   160.2143   -4.2143    -0.6655    0.9601
      6     74   162   155.6220    6.3780     1.0072    0.9097
      7     70   151   152.5604   -1.5604    -0.2464    0.9945
      8     80   156   160.2143   -4.2143    -0.6655    0.9601
      9     85   162   164.0413   -2.0413    -0.3224    0.9906
     10     72   158   154.0912    3.9088     0.6173    0.9656
     11     64   155   147.9680    7.0320     1.1104    0.8908
     12     81   160   160.9797   -0.9797    -0.1547    0.9978
     13     41   125   130.3639   -5.3639    -0.8470    0.9357
     14     61   150   145.6718    4.3282     0.6835    0.9579
     15     75   165   156.3874    8.6126     1.3601    0.8386
     16     53   135   139.5487   -4.5487    -0.7183    0.9535
     17     77   153   157.9182   -4.9182    -0.7766    0.9458
     18     60   146   144.9064    1.0936     0.1727    0.9973
     19     82   156   161.7451   -5.7451    -0.9072    0.9264
     20     55   150   141.0794    8.9206     1.4087    0.8274
     21     71   158   153.3258    4.6742     0.7381    0.9510
     22     76   158   157.1528    0.8472     0.1338    0.9984
     23     44   130   132.6601   -2.6601    -0.4201    0.9840
```

24	55	144	141.0794	2.9206	0.4612	0.9807
25	80	162	160.2143	1.7857	0.2820	0.9928
26	63	150	147.2026	2.7974	0.4417	0.9823
27	82	160	161.7451	-1.7451	-0.2756	0.9931
28	53	140	139.5487	0.4513	0.0713	0.9995
29	65	140	148.7334	-8.7334	-1.3791	0.8342
30	48	130	135.7217	-5.7217	-0.9035	0.9270

We see that for observations 2 and 4, the weights applied are respectively 0.3382 and 0 respectively which helps dampen the influential effects of these two observations. We see that the weights applied have very strong correlation with the residuals or standardized residuals.

The multiple robust regression coefficient is $R^2 = 0.5855$. The Q-Q plot and histogram plots for the standardized robust residuals are displayed in Figures 5.14 and 5.15.

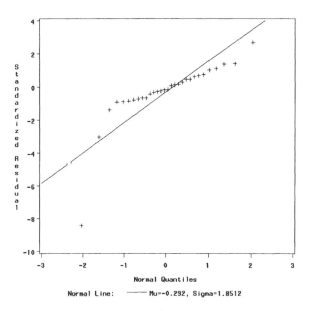

Figure 5.14: Q-Q plot of robust residuals for y

Use of the MM method

The use of the MM estimation method which is specified in the PROC statement gives the following results:

```
set one;
proc robustreg data=one method=mm;
model sbp=age/diagnostics;
run;
```

                    The ROBUSTREG Procedure

                        Model Information

        Data Set   WORK.ONE
        Dependent Variable   sbp
        Number of Independent Variables   1
        Number of Observations   30
        Method   MM Estimation

            Number of Observations Read   30
            Number of Observations Used   30

                    Summary Statistics

| | | | | | Standard | |
Variable	Q1	Median	Q3	Mean	Deviation	MAD
age	53.0000	67.5000	77.0000	64.9333	14.0931	18.5325
sbp	140.0	150.5	158.0	146.8	15.4498	12.6021

                Profile for the Initial LTS Estimate

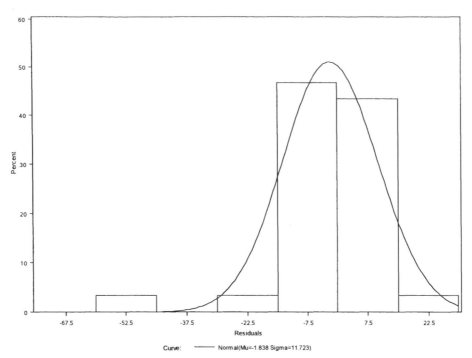

Figure 5.15: Distribution of robust residuals for y

```
Total Number of Observations              30
Number of Squares Minimized               23
Number of Coefficients                     2
Highest Possible Breakdown Value      0.2667

                    MM Profile

         Chi Function         Tukey
         K1                  3.4400
         Efficiency          0.8500

                Parameter Estimates

                    Standard   95% Confidence     Chi-
Parameter DF Estimate  Error       Limits      Square Pr > ChiSq

Intercept  1  98.9585  5.7121  87.7630 110.1540 300.13   <.0001
age        1   0.7656  0.0847   0.5995   0.9317  81.65   <.0001
Scale      0   6.7781

                   Diagnostics

                 Standardized
                    Robust
            Obs     Residual      Outlier

             4      -7.8659          *

                Diagnostics Summary

       Observation
       Type         Proportion    Cutoff

       Outlier         0.0333      3.0000

                 Goodness-of-Fit

          Statistic      Value

          R-Square       0.5063
          AICR          25.0494
          BICR          29.8860
          Deviance    1060.535
```

The method flags only observation 4 as the outlier in this case and gives the estimated regression equation as

$$\hat{y}_i = 98.9585 + 0.7656x_i, \tag{5.38}$$

with $R^2 = 0.5063$, a little lower than the R^2 from the M method. For the data in Example 5.4, we have the following SAS output. Again, we present the residuals and the weights applied or employed for the weighted least squares under the MM method.

Obs	age	sbp	pred	resid	stdr	wt
1	42	130	131.1139	-1.1139	-0.1643	0.9954
2	46	115	134.1764	-19.1764	-2.8291	0.1047
3	42	148	131.1139	16.8861	2.4913	0.2261
4	71	100	153.3165	-53.3165	-7.8659	0.0000
5	80	156	160.2069	-4.2069	-0.6207	0.9360
6	74	162	155.6133	6.3867	0.9423	0.8556
7	70	151	152.5509	-1.5509	-0.2288	0.9912
8	80	156	160.2069	-4.2069	-0.6207	0.9360
9	85	162	164.0349	-2.0349	-0.3002	0.9848
10	72	158	154.0821	3.9179	0.5780	0.9443
11	64	155	147.9572	7.0428	1.0390	0.8259
12	81	160	160.9725	-0.9725	-0.1435	0.9965
13	41	125	130.3483	-5.3483	-0.7891	0.8975
14	61	150	145.6604	4.3396	0.6402	0.9319
15	75	165	156.3789	8.6211	1.2719	0.7453
16	53	135	139.5356	-4.5356	-0.6691	0.9258
17	77	153	157.9101	-4.9101	-0.7244	0.9133
18	60	146	144.8948	1.1052	0.1631	0.9988
19	82	156	161.7381	-5.7381	-0.8466	0.8825
20	55	150	141.0668	8.9332	1.3179	0.7280
21	71	158	153.3165	4.6835	0.6910	0.9209
22	76	158	157.1445	0.8555	0.1262	0.9973
23	44	130	132.6451	-2.6451	-0.3902	0.9744
24	55	144	141.0668	2.9332	0.4327	0.9686
25	80	162	160.2069	1.7931	0.2645	0.9882
26	63	150	147.1916	2.8084	0.4143	0.9712
27	82	160	161.7381	-1.7381	-0.2564	0.9889
28	53	140	139.5356	0.4644	0.0685	0.9992
29	65	140	148.7228	-8.7228	-1.2869	0.7397
30	48	130	135.7076	-5.7076	-0.8421	0.8838

Clearly, different weights from those employed in the M estimators are employed here. Again, observation 4 receives a weight of 0, while observation 2 now receives a weight of 0.1047 and is not classified as an outlier under the MM estimation method.

5.9.2 Robust Multiple Regression Example

For this example, we consider the data previously analyzed by Daniel and Wood (1980, p. 61) and reproduced by permission of John Wiley and Sons, Inc. The data consist of 21 observations taken from a plant oxidizing ammonia to nitric acid. Here, y is the Stack loss, x_1, x_2 and x_3 are respectively, the air flow, cooling water inlet temperature and acid concentration.

Obs	y	x1	x2	x3
1	42	80	27	89
2	37	80	27	88
3	37	75	25	90
4	28	62	24	87
..	..	..	..	..
18	8	50	19	79
19	9	50	20	80
20	15	56	20	82
21	15	70	20	91

An ordinary least squares (OLS) fit to the above data produces the estimated regression equation

$$\hat{y} = -39.9197 + 0.7156x_1 + 1.2953x_2 - 0.1521x_3,$$

with R^2 and adjusted R^2 being 0.9136 and 0.8983 respectively. The diagnostic statistics are presented in the following:

Obs	y	pred	resid	stu	stud	cooks	ddfit	pre	lev
1	42	38.7654	3.2346	1.1933	1.2095	0.1537	0.7947	4.6312	0.3016
2	37	38.9175	-1.9175	-0.7158	-0.7051	0.0597	-0.4813	-2.8109	0.3178
3	37	32.4445	4.5555	1.5460	1.6179	0.1264	0.7442	5.5193	0.1746
4	28	22.3022	5.6978	1.8818	2.0518	0.1305	0.7879	6.5379	0.1285
5	18	19.7117	-1.7117	-0.5421	-0.5305	0.0040	-0.1245	-1.8060	0.0522
6	18	21.0069	-3.0069	-0.9653	-0.9632	0.0196	-0.2792	-3.2595	0.0775
7	19	21.3895	-2.3895	-0.8338	-0.8259	0.0488	-0.4377	-3.0605	0.2192
8	20	21.3895	-1.3895	-0.4848	-0.4737	0.0165	-0.2510	-1.7797	0.2192
9	15	18.1444	-3.1444	-1.0455	-1.0486	0.0446	-0.4234	-3.6570	0.1402
10	14	12.7328	1.2672	0.4368	0.4262	0.0119	0.2131	1.5841	0.2000
11	14	11.3637	2.6363	0.8843	0.8783	0.0359	0.3762	3.1200	0.1550
12	13	10.2205	2.7795	0.9686	0.9667	0.0651	0.5092	3.5506	0.2172
13	11	12.4286	-1.4286	-0.4799	-0.4687	0.0108	-0.2027	-1.6957	0.1575
14	12	12.0505	-0.0505	-0.0175	-0.0170	0.0000	-0.0086	-0.0636	0.2058
15	8	5.6386	2.3614	0.8092	0.8006	0.0385	0.3883	2.9170	0.1905
16	7	6.0949	0.9051	0.2994	0.2912	0.0034	0.1131	1.0416	0.1311
17	8	9.5200	-1.5200	-0.6112	-0.5996	0.0655	-0.5020	-2.5855	0.4121
18	8	8.4551	-0.4551	-0.1532	-0.1487	0.0011	-0.0650	-0.5422	0.1606
19	9	9.5983	-0.5983	-0.2030	-0.1972	0.0022	-0.0907	-0.7248	0.1745
20	15	13.5879	1.4121	0.4540	0.4431	0.0045	0.1308	1.5353	0.0802
21	15	22.2377	-7.2377	-2.6382	-3.3305	0.6920	-2.1003	-10.1161	0.2845

hh	cookf	dff	dfbeta
0.28571	0.82121	0.87287	0.43644

Examination of the various diagnostic statistics indicate that observations 1, 3, 4, 6, 9 and 21 have very high residual values. The leverage values on the other hand indicate that observations 1, 2, and 17 have high leverage values. The studentized values show that observations 4 and 21 have very significant values and could be considered as outliers.

Since observations 1, 3, 4, and 21 have the largest residuals, suppose we delete these from the data and refit our model. This can be accomplished in SAS with the **reweight** statement and the following results are obtained.

```
data loss;
input y x1-x3 @@;
datalines;
42 80 27 89 37 80 27 88 37 75 25 90 28 62 24 87 18 62 22 87 18 62 23 87 19 62 24 93 20 62 24 93 15 58 23 87
14 58 18 80 14 58 18 89 13 58 17 88 11 58 18 82 12 58 19 93  8 50 18 89  7 50 18 86  8 50 19 72  8 50 19 79
 9 50 20 80 15 56 20 82 15 70 20 91
;
proc reg;
model y=x1-x3;
reweight obs.=1 or obs.=3;
refit;
reweight obs.=4 or obs.=21;
print;
output out=cc p=pred r=resid;
run;
proc univariate normal plot data=cc;
var resid;
run;
```

```
                          The REG Procedure
                          Model: MODEL1.2
                      Dependent Variable: y

              Number of Observations Read        21
              Number of Observations Used        17


                          Weight: REWEIGHT

              Root MSE              1.25271    R-Square     0.9750
              Dependent Mean      14.47059    Adj R-Sq     0.9692
              Coeff Var             8.65697


                          Parameter Estimates

                      Parameter      Standard
         Variable  DF   Estimate        Error    t Value   Pr > |t|

         Intercept  1  -37.65246      4.73205      -7.96    <.0001
         x1         1    0.79769      0.06744      11.83    <.0001
         x2         1    0.57734      0.16597       3.48     0.0041
         x3         1   -0.06706      0.06160      -1.09     0.2961


                      Tests for Normality

         Test                --Statistic---    -----p Value------
```

```
Shapiro-Wilk          W     0.973986   Pr < W      0.8186
Kolmogorov-Smirnov    D     0.10749    Pr > D     >0.1500
Cramer-von Mises      W-Sq  0.036848   Pr > W-Sq  >0.2500
Anderson-Darling      A-Sq  0.253341   Pr > A-Sq  >0.2500
```

The estimated regression model with these four observations deleted is now

$$\hat{y} = -37.6525 + 0.7977x_1 + 0.5773x_2 - 0.0671x_3,$$

with R^2 and adjusted R^2 being 0.9750 and 0.9692 respectively. The reweight command put the corresponding weights of observations 1, 3, 4 and 21 to be zero, the others being 1. A test of normality, using the Anderson-Darling test, of the residuals from the model indicates that the assumption of normality holds in this case.

Let us now apply a robust regression to these data. We shall employ the three methods, M estimators, the MM estimators as well as the LTS estimators.

```
set loss;
proc robustreg data=loss method=lts;
model y=x1-x3/diagnostics leverage;;
run;
                    The ROBUSTREG Procedure

                    Model Information

        Data Set                          WORK.LOSS
        Dependent Variable                        y
        Number of Independent Variables           3
        Number of Observations                   21
        Method                      LTS Estimation

            Number of Observations Read      21
            Number of Observations Used      21

                    Summary Statistics

                                                  Standard
Variable        Q1     Median       Q3      Mean  Deviation       MAD
x1         53.0000    58.0000  62.0000   60.4286     9.1683    5.9304
x2         18.0000    20.0000  24.0000   21.0952     3.1608    2.9652
x3         82.0000    87.0000  89.5000   86.2857     5.3586    4.4478
y          10.0000    15.0000  19.5000   17.5238    10.1716    5.9304

                    LTS Profile

        Total Number of Observations         21
        Number of Squares Minimized          17
        Number of Coefficients                4
        Highest Possible Breakdown Value  0.2381

                    LTS Parameter Estimates

            Parameter       DF    Estimate

            Intercept        1    -37.6525
            x1               1      0.7977
            x2               1      0.5773
            x3               1     -0.0671
            Scale (sLTS)     0      1.6288
            Scale (Wscale)   0      1.2527

                    Diagnostics

                    Robust                  Standardized
         Mahalanobis    MCD                      Robust
Obs      Distance   Distance   Leverage        Residual   Outlier
  1        2.2536     5.5284        *            4.9634        *
  2        2.3247     5.6374        *            0.9186
  3        1.5937     4.1972        *            5.1312        *
  4        1.2719     1.5887                     6.5250        *
 21        2.1768     3.6573        *           -6.8889        *

                    Diagnostics Summary

            Observation
            Type        Proportion     Cutoff
            Outlier         0.1905     3.0000
            Leverage        0.1905     3.0575
```

```
                            R-Square for LTS
                               Estimation

                    R-Square        0.9273
```

Here, using the LTS method, the estimated regression model is given by

$$\hat{y} = -37.6525 + 0.7977x_1 + 0.5773x_2 - 0.0671x_3,$$

with $R^2 = 0.9273$. The resulting equation is exactly those provided with the OLS when the four observations were deleted from the data. The outlier observations deleted from the analysis are again, 1, 3, 4 and 21 and these are starred in the output. Note that the procedure uses only 17 observations in its analysis. The diagnostics here indicate that there are no four outliers found, however, there are also four observations (1, 2, 3, 21) displaying leverage points. For a leverage point to be considered *a bad leverage* point, its absolute value must be greater than the cut off point (here, 3.0575). Otherwise, the leverage point will be described as a *good leverage point*. We would still however, conduct tests of homoscedasticity as well as the normality of error terms.

In conclusion, robust regression fits our data better than would the OLS when we have outliers in the data. It will in all cases reliably identifies outliers in our model. However, robust regression does not protect us from the presence of heteroscedasticity in the model, or in cases when the error terms are correlated or if the structural form of the model is not as specified.

Had we used the M estimator for these data, our estimated regression model would have been

$$\hat{y} = -42.2854 + 0.9276x_1 + 0.6507x_2 - 0.1123x_3,$$

with observations 1, 2, 3 and 21 still identified as having significant leverages but only observations 4 and 21 are declared as outliers in this case. On the other hand, if we had used the MM estimators, our results would have been an estimated regression model

$$\hat{y} = -42.3209 + 0.9166x_1 + 0.6872x_2 - 0.1131x_3.$$

Again with observations 1, 2, 3 and 21 still identified as having significant leverages but only observation 21 being identified as an outlier by this method.

5.10   Quantile Regression

While robust regression is concerned with data robustness, however, what happens with model specification robustness. Quantile regression (Koenkar and Bassett, 1978) provides an alternative to fitting linear models to conditional quantiles. SAS **proc quantreg** fits quantile regressions. Quantile regression estimates the quantile of a dependent variable conditional on the explanatory variables, in much the same way as OLS estimates the mean of the dependent variable conditional on the explanatory variables. The most common quantile regression is the *median* regression which produces a line that minimizes the sum of the absolute residuals. If we define τ to be the quantile to be estimated, then, for a random sample $y_1, y_2, \ldots, y_n$ for the random variable Y, the quantile regression estimates the linear conditional quantile function $Q(\tau|X = x'\beta(\tau))$, by solving

$$\beta(\tau) = \arg\min_{\beta \in \mathbf{R}^p} \sum_{i=1}^{n} \rho_\tau(y_i - x'\beta), \quad \text{where} \quad \tau \in (0, 1).$$

We can obtain quantile regressions for $\tau = .05, .10, .25, .50, .75, .90, .95$ for instance by simply specifying τ. However, if the homogeneity assumption is satisfied, all the quantile regressions will usually have similar slopes but different intercepts. On the other hand, if the errors are not homogeneous, then, the slopes and intercepts may certainly be different for the various quantile regressions. Quantile regression therefore, provides a means of ameliorating the violation of the homogeneity of variances in regression analysis.

y	x	y	x	y	x	y	x	y	x
3.929	1790	12.866	1830	39.818	1870	91.972	1910	151.325	1950
5.308	1800	17.069	1840	50.155	1880	105.710	1920	179.323	1960
7.239	1810	23.191	1850	62.947	1890	122.775	1930	203.211	1970
9.638	1820	31.443	1860	75.994	1900	131.669	1940		

Table 5.10: USA population from 1790 to 1970 in steps of 10 years

Example 5.7

The following data are on the United States population in millions from 1790 to 1970. An ordinary least squares equation of the form (that is, quadratic model) is suggested.

$$y_i = \beta_0 + \beta_1 x + \beta_2 x^2 + \varepsilon_i. \tag{5.39}$$

In contrast, the corresponding quantile regression model is

$$Q_\tau(y) = \beta_{\tau,0} + \beta_{\tau,1} x + \beta_{\tau,2} x^2.$$

Thus, the median regression would have $\tau = 0.5$ in the above quantile regression model. We start by fitting the OLS model specified in (5.39) to the data yielding the following SAS output.

```
data quant;
input y x @@;
xsq=x*x;
datalines;
........
;
proc print noobs;
var y x;
run;
proc reg;
model y=x xsq;
output out=bb r=resid1 p=pred1;
run;
                        The REG Procedure
                          Model: MODEL1
                       Dependent Variable: y

                Number of Observations Read         19
                Number of Observations Used         19

                       Analysis of Variance

                             Sum of      Mean
     Source            DF   Squares     Square    F Value   Pr > F
     Model              2     71799      35900    4641.72   <.0001
     Error             16  123.74557    7.73410
     Corrected Total   18     71923

            Root MSE           2.78102   R-Square    0.9983
            Dependent Mean    69.76747   Adj R-Sq    0.9981
            Coeff Var          3.98613

                       Parameter Estimates

                       Parameter    Standard
     Variable    DF     Estimate       Error    t Value   Pr > |t|
     Intercept    1        20450    843.47533     24.25    <.0001
     x            1    -22.78061      0.89785    -25.37    <.0001
     xsq          1      0.00635   0.00023877     26.58    <.0001
```

A robust regression applied to the data identified two outliers (observations 16 and 17) and four observations with significant leverages, observations 1, 2, 3 and 4. A median regression applied to the data using **proc quantreg** in SAS with the options to give all **diagnostics** that otherwise we could obtain from a robust regression on the data give the following results.

```
set quant;
```

```
proc quantreg alpha=.1 ci=resampling data=quant;
model y=x xsq/quantile=.5 diagnostics leverage;
output out=aa p=pred q=quant res=resid;
run;
proc print data=aa;
run;
```

The QUANTREG Procedure

Model Information

Data Set	WORK.QUANT
Dependent Variable	y
Number of Independent Variables	2
Number of Observations	19
Optimization Algorithm	Simplex
Method for Confidence Limits	Resampling

Number of Observations Read	19
Number of Observations Used	19

Parameter Information

Parameter	Effect
Intercept	Intercept
x	x
xsq	xsq

Summary Statistics

Variable	Q1	Median	Q3	Mean	Standard Deviation	MAD
x	1830.0	1880.0	1930.0	1880.0	56.2731	74.1301
xsq	3348900	3534400	3724900	3537400	211605	275023
y	12.8660	50.1550	122.8	69.7675	63.2116	61.9980

Quantile and Objective Function

Quantile	0.5
Objective Function	14.8264
Predicted Value at Mean	70.9329

Parameter Estimates

Parameter	DF	Estimate	Standard Error	90% Confidence Limits		t Value	Pr > |t|
Intercept	1	21132.76	1008.196	19372.565	22892.951	20.96	<.0001
x	1	-23.5257	1.0880	-25.4253	-21.6262	-21.62	<.0001
xsq	1	0.0065	0.0003	0.0060	0.0071	22.31	<.0001

Diagnostics

Obs	Mahalanobis Distance	Robust MCD Distance	Leverage	Standardized Residual	Outlier
1	2.4513	4.1516	*	-1.3734	
2	1.8854	3.1897	*	0.0000	
16	1.0884	1.6792		-8.2011	*
17	1.4235	2.3606		-8.0608	*
18	1.8854	3.1897	*	-1.5914	
19	2.4513	4.1516	*	0.0000	

Diagnostics Summary

Observation Type	Proportion	Cutoff
Outlier	0.1053	3.0000
Leverage	0.2105	2.7162

Obs	y	x	xsq	pred	resid	quant
1	3.929	1790	3204100	5.455	-1.52592	0.5
2	5.308	1800	3240000	5.308	0.00000	0.5
3	7.239	1810	3276100	6.471	0.76811	0.5
4	9.638	1820	3312400	8.944	0.69451	0.5
5	12.866	1830	3348900	12.726	0.13991	0.5
6	17.069	1840	3385600	17.818	-0.74941	0.5
7	23.191	1850	3422500	24.221	-1.02953	0.5
8	31.443	1860	3459600	31.932	-0.48946	0.5
9	39.818	1870	3496900	40.954	-1.13620	0.5
10	50.155	1880	3534400	51.286	-1.13074	0.5

11	62.947	1890	3572100	62.927	0.01991	0.5
12	75.994	1900	3610000	75.878	0.11575	0.5
13	91.972	1910	3648100	90.139	1.83278	0.5
14	105.710	1920	3686400	105.710	0.00000	0.5
15	122.775	1930	3724900	122.591	0.18442	0.5
16	131.669	1940	3763600	140.781	-9.11198	0.5
17	151.325	1950	3802500	160.281	-8.95618	0.5
18	179.323	1960	3841600	181.091	-1.76818	0.5
19	203.211	1970	3880900	203.211	0.00000	0.5

When the quantile regression with $\tau = 0.5$ is applied to the data, we see that the high leverage points for observations 3 and 4 have been replaced with observations 18 and 19 and the outlier points are still left in the model. The objective function obtained is 14.8264 with the median regression. We also present the predicted and residuals under the median quadratic regression model as specified earlier. Our estimated equation therefore is

$$Q_{.5}(y) = 21132.76 - 23.5257x + 0.0065x^2.$$

The residual OLS and median residual plots against years are presented in Figure 5.16. The '+' are the median residuals, while the '*' are the corresponding OLS residuals. In Figure 5.16, the x-axis contain the years transformed as

$$x_i = \frac{(\text{Years} - 1780)}{10}, \quad i = 1790, 1800, \ldots, 1970.$$

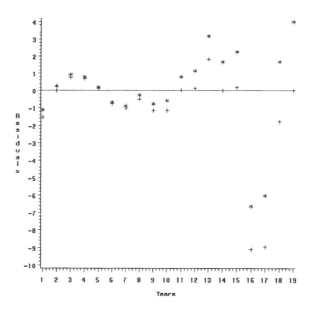

Figure 5.16: Residuals plots for the OLS and median regression

We also obtain quantile regression for $\tau = .05, .50, .95$. The predicted values and a plot of the estimated quantile regression models for these values of τ are presented in Figure 5.17 with the actual computed predicted values displayed below.

x	y	.05	.50	.95
1790	3.929	3.9290	5.4549	5.3922
1800	5.308	4.1917	5.3080	5.6864
1810	7.239	5.6335	6.4709	7.2390
1820	9.638	8.2543	8.9436	10.0499
1830	12.866	12.0542	12.7261	14.1191
1840	17.069	17.0331	17.8184	19.4466
1850	23.191	23.1910	24.2205	26.0324
1860	31.443	30.5280	31.9325	33.8765
1870	39.818	39.0440	40.9542	42.9790
1880	50.155	48.7390	51.2857	53.3398
1890	62.947	59.6131	62.9271	64.9589
1900	75.994	71.6662	75.8783	77.8363
1910	91.972	84.8983	90.1392	91.9720
1920	105.710	99.3095	105.7100	107.3660

1930	122.775	114.8997	122.5906	124.0184
1940	131.669	131.6690	140.7810	141.9291
1950	151.325	149.6173	160.2812	161.0981
1960	179.323	168.7446	181.0912	181.5254
1970	203.211	189.0510	203.2110	203.2110

The lower quantile regression models tend to predict the left lower tail of the data well as exemplified by the quantile regression for $\tau = .05$. However, both the 95th quantile and the 50th quantile regression models fit well the right tailed data. Overall the three quantile regressions did very well in predicting the data. In Figure 5.17, we see that both the 50th and 95th quantile models are very close in their predictions.

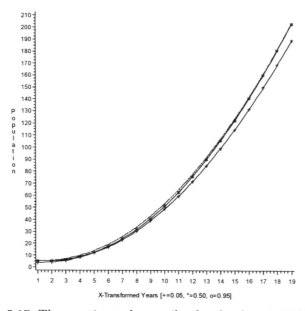

Figure 5.17: Three estimated quantiles for the data in Table 5.10

Example 5.8 (The Acorn Data):

The data in Table 5.11 relate to data on acorn biomass and suitability index of 43 oak trees. The data were originally published by Schroeder and Vangilder (1997). The biomass is the dependent variable, while the suitability index is the explanatory variable. A median quantile regression applied to the data gives the following partial SAS output.

Tree	Biomas	Index	Tree	Biomass	Index
1	43.04055	0.826136	23	11.51406	0.626498
2	4.946344	0.115109	24	67.86525	0.89861
3	60.21557	0.447214	25	80.7973	0.597913
4	81.66009	0.9	26	25.6447	0.33541
5	50.61638	0.52915	27	108.2565	0.953939
⋮	⋮	⋮	⋮	⋮	⋮
17	20.869	0.455522	39	8.10927	0.259808
18	46.28904	0.83666	40	50.85466	0.606218
19	50.56929	0.648074	41	89.92351	0.67082
20	19.39652	0.860233	42	16.45993	0.736546
21	71.11756	0.61441	43	29.18572	0.8
22	16.27327	0.316228			

Table 5.11: Acorn data from Schroeder & Vangilder

```
data quant;
input y x;
label x='suitabilityIndex'
      y='Acornbiomass';
datalines;
43.04055      0.826136
4.946344      0.115109
.....................
16.45993      0.736546
29.18572      0.8
;
run;
proc reg;
model y=x;
output out=bb r=resid1 p=pred1;
run;
proc quantreg alpha=.1 ci=resampling data=quant;
model y=x/quantile=.5 diagnostics leverage;
output out=aa p=pred q=quant res=resid leverage=lev md=mde rd=rde outlier=outl;
run;
proc print data=aa;
run;
```

The REG Procedure
Model: MODEL1
Dependent Variable: y Acornbiomass

Number of Observations Read   43
Number of Observations Used   43

Analysis of Variance

Source	DF	Sum of Squares	Mean Square	F Value	Pr > F
Model	1	13158	13158	12.07	0.0012
Error	41	44701	1090.25705		
Corrected Total	42	57858			

Root MSE	33.01904	R-Square	0.2274	
Dependent Mean	52.03640	Adj R-Sq	0.2086	
Coeff Var	63.45373			

Parameter Estimates

Variable	Label	DF	Parameter Estimate	Standard Error	t Value	Pr > |t|
Intercept	Intercept	1	-2.55629	16.50187	-0.15	0.8777
x	suitabilityIndex	1	77.54977	22.32320	3.47	0.0012

The QUANTREG Procedure

Model Information

Data Set	WORK.QUANT	
Dependent Variable	y	Acornbiomass
Number of Independent Variables	1	
Number of Observations	43	
Optimization Algorithm	Simplex	
Method for Confidence Limits	Resampling	

Number of Observations Read   43
Number of Observations Used   43

Parameter Information

Parameter	Effect
Intercept	Intercept
x	x

Summary Statistics

Variable	Q1	Median	Q3	Mean	Standard Deviation	MAD
x	0.5466	0.7365	0.9000	0.7040	0.2282	0.2423
y	20.8690	49.9990	72.8099	52.0364	37.1157	34.2936

Quantile and Objective Function

Quantile	0.5
Objective Function	467.8458
Predicted Value at Mean	52.3441

```
                              Parameter Estimates

                              Standard    90% Confidence
             Parameter DF Estimate    Error       Limits      t Value Pr > |t|
             Intercept  1  -4.3188  12.3521  -25.1059  16.4682   -0.35   0.7284
             x          1  80.4905  18.7499   48.9368 112.0443    4.29   0.0001

                                  Diagnostics

                           Robust
                             MCD
                Mahalanobis  Distance            Standardized
        Obs     Distance    Distance   Leverage    Residual      Outlier
         2       2.5801      3.0293        *        -0.0000
        11       0.3376      0.1271                  5.7742          *
        39       1.9461      2.3987        *        -0.3397

                              Diagnostics Summary

                        Observation
                        Type        Proportion    Cutoff

                        Outlier       0.0233       3.0000
                        Leverage      0.0465       2.2414

   Obs      y       x        pred     resid   outl    mde       rde     lev  quant
    1    43.041  0.82614   62.1773  -19.137    0    0.53526   0.06954    0    0.5
    2     4.946  0.11511    4.9463   -0.000    0    2.58006   3.02929    1    0.5
    3    60.216  0.44721   31.6777   28.538    0    1.12496   1.58189    0    0.5
    4    81.660  0.90000   68.1226   13.537    0    0.85889   0.39146    0    0.5
    5    50.616  0.52915   38.2727   12.344    0    0.76596   1.22479    0    0.5
   ..   ......  .......   .......   ......    .    .......   .......    .    ...
   39     8.109  0.25981   16.5932   -8.484    0    1.94607   2.39865    1    0.5
   40    50.855  0.60622   44.4760    6.379    0    0.42829   0.88891    0    0.5
   41    89.924  0.67082   49.6758   40.248    0    0.14524   0.60736    0    0.5
   42    16.460  0.73655   54.9661  -38.506    0    0.14273   0.32091    0    0.5
   43    29.186  0.80000   60.0736  -30.888    0    0.42075   0.04436    0    0.5
```

In the above SAS output, we note the following:

- **pred** contains the predicted values of y

- **lev** contains the leverages (coded 1 or 0)

- **resid** contains the residuals

- **mde** contains the Mahalanobis distances

- **rde** contains the robust MCD distances

- **outl** contains the outliers (coded 1 or 0)

Other statistics that can be requested are the standardized residuals, and standard errors of the estimated response. Some of these requests are contingent upon the simultaneous requests of some other statistics. For details, please see the **proc quantreg** documentation in SAS.

In Figure 5.18, we have the plot of the residuals against x (the suitability index), and referenced at zero. Clearly observation 11, is an outlier and the residual plot indicates this to be the case. Also observations 2 and 39 have very high leverages. We therefore sought to fit a quantile regression to the data with weights. However, how do we determine weights for the outlier and high leveraged data points? Our first choice of weights here is arbitrary. We set the three observations (2, 11, 39) to weights of zeros and the others to weights of 1.

The resulting median regression model treats the three data points as missing and we present the results from this regression in the following, with the accompanying residual plot against predicted values.

```
                        Model Information

        Data Set                        WORK.QUANT
        Dependent Variable                       y    Acornbiomass
        Weight Variable                         wt
        Number of Independent Variables          1
        Number of Observations                  40
        Optimization Algorithm             Simplex
        Method for Confidence Limits    Resampling
```

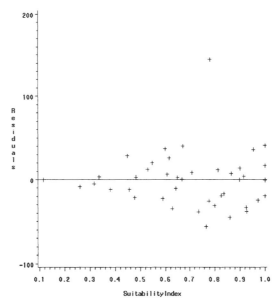

Figure 5.18: Residuals plot for the median regression on the acorn data

```
        Number of Observations Read          43
        Number of Observations Used          40
        Sum of Weights                       40
        Missing Values                        3

                      Parameter Information

                      Parameter      Effect

                      Intercept      Intercept
                      x              x

                   Summary Statistics

                                             Standard
Variable        Q1      Median        Q3      Mean    Deviation        MAD
x           0.5937     0.7515     0.9076    0.7279       0.2035      0.2340
y          27.4152    50.2842    72.4631   50.5433      28.1477     32.8826

                  Quantile and Objective Function

                  Quantile                   0.5
                  Objective Function      390.6172
                  Predicted Value at Mean  54.9969

                  Parameter Estimates

                          Standard    90% Confidence
        Parameter DF Estimate    Error       Limits     t Value Pr > |t|
        Intercept  1   0.1390  19.7909 -33.2275  33.5055    0.01   0.9944
        x          1  75.3678  28.2561  27.7292 123.0064    2.67   0.0112

                  Diagnostics Summary

        Observation
        Type          Proportion      Cutoff
        Outlier           0.0000      3.0000
        Leverage          0.0000      2.2414
```

The output indicates that there are no outlier or high leverage points in our chosen model. The plot of the residuals against the predicted under this model is presented in Figure 5.19. It should be noted that no other choice of weights for the three data points yield any fruitful results as the quantile regression under such other weights are deemed not to be robust.

We present in Figure 5.20 the plots of estimated quantile regressions for $\tau = 0.05, 0.50, 0.95$ with the observations weighted with weights of 0 and 1 as suggested earlier. The plots in Figure 5.20 corresponds respectively to the

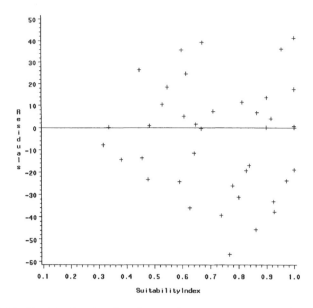

Figure 5.19: Residuals plot for the median regression on the acorn data

following estimated quantile regression models:

$$Q_{.05}(y) = -9.6139 + 33.7239x \qquad \text{(obf=86.3449)}$$
$$Q_{.50}(y) = 0.1390 + 75.3678x \qquad \text{(obf=390.6172)}$$
$$Q_{.95}(y) = 27.2803 + 89.5063x \qquad \text{(obf=86.3722)}.$$

Here, obf stands for 'objective function' from each of the models. The significant test for the suitability index is not significant when $\tau = 0.05$, 0.95 while it is significant at $\tau = 0.50$ with a p-value of less than 0.01.

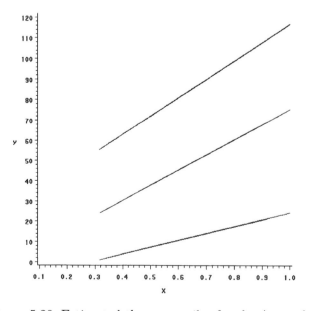

Figure 5.20: Estimated three quantiles for the Acorn data

The linear regression estimates the mean value of the response variable, biomass (y), for given levels of the independent variable, suitability index (x). A corresponding linear regression equation is

$$\hat{y}_i = -0.0411 + 69.4966x_i, \quad i = 1, 2, 3, \ldots, n.$$

The linear regression equation estimates how, on average, suitability index affect biomass. For each additional unit increase in suitability index, the biomass increases by about 69.4966. The suitability index significantly influences the biomass since its parameter estimate has a p-value of less than 0.01. Thus, the linear regression model is able to answer the question "is suitability index important to biomass?". Suppose the question of interest is "does suitability index influence biomass differently for oak trees with low biomass than oak trees with average biomass?". Unfortunately, the linear regression model cannot address the question. This is where quantile regression is also very useful. In quantile regression, the relation between the suitability index and a specific percentile (or quantile), τ, of the biomass is modeled. Since the suitability index is not significant at $\tau = 0.05, 0.95$, it influences oak trees with average ($\tau = 0.50$) biomass differently than oak tress with either low ($\tau = 0.05$) or high ($\tau = 0.95$) biomass.

5.11 Autocorrelated Error Terms

For the general linear regression model

$$y_i = \beta_0 + \sum_{j=1}^{p-1} \beta_j x_{ij} + \varepsilon_i, \quad \text{for } i = 1, 2, \ldots, n, \tag{5.40}$$

one of the basic assumptions is that the errors terms are independently distributed. In other words, we assume that the expected values of product of two error terms will be zero. That is,

$$E(\varepsilon_i \varepsilon_j) = 0, \quad \text{for } i \neq j. \tag{5.41}$$

However, if $E(\varepsilon_i \varepsilon_j) \neq 0$, then we say that the error terms are autocorrelated. Correlated data often arise in time series data due to correlation between successive times. That is for data collected over time and in this case, we say that the errors are *autocorrelated* or *serially correlated*. Longitudinal data also display serial correlations. We discuss in this section an example of a serially correlated data and how to detect and correct serial correlation in our data. The data in Table 5.12 gives the amounts of crude oil (millions of barrels) imported into the United States from the Organization of Petroleum Exporting Countries (OPEC) for the years 1974-1993.

Year	t	Imports y_t	Year	t	Imports y_t
1974	1	926	1984	11	553
1975	2	1,171	1985	12	479
1976	3	1,663	1986	13	771
1977	4	2,058	1987	14	876
1978	5	1,892	1988	15	987
1979	6	1,866	1989	16	1,232
1980	7	1,414	1990	17	1,282
1981	8	1,067	1991	18	1,233
1982	9	633	1992	19	1,247
1983	10	540	1993	20	1,339

Table 5.12: Import of crude oil to the USA from 1974-1993

The fit of the simple linear regression of the form

$$y_i = \beta_0 + \beta_1 t \quad \text{for } t = 1, 2, \ldots, 20, \tag{5.42}$$

where $t = 1$ corresponds to 1974 and $t = 20$ corresponds to 1993 is considered. Other than the very low value of $R^2 = 0.0743$, we can see immediately that the model (5.42) does not fit the data from the SAS output below.

```
data serial;
do year=1974 to 1993;
input y @@;
t=year-1973;
output;
```

```
end;
datalines;
926 1171 1663 2058 1892 1866 1414 1067 633 540
553 479 771 876 987 1232 1282 1233 1247 1339
;
proc print;
run;
proc reg;
model y=t/dw;
output out=aa r=resid p=pred;
run;
proc sort data=aa;
by t;
run;
goptions vsize=6 hsize=6;
symbol1 I=spline c=black value=none  height=.75;
symbol2 c=black value=plus  height=.75;
axis1 label=(angle=-90 rotate=90 'Residuals');
axis2 label=('Year (t)')
          offset=(2);
proc gplot data=aa;
plot resid*t resid*t/overlay vaxis=axis1 haxis=axis2 vref=0  noframe;
run;
```

The REG Procedure
Model: MODEL1
Dependent Variable: y

Number of Observations Read   20
Number of Observations Used   20

Analysis of Variance

Source	DF	Sum of Squares	Mean Square	F Value	Pr > F
Model	1	300342	300342	1.44	0.2450
Error	18	3742583	207921		
Corrected Total	19	4042925			

Root MSE	455.98384	R-Square	0.0743	
Dependent Mean	1161.45000	Adj R-Sq	0.0229	
Coeff Var	39.25988			

Parameter Estimates

Variable	DF	Parameter Estimate	Standard Error	t Value	Pr > |t|
Intercept	1	1384.59474	211.81884	6.54	<.0001
t	1	-21.25188	17.68229	-1.20	0.2450

Dependent Variable: y

Durbin-Watson D	0.327
Number of Observations	20
1st Order Autocorrelation	0.792

The plot of residuals versus time is displayed in Figure 5.21 and will be used to detect autocorrelation. The residuals have long positive and long negative runs. Long cycles indicate positive autocorrelation while for negative autocorrelation, the runs will be very short with rapid changes in signs. In this case, positive correlation implies that if the correlation is positive in year t, then the residual in year $(t + 1)$ would also be expected to be positive.

Apart from the residual diagnostics for detecting autocorrelation, there is a formal test for detecting autocorrelation. Suppose for equally spaced time intervals, the error term is governed by the following relationship

$$\varepsilon_t = \rho \varepsilon_{t-1} + u_t, \tag{5.43}$$

where

- ε_t is error term at time t for the regression model in (5.42)

- ρ is the autocorrelation coefficient

- u_t is the random error term assumed to be distributed $N(0, \sigma^2)$.

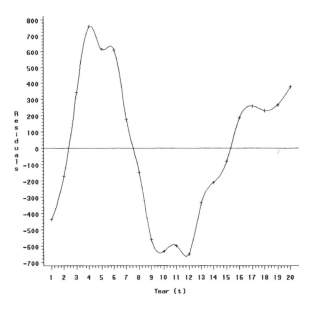

Figure 5.21: Residual time plot (residual versus time)

The expression in (5.43) when used successfully leads to the the *first-order autoregressive* model

$$\varepsilon_t = \sum_{j=0}^{\infty} \rho^j u_{t-j},$$ (5.44)

which leads to

$$\mathrm{E}(\varepsilon_t) = 0$$

$$\mathrm{Var}(\varepsilon_t) = \sigma_u^2 \sum_{j=0}^{\infty} \rho^{2j} = \frac{\sigma_u^2}{1 - \rho^2}$$ (5.45)

$$\mathrm{Cov}(\varepsilon_t, \varepsilon_{t+j}) = \frac{\rho^{|j|} \sigma_u^2}{1 - \rho^2}.$$

5.11.1   The Durbin-Watson d Statistic

The Durbin-Watson d statistic used to detect the presence of serial correlation is given by

$$d = \frac{\sum_{t=2}^{n} (e_t - e_{t-1})^2}{\sum_{t=1}^{n} e_t^2}, \quad 0 \le d \le 4,$$ (5.46)

where the e_t is the raw residuals for observation at time t. To interpret the Durbin-Watson statistic, we note the following:

1. If $d \approx 2.0$, then the residuals are uncorrelated

2. If $d < 2$, then the residuals are positively correlated, and $d \approx 0$ indicates very strong positive correlation .

3. If $d > 2$, then the residuals are negatively correlated and $d \approx 4$ indicates very strong negative correlation.

Decision Rules:

The null and alternative hypotheses for the one-tailed case can be succinctly written as

$$H_0 : \rho = 0$$
$$H_a : \rho > 0.$$ (5.47)

Durbin and Watson (1951) obtained the lower bound $d_{L,\alpha}$ and the upper bound $d_{U,\alpha}$ in order to test the null hypothesis in (5.47) against any alternative hypothesis. Table 5 in the Appendix contains these lower and upper bounds. We would reject H_0 if $d < d_{L,\alpha}$, where $d_{L,\alpha}$ is the lower bound (at significance level α) from Table 5 in the Appendix or if p-value is less than or equal to α, usually, 0.05. If the alternative hypothesis had been $H_a : \rho < 0$, then we would have rejected H_0 if $(4 - d) > d_{U,\alpha}$, where $d_{U,\alpha}$ is the upper bound from Table 5 in the Appendix at significance level α. The decision will be inconclusive for both alternatives if $d_{L,\alpha} \leq d \leq d_{U,\alpha}$ or if $d_{L,\alpha} \leq (4-d) \leq d_{U,\alpha}$ for positive and negative correlations respectively.

Consider the two-tailed hypotheses

$$H_0 : \rho = 0$$
$$H_a : \rho \neq 0. \tag{5.48}$$

The decision rules are exactly the same, except that the lower and upper bounds of the Durbin-Watson are obtained at $d_{L,\alpha/2}$ and $d_{U,\alpha/2}$. In SAS, **proc reg** can be requested to output the Durbin-Watson test statistic. In the example, this is

```
            Dependent Variable: y

Durbin-Watson D                   0.327
Number of Observations               20
1st Order Autocorrelation         0.792
```

For this data, $d = 0.327$ and $\hat{\rho} = 0.792$. From Table 5 in the Appendix, $d_{L,0.05} = 1.20$ and $d_{U,0.05} = 1.41$. In this example, $d = 0.327 < 1.20$, hence, we would reject H_0 and conclude that the errors are positively correlated.

5.11.2   Handling Serially Correlated Data

Now that we have established that the errors in the previous example are positively correlated, we examine in this section how to overcome this problem generally in multiple regression.

Autoregressive Error Models

Earlier we formulated our model as

$$y_t = \beta_0 + \beta_1 x_t + \varepsilon_t \tag{5.49a}$$
$$\varepsilon_t = \rho\varepsilon_{t-1} + \nu_t, \quad -1 < \rho < 1. \tag{5.49b}$$

The model in (5.49) assumes that residuals at times t and $(t+1)$ apart are correlated, and are described as *first-order autoregressive*, AR(1). Thus the autocorrelations for an AR(1) diminishes rapidly as the interval $\Delta_t = t - (t + \Delta_t)$ between two residuals increases. On the other hand, an m-th order AR(m), autoregressive model would have ε_t defined as

$$\varepsilon_t = \rho_1\, \varepsilon_{t-1} + \rho_2\, \varepsilon_{t-2} + \cdots + \rho_m\, \varepsilon_{t-m} + \nu_t. \tag{5.50}$$

The best approach for handling serially correlated data with auto regressive error terms is the use of transformation of the variables which would lead to the use of the Cochrane-Orcutt procedure. Consider, again the model in (5.49), and the following transformation of the response or dependent variable y:

$$y_t' = y_t - \rho y_{t-1}. \tag{5.51}$$

Substituting in (5.51), y_t and y_{t-1} in (5.49), we have

$$y_t' = \beta_0 + \beta_1 x_t - \rho(\beta_0 + \beta_1 x_{t-1} + \varepsilon_{t-1}) \tag{5.52a}$$
$$= \beta_0(1 - \rho) + \beta_1(x_t - \rho\, x_{t-1}) + (\varepsilon_t - \rho\varepsilon_{t-1}), \tag{5.52b}$$

and using the expression in (5.49b) for the third expression in (5.52b), we have

$$y'_t = \beta'_0 + \beta'_1 x'_t + \nu_t, \quad (t = 1, 2, \ldots, n) \tag{5.53}$$

where

$$y'_t = y_t - \rho\, y_{t-1}$$
$$x'_t = x_t - \rho\, x_{t-1}$$
$$\beta'_0 = \beta_0(1 - \rho)$$
$$\beta'_1 = \beta_1,$$

and the error terms in (5.53), ν_t, are independent and normally distributed with mean zero and variance σ_ν^2 which satisfies the ordinary least squares assumption. The use of the model in (5.53) requires that we have knowledge of ρ. A good estimate of ρ is obtained as follows (Cochrane-Orcutt).

$$\hat{\rho} = r = \frac{A}{B}, \tag{5.54}$$

where

$$A = \sum_{t=2}^{n}(e_t, e_{t-1}) \quad \text{and} \quad B = \sum_{t=1}^{n} e_t^2 \tag{5.55}$$

The e_t are obtained from the ordinary least squares regression. With this value of r, we form y'_t and x'_t as in (5.56) and hence use OLS regression of y'_t on x'_t and obtain estimates for β'_0 and β'_1.

$$y'_t = y_t - \rho\, y_{t-1} \tag{5.56a}$$
$$x'_t = x_t - \rho\, x_{t-1}. \tag{5.56b}$$

The estimators for the intercept and slope are therefore given by $\beta'_0/(1 - \rho)$ and β'_1.

The procedure might require more than one iteration to ensure that positive autocorrelation is no longer present in the data. We can use the Durbin-Watson statistic to achieve this. In SAS, **proc autoreg** implements this procedure and we shall use this for our data in Table 5.12.

```
data serial;
do year=1974 to 1993;
input y @@;
t=year-1973;
output;
end;
datalines;
926 1171 1663 2058 1892 1866 1414 1067 633 540
553 479 771 876 987 1232 1282 1233 1247 1339
;
proc print;
run;
proc autoreg;
model y=t/nlag=1 dwprob normal partial covb corrb coef;
run;
proc autoreg;
model y=t/nlag=1 dwprob coef method=ml;
model y=t/nlag=1 dwprob coef method=uls;
run;
```

```
                    The AUTOREG Procedure

                  Dependent Variable     y

              Ordinary Least Squares Estimates

          SSE            3742582.76  DFE                   18
          MSE                207921  Root MSE       455.98384
          SBC            305.540091  AIC            303.548626
          MAE             382.40188  AICC           304.254508
          MAPE          41.9948852   HQC            303.937381
```

```
        Durbin-Watson       0.3265    Regress R-Square      0.0743
                                      Total R-Square        0.0743

                          Miscellaneous Statistics

            Statistic         Value      Prob        Label

            Normal Test      1.1535     0.5617     Pr > ChiSq

                          Durbin-Watson Statistics

            Order          DW     Pr < DW    Pr > DW

              1          0.3265    <.0001    1.0000

NOTE: Pr<DW is the p-value for testing positive autocorrelation, and Pr>DW is the
      p-value for testing negative autocorrelation.

                                   Standard              Approx
        Variable      DF    Estimate      Error   t Value  Pr > |t|

        Intercept      1       1385    211.8188     6.54    <.0001
        t              1    -21.2519    17.6823    -1.20    0.2450

                         Estimates of Autocorrelations

 Lag   Covariance   Correlation   -1 9 8 7 6 5 4 3 2 1 0 1 2 3 4 5 6 7 8 9 1

  0      187129      1.000000    |                     |********************|
  1      148197      0.791950    |                     |***************     |

                   Preliminary MSE      69764.5

               Estimates of Autoregressive Parameters

                                    Standard
               Lag    Coefficient      Error    t Value
                1      -0.791950     0.148089    -5.35

                          Expected
                       Autocorrelations

                       Lag    Autocorr
                        0      1.0000
                        1      0.7920

               Coefficients for First NLAG Observations

                                      1
                        1     0.6105854657

                       Yule-Walker Estimates

        SSE          1134566.61    DFE                     17
        MSE               66739    Root MSE          258.33934
        SBC          285.651993    AIC               282.664796
        MAE          183.157101    AICC              284.164796
        MAPE          18.6183548    HQC               283.247928
        Durbin-Watson     0.8377    Regress R-Square     0.0016
                                    Total R-Square       0.7194

                          Durbin-Watson Statistics

            Order          DW     Pr < DW    Pr > DW

              1          0.8377    0.0006    0.9994

NOTE: Pr<DW is the p-value for testing positive autocorrelation, and Pr>DW is the
      p-value for testing negative autocorrelation.

                                   Standard              Approx
        Variable      DF    Estimate      Error   t Value  Pr > |t|

        Intercept      1       1102    394.7783     2.79    0.0125
        t              1     4.9178     30.1187     0.16    0.8722

                          Expected
                       Autocorrelations

                       Lag    Autocorr
                        0      1.0000
```

```
                              1      0.7920

             Coefficients for First NLAG Observations

                                     1
                         1     0.6105854657
```

Algorithm converged.

Maximum Likelihood Estimates

SSE	1117616.77	DFE		17
MSE	65742	Root MSE		256.40235
SBC	285.533329	AIC		282.546132
MAE	179.488931	AICC		284.046132
MAPE	18.1265318	HQC		283.129264
Durbin-Watson	0.8594	Regress R-Square		0.0035
		Total R-Square		0.7236

Durbin-Watson Statistics

Order	DW	Pr < DW	Pr > DW
1	0.8594	0.0008	0.9992

NOTE: Pr<DW is the p-value for testing positive autocorrelation, and Pr>DW is the p-value for testing negative autocorrelation.

Parameter Estimates

Variable	DF	Estimate	Standard Error	t Value	Approx Pr > |t|
Intercept	1	1067	444.6497	2.40	0.0281
t	1	8.0562	33.3188	0.24	0.8118
AR1	1	-0.8303	0.1246	-6.66	<.0001

Expected Autocorrelations

Lag	Autocorr
0	1.0000
1	0.8303

Coefficients for First NLAG Observations

```
                                     1
                         1     0.5573694081
```

The AUTOREG Procedure

Autoregressive parameters assumed given

Variable	DF	Estimate	Standard Error	t Value	Approx Pr > |t|
Intercept	1	1067	443.4831	2.41	0.0277
t	1	8.0562	32.9996	0.24	0.8101

Algorithm converged.

Unconditional Least Squares Estimates

SSE	1111904.01	DFE		17
MSE	65406	Root MSE		255.74620
SBC	285.664223	AIC		282.677026
MAE	178.306444	AICC		284.177026
MAPE	17.7855009	HQC		283.260158
Durbin-Watson	0.8735	Regress R-Square		0.0058
		Total R-Square		0.7250

Durbin-Watson Statistics

Order	DW	Pr < DW	Pr > DW
1	0.8735	0.0009	0.9991

NOTE: Pr<DW is the p-value for testing positive autocorrelation, and Pr>DW is the p-value for testing negative autocorrelation.

Parameter Estimates

```
                                      Standard                   Approx
          Variable        DF     Estimate        Error   t Value  Pr > |t|

          Intercept       1          1029     512.7556      2.01    0.0610
          t               1       11.5521      36.9615      0.31    0.7584
          AR1             1       -0.8683       0.1218     -7.13   <.0001

                                 Expected
                             Autocorrelations

                         Lag      Autocorr
                          0        1.0000
                          1        0.8683

          Coefficients for First NLAG Observations

                                        1
                   1       0.4959795038

                    The AUTOREG Procedure

          Autoregressive parameters assumed given

                                      Standard                   Approx
          Variable        DF     Estimate        Error   t Value  Pr > |t|

          Intercept       1          1029     511.7200      2.01    0.0606
          t               1       11.5521      36.6614      0.32    0.7565
```

We present in Table 5.13, the regression results for five methods of estimating the parameters of the model. We now discuss these results in turn.

Estimation Procedure	$\hat{\beta}_1$	s.e.$(\hat{\beta}_1)$	r	$\hat{\sigma}^2$ MSE	R^2
OLS	-21.2518	17.6823	0.7920	207921	0.0743
Yule-Walker I	4.9178	30.1187	0.7920	66739	0.7194
Yule-Walker II	11.1227	36.1700	0.8638	65411	0.7250
MLE	8.0562	32.9996	0.8303	65742	0.7236
ULS	11.5521	36.6614	0.8683	65406	0.7250

Table 5.13: Regression results for various forms of estimation

1. The OLS Results

The Ordinary least squares results gives us the estimate of ρ to be 0.7920 with the estimated regression model

$$\hat{y}_t = 1385 - 21.2519t, \qquad R^2 = 0.0743, \quad r = 0.7920.$$

2. Yule-Walker I

The Yule-Walker I method for estimating an autoregressive error model is very similar to the Cochrane-Orcutt method except that it retains more information from the first observation. For this method, the estimated regression model is

$$\hat{y}_t = 1102 + 4.9178\, t + \hat{\varepsilon}_t$$
$$\hat{\varepsilon}_t = 0.7920\, \hat{\varepsilon}_{t-1} + \hat{\nu}_t,$$

with estimated variance, $\mathrm{Var}(\hat{\nu}_t) = 66739$. We notice here that estimated serial correlation coefficient for this method is the same as those obtained for the ordinary least squares method. This is because there is only one iteration carried out. However, R^2 for this model is 0.7194, a considerable improvement over that of the OLS method.

3. Yule-Walker II

This is the iterated Yule-Walker method. However, the process is iterated twice and is a more efficient method than the type I or the OLS method. For this method, the estimated regression equation is

$$\hat{y}_t = 1033 + 11.1227\, t + \hat{\varepsilon}_t$$
$$\hat{\varepsilon}_t = 0.8638\, \hat{\varepsilon}_{t-1} + \hat{\nu}_t,$$

with estimated variance, $\mathrm{Var}(\hat{\nu}_t) = 65411$ and $R^2 = 0.7250$.

4. **The Maximum Likelihood Method**

The **proc autoreg** in SAS produces maximum likelihood estimates and for these data, the estimated regression equation for the AR(1), auto regressive model, is

$$\hat{y}_t = 1067 + 8.0562\, t + \hat{\varepsilon}_t$$
$$\hat{\varepsilon}_t = 0.8303\, \hat{\varepsilon}_{t-1} + \hat{\nu}_t,$$

with estimated variance, $\mathrm{Var}(\hat{\nu}_t) = 65742$ and $R^2 = 0.7236$.

5. **The Unconditional Least Squares Method**

The **proc autoreg** in SAS produces unconditional least squares estimates and for these data, the estimated regression equation for the AR(1), auto regressive model, is

$$\hat{y}_t = 1029 + 11.5521\, t + \hat{\varepsilon}_t$$
$$\hat{\varepsilon}_t = 0.8683\, \hat{\varepsilon}_{t-1} + \hat{\nu}_t,$$

with estimated variance, $\mathrm{Var}(\hat{\nu}_t) = 65406$ and $R^2 = 0.7250$.

5.11.3 ARMA Models

A first-order moving average model assumes that the error terms more than 1 interval apart are uncorrelated. That is, adjacent residuals are correlated while non-adjacent residuals (further apart by more than 1) display very little or small correlations. The *first-order moving average model* MA(1) is written as

$$y_t = \beta_0 + \beta_1\, x_t + \varepsilon_t \tag{5.57a}$$
$$\varepsilon_t = \theta\, \varepsilon_{t-1} + \nu_t, \quad -1 < \theta < 1, \tag{5.57b}$$

with ν_t being identically and independently distributed with mean zero. With autocorrelation between residuals at t and $t + k$ given by

$$\mathrm{AC}(\varepsilon_t, \varepsilon_{t+k}) = \begin{cases} \dfrac{\theta}{1 + \theta^2}, & \text{if } k = 1 \\ 0, & \text{if } k > 1. \end{cases} \tag{5.58}$$

A q-th order moving average model MA(q) is defined as

$$\varepsilon_t = \theta_1\, \varepsilon_{t-1} + \theta_2\, \varepsilon_{t-2} + \cdots + \theta_q\, \varepsilon_{t-q} + \nu_t, \tag{5.59}$$

with residuals q times apart assumed correlated and residuals more than q times apart having zero correlation. A model that combines both autoregressive and moving average models is termed an *autoregressive moving average model* and is often referred to as an (ARMA) model. An ARMA(p, q) model has its error term ε_t defined by

$$\varepsilon_t = \rho_1\, \varepsilon_{t-1} + \cdots + \rho_p\, \varepsilon_{t-p} + \theta_1\, \varepsilon_{t-1} + \cdots + \theta_q\, \varepsilon_{t-q} + \nu_t. \tag{5.60}$$

A more general combined model is the ARIMA, *the autoregressive integrated moving average model* ARIMA(p, d, q), where we combine the techniques of an AR and MA on the differenced data and d is the order of differencing. For instance, an order 1 differencing is defined as

$$x_t - x_{t-1} = \varepsilon_t - \theta\varepsilon_{t-1}.$$

The p is the order of the autoregressive process part, the d, is the order of differencing, while the q refers to the order of the moving-average process. The following equivalents are worth noting

$$\mathrm{ARIMA}(1,0,0) \equiv \mathrm{AR}(1)$$
$$\mathrm{ARIMA}(0,0,1) \equiv \mathrm{MA}(1).$$

The analysis of ARIMA models are usually divided into three stages, namely, (1) identification, (2) estimation and (3) forecasting stages. The Box-Jenkins method utilizes this approach. The **proc arima** in SAS will be used to implement both an AR(1), an MA(1) and an ARIMA(1, 0, 1) $\equiv$ ARMA (1, 1) models for the data in Table 5.12. The following SAS program implements these three models in turn. But first we present the output for the ARIMA(1, 0, 0) or the AR(1) model as implemented under the estimate $p = 1$, which assumes that, q and d are both zero.

```
set serial;
proc arima data=serial;
identify var=y nlag=1 crosscorr=t;
run;
estimate p=1;
run;
estimate q=1;
run;
estimate p=1 q=1;
run;
```

The ARIMA Procedure

Name of Variable = y

Mean of Working Series	1161.45
Standard Deviation	449.6068
Number of Observations	20

Autocorrelations

Lag	Covariance	Correlation	-1 9 8 7 6 5 4 3 2 1 0 1 2 3 4 5 6 7 8 9 1	Std Error
0	202146	1.00000	|********************|	0
1	170075	0.84134	| . |***************** |	0.223607

"." marks two standard errors

Partial Autocorrelations

Lag	Correlation	-1 9 8 7 6 5 4 3 2 1 0 1 2 3 4 5 6 7 8 9 1
1	0.84134	| . |**************** |

Correlation of y and t

Variance of input =	33.25
Number of Observations	20

Crosscorrelations

Lag	Covariance	Correlation	-1 9 8 7 6 5 4 3 2 1 0 1 2 3 4 5 6 7 8 9 1
-1	-799.839	-.30851	| . ******| . |
0	-706.625	-.27256	| . *****| . |
1	-830.236	-.32024	| . ******| . |

"." marks two standard errors

Conditional Least Squares Estimation

Parameter	Estimate	Standard Error	t Value	Approx Pr > |t|	Lag
MU	1019.2	221.61735	4.60	0.0002	0
AR1,1	0.87310	0.12157	7.18	<.0001	1

Constant Estimate	129.3361
Variance Estimate	63037.94
Std Error Estimate	251.0736
AIC	279.6802
SBC	281.6716
Number of Residuals	20

* AIC and SBC do not include log determinant.

Correlations of Parameter

```
                              Estimates

                   Parameter        MU      AR1,1

                       MU          1.000   -0.134
                      AR1,1       -0.134    1.000

                  Autocorrelation Check of Residuals

   To      Chi-          Pr >
  Lag     Square   DF   ChiSq   -----------------Autocorrelations----------------

    6      23.59    5   0.0003    0.558    0.286   -0.044   -0.291   -0.458   -0.431
   12      28.46   11   0.0027   -0.254   -0.207   -0.057    0.073    0.076    0.092
   18      30.92   17   0.0204    0.121    0.078    0.003    0.060    0.064    0.021

                         Model for variable y

                   Estimated Mean      1019.236

                     Autoregressive Factors

                 Factor 1:  1 - 0.8731 B**(1)
```

The default method of estimation of parameters in SAS **proc arima** is the conditional least squares. Other available methods of estimation are the maximum likelihood and the unconditional least squares estimations. These can be invoked by say, **estimate p method=ml** for MLE for instance. The parameters of the structural part of the model are presented under the 'Conditional Least Squares Estimation'. Here, $\hat{\beta}_0 = 1019.2$ and $\hat{\rho} = 0.8731$. Further, the AIC (Akaike Information Criterion) and the BIC (Bayesian Information Criterion) are generated to enable us compare our estimated model with other possible models. Lowest is best for each of these statistics. The goodness-of-fit (a check for the autocorrelations of the residuals or check for white noise residuals) of the model is checked under the output heading 'Autocorrelation Check of Residuals'. The hypotheses of interest in this case are

$$H_0 : \text{The error terms are uncorrelated (white noise)}$$
$$H_a : \text{The error terms are correlated.}$$

(5.61)

The p-value for the residual series is ($p = 0.0003$) for the first six lags. Since $p < .05$, we would therefore reject H_0 and conclude that the residuals are correlated. This implies that the AR(1) model does not fully fit our data. We present in Table 5.14, the goodness-of-fit test statistics and corresponding p-values for the first six lags for model combinations ($p = 0, 1, 2$) and ($q = 0, 1, 2$).

p	q	ARIMA(p,d,q)	Chi-Square	d.f.	p-value	AIC
0	0	ARIMA(0,0,0)	37.76	6	$< .000$	303.0925
1	0	ARIMA(1,0,0)	23.59	5	0.0003	279.6802
2	0	ARIMA(2,0,0)	2.35	4	0.6718	270.6027
0	1	ARIMA(0,0,1)	23.35	5	0.0003	287.4877
1	1	ARIMA(1,0,1)	10.56	4	0.0320	275.8565
2	1	ARIMA(2,0,1)	0.56	3	0.9047	271.3588
0	2	ARIMA(0,0,2)	11.60	4	0.0206	276.2293
1	2	ARIMA(1,0,2)	5.22	3	0.1563	274.7775
2	2	ARIMA(2,0,2)	0.87	2	0.6467	271.9039

Table 5.14: Fit of possible models to our data

The ARIMA(2,0,1) model has the highest p-value but the most parsimonious model is the AR(2) model which has the lowest AIC value of 270.6027. That is, the model with

$$\varepsilon_t = y_t - \rho_1\, y_{t-1} - \rho_2\, y_{t-2}.$$

(5.62)

The following SAS program implements the model in (5.62).

```
proc arima data=serial;
identify var=y nlag=1 crosscorr=t;
```

```
run;
estimate p=2 q=0;
run;
```

The ARIMA Procedure

Conditional Least Squares Estimation

Parameter	Estimate	Standard Error	t Value	Approx Pr > |t|	Lag
MU	949.13434	146.18820	6.49	<.0001	0
AR1,1	1.46717	0.18588	7.89	<.0001	1
AR1,2	-0.66474	0.18653	-3.56	0.0024	2

```
          Constant Estimate        187.514
          Variance Estimate       38360.17
          Std Error Estimate      195.8575
          AIC                     270.6027
          SBC                     273.5899
          Number of Residuals           20
  * AIC and SBC do not include log determinant.
```

Correlations of Parameter Estimates

Parameter	MU	AR1,1	AR1,2
MU	1.000	-0.110	-0.021
AR1,1	-0.110	1.000	-0.867
AR1,2	-0.021	-0.867	1.000

Autocorrelation Check of Residuals

To Lag	Chi-Square	DF	Pr > ChiSq	Autocorrelations					
6	2.35	4	0.6718	-0.170	0.169	0.026	-0.027	-0.074	-0.158
12	5.13	10	0.8825	0.175	-0.149	-0.018	0.126	-0.066	-0.004
18	9.98	16	0.8676	0.092	0.057	-0.051	0.100	0.122	0.052

Model for variable y

Estimated Mean   949.1343

Autoregressive Factors

Factor 1:   1 - 1.46717 B**(1) + 0.66474 B**(2)

The parameter estimates are $\hat{\rho}_1 = 1.4672$ and $\hat{\rho}_2 = -0.6647$, and both are highly significant. The fitted model is

$$\hat{\varepsilon}_t = \hat{y}_t - 1.4672\,\hat{y}_{t-1} + 0.6647\,\hat{y}_{t-2}, \tag{5.63}$$

with $\hat{\sigma}_\varepsilon^2 = 38360.17$.

5.12   Methods for Solving Multicollinearity Problems

Several methods are available for solving the problems created by multicollinearity. These include:

(a) Centering

(b) Centering and Scaling

(c) Use of ridge regression or the so called *biased estimation techniques*

(d) Dropping one or more of the correlated explanatory variables from the final model. This can be accomplished by a selection strategy procedure such as stepwise selection strategy (see Chapter 7).

We now consider the above procedures one at a time.

5.12.1   Centering

For $p - 1$ explanatory variables, let our regression, say, be of the form

$$y_i = \beta_0 + \beta_1 x_{i1} + \beta_2 x_{i2} + \cdots + \beta_{p-1} x_{i,p-1} + \varepsilon_i. \tag{5.64}$$

Then centering involves fitting the model

$$y_i - \bar{y} = \beta_1(x_{i1} - \bar{x}_1) + \beta_2(x_{i2} - \bar{x}_2) + \cdots + \beta_{p-1}(x_{i,p-1} - \bar{x}_{p-1}) + \varepsilon_i. \tag{5.65}$$

The least squares estimates for β_0 in (5.64) is $\hat{\beta}_0 = \bar{y} - \sum_{i=1}^{p-1} \hat{\beta}_i \bar{x}_i$, while that for (5.65) has $\hat{\beta}_0 = 0$.

Centering of data to minimize or remove collinearity is most effective for polynomial regressions. Although we shall discuss other procedures for implementing polynomial regressions at a later chapter.

Example 5.1 continued

The data in Table 5.1 represent experimental data for 10 sales trainees on number of days of training received (x) and performance score (y) in a battery of simulated sales situations. In section 5.2, we observe that the residual plot against $\hat{y}_i$ shows that the regression function is not linear. Now we consider a regression model of the form

$$y_i = \beta_0 + \beta_1 x_i + \beta_2 x_i^2 + \varepsilon_i. \tag{5.66}$$

for the data. That is, a second degree polynomial equation or a quadratic model. The SAS program and the residual plot is presented below.

```
set example;
x2=x*x;
proc reg;
model y=x x2 / vif;
output out=aa p=pred r=resid;
test1: test x2=0;
run;
proc plot data=aa vpct=50;
plot resid*pred='+'/vref=0;
run;
```

```
                          Model: MODEL1
                          Dependent Variable: y

                          Analysis of Variance

                                   Sum of          Mean
     Source            DF         Squares         Square    F Value    Pr > F

     Model              2      6932.37143     3466.18571     270.41    <.0001
     Error              7        89.72857       12.81837
     Corrected Total    9      7022.10000

              Root MSE              3.58027    R-Square    0.9872
              Dependent Mean       89.70000    Adj R-Sq    0.9836
              Coeff Var             3.99139

                          Parameter Estimates

                      Parameter    Standard                      Variance
     Variable   DF     Estimate       Error   t Value   Pr > |t|   Inflation

     Intercept   1     19.60000     5.42976      3.61     0.0086           0
     x           1     62.18571     8.27567      7.51     0.0001    26.71429
     x2          1     -8.42857     2.70643     -3.11     0.0170    26.71429

              Test TEST1 Results for Dependent Variable y

                                     Mean
              Source         DF     Square    F Value    Pr > F

              Numerator       1   124.32143       9.70    0.0170
              Denominator     7    12.81837
```

The residuals plot against the predicted in the first half of Figure 5.22 shows a pattern that is random around the zero mean-that is, it is now of type (a) in Figure 5.1. Further, the partial F test for the x^2 term gives a p-value of 0.0170 indicating that x^2 is important for the model already containing x. However, for this model, because x and x^2 are highly correlated, we would have to be careful of the problem of multicollinearity. We shall examine this example later in this chapter. For now, the estimated regression model is

$$\hat{y}_i = 19.60 + 62.1857x_i - 8.4286x_i^2,$$

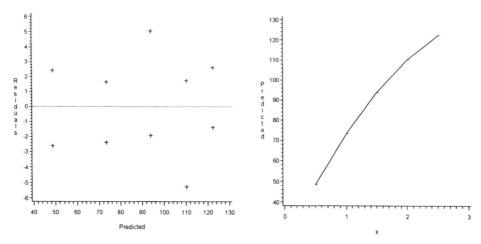

Figure 5.22: Residuals and predicted plots

and its plot is also presented in the second half of Figure 5.22.

The following SAS program will implement the quadratic model and request for multicollinearity diagnostics.

```
data ex1;
input x y @@;
x2=x*x;
datalines;
0.5 46 0.5 51 1.0 71 1.0 75
1.5 92 1.5 99 2.0 105 2.0 112
2.5 121 2.5 125
;
proc means data=ex1;
var x y; run;
proc reg data=ex1;
model y=x x2 / vif collinoint; run;
```

The MEANS Procedure

Variable	N	Mean	Std Dev	Minimum	Maximum
x	10	1.5000000	0.7453560	0.5000000	2.5000000
y	10	89.7000000	27.9326571	46.0000000	125.0000000

The REG Procedure
Model: MODEL1
Dependent Variable: y

Analysis of Variance

Source	DF	Sum of Squares	Mean Square	F Value	Pr > F
Model	2	6932.37143	3466.18571	270.41	<.0001
Error	7	89.72857	12.81837		
Corrected Total	9	7022.10000			

Root MSE	3.58027	R-Square	0.9872	
Dependent Mean	89.70000	Adj R-Sq	0.9836	
Coeff Var	3.99139			

Parameter Estimates

Variable	DF	Parameter Estimate	Standard Error	t Value	Pr > |t|	Variance Inflation
Intercept	1	19.60000	5.42976	3.61	0.0086	0
x	1	62.18571	8.27567	7.51	0.0001	26.71429
x2	1	-8.42857	2.70643	-3.11	0.0170	26.71429

Collinearity Diagnostics (intercept adjusted)

Number	Eigenvalue	Condition Index	--Proportion of Variation--	
			x	x2

```
1      1.98110      1.00000      0.00945      0.00945
2      0.01890     10.23951      0.99055      0.99055
```

From the first output, we see that VIF for both the linear and quadratic terms are both greater than 10, hence multicollinearity is present. This is expected since x^2 is highly correlated with x. Further, the x^2 term is important in the model based on its p-value of 0.0170 derived from the partial test and using $\alpha = .05$. However, the condition index of 10.2395 does not appear too high but the variance proportions are unusually high. So we would need to proceed to fitting a model where the explanatory variables would be centered. In this example, $\bar{x} = 1.50$, hence a model of the form

$$y_i = \beta_0 + \beta_1(x_i - 1.50) + \beta_2(x_i - 1.50)^2 + \varepsilon_i \tag{5.67}$$

is suggested. Note that in this example, we did not center y_i. If we want, this could also be implemented as

$$y_i - 89.70 = \beta_1'(x_i - 1.50) + \beta_2'(x_i - 1.50)^2 + \varepsilon_i'. \tag{5.68}$$

When the model in (5.67) is implemented in SAS we have the following results.

```
data new;
set ex1;
x1=x-1.50;
x21=x1*x1;
proc reg data=new;
model y=x x21 / vif collinoint; run;
```

```
                The REG Procedure
                   Model: MODEL1
               Dependent Variable: y

               Analysis of Variance

                        Sum of        Mean
Source          DF     Squares      Square    F Value   Pr > F

Model            2   6932.37143   3466.18571   270.41   <.0001
Error            7     89.72857     12.81837
Corrected Total  9   7022.10000

        Root MSE            3.58027   R-Square    0.9872
        Dependent Mean     89.70000   Adj R-Sq    0.9836
        Coeff Var           3.99139

               Parameter Estimates

             Parameter   Standard                      Variance
Variable  DF  Estimate      Error   t Value  Pr > |t|  Inflation

Intercept  1  93.91429    1.76438    53.23    <.0001          0
x1         1  36.90000    1.60115    23.05    <.0001    1.00000
x21        1  -8.42857    2.70643    -3.11    0.0170    1.00000

       Collinearity Diagnostics (intercept adjusted)

                          Condition    --Proportion of Variation-
      Number   Eigenvalue    Index          x1           x21

         1      1.00000     1.00000      1.00000            0
         2      1.00000     1.00000            0      1.00000
```

Notice that the analysis of variance table remained unchanged, but the parameter estimates changed. Further, the VIF are now 1.0 each and are much less than 10 and the condition index is 1 with variance contributions between x and x^2 now being 0. We see that the multicollinearity has been removed simply by centering the explanatory variable x and $x_2 = x^2$. In general, centering is very effective in most polynomial regressions but certainly, we should also be aware that centering alone may not be adequate in controlling the problem of collinearity in such regressions.

5.12.2 Centering and Scaling

This method is often used for polynomial regression. Here, the variables are not only centered but standardized by dividing by their respective sample standard deviations. That is, with $p - 1$ explanatory variables, our regression model in (5.64) now becomes

$$\frac{y_i - \bar{y}}{s_y} = \beta_1^* \frac{(x_{i1} - \bar{x}_1)}{s_{x1}} + \beta_2^* \frac{(x_{i2} - \bar{x}_2)}{s_{x2}} + \cdots + \beta_{p-1}^* \frac{(x_{i,p-1} - \bar{x}_{p-1})}{s_{x_{p-1}}} + \varepsilon_i^*, \tag{5.69}$$

where s_y and $s_{x_i}, i = 1, 2, \ldots, p-1$ are the sample standard deviations of y and x_i respectively. The model in (5.69) is often described as a *standardized regression model*. Thus, if we define

$$y_i = \frac{y_i - \bar{y}}{s_y} \quad \text{and}$$

$$x_{ji} = \frac{(x_{ij} - \bar{x}_j)}{s_{x_j}} \quad j = 1, 2, \ldots, p-1; \ i = 1, 2, \ldots, n,$$

then the *standardized regression* equation in (5.69) can be re-written as

$$y_i = \beta_1^* x_{i1} + \beta_2^* x_{i2} + \cdots + \beta_{p-1}^* x_{i,p-1} + \varepsilon_i^*. \tag{5.70}$$

The least squares estimates $\hat{\beta}_j^*, \ j = 1, 2, \ldots, p-1$ are referred to as *standardized regression coefficients* and are related to the original least squares estimates for the model in (5.65) by the following expression

$$\hat{\beta}_j^* = \hat{\beta}_j \frac{s_{x_j}}{s_y}, \quad j = 1, 2, \ldots, p-1. \tag{5.71}$$

Example 5.5 continued

To implement standardized regression in SAS, we shall use the option **STB** in the model line of the program for Example 5.5. The results of invoking this on the data produced the partial output below (some output have been deliberately deleted).

```
set multi;
proc means data=multi;
var y x1-x4;
run;
proc reg;
model y=x1-x4 / stb;
run;
```

Variable	N	Mean	Std Dev	Minimum	Maximum
y	13	95.4230769	15.0437226	72.5000000	115.9000000
x1	13	7.4615385	5.8823944	1.0000000	21.0000000
x2	13	48.1538462	15.5608813	26.0000000	71.0000000
x3	13	11.7692308	6.4051262	4.0000000	23.0000000
x4	13	30.0000000	16.7381799	6.0000000	60.0000000

Parameter Estimates

Variable	DF	Parameter Estimate	Standard Error	t Value	Pr > |t|	Standardized Estimate
Intercept	1	62.40537	70.07096	0.89	0.3991	0
x1	1	1.55110	0.74477	2.08	0.0708	0.60651
x2	1	0.51017	0.72379	0.70	0.5009	0.52771
x3	1	0.10191	0.75471	0.14	0.8959	0.04339
x4	1	-0.14406	0.70905	-0.20	0.8441	-0.16029

We see in the third column that the regular regression model with intercept, give the following:

$$\hat{\beta}_0 = 62.4054, \ \hat{\beta}_1 = 1.5511, \ \hat{\beta}_2 = 0.5102, \ \hat{\beta}_3 = 0.1019, \ \text{and} \ \hat{\beta}_4 = -0.1441.$$

Similarly, the estimated standardized regression coefficients are given in the last column of the output and are:

$$\hat{\beta}_1^* = 0.6065, \ \hat{\beta}_2^* = 0.5277, \ \hat{\beta}_3^* = 0.0434, \ \text{and} \ \hat{\beta}_4^* = -0.1603.$$

In this example for instance, $s_y = 15.0437226$ and $s_{x_1} = 5.8823944$. This gives the parameter estimate of β_1, the corresponding standardized parameter estimate of β_1^*, by using the expression in (5.71) as

$$\hat{\beta}_1^* = 1.5511 \times \frac{5.88239}{15.04372} = 0.60651.$$

The above as expected agrees with the results obtained from SAS. Similarly, given $\hat{\beta}_j^*$, we can again use the expression in (5.71) to obtain $\hat{\beta}_j$.

From the magnitude of the estimates of the standardized parameters, we se that x_1 has the most influence on the dependent variable y, followed by x_2, x_4 and x_3 in that order. That is, x_3 has the least effect on y.

5.12.3 Ridge Regression

For the least squares estimation procedure, we recall that the parameter estimates under the OLS are unbiased and satisfy MVUE (minimum variance unbiased estimation) theory. Ridge regression is a method where the method of least squares is modified, with the resulting parameter estimates now being biased. However it does allow for small bias in the parameter estimates but ensures higher precision of the parameter estimates. Hence, this procedure is often referred to as a *biased estimation procedure*. To motivate this discussion, let us consider a three explanatory variable situation and let us imagine we wish to fit the model

$$y_i = \beta_0 + \beta_1\,x_1 + \beta_2\,x_2 + \beta_3\,x_3 + \varepsilon_i.$$

Then the above model can be written in matrix notation as

$$\mathbf{Y} = \mathbf{X}\beta + \varepsilon. \tag{5.72}$$

The least squares solutions of the linear model in (5.72) is given by

$$\hat{\beta} = (\mathbf{X}'\mathbf{X})^{-1}\,\mathbf{X}'\mathbf{Y}.$$

Multicollinearity occurs when the determinant $|\mathbf{X}'\mathbf{X}|$ is very small resulting in inflated inverse of $\mathbf{X}'\mathbf{X}$, and this usually happens when the explanatory variables are highly correlated. It can be shown that the variances of the least squares parameter estimates are given by

$$\sum_{i=1}^{k}\left(\frac{\operatorname{Var}\hat{\beta}_i}{\sigma^2}\right) = \sum_{i=1}^{k}\frac{1}{\lambda_i}, \quad i - 1, 2, \ldots, k, \tag{5.73}$$

where λ_i are the eigenvalues of the $\mathbf{X}'\mathbf{X}$ or its equivalent correlation matrix $\mathbf{R}$. That is, the λ's are the solutions of the equation

$$|\mathbf{R} - \lambda\mathbf{I}| = 0.$$

Values of λ very close to zero indicate multicollinearity might be present. Let us consider an example involving three standardized explanatory variables and let us denote by $\mathbf{X}'\mathbf{X}$ the variance-covariance matrix (or in this case the correlation matrix $\mathbf{R}$) from the data. We consider the following hypothetical 3×3 correlation matrix.

$$\mathbf{X}'\mathbf{X} = \begin{bmatrix} 1 & 0.924 & 0.458 \\ 0.924 & 1 & 0.085 \\ 0.458 & 0.085 & 1 \end{bmatrix}$$

For the above, the eigenvalues are obtained as

$$\lambda_1 = 2.0669, \quad \lambda_2 = 0.9325, \quad \lambda_3 = 0.0006.$$

Clearly, one of these eigenvalues is very close to zero hence we suspect multicollinearity to be present. The variance inflation factors are the diagonals of the inverse of $\mathbf{X}'\mathbf{X}$, which again are obtained as

$$(\mathbf{X}'\mathbf{X})^{-1} = \begin{bmatrix} 843.0208 & -751.5624 & -322.2207 \\ -751.5624 & 671.0336 & 287.1777 \\ -322.2207 & 287.1777 & 124.1670 \end{bmatrix}.$$

Hence, the variance inflation factors of the regression coefficients are given by

$$\text{VIF}_1 = 843.0208, \quad \text{VIF}_2 = 671.0336, \quad \text{and} \quad \text{VIF}_3 = 124.1670.$$

In the three explanatory variables example, the expression in (5.73) gives

$$\sum_{i=1}^{3} \frac{1}{\lambda_i} = \frac{1}{0.0006} + \frac{1}{0.9325} + \frac{1}{2.0669} = 1667.6667 + 1.0724 + 0.4838 = 1669.2229.$$

The principle behind ridge regression is therefore based on modifying the matrix $\mathbf{X}'\mathbf{X}$ with the new matrix $\mathbf{X}^{*'}\mathbf{X}^{*} = (\mathbf{X}'\mathbf{X} + k\mathbf{I})$, where $0 \leq k < 1$ is a constant, such that $|\mathbf{X}^{*'}\mathbf{X}^{*}| \neq 0$. Specifically, suppose we choose $k = 0.05$ and $k = 0.10$, then we would now have for the case when $k = 0.05$, the modified matrix

$$\mathbf{X}^{*'}\mathbf{X}^{*} = \begin{bmatrix} 1.05 & 0.924 & 0.458 \\ 0.924 & 1.05 & 0.085 \\ 0.458 & 0.085 & 1.05 \end{bmatrix},$$

whose eigenvalues are now computed to be

$$\lambda_1 = 2.1169, \quad \lambda_2 = 0.9825, \quad \lambda_3 = 0.0506,$$

and with corresponding variance inflation factors of the coefficients derived as

$$\text{VIF}_1 = 10.4050, \quad \text{VIF}_2 = 8.4809, \quad \text{and} \quad \text{VIF}_3 = 2.3629.$$

The corresponding expression for the variances of the parameter estimates under ridge regression is given by

$$\sum_{i=1}^{k} \left(\frac{\text{Var}\,\hat{\beta}_i^R}{\sigma^2} \right) = \sum_{i=1}^{k} \frac{\lambda_i}{(\lambda_i + k)^2}. \tag{5.74}$$

For the case when $k = 0.05$, we have

$$\sum_{i=1}^{3} \frac{\lambda_i}{(\lambda_i + k)^2} = \frac{0.0506}{(0.0506 + .05)^2} + \frac{0.9825}{(0.9825 + .05)^2} + \frac{2.1169}{(2.1169 + .05)^2}$$

$$= 4.9998 + 0.9216 + 0.4508 = 6.3722.$$

Similarly, if we use $k = 0.10$, then we have

$$\mathbf{X}^{*'}\mathbf{X}^{*} = \begin{bmatrix} 1.10 & 0.924 & 0.458 \\ 0.924 & 1.10 & 0.085 \\ 0.458 & 0.085 & 1.10 \end{bmatrix},$$

whose eigenvalues are now computed to be

$$\lambda_1 = 2.1669, \quad \lambda_2 = 1.0325, \quad \lambda_3 = 0.1006,$$

and with corresponding variance inflation factors of the coefficients derived as

$$\text{VIF}_1 = 5.3433, \quad \text{VIF}_2 = 4.4435, \quad \text{and} \quad \text{VIF}_3 = 1.5825.$$

The corresponding expression for the variances of the parameter estimates are again given as in (5.74), and therefore for the case when $k = 0.10$, we have

$$\sum_{i=1}^{3} \frac{\lambda_i}{(\lambda_i + k)^2} = \frac{0.1006}{(0.1006 + .10)^2} + \frac{1.0325}{(1.0325 + .10)^2} + \frac{2.1669}{(2.1669 + .10)^2}$$

$$= 2.500 + 0.8050 + 0.4217 = 3.7267.$$

We have therefore seen from these examples of modifying with $k = .05$ and $k = .10$, the considerable reduction in the variance inflation factors as well as reduced standard errors of the parameter estimates. Thus, ridge regression is obtained by adding a constant k, where, $0 \leq k \leq 1$ to the diagonal entries of $\mathbf{X}'\mathbf{X}$ and is implemented by solving the following equations:

$$\hat{\beta}^* = (\mathbf{X}'\mathbf{X} + k\mathbf{I})^{-1}\mathbf{X}'\mathbf{Y} = (\mathbf{X}^{*'}\mathbf{X}^*)^{-1}\mathbf{X}^{*'}\mathbf{Y}. \tag{5.75}$$

The stabilizing constant k is often referred to as the *regularization parameter* or *shrinkage parameter*.

Choice of Biasing Shrinkage Parameter

Choosing the shrinkage parameter k in ridge regression has often generated a lot of controversy. We present here two methods for choosing k. The first example examines the the ridge trace output. We demonstrate this with an example. As will be shown from the SAS output in this example, as k increases from zero, the corresponding ridge regression error mean square increases while the corresponding variances of the parameter estimates decrease. There is therefore some value k at which the ridge regression estimator $\hat{\beta}^*$ has a smaller mean square error than the ordinary least squares estimator. Consequently, we often plot the *ridge trace* which is a plot of the standardized ridge regression parameters against varying values of k, $0 \leq k \leq 1$ along with observing the ridge VIF values for various k values. The best choice of k is that value where the parameter estimates first seem stable and the VIFs seem moderately small. The choice is thus highly subjective at times.

Example 5.9

We now apply **proc reg** in SAS to fit a ridge regression model to our data. The first fit specifies that the shrinkage parameter varies from 0 to .40 in increments of 0.01. We have presented a partial output from this implementation below for values of k between 0 and 0.05. The other values are presented later in our further discussion on this example.

```
set multi;
proc reg data=multi graphics outvif outest=aa ridge=0 to .4 by 0.01;
model y=x1-x4;
plot/ridgeplot vref=0 lvref=1;
run;
proc print data=aa;
run;
```

```
                        The REG Procedure
                         Model: MODEL1
                     Dependent Variable: y

                     Analysis of Variance

                                Sum of          Mean
Source               DF         Squares        Square     F Value    Pr > F

Model                 4      2667.89944     666.97486      111.48    <.0001
Error                 8        47.86364       5.98295
Corrected Total      12      2715.76308


            Root MSE            2.44601     R-Square     0.9824
            Dependent Mean     95.42308     Adj R-Sq     0.9736
            Coeff Var           2.56333


                     Parameter Estimates

                      Parameter       Standard
    Variable    DF     Estimate          Error    t Value    Pr > |t|

    Intercept    1     62.40537       70.07096       0.89      0.3991
    x1           1      1.55110        0.74477       2.08      0.0708
    x2           1      0.51017        0.72379       0.70      0.5009
    x3           1      0.10191        0.75471       0.14      0.8959
    x4           1     -0.14406        0.70905      -0.20      0.8441
```

```
                                       I
                                       n
              _    _                    t
         _    D  _ P                    e
         M    _  E R  C   _             e
```

```
        O     T    P   I  O    R       r
        D     Y    V   D  M    M       c
   O    E     P    A   G  I    S       e
   b    L     E    R   E  T    E       p      x           x      x          x
   s    _     _    _   _  _    _       t      1           2      3          4   y

   1 MODEL1 PARMS  y   .   . 2.44601 62.4054  1.5511     0.510  0.1019     -0.144 -1

   2 MODEL1 RIDGEVIF y 0.00 .   .        .    38.4962  254.423  46.8684  282.513 -1
   3 MODEL1 RIDGE    y 0.00 . 2.44601 62.4054  1.5511   0.510    0.1019   -0.144 -1

   4 MODEL1 RIDGEVIF y 0.01 .   .        .     3.1639    5.675   3.1275    5.949 -1
   5 MODEL1 RIDGE    y 0.01 . 2.46291 82.6756  1.3152    0.306  -0.1290   -0.343 -1

   6 MODEL1 RIDGEVIF y 0.02 .   .        .     2.4564    2.108   2.3227    2.015 -1
   7 MODEL1 RIDGE    y 0.02 . 2.47672 84.3596  1.2723    0.293  -0.1630   -0.354 -1

   8 MODEL1 RIDGEVIF y 0.03 .   .        .     2.1661    1.305   2.0256    1.151 -1
   9 MODEL1 RIDGE    y 0.03 . 2.49529 85.0762  1.2413    0.290  -0.1852   -0.356 -1

  10 MODEL1 RIDGEVIF y 0.04 .   .        .     1.9668    0.986   1.8332    0.820 -1
  11 MODEL1 RIDGE    y 0.04 . 2.51829 85.5145  1.2151    0.289  -0.2029   -0.356 -1

  12 MODEL1 RIDGEVIF y 0.05 .   .        .     1.8077    0.819   1.6840    0.656 -1
  13 MODEL1 RIDGE    y 0.05 . 2.54498 85.8306  1.1917    0.289  -0.2180   -0.354 -1
```

The output initially reproduces the analysis of variance table, the parameter estimates, etc. For the ridge regression analysis, the vertical labeling of the columns are self explanatory. The first ridge regression has $k = 0$, (Models labeled observations 2 and 3) which of course corresponds to the ordinary least squares analysis. The ridge variance inflation factors RIDGEVIF are presented in the first line of the ridge regression output (line numbered 2 here), while, the root mean square estimate (RMSE), and the parameter estimates are presented in the second line of the ridge analysis (that is, line 3 here). We see that these results correspond to our earlier results under the OLS procedure. For $k = .01$, the corresponding results are displayed in output numbered 4 and 5. We see immediately that the variance inflation factors are all now considerably reduced, in fact all are now less than 10, but the RMSE has now increased to 2.46291 from the OLS value of 2.44601.

Similarly, for $k = 0.02$, ridge regression output lines 6 and 7, again the VIF are further reduced, but the RMSE again increases to 2.47672. It has been suggested that possible choice of k should be motivated by the VIFs all being very close to 1.0, which in our case would correspond to the case when $k = 0.04$ (lines 10 and 11), with an RMSE value of 2.51829. Figure 5.23 presents the ridge trace of the parameter estimates for values of $.01 \leq k \leq 0.40$. We see for instance that $\hat{\beta}_3^*$ was initially positive under the ordinary least squares but becomes negative under the ridge regression procedure, indicating that initially, the parameters are not stable, but become stable as k increases. Again the trace when $k = .04$ seems to give us a stable value of the parameter estimates after careful examination of the closeness of the VIFs to 1.0.

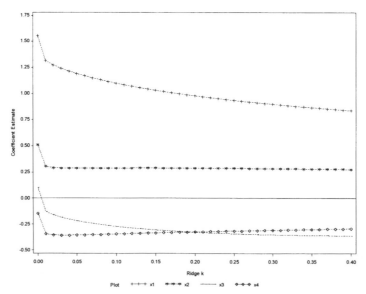

Figure 5.23: Ridge trace of estimated regression coefficients

The other ridge regressions for $.06 \leq k \leq 0.40$ are summarily presented below.

Obs	_MODEL_	_TYPE_	_DEPVAR_	_RIDGE_	_PCOMIT_	_RMSE_	Intercept	x1	x2	x3	x4	y
14	MODEL1	RIDGEVIF	y	0.06	.	.	.	1.6736	0.715	1.5600	0.560	-1
15	MODEL1	RIDGE	y	0.06	.	2.57472	86.0802	1.1704	0.289	-0.2312	-0.352	-1
16	MODEL1	RIDGEVIF	y	0.07	.	.	.	1.5575	0.644	1.4534	0.498	-1
17	MODEL1	RIDGE	y	0.07	.	2.60698	86.2883	1.1508	0.289	-0.2430	-0.350	-1
18	MODEL1	RIDGEVIF	y	0.08	.	.	.	1.4554	0.591	1.3602	0.454	-1
19	MODEL1	RIDGE	y	0.08	.	2.64132	86.4680	1.1326	0.289	-0.2535	-0.348	-1
20	MODEL1	RIDGEVIF	y	0.09	.	.	.	1.3649	0.549	1.2777	0.421	-1
21	MODEL1	RIDGE	y	0.09	.	2.67740	86.6270	1.1156	0.290	-0.2631	-0.346	-1
22	MODEL1	RIDGEVIF	y	0.10	.	.	.	1.2839	0.516	1.2041	0.396	-1
23	MODEL1	RIDGE	y	0.10	.	2.71492	86.7702	1.0996	0.290	-0.2717	-0.344	-1
24	MODEL1	RIDGEVIF	y	0.11	.	.	.	1.2111	0.488	1.1379	0.376	-1
25	MODEL1	RIDGE	y	0.11	.	2.75368	86.9007	1.0846	0.290	-0.2796	-0.342	-1
26	MODEL1	RIDGEVIF	y	0.12	.	.	.	1.1453	0.463	1.0782	0.359	-1
27	MODEL1	RIDGE	y	0.12	.	2.79346	87.0208	1.0704	0.290	-0.2868	-0.339	-1

....

....

Obs	_MODEL_	_TYPE_	_DEPVAR_	_RIDGE_	_PCOMIT_	_RMSE_	Intercept	x1	x2	x3	x4	y
60	MODEL1	RIDGEVIF	y	0.29	.	.	.	0.57260	0.27806	0.55551	0.23749	-1
61	MODEL1	RIDGE	y	0.29	.	3.54038	88.3154	0.90483	0.28512	-0.34935	-0.30873	-1
62	MODEL1	RIDGEVIF	y	0.30	.	.	.	0.55541	0.27254	0.53964	0.23384	-1
63	MODEL1	RIDGE	y	0.30	.	3.58583	88.3683	0.89783	0.28455	-0.35097	-0.30720	-1
64	MODEL1	RIDGEVIF	y	0.31	.	.	.	0.53917	0.26729	0.52463	0.23034	-1
65	MODEL1	RIDGE	y	0.31	.	3.63129	88.4196	0.89103	0.28397	-0.35246	-0.30570	-1
66	MODEL1	RIDGEVIF	y	0.32	.	.	.	0.52379	0.26229	0.51040	0.22698	-1
67	MODEL1	RIDGE	y	0.32	.	3.67674	88.4696	0.88441	0.28337	-0.35383	-0.30422	-1
68	MODEL1	RIDGEVIF	y	0.33	.	.	.	0.50923	0.25752	0.49691	0.22376	-1
69	MODEL1	RIDGE	y	0.33	.	3.72218	88.5182	0.87797	0.28275	-0.35508	-0.30275	-1
70	MODEL1	RIDGEVIF	y	0.34	.	.	.	0.49541	0.25296	0.48409	0.22065	-1
71	MODEL1	RIDGE	y	0.34	.	3.76758	88.5656	0.87169	0.28212	-0.35622	-0.30131	-1
72	MODEL1	RIDGEVIF	y	0.35	.	.	.	0.48228	0.24859	0.47190	0.21766	-1
73	MODEL1	RIDGE	y	0.35	.	3.81294	88.6118	0.86556	0.28148	-0.35725	-0.29989	-1
74	MODEL1	RIDGEVIF	y	0.36	.	.	.	0.46981	0.24440	0.46029	0.21477	-1
75	MODEL1	RIDGE	y	0.36	.	3.85824	88.6568	0.85959	0.28083	-0.35819	-0.29849	-1
76	MODEL1	RIDGEVIF	y	0.37	.	.	.	0.45793	0.24039	0.44922	0.21198	-1
77	MODEL1	RIDGE	y	0.37	.	3.90348	88.7008	0.85375	0.28016	-0.35904	-0.29711	-1
78	MODEL1	RIDGEVIF	y	0.38	.	.	.	0.44661	0.23653	0.43867	0.20928	-1
79	MODEL1	RIDGE	y	0.38	.	3.94864	88.7438	0.84805	0.27949	-0.35981	-0.29575	-1
80	MODEL1	RIDGEVIF	y	0.39	.	.	.	0.43581	0.23281	0.42858	0.20666	-1
81	MODEL1	RIDGE	y	0.39	.	3.99372	88.7859	0.84248	0.27881	-0.36050	-0.29441	-1
82	MODEL1	RIDGEVIF	y	0.40	.	.	.	0.42550	0.22924	0.41894	0.20412	-1
83	MODEL1	RIDGE	y	0.40	.	4.03871	88.8270	0.83703	0.27813	-0.36111	-0.29308	-1

For $k = .04$, the program below will implement the ridge regression and the estimated ridge regression equation is therefore

$$\hat{y}^* = 85.515 + 1.215x_1 + 0.289x_2 - 0.203x_3 - 0.356x_4.$$

```
proc reg data=multi graphics outvif outest=bb ridge=0.04;
model y=x1-x4;
run;
proc print data=bb;
run;
```

Obs	_MODEL_	_TYPE_	_DEPVAR_	_RIDGE_	_PCOMIT_	_RMSE_	Intercept	x1	x2	x3	x4	y
1	MODEL1	PARMS	y	.	.	2.44601	62.4054	1.55110	0.51017	0.10191	-0.14406	-1
2	MODEL1	RIDGEVIF	y	0.04	.	.	.	1.96677	0.98590	1.83321	0.82047	-1
3	MODEL1	RIDGE	y	0.04	.	2.51829	85.5145	1.21511	0.28868	-0.20287	-0.35571	-1

We see that the RMSE has increased from 2.44601 when $k = 0$ to 2.51829 when $k = 0.04$. Similarly, the coefficient of multiple determination has decreased from 0.9824 when $k = 0$ to $2715.76308 - 8(2.5829^2) = \dfrac{266.50288}{2715.76308} = 0.9813$

when $k = 0.04$. Both are not substantial and it appears that the ridge regression provides a good alternative to the ordinary least squares approach in this example.

Computational Selection of the Shrinkage Parameter

While the ridge trace method might be subjective, we consider here a method for a possible initial estimate of the shrinkage parameter that has been suggested by Hoerl, Kennard and Baldwin (1975). The k is computed (HKB) as

$$k = \frac{p\hat{\sigma}^2}{\hat{\boldsymbol{\beta}}'\hat{\boldsymbol{\beta}}}, \tag{5.76}$$

where p is the number of explanatory variables in the model, $\hat{\sigma}^2$ and $\hat{\boldsymbol{\beta}}$ are the error mean square and standardized regression parameters respectively under the ordinary least squares estimation with the dependent variable y also standardized. That is, we fit a model of the form in (5.69). Thus, for the four explanatory data example, we have

$$p = 4, \quad \hat{\sigma}^2 = 0.02415, \text{ and } \hat{\boldsymbol{\beta}} = \{0.7846, 0.5926, -0.2256, -1.1247\}.$$

Hence, using (5.76), we have

$$k = \frac{4(0.02415)}{0.7846^2 + \cdots + (-1.1247)^2} = \frac{0.0966}{2.2826} = 0.042.$$

We notice that the value computed here is very close to our chosen value of $k = 0.04$ from the inspection of the ridge trace plot.

Biasedness of Ridge Regression Parameters

From (5.75), the ridge estimator has the form

$$\hat{\boldsymbol{\beta}}^* = (\mathbf{X}'\mathbf{X} + k\mathbf{I})^{-1}\mathbf{X}'\mathbf{Y} = (\mathbf{X}^{*'}\mathbf{X}^*)^{-1}\mathbf{X}^{*'}\mathbf{Y}.$$

Thus,

$$\text{MSE}(\hat{\boldsymbol{\beta}}^*) = \text{E}(\hat{\boldsymbol{\beta}}^* - \boldsymbol{\beta})^2 = \text{Var}(\hat{\boldsymbol{\beta}}^*) + (\text{bias in } \hat{\boldsymbol{\beta}}^*)^2, \tag{5.77}$$

where $(\text{bias in } \hat{\boldsymbol{\beta}}^*)^2 = [\text{E}(\hat{\boldsymbol{\beta}}^*) - \boldsymbol{\beta}]^2$. Hoerl and Kennard (1970) showed that (5.77) can be written as

$$\text{MSE}(\hat{\boldsymbol{\beta}}^*) = \sigma^2 \sum_{i=1}^{p} \frac{\lambda_i}{(\lambda_i + k)^2} + k^2 \boldsymbol{\beta}'(\mathbf{X}'\mathbf{X} + k\mathbf{I})^{-2}\boldsymbol{\beta}. \tag{5.78}$$

The second term on the right hand side of (5.78) is called the square of the bias, while the first term is the sum of the variances of the parameter estimates of the ridge regression. We see that, as k increases, the bias in the estimator increases but the variance decreases. An ideal situation would be choosing k so that the variance reduction is greater than the bias term in (5.78). If this is accomplished, then the mean square error of the ridge estimator $\hat{\boldsymbol{\beta}}^*$ will be less than the variance of the corresponding OLS estimator $\hat{\boldsymbol{\beta}}$. Hoerl and Kennard (1970) had shown that such a k can be found provided that $\boldsymbol{\beta}'\boldsymbol{\beta}$ is bounded.

5.12.4   Other Remedial Methods for Multicollinearity

Most often, the problem of multicollinearity arises because we do not have enough data. **Adding** or **collecting more data** will generally result in parameter estimates being more stable and highly more precise. However, collection of additional data is often impossible because the conditions under which the initial data were collected might have changed or no longer exists or the researcher might not have enough financial or economic resources to collect additional data. Even if such data were collected, we are not guaranteed that the range of the new data will conform to those of the variables in the original data. However, it is worth trying if collection of additional data is feasible and conditions have not changed that much since the last collection.

Another procedure for combating the problem of multicollinearity is to use a **selection strategy**, for example the stepwise selection strategy which will be discussed in Chapter 7. We shall use this procedure to find the most parsimonious model for our data after our discussions of various selection strategies.

5.13 Exercises

5.1 The data in the file **sat_bk.txt** is a subset of the SAT data submitted by Guber, D.L. to the Journal of Statistics Education Data Archive (see also Guber, D.L., 1999). The data contains 25 cases and the following variables: The current expenditure per pupil (x_1), the average pupil/teacher ratio (x_2), the estimated annual salary of teachers (x_3), the percentage of students taking SAT (x_4), the average verbal SAT score (y_1), the average math SAT score (y_2), and the average total SAT score (yy). A model of the form

$$y_{2i} = \beta_0 + \beta_1 x_1 + \beta_2 x_2 + \beta_3 x_3 + \beta_4 x_4 + \varepsilon_i$$

is suggested for the average math SAT score, y_2. Analyze these data and find

(a) if there are any sample items that are influential observations. If so, what are the sample item numbers?

(b) if there is any indication of multicollinearity among the predictor variables. If so, explain what an investigator should do about it. Obtain R_j^2 for each of the variables.

5.2 Use the data in the file **sat_bk.txt** for Problem 5.1.

A model of the form $yy_i = \beta_0 + \beta_1 x_1 + \beta_2 x_2 + \beta_3 x_3 + \varepsilon_i$ is suggested for the average total SAT score, yy.

(a) Obtain standardized parameter estimates $\hat{\beta}_1^*$, $\hat{\beta}_2^*$, and $\hat{\beta}_3^*$ from the unstandardized regression coefficients. How do the relative effects of current expenditure per pupil, average pupil/teacher ratio and estimated annual salary of teachers compare for predicting average total SAT score?

(b) Is there any multicollinearity? What do the VIF's tell us?

(c) Which variables are the most important from **SS1** and **SS2**?

5.3 An investigator is interested in studying how y, the salaries (\$'000) of baseball players are related to the following variables: batting (x_1), on-base percentage (x_2), number of runs (x_3), number of hits (x_4), number of home runs (x_5), number of runs batted in (x_6), and number of errors (x_7). The data in the file **baseball.txt** is taken from baseball players' salaries data submitted by Watnik, M.R. to the Journal of Statistics Education Data Archive (see also Watnik, 1998). The data for this problem is the first 25 cases of the whole data set. A multiple regression model of the form

$$y_i = \beta_0 + \beta_1 x_1 + \beta_2 x_2 + \beta_3 x_3 + \beta_4 x_4 + \beta_5 x_5 + \beta_6 x_6 + \beta_7 x_7 + \varepsilon_i,$$

is suggested. Fit the regression model and answer the following questions:

(a) Are there any sample items that have high leverage values? If so, what are the sample item numbers?

(b) Are there any sample items that are outliers? If so, identify them.

(c) Are there any influential observations in the data? If so, identify them.

(d) Is there any indication of multicollinearity among the predictor variables? If so, explain what an investigator should do about it.

5.4 For the baseball data used in Problem 5.3, answer the following questions by using the last 40 cases of the complete data set:

(a) Obtain h_3, $\hat{y}_{12}$, and DFFITS$_{16}$.

(b) What is the largest (in absolute value) studentized deleted residual? Is this large enough to consider the corresponding observation as a possible outlier?

(c) Are there any sample items that have high leverage values? If so, what are the sample item numbers?

(d) Are there any sample items that are outliers? If so, what are the sample item numbers?

(e) Are there any sample items that are influential observations? If so, what are the sample item numbers?

(f) Write a short summary discussing unusual sample items. Explain what an investigator should do about them.

5.5 In an experiment involving 12 similar but scattered branch offices of a commercial bank, holders of checking accounts at the offices were offered gifts for setting up savings accounts. The initial deposit in the new savings account had to be for a specified minimum amount to qualify for the gift. The value of the gift was directly proportional to the specified minimum deposit. Various levels of minimum deposit and related gift values were used in the experiment in order to ascertain the relation between the specified minimum deposit and gift value on one hand and number of accounts opened at the office on the other. The data below contains the results, where x is the amount of minimum deposit and y is the number of new savings accounts that were opened and qualified for the gift during the test period. Consider a regression model of the form $y_i = \beta_0 + \beta_1 x_i + \varepsilon_i$.

Branch	1	2	3	4	5	6	7	8	9	10	11
x	125	100	200	75	150	175	75	175	125	200	100
y	160	112	124	28	152	156	42	124	150	104	136

(a) Fit the regression model and give a plot of the residuals (i) against x_i and (ii) against $\hat{y}_i$.

(b) Conduct a lack of fit test for the model.

(c) Do you think that a quadratic model will be adequate?

5.6 The number of defective items produced by a machine (y) is known to be linearly related to the speed setting of the machine (x). The data below were collected from recent quality control records.

i	1	2	3	4	5	6	7	8	9	10	11	12
x_i	200	400	300	400	200	300	300	400	200	400	200	300
y_i	28	75	37	53	22	58	40	96	46	52	30	69

(a) Fit a regression model of the form $y_i = \beta_0 + \beta_1 x_i + \varepsilon_i$ to the data and obtain the estimated regression function. Plot the residuals against x. What does the residual plot show?

(b) Calculate the sample variance s^2 of the residuals for each of the three machine speeds: $x = 200, 300, 400$. What is suggested by these three sample variances about whether or not the true error variances at the three x levels are equal?

(c) For each of the three machine speeds, calculate s^2/x, s^2/x^2, and $s^2/\sqrt{x}$. Do any of these relations appear to be stable?

(d) Using weights $\omega_i = 1/x_i^2$, obtain the weighted least squares estimates of your parameters. Are these estimates similar to the ones obtained by ordinary least squares in part (a)?

(e) Compare the estimated standard errors of the weighted least squares estimates in part (d) with those for the ordinary least squares in part (a). What do you find?

5.7 The data in the file **cars2004_bk.txt** is a subset of the 2004 new car and truck data submitted by Johnson, R.W. to the Journal of Statistics Education Data Archive. The data contains 26 cases of the whole data set and the following variables: The city miles per gallon (y_1), the highway miles per gallon (y_2), the horsepower (x_1), the weight in pounds (x_2), the wheel base in inches (x_3), the length in inches (x_4), and the width in inches (x_5).

A model of the form $y_{1i} = \beta_0 + \beta_1 x_1 + \beta_2 x_2 + \varepsilon_i$ is suggested for the response variable y_1.

(a) Write down the estimated regression parameters. Give 99% confidence intervals for these parameters.

(b) Obtain standardized parameter estimates $\hat{\beta}_1^*$ and $\hat{\beta}_2^*$, from the unstandardized regression coefficients $\hat{\beta}_1$ and $\hat{\beta}_2$. How do the relative effects of horsepower and weight compare for predicting city miles per gallon?

(c) Based on the residual analysis of the plot of residuals versus predicted, what would you say about this model?

(d) How were the studentized deleted residual (d_i^*) and COOK's D computed for the 7th observation? That is, show how d_7^* and D_7 were computed from the results presented.

(e) Do we need to worry about multicollinearity? Check for it and what do the VIF's tell us?

(f) Are there any outliers or influential observations? Base your answers on the deleted studentized residuals, the leverages and the DFFITS only.

5.8 Use the SAT data for Problem 5.1.

A model of the form $y_{1i} = \beta_0 + \beta_1 x_1 + \beta_2 x_2 + \beta_3 x_3 + \beta_4 x_4 + \varepsilon_i$ is suggested for the average verbal SAT score. Fit the regression model and answer the following questions.

(a) Write down the estimated regression parameters. Give 99% confidence intervals for these parameters.

(b) Obtain standardized parameter estimates $\hat{\beta}_1^*$ and $\hat{\beta}_2^*$, from the unstandardized regression coefficients $\hat{\beta}_1$ and $\hat{\beta}_2$. How do the relative effects of current expenditure per pupil and the average pupil/teacher ratio compare for predicting verbal SAT score?

(c) Based on the residual analysis of the plot of residuals versus predicted, what would you say about this model?

(d) How were the studentized deleted residual (d_i^*) and COOK's D computed for the 8th observation? That is, show how d_8^* and D_8 were computed from the results presented.

(e) Do we need to worry about multicollinearity? Check for it and what do the VIF's tell us?

(f) Are there any outliers or influential observations? Base your answers on the deleted studentized residuals, the leverages and the DFFITS only.

5.9 For the SAT data used in Problem 5.8,

(a) What regressor variables are involved in dependencies that are damaging to the regression coefficients?

(b) Use ridge regression for estimating the coefficients. Use k on the basis of the ridge trace.

(c) For the model containing all the explanatory variables, compute the following influence diagnostics: (i) DFBETAS (ii) Cook's D (iii) DFFITS and (iv) HAT diagonals.

Discuss your interpretation of the above results and state which observation should be reexamined or remeasured. Are there any individual observations that are exerting undue influence? Do any points exert a disproportionate amount of influence on the regression results?

(e) Compute the studentized residuals and plot them against the predicted. Does the plot suggest any need for model modification? Compute the R-student values and HAT diagonals. Does any of the outlier diagnostics procedure suggest any 'suspect' data points.

5.10 Use the SAT data for Problem 5.1.

(a) Fit a model of the form $yy_i = \beta_0 + \beta_1 x_1 + \beta_2 x_1^2 + \varepsilon_i$ to the total SAT score.

(b) Give a plot of the residuals (i) against x_i and (ii) against $\hat{y}y_i$.

(c) Do we need to worry about multicollinearity? Check for it.

(d) Center the regressors and compare your results with those in parts (a)-(c).

(e) Is there evidence that the model is useful for predicting yy? Test using $\alpha = 0.05$.

(f) If the test in (e) is significant, which of the model terms contribute to the overall significance? If none, explain the underlying problem in the analysis.

5.11 Consider the following data in Table 5.15:

y	x_1	x_2	y	x_1	x_2	y	x_1	x_2	y	x_1	x_2	y	x_1	x_2
2.4	11	35	3.2	15	37	2.6	12	33	2.9	13	30	3.0	14	36
2.8	16	38	3.3	18	40	3.7	17	36	3.7	20	40	3.5	18	39
2.9	13	34	3.1	19	42	3.8	15	38	2.5	12	35			

Table 5.15: Data for Problem 5.11

Would it be reasonable to use ridge regression for these data? If so, construct the ridge trace, using SAS, and also compute the value of the HKB value of k (see equation 5.76). What do you suggest for the choice of k?

5.12 Explain the difference between a regression outlier and a residual outlier.

5.13 The following data in Table 5.16 are randomly generated, but one value was changed considerably to represent an error.

y	x_1	x_2	y	x_1	x_2	y	x_1	x_2	y	x_1	x_2
14.3171	6.2	3.8	12.4540	5.5	3.2	14.3840	6.8	3.1	14.1014	6.2	3.8
14.1845	5.8	4.2	14.1663	5.9	4.1	13.9815	6.9	3.0	15.7612	6.9	4.5
13.8039	6.0	3.7	13.9912	6.3	3.8	12.9733	5.7	3.4	15.2926	5.1	3.3
13.5998	5.9	4.0	14.0843	6.4	3.7	13.9148	5.5	4.2	13.9870	5.9	4.2
14.3339	6.1	4.1	14.7547	6.6	3.9	14.4456	6.5	4.3	15.4053	6.6	4.4

Table 5.16: Simulated data for Problem 5.13

(a) Can you identify the bad data point?

(b) Could this identification had been made using the OLS analysis?

(c) Are there other points that appear to be questionable?

(d) Using OLS, trim all observations whose standardized residuals are greater than 2 in absolute value. Compare your results from those generated from the robust regression with the LTS.

(e) What is the R^2 with and without the bad point identified in (a)? What does the large difference between the two suggest?

5.14 Use the 2004 cars data for Problem 5.7. A model of the form

$$y_{2i} = \beta_0 + \beta_1 x_1 + \beta_2 x_2 + \beta_3 x_3 + \beta_4 x_4 + \varepsilon_i$$

is suggested for the highway miles per gallon (y_2).

Obtain the correlation matrix for the explanatory variables.

(a) Does the correlation matrix give any indication of multicollinearity?

(b) Calculate the variance inflation factors and the condition number of $\mathbf{X'X}$. Is there any evidence of multicollinearity?

(c) Use the ridge trace to select an appropriate value of k. Is this final model a good one? How much inflation in the residual sum of squares has resulted from the use of ridge regression?

(d) Estimate the parameters in the model by using ridge regression with the value of k determined by equation (5.76). Does this model differ dramatically from the one in (c) above?

(e) How much reduction in R^2 result from the use of the ridge regression?

(f) How much has the residual sum of squares increased compared to the least squares?

5.15 The following data is on the body mass index (y) and the waist girth (x) of twenty men.

Obs	y	x	Obs	y	x	Obs	y	x	Obs	y	x
1	21.67	71.5	6	22.71	82.5	11	23.64	81.9	16	23.51	82.0
2	23.36	79.0	7	25.52	82.0	12	23.45	82.6	17	25.26	88.3
3	21.55	83.2	8	23.03	76.8	13	24.41	85.0	18	23.80	73.6
4	20.87	77.8	9	20.24	68.5	14	24.08	85.6	19	24.19	78.5
5	22.49	80.0	10	24.10	77.5	15	23.45	78.0	20	27.73	87.3

(a) Draw a scatter plot of the data and discuss possible outliers that may be present.

(b) Fit a straight line model to these data using OLS. Does this fit seem satisfactory?

(c) Fit a straight line to these data with an M-estimator of your choice. Is this fit satisfactory? Discuss why the M-estimator is a poor choice for this problem?

(d) Repeat with an LTS. Is this satisfactory? Discuss the difference between the LTS results of the parameters and those of OLS and M-estimators.

5.16 The following data from ten men is a subset of body dimension measurements data used in Heinz et al. (2003). The data is on the height (y) in centimeters and shoulder girth (x).

Obs	1	2	3	4	5	6	7	8	9	10
x	106.2	110.5	115.1	104.5	107.5	119.8	123.5	120.4	111.0	119.5
y	65.6	71.8	80.7	72.6	78.8	74.8	86.4	78.4	62.0	81.6

(a) Find an OLS fit and sketch the line on the scatter plot.

(b) Find the least trimmed squares (LTS) fit and sketch the line on the scatter diagram.

(c) Discuss your findings. Does this exercise confirm the points made in the text about high-breakdown-point estimators.

5.17 The data in the file **cars2004.txt** is taken from 2004 new car and truck data submitted by Johnson, R.W. to the Journal of Statistics Education Data Archive. For this problem, use the following variables: The city miles per gallon (y), the horsepower (x_1), the weight in pounds (x_2), and the length in inches (x_3). A model of the form

$$y_i = \beta_0 + \beta_1 x_1 + \beta_2 x_2 + \beta_3 x_3 + \varepsilon_i,$$

is suggested.

(a) Fit the regression equation with city miles per gallon as the dependent variable.

(b) In the model, check for multicollinearity, influential observations, outliers as well as other relevant diagnostics.

(c) Plot the residuals from the model against the predicted values.

(d) Carry out a robust regression on the data.

5.18 Use the first 50 cases of the data for Problem 5.17 and the following variables: The highway miles per gallon (y), the horsepower (x_1), the weight in pounds (x_2), and the length in inches (x_3). A model of the form

$$y_i = \beta_0 + \beta_1 x_1 + \beta_2 x_2 + \beta_3 x_3 + \varepsilon_i,$$

is suggested. Fit the regression model and answer the following questions.

(a) Are there any sample items that are influential observations? If so, what are the sample item numbers?

(b) Is there any indication of multicollinearity among the predictor variables? If so, explain what an investigator should do about it.

5.19 Consider the data shown here with the summary statistics:

y	25	30	36	37	42	50	55
x	10	12	14	15	18	19	23

$$\sum x_i = 111, \ \sum y_i = 275, \ \sum x_i y_i = 4640, \ \sum x_i^2 = 1879, \ \sum y_i^2 = 11479, \ \text{MSE} = 3.835.$$

(a) Using the data, find the least squares estimates for the regression model $y_i = \beta_0 + \beta_1 x + \varepsilon_i$.

(b) Calculate the residuals for the model.

(c) Plot the residuals against the predicted values of y. What are your findings?

(d) Plot the residuals against x. Do you detect any trends? Explain.

5.20 The following data on 10 naval bases were used to fit a quadratic model $y_i = \beta_0 + \beta_1 x + \beta_2 x^2 + \varepsilon_i$.

% improvement(y)	18	32	9	37	6	3	30	10	25	2
Modification cost(x)	125	160	80	162	110	90	140	85	150	50

(a) Is there an outlier? If yes, which observations? Explain your results.

(b) Based on the independent variables, state the influential observations. Explain your results.

(c) Based on your analysis, is there a multicollinearity problem? Explain your answer.

(d) Is there evidence that the residuals are positively correlated at $\alpha = 0.05$?

(e) Is there evidence that the model is useful for predicting y? Test using $\alpha = 0.05$.

(f) If the test in (e) is significant, which of the model terms contribute to the overall significance? If none, explain the underlying problem in the analysis.

5.21 (a) Explain the term multicollinearity.

(b) What are the effects of multicollinearity in regression analysis?

(c) How can you detect multicollinearity?

(d) What are the solutions to the problems created by multicollinearity?

Chapter 6

Multiple and Partial Correlations

6.1 Introduction

In this chapter, we shall focus our discussion on data presented in Table 4.2 of Chapter 4, involving a dependent variable y, and three explanatory variables x_1, x_2 and x_3. For the data, the summary statistics are:

$$S_{x_1} = 38.3583, \qquad S_{x_1y} = 108.4700, \qquad S_{x_1x_2} = 155.6583,$$
$$S_{x_3} = 39.0863, \qquad S_{x_3y} = 91.6200, \qquad S_{x_2x_3} = 85.3875,$$
$$S_{x_2} = 2896.9583, \qquad S_{x_2y} = 1213.2000, \qquad S_{x_1x_3} = 12.4975,$$
$$S_{yy} = 689.2600.$$

The above summary statistics are also obtained in SAS as follows:

```
set exam2;
proc corr csscp;
var y x1 x2 x3;
run;
                  CSSCP Matrix

            y              x1              x2              x3
  y    689.260000     108.470000     1213.200000      91.620000
  x1   108.470000      38.358333      155.658333      12.497500
  x2  1213.200000     155.658333     2896.958333      85.387500
  x3    91.620000      12.497500       85.387500      39.086250
```

The sample correlation coefficient between y and x_j can be obtained from

$$r_{x_jy} = \frac{\sum x_{ji}y_i - \frac{1}{n}\sum x_{ji}\sum y_i}{\sqrt{\sum(x_{ji}-\bar{x}_j)^2\sum(y_i-\bar{y})^2}} = \frac{S_{x_jy}}{\sqrt{S_{x_j}\times S_{yy}}}. \tag{6.1}$$

Thus for r_{x_1y}, we have

$$r_{x_1y} = \frac{S_{x_1y}}{\sqrt{S_{x_1}\times S_{yy}}} = \frac{108.4700}{\sqrt{38.3583\times689.2600}} = 0.6671.$$

Similarly,

$$r_{x_1x_2} = r_{12} = \frac{S_{x_1x_2}}{\sqrt{S_{x_1}\times S_{x_2}}} = \frac{155.6583}{\sqrt{38.3583\times2896.9583}} = 0.4670.$$

The $\binom{4}{2} = 6$ all possible zero-order sample correlations between y and the three explanatory variables are presented in the form

$$
\begin{array}{c}
\quad\quad y \quad x_1 \quad x_2 \quad x_3 \\
\begin{array}{c} y \\ x_1 \\ x_2 \\ x_3 \end{array}
\begin{pmatrix}
1 & r_1 & r_2 & r_3 \\
 & 1 & r_{12} & r_{13} \\
 & & 1 & r_{23} \\
 & & & 1
\end{pmatrix}
\end{array}
=
\begin{array}{c}
\quad\quad\quad y \quad\quad x_1 \quad\quad x_2 \quad\quad x_3 \\
\begin{array}{c} y \\ x_1 \\ x_2 \\ x_3 \end{array}
\begin{pmatrix}
1.000 & 0.6671 & 0.8586 & 0.5582 \\
 & 1.0000 & 0.4670 & 0.3228 \\
 & & 1.0000 & 0.2538 \\
 & & & 1.0000
\end{pmatrix}.
$$

The above sample correlation coefficients are often referred to as the *zero-order* correlations.

141

6.2 Partial Correlation Coefficient

The *partial correlation coefficient* between two variables y and x measures the strength of the linear association between the two variables after controlling for the effects of other variables $z_1, z_2, \ldots, z_k$ and will be written as $r_{yx|z_1, z_2, \ldots, z_k}$. The partial correlation of the *first-order* assumes only one variable z_1 is being controlled and can be denoted by $r_{yx|z_1}$. That of a *second-order* assumes that variables z_1, z_2 are being controlled and has the form $r_{yx|z_1, z_2}$, etc. In general, kth-order partials have k variables being controlled and are denoted by $r_{yx|z_1, z_2, \ldots, z_k}$.

For the example at the beginning of this chapter involving three explanatory variables x_1, x_2 and x_3, the highest partial that can be obtained are second-order partials. To obtain a first-order partial, say, between y and x, controlling for z, that is, $r_{yx|z}$, we use the following expression:

$$r_{yx|z} = \frac{r_{yx} - r_{yz}r_{xz}}{\sqrt{(1 - r_{yz}^2)(1 - r_{xz}^2)}}, \tag{6.2}$$

where r_{yx}, r_{yz}, r_{xz} are the zero-order partials. Thus to compute the first-order partial between y and x_2, controlling for x_1, that is, $r_{yx_2|x_1}$, we use the expression in (6.2) to obtain

$$r_{yx_2|x_1} = \frac{r_{yx_2} - r_{yx_1}r_{x_2x_1}}{\sqrt{(1 - r_{yx_1}^2)(1 - r_{x_2x_1}^2)}} = \frac{r_2 - r_1 r_{12}}{\sqrt{(1 - r_1^2)(1 - r_{12}^2)}}$$

$$= \frac{0.8586 - (0.6671)(0.4670)}{\sqrt{1 - 0.6671^2)(1 - 0.4670^2)}}$$

$$= \frac{0.8586 - 0.3115}{\sqrt{(0.5550)(0.7819)}}$$

$$= 0.8305.$$

Similar calculations lead to the following first-order partials in Table 6.1 for the data in our example.

Order	Controlling variable	Definition form	Computed value
1	x_1	$r_{yx_2|x_1}$	0.8304
1	x_1	$r_{yx_3|x_1}$	0.4863
1	x_1	$r_{x_2x_3|x_1}$	0.1231
1	x_2	$r_{yx_1|x_2}$	0.5871
1	x_2	$r_{yx_3|x_2}$	0.6863
1	x_2	$r_{x_1x_3|x_2}$	0.2388
1	x_3	$r_{yx_1|x_3}$	0.6201
1	x_3	$r_{yx_2|x_3}$	0.8933
1	x_3	$r_{x_2x_1|x_3}$	0.4206
2	x_1, x_2	$r_{yx_3|x_1,x_2}$	0.6946
2	x_1, x_3	$r_{yx_2|x_1,x_3}$	0.8886
2	x_2, x_3	$r_{yx_1|x_2,x_3}$	0.5992

Table 6.1: All possible partials for our data example

We can also compute the second-order partials between say y and x, given or controlling for z and w as

$$r_{yx|z,w} = \frac{r_{yx|z} - (r_{yw|z})(r_{xw|z})}{\sqrt{(1 - r_{yw|z}^2)(1 - r_{xw|z}^2)}} = \frac{r_{yx|w} - (r_{yz|w})(r_{xz|w})}{\sqrt{(1 - r_{yz|w}^2)(1 - r_{xz|w}^2)}}. \tag{6.3}$$

Specifically, the second order-partial $r_{yx_2|x_1, x_3}$ can be calculated from the expression

$$r_{yx_2|x_1,x_3} = \frac{r_{yx_2|x_1} - (r_{yx_3|x1})(r_{x_2x_3|x_1})}{\sqrt{(1 - r_{yx_3|x_1}^2)(1 - r_{x_2x_3|x_1}^2)}}.$$

Hence,

$$r_{yx_2|x_1,x_3} = \frac{0.8304 - (0.4863)(0.1231)}{\sqrt{(1 - 0.4863^2)(1 - 0.1231^2)}} = 0.8886.$$

In SAS, **proc reg** can be used to obtain each of the above partials. For instance, to obtain the partials involving the dependent variable y, run a regression model with x_1, x_2 and x_3 as the explanatory variables and request SAS to output the partials by specifying **pcorr1** and **pcorr2** as options in the model statements. SAS will give the squares of the partials as in the output below.

```
set exam2;
proc reg data=exam2;
model y=x1 x2 x3/ss1 ss2 pcorr1 pcorr2;
run;
```

```
                            Parameter Estimates

                    Parameter    Standard
Variable     DF     Estimate        Error   t Value  Pr > |t|    Type I SS    Type II SS
-----------------------------------------------------------------------------------------
Intercept    1      17.84693      2.00188      8.92    <.0001        37446     244.17168
x1           1       1.10313      0.32957      3.35    0.0032    306.73233      34.41851
x2           1       0.32152      0.03711      8.66    <.0001    263.79445     230.62548
x3           1       1.28894      0.29848      4.32    0.0003     57.29022      57.29022


                              Squared            Squared
                              Partial            Partial
             Variable   DF  Corr Type I     Corr Type II
             ------------------
             Intercept  1          .                 .
             x1         1       0.44502           0.35904
             x2         1       0.68961           0.78963
             x3         1       0.48251           0.48251
```

For instance, the square of the zero-order partial between y and x_1, that is, $r_{yx_1}^2 = 0.44502$ from the output. Consequently, $r_{yx_1} = \pm\sqrt{0.44502} = \pm0.6671$. However, from the SAS output, we notice that the parameter estimate $\hat{\beta}_1$ is positive and therefore, $r_{yx_1} = 0.6671$. Similarly, we have the following from the **'Squared Partial Corr Type I'**:

$$r_{yx_2|x_1} = \sqrt{0.68961} = 0.8304$$
$$r_{yx_3|x_1,x_2} = \sqrt{0.48251} = 0.6946.$$

Both partials are positive because the parameter estimates in the regression output are positive. From the **'Squared Partial Corr Type II'** output, we can obtain the following second-order partials:

$$r_{yx_1|x_2,x_3} = \sqrt{0.35904} = 0.5992$$
$$r_{yx_2|x_1,x_3} = \sqrt{0.78963} = 0.8886$$
$$r_{yx_3|x_1,x_2} = \sqrt{0.48251} = 0.6946.$$

Alternatively, the above could also be computed from multiple correlation coefficients. In general,

$$R_{yx|z_1,z_2,\ldots,z_p}^2 = \frac{R^2(k) - R^2(p)}{1 - R^2(p)},$$

where k refers to the number of explanatory variables in the model including the controlled variables, and p is the number of controlled variables. For our example, we have presented below the multiple correlation coefficients when, one, two and all the three variables are presented in the model.

```
                 R-Square Selection Method

    Number in
      Model       R-Square      Variables in Model

         1         0.7371       x2
         1         0.4450       x1
```

```
     1        0.3116      x3
---------------------------------------------
     2        0.8609      x2 x3
     2        0.8277      x1 x2
     2        0.5763      x1 x3
---------------------------------------------
     3        0.9109      x1 x2 x3
```

Thus,

$$r^2_{yx_1} = \frac{0.4450 - 0}{1 - 0} = 0.4450$$

$$r^2_{yx_2} = \frac{0.7371 - 0}{1 - 0} = 0.7371$$

$$r^2_{yx_2|x_1} = \frac{0.8277 - 0.4450}{1 - 0.4450} = 0.6896$$

$$r^2_{yx_1|x_2} = \frac{0.8277 - 0.7371}{1 - 0.7371} = 0.3446$$

$$r^2_{yx_3|x_1,x_2} = \frac{0.9109 - 0.8277}{1 - 0.8277} = 0.4829$$

$$r^2_{yx_1|x_2,x_3} = \frac{0.9109 - 0.8609}{1 - 0.8609} = 0.3595.$$

6.2.1   Calculating Partials from Type I and Type II SS

Consider, first, the regression of y with the explanatory variables x_1, x_2 and x_3 in that order in the regression equation. Since, x_1 enters first, the zero-order squared partial correlation between y and x_1, r_{yx_1}, as well as the first and second order squared partials are computed as follows from Type I SS. First we recall from the results in Table 4.7 in Chapter 4, which is reproduced in the following table.

Source	df	SS	MS	F
x_1	1	306.7323	306.7323	99.84
$x_2|x_1$	1	263.7945	263.7945	85.87
$x_3|x_1, x_2$	1	57.2902	57.2902	18.65
Residual	20	61.4430	3.0722	
Total	23	689.2600		

Hence,

$$r^2_{yx_1} = \frac{SS(x_1)}{S_{yy}} = \frac{306.7323}{689.2600} = 0.4450,$$

$$r^2_{yx_2|x_1} = \frac{SS(x_2|x_1)}{S_{yy} - SS(x_1)}$$
$$= \frac{263.7945}{689.2600 - 306.7323} = \frac{263.7945}{382.5277}$$
$$= 0.6896,$$

and

$$r^2_{yx_3|x_1,x_2} = \frac{SS(x_3|x_1, x_2)}{S_{yy} - SS(x_1, x_2)}$$
$$= \frac{SS(x_3|x_1, x_2)}{S_{yy} - [SS(x_1) + SS(x_2|x_1)]}$$
$$= \frac{57.2902}{689.2600 - 306.7323 - 263.7945}$$
$$= \frac{57.2902}{118.7332}$$
$$= 0.4825.$$

We can also calculate $r^2_{yx_3|x_1,x_2}$ from the Type II *SS* output in SAS.

6.2.2 Alternative Calculations

The partial correlation between x and y holding a set of variables fixed will have the same sign as the multiple regression coefficient of x when y is regressed on x and the set of variables being held fixed. Also,

$$r_{xy|\text{list}} = \frac{t}{\sqrt{t^2 + \text{Error } df}}.$$

where t is the t statistic for the coefficient of x in the multiple regression of y on x and the variables in the list. For instance,

$$r_{yx_1} = \frac{3.35}{\sqrt{3.35^2 + 20}} = 0.5995.$$

6.3 Testing Hypothesis Involving Partial Correlations

Suppose we wish to test the hypotheses

$$H_0 : \rho_{yx|z_1,z_2,\cdots,z_p} = 0$$
$$H_a : \rho_{yx|z_1,z_2,\cdots,z_p} \neq 0. \tag{6.4}$$

Then the required test statistic is calculated either as

(a)

$$F^* = \frac{[R^2(k) - R^2(p)]/(k-p)}{[1 - R^2(k)]/(n-k-1)}, \tag{6.5}$$

where F^* is distributed as an F distribution with $(k-p)$ and $(n-k-1)$ degrees of freedom.

(b) Or, alternatively as

$$t^* = r_{yx|z_1,z_2,\cdots,z_p} \frac{\sqrt{n-p-2}}{\sqrt{(1 - r^2_{yx|z_1,z_2,\cdots,z_p})}}, \tag{6.6}$$

where t^* has a Student's t distribution with $n-p-2$ degrees of freedom. Thus the hypothesis is rejected when

$$t^* \geq t_{\alpha/2,n-p-2}.$$

Example 6.1

For the data in Table 4.2 of Chapter 4, suppose we wish to test the following hypotheses:

1.

$$H_0 : \rho_{yx_2|x_1} = 0$$
$$H_a : \rho_{yx_2|x_1} \neq 0. \tag{6.7}$$

Here, $k = 2$ (x_1, x_2) and $p = 1$ (x_1). Hence,

$$F^* = \frac{(0.8277 - 0.4450)/1}{(1 - 0.8277)/21} = 46.64.$$

2.

$$H_0 : \rho_{yx_3|x_1,x_2} = 0$$
$$H_a : \rho_{yx_3|x_1,x_2} \neq 0. \tag{6.8}$$

Here, $k = 3$ (x_1, x_2, x_3) and $p = 2$ (x_1, x_2). Hence,

$$F^* = \frac{(0.9109 - 0.8277)/1}{(1 - 0.9109)/20} = 18.676.$$

Using this to test the two hypotheses, we have for the hypothesis in (6.7), $r_{yx_2|x_1} = 0.8304$, $n = 24$, $p = 1$ and thus,

$$t^* = 0.8304 \frac{\sqrt{24 - 1 - 2}}{\sqrt{1 - 0.8304^2}} = 6.829828.$$

Comparing this value with the tabulated $t_{0.025,21} = 2.080$, we see that $t^* \gg 2.080$, hence we would strongly reject H_0. That is, given that x_1 is already in the model, x_2 is very important to be added to the model. Also for the second hypotheses in (6.8), $r_{yx_3|x_1,x_2} = 0.6946$, $n = 24$, $p = 2$ and again, we have

$$t^* = 0.6946 \frac{\sqrt{24 - 2 - 2}}{\sqrt{1 - 0.6946^2}} = 4.31816.$$

Again, comparing this with the tabulated $t_{0.025,20} = 2.086$, we would strongly reject H_0, indicating that the model that already contains x_1 and x_2 deserved the inclusion of x_3. We may note that $6.829828^2 = 46.647$ as well as $4.31816^2 = 18.6465$ are very close to the F^* values for the method in (a). As a matter of fact, barring rounding errors, both results should be the same, that is, give the same exact results.

Obtaining Partial Correlations from Residuals

To obtain, for instance, the sample first-order partial correlation $r_{yx_2|x_1}$ from regression analysis residuals, we would need to do the following:

1. Regress y on x_1. That is, fit the model

$$y_i = \beta_0 + \beta_1 x_1 + \varepsilon_i. \tag{6.9}$$

   From this model, obtain the residuals $\hat{\varepsilon}_{i1} = y_i - \hat{y}_i = (y_i - \bar{y}) + \hat{\beta}_1(x_{i1} - \bar{x}_1)$.

2. Regress x_2 on x_1. That is, fit the model

$$x_{i2} = \beta_0^* + \beta_1^* x_{i1} + \varepsilon_i. \tag{6.10}$$

   From this model, again obtain the residuals $\hat{\varepsilon}_{i2} = x_{i2} - \hat{x}_{i2} = (x_{i2} - \bar{x}_2) + \hat{\beta}_1^*(x_{i1} - \bar{x}_1)$.

3. Now obtain the sample correlation coefficient between $y_i - \hat{y}_i$ and $x_{i2} - \hat{x}_{i2}$, that is, obtain the sample correlation coefficient between $\hat{\varepsilon}_{i1}$ and $\hat{\varepsilon}_{i2}$.

4. Then, compute

$$r_{yx_2|x_1} = r_{y-\hat{y},x_2-\hat{x}_2} = r_{\hat{\varepsilon}_1,\hat{\varepsilon}_2}. \tag{6.11}$$

The residuals approach is implemented in SAS with the following SAS program and output for the data in Table 4.2 in Chapter 4.

```
set exam2;
proc reg data=exam2;
model y=x1/noprint;
output out=aa r=residy;
model x2=x1/noprint;
output out=bb r=residx;
run;
data comb;
merge aa bb;
run;
proc corr data=comb nosimple;
var residy residx;
run;
```

```
                        The CORR Procedure

              2 Variables:    residy   residx


                  Pearson Correlation Coefficients, N = 24
                     Prob > |r| under H0: Rho=0

                            residy         residx

            residy         1.00000        0.83043
            Residual                       <.0001

            residx         0.83043        1.00000
            Residual       <.0001
```

The computed value agrees with that obtained from our earlier calculations.

6.4   Multiple Partial Correlation

These are of the form $r_{y(x_1,x_2,\ldots,x_k)|z_1,z_2,\ldots,z_p}$, where we wish to find the correlation coefficient of y with two or more independent variables $x_1, x_2, \ldots, x_k$ after controlling for the effects of $z_1, z_2, \ldots, z_p$. For example, for our three variable explanatory variables x_1, x_2 and x_3, we may wish to calculate the sample multiple partial correlation coefficient $r_{y(x_2,x_3)|x_1}$. This will be computed from

$$r^2_{y(x_2,x_3)|x_1} = \frac{R^2_{y|x_1,x_2,x_3} - R^2_{y|x_1}}{1 - R^2_{y|x_1}}, \tag{6.12}$$

and the hypotheses,

$$\begin{aligned} H_0 &: \rho_{y(x_2,x_3)|x_1} = 0 \\ H_a &: \rho_{y(x_2,x_3)|x_1} \neq 0, \end{aligned} \tag{6.13}$$

is equivalent to the hypotheses

$$\begin{aligned} H_0 &: \beta_2 = \beta_3 = 0 \\ H_a &: \text{at least one of } \beta_2 \text{ or } \beta_3 \neq 0. \end{aligned} \tag{6.14}$$

The appropriate test statistic for testing the above equivalent hypotheses (6.13) and (6.14) is

$$F^* = \frac{[SS(x_1,x_2,x_3) - SS(x_1)]/2}{MSE(x_1,x_2,x_3)}. \tag{6.15}$$

The above F^* will be compared with an F distribution with 2 and $(n - k - 1)$ degrees of freedom. For our data, we have

$$r^2_{y(x_2,x_3)|x_1} = \frac{0.9109 - 0.4450}{1 - 0.4450} = 0.8395, \text{ and therefore}$$
$$r_{y(x_2,x_3)|x_1} = \sqrt{0.8395}$$
$$= 0.9162.$$

The corresponding test statistic based on the regression sum of squares is computed as

$$F^* = \frac{(627.8170 - 306.7323)/2}{3.0722} = 52.26,$$

since $SS(x_1, x_2, x_3) = 627.8170$, $SS(x_1) = 306.7323$ and MSE is 3.07215. An alternative test statistic that is based on squared multiple correlations is given by

$$F^* = \frac{[R^2(\text{full model}) - R^2(\text{reduced model})]/df_1}{[1 - R^2(\text{full model})]/df_2}, \tag{6.16}$$

where,

- df_1 = Regression degrees of freedom (full model) − Regression degrees of freedom (reduced model).

- df_2 = Residual degrees of freedom (full model).

Again for our data, using this alternative procedure, we have

$$F^* = \frac{(0.9109 - 0.4450)/(3 - 1)}{(1 - 0.9109)/20} = \frac{0.23295}{0.00446} = 52.29.$$

Both approaches give same results barring round off errors. Our decision rule rejects H_0 if $F^* \geq f_{0.025,2,20} = 4.46$. Since $52.29 \ggg 4.46$, we would therefore strongly reject H_0 and conclude that at least one of β_2 or β_3 is not zero. The above tests for multiple partials can be carried out in the same way for a much larger set of variables in both the reduced and full models.

6.5 Semi-partial Correlation Coefficients

The semi-partial correlation between a dependent variable y and two explanatory variables x and z is defined as the correlation between two variables when only one of the two variables has been adjusted or controlled for a third variable. In our example, the semi-partial correlation between y and x for instance can be considered as the partial correlation when either only x has been adjusted for z or with only y adjusted for z. Thus the semi-partial correlation $r_{y(x|z)}$ is the semi-partial correlation between y and x, when x has been adjusted for z. Similarly, the semi-partial correlation $r_{x(y|z)}$ is the semi-partial correlation between x and y when only y has been adjusted for z. Both can be computed as follows:

$$r_{y(x|z)} = \frac{r_{yx} - r_{yz}r_{xz}}{\sqrt{1 - r_{xz}^2}}, \quad \text{and} \tag{6.17a}$$

$$r_{x(y|z)} = \frac{r_{yx} - r_{yz}r_{xz}}{\sqrt{1 - r_{yz}^2}}. \tag{6.17b}$$

For our data, for instance, suppose we wish to find the semi-partial correlations $r_{y(x_1|x_2)}$ and $r_{y(x_2|x_1)}$. Then we would have

$$r_{y(x_1|x_2)} = \frac{r_{yx_1} - r_{yx_2}r_{x_1x_2}}{\sqrt{1 - r_{x_1x_2}^2}} = \frac{0.6671 - (0.8586)(0.4670)}{\sqrt{1 - 0.4670^2}} = 0.3010.$$

Similarly,

$$r_{y(x_2|x_1)} = \frac{r_{yx_2} - r_{yx_1}r_{x_1x_2}}{\sqrt{1 - r_{x_1x_2}^2}} = \frac{0.8586 - (0.6671)(0.4670)}{\sqrt{1 - 0.4670^2}} = 0.6187.$$

As in the previous section, the semi-partial correlations can also be obtained from regression residuals. Thus, $r_{y(x_1|x_2)}$ can be obtained as $r_{y\hat{x}_1}$, while $r_{y(x_2|x_1)}$ is obtained as $r_{y\hat{x}_2}$.

6.6   Exercises

6.1 The data in the file **baseball.txt** is taken from baseball players' salaries data submitted by Watnik, M.R. to the Journal of Statistics Education Data Archive (see also Watnik, 1998). For this problem, use the first 30 cases of the whole data set and the following variables: The 1992 salary in thousands of dollars (y), the batting average (x_1), the number of hits (x_2), and the number of home runs (x_3). A model of the form

$$y_i = \beta_0 + \beta_1 x_1 + \beta_2 x_2 + \beta_3 x_3 + \varepsilon_i,$$

is suggested. Analyze the data and find:

  (i) $SSR(x_2)$.

  (ii) $SSR(x_3 \mid x_2)$.

  (iii) $SSR(x_1 \mid x_2, x_3)$.

  (iv) $SSR(x_3, x_1 \mid x_2)$.

where SSR refers to the extra sum of squares.

  (a) Test whether there is a regression relation of the form suggested above. Use $\alpha = 0.05$. State the null and the alternative hypotheses, the decision rule, and conclusion. Interpret R^2 for the model. How is it calculated? Discuss the adequacy of this model based on the residual plots and the normal plot of the residuals.

  (b) Test whether x_1 can be dropped from the model, given that both x_2 and x_3 are already in the model. Use $\alpha = 0.01$. State the null and the alternative hypotheses, the decision rule and conclusion.

  (c) Test whether both x_1 and x_3 can be dropped from the model, given that x_2 is already in the model. State the null and the alternative hypotheses, the decision rule and conclusion.

  (d) Test whether $\beta_1 = \beta_3$; use $\alpha = 0.01$. State the null and the alternative hypotheses, the decision rule and conclusion.

  (e) Suppose we wish to obtain the partial multiple regression coefficient $r_{y1.23}$ from residual analysis, describe briefly, how this can be accomplished.

6.2 The data in the file **sat_bk.txt** is a subset of the SAT data submitted by Guber, D.L. to the Journal of Statistics Education Data Archive (see also Guber, D.L., 1999). The data contains 25 cases and the following variables: The current expenditure per pupil (x_1), the average pupil/teacher ratio (x_2), the estimated annual salary of teachers (x_3), the percentage of students taking SAT (x_4), the average verbal SAT score (y_1), the average math SAT score (y_2), and the average total SAT score (yy). A model of the form

$$y_{1i} = \beta_0 + \beta_1 x_1 + \beta_2 x_2 + \beta_3 x_3 + \beta_4 x_4 + \epsilon_i, \tag{6.18}$$

is suggested, and the director of admissions decides to estimate the response model in (6.18). Use a statistical software of your choice to answer the following questions:

  (a) Obtain the following sample partial correlation coefficients. What orders are these?

    (i) $r_{yx_2 \mid x_1}$.

    (ii) $r_{yx_1 \mid x_2, x_3, x_4}$.

    (iii) $r_{yx_4 \mid x_1, x_2, x_3}$.

  (b) Test whether x_4 can be dropped from the regression model given that x_1, x_2 and x_3 are retained. State the null and alternative hypotheses for this test.

6.3 For the baseball players' salaries data (see Problem 5.3 in Chapter 5), obtain $r_{yx_2 \mid x_1, x_3}$.

6.4 Use the data for Problem 5.11 in Chapter 5.

  (a) Obtain $r^2_{yx_1 \mid x_2}$.

(b) Test whether x_2 can be dropped from the model already containing x_1. Use $\alpha = 0.05$. State the null and alternative hypotheses, the decision rule and conclusion.

6.5 Use the baseball players' salaries data (see Problem 5.3 in Chapter 5).

   (a) Obtain $r_{y,x_5 | x_4, x_6}$.

   (b) Test whether x_6 can be dropped from the model already containing x_4 and x_5. Use $\alpha = 0.05$. State the null and alternative hypotheses, the decision rule and conclusion.

6.6 The data in the file **sat_bk.txt** is a subset of the SAT data submitted by Guber, D.L. to the Journal of Statistics Education Data Archive (see also Guber, D.L., 1999). The data contains 25 cases and the following variables: The current expenditure per pupil (x_1), the average pupil/teacher ratio (x_2), the estimated annual salary of teachers (x_3), the percentage of students taking SAT (x_4), the average verbal SAT score (y_1), the average math SAT score (y_2), and the average total SAT score (yy). A model of the form

$$y_{2i} = \beta_0 + \beta_1 x_1 + \beta_2 x_2 + \beta_3 x_3 + \varepsilon_i,$$

is suggested. Analyze the data and answer the following questions:

   (a) Find the following sum of squares:

      (i) $SS(x_1 \mid x_2, x_3)$.
      (ii) $SS(x_2 \mid x_1, x_3)$.
      (iii) $SS(x_2 \mid x_1)$.
      (iv) $SS(x_3 \mid x_1, x_2)$, where SS refers to the sum of squares.

   (b) Based on your results in (a), which of the variables x_1, x_2, x_3 would you consider to be the most important predictor of y_2?

   (c) Obtain the partial multiple correlation coefficients $r_{y_2 x_2 | x_1}$, $r_{y_2 x_2 | x_1, x_3}$, and $r_{y_2 x_3 | x_1, x_2}$.

   (d) Test $H_0 : \rho_{y_2 x_3 | x_1, x_2} = 0$ versus $H_a : \rho_{y_2 x_3 | x_1, x_2} \neq 0$. State the decision rule, conclusion and interpretation.

6.7 The data in the file **cars2004.txt** is taken from 2004 new car and truck data submitted by Johnson, R.W. to the Journal of Statistics Education Data Archive. For this problem, use the following variables: The dealer cost in US dollars (y), the horsepower (x_1), the weight in pounds (x_2), the wheel base in inches (x_3), the length in inches (x_4), and the width in inches (x_5). Assume the regression model

$$y_i = \beta_0 + \beta_1 x_1 + \beta_2 x_2 + \beta_3 x_3 + \beta_4 x_4 + \beta_5 x_5 + \varepsilon_i.$$

   (a) Obtain the following sample partial correlation coefficients. What orders are these?

      (i) $r_{y x_2 | x_1}$.
      (ii) $r_{y x_1 | x_2, x_3, x_4, x_5}$.
      (iii) $r_{y x_4 | x_1, x_2, x_3}$.

   (b) Test whether x_5 can be dropped from the regression model given that x_3 and x_4 are retained. State the null and alternative hypotheses for this test.

   (c) Regress y on x_2 using the simple linear regression model and obtain the residuals. Regress x_1 on x_2 using the simple linear regression model and obtain the residuals. Calculate the coefficient of simple correlation between the two sets of residuals and show that it equals $r_{y x_1 | x_2}$. Explain the meaning of this correlation coefficient.

6.8 The data in the file **cars2004.txt** is taken from 2004 new car and truck data submitted by Johnson, R.W. to the Journal of Statistics Education Data Archive. For this problem, use the following variables: The city miles per gallon (y), the horsepower (x_1), the weight in pounds (x_2), and the length in inches (x_3). Assume the regression model

$$y_i = \beta_0 + \beta_1 x_1 + \beta_2 x_2 + \beta_3 x_3 + \varepsilon_i.$$

(a) Obtain the following sample partial correlation coefficients. What orders are these?

   (i) $r_{yx_2|x_1}$.

   (ii) $r_{yx_1|x_2,x_3}$.

   (iii) $r_{yx_3|x_1,x_2}$.

(b) Test whether x_3 can be dropped from the regression model given that x_1 and x_2 are retained. State the null and alternative hypotheses for this test.

(c) Regress y on x_2 using the simple linear regression model and obtain the residuals. Regress x_1 on x_2 using the simple linear regression model and obtain the residuals. Calculate the coefficient of simple correlation between the two sets of residuals and show that it equals $r_{yx_1|x_2}$. Explain the meaning of this correlation coefficient.

6.9 The data in the following table from 20 women is a subset of body dimension measurements data used in Heinz et al. (2003). The dependent variable is weight (y, in kg) and the three independent variables are chest diameter (x_1, in cm), age (x_2, in years) and height (x_3, in cm). The data is as follows:

Obs	y	x_1	x_2	x_3	Obs	y	x_1	x_2	x_3
1	51.6	24.9	22	161.2	11	55.2	24.3	32	172.5
2	59.0	24.5	20	167.5	12	54.2	25.5	25	170.9
3	49.2	24.2	19	159.5	13	62.5	27.3	25	172.9
4	63.0	24.9	25	157.0	14	42.0	22.6	29	153.4
5	53.6	26.2	21	155.8	15	50.0	23.7	22	160.0
6	59.0	25.7	23	170.0	16	49.8	25.7	25	147.2
7	47.6	24.5	26	159.1	17	49.2	23.9	23	168.2
8	69.8	26.6	22	166.0	18	73.2	26.7	37	175.0
9	66.8	25.0	28	176.2	19	47.8	22.6	19	157.0
10	75.2	29.4	40	160.2	20	68.8	28.8	23	167.6

Assume the regression model $y_i = \beta_0 + \beta_1 x_1 + \beta_2 x_2 + \beta_3 x_3 + \varepsilon_i$.

(a) Obtain the following sample partial correlation coefficients. What orders are these?

   (i) $r_{yx_2|x_1}$.

   (ii) $r_{yx_1|x_2,x_3}$.

   (iii) $r_{yx_3|x_1,x_2}$.

   (iv) $r_{yx_2|x_1,x_3}$.

(b) Test whether x_3 can be dropped from the regression model given that x_1 and x_2 are retained. State the null and alternative hypotheses for this test.

(c) Regress y on x_2 using the simple linear regression model and obtain the residuals. Regress x_1 on x_2 using the simple linear regression model and obtain the residuals. Calculate the coefficient of simple correlation between the two sets of residuals and show that it equals $r_{yx_1|x_2}$. Explain the meaning of this correlation coefficient.

Chapter 7

Model Selection Strategies

7.1 Introduction

Given $p-1$ independent explanatory variables, our goal in this chapter is to find an appropriate set or subset of the $p-1$ independent variables to describe or predict values of the dependent variable. There are various approaches or selection strategies for doing this and we present in the following sections these approaches which are also illustrated with examples.

7.2 Approaches

Approach I: All Possible Regression Methods

- We wish to select the subset(s) of models that optimize certain criteria.

- The method considers all possible models that can be fitted with the given set of $p-1$ independent variables ($2^{p-1}-1$ of them with at least one independent variable). These independent variables can include squared terms and interaction terms as well.

- Allows you to select a group of good candidate models not just one.

- It can take a considerable amount of computer time when $p-1$ is very large and it might also be time consuming to look at all the various models trying to find the best ones. If $p-1=10$ for instance, then there are $2^{10}-1=1023$ possible models. If $p-1=20$, then there are 1,048,575 possible models. If $p-1=50$, then there are 1.1259×10^{15} possible models. We see that as $p-1$ increases, the number of all possible models increases, and this may task considerably, our ability to implement this approach. We shall discuss available SAS mechanisms for reducing the number of possible regression models to be examined.

Approach II: Stepwise Regression Methods

- Systematically add or delete variables from the model in an attempt to find an adequate model.

- Examine only a subset of all the possible models.

- Generally select only one model from the set of all possible models.

- Do not provide models that are almost as good as the selected model.

- Requires less computer time than all possible regression methods.

NOTE: Neither approach considers model diagnostics, and they cannot detect data problems or violations of the various model assumptions.

7.3   Guidelines When Selecting Candidate Models

- Simple regression models are preferred. Therefore, consider models with fewer variables unless a more complex model contains additional important variables.

- Practical considerations are important. For example, certain variables may be expensive or impractical to measure.

- For each candidate model, you should examine the data for unusual observations and verify inference assumptions, which can help eliminate certain candidate models. If a particular model of a certain size is not appropriate, you can consider other models of the same size.

- With model-selection methods, there is a tendency to over specify the model. Recall that a model is overspecified if the model, on the average, fits the data better than the true regression curve.

To implement a model selection strategy in SAS, we would need for a multiple regression for instance, the following SAS statement:

$$\text{proc reg;}$$
$$\text{model } y = x_1 \ x_2 \ldots x_{p-1}/\text{selection} = \text{options,}$$

where the **options** available include

- STEPWISE: stepwise selection procedure

- FORWARD: forward selection procedure

- BACKWARD: backward selection procedure

- MAXR: maximum R^2-selection

- MINR: minimum R^2-selection

- ADJRSQ: Adjusted R^2-selection

- CP: Mallows' C_p-selection

These selection criteria order the models according to certain rules and they are invoked in SAS **proc reg** by specifying them using the **selection**=option. The selection procedure also employs certain **selection** or **optimality** criteria, namely,

- RMSE: prints Root Mean Square Error for each model

- AIC: prints Akaike Information Criterion for each model

- BC: prints Schwarz's Bayesian Criterion

- BIC: prints Bayesian Information Criterion

The Rsquare method lists all models for each subset size in order of the R^2 values, whereas the **adjrsq** and **CP** methods list all models in order of the magnitude of the requested statistics. The command **include** $= n$ includes the first n variables listed in the **model** statement in all candidate models. We now describe some of the selection or optimality criteria enumerated above.

1. **RSQUARE Method:**

   - Finds the subset with the largest R^2 value.

$$R_p^2 = \frac{\text{SSR}_p}{\text{SST}} = 1 - \frac{\text{SSE}_p}{\text{SST}},$$

   where p is the total number of parameters in the regression model.

- Lists all models for each subset size in order of the R^2 values.
- Objective: Account for as much variability in the dependent variable as is practical.
- Because R^2 increases as you add more variables to the model, you must judge whether the increase in R^2 is worth the increase in model complexity.
- R^2 always is largest for the full model.

2. **ADJRSQ Method:**

  - Attempts to account for the difficulty in using R^2 as a model-selection criterion

$$R^2_{\text{adj},p} = 1 - \frac{\text{SSE}_p/(n-p)}{\text{SST}/(n-1)} = 1 - \frac{(n-1)\text{MSE}_p}{\text{SST}}$$

  - Lists all models in order of the magnitude of adjusted R^2 values.
  - Objective: Your objective is to maximize the value of the adjusted R^2.
  - The adjusted R^2 tends to increase as you add significant variables to the model, but it tends to decrease when non-significant variables are added to the model.

3. **CP Method:**

  - The CP method finds models that minimizes Mallows' C_p statistic

$$C_p = \frac{\text{SSE}_p}{\text{MSE}} + 2p - n,$$

  where MSE = mean square error under the full model, SSE_p is the error sum of squares and p is the total number of parameters in the regression model.

  - Lists all models in order of the C_p value.
  - Mallows' C_p statistic consists of a variance component plus a bias component.
  - If an important variable has been left out of the model, then Mallows' C_p statistic usually is larger than p.
  - If all important variables are in the model, then Mallows' C_p statistic is approximately equal to p.
  - For full model, $C_p = p$.
  - Mallows' C_p assumes that the full model has the correct form and that error terms have constant variance.
  - Thus you should choose candidate models such that C_p is close to p and C_p is small.

4. **Akaike's Information Criterion (AIC):** Small AIC is good.

  - The AIC is defined as
$$\text{AIC} = n\log\left(\text{SSE}/n\right) + 2p.$$

  - It takes into account both the goodness of fit (SSE/n) and the number of parameters (p).
  - If AIC is specified as a model option and the selection method is RSQUARE, ADJRSQ or CP, the AIC is included in the printout.

5. **Schwarz's Bayesian Criterion (SBC):** Small SBC is good.

  - The SBC is defined as
$$\text{SBC} = n\log\left(\text{SSE}/n\right) + p\log(n).$$

  - It takes into account both the goodness of fit (SSE/n) and the number of parameters (p).
  - If SBC is specified as a model option and the selection method is RSQUARE, ADJRSQ or CP, the SBC is included in the printout.

7.4   All Possible Regression

To illustrate the use of the model selection strategies discussed in the preceding sections, we consider the following example involving four explanatory variables $x_1 - x_4$ and a response variable y. The data was used in Example 5.5 and it is presented in the following example.

Example 7.1

The data in Table 7.1 relate to the heat evolved in calories during hardening of cement on a per gram basis (y) along with the percentages of four ingredients: tricalcium aluminate (x_1), tricalcium silicate (x_2), tetracalcium alumino ferrite (x_3), and decalcium silicate (x_4). A regression model of the form

$$y_i = \beta_0 + \beta_1 x_1 + \beta_2 x_2 + \beta_3 x_3 + \beta_4 x_4 + \varepsilon_i \tag{7.1}$$

is to be fitted.

No	x_1	x_2	x_3	x_4	y
1	7	26	6	60	78.5
2	1	29	15	52	74.3
3	11	56	8	20	104.3
4	11	31	8	47	87.6
5	7	52	6	33	95.9
6	11	55	9	22	109.2
7	3	71	17	6	102.7
8	1	31	22	44	72.5
9	2	54	18	22	93.1
10	21	47	4	26	115.9
11	1	40	23	34	83.8
12	11	66	9	12	113.3
13	10	68	8	12	109.4

Table 7.1: Heat evolved in calories during hardening of cement

In this example, we would use SAS to analyze all possible regression models. There would be a total of $2^4 - 1 = 15$ such possible regression models comprising of the following:

1. $\binom{4}{1} = 4$ models with one explanatory variable. That is, each model is of the form

$$y_i = \beta_0 + \beta_1 x_i + \varepsilon_i.$$

2. $\binom{4}{2} = 6$ models with two explanatory variables. That is, each model is of the form

$$y_i = \beta_0 + \beta_1 x_{ih} + \beta_2 x_{ij} + \varepsilon_i, \quad h \neq j = 1, 2, 3, 4.$$

3. $\binom{4}{3} = 4$ models with three explanatory variables. That is, each model is of the form

$$y_i = \beta_0 + \beta_1 x_{ih} + \beta_2 x_{ij} + \beta_3 x_{ik} + \varepsilon_i.$$

4. $\binom{4}{4} = 1$ model with all the four explanatory variables. That is, a model of the form in (7.1).

We can use any of the optimality criteria discussed earlier. In the SAS program for this example, we have used both Mallows' C_p and Akaike Information Criterion (AIC) separately. We could have used both in the same **option** statement in SAS. The results are displayed below:

```
data one;
infile 'C:\Research\Book\Data\table611.txt' truncover;
input id x1 x2 x3 x4 y;
run;
proc reg;
model y=x1-x4/selection=rsquare cp;
model y=x1-x4/selection=rsquare aic;
run;
```

The REG Procedure
Model: MODEL1
Dependent Variable: y

R-Square Selection Method

Number in Model	R-Square	C(p)	Variables in Model
1	0.6745	138.7308	x4
1	0.6663	142.4864	x2
1	0.5339	202.5488	x1
1	0.2859	315.1543	x3
2	0.9787	2.6782	x1 x2
2	0.9725	5.4959	x1 x4
2	0.9353	22.3731	x3 x4
2	0.8470	62.4377	x2 x3
2	0.6801	138.2259	x2 x4
2	0.5482	198.0947	x1 x3
3	0.9823	3.0182	x1 x2 x4
3	0.9823	3.0413	x1 x2 x3
3	0.9813	3.4968	x1 x3 x4
3	0.9728	7.3375	x2 x3 x4
4	0.9824	5.0000	x1 x2 x3 x4

Model: MODEL2
Dependent Variable: y

R-Square Selection Method

Number in Model	R-Square	AIC	Variables in Model
1	0.6745	58.8516	x4
1	0.6663	59.1780	x2
1	0.5339	63.5195	x1
1	0.2859	69.0674	x3
2	0.9787	25.4200	x1 x2
2	0.9725	28.7417	x1 x4
2	0.9353	39.8526	x3 x4
2	0.8470	51.0371	x2 x3
2	0.6801	60.6293	x2 x4
2	0.5482	65.1167	x1 x3
3	0.9823	24.9739	x1 x2 x4
3	0.9823	25.0112	x1 x2 x3
3	0.9813	25.7276	x1 x3 x4
3	0.9728	30.5759	x2 x3 x4
4	0.9824	26.9443	x1 x2 x3 x4

In SAS, the option **selection=rsquare** is given in order to generate all possible regressions. The optimality criterion of choice follows. Results from our use of both C_p and AIC indicate the following:

1. Explanatory variable x_4 gives the best C_p and AIC criteria among the 4 one-variable models.

2. Explanatory variables x_1, x_2 give the best two-variable subset among the six possible models in this category.

3. Explanatory variables x_1, x_2, x_4 give the best three-variable subset among the 4 possible models in this category.

4. We will not discuss the model containing all the four explanatory variables since our goal in this chapter is to select parsimonious models with subsets of the 4 explanatory variables. In this case, we are looking for models with $p - 1$ explanatory variables such that $p - 1 < 4$.

The all possible regression analysis does not in itself give us the most parsimonious model. We would have to consider this in the context of optimal selection criteria. In this example, best model is provided with the subset model with the lowest values of C_p and AIC. For the C_p, the subset would be (x_1, x_2) with $C_p = 2.6782$. However in terms of the criterion $C_p - p \simeq 0$, the choice would have to be between (x_1, x_2) and (x_1, x_2, x_4) with a C_p value of 3.0182. We shall examine this case later.

On the other hand, in terms of AIC, the most parsimonious model (model with lowest AIC) is the subset (x_1, x_2, x_4) with AIC value of 24.9739.

7.5   The Forward Selection Strategy

For the forward selection strategy, we need to define upfront the criterion for allowing a variable in, usually denoted as **sle** (significance level for entry) in SAS. The default in SAS is 0.50. The following are the steps necessary to accomplish the forward selection strategy for the data in Example 7.1.

Step 1: The first variable to enter the regression equation is obtained by examining which of the variables has the highest absolute zero-order correlation coefficient with the dependent variable y. In our case, this variable is x_4, since

$$r_{yx_1} = 0.73072, \quad r_{yx_2} = 0.81625, \quad r_{yx_3} = -0.53467 \quad \text{and} \quad r_{yx_4} = -0.82131.$$

Thus variable x_4 is entered first in the model. Alternatively, if we choose an optimality criterion, the first variable to enter is that variable that satisfies best the optimality criterion. For instance, to select the first variable to enter, suppose we employ the C_p optimality criterion. Then the variable with the smallest value of C_p is first chosen to go into the model. In this example the C_p values for the four variables are respectively $(x_4, x_2, x_1, x_3) \equiv (138.7308, 142.4864, 202.5488, 315.1543)$. Hence, variable x_4 would again enter the model first. A similar situation occurs if our chosen criterion had been SSE (the error sum of squares).

Step 2: Using the C_p criterion, we next fit the three models (x_4, x_1), (x_4, x_2), and (x_4, x_3). That is, models that already has x_4 in the model. The results of these are displayed below in the SAS program and a partial output. Clearly, of the three models, the model with x_1 has the lowest C_p criterion, given that x_4 is already in the model. Hence variable x_1 is next selected to go into the model, since it has the lowest C_p criterion value of 5.4959 among this group.

```
proc reg data=one;
model y=x4 x1 x2 x3/selection=rsquare cp include=1;
run;
                        Model: MODEL1
                   Dependent Variable: y

   NOTE: The variables in the 1 variable model are included in all models.

       Number in
         Model      R-Square       C(p)      Variables in Model

           1         0.6745      138.7308    x4
       ----------------------------------------------------------------
           2         0.9725        5.4959    x1
           2         0.9353       22.3731    x3
           2         0.6801      138.2259    x2
       ----------------------------------------------------------------
```

Next we would want to conduct a partial F-test on the newly included variable from the model

$$y_i = \beta_0 + \beta_4\,x_4 + \beta_1\,x_1 + \varepsilon_i.$$

That is, we wish to test the hypothesis that

$$\begin{aligned} H_0 &: \beta_1 | \beta_4 = 0 \\ H_a &: \beta_1 | \beta_4 \neq 0, \end{aligned} \tag{7.2}$$

and the result is given in the following SAS statements and output.

```
proc reg data=one;
model y = x4 x1;
testa: test x1;
run;
```

Parameter Estimates

Variable	DF	Parameter Estimate	Standard Error	t Value	Pr > |t|
Intercept	1	103.09738	2.12398	48.54	<.0001
x4	1	−0.61395	0.04864	−12.62	<.0001
x1	1	1.43996	0.13842	10.40	<.0001

Test testa Results for Dependent Variable y

Source	DF	Mean Square	F Value	Pr > F
Numerator	1	809.10480	108.22	<.0001
Denominator	10	7.47621		

The partial F-test or t-test of the hypotheses in (7.2) gives a p-value of $< 0.0001 \lll 0.50$(**sle**). Hence, variable x_1 satisfies our entry requirement into the model.

Step 3: Next, using the C_p criterion, we fit the two models (x_4, x_1, x_2) and (x_4, x_1, x_3). That is, models that already contained variables x_4 and x_1, since the remaining variables left to be considered are x_2 and x_3. This is again accomplished in SAS with the following statements and partial output.

```
proc reg data=one;
model y=x4 x1 x2 x3/selection=rsquare cp include=2;
run;
```

Dependent Variable: y

R-Square Selection Method

NOTE: The variables in the 2 variable model are included in all models.

Number in Model	R-Square	C(p)	Variables in Model
2	0.9725	5.4959	x4 x1
3	0.9823	3.0182	x2
3	0.9813	3.4968	x3

Given that variables x_4 and x_1 are already in the model, by the C_p criterion, variable x_2 is next to be included, because that would be the model with the lowest C_p of the two competing models. That is, we now have the model

$$y_i = \beta_0 + \beta_4\, x_4 + \beta_1\, x_1 + \beta_2\, x_2 + \varepsilon_i.$$

Again, we would like to test whether x_2 meets our entry criterion of **sle=0.50**. We can do this from a partial F or t-test as in (7.3).

$$H_0 : \beta_2|(\beta_4, \beta_1) = 0$$
$$H_a : \beta_2|(\beta_4, \beta_1) \neq 0. \tag{7.3}$$

```
proc reg data=one;
model y = x4 x1 x2;
testb: test x2;
run;
```

Parameter Estimates

Variable	DF	Parameter Estimate	Standard Error	t Value	Pr > |t|
Intercept	1	71.64831	14.14239	5.07	0.0007
x4	1	−0.23654	0.17329	−1.37	0.2054
x1	1	1.45194	0.11700	12.41	<.0001
x2	1	0.41611	0.18561	2.24	0.0517

Test testb Results for Dependent Variable y

```
                                    Mean
          Source           DF      Square     F Value    Pr > F

          Numerator         1     26.78938       5.03     0.0517
          Denominator       9      5.33030
```

Again, the p-value for the hypotheses in (7.3) is $0.0517 \lll 0.50$. Hence, variable x_2 can be admitted into the model.

Step 4: Now that there is only one explanatory variable left to be considered, there is no need to obtain the C_p criterion first, because this is the last variable. However, we would like to test the hypothesis whether x_3 would be important, given a model that already contained variables x_4, x_1 and x_2. The result is given in the following SAS statements and partial output.

```
proc reg data=one;
model y = x4 x1 x2 x3;
testc: test x3;
run;
                            Parameter Estimates

                         Parameter     Standard
          Variable    DF   Estimate        Error    t Value    Pr > |t|

          Intercept    1   62.40537     70.07096       0.89      0.3991
          x4           1   -0.14406      0.70905      -0.20      0.8441
          x1           1    1.55110      0.74477       2.08      0.0708
          x2           1    0.51017      0.72379       0.70      0.5009
          x3           1    0.10191      0.75471       0.14      0.8959
```

The partial F-test from the hypotheses

$$H_0 : \beta_3|(\beta_4, \beta_1, \beta_2) = 0$$
$$H_a : \beta_3|(\beta_4, \beta_1, \beta_2) \neq 0, \tag{7.4}$$

gives a computed p-value of $0.8958 > 0.50$ (the **sle**), hence variable x_3 can not enter the model that already contained x_4, x_1 and x_2 and we would have to stop at step 3. That is, the model with the subset variables (x_4, x_1, x_2) is obtained. We also note that to two decimal places, the p-values of x_4 and x_2 are 0.84 and 0.50 respectively at step 4. Thus while x_2 would still qualify for inclusion in our model (with **sle** $= 0.50$), however, x_4, the first favored variable would no longer qualify with such a high p-value. Unfortunately, we could not remove variable x_4 at this stage. In fact, once a variable enters at a preceding step, we can not remove that variable at subsequent steps even if we now find its p-value unacceptable for entry. This dilemma highlights the problem with the forward selection strategy. Once a variable is in, it is in permanently.

The entire forward selection strategy process can be implemented in SAS with the following program. Here we have specified that the selection strategy is the forward selection approach and that the optimality criteria be CP (We note here that R^2 is given in SAS). Only step 3 and the summary of the forward selection procedure are included with the program.

```
proc reg data=one;
model y=x1-x4/selection=forward cp;
run;
                            Model: MODEL1
                       Dependent Variable: y


                       Forward Selection: Step 3

                     Parameter    Standard
          Variable    Estimate       Error    Type II SS   F Value   Pr > F

          Intercept   71.64831    14.14239    136.81003     25.67    0.0007
          x1           1.45194     0.11700    820.90740    154.01   <.0001
          x2           0.41611     0.18561     26.78938      5.03    0.0517
          x4          -0.23654     0.17329      9.93175      1.86    0.2054

          Bounds on condition number: 18.94, 116.36
--------------------------------------------------------------------------------
```

No other variable met the 0.5000 significance level for entry into the model.

Summary of Forward Selection

Step	Variable Entered	Number Vars In	Partial R-Square	Model R-Square	C(p)	F Value	Pr > F
1	x4	1	0.6745	0.6745	138.731	22.80	0.0006
2	x1	2	0.2979	0.9725	5.4959	108.22	<.0001
3	x2	3	0.0099	0.9823	3.0182	5.03	0.0517

We observe from the SAS output that the forward selection strategy stops at step 3, with estimated regression equation

$$\hat{y}_i = 71.6483 + 1.4519\,x_1 + 0.4161\,x_2 - 0.2365\,x_4.$$

It is necessary to conduct the regression diagnostics on this final model. The chosen final model indicates that we have a condition number between 18.94 and $116.36 \gg 30$. Hence, this final model exhibits multicollinearity (discussed in Chapter 5) and we would need to fix this. We shall come back to this problem later. If we had decided to use a more stringent entry criterion, we could specify **sle**= 0.15 for instance as:

```
proc reg data=one;
model y=x1-x4/selection=forward cp sle=0.15;
run;
```

7.6 The Backward Selection Strategy

As in the forward selection strategy, we also need to define the criterion for allowing a variable to stay or remain in the model. This is denoted in SAS as **sls** (significance level for staying). The default in SAS is 0.10. The following are the steps necessary to accomplish the backward selection strategy for the data in Example 7.1.

Step 0: The model containing all the explanatory variables is fitted to the data. For our example, we present a summary output from such a model

Variable	Parameter Estimate	Standard Error	Type II SS	F Value	Pr > F
Intercept	62.40537	70.07096	4.74552	0.79	0.3991
x1	1.55110	0.74477	25.95091	4.34	0.0708
x2	0.51017	0.72379	2.97248	0.50	0.5009
x3	0.10191	0.75471	0.10909	0.02	0.8959
x4	-0.14406	0.70905	0.24697	0.04	0.8441

The p-values for the explanatory variables are $0.0708, 0.5009, 0.8959, 0.8441$ for x_1, x_2, x_3 and x_4 respectively. If we employ the default **sls=0.10** in this example, then variables x_2, x_3 and x_4 have p-values that are each greater than 0.10. Hence, each of these variables is a candidate for removal from our model. However, of the three variables, the p-value 0.8959 for x_3 is the largest. Hence variable x_3 will be the first to be eliminated from the model.

Step 1: Variable x_3 is eliminated from the model and we are now left with a model containing (x_1, x_2, x_4). The result from the analysis involving the variables x_1, x_2 and x_4 is presented below.

Variable	Parameter Estimate	Standard Error	Type II SS	F Value	Pr > F
Intercept	71.64831	14.14239	136.81003	25.67	0.0007
x1	1.45194	0.11700	820.90740	154.01	<.0001
x2	0.41611	0.18561	26.78938	5.03	0.0517
x4	-0.23654	0.17329	9.93175	1.86	0.2054

Again, since for a variable to remain in the model, the p-value must be less than 0.10, we see that variable x_4 has a p-value of $0.2054 > 0.10$. Hence, variable x_4 will be removed from the model, leaving a model containing only variables x_1 and x_2.

Step 2: At this stage, we are now left with variables x_1 and x_2 for further consideration for our parsimonious model. The model containing these two variables have the following results:

Variable	Parameter Estimate	Standard Error	Type II SS	F Value	Pr > F
Intercept	52.57735	2.28617	3062.60416	528.91	<.0001
x1	1.46831	0.12130	848.43186	146.52	<.0001
x2	0.66225	0.04585	1207.78227	208.58	<.0001

Clearly now, the p-values for both variables are much less than the **sls** value of 0.10. Hence, no other variable in the model will be removed at this stage and therefore our final model is a model having x_1 and x_2 as the explanatory variables with estimated regression equation

$$\hat{y}_i = 52.57774 + 1.4683\,x_1 + 0.6623\,x_2.$$

Once again, we would have to subject this final model to the diagnostics procedures discussed in Chapter 5.

The entire backward selection strategy can be implemented in SAS with the following SAS program and the partial output have been presented under steps 0 through 2. Here, we are employing the C_p criterion for the selection procedure.

```
proc reg data=one;
model y=x1-x4 / selection=backward cp;
run;
```

A major disadvantage of the backward selection strategy is that once a variable is eliminated at an earlier step of the analysis, that variable is lost completely to the model and cannot be brought back, even when it might be desirable.

Because of the shortcomings of both the forward and backward selection strategies (one adds variables permanently, while the other eliminates variables permanently), we consider in the next section a selection strategy that may add and remove variables at any stage of the analysis.

7.7   The Stepwise Regression Procedure

For this procedure we require both the significance level for entry (**sle**) and the significant level for staying (**sls**). The default in SAS for both are 0.15. We could of course use other values of our choice. For now, we would use the default values in SAS.

The stepwise procedure initially behaves like a forward selection procedure, admitting variables that satisfy the entry requirement at each step. However, once a variable is admitted, the procedure obtains partial p-values and checks these against the **sls** of 0.15 in our case. Any partial p-value not satisfying this requirement to remain in the model is then removed and the backward elimination procedure kicks in. This process continues until a parsimonious model is attained. For the example, we have the following steps:

Step 1: A variable having the highest absolute zero-order correlation or best optimal criterion with the dependent variable is first admitted into the model. In our case, variable x_4 is first admitted.

Variable	Parameter Estimate	Standard Error	Type II SS	F Value	Pr > F
Intercept	117.56793	5.26221	40108	499.16	<.0001
x4	-0.73816	0.15460	1831.89616	22.80	0.0006

Step 2: As in the case for the forward selection strategy, variable x_1 is next admitted, giving the following:

Variable	Parameter Estimate	Standard Error	Type II SS	F Value	Pr > F
Intercept	103.09738	2.12398	17615	2356.10	<.0001
x1	1.43996	0.13842	809.10480	108.22	<.0001
x4	-0.61395	0.04864	1190.92464	159.30	<.0001

The p-values for both variables are now cross-checked against **sls** of 0.15. Both p-values are admissible.

Step 3: Next, the procedure admits variable x_2. We see that this produces exactly the same forward selection procedure results we discussed earlier.

Variable	Parameter Estimate	Standard Error	Type II SS	F Value	Pr > F
Intercept	71.64831	14.14239	136.81003	25.67	0.0007
x1	1.45194	0.11700	820.90740	154.01	<.0001
x2	0.41611	0.18561	26.78938	5.03	0.0517
x4	-0.23654	0.17329	9.93175	1.86	0.2054

Step 4: We again compare the p-values for each variable with **sls** of 0.15. In this case, the p-value for variable x_4 is $0.2054 > 0.15$. Hence, variable x_4 is removed, and we now have the following results:

Variable	Parameter Estimate	Standard Error	Type II SS	F Value	Pr > F
Intercept	52.57735	2.28617	3062.60416	528.91	<.0001
x1	1.46831	0.12130	848.43186	146.52	<.0001
x2	0.66225	0.04585	1207.78227	208.58	<.0001

The estimated final regression equation is

$$\hat{y}_i = 52.57774 + 1.4683\,x_1 + 0.6623\,x_2.$$

The entire stepwise selection procedure can be implemented in SAS with the following statements and a partial output have been presented under steps 1 through 4. The output from the final step is included with the program.

```
proc reg data=one;
model y=x1-x4 / selection=stepwise cp;
run;
```

                        Stepwise Selection: Step 4

        Variable x4 Removed: R-Square = 0.9787 and C(p) = 2.6782

                          Analysis of Variance

Source	DF	Sum of Squares	Mean Square	F Value	Pr > F
Model	2	2657.85859	1328.92930	229.50	<.0001
Error	10	57.90448	5.79045		
Corrected Total	12	2715.76308			

Variable	Parameter Estimate	Standard Error	Type II SS	F Value	Pr > F
Intercept	52.57735	2.28617	3062.60416	528.91	<.0001
x1	1.46831	0.12130	848.43186	146.52	<.0001
x2	0.66225	0.04585	1207.78227	208.58	<.0001

            Bounds on condition number: 1.0551, 4.2205

    All variables left in the model are significant at the 0.1500 level.

  No other variable met the 0.1500 significance level for entry into the model.

The condition number for the final model indicates 'no multicollinearity'. Of course, we would still need to conduct other diagnostics tests such as normality of error terms, checking for influential observations or outliers as well as testing for the adequacy of the model.

Use of Selection Strategy to Solve Multicollinearity Problem

One of the best ways of combating multicollinearity is to conduct a stepwise selection strategy. In the previous data analysis, the stepwise regression strategy indicates that only variables x_1 and x_2 are important to be included in our estimated regression equation. Variable x_3 was not at any stage included in the model. Based on this result, the estimated regression equation is $\hat{y} = 52.5774 + 1.4683x_1 + 0.6623x_2$. The condition numbers for this model are 1.0551 and 4.2205, indicating no presence of multicollinearity in the estimated regression equation. The advantage of this procedure is that the parameter estimates are OLS and are unbiased.

Example 7.2

An investigator is interested in understanding the relationship, if any, between the analytical skills of young gifted children and the following variables:

$x_1 = $ father's IQ

$x_2 = $ mother's IQ

$x_3 = $ age in months when the child first said 'mommy' or 'daddy'

$x_4 = $ age in months when the child first counted to 10 successfully

$x_5 = $ average number of hours per week the child's mother or father read to the child

$x_6 = $ average number of hours per week the child watched an educational program on TV during the past 3 months

$x_7 = $ average number of hours per week the child watched cartoons on TV during the past 3 months

Data were collected on 36 children who were identified as gifted children soon after they reached the age of four. The data for the first two children in the sample as well as the last two children are displayed. The complete data can be found on the CD.

Gifted Children Data

Child	y	x_1	x_2	x_3	x_4	x_5	x_6	x_7
1	159	115	117	18	26	1.9	3.00	2.00
2	164	117	113	20	37	2.5	1.75	3.25
...	...	...	...	...	...	...	...	...
35	164	111	121	18	36	2.3	1.00	4.50
36	159	114	123	20	30	2.2	1.75	3.25

A multiple regression model of the form

$$y_i = \beta_0 + \beta_1 x_1 + \beta_2 x_2 + \beta_3 x_3 + \beta_4 x_4 + \beta_5 x_5 + \beta_6 x_6 + \beta_7 x_7 + \varepsilon_i,$$

is suggested and a subset regression analysis is envisaged. Suppose we work through this example by asking and answering the following questions based on our analysis.

(a) Find the predictor variable that results in the best prediction function among all subset models containing only *one* predictor.

(b) Find the two predictor variables that result in the best prediction function among all subset models containing exactly *two* predictors.

(c) Find the three predictor variables that result in the best prediction function among all subset models containing exactly *three* predictors.

(d) Find a short list of the three best models using the R^2 criterion.

(e) Find a short list of the three best models using the R^2_{adj} criterion.

(f) Find a short list of the three best models using the C_p criterion.

(g) Find a short list of the three best models using the C_p and $C_p - p$ together.

(h) Run a stepwise regression model with **sle** $= 0.50$ and **sls** $= 0.10$. List the final model.

An all possible regression will generate $2^7 - 1 = 127$ total regressions comprising of subsets of one, two, three, four, five, six and seven explanatory variables with

$$\binom{7}{1} = 7, \quad \binom{7}{2} = 21, \quad \binom{7}{3} = 35, \quad \binom{7}{4} = 35, \quad \binom{7}{5} = 21, \quad \binom{7}{6} = 7, \quad \binom{7}{7} = 1,$$

respectively. In SAS, we can generate only the top two or three for each subset by specifying an option **best** $= k$, where k is 2 or 3 in this case. As an example, suppose, we let $k = 3$, then we have the following all possible regressions based on the C_p criterion.

```
data one;
infile 'C:\Research\Book\Data\example72.txt' truncover;
input y x1-x7;
run;
proc reg data=one;
model y=x1-x7/selection=rsquare cp best=3;
run;
```

                        Dependent Variable: y

                    R-Square Selection Method

Number in Model	R-Square	C(p)	Variables in Model
1	0.3263	43.3217	x2
1	0.2962	46.6934	x4
1	0.2758	48.9662	x5
2	0.6131	11.1050	x2 x5
2	0.6078	13.8557	x2 x4
2	0.3780	39.5478	x2 x3
3	0.6872	6.9673	x1 x2 x5
3	0.6666	9.2811	x1 x2 x4
3	0.6465	11.5240	x2 x3 x4
4	0.7077	6.6799	x1 x2 x3 x4
4	0.7069	6.7730	x2 x5 x6 x7
4	0.7040	7.0929	x1 x2 x3 x5
5	0.7338	5.7678	x1 x2 x5 x6 x7
5	0.7179	7.5400	x1 x2 x3 x4 x5
5	0.7160	7.7482	x2 x3 x5 x6 x7
6	0.7442	6.6012	x1 x2 x3 x5 x6 x7
6	0.7351	7.6147	x1 x2 x4 x5 x6 x7
6	0.7333	7.8237	x1 x2 x3 x4 x6 x7
7	0.7496	8.0000	x1 x2 x3 x4 x5 x6 x7

In this case, only 19 of the possible 127 regressions are displayed by use of the option **best** $= 3$. Similarly, an all possible regression using the SSE (error sum of squares) as the optimality criterion is presented below, this time with the option **best** $= 2$.

```
proc reg data=one;
model y=x1-x7/selection=rsquare sse best=2;
run;
```

                        Dependent Variable: y

                    R-Square Selection Method

Number in Model	R-Square	SSE	Variables in Model
1	0.3263	505.46781	x2
1	0.2962	528.09446	x4
2	0.6291	278.26264	x2 x5
2	0.6078	294.30597	x2 x4
3	0.6872	234.65806	x1 x2 x5
3	0.6666	250.18560	x1 x2 x4
4	0.7077	219.30784	x1 x2 x3 x4
4	0.7069	219.93248	x2 x5 x6 x7
5	0.7338	199.76508	x1 x2 x5 x6 x7
5	0.7179	211.65832	x1 x2 x3 x4 x5

```
------------------------------------------------------------------
      6      0.7442      191.93650      x1 x2 x3 x5 x6 x7
      6      0.7351      198.73806      x1 x2 x4 x5 x6 x7
------------------------------------------------------------------
      7      0.7496      187.90202      x1 x2 x3 x4 x5 x6 x7
```

We present in Table 7.2 the results of the best prediction function among all subset models containing only 1, 2, 3, 4 and 5 predictors. We have also presented the corresponding values of the optimal selection criteria. These are presented in brackets underneath.

| No of | Best Subsets | | | |
predictors	R^2	C_p	SSE	AIC
1	(x_2)	(x_2)	(x_2)	(x_2)
	(0.3263)	(43.3217)	(505.4678)	(99.1108)
2	(x_2, x_5)	(x_2, x_5)	(x_2, x_5)	(x_2, x_5)
	(0.6291)	(11.4650)	(278.2626)	(79.6217)
3	(x_1, x_2, x_5)	(x_1, x_2, x_5)	(x_1, x_2, x_5)	(x_1, x_2, x_5)
	(0.6872)	(6.9673)	(234.6581)	(75.4860)
4	(x_1, x_2, x_3, x_4)	(x_1, x_2, x_3, x_4)	(x_1, x_2, x_3, x_4)	(x_1, x_2, x_3, x_4)
	(0.7077)	(6.6799)	(219.3078)	(75.0505)
5	$(x_1, x_2, x_5, x_6, x_7)$	$(x_1, x_2, x_5, x_6, x_7)$	$(x_1, x_2, x_5, x_6, x_7)$	$(x_1, x_2, x_5, x_6, x_7)$
	(0.7338)	(5.7678)	(199.7651)	(73.6904)

Table 7.2: Best subsets based on four optimal selection criteria

A short list of four models based on the C_p criterion would be

Subset	C_p value	$C_p - p$
x_1, x_2, x_5, x_6, x_7	5.7678	-0.232
$x_1, x_2, x_3, x_5, x_6, x_7$	6.6012	-0.399
x_1, x_2, x_3, x_4	6.6799	1.680
x_2, x_5, x_6, x_7	6.7730	1.77

Of the four models with the lowest C_p value, the subset x_1, x_2, x_5, x_6, x_7 has the closest $C_p - p$ value to zero. Hence, this will be the best subset model selected by Mallows' C_p criterion.

Using Forward, Backward and Stepwise Selection Strategies

The forward selection procedure that sets the significance level for entry at 0.15 is considered in this case. For this, the selected subset model is (x_1, x_2, x_5), with estimated regression equation

$$\hat{y}_i = 44.3656 + 0.3215\,x_1 + 0.4283\,x_2 + 12.7663\,x_5.$$

The model has an overall $R^2 = 0.6872$. The SAS program and a partial output are presented below.

```
proc reg data=one;
model y=x1-x7/selection=forward sle=0.15;
run;
                          The REG Procedure
                            Model: MODEL1
                        Dependent Variable: y

                     Forward Selection: Step 3

                  Parameter    Standard
      Variable     Estimate      Error    Type II SS  F Value  Pr > F

      Intercept    44.36557    18.48732    42.23081     5.76   0.0224
      x1            0.32148     0.13183    43.60458     5.95   0.0205
      x2            0.42825     0.07046   270.89628    36.94   <.0001
      x5           12.76632     2.23107   240.09817    32.74   <.0001

          Bounds on condition number: 1.0067, 9.0444
--------------------------------------------------------------------------

    No other variable met the 0.1500 significance level for entry into the model.

                   Summary of Forward Selection
```

Step	Variable Entered	Number Vars In	Partial R-Square	Model R-Square	C(p)	F Value	Pr > F
1	x2	1	0.3263	0.3263	43.3217	16.47	0.0003
2	x5	2	0.3028	0.6291	11.4650	26.94	<.0001
3	x1	3	0.0581	0.6872	6.9673	5.95	0.0205

A backward selection strategy which also sets the significance level for staying in the model at 0.15 is employed. For this, the selected subset model is $(x_1, x_2, x_5, x_6, x_7)$, with estimated regression equation

$$\hat{y}_i = 79.0468 + 0.2374\,x_1 + 0.4121\,x_2 + 11.9013\,x_5 - 4.8881\,x_6 - 3.8202\,x_7.$$

The model has an overall $R^2 = 0.7338$. We envisage a multicollinearity problem with this model as the bounds for condition index for the model are between 8.0214 and 95.388. The latter being greater than 30. The SAS program and a partial output are presented below.

```
proc reg data=one;
model y=x1-x7/selection=backward sls=0.15;
run;
                    The REG Procedure
                     Model: MODEL1
                  Dependent Variable: y


            Backward Elimination: Step 2
```

Variable	Parameter Estimate	Standard Error	Type II SS	F Value	Pr > F
Intercept	79.04678	23.73961	73.82773	11.09	0.0023
x1	0.23736	0.13639	20.16741	3.03	0.0921
x2	0.41211	0.07177	219.57291	32.97	<.0001
x5	11.90132	2.17231	199.86868	30.02	<.0001
x6	-4.88814	2.16155	34.05314	5.11	0.0311
x7	-3.82023	1.96759	25.10199	3.77	0.0616

```
         Bounds on condition number: 8.0214, 95.388
-------------------------------------------------------------------------------

    All variables left in the model are significant at the 0.1500 level.

            Summary of Backward Elimination
```

Step	Variable Removed	Number Vars In	Partial R-Square	Model R-Square	C(p)	F Value	Pr > F
1	x4	6	0.0054	0.7442	6.6012	0.60	0.4446
2	x3	5	0.0104	0.7338	5.7678	1.18	0.2857

For the stepwise procedure applied to this example, we have chosen to use **sls** and **sle** values of 0.15 (the default in SAS). The stepwise procedure selects the subset model (x_1, x_2, x_5), with estimated regression equation

$$\hat{y}_i = 44.3656 + 0.3215\,x_1 + 0.4283\,x_2 + 12.7663\,x_5.$$

The model has an overall $R^2 = 0.6872$. This is the same model from the forward procedure. As we can see, the best models from forward, backward and stepwise need not be the same. The SAS program and a partial output for the stepwise procedure are presented below.

```
proc reg data=one;
model y=x1-x7/selection=stepwise sle=0.15 sls=0.15;
run;
                    The REG Procedure
                     Model: MODEL1
                  Dependent Variable: y


              Stepwise Selection: Step 3
```

Variable	Parameter Estimate	Standard Error	Type II SS	F Value	Pr > F
Intercept	44.36557	18.48732	42.23081	5.76	0.0224
x1	0.32148	0.13183	43.60458	5.95	0.0205
x2	0.42825	0.07046	270.89628	36.94	<.0001
x5	12.76632	2.23107	240.09817	32.74	<.0001

```
                    Bounds on condition number: 1.0067, 9.0444
-------------------------------------------------------------------------------
```

All variables left in the model are significant at the 0.1500 level.

No other variable met the 0.1500 significance level for entry into the model.

Summary of Stepwise Selection

Step	Variable Entered	Variable Removed	Number Vars In	Partial R-Square	Model R-Square	C(p)	F Value	Pr > F
1	x2		1	0.3263	0.3263	43.3217	16.47	0.0003
2	x5		2	0.3028	0.6291	11.4650	26.94	<.0001
3	x1		3	0.0581	0.6872	6.9673	5.95	0.0205

7.8   Model Validation

In the previous sections of this chapter, we discuss the various model selection strategies needed to select an appropriate model. In Chapter 5, we discuss the necessary remedial measures or diagnostics needed on a chosen model. If such remedial measures have been accomplished for the chosen model, the next phase in model building in regression analysis is **model validation**. We distinguish here the difference between model validation and model adequacy which relates to such things as residual analysis, influential observations and such issues that are raised in a normal diagnostics procedures enumerated in Chapter 5. Validation on the other hand relates to how robust our model will hold or perform in post sample situations. However, model selection has its difficulties. There are usually two approaches to model selection, namely,

1. You can select a model or a set of potential models based upon theoretical and practical considerations of the applied problem.

2. You can use the data to select a model. This is often the approach we have used to date.

The difficulty with selecting a model from the data is that bias can be created in the estimated regression equation. This can have a variety of effects on our analysis:

- The estimated regression equation may fit the data better than the actual, unknown regression model.

- The estimate of the error variance can be too small, thus invalidating parameter tests and confidence intervals.

In most applied situations, some information about the problem is available before the data are collected. The more background information that we use in the model-selection process, the less likely that bias will appear in the model. The process of verifying that we have selected an appropriate model for the populations we are studying is called *model validation*. Model validation are usually conducted in either one of three different ways:

1. Collection of new or fresh data to enable us investigate the model's predictive ability.

2. Detailed examination of the model's coefficients and predictive values with theoretical expectations, and other analytic models or simulation results.

3. Use of **data splitting**, that is, use a holdout sample or set aside a sample of the original data to investigate the model's predictive ability.

Data splitting has proved to be the most popular for model validation although collection of new data is also the most difficult to accomplish. However, if the data is large, we can split the data into two data sets. The first data set is often called the **estimation data** which are used to build the model (including the carrying out of all the diagnostics), while the second set of data is called the **prediction data** which will be used to study the predictive and appropriateness of the model. Often the data are split equally, but let us assume that both the estimation and prediction data sets have respectively n and n^* observations.

   The mean square prediction error (MSPR) which measures the predictive ability of the model is computed from the prediction data set as

$$\text{MSPR} = \frac{1}{n^*} \sum_{i=1}^{n^*} (y_i - \hat{y}_i)^2. \tag{7.5}$$

The computed MSPR which are usually larger than the residual mean square error from the estimation data can then be compared to the MSE from the estimation data. If both are very close, then, the MSE from the model building data is perhaps not biased and would have a fairly good predictive capability. However, if the MSE is much smaller than the MSPR, then we need to be wary of the predictive capability of our chosen model.

Another means of validating the model based on split data approach is to compare the multiple R^2 obtained from the chosen model to the percentage of variability in the prediction data. Again this is defined as

$$R^2_{\text{prediction}} = 1 - \frac{\sum\limits_{i=1}^{n^*}(y_i - \hat{y}_i)^2}{\sum\limits_{i=1}^{n^*}(y_i - \bar{y})^2}. \tag{7.6}$$

We can relate the $R^2_{\text{prediction}}$ defined in (7.6) to the **PRESS** statistic defined in Chapter 5. Recall that

$$\text{PRESS} = \sum_{i=1}^{n}\left(y_i - \hat{y}_{(i)}\right)^2 = \sum_{i=1}^{n}\left(\frac{e_i}{1 - h_i}\right)^2, \tag{7.7}$$

and a prediction like statistic based on the PRESS statistic can be defined as

$$R^2_{\text{prediction}} = 1 - \frac{\text{PRESS}}{\text{SST}}, \tag{7.8}$$

where PRESS is as defined in (7.7) and SST is the total sum of squares from the estimation data.
Note here that the PRESS statistic is defined for the estimation data and thus the PRESS statistic can be used as a model selection criterion as well as a validation statistic. Thus, the PRESS statistic

- allows a model to "predict itself"

- measures the overall influence of all observations in the regression

- is always larger than the error sum of squares.

For the PRESS statistic; smaller values are preferred.

Example 7.3

The following example relating to **job proficiency** was adapted from Kutner et al. (2005, p. 377) and reproduced with permission of The McGraw-Hill Companies. The data relate to the proficiency after a probationary period for 25 applicants admitted to entry-level clerical positions in a governmental agency. The proficiency aptitude scores on four tests (x_1, x_2, x_3, x_4) and the job proficiency score y are displayed in Table 7.3 for the first and last three applicants.

In Table 7.4 are the data on additional 25 applicants for the entry-level positions within the governmental agency who were similarly tested and hired irrespective of their test scores. We shall use this second data set as our prediction data.

We present in Figure 7.1 the scatter plot matrix of the dependent variable y with the four explanatory variables $x_1 - x_4$ for the estimation data in Table 7.3. These plot reveals that x_3, x_4 and x_1 are linearly related to the dependent variable y. Further, it also shows that x_3 and x_4 are similarly highly correlated.

The sample correlation coefficients between the dependent and all explanatory variables are presented in (7.9). The results confirm what we surmise from the scatter plot matrix in Figure 7.1.

$$\begin{array}{c} \\ y \\ x_1 \\ x_2 \\ x_3 \\ x_4 \end{array} \begin{pmatrix} y & x_1 & x_2 & x_3 & x_4 \\ 1 & 0.5144 & 0.4970 & 0.8971 & 0.8694 \\ & 1 & 0.1023 & 0.1808 & 0.3267 \\ & & 1 & 0.5190 & 0.3967 \\ & & & 1 & 0.7820 \\ & & & & 1 \end{pmatrix} \tag{7.9}$$

Subject	Test Scores				Job Proficiency Score
i	x_{i1}	x_{i2}	x_{i3}	x_{i4}	y_i
1	86	110	100	87	88
2	62	97	99	100	80
3	110	107	103	103	96
⋮	⋮	⋮	⋮	⋮	⋮
23	104	73	93	80	78
24	94	121	115	104	115
25	91	129	97	83	83

Table 7.3: Job proficiency estimation data set for 25 applicants

Subject	Test Scores				Job Proficiency Score
i	x_{i1}	x_{i2}	x_{i3}	x_{i4}	y_i
26	65.0	109.0	88.0	84.0	58
27	85.0	90.0	104.0	98.0	92
28	93.0	73.0	91.0	82.0	71
⋮	⋮	⋮	⋮	⋮	⋮
48	115.0	119.0	102.0	94.0	95
49	129.0	70.0	94.0	95.0	81
50	136.0	104.0	106.0	104.0	109

Table 7.4: Job proficiency prediction data set for 25 applicants

We fit a stepwise first-order regression model to the data. The result of the stepwise regression with **sle**=0.05 and **sls**=0.10 is presented in the following SAS program and a partial output.

```
data valid;
input y x1-x4;
datalines;
......
   ;
  proc reg data=valid;
  model y=x1-x4/selection=stepwise sse cp sle=0.05 sls=0.10;
  run;
                         Model: MODEL1
                     Dependent Variable: y

               Stepwise Selection: Step 3

         Variable x4 Entered: R-Square = 0.9615 and C(p) = 3.7274

                       Analysis of Variance

                               Sum of          Mean
     Source            DF      Squares        Square    F Value   Pr > F

     Model              3    8705.80299    2901.93433    175.02   <.0001
     Error             21     348.19701      16.58081
     Corrected Total   24    9054.00000

                     Parameter    Standard
            Variable  Estimate      Error    Type II SS  F Value  Pr > F

            Intercept -124.20002    9.87406  2623.35826  158.22   <.0001
            x1          0.29633     0.04368   763.11559   46.02   <.0001
            x3          1.35697     0.15183  1324.38825   79.87   <.0001
            x4          0.51742     0.13105   258.46044   15.59   0.0007

               Bounds on condition number: 2.8335, 19.764
-------------------------------------------------------------------------------
```

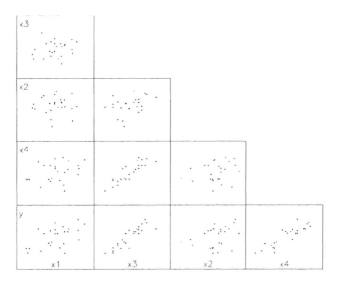

Figure 7.1: Bivariate scatter plot matrix for the five variables

All variables left in the model are significant at the 0.1000 level

No other variable met the 0.0500 significance level for entry into the model.

Summary of Stepwise Selection

Step	Variable Entered	Variable Removed	Number Vars In	Partial R-Square	Model R-Square	C(p)	F Value	Pr > F
1	x3		1	0.8047	0.8047	84.2465	94.78	<.0001
2	x1		2	0.1283	0.9330	17.1130	42.12	<.0001
3	x4		3	0.0285	0.9615	3.7274	15.59	0.0007

The predicted model chosen by the stepwise selection method is therefore

$$\hat{y}_i = -124.2000 + 0.2963x_1 + 1.3570x_3 + 0.5174x_4.$$

7.8.1 Diagnostics for the Chosen Model

The variance inflation factors are small, indicating no presence of multicollinearity. Both the Cook's distances and leverages indicate no influential or outlying observations in the data when these are compared with the cut-off points for significance generated at the end of the output.

```
proc reg data=valid;
model y=x1 x3 x4/vif influence;
output out=aa r=resid p=pred student=stu rstudent=stud h=lev cookd=cooks dffits=ddfit press=press;
run;
proc print data=aa;
var y pred resid stu stud lev cooks ddfit press;
format pred resid stu stud lev cooks ddfit press 8.3;
data new;
set aa;
hh=(2*4)/25; cookf=finv(0.5, 3, 20); dff = 2*sqrt(4/25); u=press*press;
run;
proc print data=new;
run;

proc univariate data=new normal plot;
var resid u;
run;
goptions vsize=5in hsize=5in;

proc gplot data=aa;
axis1 label=(a=-90 rotate=90 'Standardized Residuals');
axis2 label=('Predicted');
plot stu*pred='plus'/vaxis=axis1 haxis=axis2 noframe vref=0;
run;
```

```
                        The REG Procedure
                         Model: MODEL1
                     Dependent Variable: y

             Number of Observations Read        25
             Number of Observations Used        25

                      Analysis of Variance

                               Sum of         Mean
     Source             DF    Squares       Square    F Value    Pr > F

     Model               3   8705.80299   2901.93433   175.02    <.0001
     Error              21    348.19701     16.58081
     Corrected Total    24   9054.00000


             Root MSE              4.07195    R-Square    0.9615
             Dependent Mean       92.20000    Adj R-Sq    0.9560
             Coeff Var             4.41644

                      Parameter Estimates

                      Parameter    Standard                          Variance
     Variable    DF    Estimate      Error    t Value   Pr > |t|    Inflation

     Intercept    1   -124.20002    9.87406    -12.58    <.0001            0
     x1           1      0.29633    0.04368      6.78    <.0001      1.13775
     x3           1      1.35697    0.15183      8.94    <.0001      2.61664
     x4           1      0.51742    0.13105      3.95    0.0007      2.83349

    Obs   y     pred    resid     stu     stud     lev    cooks    ddfit    press
     1    88   81.996    6.004    1.553    1.611   0.099   0.066    0.533    6.662
     2    80   80.254   -0.254   -0.077   -0.075   0.342   0.001   -0.054   -0.386
     3    96  101.458   -5.458   -1.397   -1.431   0.079   0.042   -0.420   -5.928
     4    76   81.082   -5.082   -1.340   -1.367   0.132   0.068   -0.533   -5.855
     5    80   79.878    0.122    0.031    0.030   0.059   0.000    0.008    0.130
     6    73   71.289    1.711    0.448    0.440   0.122   0.007    0.164    1.948
     7    58   58.206   -0.206   -0.063   -0.061   0.350   0.001   -0.045   -0.316
     8   116  117.099   -1.099   -0.304   -0.297   0.209   0.006   -0.153   -1.391
     9   104  107.156   -3.156   -0.809   -0.802   0.082   0.015   -0.240   -3.439
    10    99  104.031   -5.031   -1.290   -1.312   0.083   0.038   -0.394   -5.485
    11    64   69.240   -5.240   -1.387   -1.421   0.140   0.078   -0.572   -6.091
    12   126  124.430    1.570    0.426    0.418   0.183   0.010    0.198    1.921
    13    94   93.605    0.395    0.111    0.108   0.235   0.001    0.060    0.516
    14    71   74.033   -3.033   -0.851   -0.845   0.233   0.055   -0.466   -3.955
    15   111  111.519   -0.519   -0.136   -0.133   0.122   0.001   -0.049   -0.591
    16   109  102.392    6.608    1.797    1.907   0.185   0.183    0.907    8.104
    17   100   95.092    4.908    1.275    1.296   0.106   0.048    0.447    5.492
    18   127  122.361    4.639    1.352    1.381   0.290   0.186    0.882    6.532
    19    99  100.547   -1.547   -0.402   -0.394   0.108   0.005   -0.137   -1.734
    20    82   86.839   -4.839   -1.251   -1.269   0.097   0.042   -0.416   -5.359
    21    67   65.193    1.807    0.485    0.476   0.163   0.011    0.210    2.158
    22   109  112.236   -3.236   -0.866   -0.861   0.158   0.035   -0.373   -3.844
    23    78   74.210    3.790    1.001    1.001   0.136   0.039    0.397    4.385
    24   115  113.518    1.482    0.402    0.393   0.178   0.009    0.183    1.803
    25    83   77.337    5.663    1.475    1.520   0.111   0.068    0.536    6.368

                    Obs    hh     cookf    dff
                     1    0.32   0.81621   0.8
```

The normality test of the residuals gives a p-value > 0.2500 for the Anderson-Darling test, indicating that we can assume that the error terms are normally distributed. The plot of the standardized residuals versus the predicted values is presented in Figure 7.2. The residuals are randomly distributed around zero indicating that the variances are homogeneous. That is, there is no heteroscedasticity in these data. Thus, this model will be acceptable to us as the final model. The PRESS statistic obtained for this model is 471.4520.

7.8.2   Validation Analysis

First we present in Figure 7.3 the scatter plot matrix for the five variables $y, x_1 - x_4$ in the validation data. Again we observe here strong linear association between explanatory variables x_3 and x_4 and between the dependent variable y and x_3, x_4. This pattern of associations are consistent with what we had for the estimation data set for building the model.

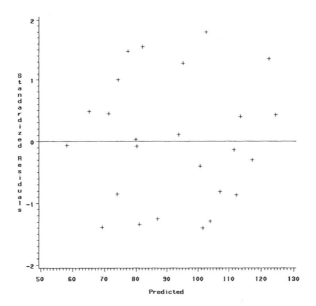

Figure 7.2: Plot of standardized residuals against predicted values

The chosen model has the estimated regression equation

$$\hat{y}_i = -124.2000 + 0.2963x_1 + 1.3570x_3 + 0.5174x_4,$$

and we would now use the validation data in Table 7.4 to obtain the predicted values by substituting the explanatory variables in the estimated regression equation for x_1, x_3 and x_4. That is, we need to obtain $\hat{y}_i$, $i = 26, \ldots, 50$. The corresponding residuals $(y_i - \hat{y}_i)$ are similarly obtained. These are presented in the following output.

```
data new;
set valid2;
yhat=-124.20002+0.29633*x1+1.35697*x3+0.51742*x4;
resid=y-yhat;
u=resid*resid;
run;
proc print data=new;
var y yhat resid;
format yhat resid 8.3;
run;
```

Obs	y	yhat	resid
1	58	57.938	0.062
2	92	92.820	-0.820
3	71	69.271	1.729
4	77	76.844	0.156
5	92	90.682	1.318
6	66	64.130	1.870
7	61	58.540	2.460
8	57	58.575	-1.575
9	66	72.100	-6.100
10	75	76.498	-1.498
11	98	92.764	5.236
12	100	103.236	-3.236
13	67	69.274	-2.274
14	111	107.375	3.625
15	97	97.905	-0.905
16	99	95.416	3.584
17	74	72.239	1.761
18	117	110.990	6.010
19	92	85.596	6.404
20	95	92.832	2.168
21	104	104.071	-0.071
22	100	107.742	-7.742
23	95	96.926	-1.926
24	81	90.737	-9.737
25	109	113.751	-4.751

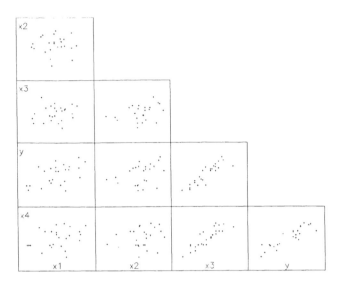

Figure 7.3: Bivariate scatter plot matrix for the validation data set

In the output, yhat $= \hat{y}_i$, and resid $= y_i - \hat{y}_i$. Consequently, for these validation data, we have

$$\sum_{i=26}^{50} (y_i - \hat{y}_i)^2 = 392.7473, \quad \sum_{i=26}^{50} (y_i - \bar{y})^2 = 7541.36.$$

The percentage of variation accounted for by our chosen model in the validation data is therefore computed as

$$R^2_{\text{prediction}} = 1 - \frac{\displaystyle\sum_{i=26}^{50} (y_i - \hat{y}_i)^2}{\displaystyle\sum_{i=26}^{50} (y_i - \bar{y})^2} = 1 - \frac{392.7473}{7541.36} = 0.9479.$$

This prediction R^2 compares favorably with the R^2 of 0.9615 obtained for the choice model based on our estimation data.

Since we also indicated that the PRESS statistic can be used for data splitting from the original data, therefore we have PRESS = 471.4520 and the total sum of squares for the original data is 9054.0. Hence, $R^2_{\text{prediction}}$ from the PRESS statistic is

$$R^2_{\text{prediction}} = 1 - \frac{\text{PRESS}}{\text{SST}} = 1 - \frac{471.4520}{9054} = 0.9479.$$

Even though, we have used the 'estimate' data, the $R^2_{\text{prediction}}$ generated from the use of the PRESS statistic is amazingly very good, when compared with that obtained through the validation data. Again, the prediction R^2 here, is very close to the 0.9615 obtained from the regression model. Hence, the estimated regression equation obtained will be very suitable for predictive analysis. Clearly, the model predicts very well the dependent variable from the validation data. Further, the MSPR for the validation data is 392.7473/25 = 15.7099, a value that is very close to the MSE of 16.5808, indicating that the chosen model has very strong predictive capability.

Alternative Model Example 7.3

It has been suggested that the two-variable subset model for the estimation data based on adjusted R^2 criterion has explanatory variables x_1 and x_3 as predictors. The selection strategy employing this criterion for the estimation data has the displayed output below with x_1, x_3 having adjusted $R^2 = 0.9269$.

```
                    The REG Procedure
                       Model: MODEL1
                  Dependent Variable: y

                 R-Square Selection Method

        Number of Observations Read        25
        Number of Observations Used        25
```

```
 Number in            Adjusted
   Model    R-Square  R-Square   Variables in Model
      1      0.8047    0.7962    x3
      1      0.7558    0.7452    x4
      1      0.2646    0.2326    x1
      1      0.2470    0.2143    x2
-------------------------------------------------------
      2      0.9330    0.9269    x1 x3
      2      0.8773    0.8661    x3 x4
      2      0.8153    0.7985    x1 x4
      2      0.8061    0.7884    x2 x3
      2      0.7833    0.7636    x2 x4
      2      0.4642    0.4155    x1 x2
-------------------------------------------------------
      3      0.9615    0.9560    x1 x3 x4
      3      0.9341    0.9247    x1 x2 x3
      3      0.8790    0.8617    x2 x3 x4
      3      0.8454    0.8233    x1 x2 x4
-------------------------------------------------------
      4      0.9629    0.9555    x1 x2 x3 x4
```

The proposed model therefore is

$$y_i = \beta_0 + \beta_1 x_{i1} + \beta_2 x_{i3} + \varepsilon_i. \tag{7.10}$$

This model when implemented has the following results:

```
                  Analysis of Variance

                           Sum of        Mean
 Source            DF      Squares      Square   F Value   Pr > F

 Model              2   8447.34255  4223.67128    153.17   <.0001
 Error             22    606.65745    27.57534
 Corrected Total   24   9054.00000


         Root MSE            5.25122    R-Square   0.9330
         Dependent Mean     92.20000    Adj R-Sq   0.9269
         Coeff Var           5.69547
```

```
                    Parameter Estimates

                    Parameter   Standard
 Variable    DF      Estimate      Error   t Value   Pr > |t|

 Intercept    1    -127.59569   12.68526    -10.06    <.0001
 x1           1       0.34846    0.05369      6.49    <.0001
 x3           1       1.82321    0.12307     14.81    <.0001
```

Diagnostics of this model indicate that the error terms can be assumed to follow a normal distribution. The plot of the standardized residuals against predicted values are presented in Figure 7.4. Although the residuals are randomly distributed, there is however a standardized residual that is greater than 2.0 indicating a suspect outlier.

Examinations of the leverages, Cook's distances and DFFITS indicate that observations 7 and 18 are possible outliers with respect to their x values while PRESS statistic for observation 16 is 13.222 indicating possible outlier with its y value. It would therefore be better to fit a robust regression to these data for comparison. The robust regression output is presented below:

```
proc robustreg data=valid;
model y=x1 x3 / diagnostics leverage;
output out=out1 r=resid p=pred sresidual=stdr weight=wt;
run;
proc print data=out1;
var y pred resid stdr wt;
run;
                    Parameter Estimates
```

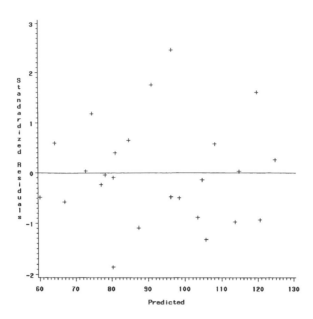

Figure 7.4: Plot of standardized residuals against predicted values

Parameter	DF	Estimate	Standard Error	95% Confidence Limits		Chi-Square	Pr > ChiSq
Intercept	1	-125.164	12.6754	-150.007	-100.321	97.51	<.0001
x1	1	0.3250	0.0537	0.2198	0.4301	36.69	<.0001
x3	1	1.8186	0.1230	1.5776	2.0596	218.68	<.0001
Scale	1	4.1242					

Diagnostics

Obs	Mahalanobis Distance	Robust MCD Distance	Leverage	Standardized Robust Residual	Outlier
7	2.6680	3.7120	*	-0.3202	
13	2.0095	3.1201	*	-0.2224	
16	0.2911	0.6608		3.2117	*
18	2.3160	3.1862	*	2.1406	

Diagnostics Summary

Observation Type	Proportion	Cutoff
Outlier	0.0400	3.0000
Leverage	0.1200	2.7162

Obs	y	pred	resid	stdr	wt
1	88	84.643	3.3574	0.81407	0.94053
2	80	75.024	4.9756	1.20644	0.87177
3	96	97.898	-1.8979	-0.46019	0.98080
4	76	76.787	-0.7873	-0.19089	0.99668
5	80	80.099	-0.0994	-0.02411	0.99995
6	73	72.950	0.0502	0.01217	0.99999
7	58	59.320	-1.3204	-0.32016	0.99068
8	116	119.915	-3.9146	-0.94917	0.91959
9	104	104.004	-0.0036	-0.00088	1.00000
10	99	104.785	-5.7849	-1.40267	0.82876
11	64	66.782	-2.7817	-0.67449	0.95898
12	126	123.558	2.4416	0.59202	0.96832
13	94	94.917	-0.9174	-0.22244	0.99550
14	71	80.355	-9.3554	-2.26842	0.58609
15	111	107.509	3.4905	0.84634	0.93580
16	109	95.754	13.2456	3.21167	0.28096
17	100	90.492	9.5077	2.30534	0.57437
18	127	118.172	8.8284	2.14062	0.62605
19	99	103.091	-4.0910	-0.99195	0.91235
20	82	86.599	-4.5991	-1.11515	0.88990

```
21      67     64.376    2.6245    0.63635   0.96344
22     109    113.353   -4.3526   -1.05536   0.90109
23      78     77.762    0.2378    0.05766   0.99970
24     115    114.521    0.4788    0.11610   0.99877
25      83     80.812    2.1882    0.53058   0.97451
```

The analysis reveals as before that observations 7, 18 and now 13 have high leverages, while observation 16 is an outlier. We now employ these two estimated chosen models for the validation analysis using the validation data to predict corresponding dependent values. The accompanying results give the predicted and residuals for both the ordinary least squares regression and the robust regression methods for the two-variable models. The label yhat1 and resid1 relate to the OLS, while yhat2 and resid2 similarly relate to the robust regression results.

```
data new;
set valid2;
yhat1=-127.59569+0.34846*x1+1.82321*x3;
yhat2=-125.164+0.3250*x1+1.8186*x3;
resid1=y-yhat1;
resid2=y-yhat2;
u=resid1*resid1;
u1=resid2*resid2;
run;
proc print data=new;
var y yhat1 resid1 yhat2 resid2;
format yhat1 resid1 yhat2 resid2 7.2;
run;
```

Obs	y	yhat1	resid1	yhat2	resid2
1	58	55.50	2.50	56.00	2.00
2	92	91.64	0.36	91.60	0.40
3	71	70.72	0.28	70.55	0.45
4	77	78.71	-1.71	78.48	-1.48
5	92	92.09	-0.09	91.66	0.34
6	66	63.92	2.08	64.44	1.56
7	61	61.07	-0.07	61.20	-0.20
8	57	62.41	-5.41	61.85	-4.85
9	66	74.43	-8.43	74.83	-8.83
10	75	79.14	-4.14	79.00	-4.00
11	98	95.68	2.32	94.66	3.34
12	100	100.62	-0.62	99.85	0.15
13	67	69.44	-2.44	69.00	-2.00
14	111	105.63	5.37	105.24	5.76
15	97	98.07	-1.07	97.83	-0.83
16	99	96.52	2.48	96.15	2.85
17	74	76.65	-2.65	76.08	-2.08
18	117	113.91	3.09	112.84	4.16
19	92	81.93	10.07	81.60	10.40
20	95	95.23	-0.23	94.59	0.41
21	104	100.00	4.00	99.40	4.60
22	100	109.25	-9.25	109.33	-9.32
23	95	98.44	-3.44	97.71	-2.71
24	81	88.74	-7.74	87.71	-6.71
25	109	113.06	-4.06	111.81	-2.81

For the validation data therefore, we have the following tabulated results:

	OLS Regression	Robust Regression
PRESS	760.9744	-
$\sum_{i=26}^{50} (y_i - \hat{y}_i)^2$	486.7914	482.3081
SST	7541.36	7541.36
$R^2_{prediction}$	0.9355	0.9360

The $R^2_{prediction}$ based on the PRESS statistic for this model is computed as 0.9160 which compares very well with the R^2 of 0.9330 obtained from the ordinary least squares model. Similarly, MSPR is computed as 19.4716 and 19.2923 respectively for both the OLS and robust regression models. Clearly, our chosen model is not as good a fit as the original model utilizing three explanatory variables. Nonetheless, the results suggest here that the two-variable model would also be a good predictive model based on our validation analysis.

7.9 Exercises

7.1 The data in the file **baseball.txt** is taken from baseball players' salaries data submitted by Watnik, M.R. to the Journal of Statistics Education Data Archive (see also Watnik, 1998). For this problem, use the first 50 cases of the whole data set and the following variables: The 1992 salary in thousands of dollars (y), the batting average (x_1), the number of runs (x_2), the number of hits (x_3), and the number of home runs (x_4). A full model or a subset regression model using the predictor variables x_1-x_4 is envisaged.

(a) Carry out a variable selection analysis by the method of all possible regression models using each of the following criteria: AIC, R^2, adjusted R^2, and the C_p. Find a short list of three best subset models in each case.

(b) State the best subset model according to the C_p criterion.

(c) State any other models that are almost as good as the best model.

(d) Conduct a stepwise regression analysis with **sle** $= 0.50$ and **sls** $= 0.10$. List the variables that are in the model at each step.

(e) Summarize the results in (a)-(d). What further analysis do you think might be necessary after your choice of best model?

7.2 The data in the file **body.txt** is taken from body dimensions data submitted by Heinz, G., Peterson, L.J., Johnson, R.W. and Kerk, C.J. to the Journal of Statistics Education Data Archive (see also Heinz, et al. 2003). For this problem, use the first 100 cases of the whole data set and the following variables: The weight in pounds (y), the biacromial diameter (x_1), the biiliac diameter (x_2), the bitrochanteric diameter (x_3), the chest diameter (x_4), the elbow diameter (x_5), the wrist diameter (x_6), the knee diameter (x_7), and the ankle diameter (x_8). An investigator is interested in understanding the relationship, if any, between the weight and the eight predictor variables.

(a) Find the predictor that results in the best one-variable subset model.

(b) Find the two predictor variables that result in the best two-variable subset model.

(c) Find the three predictor variables that result in the best three-variable subset model.

(d) Find a short list of the three best models using the R^2 criterion.

(e) Find a short list of the three best models using the C_p criterion.

(f) Find a short list of three models using C_p and $C_p - p$ together.

(g) Conduct a stepwise regression analysis with **sle** $= 0.15$ and **sls** $= 0.15$. List the variables that are in the model at each step.

(h) Summarize the results in (a)-(g). What further analysis do you think might be necessary after your choice of best model?

7.3 The data in the file **sat.txt** is taken from the SAT data submitted by Guber, D.L. to the Journal of Statistics Education Data Archive (see also Guber, D.L., 1999). For this problem, use the following variables: The current expenditure per pupil (x_1), the average pupil/teacher ratio (x_2), the estimated annual salary of teachers (x_3), the percentage of students taking SAT (x_4), and the average math SAT score (y). A model of the form

$$y_i = \beta_0 + \beta_1 x_1 + \beta_2 x_2 + \beta_3 x_3 + \beta_4 x_4 + \varepsilon_i \tag{7.11}$$

is suggested, and a researcher decides to estimate the response model in (7.11).

(a) Test whether there is a regression relation, use $\alpha = 0.01$. State the null and alternative hypotheses, the decision rule, and your conclusion. What does your test imply about β_1, β_2, β_3 and β_4?

(b) Obtain the analysis of variance table that decomposes the regression sum of squares into extra sums of squares

- $SS(x_2 \mid x_1)$,

- $SS(x_3 \mid x_1, x_2)$,
- $SS(x_4 \mid x_1, x_2, x_3)$, and
- $SS(x_1 \mid x_2, x_3, x_4)$.

(c) Test whether x_4 can be dropped from the regression model given that x_1, x_2 and x_3 are retained. Use the F-test statistic at $\alpha = .025$. What is the p-value of the test?

(d) Test whether both x_3 and x_4 can be dropped from the regression model given that x_1 and x_2 are retained (Use $\alpha = .05$). State the null and alternative hypotheses, decision rule, and conclusion. What is the p-value for this test?

(e) Obtain the following sample partial multiple correlation coefficients. What orders are these?

$$r_{yx_2 \mid x_1} \quad \text{and} \quad r_{yx_1 \mid x_2, x_3, x_4}.$$

(f) Test whether $\beta_1 = \beta_2$; use $\alpha = .025$. State the null and the alternative hypotheses, full and reduced models, decision rule, and conclusion.

(g) Determine the most parsimonious model from the above results.

7.4 The data in the file **body.txt** is taken from body dimensions data submitted by Heinz, G., Peterson, L.J., Johnson, R.W. and Kerk, C.J. to the Journal of Statistics Education Data Archive (see also Heinz, et al. 2003). For this problem, use the first 100 cases of the whole data set and the following variables: The weight in pounds (y), the shoulder girth (x_1), the chest girth (x_2), the waist girth (x_3), the navel girth (x_4), the hip girth (x_5), the thigh girth (x_6), and the bicep girth (x_7). An investigator is interested in studying how the weight y is related to the seven predictor variables. A subset regression analysis is envisaged.

(a) Find the predictor that results in the best one-variable subset model.

(b) Find the two predictor variables that result in the best two-variable subset model.

(c) Find the three predictor variables that result in the best three-variable subset model.

(d) Find a short list of the three best models using the R^2 criterion.

(e) Find a short list of the three best models using the C_p criterion.

(f) Carry out a stepwise regression analysis with **sle** $= 0.15$ and **sls** $= 0.15$. List the variables that are in the model at each step.

(g) Summarize the results in (a)-(f). What further analysis do you think might be necessary after your choice of best model?

7.5 Refer to the SAT data for Problem 7.3. Use the average verbal SAT score as the response variable y. A subset regression analysis is envisaged.

(a) Carry out a variable selection analysis by the method of all possible regression models using each of the following criteria: AIC, R^2, and the C_p. Find a short list of two best subset models in each case.

(b) State the best subset model according to the C_p criterion.

(c) State any other models that are almost as good as the best model.

(d) Conduct a stepwise regression analysis with **sle** $= 0.50$ and **sls** $= 0.10$. List the variables that are in the model at each step.

7.6 Refer to the SAT data for Problem 6.6 in Chapter 6. Use the average total SAT score as the response variable y and the four predictor variables in Problem 6.6.

(a) Find the predictor that results in the best one-variable subset model.

(b) Find the two predictor variables that result in the best two-variable subset model.

(c) Find the three predictor variables that result in the best three-variable subset model.

(d) Find a short list of the three best models using the R^2 criterion.

(e) Find a short list of the three best models using the C_p criterion.

(f) Find a short list of three models using C_p and $C_p - p$ together.

(g) Conduct a stepwise regression analysis with **sle** $= 0.15$ and **sls** $= 0.15$. List the variables that are in the model at each step.

(h) Summarize the results in (a)-(g). What further analysis do you think might be necessary after your choice of best model?

7.7 The data in the file **baseball.txt** is taken from baseball players' salaries data submitted by Watnik, M.R. to the Journal of Statistics Education Data Archive (see also Watnik, 1998). For this problem, use the first 100 cases of the whole data set and the following variables: The 1992 salary in thousands of dollars (y), the batting average (x_1), the on-base percentage (x_2), the number of runs (x_3), the number of hits (x_4), the number of doubles (x_5), the number of triples (x_6), and the number of home runs (x_7). An investigator is interested in understanding the relationship, if any, between the 1992 salary and the seven predictor variables.

A regression of the form

$$y_i = \beta_0 + \beta_1 x_1 + \beta_2 x_2 + \beta_3 x_3 + \beta_4 x_4 + \beta_5 x_5 + \beta_6 x_6 + \beta_7 x_7 + \varepsilon_i$$

is suggested. A subset regression analysis is also envisaged.

(a) Find the predictor variable that results in the best prediction equation among all subset models containing only one predictor.

(b) Find the two predictor variables that result in the best prediction equation among all subset models containing exactly two predictors.

(c) Find the three predictor variables that result in the best prediction equation among all subset models containing exactly three predictors.

(d) Find a short list of the three best models using the criterion R^2.

(e) Find a short list of the three best models using the criterion R^2_{adj}.

(f) Find a short list of the three best models using the C_p criterion.

(g) Find a short list of the three best models using the C_p and $C_p - p$ together.

(h) Run a stepwise regression model with **sle** $= 0.50$ and **sls** $= 0.10$. List the final model.

(i) Write a short report summarizing the results of parts (a) through (h).

7.8 Refer to the baseball data for Problem 6.1 in Chapter 6.

(a) Show that in multiple linear regression analysis, a ranking of competing models using adjusted R^2 is equivalent to a ranking of competing models with the use of the error mean square, MSE.

(b) Compute s^2, C_p, PRESS, and the sum of the absolute PRESS residuals for the model involving all the three predictor variables with the response variable y.

(c) For all possible regression models for the data, compute R^2, PRESS, $\sum_{i=1}^{n} | y_i - \hat{y}_{i-1} |$, s^2, and C_p. On the basis of this information, reduce your pool of models from all x's to a small subset for further consideration.

7.9 The data below represents the sales price $(y, \$'000)$, square footage x_1, number of rooms x_2, number of bedrooms x_3, age x_4, and number of bathrooms x_5 for each of 23 single-family residences sold during a past year in a Michigan city.

Assume the regression model $y_i = \beta_0 + \beta_1 x_1 + \beta_2 x_2 + \beta_3 x_3 + \beta_4 x_4 + \beta_5 x_5 + \varepsilon_i$. An analyst claims that not all the predictor variables are important in predicting y. Based on this claim, a stepwise selection procedure was applied to the data.

(a) Obtain the best regression model based on this selection method.

y	x_1	x_2	x_3	x_4	x_5	y	x_1	x_2	x_3	x_4	x_5
93.5	1008	5	2	35	1.0	89.5	1008	6	3	35	2.0
89.0	1290	6	3	36	1.0	145.0	1950	8	3	52	1.5
90.5	860	8	2	36	1.0	192.5	2086	7	3	12	2.0
89.9	912	5	3	41	1.0	125.0	2011	9	4	76	1.5
92.0	1204	6	3	40	1.0	100.0	1465	6	3	102	1.0
95.0	1204	5	3	10	1.5	98.5	1232	5	2	69	1.5
120.5	1764	8	4	64	1.5	141.0	1736	7	3	67	1.0
126.0	1600	7	3	19	2.0	119.4	1296	6	3	11	1.5
109.0	1255	5	3	16	2.0	165.0	1996	7	3	9	2.5
189.0	3600	10	5	17	2.5	127.9	1874	5	2	14	2.0
86.0	864	5	3	37	1.0	120.0	1580	5	3	11	1.0
78.0	720	4	2	41	1.0						

(b) Comment on the claim by the analyst and your results in (a).

7.10 A fisheries commission wants to estimate the number of bass caught in a given lake during a season in order to restock the lake with appropriate number of young fish. The commission samples a number of lakes and records y, the seasonal catch (thousands of bass per square mile of lake area); x_1, the number of lake shore residences per square mile of lake area; x_2, the size of lake in square miles; $x_3 = 1$, if the lake has public access, 0 if not; and x_4, a structure index. Assume the regression model $y_i = \beta_0 + \beta_1 x_1 + \beta_2 x_2 + \beta_3 x_3 + \beta_4 x_4 + \varepsilon_i$.

Obs	y	x_1	x_2	x_3	x_4	Obs	y	x_1	x_2	x_3	x_4
1	3.6	92.2	0.21	0	81	11	2.4	64.6	0.91	1	40
2	0.8	86.7	0.30	0	26	12	1.9	50.0	1.10	1	22
3	2.5	80.2	0.31	0	52	13	2.0	50.0	1.24	1	50
4	2.9	87.2	0.40	0	64	14	1.9	51.2	1.47	1	37
5	1.4	64.9	0.44	0	40	15	3.1	40.1	2.21	1	61
6	0.9	90.1	0.56	0	22	16	2.6	45.0	2.46	1	39
7	3.2	60.7	0.78	0	80	17	3.4	50.0	2.80	1	53
8	2.7	50.9	1.21	0	60	18	3.6	70.0	0.78	1	61
9	2.2	86.1	0.34	1	30	19	2.9	75.0	0.66	1	50
10	5.9	90.0	0.40	1	90	20	3.3	80.4	0.52	1	74

(a) Which independent variable has the least correlation with y?

(b) Is multicollinearity a problem in this regression study? Comment, referring to appropriate results in your computer output.

(c) State the estimated regression function.

(d) Obtain a point estimate of the change in mean response when x_3 increases by two units while other variables are held constant.

(e) Obtain a 95% confidence interval in place of the point estimate in (d). Interpret your interval.

(f) Test whether or not $\beta_4 = 0$ controlling the α risk at 0.025 when $\beta_4 = 0$. State the null and the alternative hypotheses, the decision rule, the value of the test statistic, and your conclusion. What can you say concerning the vector whose elements are denoted by x_4?

(g) The commission claims that x_1 and x_2 are important variables in predicting y because they both reflect how intensive the lake has been fished. There is some question as to whether x_3 and x_4 are useful as additional predictors variables. Based on this claim, a forward selection procedure was applied to the data.

(i) Obtain the best regression model based on this selection method.

(ii) Comment on the claim by the analyst and your results in (i).

Chapter 8

Use of Dummy Variables in Regression Analysis

8.1 Introduction

Often times, a regression model might include **qualitative** or **categorical** variable(s). Unlike quantitative variables, these are variables that cannot be measured on a continuum or numerical scale. To incorporate this type of variables into our model, we usually have to create dummy or indicator variables which essentially implies coding the qualitative variables into **levels** which are numeric (usually $\ldots, \pm 2, \pm 1, 0$). For a categorical variable with k categories, we would need to create $k - 1$ dummy variables, say, $x_1, x_2, \ldots, x_{k-1}$. Two types of coding schemes are often employed. The first is **the cell reference** coding scheme and the second is the **effect coding scheme**. We illustrate both with the following example:

Example 8.1

The data in Table 8.1 is from Daniel (1999, p. 522) reproduced by permission of John Wiley and Sons, Inc. The data relate to factors thought to be associated with birth weight in grams (y) and the data consist of a random sample of 32 birth records. The factors of interest are length of gestation in weeks (x_1), and smoking status of mothers (S for smoker, and N for non-smoker).

Case	Birth weight (grams) y	Gestation (weeks) x_1	Smoking Status of mother x_2	Case	Birth weight (grams) y	Gestation (weeks) x_1	Smoking Status of mother x_2
1	2940	38	S	17	3523	41	N
2	3130	38	N	18	3446	42	S
3	2420	36	S	19	2920	38	N
4	2450	34	N	20	2957	39	S
5	2760	39	S	21	3530	42	N
6	2440	35	S	22	2580	38	S
7	3226	40	N	23	3040	37	N
8	3301	42	S	24	3500	42	S
9	2729	37	N	25	3200	41	S
10	3410	40	N	26	3322	39	N
11	2715	36	S	27	3459	40	N
12	3095	39	N	28	3346	42	S
13	3130	39	S	29	2619	35	N
14	3244	39	N	30	3175	41	S
15	2520	35	N	31	2740	38	S
16	2928	39	S	32	2841	36	N

Table 8.1: Data on a random sample of 32 birth records

Since status is categorical with two levels (N and S), we would therefore need to create $2 - 1 = 1$ dummy variable, say, x_2.

8.2   Cell Reference Scheme

For the cell reference coding scheme, one of the categories of the categorical variable is adopted as the reference cell. This is usually the last category of the variable (it could also be the first category of the variable). We adopt here the last category as the reference cell. In this case, the dummy variables generated would be

$$x_2 = \begin{cases} 1, & \text{if status is N} \\ 0, & \text{elsewhere,} \end{cases} \tag{8.1}$$

which results into the following:

status	x_2
N	1
S	0

In this coding scheme, we notice that status S is the reference cell for the categorical variable 'status'. If we assume that the error variances for the two status categories are the same and that a regression model of the form

$$y = \beta_0 + \beta_1 x_1 + \beta_2 x_2 + \beta_3 x_3 + \varepsilon, \tag{8.2}$$

would be appropriate, where x_1 relates to the effect of gestation, and x_2 relates to the effect of the status, and $x_3 = x_1 \times x_2$ relates to the interaction effects of gestation period and smoking status.
For smoking mothers (S), $x_2 = 0$ in equation (8.2) and we have therefore,

$$y = \beta_0 + \beta_1 x_1 + \varepsilon. \tag{8.3}$$

For non-smoking mothers (N), equation (8.2) has $x_2 = 1$ and we have therefore,

$$y = \beta_0 + \beta_1 x_1 + \beta_2(1) + \beta_3 x_1(1) + \varepsilon$$
$$y = (\beta_0 + \beta_2) + (\beta_1 + \beta_3)x_1 + \varepsilon. \tag{8.4}$$

8.2.1   Test of Parallelism

A test of parallelism of the two regression lines in (8.3) and (8.4) will be accomplished by setting the slopes of the two models equal. That is, for this to happen, we must have

$$\beta_1 \equiv \beta_1 + \beta_3.$$

The above can only be true if $\beta_3 = 0$. Hence, a test of parallelism will be accomplished by testing the following hypotheses from the general model in (8.2).

$$H_0 : \beta_3 = 0$$
$$H_a : \beta_3 \neq 0. \tag{8.5}$$

The above test is equivalent to saying that the interaction term $(x_1 x_2)$ in the model in (8.2) is not significant. Thus a test for parallelism is equivalent to a test of no interaction in the model. Under the null hypothesis, the model in (8.2) reduces, for smoking and non-smoking mothers respectively, to

$$y = \beta_0 + \beta_1 x_1 + \varepsilon \tag{8.6a}$$
$$y = (\beta_0 + \beta_2) + \beta_1 x_1 + \varepsilon. \tag{8.6b}$$

8.2.2 Test for a Single Straight Line

Can we completely ignore the effect of smoking status of mothers and believe that a single straight line will be appropriate for the response profiles of the two status categories? For this to happen, we must have the following:

$$\beta_0 \equiv \beta_0 + \beta_2$$
$$\beta_1 \equiv \beta_1 + \beta_3.$$

That is, we must have

$$\beta_2 = \beta_3 = 0.$$

That is, a test of coincidence as is often called can only be accomplished by testing the hypotheses

$$\begin{aligned} H_0 &: \beta_2 = \beta_3 = 0 \\ H_a &: \text{at least one of these} \neq 0. \end{aligned} \tag{8.7}$$

That is, under the null hypothesis, the model in (8.2) reduces in this case to

$$y = \beta_0 + \beta_1 x_1 + \varepsilon. \tag{8.8}$$

8.2.3 Test for a Common Intercept

Often times, there is a need to test for a common intercept. In these case, we would expect the following to be equivalent from model (8.2)

$$\beta_0 \equiv \beta_0 + \beta_2.$$

This can only happen only if $\beta_2 = 0$. A test of common intercept will be conducted from the following hypotheses relating to the model in (8.2):

$$\begin{aligned} H_0 &: \beta_2 = 0 \\ H_a &: \beta_2 \neq 0. \end{aligned} \tag{8.9}$$

Under the null hypothesis, the model in (8.2) reduces to the following:

$$y - \beta_0 + \beta_1 x_1 + \varepsilon \tag{8.10a}$$
$$y = \beta_0 + (\beta_1 + \beta_3) x_1 + \varepsilon. \tag{8.10b}$$

We illustrate these tests with the example data in Table 8.1. The following are the SAS codes required to accomplish these tests and the corresponding partial output.

```
data one;
input y x1 status$ @@;
x2=(status='N'); x3=x1*x2;
datalines;
.........
;
run;

proc reg data=one;
model y=x1 x2 x3 / clb;
Test1: test x3;      /* Test for parallelism */
Test2: test x2, x3;  /* Test for coincidence */
Test3: test x2;      /* Test for common intercept */
run;
                         The REG Procedure
                          Model: MODEL1
                     Dependent Variable: y

                      Analysis of Variance

                              Sum of       Mean
        Source        DF      Squares      Square    F Value    Pr > F

        Model          3      3351398     1117133     81.37     <.0001
        Error         28       384391       13728
```

```
Corrected Total          31         3735790

                 Root MSE          117.16768    R-Square     0.8971
                 Dependent Mean   3019.87500    Adj R-Sq     0.8861
                 Coeff Var           3.87989

                          Parameter Estimates

                Parameter    Standard
Variable   DF   Estimate      Error    t Value   Pr > |t|     95% Confidence Limits
-----------------------------------------------------------------------------------
Intercept  1  -2474.56410   512.78563   -4.83    <.0001   -3524.95785  -1424.17036
x1         1    139.02875    13.06407   10.64    <.0001     112.26821    165.78929
x2         1    -71.57383   716.94989   -0.10    0.9212   -1540.17910   1397.03145
x3         1      8.17815    18.51516    0.44    0.6621     -29.74844     46.10474

              Test Test1 Results for Dependent Variable y

                                      Mean
          Source            DF       Square     F Value    Pr > F
          ------------------------------------------------------------
          Numerator          1     2678.36964     0.20     0.6621
          Denominator       28        13728

              Test Test2 Results for Dependent Variable y

                                      Mean
          Source            DF       Square     F Value    Pr > F
          ------------------------------------------------------------
          Numerator          2      227780       16.59    <.0001
          Denominator       28        13728

              Test Test3 Results for Dependent Variable y

                                      Mean
          Source            DF       Square     F Value    Pr > F
          ------------------------------------------------------------
          Numerator          1      136.81906     0.01     0.9212
          Denominator       28        13728
```

The following conclusions are obtained from the above analysis:

1. The hypotheses of parallelism in (8.5), which is the result in Test1 indicates that we would fail to reject H_0, since p-value $= 0.6611$. Hence, the lines are parallel and interaction is therefore not present.

2. The hypotheses of coincidence from the Test2 partial F-test result indicates that this hypothesis is not tenable. Thus the effect of smoking status cannot be ignored in our model.

3. The hypothesis of common intercept does not make sense here in view of our results from the parallelism test.

Since there is no interaction in the model, we would now fit a revised model

$$y = \beta_0 + \beta_1 x_1 + \beta_2 x_2 + \varepsilon. \tag{8.11}$$

```
proc reg data=one;
model y=x1 x2 / clb;
Test2b: test x2; /* Test for coincidence */
run;
                       Dependent Variable: y

                       Analysis of Variance

                             Sum of       Mean
Source            DF         Squares      Square    F Value   Pr > F
--------------------------------------------------------------------
Model              2         3348720     1674360    125.45    <.0001
Error             29          387070       13347
Corrected Total   31         3735790

                 Root MSE          115.53024    R-Square     0.8964
                 Dependent Mean   3019.87500    Adj R-Sq     0.8892
                 Coeff Var           3.82566

                       Parameter Estimates

                Parameter    Standard
```

```
Variable   DF    Estimate     Error  t Value  Pr > |t|    95% Confidence Limits
------------------------------------------------------------------------------
Intercept   1  -2634.11695  358.87235   -7.34   <.0001  -3368.09332  -1900.14058
x1          1    143.10027    9.12812   15.68   <.0001    124.43117    161.76938
x2          1    244.54404   41.98176    5.83   <.0001    158.68171    330.40637

            Test Test2b Results for Dependent Variable y

                             Mean
          Source        DF   Square   F Value   Pr > F
          ------------------------------------------------
          Numerator      1   452881    33.93    <.0001
          Denominator   29    13347
```

Thus the estimated regression equation for the data is

$$\hat{y} = -2634.1170 + 143.1003x_1 + 244.5440x_2. \tag{8.12}$$

8.3 Effect Coding Scheme

In this case, the dummy variables generated would be

$$x_2 = \begin{cases} 1, & \text{if status is N} \\ -1, & \text{if status is S,} \end{cases} \tag{8.13}$$

which results into the following:

Status	x_2
N	1
S	-1

For the model in equation (8.2), we have for smoking status S, $x_2 = -1$ and we have therefore,

$$\begin{aligned} y &= \beta_0 + \beta_1 x_1 + \beta_2(-1) + \beta_3 x_1(-1) + \varepsilon \\ y &= (\beta_0 - \beta_2) + (\beta_1 - \beta_3)\, x_1 + \varepsilon. \end{aligned} \tag{8.14}$$

For smoking status N, equation (8.2) has $x_2 = 1$ and we have therefore,

$$\begin{aligned} y &= \beta_0 + \beta_1 x_1 + \beta_2(1) + \beta_3 x_1(1) + \varepsilon \\ y &= (\beta_0 + \beta_2) + (\beta_1 + \beta_3)x_1 + \varepsilon. \end{aligned} \tag{8.15}$$

Tests of Parallelism, Coincidence and Common Intercept

A test of parallelism of the two regression lines in (8.14) and (8.15) will be accomplished by setting the slopes of the two models equal. That is, for this to happen, we must have

$$\beta_1 - \beta_3 \equiv \beta_1 + \beta_3.$$

The above can only be true if $\beta_3 = 0$. Hence, a test of parallelism will be accomplished by testing the following hypotheses from the general model in (8.2).

$$\begin{aligned} H_0 &: \beta_3 = 0 \\ H_a &: \beta_3 \neq 0. \end{aligned} \tag{8.16}$$

Similarly, the tests of coincidence can be conducted with the hypotheses in (8.17).

$$\begin{aligned} H_0 &: \beta_2 = \beta_3 = 0 \\ H_a &: \text{at least one of these} \neq 0, \end{aligned} \tag{8.17}$$

while the test of common intercept can similarly be conducted with the set of hypotheses in (8.18).

$$H_0 : \beta_2 = 0$$
$$H_a : \beta_2 \neq 0. \tag{8.18}$$

Again, the SAS code for implementing the effect coding scheme is presented below together with the required tests and partial outputs.

```
data new;
set one;
if status = 'N' then x2=1; else x2=-1;
x3=x1*x2;
proc reg data=new;
model y=x1 x2 x3 / clb;
Test1: test x3;     /* Test for parallelism */
Test2: test x2, x3; /* Test for coincidence */
Test3: test x2;     /* Test for common intercept */
run;
```

                            The REG Procedure
                             Model: MODEL1
                         Dependent Variable: y

                          Analysis of Variance

                                    Sum of   Mean
Source   DF   Squares   Square   F Value   Pr > F

Model   3   3351398   1117133   81.37   <.0001
Error   28   384391   13728
Corrected Total   31   3735790

              Root MSE   117.16768   R-Square   0.8971
              Dependent Mean   3019.87500   Adj R-Sq   0.8861
              Coeff Var   3.87989

                          Parameter Estimates

              Parameter   Standard
Variable   DF   Estimate   Error   t Value   Pr > |t| 95% Confidence Limits

Intercept   1 -2510.35102 358.47494   -7.00   <.0001   -3244.65365 -1776.04838
x1   1   143.11782   9.25758   15.46   <.0001   124.15453   162.08112
x2   1   -35.78691 358.47494   -0.10   0.9212   -770.08955   698.51572
x3   1   4.08907   9.25758   0.44   0.6621   -14.87422   23.05237

                 Test Test1 Results for Dependent Variable y

                                        Mean
        Source   DF   Square   F Value   Pr > F

        Numerator   1   2678.36964   0.20   0.6621
        Denominator   28   13728

                 Test Test2 Results for Dependent Variable y

                                        Mean
        Source   DF   Square   F Value   Pr > F

        Numerator   2   227780   16.59   <.0001
        Denominator   28   13728

                 Test Test3 Results for Dependent Variable y

                                        Mean
        Source   DF   Square   F Value   Pr > F

        Numerator   1   136.81906   0.01   0.9212
        Denominator   28   13728

We observe that the three tests give identical results to those obtained under the 'cell reference coding scheme'. In this case, the hypotheses of parallelism is tenable and we can again drop x_3 from our model. Furthermore, the following table that summarizes the parameter estimates involving the categorical variable, viz, smoking status (x_2), under the full model in (8.2) and under both coding schemes shows that the parameter estimates under the effect coding scheme are one-half those under the cell reference scheme. This is also true for their standard errors.

Parameter	Cell Reference Scheme		Effect Coding Scheme	
	Estimate	Standard Error	Estimate	Standard Error
β	$\hat{\beta}$	$S_{\hat{\beta}_0}$	$\hat{\beta}_0$	$S_{\hat{\beta}_0}$
β_0	-2474.5641	512.7856	-2510.3510	358.4749
β_1	139.0288	13.0641	143.1178	9.2576
β_2	-71.5738	716.9499	-35.7869	358.4749
β_3	8.1782	18.5152	4.0891	9.2576

The revised model when x_3 is dropped under the effect coding scheme is implemented in SAS with the following results. The corresponding estimated regression profiles for both smoking status categories are presented in Figure 8.1.

```
proc reg data=new;
model y=x1 x2 / clb;
Test2b: test x2; /* Test for coincidence */
output out=aa p=pred r=resid;
run;
proc sort data=aa; by status;
run;
goptions vsize=5in hsize=5in;
symbol1 c=black i=join v=plus; symbol2 c=black i=join v=none;
axis1 label=(angle=-90 rotate=90 'PREDICTED'); axis2 label=('GESTATION IN WEEKS (X1)');
proc gplot data=aa;
plot pred*x1=status / vaxis=axis1 haxis=axis2 noframe;
run;
```

```
                        The REG Procedure
                        Model: MODEL1
                    Dependent Variable: y

                     Analysis of Variance

                                Sum of         Mean
Source                DF        Squares       Square    F Value   Pr > F

Model                  2        3348720       1674360    125.45   <.0001
Error                 29         387070         13347
Corrected Total       31        3735790

            Root MSE              115.53024    R-Square     0.8964
            Dependent Mean       3019.87500    Adj R Sq     0.8892
            Coeff Var               3.82566

                     Parameter Estimates

                 Parameter      Standard
Variable    DF    Estimate         Error   t Value  Pr > |t|     95% Confidence Limits

Intercept    1  -2511.84493     353.44943    -7.11   <.0001   -3234.73019  -1788.95968
x1           1    143.10027       9.12812    15.68   <.0001     124.43117    161.76938
x2           1    122.27202      20.99088     5.83   <.0001      79.34085    165.20319

                        The REG Procedure
                        Model: MODEL1

            Test Test2b Results for Dependent Variable y

                                 Mean
            Source        DF     Square    F Value   Pr > F

            Numerator      1     452881      33.93    <.0001
            Denominator   29      13347
```

Here, however, the estimated regression equation under this coding scheme is

$$\hat{y} = -2511.8449 + 143.1003x_1 + 122.2720x_2. \qquad (8.19)$$

Homogeneity Test in Regression Models

From our discussions in the previous section and this section, homogeneity tests for regression models comprise of testing for:

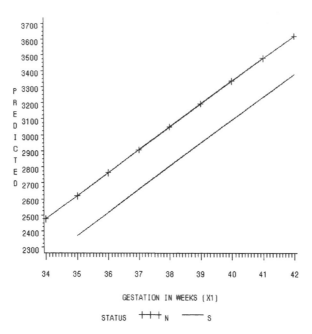

Figure 8.1: Plot of estimated smoking status profiles

- Fitting a model with different slopes and intercepts and testing whether the slopes or intercepts are equal.

- Fitting regression models with common slope and different intercept and testing for equality of intercepts.

- Fitting separate regression lines to the data.

- We can of course also fit a single regression line to the data (A model that ignores the effect of status in this case).

We can employ **proc glm** to implement all the above tests. The following SAS codes will implement the first three of the above situations.

```
proc glm data=one;
class status;
model y=x1|status; /* Fits different slopes and intercepts */
run;
proc glm data=one;
class status;
model y=x1 status; /* Fits common slope and different intercepts */
run;
proc glm data=one;
class status;
model y=status x1(status) / solution; /* Fits separate regression lines */
output out=aa p=pred r=resid;
run;
goptions vsize=5in hsize=5in;
symbol1 c=black v=plus; symbol2 c=black v=star;
axis1 label=(angle=-90 rotate=90 'Residuals'); axis2 label=('Gestation in weeks (x1)');
proc gplot data=aa;
plot resid*x1=status / vaxis=axis1 haxis=axis2 vref=0 noframe;
run;
```

We present in Figure 8.2 the residual plots under the separate regression models.

8.4 Categorical Variable with more than Two Levels: Cell Reference Scheme

We will illustrate the use of cell reference scheme for a categorical variable with more than two levels by using the following example.

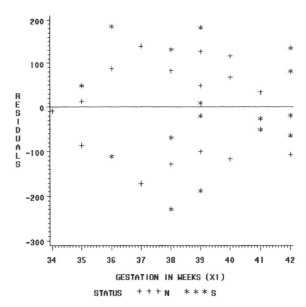

Figure 8.2: Plot of residuals from the separate regression lines

Example 8.2

We consider in this example, the case of a categorical variable having more than two levels. In this example, the categorical variable has three levels and we again show how what we have done in the previous example can be applied to this case. The following data in Table 8.2 relate to whether or not y, the number of days of prison sentence for thefts whose total value is under \$1,000 for first time offenders, is related to x_1, the amount of money stolen. Sample data from thefts falling in this category for a particular year in three different states in the United States, denoted as states A, B, and C, are obtained from police records in each state.

y	x_1	State	y	x_1	State	y	x_1	State
44	367	A	43	538	B	50	732	C
81	855	A	32	290	B	52	556	C
43	284	A	53	759	B	53	960	C
40	305	A	55	734	B	39	134	C
38	215	A	40	499	B	55	826	C
44	308	A	44	541	B	53	738	C
49	433	A	42	474	B	37	403	C
51	455	A	51	940	B	46	511	C
49	454	A	35	314	B	45	699	C
57	429	A	39	351	B	49	778	C
47	345	A	51	703	B	37	530	C
37	167	A	37	459	B	35	140	C
67	689	A				38	429	C
55	499	A				48	554	C
						50	672	C
						30	125	C
						29	124	C

Table 8.2: Number of days as a function of amount of money and state

Since State is categorical with three levels, we would therefore need to create $3 - 1 = 2$ dummy variables, say, x_2 and x_3. For the cell reference coding scheme, one of the categories of the categorical variable is adopted as the reference cell. We adopt here the last category as the reference cell. In this case, the dummy variables generated would be

$$x_2 = \begin{cases} 1, & \text{if State is A} \\ 0, & \text{elsewhere,} \end{cases} \qquad x_3 = \begin{cases} 1, & \text{if State is B} \\ 0, & \text{elsewhere,} \end{cases} \qquad (8.20)$$

which results into the following:

State	x_2	x_3
A	1	0
B	0	1
C	0	0

In this coding scheme, we notice that State C is the reference cell for the categorical variable 'state'. If we assume that the error variances for the three states are the same and that a regression model of the form

$$y = \beta_0 + \beta_1 x_1 + \beta_2 x_2 + \beta_3 x_3 + \beta_4 x_4 + \beta_5 x_5 + \varepsilon, \qquad (8.21)$$

would be appropriate, where x_1 relates to the effect of amount stolen, x_2 and x_3 relate to the effect of the states, and $x_4 = x_1 \times x_2$ and $x_5 = x_1 \times x_3$ relate to the interaction effects (two components) of amount stolen and states. For state C, equation (8.21) has $x_2 = 0$ and $x_3 = 0$ and we have

$$y = \beta_0 + \beta_1 x_1 + \varepsilon. \qquad (8.22)$$

For state A, equation (8.21) has $x_2 = 1$ and $x_3 = 0$ and we have

$$\begin{aligned} y &= \beta_0 + \beta_1 x_1 + \beta_2(1) + \beta_3(0) + \beta_4 x_1(1) + \beta_5 x_1(0) + \varepsilon \\ y &= (\beta_0 + \beta_2) + (\beta_1 + \beta_4)x_1 + \varepsilon. \end{aligned} \qquad (8.23)$$

Similarly for state B, equation (8.21) has $x_2 = 0$ and $x_3 = 1$ and we have,

$$\begin{aligned} y &= \beta_0 + \beta_1 x_1 + \beta_2(0) + \beta_3(1) + \beta_4 x_1(0) + \beta_5 x_1(1) + \varepsilon \\ y &= (\beta_0 + \beta_3) + (\beta_1 + \beta_5)x_1 + \varepsilon. \end{aligned} \qquad (8.24)$$

8.4.1   Test of Parallelism

A test of parallelism of the three regression lines in (8.22), (8.23) and (8.24) will be accomplished by setting the slopes of the three models equal. That is, for this to happen, we must have

$$\beta_1 \equiv \beta_1 + \beta_4 \equiv \beta_1 + \beta_5.$$

The above can only be true if $\beta_4 = \beta_5 = 0$. Hence, a test of parallelism will be accomplished by testing the following hypotheses from the general model in (8.21).

$$\begin{aligned} H_0 &: \beta_4 = \beta_5 = 0 \\ H_a &: \text{at least one of these} \neq 0. \end{aligned} \qquad (8.25)$$

The above test is equivalent to saying that the interaction terms in the model in (8.21) are not significant. Thus a test for parallelism is equivalent to a test of no interaction in the model. Under the null hypothesis, the model in (8.21) reduces for states C, A and B respectively to:

$$y = \beta_0 + \beta_1 x_1 + \varepsilon \qquad (8.26a)$$
$$y = (\beta_0 + \beta_2) + \beta_1 x_1 + \varepsilon \qquad (8.26b)$$
$$y = (\beta_0 + \beta_3) + \beta_1 x_1 + \varepsilon. \qquad (8.26c)$$

8.4.2 Test for a Single Straight Line

Can we completely ignore the effect of State and believe that a single straight line will be appropriate for the response profiles of the three states? For this to happen, we must have the following

$$\beta_0 \equiv \beta_0 + \beta_2 \equiv \beta_0 + \beta_3$$
$$\beta_1 \equiv \beta_1 + \beta_4 \equiv \beta_1 + \beta_5.$$

For these to happen, we must have

$$\beta_2 = \beta_3 = \beta_4 = \beta_5 = 0.$$

That is, a test of coincidence as is often called can only be accomplished by testing the hypotheses:

$$H_0 : \beta_2 = \beta_3 = \beta_4 = \beta_5 = 0$$
$$H_a : \text{at least one of these} \neq 0. \tag{8.27}$$

That is, under the null hypothesis, the model in (8.21) reduces in this case to

$$y = \beta_0 + \beta_1 x_1 + \varepsilon. \tag{8.28}$$

8.4.3 Test for a Common Intercept

Often times, there is a need to test for a common intercept. In these case, we would expect the following to be equivalent from model (8.21)

$$\beta_0 \equiv \beta_0 + \beta_2 \equiv \beta_0 + \beta_3,$$

which can only happen only if $\beta_2 = \beta_3 = 0$. A test of common intercept will be conducted from the following hypotheses relating to the model in (8.21):

$$H_0 : \beta_2 = \beta_3 = 0$$
$$H_a : \text{at least one of these} \neq 0. \tag{8.29}$$

Under the null hypothesis, the model in (8.21) reduces to the following

$$y = \beta_0 + \beta_1 x_1 + \varepsilon \tag{8.30a}$$
$$y = \beta_0 + (\beta_1 + \beta_4) x_1 + \varepsilon \tag{8.30b}$$
$$y = \beta_0 + (\beta_1 + \beta_5) x_1 + \varepsilon. \tag{8.30c}$$

The SAS implementation is accomplished by first creating dummy variables x_2 and x_3 and hence, their interaction terms x_4 and x_5. The results are presented as follows:

```
/* Program to analyze the data for Example 8.2  */
data one;
input y x1 state$ @@;
x2=(state='A'); x3=(state='B'); x4=x1*x2; x5=x1*x3;
datalines;
.........
;
run;
proc reg data=one;
model y=x1 x2 x3 x4 x5 / clb;
Test1: test x4, x5;          /* Test for parallelism */
Test2: test x2, x3, x4, x5; /* Test for coincidence */
Test3: test x2, x3;          /* Test for common intercept  */
output out=aa p=pred r=resid;
run;
proc sort data=aa; by state;
run;
goptions vsize=5in hsize=5in;
symbol1 c=black i=join v=plus; symbol2 c=black i=join v=diamond; symbol3 c=black i=join v=none;
axis1 label=(angle=-90 rotate=90 'PREDICTED'); axis2 label=('AMOUNT STOLEN (X1)');
proc gplot data=aa;
plot pred*x1=state / vaxis=axis1 haxis=axis2 noframe;
run;
                        The REG Procedure
```

```
                    Model: MODEL1
                 Dependent Variable: y

                 Analysis of Variance

                          Sum of        Mean
Source              DF    Squares       Square     F Value   Pr > F

Model                5   3555.46742   711.09348     58.85    <.0001
Error               37    447.04421    12.08228
Corrected Total     42   4002.51163


        Root MSE              3.47596   R-Square    0.8883
        Dependent Mean       45.81395  Adj R-Sq    0.8732
        Coeff Var             7.58711

                 Parameter Estimates

            Parameter    Standard
Variable  DF  Estimate     Error   t Value  Pr > |t|    95% Confidence Limits

Intercept  1  29.30934   1.90879    15.35   <.0001    25.44177    33.17691
x1         1   0.02780   0.00327     8.51   <.0001     0.02118     0.03442
x2         1  -5.83727   3.05795    -1.91   0.0641   -12.03326     0.35872
x3         1  -4.81362   3.61935    -1.33   0.1917   -12.14712     2.51988
x4         1   0.03652   0.00623     5.86   <.0001     0.02389     0.04915
x5         1   0.00674   0.00621     1.09   0.2849    -0.00585     0.01933

          Test Test1 Results for Dependent Variable y

                             Mean
Source              DF      Square    F Value   Pr > F

Numerator            2   209.22606    17.32    <.0001
Denominator         37    12.08228

          Test Test2 Results for Dependent Variable y

                             Mean
Source              DF      Square    F Value   Pr > F

Numerator            4   368.80828    30.52    <.0001
Denominator         37    12.08228

          Test Test3 Results for Dependent Variable y

                             Mean
Source              DF      Square    F Value   Pr > F

Numerator            2    25.34783     2.10    0.1371
Denominator         37    12.08228
```

The following conclusions are obtained from the above analysis:

1. The hypotheses of parallelism in (8.25), which is the result in Test1 indicates that we would strongly reject H_0, since p-value $\leq .0001$. Hence, the lines are not parallel and interaction is present.

2. The hypotheses of coincidence from Test2 partial F-test results also indicate that this hypothesis is not tenable. Thus the effect of states cannot be ignored in our model.

3. The hypothesis of common intercept seems plausible here. The p-value is 0.1371, which indicates that we would fail to reject the null hypothesis in (8.29).

From the cell reference coding scheme therefore, the estimated regression equation is

$$\hat{y}_i = 29.3093 + 0.0278x_1 - 5.8373x_2 - 4.8136x_3 + 0.0365x_4 + 0.0067x_5. \tag{8.31}$$

Hence, from (8.22), (8.23) and (8.24), the estimated regression equations for states C, A and B are given respectively by

$$\hat{y}_i = 29.3093 + 0.0278x_1,$$
$$\hat{y}_i = (29.3093 - 5.8373) + (0.0278 + 0.0365)x_1,$$
$$\hat{y}_i = (29.3093 - 4.8136) + (0.0278 + 0.0067)x_1.$$

That is, the estimated equations for states A, B and C are given respectively as

$$\hat{y}_i = 23.4720 + 0.0643x_1 \tag{8.32a}$$
$$\hat{y}_i = 24.4957 + 0.0345x_1 \tag{8.32b}$$
$$\hat{y}_i = 29.3093 + 0.0278x_1. \tag{8.32c}$$

These estimated regression equations for the three states are presented in Figure 8.3.

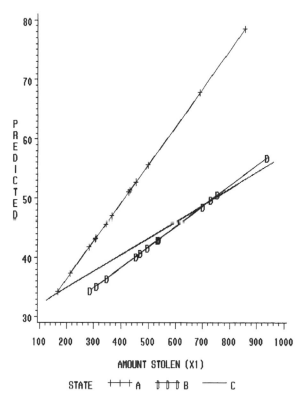

Figure 8.3: Plot of estimated state profiles

8.5 Categorical Variable with more than Two Levels: Effect Coding Scheme

In this case, the dummy variables generated would be

$$x_2 = \begin{cases} 1, & \text{if State A} \\ -1, & \text{if State C} \\ 0, & \text{elsewhere,} \end{cases} \qquad x_3 = \begin{cases} 1, & \text{if State B} \\ -1, & \text{if State C} \\ 0, & \text{elsewhere,} \end{cases} \tag{8.33}$$

which results into the following:

State	x_2	x_3
A	1	0
B	0	1
C	-1	-1

For the model in equation (8.21), for state C, $x_2 = -1$ and $x_3 = -1$ and we have

$$y = \beta_0 + \beta_1 x_1 + \beta_2(-1) + \beta_3(-1) + \beta_4 x_1(-1) + \beta_5 x_1(-1) + \varepsilon$$
$$y = (\beta_0 - \beta_2 - \beta_3) + (\beta_1 - \beta_4 - \beta_5)x_1 + \varepsilon. \tag{8.34}$$

For state A, equation (8.21) has $x_2 = 1$ and $x_3 = 0$ and we have

$$y = \beta_0 + \beta_1 x_1 + \beta_2(1) + \beta_3(0) + \beta_4 x_1(1) + \beta_5 x_1(0) + \varepsilon$$
$$y = (\beta_0 + \beta_2) + (\beta_1 + \beta_4)x_1 + \varepsilon.$$

(8.35)

Similarly for state B, equation (8.21) has $x_2 = 0$ and $x_3 = 1$ and we have

$$y = \beta_0 + \beta_1 x_1 + \beta_2(0) + \beta_3(1) + \beta_4 x_1(0) + \beta_5 x_1(1) + \varepsilon$$
$$y = (\beta_0 + \beta_3) + (\beta_1 + \beta_5)x_1 + \varepsilon.$$

(8.36)

Tests of Parallelism, Coincidence and Common Intercept

A test of parallelism for the three regression lines in (8.34), (8.35) and (8.36) will be accomplished by setting the slopes of the three models equal. That is, for this to happen, we must have

$$\beta_1 - \beta_4 - \beta_5 \equiv \beta_1 + \beta_4 \equiv \beta_1 + \beta_5.$$

The above can only be true if $\beta_4 = \beta_5 = 0$. Hence, a test of parallelism will be accomplished by testing the following hypotheses from the general model in (8.21).

$$H_0 : \beta_4 = \beta_5 = 0$$
$$H_1 : \text{at least one of these} \neq 0.$$

(8.37)

It is not difficult to show that the test of coincidence leads to the null hypothesis $H_0 : \beta_2 = \beta_3 = \beta_4 = \beta_5 = 0$ while the test of common intercept leads to the null hypothesis $H_0 : \beta_2 = \beta_3 = 0$. The following SAS codes will implement the effect coding scheme method with corresponding tests and partial outputs.

```
data new;
set one;
if state='A' then x2=1; else if state='C' then x2=-1; else x2=0;
if state='B' then x3=1; else if state='C' then x3=-1; else x3=0;
x4=x1*x2; x5=x1*x3;
run;
proc reg data=new;
model y=x1 x2 x3 x4 x5 / clb;
Test1: test x4, x5;        /* Test for parallelism */
Test2: test x2, x3, x4, x5; /* Test for coincidence */
Test3: test x2, x3;        /* Test for common intercept  */
run;
                      Model: MODEL1
                  Dependent Variable: y

                    Analysis of Variance

                              Sum of         Mean
    Source            DF      Squares       Square    F Value   Pr > F
    ----------------------------------------------------------------------
    Model              5    3555.46742    711.09348     58.85   <.0001
    Error             37     447.04421     12.08228
    Corrected Total   42    4002.51163

              Root MSE              3.47596    R-Square    0.8883
              Dependent Mean       45.81395    Adj R-Sq    0.8732
              Coeff Var             7.58711

                    Parameter Estimates

               Parameter    Standard
    Variable DF  Estimate      Error   t Value  Pr > |t|   95% Confidence Limits
    ----------------------------------------------------------------------------
    Intercept 1   25.75904    1.44558    17.82    <.0001    22.83002    28.68806
    x1        1    0.04222    0.00272    15.50    <.0001     0.03670     0.04774
    x2        1   -2.28697    1.99805    -1.14    0.2597    -6.33541     1.76147
    x3        1   -1.26333    2.28949    -0.55    0.5844    -5.90228     3.37563
    x4        1    0.02210    0.00410     5.39    <.0001     0.01379     0.03041
    x5        1   -0.00768    0.00409    -1.88    0.0683    -0.01597  0.00060647

           Test Test1 Results for Dependent Variable y
```

Source	DF	Mean Square	F Value	Pr > F
Numerator	2	209.22606	17.32	<.0001
Denominator	37	12.08228		

Test Test2 Results for Dependent Variable y

Source	DF	Mean Square	F Value	Pr > F
Numerator	4	368.80828	30.52	<.0001
Denominator	37	12.08228		

Test Test3 Results for Dependent Variable y

Source	DF	Mean Square	F Value	Pr > F
Numerator	2	25.34783	2.10	0.1371
Denominator	37	12.08228		

The tests of coincidence and common intercept also give p-values of < 0.0001 and 0.1371 respectively. These are exactly the same results obtained with the cell reference coding scheme. Thus the two coding schemes as expected to give the same results.

8.6 Models with Two or More Categorical Independent Variables

In this section, we consider the case with two categorical variables. Let us suppose we have an independent quantitative variable x_1 and two categorical variables A at three levels (A_1, A_2, A_3) and B also at three levels (B_1, B_2, B_3). Then, the response model in this case (with only first-order consideration) will be:

$$y = \beta_0 + \beta_1 x_1 + \underbrace{\beta_2 x_2 + \beta_3 x_3}_{\text{ME of A}} + \underbrace{\beta_4 z_1 + \beta_5 z_2}_{\text{ME of B}}$$

$$+ \underbrace{\beta_6 x_1 x_2 + \beta_7 x_1 x_3 + \beta_8 x_1 z_1 + \beta_9 x_1 z_2}_{\text{Interaction Effects}} + \underbrace{\beta_{10} x_2 z_1 + \beta_{11} x_2 z_2 + \beta_{12} x_3 z_1 + \beta_{13} x_3 z_2}_{\text{AB interaction terms}} \quad (8.38)$$

$$+ \underbrace{\beta_{14} x_1 x_2 z_1 + \beta_{15} x_1 x_2 z_2 + \beta_{16} x_1 x_3 z_1 + \beta_{17} x_1 x_3 z_2}_{\text{Interaction of AB with quantitative term}} + \varepsilon,$$

where if we use the cell reference coding scheme, we have

$$x_2 = \begin{cases} 1, & \text{if level is } A_1 \\ 0, & \text{elsewhere,} \end{cases} \qquad x_3 = \begin{cases} 1, & \text{if level is } A_2 \\ 0, & \text{elsewhere,} \end{cases}$$

$$z_1 = \begin{cases} 1, & \text{if level is } B_1 \\ 0, & \text{elsewhere,} \end{cases} \qquad z_2 = \begin{cases} 1, & \text{if level is } B_2 \\ 0, & \text{elsewhere,} \end{cases}$$

and 'ME of A' for instance stands for main effects of A. We have assumed in the above model that there is interaction between independent variables A and B. This model provides nine response profiles corresponding namely to

Models	Levels of A and B	x_2	x_3	z_1	z_2
1	$A_1 B_1$	1	0	1	0
2	$A_1 B_2$	1	0	0	1
3	$A_1 B_3$	1	0	0	0
4	$A_2 B_1$	0	1	1	0
5	$A_2 B_2$	0	1	0	1
6	$A_2 B_3$	0	1	0	0
7	$A_3 B_1$	0	0	1	0
8	$A_3 B_2$	0	0	0	1
9	$A_3 B_3$	0	0	0	0

Substituting these values in (8.38), we have respectively,

$$y = (\beta_0 + \beta_2 + \beta_4 + \beta_{10}) + (\beta_1 + \beta_6 + \beta_8 + \beta_{14}) x_1 + \varepsilon \qquad \text{(8.39a)}$$

$$y = (\beta_0 + \beta_2 + \beta_5 + \beta_{11}) + (\beta_1 + \beta_6 + \beta_9 + \beta_{15}) x_1 + \varepsilon \qquad \text{(8.39b)}$$

$$y = (\beta_0 + \beta_2) + (\beta_1 + \beta_6) x_1 + \varepsilon \qquad \text{(8.39c)}$$

$$y = (\beta_0 + \beta_3 + \beta_4 + \beta_{12}) + (\beta_1 + \beta_7 + \beta_8 + \beta_{16}) x_1 + \varepsilon \qquad \text{(8.39d)}$$

$$y = (\beta_0 + \beta_3 + \beta_5 + \beta_{13}) + (\beta_1 + \beta_7 + \beta_9 + \beta_{17}) x_1 + \varepsilon \qquad \text{(8.39e)}$$

$$y = (\beta_0 + \beta_3) + (\beta_1 + \beta_7) x_1 + \varepsilon \qquad \text{(8.39f)}$$

$$y = (\beta_0 + \beta_4) + (\beta_1 + \beta_8) x_1 + \varepsilon \qquad \text{(8.39g)}$$

$$y = (\beta_0 + \beta_5) + (\beta_1 + \beta_9) x_1 + \varepsilon \qquad \text{(8.39h)}$$

$$y = \beta_0 + \beta_1 x_1 + \varepsilon. \qquad \text{(8.39i)}$$

The following hypotheses are of interest.

1. The hypothesis that all the nine regression lines are parallel can be tested with the test

$$H_0 : \beta_6 = \beta_7 = \beta_8 = \beta_9 = \beta_{14} = \beta_{15} = \beta_{16} = \beta_{17} = 0.$$

2. The hypothesis that all the nine lines have a common intercept is tested by the test

$$H_0 : \beta_2 = \beta_3 = \beta_4 = \beta_5 = \beta_{10} = \beta_{11} = \beta_{12} = \beta_{13} = 0.$$

3. The hypothesis that all the nine lines are coincident is tested by

$$H_0 : \beta_2 = \beta_3 = \beta_4 = \cdots = \beta_{16} = \beta_{17} = 0.$$

4. The hypothesis that that all three categories of factor A lines are parallel (controlling for variable B) is tested with:

$$H_0 : \beta_6 = \beta_7 = \beta_{14} = \beta_{15} = \beta_{16} = \beta_{17} = 0,$$

leading to $A_1 B_1$, $A_2 B_1$, and $A_3 B_1$ respectively,

$$y = (\beta_0 + \beta_2 + \beta_4 + \beta_{10}) + (\beta_1 + \beta_8) x_1 + \varepsilon \qquad A_1 B_1$$

$$y = (\beta_0 + \beta_3 + \beta_4 + \beta_{12}) + (\beta_1 + \beta_8) x_1 + \varepsilon \qquad A_2 B_1$$

$$y = (\beta_0 + \beta_4) + (\beta_1 + \beta_8) x_1 + \varepsilon \qquad A_3 B_1$$

$$y = (\beta_0 + \beta_2 + \beta_5 + \beta_{11}) + (\beta_1 + \beta_9) x_1 + \varepsilon \qquad A_1 B_2$$

$$y = (\beta_0 + \beta_3 + \beta_5 + \beta_{13}) + (\beta_1 + \beta_9) x_1 + \varepsilon \qquad A_2 B_2$$

$$y = (\beta_0 + \beta_5) + (\beta_1 + \beta_9) x_1 + \varepsilon \qquad A_3 B_2$$

$$y = (\beta_0 + \beta_2) + \beta_1 x_1 + \varepsilon \qquad A_1 B_3$$

$$y = (\beta_0 + \beta_3) + \beta_1 x_1 + \varepsilon \qquad A_2 B_3$$

$$y = \beta_0 + \beta_1 x_1 + \varepsilon \qquad A_3 B_3$$

We see that within and given B factor level, all the three equations have the same coefficients for x_1 and are therefore parallel. Similar results can be obtained for the lines relating to variable B (controlling for variable A).

5. The hypothesis of no interaction effects between factors A and B can be tested by

$$H_0 : \beta_{10} = \beta_{11} = \beta_{12} = \beta_{13} = \beta_{14} = \beta_{15} = \beta_{16} = \beta_{17} = 0.$$

8.7   Use of Dummy Variables in Piece-wise Regression

Sometimes the regression model may be different or not adequately described over the entire range of the explanatory variable. Suppose $[a, b]$ for instance denotes the range of the explanatory variable x. That is, $a \leq x \leq b$. Then it may happen that for the interval $a \leq x \leq q$, where $a < q < b$, the regression function is a straight line or a certain function and is a different line or function between $q < x \leq b$, such that the two lines intersect at q. The point q at which the lines meet are often described as the **knot-point** and the graph consisting of the lines are often called *piece-wise regression* or *linear spline with a knot-point*. We consider the data in Table 8.3 in the following example.

Example 8.3

An investigator notices that for small companies, the volume of sales (y in thousands of dollars) tends to increase as a function of advertising budget (x_1 in thousands of dollars). The rate of increase in sales is rapid for the first several thousand dollars spent on advertising, but it slows down at some point. We wish to model the relationship between the average yearly sales and dollars spent on advertising. This data is from Graybill and Iyer (1994, p. 471) reproduced by permission of Brooks/Cole, a part of Cengage Learning, Inc.

Obs.	y	x_1	Obs.	y	x_1	Obs.	y	x_1	Obs.	y	x_1
1	260	12	6	399	41	11	439	47	16	462	66
2	328	25	7	404	41	12	452	47	17	472	73
3	376	30	8	414	44	13	465	55	18	450	74
4	356	35	9	428	45	14	461	59	19	490	83
5	404	41	10	436	46	15	475	64	20	496	87

Table 8.3: Advertising and sales data

A scatter plot of the dependent variable against the explanatory variable x_1 is presented in Figure 8.4.

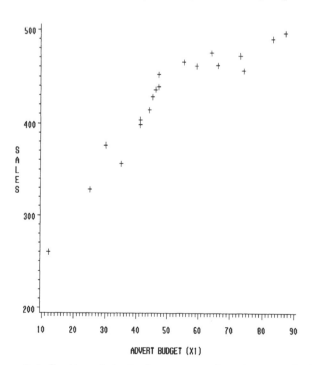

Figure 8.4: Scatter plot of sales versus advertisement budget

We notice from Figure 8.4 that there is a change of slope between $45 < x_1 < 55$. Hence, we would expect our knot-point q to lie between 45 and 55 and an appropriate model would be

$$y = \begin{cases} \alpha_1 + \beta_1\, x_1 & \text{for } a \le x_1 \le q \\ \alpha_2 + \beta_2\, x_1 & \text{for } q \le x_1 \le b, \end{cases} \tag{8.40}$$

and equation (8.40) can be written alternatively as

$$y = \beta_0 + \beta_1\, x_1 + \beta_2\, (x_1 - q)x_2 \quad \text{for } a \le x_1 \le b, \tag{8.41}$$

where

$$x_2 = \begin{cases} 1, & \text{for } q \le x_1 \le b \\ 0, & \text{for } a \le x_1 \le q. \end{cases}$$

For our data, the model in (8.41) becomes

$$y = \beta_0 + \beta_1\, x_1 + \beta_2\, (x_1 - q)x_2 \quad \text{for } 12 \le x_1 \le 87, \tag{8.42}$$

where

$$x_2 = \begin{cases} 1, & \text{for } q \le x_1 \le 87 \\ 0, & \text{for } 12 \le x_1 \le q. \end{cases}$$

How do we find q? It has been suggested that, first we fit a model of the form $y = \beta_0 + \beta_1\, x_1$ to the entire data, that is for $12 \le x_1 \le 87$ and plot the residuals versus the explanatory variable. The knot-point will usually be revealed in such a plot. The plot of residuals against x_1 in Figure 8.5 indicates that a possible knot-point would be $x_1 = 50$. Hence, we would consider a knot-point at $q = 50$. Thus equation (8.42) now becomes

$$y = \beta_0 + \beta_1\, x_1 + \beta_2\, (x_1 - 50)x_2 \quad \text{for } 12 \le x_1 \le 87, \tag{8.43}$$

where

$$x_2 = \begin{cases} 1, & \text{for } 50 \le x_1 \le 87 \\ 0, & \text{for } 12 \le x_1 \le 50. \end{cases}$$

The piecewise regression for these data is implemented in SAS. The SAS program and a partial output are presented below.

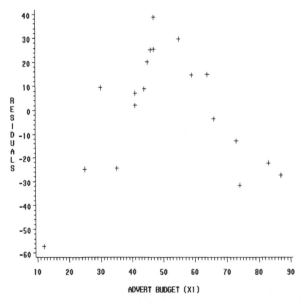

Figure 8.5: Plot of residuals against x_1 over the entire range of x_1

```
/* Program to analyze the data in Example 8.3 */
data one;
input y x1 @@;
if x1 > 50 then x2=1; else x2=0; x3=(x1-50)*x2;
datalines;
........
;
run;
proc reg data=one;
model y=x1 x3 / covb clb;
output out=aa p=pred r=resid;
run;

proc sort data=aa; by x1;
run;
goptions vsize=5in hsize=5in;
axis1 label=(angle=-90 rotate=90 'SALES'); axis2 label=('ADVERT BUDGET (X1)');
proc gplot data=one;
plot y*x1='plus' / vaxis=axis1 haxis=axis2 noframe;
run;
```

                            The REG Procedure
                            Model: MODEL1
                          Dependent Variable: y

                          Analysis of Variance

| | | Sum of | Mean | | |
Source	DF	Squares	Square	F Value	Pr > F
Model	2	63619	31810	260.57	<.0001
Error	17	2075.30947	122.07703		
Corrected Total	19	65695			

Root MSE	11.04885	R-Square	0.9684
Dependent Mean	423.65000	Adj R-Sq	0.9647
Coeff Var	2.60801		

                          Parameter Estimates

| | | Parameter | Standard | | | |
Variable	DF	Estimate	Error	t Value	Pr > |t|	95% Confidence Limits
Intercept	1	201.44537	11.69917	17.22	<.0001	176.76228 226.12845
x1	1	5.02178	0.28747	17.47	<.0001	4.41527 5.62829
x3	1	-4.05602	0.45708	-8.87	<.0001	-5.02037 -3.09167

                        Covariance of Estimates

Variable	Intercept	x1	x3
Intercept	136.87053757	-3.251932308	4.2570035371
x1	-3.251932308	0.0826394305	-0.117020967
x3	4.2570035371	-0.117020967	0.2089205658

The estimated regression parameters are

$$\hat{\beta}_0 = 201.4454, \quad \hat{\beta}_1 = 5.0218, \text{ and } \hat{\beta}_3 = -4.0560.$$

The predicted equation is therefore

$$\hat{y} = 201.4454 + 5.0218\,x_1 - 4.0560\,(x_1 - 50)\,x_2, \text{ for } 12 < x_1 < 87. \tag{8.44}$$

Therefore, when $x_2 = 0$, that is for $12 \leq x_1 \leq 50$, we have the estimated equation

$$\hat{y} = 201.4454 + 5.0218\,x_1. \tag{8.45}$$

Similarly, when $x_2 = 1$, that is for $50 \leq x_1 \leq 87$, we have the estimated equation

$$\hat{y} = 201.4454 + 5.0218\,x_1 - 4.0560\,(x_1 - 50), \text{ that is,}$$
$$\hat{y} = 404.2454 + 0.9658\,x_1. \tag{8.46}$$

The graph of the predicted model in (8.44) is presented in Figure 8.6.

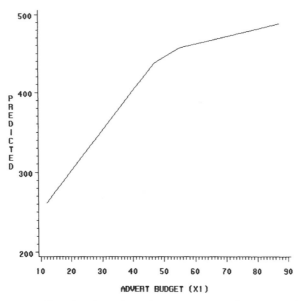

Figure 8.6: Graph of predicted versus x_1 under model (8.44)

Alternatively, we might wish to model the data in this example with a quadratic model of the form

$$y = \beta_0 + \beta_1 x_1 + \beta_2 x_1^2 + \varepsilon, \quad 12 \le x \le 87. \tag{8.47}$$

The estimated regression model is

$$\hat{y} = 172.9083 + 7.6651 \, x_1 - 0.0472 \, x_1^2.$$

The error mean square under this model is also based on 17 degrees of freedom with corresponding MSE = 190.234 and $R^2 = 0.9508$. The plot of the estimated quadratic response is presented in Figure 8.7.

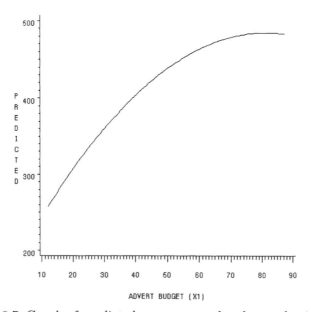

Figure 8.7: Graph of predicted versus x_1 under the quadratic model

Of the two models, the piecewise regression model is better because its MSE = 122.0770 which is much smaller than the 190.234 under the quadratic model. Further, R^2 under the piecewise and quadratic models are respectively, 0.9684 and 0.9508, again indicating that the former model is better. Thus, the piecewise regression model would be recommended in this case.

In the above example, while we have considered the case in which we have only one knot-point, it is not uncommon to have situations where there are more than one-knot points. Further, the structure of the model function may also be polynomial rather than the simple straight line we consider in this example.

8.8 Exercises

8.1 The Tri-City Office Equipment Corporation sells an imported calculator on a franchise basis and performs preventive maintenance and repair on this calculator. The data below has been collected by 18 recent calls on users to perform routine preventive maintenance service. For each call, x_1 is the number of machines serviced, y is the total number of minutes spent by the service person, and type of calculator model serviced (S is student and C is commercial).

x_1	y	Type	x_1	y	Type	x_1	y	Type	x_1	y	Type
7	97	C	4	62	S	8	118	C	1	17	C
6	86	S	7	101	S	5	65	S	4	49	C
5	78	C	3	39	C	2	25	S	5	68	C
1	10	C	4	53	C	5	71	C			
5	75	C	2	33	C	7	105	S			

Suppose we define the variable x_2 to be:

$$x_2 = \begin{cases} 1, & \text{if type is student model} \\ 0, & \text{if type is commercial model.} \end{cases}$$

(a) Define the first-order multiple regression model with interaction for the response variable and the predictor variables. Define the intercepts and slopes for the expected response models for both student and commercial type calculators in terms of the regression coefficients of the single regression model.

(b) Define separate regression lines of y on x_1 for $x_2 = 1$ and 0 respectively.

(c) Using a statistical software, write down the estimated regression lines obtained from the fit in (b).

(d) Test the null hypothesis that the two lines are parallel.

(e) Test the null hypothesis that the two lines are coincident. What does this mean?

(f) State the predicted equation based on your result in (e).

8.2 It is known that the compressive strength of concrete depends on the proportion of water mixed with the cement. A certain type of concrete, when mixed in batches with varying water/cement ratios (measured as a percentage), may yield compressive strengths that follow a pattern where the compressive strength decreases at a much faster rate for batches with water/cement ratios greater than $p\%$. The data below on compressive strength (y) and water ratio (x) for 18 batches of concrete are recorded. Suppose that the assumptions for the linear regression are valid where the data were obtained by simple random sampling. A first order regression function is proposed for the data.

Batch	y	x	Batch	y	x	Batch	y	x
1	4.67	47	7	0.76	82	13	1.67	80
2	3.54	68	8	3.01	72	14	1.59	75
3	2.25	75	9	4.29	52	15	3.91	63
4	3.82	65	10	2.21	73	16	3.15	70
5	4.50	50	11	4.10	60	17	4.37	50
6	4.07	55	12	1.13	85	18	3.75	57

(a) Plot a scatter diagram of y against x.

(b) Fit a simple linear regression model $y_i = \beta_0 + \beta_1 x_i + \varepsilon_i$ and plot the residual against x.

(c) What is $p\%$, the knot value?

(d) Write down the regression equation in terms of the population parameters? You must define x_2 and x_3.

(e) Write down the equations relating to the following in terms of the parameters: for $47 < x \le p$ and $x > p$.

(f) Write down the estimated regression equation and find an estimate for σ.

(g) In terms of the parameters, give the y-intercept and the slope for observations with $x \le p$ and for $x > p$.

(h) Predict the compressive strength at $x = 60$ and 80 respectively.

8.3 The data in the file **cars2004.txt** is taken from 2004 new car and truck data submitted by Johnson, R.W. to the Journal of Statistics Education Data Archive. For this problem, use the last 100 cases and the following variables: The highway miles per gallon (y), the length in inches (x_1), and whether wagon or not (x_2). We are interested in whether there are differences between the highway miles per gallon of wagon and no-wagon vehicles. The variable wagon has been dichotomized as follows:

$$x_2 = \begin{cases} 1, & \text{if wagon} \\ 0, & \text{if non-wagon.} \end{cases}$$

(a) Define a first-order multiple regression model with interaction for the data for both wagon and non-wagon, and which defines straight-line models for each group with possibly different intercepts and slopes. Define the intercept and slopes for the expected response models for each group in terms of the regression coefficients of the single regression model.

(b) Define separate regression lines of highway miles per gallon (y) on length (x_1) for wagon and non-wagon respectively.

(c) Using a statistical package, state the two estimated regression lines obtained from the fit.

(d) Test the null hypothesis that the two lines are parallel.

(e) Test the null hypothesis that the two lines are coincident.

(f) What is the difference between the average highway miles per gallon for wagon and non-wagon vehicles with the same length? Express your answer in terms of population parameters.

(g) What is the difference between the average highway miles per gallon of wagon and non-wagon vehicles? Express your answer in terms of parameter estimates.

8.4 The weight $(y$ in pounds$)$ of a newborn baby increases quite rapidly during the first 100 days after birth, and after that it slows down somewhat. Data for seventeen babies were obtained from the past 3 years' records for babies born at a certain hospital. Suppose that the assumptions for the linear regression between y and age $(x$ in days$)$ are valid where the data were obtained by simple random sampling.

Baby	y	x	Baby	y	x	Baby	y	x
1	7.5	7	7	13.4	92	13	16.1	156
2	7.9	12	8	13.9	99	14	16.5	159
3	8.4	18	9	13.5	105	15	16.9	167
4	10.1	45	10	14.5	108	16	17.3	183
5	11.5	67	11	14.7	120	17	17.6	195
6	12.8	88	12	15.8	147			

A first order regression function is proposed for the data.

(a) Plot a scatter diagram of y against x.

(b) Fit the regression model and plot the residuals against x.

(c) Fit a piece-wise regression model of the form $y = \beta_0 + \beta_1 x + \beta_2 (x - q) x_2$ to the data. Write down the regression equation in terms of the population parameters? You must define x_2.

(d) Write down the equations relating to the following in terms of the parameters: for $0 < x \le 100$ and $100 < x \le 200$.

(e) Write down the estimated regression equation and find an estimate for σ.

(f) What are the population quantities needed to determine the difference between the average growth rate of newborn babies during the first 100 days and the average growth rate during the next 100 days.

(g) Instead of using the equation above, suppose a quadratic model of the form

$$y_i = \beta_0 + \beta_1 x + \beta_2 x^2,$$

holds for $0 \leq x \leq 200$. Estimate the regression equation. From your computer output, compare the fits, residuals and standardized residuals for both models. Compare the residual plots in both models. Which model is better for this problem? Give reasons for your answer.

(h) Predict the weights of newborn babies at 84 and 160 days.

8.5 The following data relate to whether or not y, the number of days of prison sentence for thefts whose total value is under \$1,000 for first time offenders, is related to x, the amount of money stolen. Sample data from thefts falling in this category in three different states in the United States, denoted as states 1, 2, and 3, are obtained from last year's police records in each state. If we assume that the error variances for the three states are the same and that a regression model of the form

$$y = \beta_0 + \beta_1 x_1 + \beta_2 x_2 + \beta_3 x_3 + \beta_4 x_1 x_2 + \beta_5 x_1 x_3 + \varepsilon,$$

would be appropriate where

$$x_2 = \begin{cases} 1, & \text{if State is 1} \\ 0, & \text{elsewhere,} \end{cases} \qquad x_3 = \begin{cases} 1, & \text{if State is 2} \\ 0, & \text{elsewhere.} \end{cases}$$

y	x_1	State	y	x_1	State	y	x_1	State
44	367	1	43	538	2	50	732	3
81	855	1	32	290	2	52	556	3
43	284	1	53	759	2	53	960	3
40	305	1	55	734	2	39	134	3
38	215	1	40	499	2	55	826	3
44	308	1	44	541	2	53	738	3
49	433	1	42	474	2	37	403	3
51	455	1	51	940	2	46	511	3
49	454	1	35	314	2	45	699	3
57	429	1	39	351	2	49	778	3
47	345	1	51	703	2	37	530	3
37	167	1	37	459	2	35	140	3
67	689	1				38	429	3
55	499	1				48	554	3
						50	672	3
						30	125	3
						29	124	3

(a) Define separate regression lines of y on x_1 for each State. You must indicate the State your equation refers to precisely.

(b) For the general model above, define the intercepts and slopes for the expected response models for each State in terms of the regression coefficients of the single regression model.

(c) Using a statistical software, obtain the three estimated regression lines for each State.

(d) Test the hypotheses of parallelism, coincident and equal intercept. Are all of the tests relevant? Conduct the relevant tests by including the hypotheses, the decision and appropriate conclusion.

8.6 The following data relate the operating cost per mile (y) to cruising speed (x_1) for a laboratory tire testing with equipment that stimulates highway driving for two makes (A, B) of a certain type of tire. An engineer wishes to study whether or not the regression of operating cost on cruising speed is the same for the two makes. If we assume that the error variances for the two makes are the same and that a regression model of the form

$$y_i = \beta_0 + \beta_1 x_{i1} + \beta_2 x_{i2} + \beta_3 x_{i1} x_{i2} + \varepsilon_i,$$

would be appropriate where,

$$x_2 = \begin{cases} 1, & \text{if make A} \\ 0, & \text{if make B.} \end{cases}$$

Make A		Make B	
x_1	y	x_1	y
10	9.8	10	15.0
20	12.5	20	14.5
20	14.2	20	16.1
30	14.9	30	16.5
40	19.0	40	16.4
40	16.5	40	19.1
50	20.9	50	20.9
60	22.4	60	22.3
60	24.1	60	19.8
70	25.8	70	21.4

(a) Define separate regression lines of y on x_1 for $x_2=1$ and 0 respectively.

(b) Interpret the model specified above and define the intercepts and slopes for the expected response models for each makes in terms of the regression coefficients of the single regression model.

(c) Using a statistical software to fit the data, state the two estimated regression lines obtained from the fit.

(d) Test the null hypothesis that the two lines are parallel.

(e) Test the null hypothesis that the two lines are coincident (that is, $\beta_2 = \beta_3 = 0$).

8.7 The data in the file **cars2004.txt** is taken from 2004 new car and truck data submitted by Johnson, R.W. to the Journal of Statistics Education Data Archive. For this problem, use the last 200 cases of the whole data set where the number of cylinders is 4, 6 or 8. The response variable is the city miles per gallon (y) and the independent variables are the weight in pounds (x_1) and the number of cylinders.

(a) Initially, the researchers fitted the first-order, main effects model

$$y = \beta_0 + \beta_1 x_1 + \beta_2 x_2 + \beta_3 x_3 + \varepsilon,$$

where

$$y = \text{city miles per gallon}$$
$$x_1 = \text{weight in pounds}$$
$$x_2 = \begin{cases} 1, & \text{if number of cylinders is 4} \\ 0, & \text{if not,} \end{cases}$$
$$x_3 = \begin{cases} 1, & \text{if number of cylinders is 6} \\ 0, & \text{if not.} \end{cases}$$

What can you say about the fitted regression model?

(b) The interaction model

$$y = \beta_0 + \beta_1 x_1 + \beta_2 x_2 + \beta_3 x_3 + \beta_4 x_1 x_2 + \beta_5 x_1 x_3 + \varepsilon,$$

was also fitted.

    (i) Define separate regression lines of y on x_1 for each type of fuel in terms of the parameters of the model. You must state the number of cylinders your equation refers to precisely.

    (ii) For the general model above, define the intercepts and slopes for the expected response models for each number of cylinders in terms of the regression coefficients of the single regression model.

    (iii) Using a statistical software, obtain the three estimated regression lines for each number of cylinders.

8.8 The data in the file **cars2004.txt** is taken from 2004 new car and truck data submitted by Johnson, R.W. to the Journal of Statistics Education Data Archive. For this problem, use the last 30 cases and the following variables: The highway miles per gallon (y) and engine size (x). A researcher wants to model the highway miles per gallon (y) as a linear function of engine size x.

(a) Specify the appropriate piecewise linear model for y.

(b) Give the least squares prediction equation.

(c) Is the model adequate for predicting highway miles per gallon (y)? Test using $\alpha = 0.10$.

(d) Give a 90% confidence interval for the mean increase in highway miles per gallon for every unit increase in engine size.

8.9 A study was conducted to study the relationship between an employee's work involvement (y, a composite behavioral rating on a 100-point scale) and the employee's age (x_1, in years), marital status (never married, divorced, and married), and gender (male and female). Data were collected for 60 employees of the same firm.

(a) How many dummy variables will be needed to describe marital status?

(b) Using never married as the base level, define the dummy variables for marital status.

(c) Using the dummy variables in (b), write a model for $E(Y)$ as a function of marital status.

(d) Interpret all the β parameters in the regression model in (c).

(e) Explain how to test the difference between the mean work involvement of married and never married employees.

(f) Using male as the base level, define the dummy variable(s) for gender.

(g) Write a first order model for $E(Y)$ as a function of age, marital status, and gender.

(h) Write a second order model for $E(Y)$ as a function of age, marital status, and gender.

8.10 A Science and Technology study investigated the variables that affect the absorption of organic vapors on clay minerals. The independent variables are temperature (x_1 in degrees), relative humidity (x_2 in percent), and organic compound (with three levels benzene, chloroform and methanol.) The dependent variable y is the retention coefficient.

(a) How many dummy variables will be needed to describe organic compound?

(b) Using benzene as the base level, define the dummy variables for organic compound.

(c) Write a first-order, main effects model for E(Y) as a function of temperature and organic compound.

(d) Interpret all the β parameters in the regression model in (c).

(e) Explain how to test the difference between the mean retention coefficients of methanol and benzene.

(f) Write a model for E(Y) as a function of relative humidity and organic compound that hypothesizes different retention-relative humidity slopes for the three compounds.

(g) Give the slopes of the three compounds (in terms of the β's) for the model in (f).

(h) Write a complete second-order model for E(Y) as a function of temperature and relative humidity.

Chapter 9

Polynomial Regression

9.1 Introduction

We have seen from the previous chapters that the general multiple regression is of the form

$$y_i = \beta_1 x_1 + \beta_2 x_2 + \beta_3 x_3 + \cdots + \varepsilon_i. \tag{9.1}$$

If however, $x_2, x_3, \cdots$, are powers of x_1, a single predictor or explanatory variable, then we have a polynomial regression. In other words, polynomial regressions are generated by the addition of explanatory variables that are successive powers of the independent variable(s). For example, we present in (9.2), (9.3) and (9.4) respectively, *first-order*, *second-order* and *third-order* degree models (with linear, quadratic and cubic response functions respectively):

$$\text{linear model} \quad y_i = \quad \beta_0 + \beta_1 x_i + \varepsilon_i \tag{9.2}$$

$$\text{quadratic model} \quad y_i = \quad \beta_0 + \beta_1 x_i + \beta_2 x_i^2 + \varepsilon_i \tag{9.3}$$

$$\text{cubic model} \quad y_i = \quad \beta_0 + \beta_1 x_i + \beta_2 x_i^2 + \beta_3 x_i^3 + \varepsilon_i. \tag{9.4}$$

Polynomial equations are often referred to by their degrees, which is the number of the largest exponent. In our example above, the linear model is a first degree polynomial, the quadratic is a second degree, while the cubic model is of the third degree, and so on. In general, a k-th order polynomial model in one variable is of the form

$$y_i = \beta_0 + \beta_1 x_i + \beta_2 x_i^2 + \beta_3 x_i^3 + \cdots + \beta_k x_i^k + \varepsilon_i.$$

The model is linear and we present in Figure 9.1 graphical plots of the cubic polynomial $y = -6(x+2)(x-1)x$ and quadratic function $y = -12 - x + x^2$ for $-2 \leq x \leq 2$. The corresponding y values are on the vertical axes to the left.

Example 9.1

The following data are from Kleinbaum et al. (2008, p. 377) and reproduced by permission of Brooks/Cole, a part of Cengage Learning, Inc.

x	y	x	y	x	y	x	y	x	y	x	y
0.5	13.90	1.0	14.08	1.5	13.75	2.0	13.32	2.5	13.45	3.0	13.59
0.5	13.81	1.0	13.99	1.5	13.60	2.0	13.39	2.5	13.53	3.0	13.64

Table 9.1: Skin response (y) as a function of concentration (x) in ml/l

In this example, there are $n = 12$ observations of which there are 6 distinct values of x. Hence, we can at most fit a $6 - 1 = 5$th degree polynomial to the data. A scatter plot of the data overlaid with a possible curvilinear plot is presented in Figure 9.2.

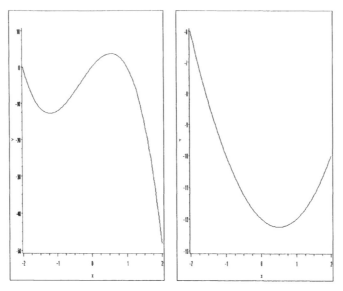

Figure 9.1: Plots of cubic and quadratic polynomials

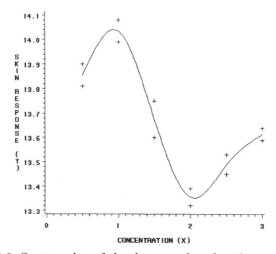

Figure 9.2: Scatter plot of the data overlayed with a possible plot

To fit a fifth degree polynomial, we need to generate powers of x, namely, x^2, x^3, x^4 and x^5, resulting in the model

$$y_i = \beta_0 + \beta_1 x_i + \beta_2 x_i^2 + \beta_3 x_i^3 + \beta_4 x_i^4 + \beta_5 x_i^5 + \varepsilon_i.$$

But from our discussion in the earlier chapter under multicollinearity, since the explanatory variables in this case are highly correlated, fitting even a second-degree polynomial to the data will result in multicollinearity problems. We shall examine how to overcome this problem in polynomial regressions at a later section. Ignoring multicollinearity for now, a fifth degree polynomial fitted to the above data gives the following SAS results.

```
/* Program to analyze the data for Example 9.1  */
data poly1;
input x y @@;
label x='Conc' y='Resp';
x2=x**2; x3=x**3; x4=x**4; x5=x**5; x1=x;
datalines;
.........
;
proc reg data=poly1;
model y = x x2 x3 x4 x5 / vif ss1 ss2;
test1: test x5=0;
run;
```

```
                              The REG Procedure
                               Model: MODEL1
                          Dependent Variable: y Resp

                            Analysis of Variance

                                   Sum of        Mean
         Source            DF      Squares      Square    F Value   Pr > F

         Model              5      0.60424     0.12085     27.62    0.0005
         Error              6      0.02625     0.00437
         Corrected Total   11      0.63049

                  Root MSE            0.06614    R-Square    0.9584
                  Dependent Mean     13.67083    Adj R-Sq    0.9237
                  Coeff Var           0.48383

                          Parameter Estimates

                          Parameter    Standard
    Variable   Label    DF   Estimate      Error   t Value  Pr > |t|   Type I SS

    Intercept  Intercept  1   13.10500    1.42094    9.22    <.0001    2242.70021
    x          Conc       1    1.52900    5.91595    0.26    0.8047       0.28440
    x2                    1    1.00000    8.63161    0.12    0.9115       0.06921
    x3                    1   -2.72833    5.69426   -0.48    0.6488       0.16857
    x4                    1    1.32000    1.73437    0.76    0.4754       0.07800
    x5                    1   -0.19067    0.19799   -0.96    0.3727       0.00406

                          Parameter Estimates

                                                    Variance
         Variable   Label    DF   Type II SS       Inflation

         Intercept  Intercept  1     0.37214               0
         x          Conc       1     0.00029224        69997
         x2                    1     0.00005872      1904836
         x3                    1     0.00100         8049982
         x4                    1     0.00253         6875506
         x5                    1     0.00406          814270

                              The REG Procedure
                               Model: MODEL1

                   Test TEST1 Results for Dependent Variable y

                                    Mean
         Source            DF      Square    F Value   Pr > F

         Numerator          1     0.00406     0.93     0.3727
         Denominator        6     0.00437
```

The first model in the SAS program above fits a fifth degree polynomial and the hypotheses of interest are:

$$H_0 : \beta_5|(\beta_1, \beta_2, \beta_3, \beta_4) = 0$$
$$H_a : \beta_5|(\beta_1, \beta_2, \beta_3, \beta_4) \neq 0. \tag{9.5}$$

The result of these hypotheses is presented in the test1 result with an F-value of 0.93 on 1 and 6 degrees of freedom. The corresponding p-value is 0.3727, which indicates that we would fail to reject H_0 and therefore, x^5 is not important in our model. Thus, x^5 can be removed from our model. We now fit a reduced model, that is, a fourth degree polynomial

$$y_i = \beta_0 + \beta_1 x_i + \beta_2 x_i^2 + \beta_3 x_i^3 + \beta_4 x_i^4 + \varepsilon_i,$$

and again conduct the following hypotheses:

$$H_0 : \beta_4|(\beta_1, \beta_2, \beta_3) = 0$$
$$H_a : \beta_4|(\beta_1, \beta_2, \beta_3) \neq 0. \tag{9.6}$$

The SAS output from fitting this model and conducting the test specified in (9.6) is presented below.

```
proc reg data=poly1;
model y = x x2 x3 x4 / vif ss1 ss2;
```

```
test2: test x4=0;
run;
```

<pre>
                          The REG Procedure
                           Model: MODEL1
                     Dependent Variable: y Resp

                      Analysis of Variance

                              Sum of   Mean
      Source   DF   Squares   Square   F Value   Pr > F

      Model   4   0.60018   0.15005   34.66   0.0001
      Error   7   0.03031   0.00433
      Corrected Total 11   0.63049

             Root MSE   0.06580   R-Square   0.9519
             Dependent Mean   13.67083   Adj R-Sq   0.9245
             Coeff Var   0.48132

                      Parameter Estimates

                           Parameter   Standard
      Variable Label   DF   Estimate   Error t Value Pr > |t| Type I SS

      Intercept Intercept 1 11.79417   0.40562   29.08   <.0001   2242.70021
      x   Conc   1   7.07914   1.32824   5.33   0.0011   0.28440
      x2   1 -7.20264   1.39045   -5.18   0.0013   0.06921
      x3   1   2.72685   0.57633   4.73   0.0021   0.16857
      x4   1 -0.34833   0.08207   -4.24   0.0038   0.07800

                      Parameter Estimates
                                               Variance
            Variable Label   DF Type II SS   Inflation

            Intercept Intercept 1   3.66059   0
            x   Conc   1   0.12299   3565.42052
            x2   1   0.11618   49948
            x3   1   0.09693   83327
            x4   1   0.07800   15555

                          The REG Procedure
                           Model: MODEL1

             Test TEST2 Results for Dependent Variable y

                                     Mean
            Source   DF   Square   F Value   Pr > F

            Numerator   1   0.07800   18.02   0.0038
            Denominator   7   0.00433
</pre>

Again the results of this test, in test2 above gives a computed F-value of 18.02 on 1 and 7 degrees of freedom respectively. The p-value for this test is 0.0038, indicating that we would strongly reject the null hypothesis and can consequently conclude that we would need a fourth degree polynomial for the above data. The predicted equation under the estimated fourth-order model is

$$\hat{y}_i = 11.7942 + 7.0791x_i - 7.2026x_i^2 + 2.7269x_i^3 - 0.3483x_i^4,$$

and this response function is plotted and displayed in Figure 9.3 with the observed values superimposed over it.

9.2   Lack of Fit Tests

Suppose we had assumed that the true relationship between y and x to the above data is of the form $y = \beta_0 + \beta_1 x$. This is an assumption we should not blindly accept but should tentatively entertain. We can use the lack of fit test to examine if indeed this simple model is adequate. The lack of fit test is conducted when there are genuine replicated observations. Consider the data in our example. There, we have each value of x being replicated twice (does not have to be the same number of times nor does all values of x need necessarily be replicated). These repeated observations and their corresponding response values (y_i) are presented in Table 9.2. For the data in Table 9.1 therefore, it would be possible to conduct a test of lack of fit by breaking the Error SS into two components:

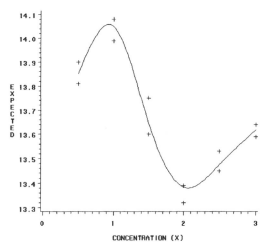

Figure 9.3: Plot of predicted versus x under the estimated model

(a) Lack of Fit SS

(b) Pure Error SS

x_i	y_i's
0.5	13.90, 13.81
1.0	14.08, 13.99
1.5	13.75, 13.60
2.0	13.32, 13.39
2.5	13.45, 13.53
3.0	13.59, 13.64

Table 9.2: Replicated observations and their corresponding response values

We proceed with the calculations to conduct a test of adequacy, by first calculating the *Pure Error SS* at these replicated points, viz:

Calculation of Pure Error SS

$$\text{At } x = 0.5, \text{ SS} = \frac{(13.90 - 13.81)^2}{2} = 0.00405 \quad \text{on 1 d.f}$$

$$\text{At } x = 1.0, \text{ SS} = \frac{(14.08 - 13.99)^2}{2} = 0.00405 \quad \text{on 1 d.f}$$

$$\text{At } x = 1.5, \text{ SS} = \frac{(13.75 - 13.60)^2}{2} = 0.01125 \quad \text{on 1 d.f}$$

$$\text{At } x = 2.0, \text{ SS} = \frac{(13.32 - 13.39)^2}{2} = 0.00245 \quad \text{on 1 d.f}$$

$$\text{At } x = 2.5, \text{ SS} = \frac{(13.45 - 13.53)^2}{2} = 0.00320 \quad \text{on 1 d.f}$$

$$\text{At } x = 3.0, \text{ SS} = \frac{(13.59 - 13.64)^2}{2} = 0.00125 \quad \text{on 1 d.f}$$

Note that the SS at each of the above values of x can also be obtained alternatively. For example, at $x = 0.5$, the SS can be computed as

$$13.90^2 + 13.81^2 - \frac{(13.90 + 13.81)^2}{2} = 383.9261 - \frac{27.71^2}{2} = 0.00405.$$

Hence, Total Pure Error SS $= 0.00405 + \cdots + 0.00125 = 0.02625$ and is based on $1+1+1+1+1+1 = 6$ degrees of freedom. We therefore have the revised analysis of variance for these data in Table 9.3.

Source of Variation	d.f.	SS	MS	F
Regression	1	0.2844	0.28440	0.22
Residual (Error)	10	0.34609	0.03461	
Lack of Fit	4	0.31984	0.07996	18.256
Pure Error	6	0.02625	$0.00438 = s_e^2$	
Total	11	0.63049		

Table 9.3: Revised analysis of variance table

Since the lack of fit is highly significant, there is therefore a strong reason to doubt the adequacy of the model. With this significant lack of fit, it means that the proposed simple linear regression model is not adequate and we must therefore seek ways to improve the model by examining the residuals. Any test of significance carried out with this type of model will not be valid. We can implement the lack of fit test in SAS with the following statements and corresponding partial output.

```
/* Create a new variable, lackfit, as lackfit=x. This variable is used in class statement
where it will be treated as a categorical variable */
data new;
set poly1;
lackfit=x;
run;
proc glm data=new;
class lackfit;
model y=x1 lackfit / ss1;
title 'Testing for Lack of Fit';
run;
```

```
                    Testing For Lack of Fit

                    The GLM Procedure

                 Class Level Information

        Class        Levels     Values

        lackfit          6      0.5 1 1.5 2 2.5 3

                    The GLM Procedure

Dependent Variable: y   Resp

                                  Sum of
Source                  DF        Squares      Mean Square    F Value    Pr > F

Model                    5      0.60424167     0.12084833      27.62     0.0005

Error                    6      0.02625000     0.00437500

Corrected Total         11      0.63049167

                R-Square    Coeff Var     Root MSE       y Mean

                0.958366    0.483831      0.066144      13.67083

Source                  DF      Type I SS     Mean Square    F Value    Pr > F

x1                       1     0.28440071     0.28440071      65.01     0.0002
lackfit                  4     0.31984095     0.07996024      18.28     0.0016
```

The lack of fit test can also be easily implemented with **proc rsreg**. We present the partial output from the implementation in **proc rsreg** in what follows.

```
proc rsreg data=poly1;
model y=x1 /lackfit covar=1;
run;
```

```
                    The RSREG Procedure
```

```
              Response Surface for Variable y: Resp

         Response Mean               13.670833
         Root MSE                     0.186035
         R-Square                     0.4511
         Coefficient of Variation     1.3608

                         Type I Sum
   Regression     DF     of Squares   R-Square   F Value   Pr > F

   Covariates      1      0.284401     0.4511      8.22     0.0168
   Linear          0             0     0.0000       .        .
   Quadratic       0             0     0.0000       .        .
   Crossproduct    0             0     0.0000       .        .
   Total Model     1      0.284401     0.4511      8.22     0.0168

                          Sum of
   Residual       DF     Squares    Mean Square   F Value   Pr > F

   Lack of Fit     4     0.319841     0.079960     18.28    0.0016
   Pure Error      6     0.026250     0.004375
   Total Error    10     0.346091     0.034609
```

The lack of fit results agree with our earlier results, with a p-value of 0.0016. Had we assumed a quadratic model to begin with, the revised analysis of variance would have looked like the one in Table 9.4. Again this indicates lack of fit. We can do the same for a third-degree polynomial model applied to the data in Table 9.1.

Source of Variation	d.f.	SS	MS	F
Regression	2	0.35362	0.17681	5.75
Residual (Error)	9	0.27688	0.03076	
Lack of Fit	3	0.25067	0.08354	19.10
Pure Error	6	0.02625	$0.00438=s_e^2$	
Total	11	0.63049		

Table 9.4: Revised analysis of variance table

Both the quadratic and third degree polynomial models lack of fit results in SAS using **proc rsreg** are presented in the following. Notice that the computed F values agree with our results in Table 9.4 for the quadratic model.

```
/* Lack of fit test for the quadratic model */
proc rsreg data=poly1;
model y=x1 x2 /lackfit covar=2;
run;
                      The RSREG Procedure

              Response Surface for Variable y: Resp

         Response Mean               13.670833
         Root MSE                     0.175397
         R-Square                     0.5609
         Coefficient of Variation     1.2830

                          Sum of
   Residual       DF     Squares    Mean Square   F Value   Pr > F

   Lack of Fit     3     0.250626     0.083542     19.10    0.0018
   Pure Error      6     0.026250     0.004375
   Total Error     9     0.276876     0.030764

/* Lack of fit test for the third degree polynomial model */
proc rsreg data=poly1;
model y=x1 x2 x3 /lackfit covar=3;
run;
                          Sum of
   Residual       DF     Squares    Mean Square   F Value   Pr > F

   Lack of Fit     2     0.082059     0.041030      9.38    0.0142
   Pure Error      6     0.026250     0.004375
   Total Error     8     0.108309     0.013539
```

Source of Variation	d.f.	SS	MS	F
Regression	4	0.60018	0.15005	34.66
Residual (Error)	7	0.03031	0.00433	
Lack of Fit	1	0.00406	0.00406	0.93
Pure Error	6	0.02625	$0.00438=s_e^2$	
Total	11	0.63049		

Table 9.5: Revised analysis of variance table

For the fourth degree polynomial model, the corresponding lack of fit test is displayed in Table 9.5.

The results from Table 9.5 indicate that the model displays adequacy as the lack of fit test is no longer significant. The SAS program and the partial output from the fourth degree polynomial model are presented below. The implementation in **proc rsreg** is displayed after the partial output. The results from using **proc rsreg** agree with the results using **proc glm**.

```
/* Lack of fit test for the fourth degree polynomial model */
proc glm data=new;
class lackfit;
model y=x1 x2 x3 x4 lackfit / ss1;
title 'Testing for Lack of Fit';
run;
                    The GLM Procedure

Dependent Variable: y   Resp

                            Sum of
Source                DF    Squares      Mean Square   F Value   Pr > F

Model                  5    0.60424167   0.12084833     27.62    0.0005

Error                  6    0.02625000   0.00437500

Corrected Total       11    0.63049167

           R-Square    Coeff Var    Root MSE      y Mean

           0.958366    0.483831     0.066144      13.67083

Source                DF    Type I SS    Mean Square   F Value   Pr > F

x1                     1    0.28440071   0.28440071     65.01    0.0002
x2                     1    0.06921488   0.06921488     15.82    0.0073
x3                     1    0.16856694   0.16856694     38.53    0.0008
x4                     1    0.07800179   0.07800179     17.83    0.0055
lackfit                1    0.00405734   0.00405734      0.93    0.3727

/* Lack of fit test for the fourth degree polynomial model: proc rsreg */
proc rsreg data=poly1;
model y=x1 x2 x3 x4 /lackfit covar=4;
run;
```

9.3   Model with Two Explanatory Variables

With two explanatory variables, say, x_1 and x_2, then, a *first-order* or *first-degree* polynomial regression involving both variables can be written as

$$y_i = \beta_0 + \beta_1 x_1 + \beta_2 x_2 + \varepsilon_i. \tag{9.7}$$

Similarly, a full *second-degree* polynomial equation can be written as

$$y_i = \beta_0 + \beta_1 x_1 + \beta_2 x_2 + \beta_{12} x_1 x_2 + \beta_{11} x_1^2 + \beta_{22} x_2^2 + \varepsilon_i. \tag{9.8}$$

The x_1x_2 term in equation (9.8) represents the interaction between the explanatory variables x_1 and x_2 and it is a *second-degree* polynomial term. We may also note that we only need at least one of the terms involving x_1^2, x_2^2 and x_1x_2 to have a second-degree polynomial regression equation.

9.3.1 Interpretation of Second Degree Polynomial Regression Parameters

- The model in (9.8) will produce what we call a response surface.

- β_0: y-intercept; the value of $E(Y_i)$ when $x_1 = x_2 = 0$.

- β_1 and β_2: Changing β_1 and β_2 causes the surface to shift along the x_1 and x_2 axes.

- β_{12}: The value of β_{12} controls the rotation of the surface.

- β_{11} and β_{22}: Signs and values of these parameters control the type of surface and the rates of curvature.

- Three types of surfaces may be produced by a second-order model.

  1. A paraboloid that opens upward if $\beta_{11} + \beta_{22} > 0$ for $\beta_{12}^2 < 4\beta_{11}\beta_{22}$.
  2. A paraboloid that opens downward if $\beta_{11} + \beta_{22} < 0$ for $\beta_{12}^2 < 4\beta_{11}\beta_{22}$.
  3. A saddle-shaped surface when $\beta_{12}^2 > 4\beta_{11}\beta_{22}$.

Example 9.2

Many companies manufacture products that are at least partially produced using chemicals (e.g. stell, paint, gasoline). In many instances, the quality of the finished product is a function of the temperature and pressure at which the chemical reactions take place. Suppose, we wish to model the quality, y, of a product as a function of the temperature, x_1, and the pressure, x_2, at which it is produced. Four inspectors independently assign a quality score between 0 and 100 to each product, and then the quality, y, is calculated by averaging the four scores. An experiment is conducted by varying temperature between 80^0 and 100^0F and pressure between 50 and 60 pounds per square inch(psi). The resulting data($n = 27$) are given in Table 9.6. We wish to fit the complete second-order model to the data and sketch the response surface.

Obs	x_1	x_2	y	Obs	x_1	x_2	y	Obs	x_1	x_2	y
1	80	50	50.8	10	90	50	63.4	19	100	50	46.6
2	80	50	50.7	11	90	50	61.6	20	100	50	49.1
3	80	50	49.4	12	90	50	63.4	21	100	50	46.4
4	80	55	93.7	13	90	55	93.8	22	100	55	69.8
5	80	55	90.9	14	90	55	92.1	23	100	55	72.5
6	80	55	90.9	15	90	55	97.4	24	100	55	73.2
7	80	60	74.5	16	90	60	70.9	25	100	60	38.7
8	80	60	73.0	17	90	60	68.8	26	100	60	42.5
9	80	60	71.2	18	90	60	71.3	27	100	60	41.4

Table 9.6: Data on product quality

```
/* Program to analyze the data in Example 9.2 */
data ta96;
input x1 x2 y @@;
x11=x1**2; x22=x2**2; x12=x1*x2;
label x1='Temp' x2='Press' y='Quality';
datalines;
........
;
proc reg data=ta96;
model y = x1 x2 x12 x11 x22;
run;

data cont;
```

```
set ta96;
do x1=80 to 100 by 0.1;
do x2=50 to  60 by 0.1;
y=-5127.899+(31.096*x1)+(139.747*x2)-0.146*(x1*x2)-0.133*(x1*x1)-1.144*(x2*x2);
output;
end; end;
proc plot vpct=50;
plot x1*x2=y /contour=8;
run;
goptions hsize=5in vsize=5in;
proc g3d data=cont;
plot x1*x2=y;
run;
```

The REG Procedure
Model: MODEL1
Dependent Variable: y Quality

Analysis of Variance

Source	DF	Sum of Squares	Mean Square	F Value	Pr > F
Model	5	8402.26454	1680.45291	596.32	<.0001
Error	21	59.17843	2.81802		
Corrected Total	26	8461.44296			

Root MSE	1.67870	R-Square	0.9930	
Dependent Mean	66.96296	Adj R-Sq	0.9913	
Coeff Var	2.50690			

Parameter Estimates

Variable	Label	DF	Parameter Estimate	Standard Error	t Value	Pr > |t|
Intercept	Intercept	1	-5127.89907	110.29601	-46.49	<.0001
x1	Temp	1	31.09639	1.34441	23.13	<.0001
x2	Press	1	139.74722	3.14005	44.50	<.0001
x12		1	-0.14550	0.00969	-15.01	<.0001
x11		1	-0.13339	0.00685	-19.46	<.0001
x22		1	-1.14422	0.02741	-41.74	<.0001

The estimated full second-order model for the data is

$$\hat{y}_i = -5127.899 = 31.096x_1 + 139.747x_2 - 0.146x_1x_2 - 0.133x_1^2 - 1.144x_2^2.$$

A contour plot of x_2 versus x_1 at various values of the response function is presented below as well as the response surface plot in Figure 9.4.

Contour plot of X1*X2.

```
        99.5  '''++++OOOOOXXXXXXXXWWWWWWWWWXXXXXXXXXOOOOO++++'''.
              '++++OOOOOXXXXXWWWWWWWWWWWWWWWWWWWWWXXXXXOOOOO+++''
              ++++OOOXXXXXWWWWWWWW**********WWWWWWWWWXXXXXOOOO+++
              ++OOOOXXXWWWWWW********************WWWWWWWXXXXOOOO+
              ++OOOXXXXWWWWW***********##**********WWWWWXXXXOOO
     T 93.0   +OOOXXXXWWWWW********###############*******WWWWWXXXOO
     e        OOOOXXXWWWWW******###################*****WWWWXXXX
     m        OOOOXXXWWWW*****#######################*****WWWWXX
     p        OOOOXXXWWW*****#########################*****WWWWX
              OOOOXXXWWW****##########################*****WWWWX
        86.5  +OOOXXXWWWW****##########################*****WWWWX
              ++OOOXXXWWW*****#########################*****WWWW
              +++OOOXXXWWW*****#########################*****WWWWW
              ++++OOOXXXWWW*****#######################*****WWWWX
              ''+++OOOXXXWWWW******####################*****WWWWWX
        80.0  '''++OOOOXXXWWWWW*******###############*****WWWWXX

                 50       52       54       56       58       60

                                  Press
```

Symbol		y	Symbol		y	Symbol		y
.....		42.121 - 49.107	OOOOO		63.080 - 70.067	*****		84.040 - 91.026
'''''		49.107 - 56.094	XXXXX		70.067 - 77.053	#####		91.026 - 98.012

```
+++++    56.094 - 63.080      WWWWW   77.053 - 84.040

NOTE: 547311 obs hidden.
```

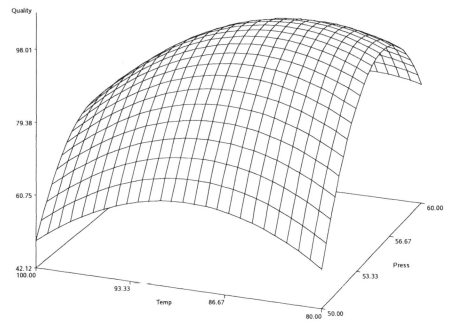

Figure 9.4: Response surface plot for the estimated model

Alternatively, we could use **proc score** in SAS to obtain the estimated response values (predicted values) from estimated parameters of a given model from a different data set. In order to do this, we create values of x_1 and x_2 as above in SAS data file called **cont** and use its contents to predict the estimated response used in plotting the contour. This is accomplished as follows:

```
proc reg data=ta96 outest=aa;
model y=x1 x2 x12 x11 x22;
run;
data cont1;
do x1=80 to 100 by 0.1;
do x2=50 to  60 by 0.1;
x12=x1*x2;
x11=x1**2;
x22=x2**2;
output;
end; end;
run;
proc score data=cont1 score=aa out=bb type=parms predict;
var x1 x2 x12 x11 x22;
run;
proc plot data=bb vpct=50;
plot x1*x2=model1 / contour=8;
run;

goptions hsize=5in vsize=5in;
symbol color=black;
proc g3d data=bb;
plot x1*x2=model1;
run;
```

In the above SAS codes, first, the parameters of the model are estimated from the original data as before with **proc reg**, but we ask that the parameter estimates be written to an output file named 'aa' with the **outest** option. Next, we create our plotting data and put these in SAS data file **cont1**. We next use **proc score**, specifying the following:

- **data=cont1** where **cont1** is the data set containing variable values to be used for prediction.

- **score=aa** gives the **outest** data set from **proc reg** used on the original data set (this contains the parameter estimates).

- **out=bb** contains the output data set with the predicted values that are labeled as **model1** in the output data set.

- **type=parms** specifies that the parameter of the estimated model from **proc reg** be used.

- **predict** specifies that predicted values be computed.

- **vars** specifies the predictor variables in the new data set.

We note here that the predicted values are contained in variable labeled **model1**.

9.3.2 Coping with Multicollinearity in Polynomial Regressions

For the data in Table 9.1 used in Example 9.1, we employed the fourth-degree polynomial or quartic model

$$y_i = \beta_0 + \beta_1 x_i + \beta_2 x_i^2 + \beta_3 x_i^3 + \beta_4 x_i^4 + \varepsilon_i.$$

Implementing this model in SAS leads to the following output. The variance inflation factors here are very large indicating very strong multicollinearity. We can ameliorate this by centering (see section 5.12) the explanatory variables. That is, fit the model

$$y_i = \beta_0 + \beta_1(x_i - \bar{x}) + \beta_2(x_i - \bar{x})^2 + \beta_3(x_i - \bar{x})^3 + \beta_4(x_i - \bar{x})^4 + \varepsilon_i,$$

where $\bar{x} = 1.75$. The resulting implementation in SAS with a partial output is provided.

```
/* Program to analyze the data in Table 9.1 for section 9.3.2 */
proc reg data=poly1;
model y=x1 x2 x3 x4 / vif;
run;
```
```
                        The REG Procedure
                          Model: MODEL1
                    Dependent Variable: y Resp


                 Number of Observations Read        12
                 Number of Observations Used        12


                        Analysis of Variance

                               Sum of        Mean
     Source            DF     Squares       Square    F Value    Pr > F

     Model              4     0.60018      0.15005      34.66    0.0001
     Error              7     0.03031      0.00433
     Corrected Total   11     0.63049


              Root MSE             0.06580    R-Square     0.9519
              Dependent Mean      13.67083    Adj R-Sq     0.9245
              Coeff Var            0.48132


                        Parameter Estimates

                         Parameter    Standard                        Variance
     Variable  Label  DF   Estimate      Error  t Value  Pr > |t|    Inflation

     Intercept Intercept 1  11.79417    0.40562    29.08   <.0001           0
     x1              1      7.07914     1.32824     5.33   0.0011  3565.42052
     x2              1     -7.20264     1.39045    -5.18   0.0013     49948
     x3              1      2.72685     0.57633     4.73   0.0021     83327
     x4              1     -0.34833     0.08207    -4.24   0.0038     15555
```
```
data poly2;
set poly1;
x1=x1-1.75; x2=x1**2; x3=x1**3; x4=x1**4;
proc reg data=poly2;
model y=x1 x2 x3 x4 / vif;
run;
```
```
                        The REG Procedure
                          Model: MODEL1
                    Dependent Variable: y Resp


                 Number of Observations Read        12
```

```
                    Number of Observations Used         12

                      Analysis of Variance

                              Sum of          Mean
        Source          DF    Squares        Square     F Value    Pr > F

        Model            4    0.60018        0.15005      34.66    0.0001
        Error            7    0.03031        0.00433
        Corrected Total  11   0.63049

                Root MSE            0.06580    R-Square    0.9519
                Dependent Mean     13.67083    Adj R-Sq    0.9245
                Coeff Var           0.48132

                      Parameter Estimates

                          Parameter    Standard                      Variance
        Variable  Label   DF  Estimate     Error   t Value  Pr > |t|  Inflation

        Intercept Intercept 1  13.47182   0.03909   344.60   <.0001        0
        x1                  1  -0.54454   0.06247    -8.72   <.0001   7.88725
        x2                  1   0.71271   0.14251     5.00   0.0016  21.89120
        x3                  1   0.28852   0.04624     6.24   0.0004   7.88725
        x4                  1  -0.34833   0.08207    -4.24   0.0038  21.89120
```

While the analysis of variance table remains the same, as expected there are changes in the parameter estimates, but the dramatic reduction in the VIF values is worth noting. Although centering did not completely remove the effect of multicollinearity (some VIF > 10), the procedure has nonetheless enhanced the reduction of the VIF considerably and the problem is not as well pronounced as in the original case. Thus centering has helped. Similarly, for the full second-order model in (9.8) for Example 9.2, the VIF generated from this model are again very high indicating multicollinearity.

```
proc reg data=ta96;
model y=x1 x2 x12 x11 x22 / vif;
run;
                      Analysis of Variance

                              Sum of          Mean
        Source          DF    Squares        Square     F Value    Pr > F

        Model            5   8402.26454   1680.45291    596.32    <.0001
        Error           21     59.17843      2.81802
        Corrected Total 26   8461.44296

                Root MSE            1.67870    R-Square    0.9930
                Dependent Mean     66.96296    Adj R-Sq    0.9913
                Coeff Var           2.50690

                      Parameter Estimates

                          Parameter    Standard                      Variance
        Variable  Label   DF  Estimate     Error   t Value  Pr > |t|  Inflation

        Intercept Intercept 1 -5127.89907 110.29601  -46.49   <.0001        0
        x1        Temp      1    31.09639   1.34441   23.13   <.0001 1154.50000
        x2        Press     1   139.74722   3.14005   44.50   <.0001 1574.50000
        x12                 1    -0.14550   0.00969  -15.01   <.0001  304.00000
        x11                 1    -0.13339   0.00685  -19.46   <.0001  973.00000
        x22                 1    -1.14422   0.02741  -41.74   <.0001 1453.00000
```

However, a model that is based on centered values of x_1 and x_2, namely,

$$y_i = \beta_0 + \beta_1(x_1 - \bar{x}_1) + \beta_2(x_2 - \bar{x}_2) + \beta_{12}(x_1 - \bar{x}_1)(x_2 - \bar{x}_2) + \beta_{11}(x_1 - \bar{x}_1)^2 + \beta_{22}(x_2 - \bar{x}_2)^2 + \varepsilon_i,$$

where $\bar{x}_1 = 90.0$ and $\bar{x}_2 = 55.0$ is considered.

```
data ta96a;
set ta96;
x1=x1-90.0; x2=x2-55.0; x12=x1*x2; x11=x1**2; x22=x2**2;
proc reg data=ta96a;
model y=x1 x2 x12 x11 x22 / vif;
run;
                      Analysis of Variance
```

Source		DF	Sum of Squares	Mean Square	F Value	Pr > F
Model		5	8402.26454	1680.45291	596.32	<.0001
Error		21	59.17843	2.81802		
Corrected Total		26	8461.44296			

Root MSE		1.67870	R-Square	0.9930	
Dependent Mean		66.96296	Adj R-Sq	0.9913	
Coeff Var		2.50690			

Parameter Estimates

Variable	Label	DF	Parameter Estimate	Standard Error	t Value	Pr > |t|	Variance Inflation
Intercept	Intercept	1	94.92593	0.72240	131.40	<.0001	0
x1	Temp	1	-0.91611	0.03957	-23.15	<.0001	1.00000
x2	Press	1	0.78778	0.07913	9.95	<.0001	1.00000
x12		1	-0.14550	0.00969	-15.01	<.0001	1.00000
x11		1	-0.13339	0.00685	-19.46	<.0001	1.00000
x22		1	-1.14422	0.02741	-41.74	<.0001	1.00000

Again, we see that the VIF have all been reduced to 1.0 and the effect of multicollinearity seems to have been removed. Thus centering of explanatory variables can sometimes be very helpful toward removing the problem of multicollinearity in multiple regression analysis.

9.4 Orthogonal Polynomials Approach

Consider the following k-th order polynomial model

$$y_i = \beta_0 + \beta_1 x_i + \beta_2 x_i^2 + \beta_3 x_i^3 + \cdots + \beta_k x_i^k + \varepsilon_i.$$

Because of the problem of multicollinearity, suppose we wish to replace the above with the following alternative response function:

$$y_i = \alpha_0 P_0(x_i) + \alpha_1 P_1(x_i) + \alpha_2 P_2(x_i) + \cdots + \alpha_k P_k(x_i) + \varepsilon_i,$$

where $P_k(x_i)$ is a k-th *order* orthogonal polynomial and

$$\sum_{i=1}^{n} P_r(x_i) = 0 \quad \text{for all } r = 0, 1, \ldots, k \tag{9.9a}$$

$$\sum_{i=1}^{n} P_r(x_i) P_s(x_i) = 0 \quad r \neq s, \quad r, s = 0, 1, \ldots, k \tag{9.9b}$$

$$P_0(x_i) = 1. \tag{9.9c}$$

The OLS estimates of the α's are given in this case by

$$\hat{\alpha}_j = \frac{\sum P_j(x_i)\, y_i}{\sum P_j^2(x_i)}, \quad j = 1, 2, 3, \ldots, k,$$

$$\hat{\alpha}_0 = \bar{y},$$

and the corresponding regression SS is computed as

$$\text{SS}_R(\alpha_j) = \hat{\alpha}_j \sum P_j(x_i)\, y_i. \tag{9.10}$$

For x equally spaced with distance d, the first four orthogonal polynomials are

$$P_0(x_i) = 1,$$

$$P_1(x_i) = \lambda_1 \left[\frac{x_i - \bar{x}}{d} \right],$$

$$P_2(x_i) = \lambda_2 \left[\left(\frac{x_i - \bar{x}}{d} \right)^2 - \left(\frac{n^2 - 1}{12} \right) \right],$$

$$P_3(x_i) = \lambda_3 \left[\left(\frac{x_i - \bar{x}}{d} \right)^3 - \left(\frac{x_i - \bar{x}}{d} \right) \left(\frac{3n^2 - 7}{20} \right) \right],$$

where λ_i and the $P_j(x_i)$ are obtained from standard table of orthogonal polynomials. Table 6 in the Appendix contains the orthogonal polynomial coefficients.

Example 9.3

Let us analyze the data in Table 9.7, where the x's are equally spaced, that is, $d = 1$. Suppose we wish to fit a second-order polynomial, that is, a parabolic model, then we need $P_1(x_i)$ and $P_2(x_i)$ from Table 6 in the Appendix. Since $n = 8$, columns 2, 3, and 4 in Table 9.7 gives these coefficients, extracted from Table 6 in the Appendix.

x_i	$P_0(x_i)$	$P_1(x_i)$	$P_2(x_i)$	y_i	$P_0(x_i)\, y_i$	$P_1(x_i)\, y_i$	$P_2(x_i)\, y_i$
1	1	-7	7	1.0	1.0	-7.0	7.0
2	1	-5	1	1.2	1.2	-6.0	1.2
3	1	-3	-3	1.8	1.8	-5.4	-5.4
4	1	-1	-5	2.5	2.5	-2.5	-12.5
5	1	1	-5	3.6	3.6	3.6	-18.0
6	1	3	-3	4.7	4.7	14.1	-14.1
7	1	5	1	6.6	6.6	33.0	6.6
8	1	7	7	9.1	9.1	63.7	63.7
36	8	168	168	30.5	30.5	93.5	28.5
		$\lambda_1 = 2$	$\lambda_2 = 1$				

Table 9.7: Set-up for the calculations

Here,

$$\sum_{i=1}^{n} P_0(x_i)\, y_i = 30.5, \quad \sum_{i=1}^{n} P_1(x_i)\, y_i = 93.5, \quad \sum_{i=1}^{n} P_2(x_i)\, y_i = 28.5$$

$$\sum_{i=1}^{n} P_0^2(x_i) = 8, \quad \sum_{i=1}^{n} P_1^2(x_i) = 168, \quad \sum_{i=1}^{n} P_2^2(x_i) = 168.$$

Hence,

$$\hat{\alpha}_0 = \frac{30.5}{8} = 3.8125, \quad \hat{\alpha}_1 = \frac{93.5}{168} = 0.5565 \quad \text{and} \quad \hat{\alpha}_2 = \frac{28.5}{168} = 0.16964.$$

The regression sums of squares (Reg SS) is computed as

$$\hat{\alpha}_j \sum P_j(x_i)\, y_i = (0.5565 * 93.5) + (0.16964 * 28.5) = 56.8675.$$

The fitted equation in terms of the orthogonal polynomials is therefore given by

$$\hat{y} = 3.8125 + 0.5565 P_1(x) + 0.16964 P_2(x).$$

Writing the above in terms of the original x variable, we have, since $\lambda_1 = 2$ and $\lambda_2 = 1$,

$$\hat{y} = 3.8125 + 0.5565(2)\left(\frac{x-4.5}{1}\right) + 0.16964(1)\left[\left(\frac{x-4.5}{1}\right)^2 - \left(\frac{8^2-1}{12}\right)\right]$$

$$= 3.8125 + 1.113(x-4.5) + 0.16964[x^2 - 9X + 20.5 - 5.25]$$

$$= 3.8125 + 1.113x - 5.0085 + 0.16964x^2 - 1.52676x + 2.5446, \quad \text{that is,}$$

$$\hat{y} = 1.3486 - 0.41376x + 0.16964x^2.$$

The above predicted equation agrees with those given by **proc orthoreg** in SAS for the same data. Note that $\bar{x} = 36/8 = 4.5$.

```
/* program to analyze the data in Example 9.3 */
data poly;
title h=1 'Polynomial Regression';
input x y @@;
label x='Dosage' y='Eight';
x2=x**2; x3=x**3;
datalines;
1 1 2 1.2 3 1.8 4 2.5 5 3.6 6 4.7 7 6.6 8 9.1
;
goptions vsize=5in hsize=5in;
axis1 label=(angle=-90 rotate=90 'Eight (y)'); axis2 label=('Dosage (x)');
proc gplot data=poly;
plot y*x='plus' / vaxis=axis1 haxis=axis2 noframe;
run;
proc orthoreg;
model y=x x2;
run;
```

                       The ORTHOREG Procedure

                   Dependent Variable: y   Eight

                              Sum of
    Source   DF   Squares   Mean Square   F Value   Pr > F

    Model   2   56.87202381   28.436011905   722.73   <.0001
    Error   5   0.1967261905   0.0393452381
    Corrected Total 7   57.06875

                    Root MSE   0.1983563412
                    R-Square   0.9965528211

                                         Standard
    Variable   DF   Parameter Estimate   Error   t Value   Pr > |t|

    Intercept   1   1.34821428571428   0.2767357866   4.87   0.0046
    x   1   -0.41369047619047   0.1410915196   -2.93   0.0326
    x2   1   0.16964285714285   0.015303524   11.09   0.0001

9.5　Non-parametric Regression

We consider in this section a nonparametric regression. By definition, *nonparametric regression is a form of regression analysis in which the predictor does not take a predetermined form but is constructed according to information derived from the data. Nonparametric regression requires larger sample sizes than regression based on parametric models because the data must supply the model structure as well as the model estimates.*

Locally Weighted Regression (LOESS)

Locally weighted regression, often referred to as **loess** is a nonparametric regression model which predicts the response variable without any model assumption. The approach is closely related to the piece-wise polynomial regression methods that we described in Chapter 8. It is very useful when the functional form of the model is either unknown or complicated. The method fits simple models to localized subsets of the data. This localized neighborhood, often described as the **span** is usually a fraction of the total points used to form such neighborhoods. The **loess** procedure uses weighted least-squares procedure to estimate the function of the response based on the points in the neighborhood set. Thus the procedure can be used to fit models to segments of the data rather than specifying a single response function for the entire data-hence the similarity with the piece-wise regression. Thus at each point in the data, a low-degree polynomial (usually simple linear or quadratic) model will be fitted to a subset of the data. The weights for the weighted least-squares part of the model estimation are chosen such that more weights are assigned to points near the point whose response value is being estimated and less to points further away. This process is carried out for each of the n data points.

Most statistical software used to fit the **loess** regression uses the nearest neighbor algorithm. Basically, we use a 'smoothing parameter' q, where usually, $(d+1)/n \leq q \leq 1$, with d being the degree of the local polynomial, and q represents the proportion of data used in each model fit. Generally, $0.25 < q < 0.5$. Smaller q values are of course preferred. Furthermore, many algorithms use the **tri-cube** weighting function defined as

$$W(x) = \begin{cases} (1- \mid x \mid^3)^3, & \text{for } \mid x \mid < 1 \\ 0, & \text{for } \mid x \mid \geq 1. \end{cases}$$

The above can of course be alternatively defined as

$$W(x) = \begin{cases} (1 - x^3)^3, & \text{for } 0 \leq x < 1 \\ 0, & \text{elsewhere.} \end{cases}$$

Example 9.4

The following SAS program and partial output are from fitting the **loess** regression to the data in Table 9.2.

```
/* Program for Non-parametric regression example in section 9.5 */
proc loess data=poly1;
model y=x/smooth=0.4 0.5 0.6 0.7 residual;
ods output OutputStatistics = aa;
run;
proc print data=aa;
id obs;
run;
symbol1 color=black value=dot;
symbol2 color=black i=spline line=1 value=none;
axis1 label=(angle=90 rotate=0);

goptions nodisplay;
proc gplot data=aa;
by SmoothingParameter;
plot DepVar*x=1 Pred*x/&opts name='fit';
run; quit;

goptions display;
proc greplay nofs tc=sashelp.templt template=l2r2;
igout gseg;
treplay 1:fit 2:fit2 3:fit1 4:fit3;
run;
```

```
                        The LOESS Procedure
                     Smoothing Parameter: 0.4
                     Dependent Variable: y

                         Fit Summary

            Fit Method                      kd Tree
            Blending                        Linear
            Number of Observations               12
            Number of Fitting Points              6
            kd Tree Bucket Size                   1
            Degree of Local Polynomials           1
            Smoothing Parameter             0.40000
            Points in Local Neighborhood          4
            Residual Sum of Squares         0.02625
```

Obs	Smoothing Parameter	x	Dep Var	Pred	Residual
1	0.4	0.5	13.90	13.85500	0.04500
2	0.4	0.5	13.81	13.85500	-0.04500
3	0.4	1.0	14.08	14.03500	0.04500
4	0.4	1.0	13.99	14.03500	-0.04500
5	0.4	1.5	13.75	13.67500	0.07500
6	0.4	1.5	13.60	13.67500	-0.07500
7	0.4	2.0	13.32	13.35500	-0.03500

8	0.4	2.0	13.39	13.35500	0.03500
9	0.4	2.5	13.45	13.49000	-0.04000
10	0.4	2.5	13.53	13.49000	0.04000
11	0.4	3.0	13.59	13.61500	-0.02500
12	0.4	3.0	13.64	13.61500	0.02500
1	0.5	0.5	13.90	13.85500	0.04500
2	0.5	0.5	13.81	13.85500	-0.04500
3	0.5	1.0	14.08	14.03500	0.04500
4	0.5	1.0	13.99	14.03500	-0.04500
5	0.5	1.5	13.75	13.67500	0.07500
6	0.5	1.5	13.60	13.67500	-0.07500
7	0.5	2.0	13.32	13.35500	-0.03500
8	0.5	2.0	13.39	13.35500	0.03500
9	0.5	2.5	13.45	13.49000	-0.04000
10	0.5	2.5	13.53	13.49000	0.04000
11	0.5	3.0	13.59	13.61500	-0.02500
12	0.5	3.0	13.64	13.61500	0.02500
1	0.6	0.5	13.90	13.91968	-0.01968
2	0.6	0.5	13.81	13.91968	-0.10968
3	0.6	1.0	14.08	13.88039	0.19961
4	0.6	1.0	13.99	13.88039	0.10961
5	0.6	1.5	13.75	13.68645	0.06355
6	0.6	1.5	13.60	13.68645	-0.08645
7	0.6	2.0	13.32	13.48527	-0.16527
8	0.6	2.0	13.39	13.48527	-0.09527
9	0.6	2.5	13.45	13.48714	-0.03714
10	0.6	2.5	13.53	13.48714	0.04286
11	0.6	3.0	13.59	13.61620	-0.02620
12	0.6	3.0	13.64	13.61620	0.02380
1	0.7	0.5	13.90	13.91968	-0.01968
2	0.7	0.5	13.81	13.91968	-0.10968
3	0.7	1.0	14.08	13.88039	0.19961
4	0.7	1.0	13.99	13.88039	0.10961
5	0.7	1.5	13.75	13.68645	0.06355
6	0.7	1.5	13.60	13.68645	-0.08645
7	0.7	2.0	13.32	13.48527	-0.16527
8	0.7	2.0	13.39	13.48527	-0.09527
9	0.7	2.5	13.45	13.48714	-0.03714
10	0.7	2.5	13.53	13.48714	0.04286
11	0.7	3.0	13.59	13.61620	-0.02620
12	0.7	3.0	13.64	13.61620	0.02380

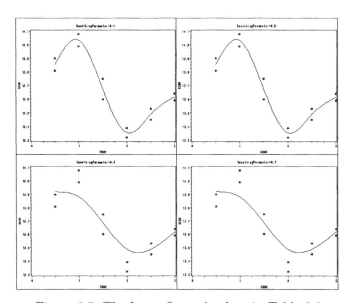

Figure 9.5: The **loess** fit to the data in Table 9.2

9.6   Exercises

9.1 The following data relate to height in inches and weight in pounds of 16 students in a community college. Carry out a lack of fit test on the data.

x_i	y_i	x_i	y_i	x_i	y_i	x_i	y_i
60	90	62	120	68	130	68	135
60	100	62	125	68	117	70	160
60	110	66	140	68	155	70	145
62	116	66	170	68	140	70	148

9.2 An operations analyst in a multinational electronics firm studied factors affecting production in a piecework operation where earnings are based on the number of pieces produced. Two employees each were selected from various age groups and data on their productivity last year were obtained (X is age of employee, in years) and y is employees's productivity. The analyst wants to fit a third degree polynomial to the data in order to predict the mean response. A model of the form below is proposed.

$$y_i = \beta_0 + \beta_1 x_i + \beta_2 x_i^2 + \beta_3 x_i^3 + \varepsilon_i, \quad \text{where} \quad x_i = X_i - \bar{X}$$

i	X_i	y_i	i	X_i	y_i	i	X_i	y_i	i	X_i	y_i
1	20	97	6	30	106	11	45	97	16	55	109
2	20	93	7	35	109	12	45	101	17	60	112
3	25	99	8	35	111	13	50	105	18	60	110
4	25	105	9	40	100	14	50	103			
5	30	109	10	40	105	15	55	105			

Fit the model and answer the following questions:

(a) Test whether or not there is a regression relation, use $\alpha = 0.01$. State the null and the alternative hypotheses, the decision rule, and conclusion. What is the p-value of this test?

(b) Write down the estimated regression equation in terms of x and X respectively.

(c) Express the fitted regression function obtained in (b) in terms of the original variable X.

(d) Test whether both the quadratic and cubic terms can be dropped from the model. Use $\alpha = 0.01$. State the null and the alternative hypotheses, the decision rule and conclusion.

(e) Test whether the cubic term can be dropped from the model. Use $\alpha = 0.01$. State the null and the alternative hypotheses, the decision rule and conclusion.

(f) The residuals from the above model is plotted first against x and against fitted or predicted values on separate graphs. What do the plots show?

(g) Is the normality assumption for the regression model satisfied in this case?

(h) Is the orthogonal regression necessary? Why? Compare the parameter estimates in this case with those obtained in (c) above.

(i) Using computer output, complete the table below and formally test for lack of fit.

Source	d.f.	SS	MS	F
Regression	3			
Degree 1 (x)	1			
Degree 2 ($x^2 \mid x$)	1			
Degree 3 ($x^3 \mid x^2, x$)	1			
Residual				
Lack of Fit				
Pure Error				

9.3 The data in the file **sat.txt** is taken from the SAT data submitted by Guber, D.L. to the Journal of Statistics Education Data Archive (see also Guber, D.L., 1999). For this problem, use the first 50 cases and the following variables: The current expenditure per pupil (x_1), the average pupil/teacher ratio (x_2), the estimated annual salary of teachers (x_3), the percentage of students taking SAT (x_4), and the average total SAT score (y).

Suppose a researcher is interested in investigating how well, y, the average total SAT score can be predicted by using the predictor variables x_1 - x_4. A model of the following form is suggested.

$$y_i = \beta_0 + \beta_1 x_1 + \beta_2 x_2 + \beta_3 x_3 + \beta_4 x_4 + \beta_5 x_5 + \beta_6 x_6 + \varepsilon_i,$$

where, $x_5 = x_1 * x_2$ and $x_6 = x_3 * x_4$.

However, two forms of the above model are employed in the analysis. One form uses the x_i as predictor variables, while the second form uses $(x_i - \bar{x}_i)$ as predictor variables. Use a statistical software to fit both forms and answer the following questions:

(a) Why was the second form of the predictor variables necessary?

(b) Which of the form will you suggest that we use and why?

(c) What would you attribute to the problem in this exercise?

(d) With your choice, which of the variables would you consider to be important predictors of y, and what would be the order of importance?

(e) Test the hypothesis that both x_5 and x_6 are not needed in the model. Conduct the test at $\alpha = 0.05$.

(f) What would you suggest as the best model? Is this final?

(g) What other solutions can you suggest for solving the problem encountered in this problem from the first regression model?

9.4 The data in the file **tryptone.txt** is taken from the bacteria counts data submitted by Binnie, N.S. to the Journal of Statistics Education Data Archive (see also Binnie, N., 2004). For this problem, use the following variables: The bacteria counts for strain 1 (y) and the temperature (x). Fit the regression model

$$y_i = \beta_0 + \beta_1 x_i + \beta_2 x_i^2 + \varepsilon_i$$

to the data. Test at $\alpha = 0.05$ whether the quadratic term can be dropped from the model.

9.5 The data in the file **tryptone.txt** is taken from the bacteria counts data submitted by Binnie, N.S. to the Journal of Statistics Education Data Archive (see also Binnie, N., 2004). For this problem, use the following variables: The bacteria counts for strain 4 (y) and the temperature (x). Consider the regression model

$$y_i = \beta_0 + \beta_1 x_i + \beta_2 x_i^2 + \varepsilon_i$$

for the data.

(a) Fit the regression model to the data.

(b) Test at $\alpha = 0.05$ whether the quadratic term can be dropped from the model.

(c) Plot the residuals from the model against the predicted values.

(d) In the model, check for multicollinearity, influential observations, outliers as well as other relevant diagnostics.

(e) Explains your findings in (d) and suggest remedial measures for the violation(s).

9.6 (a) Consider the regression model
$$y_i = \beta_0 + \beta_1 x_i + \beta_2 x_i^2 + \varepsilon_i.$$

(i) How many levels of x are required to fit the regression model?

(ii) How large a sample size is required to have sufficient degrees of freedom for estimating the error variance?

(b) Consider the regression model

$$y_i = \beta_0 + \beta_1 x_1 + \beta_2 x_2 + \beta_3 x_1 x_2 + \varepsilon_i.$$

(i) How many levels of x_1 are required to fit the regression model?

(ii) How many levels of x_2 are required to fit the regression model?

(iii) How large a sample size is required to have sufficient degrees of freedom for estimating the error variance?

Chapter 10

Logistic Regression

10.1　Introduction

In the previous chapters, we have built regression models based on the response or dependent variable(s) being of the continuous type. As a matter of fact, all the regression models employed to data so far, have all been on continuous type dependent variables. What happens if the dependent or response variable is not continuous but categorical? For example, given several explanatory variables, suppose the outcome variable is 'yes' or 'no'; survived or dead; defective or not defective etc. In all these cases, the outcome variables are binary (or dichotomous) and categorical. In such a case the classical regression analysis that we have employed so far would be inadequate to model the response variable as function(s) of the given explanatory variables. We examine below various other situations that we may have with categorical response variable.

1. The dependent variable has a binary outcome as in the examples above.

2. The response variable has more than two outcome categories. For example, suppose n subjects were jointly classified by gender, income with the response being party affiliation of the respondents which have say, three categories 'democrat', 'independent', and 'republican'. Clearly, the response variable is not only categorical but also nominal.

3. Consider the example in (2) above, but the response variable is the respondents' political ideology, namely, 'liberal', 'moderate' and 'conservative'. Similarly, for example, patients classified by the severity of nausea of patients undergoing certain chemotherapy could be (none, moderate, severe). Other examples of this type are the Likert type variables, for example, the attitudinal response toward abortion can be classified as (disapprove, middle, approve).

It should be quite obvious that the form of analysis we employ depend on the type of categorical response variable we have. For the binary response variable in (1), the logistic regression models will be employed and we shall develop these further later in this chapter. For the response variable in (2) having more than two categories and nominal, the *baseline category* model (Lawal, 2003) can readily be employed in this case.

For the type of response variables in (3), we observe that the response variables not only have more than two categories but there is intrinsic ordering about the category levels. In other words, the response variables can be considered ordinal. In this case the class of *cumulative logit* models, *proportional odds, adjacent-category logit, continuation ratio* models, as well as *mean response* models will be appropriate. For a fuller explanation of these models and their implementations see Lawal (2003).

In this chapter however, our goal is to concentrate on the first case, namely, when the response variable is binary and factor or explanatory variables can be categorical, continuous or both. That is, the explanatory variables in logistic regression can be discrete or continuous. We discuss this in the next section.

10.2   Binary Response Model

Logistic regression is used to model the probability π of occurrence of a binary or dichotomous outcome. Many areas of research these days have outcomes that are binary. Examples of such outcomes could be mortality (dead or alive). Binary logistic regression therefore is a type of regression when the dependent variable is categorical. Here, the theory of ordinary least squares (OLS) does not work well. A model that utilizes the OLS has been described as the *linear probability model* (Lawal, 2003). We now consider the linear probability model.

The Linear Probability Model

Consider a binary outcome variable y_i, $(i = 1, 2, \ldots, n)$ such that

$$y_i = \begin{cases} 1, & \text{if event occurs} \\ 0, & \text{if event does not occur.} \end{cases}$$

Let the probability that y_i takes the value 1 be π_i and 0 with probability $1 - \pi_i$. If we let $\mathbf{X}$ be $p - 1$ factor or explanatory variables, then, a linear probability model (Grizzle et al., 1969) is of the form

$$y_i = \beta_0 + \beta_0 + \beta_1 x_{i1} + \beta_2 x_{i2} + \ldots + \beta_{p-1} x_{i,p-1} + \varepsilon_i, \quad i = 1, 2, \ldots, n. \tag{10.1}$$

Example 10.1

Consider the following bio-assay example from Lawal (2003). The data in Table 10.1 give the effect of different concentrations of nicotine sulphate in a 1% saponin solution on a certain insect *Drosophila Melanogesta-the fruit fly.*

Nicotine Sulphate gm/100cc x_i	Number killed r_i	Number of insects n_i	Observed proportions π_i
0.10	8	47	0.1702
0.15	14	53	0.2642
0.20	24	55	0.4384
0.30	32	52	0.6154
0.50	38	46	0.8261
0.70	50	54	0.9259
0.95	50	52	0.9615

Table 10.1: Effect of different concentrations of nicotine sulphate on DM, Hubert (1992)

We usually employ the logarithm to base e of the explanatory variable, that is, $\log(x_i)$. If we let π_i be the observed proportion of insects killed at dose level $i = 1, 2, 3, 4, 5, 6, 7$, then the model of interest is

$$\pi_i = \beta_0 + \beta_1 \log x_i + \varepsilon_i, \quad i = 1, 2, \ldots, 7. \tag{10.2}$$

We employ SAS to fit the linear probability model using OLS to the data. The resulting estimated linear probability model is

$$\hat{\pi}_i = 1.04170 + 0.3810 \log x_i.$$

```
data new;
input x1 r n @@;
x=log(x1);
p=r/n;
datalines;
.1 8 47 .15 14 53 .2 24 55
.3 32 52 .5 38 46 .7 50 54 .95 50 52
;
run;
proc print;
proc reg;
model p=x;
```

```
output out=aa p=pred r=resid;
run;
proc print data=aa;
run;
```

The REG Procedure
Model: MODEL1
Dependent Variable: p

```
            Number of Observations Read          7
            Number of Observations Used          7
```

Analysis of Variance

Source	DF	Sum of Squares	Mean Square	F Value	Pr > F
Model	1	0.60201	0.60201	284.73	<.0001
Error	5	0.01057	0.00211		
Corrected Total	6	0.61258			

Root MSE	0.04598	R-Square	0.9827	
Dependent Mean	0.59995	Adj R-Sq	0.9793	
Coeff Var	7.66418			

Parameter Estimates

Variable	DF	Parameter Estimate	Standard Error	t Value	Pr > |t|
Intercept	1	1.04170	0.03142	33.15	<.0001
x	1	0.38109	0.02258	16.87	<.0001

x1	r	n	x	p	pred	resid
0.10	8	47	-2.30259	0.17021	0.16421	0.005999
0.15	14	53	-1.89712	0.26415	0.31873	-0.054580
0.20	24	55	-1.60944	0.43636	0.42836	0.008001
0.30	32	52	-1.20397	0.61538	0.58288	0.032504
0.50	38	46	-0.69315	0.82609	0.77755	0.048538
0.70	50	54	-0.35667	0.92593	0.90577	0.020152
0.95	50	52	-0.05129	0.96154	1.02215	-0.060613

One of the main problems of using the probability model is in its inability to predict or guarantee expected or predicted probabilities or proportions within the range $[0, 1]$. That is, the model can give predicted probabilities that are either < 0 or > 1.0 or both. In our example above, we see from the SAS output, that the predicted probability or proportion for the sample data when $x_7 = 0.95$ has a predicted proportion $\hat{\pi}_7 = 1.0222 > 1$. Other violations of the OLS assumptions when using the linear probability model include the fact that the dependent variable is binomially distributed rather than the normality assumption under the general linear model. Since y_i in (10.1) takes the value 1 with probability π_i and 0 with probability $1 - \pi_i$, then $\text{Var}(Y_i) = \pi_i(1 - \pi_i)$. Thus, the variance of the error term is a function of π, that is, $\text{Var}(\varepsilon_i) = \pi_i(1 - \pi_i)$, where π_i is a function of x_i. The variance is not constant across the values of x_i (as x_i changes) and this violates the assumption of constant variance or homoscedasticity.

10.3   Linear Logistic Regression

For a simple logistic regression having only one explanatory variable, the logistic regression model has the form

$$\ln\left(\frac{\pi_i}{1 - \pi_i}\right) = \beta_0 + \beta_1 x_i. \tag{10.3}$$

The expression in (10.3) can sometimes be written as

$$\frac{\pi_i}{1 - \pi_i} = e^{\beta_0 + \beta_1 x_i}, \tag{10.4}$$

where

- $e = 2.71828$, is the base of the natural logarithm.

- $\frac{\pi_i}{1-\pi_i}$ is the odds ratio.

- $\ln\left(\frac{\pi_i}{1-\pi_i}\right)$ is the log odds ratio or simply the logit, and

- β_0 and β_1 are the parameters of the model to be estimated.

The logistic regression in (10.3) does not assume normality of error terms nor does it assume homoscedastic of error variances. The result in (10.4) can be re-written as

$$\pi_i = \frac{e^{\beta_0+\beta_1 x_i}}{1 + e^{\beta_0+\beta_1 x_i}} \quad \text{or} \quad \pi_i = \frac{1}{1 + e^{-(\beta_0+\beta_1 x_i)}}.$$

In Figure 10.1, we have the graphs of the logistic function

$$\pi_i = \frac{1}{1 + e^{-(\beta_0+\beta_1 x_i)}},$$

for fixed $\beta_1 = 0.1$, varying $\beta_0 = -2.5, -1.5, -1.0, 2.0$ values and values of x ranging from -40 to 80. That is, for β_0 taking values $\{-2.5, -1.5, -1.0, 2.0\}$, we plot the graph of π_i against x for a constant value of β_0 at 0.1. The plot is presented in Figure 10.1 where

$$\pi_i = \frac{1}{1 + e^{-(\beta_0+0.1 x_i)}}; \quad x = -40, \ldots, 80.$$

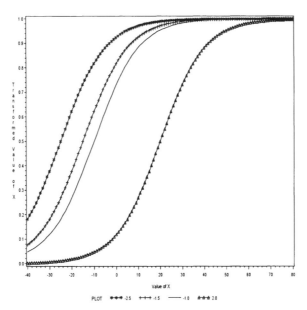

Figure 10.1: Plot of logistic function

10.3.1   Estimating the Parameters of a Logistic Regression Model

The parameters are estimated by maximum likelihood estimation (MLE) where the likelihood equations are derived by using the Fisher's scoring algorithm. SAS employs this procedure to estimate the parameters of the logistic model. A more detailed explanation of Fisher's scoring algorithm can be found in Lawal (2003) or in similar books on categorical data analysis. SAS **proc logistic**, **genmod** and **catmod** can be employed to fit logistic models to a data set. However, before using SAS it is important to discuss two important forms in which data may arise. The data may be in grouped form like the data in Table 10.1 or the data may come as ungrouped. We shall consider both cases and how to use any of the above SAS procedures to analyze each case.

10.3.2 Grouped Data

The form of the data in Table 10.1 is an example of grouped data where we do not have individual observations but rather grouped data at a given design point. For instance when the nicotine Sulphate gm/100cc is 0.10 gm/100cc, we have a total of 47 flies exposed at this concentration and a total of 8 as the frequency of success, 'died' in this case. Thus, we have frequency of successes to be 8 and frequency of failure 'alive' to be $47 - 8 = 39$. We usually denote the frequency of successes as r_i and number of trials as n_i. At this design point, both r_i and n_i are respectively 8 and 47. In this case, the simple logistic regression model becomes

$$\ln\left(\frac{r_i}{n_i - r_i}\right) = \beta_0 + \beta_1 x_i. \tag{10.5}$$

The goodness-of-fit test statistics often used are the Pearson's chi-square (X^2) and the deviance (G^2) which are defined in this case as

$$X^2 = \sum_{i=1}^{k} \frac{(r_i - \hat{r}_i)^2}{\hat{r}_i},$$

$$G^2 = 2\sum_{i=1}^{k} \left[r_i \ln\left(\frac{r_i}{\hat{r}_i}\right) + (n_i - r_i)\ln\left(\frac{n_i - r_i}{n_i - \hat{r}_i}\right)\right],$$

where $\hat{r}_i = n_i \hat{\pi}_i$, and $\hat{\pi}_i$ is the estimated probability from the model in (10.5) and we set the case when $0\ln(0)$ to be zero. Implementation of the grouped logistic regression in **proc logistic** in SAS is often referred to as the *events/trials* modeling. We now implement this in SAS with **proc logistic**. The events/trial is implemented in the model statement followed after the equality sign with the explanatory variable(s). In this case, the explanatory variable is $\log 10(x_1)$. The options in the model statement in **proc logistic** states, **scale=none, aggregate covb lackfit**.

```
data new;
input x1 r n @@;
x=log10(x1);
datalines;
.1 8 47 .15 14 53 .2 24 55
.3 32 52 .5 38 46 .7 50 54 .95 50 52
;
run;
proc print;
proc logistic data=new;
model r/n=x/scale=none aggregate covb lackfit;
output out=aa p=pred stdxbeta=self h=lev;
run;
proc print data=aa noobs;
var r n x1 x pred self lev;
format pred self lev 10.4;
run;
```

The option **aggregate** must accompany the **scale=none** option in the logistic model statement. The **aggregate** option allows the goodness-of-fit test statistics (deviance or G^2 and the Pearson's X^2 to be presented.

```
                    The LOGISTIC Procedure

                    Model Information

        Data Set                    WORK.NEW
        Response Variable (Events)   r
        Response Variable (Trials)   n
        Model                       binary logit
        Optimization Technique      Fisher's scoring

            Number of Observations Read      7
            Number of Observations Used      7
            Sum of Frequencies Read        359
            Sum of Frequencies Used        359

                    Response Profile

            Ordered    Binary        Total
            Value      Outcome     Frequency
```

```
              1      Event           216
              2      Nonevent        143

              Model Convergence Status

       Convergence criterion (GCONV=1E-8) satisfied.

       Deviance and Pearson Goodness-of-Fit Statistics

   Criterion        Value      DF    Value/DF    Pr > ChiSq
   -------------------------------------------------------------
   Deviance        0.7336       5     0.1467        0.9811
   Pearson         0.7351       5     0.1470        0.9810

            Number of unique profiles: 7
```

The output tells us that there are 7 observations, and we are modeling the binary logit using Fisher's scoring algorithm as the method of optimization. The model fitted gives a deviance or likelihood ratio test statistic G^2 value of 0.7336 on 5 d.f. (p-value= 0.9811). Thus the model fits the data very well.

Next, the output considers fitting a model of the form involving intercept only and the one involving intercept and the covariate $\log 10(x_1)$. The difference between the two values of $-2\log L$ ($482.732 - 337.443 = 145.289$) gives the extra contribution toward the overall value by the covariate given that the intercept is already in the model. This value and those generated from SAS using either the score or Wald's test are presented in the following.

```
           Model Fit Statistics

                                     Intercept
                        Intercept       and
        Criterion         Only       Covariates

        AIC              484.732       341.443
        SC               488.615       349.210
        -2 Log L         482.732       337.443
```

This value and those generated from using either the score or Wald's test are presented in the test of hypothesis concerning the parameter of the covariate. That is, for the hypotheses

$$H_0 : \beta_1 = 0 \quad \text{versus} \quad H_a : \beta_1 \neq 0.$$

The output below from **proc logistic** implementation allows us to test the null hypothesis against the alternative hypothesis.

```
        Testing Global Null Hypothesis: BETA=0

   Test              Chi-Square     DF    Pr > ChiSq

   Likelihood Ratio   145.2882       1      <.0001
   Score              127.2173       1      <.0001
   Wald                92.3620       1      <.0001
```

The p-values from each of the three test criteria indicate that we would strongly reject H_0. Indicating that the covariate is very important for our model specification. The parameters of the model from the SAS output are again displayed in the following.

```
            The LOGISTIC Procedure

       Analysis of Maximum Likelihood Estimates

                          Standard      Wald
   Parameter  DF  Estimate   Error   Chi-Square  Pr > ChiSq

   Intercept   1   3.1236    0.3349   86.9809      <.0001
   x           1   4.8995    0.5098   92.3620      <.0001

              Odds Ratio Estimates

                  Point        95% Wald
         Effect  Estimate   Confidence Limits
           x     134.228    49.419    364.582

   Association of Predicted Probabilities and Observed Responses
```

```
Percent Concordant      80.0    Somers' D    0.692
Percent Discordant      10.8    Gamma        0.762
Percent Tied             9.2    Tau-a        0.333
Pairs                  30888    c            0.846
```

The parameter estimates are $\hat{\beta}_0 = 3.1236$ and $\hat{\beta}_1 = 4.8995$ and are both significantly different from zero by using the *p*-values. The variance-covariance matrix of the parameter estimates are obtained with the option **covb** in the model statement. These estimates are

```
           Estimated Covariance Matrix

    Parameter      Intercept            x

    Intercept      0.112172      0.15627
    x              0.15627       0.259907
```

With each increase in $\log_{10}(\text{dose})$, the odds of a fruit fly dying increases by $e^{4.8995} = 134.228$ times. This odd is also presented in the SAS output along with its 95% confidence limits. The estimated killing probability as a function of the $\log_{10}(\text{dose})$ is estimated as

$$\hat{\pi} = \frac{e^{3.1236 + 4.8995\,\log_{10}(\text{dose})}}{1 + e^{3.1236 + 4.8995\,\log_{10}(\text{dose})}},$$

which for the first dose becomes

$$\hat{\pi}_1 = \frac{e^{3.1236 + 4.8995(-1)}}{1 + e^{3.1236 + 4.8995(-1)}} = \frac{0.1693}{1.1693} = 0.1448.$$

The expected number of deaths for this dose level is $n_1 * \hat{\pi}_1 = 47 * 0.1448 = 6.8056$ or about 7 insects. These and other relevant parameters are displayed below.

```
Obs    r    n     yhat     phat     selp     lev     resid
-----------------------------------------------------------
 1     8    47    6.8056   0.1448   0.2440   0.3465   0.4838
 2    14    53   15.1792   0.2864   0.1763   0.3368  -0.3604
 3    24    55   23.3915   0.4253   0.1439   0.2782   0.1657
 4    32    52   33.1188   0.6369   0.1408   0.2382  -0.3206
 5    38    46   38.5802   0.8387   0.2041   0.2591  -0.2305
 6    50    54   49.3614   0.9141   0.2646   0.2968   0.3172
 7    50    52   49.5664   0.9532   0.3246   0.2442   0.2926
-----------------------------------------------------------
```

The final model based on the logistic regression is given by

$$\ln\left(\frac{\hat{\pi}_i}{1 - \hat{\pi}_i}\right) = 3.1236 + 4.8995\,\log_{10}(\text{dose}_i), \quad i = 1, 2, \ldots, 7. \tag{10.6}$$

The plot of the estimated logistic regression probabilities are presented in Figure 10.2.

```
proc logistic data=new;
model r/n=x/scale=none aggregate covb lackfit;
output out=aa p=pred stdxbeta=self h=lev;
run;
proc print data=aa noobs;
var r n x1 x pred self lev;
format pred self lev 10.4;
run;
proc gslide;
run;
goptions vsize=6in
        hsize=6in;
 proc sort data=aa;
 by x1;
 run;
symbol i=spline value=none height=.75;
axis1 label=(angle=-90 rotate=90 'predicted probs');
axis2 label=('log10 of dose');
proc gplot data=aa;
plot pred*x1/vaxis=axis1 haxis=axis2 noframe vref=1.0;
run;
```

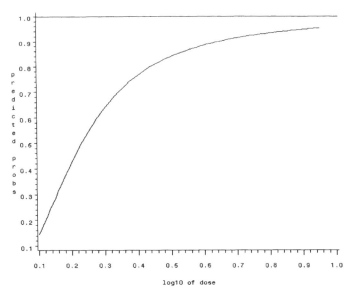

Figure 10.2: Fitted logistic model of logistic function

```
                    Partition for the Hosmer and Lemeshow Test

                               Event              Nonevent
          Group     Total   Observed  Expected  Observed  Expected
            1        47        8        6.81       39      40.19
            2        53       14       15.18       39      37.82
            3        55       24       23.39       31      31.61
            4        52       32       33.12       20      18.88
            5        46       38       38.58        8       7.42
            6        54       50       49.36        4       4.64
            7        52       50       49.57        2       2.43

            Hosmer and Lemeshow Goodness-of-Fit Test

            Chi-Square      DF      Pr > ChiSq
             0.7351          5        0.9810
```

10.3.3   PROC CATMOD Implementation

The logistic model for the data in Table 10.1 can also be implemented with **proc catmod**. We present below the SAS program required to implement this together with a partial SAS output.

```
data logit;
set new;
count=r;
y=0;
output;
count=n-r;
y=1;
output;
drop n r;
proc catmod data=logit;
weight count;
direct x;
model y=x/freq ml nogls;
run;
ml nogls;
run;
                         The CATMOD Procedure

                            Data Summary

             Response            y      Response Levels    2
             Weight Variable     count  Populations        7
             Data Set            logit  Total Frequency   359
             Frequency Missing   0      Observations       14

                         Population Profiles
```

```
Sample    x                  Sample Size
----------------------------------------
  1              -1               47
  2      -0.823908741            53
  3      -0.698970004            55
  4      -0.522878745            52
  5      -0.301029996            46
  6       -0.15490196            54
  7      -0.022276395            52
```

```
             Response Profiles

          Response    y
          -------------
             1        0
             2        1
```

```
          Response Frequencies

              Response Number
       Sample     1      2
       -----------------------
          1       8      39
          2      14      39
          3      24      31
          4      32      20
          5      38       8
          6      50       4
          7      50       2
```

```
          Maximum Likelihood Analysis

    Maximum likelihood computations converged.
```

```
     Maximum Likelihood Analysis of Variance

   Source        DF    Chi-Square    Pr > ChiSq
   ------------------------------------------------
   Intercept      1       86.98        <.0001
   x              1       92.36        <.0001

   Likelihood Ratio  5     0.73        0.9811
```

```
     Analysis of Maximum Likelihood Estimates

                       Standard      Chi-
   Parameter  Estimate    Error     Square   Pr > ChiSq
   ----------------------------------------------------
   Intercept   3.1236    0.3349     86.98      <.0001
   x           4.8996    0.5098     92.36      <.0001
```

Results from the **catmod** procedure agree with those obtained from **proc logistic**.

10.3.4 PROC GENMOD Implementation

SAS **proc genmod** can also be employed to fit the logistic model to the data in Table 10.1. The following SAS program will fit this model and output some relevant statistics of interest.

```
set new;
proc genmod;
make 'obstats' out=aa;
model r/n=x/dist=bin link=logit obstats;
run;
proc print data=aa noobs;
var r n pred xbeta std resraw reschi resdev stresdev streschi;
format pred xbeta std resraw reschi resdev stresdev streschi 8.3;
run;
                        The GENMOD Procedure

                        Model Information

              Data Set                    WORK.NEW
              Distribution                Binomial
              Link Function               Logit
              Response Variable (Events)         r
              Response Variable (Trials)         n
```

```
Number of Observations Read       7
Number of Observations Used       7
Number of Events                216
Number of Trials                359

Criteria For Assessing Goodness Of Fit

Criterion              DF       Value      Value/DF

Deviance                5      0.7336        0.1467
Scaled Deviance         5      0.7336        0.1467
Pearson Chi-Square      5      0.7351        0.1470
Scaled Pearson X2       5      0.7351        0.1470
Log Likelihood               -168.7217
```

Algorithm converged.

```
                  Analysis Of Parameter Estimates

                       Standard   Wald 95% Confidence    Chi-
Parameter   DF  Estimate   Error        Limits          Square  Pr > ChiSq

Intercept    1   3.1236   0.3349    2.4672   3.7800      86.98     <.0001
x            1   4.8996   0.5098    3.9003   5.8988      92.36     <.0001
Scale        0   1.0000   0.0000    1.0000   1.0000
```

NOTE: The scale parameter was held fixed.

```
             r         n        Pred      Xbeta       Std
             8        47       0.145     -1.776      0.244
            14        53       0.286     -0.913      0.176
            24        55       0.425     -0.301      0.144
            32        52       0.637      0.562      0.141
            38        46       0.839      1.649      0.204
            50        54       0.914      2.365      0.265
            50        52       0.953      3.014      0.325

     Resraw    Reschi    Resdev    Stresdev   Streschi
      1.194     0.495     0.484     0.599      0.612
     -1.177    -0.357    -0.360    -0.443     -0.439
      0.608     0.166     0.166     0.195      0.195
     -1.116    -0.322    -0.321    -0.367     -0.369
     -0.581    -0.233    -0.230    -0.268     -0.271
      0.639     0.310     0.317     0.378      0.370
      0.432     0.284     0.293     0.337      0.327
```

Again the results from **proc genmod** agree with those obtained from **proc logistic** and **proc catmod**.

Alternative Presentation of Data in Table 10.1

We give in Table 10.2 another presentation of the data in Table 10.1 in terms of status of the insects 'dead' or 'alive'. Again we would want to model the 'dead' status in SAS. The SAS program as well as the output are presented in what follows.

dose	count	status	dose	count	status
0.10	8	died	0.10	39	alive
0.15	14	died	0.15	39	alive
0.20	24	died	0.20	31	alive
0.30	32	died	0.30	20	alive
0.50	38	died	0.50	8	alive
0.70	50	died	0.70	4	alive
0.95	50	died	0.95	2	alive

Table 10.2: Died versus alive for the data in Table 10.1

```
data fruit;
input dose count status $@@;
ldose=log10(dose);
datalines;
.10  8 died .10 39 alive
```

```
.15 14 died .15 39 alive
.20 24 died .20 31 alive
.30 32 died .30 20 alive
.50 38 died .50  8 alive
.70 50 died .70  4 alive
.95 50 died .95  2 alive
;
proc logistic data=fruit descending;
freq count;
model status=ldose /scale=none aggregate;
run;
```

The SAS System

The LOGISTIC Procedure

Model Information

Data Set	WORK.FRUIT
Response Variable	status
Number of Response Levels	2
Frequency Variable	count
Model	binary logit
Optimization Technique	Fisher's scoring

Number of Observations Read	14
Number of Observations Used	14
Sum of Frequencies Read	359
Sum of Frequencies Used	359

Response Profile

Ordered Value	status	Total Frequency
1	died	216
2	alive	143

Probability modeled is status='died'.

Model Convergence Status

Convergence criterion (GCONV=1E-8) satisfied.

Deviance and Pearson Goodness-of-Fit Statistics

Criterion	Value	DF	Value/DF	Pr > ChiSq
Deviance	0.7336	5	0.1467	0.9811
Pearson	0.7351	5	0.1470	0.9810

Number of unique profiles: 7

Model Fit Statistics

Criterion	Intercept Only	Intercept and Covariates
AIC	484.732	341.443
SC	488.615	349.210
-2 Log L	482.732	337.443

Testing Global Null Hypothesis: BETA=0

Test	Chi-Square	DF	Pr > ChiSq
Likelihood Ratio	145.2882	1	<.0001
Score	127.2173	1	<.0001
Wald	92.3620	1	<.0001

Analysis of Maximum Likelihood Estimates

Parameter	DF	Estimate	Standard Error	Wald Chi-Square	Pr > ChiSq
Intercept	1	3.1236	0.3349	86.9809	<.0001
ldose	1	4.8995	0.5098	92.3620	<.0001

Odds Ratio Estimates

Effect	Point Estimate	95% Wald Confidence Limits	

```
ldose       134.228      49.419      364.582
```

```
Association of Predicted Probabilities and Observed Responses
```

```
Percent Concordant     80.0    Somers' D    0.692
Percent Discordant     10.8    Gamma        0.762
Percent Tied            9.2    Tau-a        0.333
Pairs                 30888    c            0.846
```

The results produced with **proc logistic** are exactly the same when we use the *events/trials* model. The **descending** option in the **proc** command is needed to tell SAS to model the status 'dead' rather than 'alive' since 'a' comes before 'd' alphabetically.

Example 10.2: Bar Examination

The data in Table 10.3 relate to the examination for admission to the bar. Interest centers on how performance of students in a certain professional school's conduct course might explain passing or failing of the ethics examination. The original data was presented in Lunneborg (2000, p. 421) and reproduced by permission.

Subject	grade	Result	Subject	grade	Result	Subject	grade	Result
1	70	fail	10	76	fail	19	84	pass
2	70	fail	11	78	fail	20	86	fail
3	72	fail	12	78	fail	21	88	pass
4	72	fail	13	78	fail	22	90	pass
5	74	fail	14	80	pass	23	94	fail
6	74	fail	15	82	fail	24	96	pass
7	74	fail	16	82	pass	25	100	fail
8	76	fail	17	82	pass	26	100	pass
9	76	fail	18	84	fail	27	100	pass

Table 10.3: Course grade and ethics examination performance for 27 subjects

In this example, the outcome variable is result (pass or fail) in the ethics examination. We wish to model the 'pass' in this example. Again, we are interested in a model of the form

$$\ln\left(\frac{\pi_i}{1 - \pi_i}\right) = \beta_0 + \beta_1 x_i + \varepsilon_i, \tag{10.7}$$

where π_i represents the probability that an individual will pass the ethics examination and x_i represents the grade score for subject i. The following SAS program and partial output fits the logistic model in (10.7) to the data in Table 10.3. We notice here that the model statement does not have the *events/trials* format. Again notice the **descending** in the proc logistic line. This again tells SAS that we wish to model the 'pass' rather than 'failure'. Again since failure starts with an 'f' and passes with a 'p', we need to invoke descending, that is, go from 'z' to 'a'; rather than from 'a' to 'z' which is in alphabetic order (or the default in SAS). The partial results are again presented below.

```
data logitt;
input grade ethics $ @@;
resp=ethics;
datalines;
70 fail 70 fail 72 fail 72 fail 74 fail
74 fail 74 fail 76 fail 76 fail 76 fail
78 fail 78 fail 78 fail 80 pass 82 fail
82 pass 82 pass 84 fail 84 pass 86 fail
88 pass 90 pass 94 fail 96 pass 100 fail
100 pass 100 pass
;
proc print noobs;
proc logistic descending;
model resp=grade/scale=none aggregate lackfit;
output out=aa predicted=probs;
run;
proc print data=aa noobs;
```

```
run;
```

The LOGISTIC Procedure

Model Information

Data Set	WORK.LOGITT
Response Variable	resp
Number of Response Levels	2
Model	binary logit
Optimization Technique	Fisher's scoring

Number of Observations Read	27
Number of Observations Used	27

Response Profile

Ordered Value	resp	Total Frequency
1	pass	9
2	fail	18

Probability modeled is resp='pass'.

Model Convergence Status

Convergence criterion (GCONV=1E-8) satisfied.

Deviance and Pearson Goodness-of-Fit Statistics

Criterion	Value	DF	Value/DF	Pr > ChiSq
Deviance	15.6622	12	1.3052	0.2072
Pearson	13.3333	12	1.1111	0.3453

Number of unique profiles: 14

Model Fit Statistics

Criterion	Intercept Only	Intercept and Covariates
AIC	36.372	30.073
SC	37.668	32.665
-2 Log L	34.372	26.073

Testing Global Null Hypothesis: BETA=0

Test	Chi-Square	DF	Pr > ChiSq
Likelihood Ratio	8.2988	1	0.0040
Score	7.9311	1	0.0049
Wald	5.9594	1	0.0146

Analysis of Maximum Likelihood Estimates

Parameter	DF	Estimate	Standard Error	Wald Chi-Square	Pr > ChiSq
Intercept	1	-12.7587	4.9926	6.5307	0.0106
grade	1	0.1449	0.0593	5.9594	0.0146

Odds Ratio Estimates

Effect	Point Estimate	95% Wald Confidence Limits	
grade	1.156	1.029	1.298

Partition for the Hosmer and Lemeshow Test

Group	Total	resp = pass Observed	resp = pass Expected	resp = fail Observed	resp = fail Expected
1	4	0	0.31	4	3.69
2	3	0	0.35	3	2.65
3	3	0	0.44	3	2.56
4	3	0	0.57	3	2.43
5	4	3	1.12	1	2.88
6	3	1	1.14	2	1.86
7	3	2	1.77	1	1.23

8	4	3	3.31	1	0.69

```
          Hosmer and Lemeshow Goodness-of-Fit Test

              Chi-Square        DF      Pr > ChiSq
                6.6218           6         0.3572

          grade    ethics    resp    _LEVEL_     probs
            70      fail      fail     pass      0.06797
            70      fail      fail     pass      0.06797
            72      fail      fail     pass      0.08879
            72      fail      fail     pass      0.08879
            74      fail      fail     pass      0.11519
          ....................................
            94      fail      fail     pass      0.70234
            96      pass      pass     pass      0.75918
           100      fail      fail     pass      0.84911
           100      pass      pass     pass      0.84911
           100      pass      pass     pass      0.84911
```

The model in (10.7) fits the data well, with a *deviance* of 15.6622 on 12 d.f. and a *p*-value of 0.2072. The parameter estimates for this model are $\hat{\beta}_0 = -12.7587$ and $\hat{\beta}_1 = 0.1449$. Thus, a unit increase in the grade results in an odds-ratio increase of $e^{0.1449} = 1.156$. That is, the chances of a subject passing the exam increases by almost 16% for a unit increase in course grade.

10.4   Multiple Logistic Regression

In the previous examples, we have assumed that there is only one explanatory variable. Most often however, we have more than one explanatory variables, $x_1, x_2, \ldots, x_{p-1}$, which may be individually distinct or some could be derivatives of others. For instance, we could have x_1 and $x_2 = x_1^2$, that is the square of x_1 and we may wish to explore a quadratic model. In general however, we have,

$$\text{Prob}(Y = 1, |x_1, x_2, \ldots, x_{p-1}) = \frac{\exp(\beta_0 + \beta_1 x_1 + \cdots + \beta_{p-1} x_{p-1})}{1 + \exp(\beta_0 + \beta_1 x_1 + \cdots + \beta_{p-1} x_{p-1})}, \tag{10.8}$$

where $\beta_0, \beta_1, \ldots, \beta_{p-1}$ are parameters to be estimated from available data. If $\pi_i, i = 1, 2, \ldots, n$, is the probability in (10.8), then this can be re-written as

$$\ln\left(\frac{\pi_i}{1 - \pi_i}\right) = \beta_0 + \beta_1 x_{i1} + \beta_2 x_{i2} + \cdots + \beta_{p-1} x_{i,p-1} = \beta_0 + \sum_{j=1}^{p-1} \beta_j x_{ij}. \tag{10.9}$$

10.4.1   Applications

Suppose we wish to fit a quadratic response model to the data in Table 10.3. That is, a model of the form

$$\ln\left(\frac{\pi_i}{1 - \pi_i}\right) = \beta_0 + \beta_1 x_i + \beta_2 x_i^2. \tag{10.10}$$

The SAS program, a partial output as well as the resulting estimated logistic function under the quadratic model are presented below. The model fits the data with a deviance or $G^2 = 11.8131$ on 11 degrees of freedom, with a corresponding *p*-value of 0.3779. While the quadratic model fits better, however, the test of the hypotheses

$$H_0 : \beta_2 = 0 \quad \text{versus} \quad H_a : \beta_2 \neq 0,$$

provided by the **test** statement in the SAS program is not significant.

```
data new;
set logitt;
x1=grade;
x2=x1*x1;
proc logistic descending;
model resp=x1 x2/scale=none aggregate;
test: test x2=0;
output out=aa predicted=probs;
```

```
run;

goptions vsize=6in hsize=6in;
proc sort data=aa;
by grade;
run;
symbol i=spline value=none line=1 height=.75;
axis1 label=(angle=-90 rotate=90 'Expected probs') order=(0 to 1.0 by 0.1);
axis2 label=('Grades');
proc gplot data=aa;
plot probs*grade/vaxis=axis1 haxis=axis2 noframe vref=1.0;
run;
```

                          The LOGISTIC Procedure

              Deviance and Pearson Goodness-of-Fit Statistics

Criterion	Value	DF	Value/DF	Pr > ChiSq
Deviance	11.8131	11	1.0739	0.3779
Pearson	10.9933	11	0.9994	0.4438

                    Number of unique profiles: 14

                          The LOGISTIC Procedure

                Analysis of Maximum Likelihood Estimates

Parameter	DF	Estimate	Standard Error	Wald Chi-Square	Pr > ChiSq
Intercept	1	-134.3	74.7311	3.2275	0.0724
x1	1	2.9400	1.0020	0.0000	0.0010
x2	1	-0.0160	0.00951	2.8280	0.0926

                          Odds Ratio Estimates

Effect	Point Estimate	95% Wald Confidence Limits	
x1	19.029	0.690	525.002
x2	0.984	0.966	1.003

The results of the test are displayed as

                    Linear Hypotheses Testing Results

Label	Wald Chi-Square	DF	Pr > ChiSq
test	2.8280	1	0.0926

Results from the test gives a *p*-value of 0.0926 which is not significant at the 5% level of significance, indicating that given the linear component of grade in the model, the quadratic term is not needed. Further, the quadratic model does not make much sense, because as we see in Figure 10.3, a subject scoring a grade of 98 for instance has a reduced probability of passing the examination than a subject scoring 90, which seems unreasonable.

As a result of the above, Lunneborg (2000, Chapter 16) considers fitting a reciprocal model of the form

$$\ln\left(\frac{\pi_i}{1 - \pi_i}\right) = \beta_0 + \beta_1\left(1 - \frac{1}{x_i - 68}\right) + \varepsilon_i, \tag{10.11}$$

to the bar examination data. The reciprocal model in (10.11) is implemented in SAS with the following SAS program, partial output and resulting predicted plot of expected probabilities versus grade scores.

```
set logitt;
x1=1-1/(grade-68);
proc logistic descending;
model resp=x1/scale=none aggregate lackfit;
output out=aa predicted=probs;
run;

goptions vsize=6in hsize=6in;
proc sort data=aa;
by grade;
run;
symbol i=spline value=none line=1 height=.75;
axis1 label=(angle=-90 rotate=90 'Expected probs') order=(0 to 1.0 by 0.1);
```

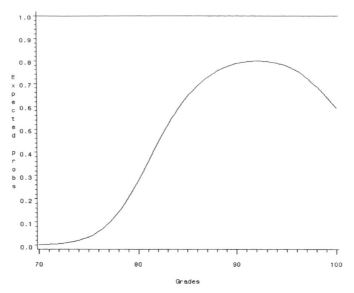

Figure 10.3: Estimated quadratic logistic model

```
axis2 label=('Grades');
proc gplot data=aa;
plot probs*grade/vaxis=axis1 haxis=axis2 noframe vref=1.0;
run;
```

The LOGISTIC Procedure

Deviance and Pearson Goodness-of-Fit Statistics

Criterion	Value	DF	Value/DF	Pr > ChiSq
Deviance	11.7537	12	0.9795	0.4657
Pearson	9.7981	12	0.8165	0.6337

Number of unique profiles: 14

Analysis of Maximum Likelihood Estimates

Parameter	DF	Estimate	Standard Error	Wald Chi-Square	Pr > ChiSq
Intercept	1	-38.7637	17.6447	4.8264	0.0280
x1	1	41.3476	18.8649	4.8039	0.0284

Odds Ratio Estimates

Effect	Point Estimate	95% Wald Confidence Limits	
x1	>999.999	79.290	>999.999

Association of Predicted Probabilities and Observed Responses

Percent Concordant	84.0	Somers' D	0.710
Percent Discordant	13.0	Gamma	0.732
Percent Tied	3.1	Tau-a	0.328
Pairs	162	c	0.855

Partition for the Hosmer and Lemeshow Test

		resp = pass		resp = fail	
Group	Total	Observed	Expected	Observed	Expected
1	4	0	0.00	4	4.00
2	3	0	0.04	3	2.96
3	3	0	0.21	3	2.79
4	3	0	0.52	3	2.48
5	4	3	1.52	1	2.48
6	3	1	1.57	2	1.43
7	3	2	2.03	1	0.97
8	4	3	3.10	1	0.90

Hosmer and Lemeshow Goodness-of-Fit Test

```
Chi-Square        DF      Pr > ChiSq
   3.6697          6        0.7213
```

The deviance for this model is 11.7537 on 12 degrees of freedom (p-value $= 0.4657$), a much better fit than the two previous models to these data. The Akaike information criterion (AIC) defined as *Deviance* $- 2$(d.f.) for the three models are presented in the following table.

Model	Deviance	d.f.	AIC
Linear (10.7)	15.6622	12	-8.3378
Quadratic (10.10)	11.8131	11	-10.1869
Reciprocal (10.11)	11.7537	12	-12.2463

Clearly, the reciprocal model is the most parsimonious based on AIC-it is the smallest of the three. The graph of the predicted probabilities for values of x_i, the grade scores, are presented in Figure 10.4.

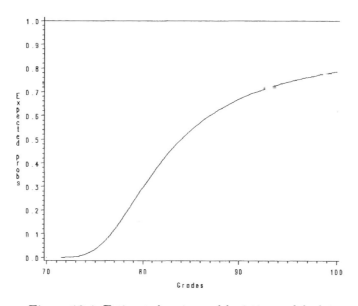

Figure 10.4: Estimated reciprocal logistic model plot

Example 10.3: Beetle Mortality Data

The following data on beetle mortality (Bliss, 1935) were analyzed in Lawal (2003) and relate to the numbers of insects dead after 5 hours of exposure to gaseous carbon disulphide at various concentrations. The data have also been analyzed by Dobson (1990) and Agresti (1990). We employ both the linear logdose (to base 10) and its quadratic counterpart to the data in the example.

logdose	n	r
1.6907	59	6
1.7242	60	13
1.7552	62	18
1.7842	56	28
1.8113	63	52
1.8369	59	53
1.8610	62	61
1.8839	60	60

Thus, we fit the following models:

$$\ln\left(\frac{\pi_i}{1-\pi_i}\right) = \beta_0 + \beta_1 \text{logdose}, \tag{10.12a}$$

$$\ln\left(\frac{\pi_i}{1-\pi_i}\right) = \beta_0 + \beta_1 \text{logdose} + \beta_2 \text{logdose}^2. \tag{10.12b}$$

We present below extracts from the SAS output for the linear and quadratic models from **proc logistic**.

```
data beet;
input logdose n r @@;
x2=logdose*logdose;
datalines;
1.6907 59 6 1.7242 60 13 1.7552 62 18
1.7842 56 28 1.8113 63 52 1.8369 59 53
1.8610 62 61 1.8839 60 60
;
proc logistic data=beet;
model r/n = logdose / scale=none aggregate;
output out=aa predicted=probs;
run;

proc logistic data=beet;
model r/n = logdose x2 / scale=none aggregate;
output out=bb predicted=probs;
test: test x2=0;
run;

goptions cback=white colors=(black) vsize=6 hsize=6;
proc sort data=aa;
by logdose; run;
symbol2 i=spline value=none height=.75;
axis1 label=(angle=-90 rotate=90 'EXPECTED PROBS');
axis2 label=('LOG OF DOSAGE');
proc gplot data=aa;
plot probs*logdose/vaxis=axis1 haxis=axis2 noframe;
run;
```

```
              Deviance and Pearson Goodness-of-Fit Statistics

        Criterion          Value      DF     Value/DF     Pr > ChiSq

        Deviance          11.2322      6      1.8720        0.0815
        Pearson           10.0253      6      1.6709        0.1236

                    Number of unique profiles: 8

                Analysis of Maximum Likelihood Estimates

                                    Standard       Wald
        Parameter    DF   Estimate    Error    Chi-Square    Pr > ChiSq

        Intercept     1   -60.7114    5.1802    137.3557       <.0001
        logdose       1    34.2669    2.9118    138.4876       <.0001

              Deviance and Pearson Goodness-of-Fit Statistics

        Criterion          Value      DF     Value/DF     Pr > ChiSq

        Deviance           3.1949      5      0.6390        0.6700
        Pearson            3.0033      5      0.6007        0.6995

                    Number of unique profiles: 8

                Analysis of Maximum Likelihood Estimates

                                    Standard       Wald
        Parameter    DF   Estimate    Error    Chi-Square    Pr > ChiSq

        Intercept     1     430.9     180.6     5.6916         0.0170
        logdose       1    -520.4     204.5     6.4764         0.0109
        x2            1     156.4    57.8560    7.3035         0.0069

                 Linear Hypotheses Testing Results

                             Wald
              Label     Chi-Square     DF    Pr > ChiSq
```

```
        test        7.3035      1       0.0069
```

Both the linear and quadratic give deviance values of 11.2322 and 3.1949 on 6 and 5 degrees of freedom respectively. Obviously, the quadratic model fits much better.

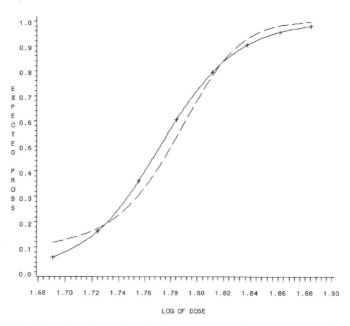

Figure 10.5: Overlay plot of both linear and quadratic estimated functions

Lawal (2003) fitted the complimentary log-log model to the beetle data and found the deviance to be 3.4464 on 6 degrees of freedom, a fit as good as the quadratic model and more parsimonious as is based on 6 rather than 5 degrees of freedom. Based on the results, the estimated logistic functions are presented in (10.13).

$$\ln\left(\frac{\hat{\pi}_i}{1 - \hat{\pi}_i}\right) = -60.7114 + 34.2669\, x_i, \tag{10.13a}$$

$$\ln\left(\frac{\hat{\pi}_i}{1 - \hat{\pi}_i}\right) = 430.90 - 520.4\, x_i + 156.4\, x_i^2, \tag{10.13b}$$

$$\ln\left[-\ln\left(1 - \hat{\pi}_i\right)\right] = -39.5725 + 22.0412\, x_i, \tag{10.13c}$$

where x_i = logdose to base 10. The estimated equations in (10.13a), (10.13b), and (10.13c) refer respectively to the linear, quadratic and complimentary log-log models. To implement the complimentary log-log model, we need to invoke **link=cloglog** in the model options in both **proc logistic** and **proc genmod**. The results from both procedures are presented in the partial SAS output below.

```
        Deviance and Pearson Goodness-of-Fit Statistics

Criterion        DF        Value      Value/DF     Pr > ChiSq
----------------------------------------------------------------
Deviance         6         3.4464     0.5744       0.7511
Pearson          6         3.2947     0.5491       0.7711

Number of unique profiles: 8

        Analysis of Maximum Likelihood Estimates

                            Standard
Parameter   DF   Estimate   Error      Chi-Square    Pr > ChiSq
----------------------------------------------------------------
Intercept   1    -39.5725   3.2403     149.1487      <.0001
dose        1     22.0412   1.7994     150.0498      <.0001

Using proc genmod:

data gen;
```

```
set tab86;
proc genmod data=tab86 descending;
model r/n=dose/link=cloglog;
run;
            Criteria For Assessing Goodness Of Fit

Criterion                   DF        Value      Value/DF
--------------------------------------------------------
Deviance                     6        3.4464       0.5744
Pearson Chi-Square           6        3.2947       0.5491

            Analysis Of Parameter Estimates

                          Standard     Chi-
Parameter   DF  Estimate     Error   Square   Pr > ChiSq
--------------------------------------------------------
Intercept    1  -39.5723    3.2290   150.19      <.0001
dose         1   22.0412    1.7931   151.10      <.0001
```

The plot of the estimated complimentary log-log model for the data in the example and the corresponding quadratic model are overlaid in Figure 10.6. The quadratic is given by the solid line, while the complimentary log-log is given by the dashed lines.

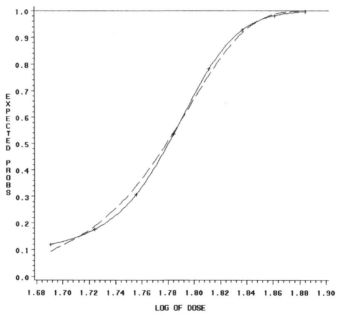

Figure 10.6: Plot of quadratic and CLL models

Example 10.4

The following data from Finney (1947) and Pregibon (1981) relate to the occurrence of vasoconstriction (y; occurrence = 1 and non-occurrence = 0) in the skin of the fingers as a function of the rate (x_2) and volume (x_1) of air breathed. In times of stress, vasoconstriction restricts blood flow to the extremities (such as fingers and toes), forcing blood to the central vital organs. The data are reproduced in Table 10.4. A constriction value of 1 indicates that constriction occurred. Fit a logistic regression to the data with rate and volume as explanatory variables. Discuss your results.

If we assume that the probability π of the occurrence of vasoconstriction in the skin of the finger depends on the volume (x_1), the rate (x_2) and their interaction term ($x_1 \times x_2$), then our logistic model would be as described in (10.14), while the corresponding probability is as defined in (10.15).

$$\ln\left(\frac{\pi_i}{1-\pi_i}\right) = \beta_0 + \beta_1 x_1 + \beta_2 x_2 + \beta_3\, x_1 * x_2, \tag{10.14}$$

y	x_1	x_2	y	x_1	x_2	y	x_1	x_2	y	x_1	x_2
1	0.825	3.7	0	0.57	0.8	0	2.0	0.4	1	0.75	2.7
1	1.09	3.5	0	2.75	0.55	0	1.36	0.95	0	0.03	2.35
1	2.5	1.25	0	3.0	0.6	0	1.35	1.35	0	1.83	1.1
1	1.5	0.75	1	2.33	1.4	0	1.36	1.5	1	2.2	1.1
1	3.2	0.8	1	3.75	0.75	1	1.78	1.6	1	2.0	1.2
1	3.5	0.7	1	1.64	2.3	0	1.5	0.6	1	3.33	0.8
0	0.75	0.6	1	1.6	3.2	1	1.5	1.8	0	1.9	0.95
0	1.7	1.1	1	1.415	0.85	0	1.9	0.95	0	1.9	0.75
0	0.75	0.9	0	1.06	1.7	1	0.95	1.9	1	1.625	1.3
0	0.45	0.9	1	1.8	1.8	0	0.4	1.6			

Table 10.4: Occurrence of vasoconstriction in the skin of fingers

$$\text{Prob}(Y = 1, |x_1, x_2) = \frac{\exp(\beta_0 + \beta_1 x_1 + \beta_2 x_2 + \beta_3\, x_1 * x_2)}{1 + \exp(\beta_0 + \beta_1 x_1 + \beta_2 x_2 + \beta_3\, x_1 * x_2)}. \tag{10.15}$$

The implementation of the model in (10.14) in SAS is carried out with the following SAS program and the corresponding partial output.

```
data vol;
input cont x1 x2 @@;
x12=x1*x2;
datalines;
..........
;
proc print;
run;
;
proc logistic data=vol descending;
model cont=x1 x2 x12/scale=none aggregate;
output out=aa predicted=probs;
test: test x12;
run;
```

```
                          The LOGISTIC Procedure

                      Model Information

          Data Set                      WORK.VOL
          Response Variable             cont
          Number of Response Levels     2
          Model                         binary logit
          Optimization Technique        Fisher's scoring

             Number of Observations Read        39
             Number of Observations Used        39

                      Response Profile

          Ordered                        Total
          Value             cont       Frequency
             1                1             20
             2                0             19

          Probability modeled is cont=1.

                   Model Convergence Status

          Convergence criterion (GCONV=1E-8) satisfied.

          Deviance and Pearson Goodness-of-Fit Statistics

     Criterion        Value      DF     Value/DF     Pr > ChiSq

     Deviance        26.5799     34      0.7818        0.8140
     Pearson         36.5097     34      1.0738        0.3529

              Number of unique profiles: 38
```

```
                        The LOGISTIC Procedure

                      Model Fit Statistics

                                               Intercept
                                  Intercept       and
              Criterion             Only       Covariates

              AIC                  56.040        34.580
              SC                   57.703        41.234
              -2 Log L             54.040        26.580

           Testing Global Null Hypothesis: BETA=0

      Test                 Chi-Square      DF     Pr > ChiSq

      Likelihood Ratio       27.4599        3       <.0001
      Score                  19.1171        3       0.0003
      Wald                    9.1214        3       0.0277

            Analysis of Maximum Likelihood Estimates

                             Standard        Wald
      Parameter   DF  Estimate   Error   Chi-Square   Pr > ChiSq

      Intercept    1   -7.1142   3.3482     4.5145       0.0336
      x1           1    0.5112   1.5081     0.1149       0.7346
      x2           1    1.2636   2.1469     0.3464       0.5561
      x12          1    2.4076   1.6616     2.0996       0.1473

                      Odds Ratio Estimates

                        Point         95% Wald
              Effect   Estimate   Confidence Limits

              x1        1.667      0.087      32.039
              x2        3.538      0.053     237.817
              x12      11.107      0.428     288.351

                     The LOGISTIC Procedure

     Association of Predicted Probabilities and Observed Responses

         Percent Concordant    93.9    Somers' D    0.879
         Percent Discordant     6.1    Gamma        0.879
         Percent Tied           0.0    Tau-a        0.451
         Pairs                  380    c            0.939

              Linear Hypotheses Testing Results

                             Wald
              Label      Chi-Square     DF    Pr > ChiSq

              test         2.0996        1       0.1473
```

The global hypothesis

$$H_0 : \beta_1 = \beta_2 = \beta_3 = 0 \quad \text{versus}$$
$$H_a : \text{at least one of these } \beta \text{ parameters is not zero,}$$

(10.16)

is tested by either the likelihood ratio test statistic, the score test or the Wald test. Both likelihood and Wald tests gives p-values of < 0.0001 and 0.0277 respectively, indication that at $\alpha = 0.05$ level of significance, we would have to reject H_0 and therefore conclude that at least one of the parameters $\beta_1, \beta_2, \beta_3$ is not zero. The model fits the data well with deviance being 26.5798 on 34 degrees of freedom (p-value $= 0.8140$). The resulting estimated model is therefore given by

$$\ln\left(\frac{\hat{\pi}_i}{1 - \hat{\pi}_i}\right) = -7.1138 + 0.5110\, x_i + 1.2633\, x_2 + 2.4077 x_1 x_2.$$

Analysis of the results however indicate that the interaction effect is not significant. This result is obtained either from the analysis of maximum likelihood estimates which gives a p-value of the hypotheses

$$H_0 : \beta_3|\beta_1, \beta_2 = 0 \quad \text{versus} \quad H_a : \beta_3|\beta_1, \beta_2 \neq 0,$$

of 0.1473 which indicates that we would fail to reject H_0. That is, the interaction term $x_{12} = x_1 \times x_2$ is not important in the model given that x_1 and x_2 are already in the model. We can therefore fit the reduced model involving only x_1 and x_2 as explanatory variables. That is, the model

$$\ln\left(\frac{\pi_i}{1 - \pi_i}\right) = \beta_0 + \beta_1 x_1 + \beta_2 x_2, \tag{10.17}$$

with corresponding probability of transient occurrence

$$\text{Prob}(Y = 1, |x_1, x_2) = \frac{\exp(\beta_0 + \beta_1 x_1 + \beta_2 x_2)}{1 + \exp(\beta_0 + \beta_1 x_1 + \beta_2 x_2)}. \tag{10.18}$$

The following SAS program with the partial output implements the model in (10.17).

```
set vol;
proc logistic data=vol descending;
model cont=x1 x2/scale=none aggregate;
output out=aa predicted=probs;
run;
                    The LOGISTIC Procedure

          Deviance and Pearson Goodness-of-Fit Statistics

     Criterion        Value     DF     Value/DF     Pr > ChiSq

     Deviance        29.7723     35     0.8506        0.7184
     Pearson         39.0106     35     1.1146        0.2942

                 Number of unique profiles: 38

          Analysis of Maximum Likelihood Estimates

                              Standard       Wald
     Parameter   DF   Estimate    Error   Chi-Square   Pr > ChiSq

     Intercept    1    -9.5293    3.2331      8.6873       0.0032
     x1           1     2.6490    0.9142      8.3966       0.0038
     x2           1     3.8820    1.4286      7.3844       0.0066

                    Odds Ratio Estimates

                      Point          95% Wald
           Effect   Estimate     Confidence Limits

            x1       14.140      2.357       84.844
            x2       48.522      2.951      797.865

     Association of Predicted Probabilities and Observed Responses

            Percent Concordant   91.3    Somers' D   0.826
            Percent Discordant    8.7    Gamma       0.826
            Percent Tied          0.0    Tau-a       0.424
            Pairs                 380    c           0.913
```

The model in (10.17) when implemented fits the data very well with a deviance value of 29.7723 on 35 degrees of freedom and p-value of 0.7184. The parameter estimates are very important in the model and the estimated regression equation is

$$\ln\left(\frac{\hat{\pi}_i}{1 - \hat{\pi}_i}\right) = -9.5292 + 2.6490 x_{i1} + 3.8820 x_{i2}, \tag{10.19}$$

with a corresponding estimated probability

$$\hat{\pi}_i = \frac{\exp(-9.5292 + 2.6490 x_{i1} + 3.8820 x_{i2})}{1 + \exp(-9.5292 + 2.6490 x_{i1} + 3.8820 x_{i2})}. \tag{10.20}$$

10.4.2   Interpretation of Parameters

We have

$$\exp(\beta_0) = \frac{\text{Prob}(Y = 1 | x_1 = x_2 = 0)}{\text{Prob}(Y = 0 | x_1 = x_2 = 0)}$$

$$= \text{Odds of an occurrence of TV in the skin of the fingers,}$$

$$\exp(\beta_1) = \frac{\text{Odds of occurrence of TV when } x_1 = 1, x_2 = 0}{\text{Odds of occurrence of TV at baseline}},$$

$$\exp(\beta_2) = \frac{\text{Odds of occurrence of TV when } x_1 = 0, x_2 = 1}{\text{Odds of occurrence of TV at baseline}},$$

where $\exp(\beta_0)$ is often referred to as the odds of occurrence of TV (transient vasoconstriction) at the baseline. That is, at $(x_1 = x_2 = 0)$.

Thus, for a unit increase in x_1 (keeping x_2 constant), the odds of the occurrence of TV increases by $\exp(\beta_1)$. For these data, this is estimated to be $\exp(2.6490) = 14.143$. Similarly, for a unit increase in x_2 (keeping x_1 constant), the odds of the occurrence of TV increases by $\exp(\beta_2)$. Again for these data, this is estimated to be $\exp(3.8820) = 48.521$.

In general, the estimated odds of the occurrence of TV for any given x_1, x_2 is

$$= \exp(\hat\beta_0) \times \exp(\hat\beta_1 x_1) \times \exp(\hat\beta_2 x_2)$$

$$= \left\{ \begin{matrix} \text{odds} \\ \text{for} \\ \text{baseline} \end{matrix} \right\} \times \left\{ \begin{matrix} \text{Factor} \\ \text{due} \\ \text{to } x_1 \end{matrix} \right\} \times \left\{ \begin{matrix} \text{Factor} \\ \text{due} \\ \text{to } x_2 \end{matrix} \right\}.$$

Example 10.5

The following example in Table 10.5 gives the data relating to fifty years survival after graduation of men and women who graduated in the years 1938 to 1947 at the University of Adelaide. The original data had four faculties for men and two for women. Here, we have used data from both faculties of Arts and Science for both men and women. Column labeled S refers to graduates who survived, while column labeled T contains the total number of graduates for that year.

Faculty of	Men Arts		Men Science		Women Arts		Women Science	
Year of Graduation	S	T	S	T	S	T	S	T
1938	16	30	9	14	14	19	1	1
1939	13	22	9	12	11	16	4	4
1940	11	25	12	19	15	18	6	7
1941	12	14	12	15	15	21	3	3
1942	8	12	20	28	8	9	4	4
1943	11	20	16	21	13	13	8	9
1944	4	10	25	31	18	22	5	5
1945	4	12	32	38	18	22	16	17
1946			4	5	1	1	1	1
1947	13	23	25	31	13	16	10	10

Table 10.5: 50 years survival for men and women after graduation

If we let x_1 represent gender, x_2 to represent faculty and x_3 to represent years, then we have

$$x_1 \equiv (\text{Gender: Men} = 1, \text{Women} = 0)$$
$$x_2 \equiv (\text{Faculty: Arts} = 1, \text{Science} = 0)$$
$$x_3 \equiv (\text{Years: Can be continuous or categorical})$$

and the model which assumes no interaction between the factor variables is presented in (10.21) as

$$\ln\left(\frac{\pi_i}{1-\pi_i}\right) = \beta_0 + \beta_1 x_1 + \beta_2 x_2 + \beta_3 x_3. \tag{10.21}$$

The SAS program for implementing the model in (10.21) is presented below together with a partial output. In model (10.21), we have assumed that the years of graduation is a continuous type variable.

```
data new;
do gender=1 to 2;
do year=1938 to 1947;
do fac=1 to 2;
input  r n @@;
output;
end; end; end;
datalines;
...........
 ;
proc print;
run;
proc logistic;
class gender(ref=last) fac (ref=last)/param=ref;
model r/n=year gender fac/scale=none aggregate;
run;
```

```
                  Model Convergence Status

          Convergence criterion (GCONV=1E-8) satisfied.

          Deviance and Pearson Goodness-of-Fit Statistics

     Criterion          Value      DF    Value/DF    Pr > ChiSq

     Deviance          29.2167     35     0.8348       0.7430
     Pearson           24.2741     35     0.6935       0.9132

              Number of unique profiles: 39

              Type 3 Analysis of Effects

                             Wald
         Effect      DF   Chi-Square    Pr > ChiSq

         year         1     1.9290       0.1649
         gender       1    31.2880       <.0001
         fac          1    22.6027       <.0001

          Analysis of Maximum Likelihood Estimates

                          Standard     Wald
    Parameter     DF   Estimate   Error   Chi-Square   Pr > ChiSq

    Intercept      1   -89.7536  66.4413    1.8248       0.1767
    year           1     0.0475   0.0342    1.9290       0.1649
    gender    1    1    -1.2879   0.2302   31.2880       <.0001
    fac       1    1    -1.0080   0.2120   22.6027       <.0001

                  Odds Ratio Estimates

                     Point        95% Wald
         Effect    Estimate   Confidence Limits

         year        1.049      0.981     1.121
         gender 1 vs 2  0.276   0.176     0.433
         fac    1 vs 2  0.365   0.241     0.553
```

The model fits the data with a deviance of 29.2167 on 45 degrees of freedom (p-value $= 0.7430$). However, examination of the type III analysis of effects indicate that the years' effect is not at all significant and can be ignored or removed from the model. Thus, removing the years' effect from the model leads to a reduced model

$$\ln\left(\frac{\pi_i}{1-\pi_i}\right) = \beta_0 + \beta_1 x_1 + \beta_2 x_2, \tag{10.22}$$

and the following corresponding SAS program and partial output.

```
proc logistic;
class gender(ref=last) fac (ref=last)/param=ref;
model r/n=gender fac/scale=none aggregate=(gender fac year);
run;
```

Deviance and Pearson Goodness-of-Fit Statistics

Criterion	Value	DF	Value/DF	Pr > ChiSq
Deviance	31.1533	36	0.8654	0.6983
Pearson	26.5053	36	0.7363	0.8760

Number of unique profiles: 39
Analysis of Maximum Likelihood Estimates

Parameter		DF	Estimate	Standard Error	Wald Chi-Square	Pr > ChiSq
Intercept		1	2.5344	0.2583	96.2430	<.0001
gender	1	1	−1.3075	0.2296	32.4205	<.0001
fac	1	1	−1.0714	0.2072	26.7470	<.0001

Odds Ratio Estimates

Effect			Point Estimate	95% Wald Confidence Limits	
gender	1 vs 2		0.271	0.172	0.424
fac	1 vs 2		0.343	0.228	0.514

We have requested SAS to use reference coding in this example, where

$$x_1 = \begin{cases} 1, & \text{if gender = men} \\ 0, & \text{if gender = women,} \end{cases} \qquad x_2 = \begin{cases} 1, & \text{if faculty = arts} \\ 0, & \text{if faculty = science.} \end{cases}$$

This is accomplished in the **class** statement in SAS. Other dummy variable procedures can be used within SAS. For this model, the deviance or G^2 is 31.1533 on 36 degrees of freedom (p-value = 0.6983), giving a very good fit. The estimated logistic regression equation is

$$\ln\left(\frac{\hat{\pi}_i}{1-\hat{\pi}_i}\right) = 2.5344 - 1.3075\, x_1 - 1.0714\, x_2. \tag{10.23}$$

The model indicates that the effect of the year of graduation can be ignored. The odds of men surviving longer than 50 years over women (assuming that both graduated from the same faculty and ignoring the year of graduation) is 0.271. Put differently, the odds of surviving for at least 50 years after graduation increases $\frac{1}{0.271} = 3.7$ times for women than men graduating from the same faculty (again ignoring the year of graduation). In other words, women are 3.7 times more likely to live longer than 50 years after graduation than men (assuming both graduated from the same faculty).

Similarly, the odds of surviving longer than 50 years is $\frac{1}{0.343} = 2.9$ times higher for those who graduated from the faculty of science than those who graduated from the faculty of Arts assuming both are of the same gender, and again ignoring the effect of year of graduation. On the other hand, if we were to use the effect coding scheme, we would have

$$x_1 = \begin{cases} 1, & \text{if gender = men} \\ -1, & \text{if gender = women,} \end{cases} \qquad x_2 = \begin{cases} 1, & \text{if faculty = arts} \\ -1, & \text{if faculty = science.} \end{cases}$$

The effect coding scheme is the default in **proc logistic** and the SAS program below implements the model in (10.22). While the deviance values and the estimated odds ratios are the same as those obtained for the reference cell scheme, we notice that the parameter estimates of β_1 and β_2 are half those of the reference cell estimates. Consequently, the odds ratios are computed in this case, for gender, for instance as $e^{-0.6537-0.6537} = 0.271$.

```
proc logistic;
class gender fac;
model r/n=gender fac/scale=none aggregate=(gender fac year);
run;
```

Analysis of Maximum Likelihood Estimates

Standard   Wald

Parameter		DF	Estimate	Error	Chi-Square	Pr > ChiSq
Intercept		1	1.3449	0.1166	133.1505	<.0001
gender	1	1	-0.6537	0.1148	32.4205	<.0001
fac	1	1	-0.5357	0.1036	26.7470	<.0001

Odds Ratio Estimates

Effect		Point Estimate	95% Wald Confidence Limits	
gender	1 vs 2	0.271	0.172	0.424
fac	1 vs 2	0.343	0.228	0.514

The model in (10.21) does not include the various interactions between say, years and gender, years and faculty, gender and faculty as well as the three factor interaction of years, gender and faculty. We fitted this model with all the three selection strategies (backward, forward, stepwise) for the most parsimonious model. All the three selection strategies return the reduced model in (10.22) that was previously examined. That is, none of the interaction terms is significant enough to be included in the model.

If we decide to consider year of graduation as a categorical variable, then, the model without interaction between the factors become

$$\ln\left(\frac{\hat{\pi}_i}{1-\hat{\pi}_i}\right) = \beta_0 + \beta_2\, x_1 + \beta_3\, x_2 + \sum_{j=1}^{9} \alpha_j z_j, \tag{10.24}$$

where again, for the cell reference scheme, x_1 and x_2 are as previously described, but

$$z_1 = \begin{cases} 1, & \text{if year} = 1938 \\ 0, & \text{elsewhere,} \end{cases} \quad z_2 = \begin{cases} 1, & \text{if year} = 1939 \\ 0, & \text{elsewhere,} \end{cases} \cdots, \quad z_9 = \begin{cases} 1, & \text{if year} = 1946 \\ 0, & \text{elsewhere.} \end{cases}$$

Since years has ten categories, thus 9 indicator variables, $z_1, \ldots, z_9$ will be created, with 1947 being used as a reference category. Again the implementation of this in SAS is presented with the resulting output.

```
set ade;
proc logistic;
class gender(ref=last) fac (ref=last) year (ref=last)/param=ref;
model r/n=gender fac year/scale=none aggregate=(gender fac year);
run;
```

Deviance and Pearson Goodness-of-Fit Statistics

Criterion	Value	DF	Value/DF	Pr > ChiSq
Deviance	26.0032	27	0.9631	0.5184
Pearson	22.1640	27	0.8209	0.7290

Type 3 Analysis of Effects

Effect	DF	Wald Chi-Square	Pr > ChiSq
gender	1	30.5577	<.0001
fac	1	21.7634	<.0001
year	9	5.1601	0.8201

Analysis of Maximum Likelihood Estimates

Parameter		DF	Estimate	Standard Error	Wald Chi-Square	Pr > ChiSq
Intercept		1	2.6537	0.3720	50.9005	<.0001
gender	1	1	-1.2861	0.2327	30.5577	<.0001
fac	1	1	-1.0090	0.2163	21.7634	<.0001
year	1938	1	-0.4357	0.3863	1.2722	0.2593
year	1939	1	-0.2622	0.4129	0.4032	0.5255
year	1940	1	-0.5649	0.3808	2.2004	0.1380
year	1941	1	0.1644	0.4455	0.1363	0.7120
year	1942	1	-0.0729	0.4290	0.0289	0.8650
year	1943	1	0.0176	0.4127	0.0018	0.9661
year	1944	1	-0.1527	0.4028	0.1438	0.7046
year	1945	1	-0.1237	0.3852	0.1031	0.7482
year	1946	1	0.2621	1.1249	0.0543	0.8157

```
                   Odds Ratio Estimates

                          Point           95% Wald
          Effect         Estimate     Confidence Limits

          gender 1 vs 2     0.276      0.175      0.436
          fac    1 vs 2     0.365      0.239      0.557
          year   1938 vs 1947  0.647   0.303      1.379
          year   1939 vs 1947  0.769   0.342      1.728
          year   1940 vs 1947  0.568   0.269      1.199
          year   1941 vs 1947  1.179   0.492      2.822
          year   1942 vs 1947  0.930   0.401      2.155
          year   1943 vs 1947  1.018   0.453      2.285
          year   1944 vs 1947  0.858   0.390      1.890
          year   1945 vs 1947  0.884   0.415      1.880
          year   1946 vs 1947  1.300   0.143     11.785
```

Our results again indicate that the following hypotheses from model (10.24),

$$H_0 : \alpha_1 = \alpha_2 = \cdots = \alpha_9 = 0 \quad \text{versus} \quad H_a : \text{at least on of these} \neq 0,$$

has a Wald test statistic of 5.1601 on 9 degrees of freedom and p-value of 0.8201. Indicating that the null hypothesis is tenable and hence the year effects are not again significant, which leads us to the reduced model in (10.22) as analyzed earlier.

10.5   A More Complex Example

The following arthritis treatment data was presented in Friendly (2000, Appendix B.1) and reproduced by permission of the author. The data originally came from Koch and Edwards (1988) and relate to arthritis treatment response on 84 patients randomly assigned to two treatments. The response in this example has three categories (no improvement, some improvement, marked improvement). These are coded as 0, 1 and 2 respectively in the data. The other explanatory variables are

- Treat: Treatments with two levels Treatment and Placebo.

- Gender: with two levels Males and Females denoted by M and F respectively.

- Age: A continuous variable.

- Improve: The response variable with 3 categories (no improvement, some improvement, marked improvement).

The response variable in this case has three categories and moreover these categories seem to be intrinsically ordered. We shall look at analysis of data in this example with more than two response levels. However, we can run our usual logistic regression analysis by dichotomizing the response variable into Improved versus Not improved. That is, we would have a response variable defined as

$$\text{Better} = \begin{cases} 1, & \text{if improve} > 0 \\ 0, & \text{if improve} = 0. \end{cases}$$

In this case the response variable 'Better' is now binary and the logistic regression can be employed with treat, gender and age as explanatory or predictor variables. In the following section, the response variable (with three levels) will be analyzed by using proportional odds model.

```
id  treat  gender age improve     id  treat  gender age improve     id  treat  gender age improve
-------------------------------    -------------------------------    -------------------------------
57  Treat   M     27   1          66  Treat   F     23   0           2  Treat   F     59   2
 9  Placbo  M     37   0          50  Placbo  F     31   1          70  Placbo  F     55   2
46  Treat   M     29   0          40  Treat   F     32   0          59  Treat   F     59   2
14  Placbo  M     44   0          38  Placbo  F     32   0          49  Placbo  F     57   0
77  Treat   M     30   0           6  Treat   F     37   1          62  Treat   F     60   2
73  Placbo  M     50   0          35  Placbo  F     33   2          10  Placbo  F     57   1
17  Treat   M     32   2           7  Treat   F     41   0          84  Treat   F     61   2
74  Placbo  M     51   0          51  Placbo  F     37   0          47  Placbo  F     58   1
36  Treat   M     46   2          72  Treat   F     41   2          64  Treat   F     62   1
```

25	Placbo	M	52	0		54	Placbo	F	44	0		44	Placbo	F	59	1
23	Treat	M	58	2		37	Treat	F	48	0		34	Treat	F	62	2
18	Placbo	M	53	0		76	Placbo	F	45	0		24	Placbo	F	59	2
75	Treat	M	59	0		82	Treat	F	48	2		58	Treat	F	66	2
21	Placbo	M	59	0		16	Placbo	F	46	0		48	Placbo	F	61	0
39	Treat	M	59	2		53	Treat	F	55	2		13	Treat	F	67	2
52	Placbo	M	59	0		69	Placbo	F	48	0		19	Placbo	F	63	1
33	Treat	M	63	0		79	Treat	F	55	2		61	Treat	F	68	1
45	Placbo	M	62	0		31	Placbo	F	49	0		3	Placbo	F	64	0
55	Treat	M	63	0		26	Treat	F	56	2		65	Treat	F	68	2
41	Placbo	M	62	0		20	Placbo	F	51	0		67	Placbo	F	65	2
30	Treat	M	64	0		28	Treat	F	57	2		11	Treat	F	69	0
8	Placbo	M	63	2		68	Placbo	F	53	0		32	Placbo	F	66	0
5	Treat	M	64	1		60	Treat	F	57	2		56	Treat	F	69	1
80	Placbo	F	23	0		81	Placbo	F	54	0		42	Placbo	F	66	0
63	Treat	M	69	0		22	Treat	F	57	2		43	Treat	F	70	1
12	Placbo	F	30	0		4	Placbo	F	54	0		15	Placbo	F	66	1
83	Treat	M	70	2		27	Treat	F	58	0		71	Placbo	F	68	1
29	Placbo	F	30	0		78	Placbo	F	54	2		1	Placbo	F	74	2

Our model of interest in this case would be

$$\text{logit}(p_{ij}) = \beta_0 + \beta_{1i}\,\text{trt}_i + \beta_{2j}\,\text{sex}_j + \beta_3\,\text{age} + \beta_{4ij}\,(\text{trt}*\text{sex})_{ij}$$
$$+ \beta_{5i}\,(\text{trt}*\text{age})_i + \beta_{6j}\,(\text{sex}*\text{age})_j + \text{higher terms}, \tag{10.25}$$

where

- trt$_i$ is the effect of the i th treatment, $(i = 1, 2)$

- sex$_j$ is the effect of the j-th sex, where sex $= 1$ if gender is F and sex $= 0$ if gender $=$ M.

- age is the effect of age.

The last three terms in (10.25) are the interaction terms. We give below the SAS statements and the corresponding partial output for a forward selection procedure that forces SAS to include the first three terms in the model regardless (invoked by **start = 3**).

```
title 'Arthritis treatment data: Logistic regression';
data arthrit;
   input id treat $ gender $ age improve @@ ;
   better  = (improve > 0);
   trt = (treat ='Treat') ;
   sex  = (gender = 'F');
   sexage  = age*sex ;
   trtage  = age*trt;
   trtsex  = sex*trt;
datalines ;
57 Treat M   27 1   9 Placbo M   37 0
46 Treat M   29 0  14 Placbo M   44 0
.....................................
;
proc print noobs (obs=10);
run;
proc logistic descending;
 model  better =trt sex age  trtsex trtage sexage /selection=forward details start=3;
run;
```

```
                     The LOGISTIC Procedure

                    Model Information

       Data Set               WORK.ARTHRIT
       Response Variable      better
       Number of Response Levels  2
       Model                  binary logit
       Optimization Technique Fisher's scoring

            Number of Observations Read      84
            Number of Observations Used      84

                    Response Profile

            Ordered                 Total
            Value       better      Frequency

              1           1           42
```

```
                        2          0          42
                Probability modeled is better=1.

                   Forward Selection Procedure

Step  0. The following effects were entered:

Intercept  trt  sex  age

                      Model Convergence Status

            Convergence criterion (GCONV=1E-8) satisfied.

                      Model Fit Statistics

                                        Intercept
                            Intercept        and
              Criterion       Only       Covariates

              AIC            118.449       100.063
              SC             120.880       109.786
              -2 Log L       116.449        92.063

            Testing Global Null Hypothesis: BETA=0

      Test                Chi-Square    DF     Pr > ChiSq

      Likelihood Ratio       24.3859     3       <.0001
      Score                  22.0051     3       <.0001
      Wald                   17.5147     3       0.0006

            Analysis of Maximum Likelihood Estimates

                                 Standard      Wald
      Parameter   DF   Estimate    Error    Chi-Square   Pr > ChiSq

      Intercept    1    -4.5033    1.3074     11.8649       0.0006
      trt          1     1.7598    0.5365     10.7596       0.0010
      sex          1     1.4878    0.5948      6.2576       0.0124
      age          1     0.0487    0.0207      5.5655       0.0183

                      Odds Ratio Estimates

                         Point        95% Wald
              Effect    Estimate    Confidence Limits

              trt         5.811     2.031      16.632
              sex         4.427     1.380      14.204
              age         1.050     1.008       1.093

      Association of Predicted Probabilities and Observed Responses

              Percent Concordant    78.9    Somers' D    0.583
              Percent Discordant    20.5    Gamma        0.587
              Percent Tied           0.6    Tau-a        0.295
              Pairs                 1764    c            0.792

                      Residual Chi-Square Test

              Chi-Square       DF      Pr > ChiSq

                3.9744          3        0.2642

            Analysis of Effects Eligible for Entry

                               Score
              Effect    DF   Chi-Square   Pr > ChiSq

              trtsex     1      0.4043      0.5249
              trtage     1      0.5951      0.4405
              sexage     1      3.6874      0.0548

NOTE: No (additional) effects met the 0.05 significance level for entry into the model.
```

The SAS partial output indicates that none of the interaction terms in model (10.25) is significant, given that the first three main terms are in the model. Hence our revised or reduced model would be

$$\text{logit}(p_{ij}) = \beta_0 + \beta_{1i}\,\text{trt}_i + \beta_{2j}\,\text{sex}_j + \beta_3\,\text{age}. \tag{10.26}$$

The model in (10.26) is implemented with the following SAS codes and partial output are also displayed.

```
set athrit;
proc logistic descending;
 model  better =trt sex age/scale=none aggregate lackfit
                        rsq plcl plrl waldcl waldrl;
units trt=1 age=1 2 3 sex=1;
output out=aa p=phat;
run;
data new;
set aa;
predicts=(phat ge 0.5);
proc freq data=new;
tables better*predicts/norow nocol nopercent agree;
run;
```

                        The LOGISTIC Procedure

                Number of Observations Read   84
                Number of Observations Used   84

                        Response Profile

                Ordered   Total
                 Value   better   Frequency
                   1   1   42
                   2   0   42

                Probability modeled is better=1.

                Model Convergence Status

        Convergence criterion (GCONV=1E 8) satisfied.

        Deviance and Pearson Goodness-of-Fit Statistics

    Criterion   Value   DF   Value/DF   Pr > ChiSq

    Deviance   67.7891   59   1.1490   0.2025
    Pearson   60.3906   59   1.0236   0.4253

                Number of unique profiles: 63

                Model Fit Statistics

                                        Intercept
                          Intercept   and
          Criterion   Only   Covariates

            AIC   118.449   100.063
            SC   120.880   109.786
            -2 Log L   116.449   92.063

      R-Square   0.2520   Max-rescaled R-Square   0.3360

        Testing Global Null Hypothesis: BETA=0

    Test   Chi-Square   DF   Pr > ChiSq

    Likelihood Ratio   24.3859   3   <.0001
    Score   22.0051   3   <.0001
    Wald   17.5147   3   0.0006

        Analysis of Maximum Likelihood Estimates

                          Standard   Wald
    Parameter   DF   Estimate   Error   Chi-Square   Pr > ChiSq

    Intercept   1   -4.5033   1.3074   11.8649   0.0006
    trt   1   1.7598   0.5365   10.7596   0.0010
    sex   1   1.4878   0.5948   6.2576   0.0124
    age   1   0.0487   0.0207   5.5655   0.0183

                Odds Ratio Estimates

                  Point   95% Wald
        Effect   Estimate   Confidence Limits

        trt   5.811   2.031   16.632
        sex   4.427   1.380   14.204
        age   1.050   1.008   1.093

```
          Association of Predicted Probabilities and Observed Responses

              Percent Concordant     78.9    Somers' D    0.583
              Percent Discordant     20.5    Gamma        0.587
              Percent Tied            0.6    Tau-a        0.295
              Pairs                  1764    c            0.792

                      Profile Likelihood Confidence
                         Interval for Parameters

              Parameter    Estimate    95% Confidence Limits

              Intercept    -4.5033     -7.3033      -2.1115
              trt           1.7598      0.7508       2.8750
              sex           1.4878      0.3724       2.7296
              age           0.0487      0.00995      0.0919

                 Wald Confidence Interval for Parameters

              Parameter    Estimate    95% Confidence Limits

              Intercept    -4.5033     -7.0657      -1.9409
              trt           1.7598      0.7083       2.8113
              sex           1.4878      0.3221       2.6536
              age           0.0487      0.00825      0.0892

      Profile Likelihood Confidence Interval for Adjusted Odds Ratios

         Effect       Unit      Estimate     95% Confidence Limits

         trt         1.0000       5.811        2.119       17.726
         sex         1.0000       4.427        1.451       15.327
         age         1.0000       1.050        1.010        1.096
         age         2.0000       1.102        1.020        1.202
         age         3.0000       1.157        1.030        1.318

              Wald Confidence Interval for Adjusted Odds Ratios

         Effect       Unit      Estimate     95% Confidence Limits

         trt         1.0000       5.811        2.031       16.632
         sex         1.0000       4.427        1.380       14.204
         age         1.0000       1.050        1.008        1.093
         age         2.0000       1.102        1.017        1.195
         age         3.0000       1.157        1.025        1.307

             Hosmer and Lemeshow Goodness-of-Fit Test

              Chi-Square      DF      Pr > ChiSq
                5.5549         8        0.6970

             Statistics for Table of better by predicts

                         McNemar's Test
                    ----------------------
                    Statistic (S)    0.0400
                    DF                    1
                    Pr > S           0.8415

                  Simple Kappa Coefficient
              ---------------------------------
              Kappa                    0.4048
              ASE                      0.0997
              95% Lower Conf Limit     0.2093
              95% Upper Conf Limit     0.6003

                    Sample Size = 84
```

The SAS output indicates that there are 63 unique profiles in the data. The model in (10.26) fits the data well with deviance or G^2 being 67.7891 on 59 degrees of freedom and a p-value of 0.2025. The ratio of Value/df is 1.1490 which indicates not much over-dispersion in the data. Both the 95% profile likelihood and Wald's confidence intervals are also produced in the output. The Hosmer and Lemeshow Goodness-of-fit test gives a p-value of 0.6970 indicating a very good fit of the model to the data.

We also wish to see to what degree does the observed outcome agree with the predicted probabilities of an outcome. To achieve this, we classify a patient to be 'better' the desired outcome, if his or her predicted probability is greater than or equal to 0.5 (this is subjective). We present in the table below the result of the McNemar's test of agreement using **proc freq** in SAS. The estimate of κ, the agreement statistic and the agreement test clearly indicate that there is very strong agreement between the observed outcome and the expected desired outcome based on the estimated logistic regression model. In Figures 10.7 to 10.10 are the plots of predicted probabilities against age for sex = 0 (or male) and sex = 1 (or female); and for trt = 0 and trt = 1 respectively.

Table of Observed Outcome 'Better' by PREDICTS

Better	PREDICTS		Total
	0	1	
0	29	13	42
1	12	30	42
Total	41	43	84

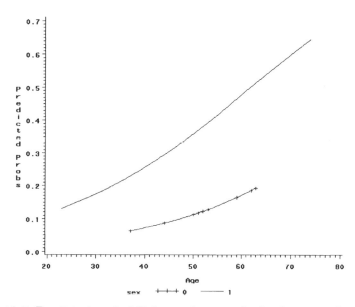

Figure 10.7: Predicted probabilities against age for both sexes when trt = 0

The plot indicates that for those patients on placebo, Figure 10.7, the probability of of getting better increases with age for both sexes, although females tend to have higher probabilities than men. Moreover, the male patients that were on placebo are aged between 35 and 68 as compared with females, who are aged between 20 and 80.

Figure 10.8 presents a similar result for patients assigned to the treatment. The females still have higher improvement probabilities than males as age increases, although with older patients, the probabilities tend to be converging with increasing age of patients, though the female patients still have much higher improvement probabilities than male.

The plots in Figures 10.9 and 10.10 give the plot of estimated improvement probabilities against age for the two treatment groups for men and women respectively. Again, the treatment group had much higher probabilities irrespective of gender and therefore the treatment is very beneficial in improving the condition of arthritis. In general, those on the treatment are 5.811 times likely to be better than those on placebo given the same age and sex of the patients.

Similarly, the odds are 4.427 times higher among females than males to feel much better for a given age of patients and assuming patients are on the same treatment. For a unit increase in age (say, by 1 year), the odds is 1.050 times higher for a better improvement (keeping sex and treatment constant). The odds go up to 1.102 for 2 unit increases and up to 1.157 for a three year increase.

10.6 Multi Category Response Variable

In our complex example data in section 10.6, we recall that the response variable has three categories, namely, 'No Improvement', 'Some improvement', 'Marked Improvement'. These categories have intrinsic ordering about them and

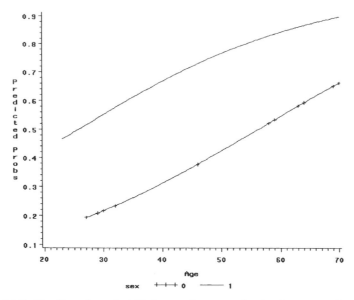

Figure 10.8: Predicted probabilities against age for both sexes when trt = 1

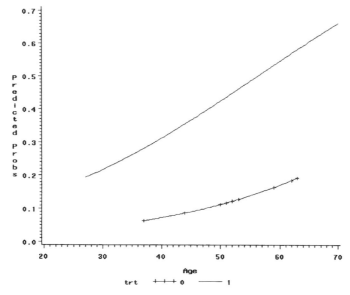

Figure 10.9: Predicted probabilities against age for both treatments when sex = 0

we can therefore consider fitting models that exploit the ordinal nature of the response variable. These models are

- The cumulative logit model

- The proportional odds model

- The adjacent-category model

- The continuation ratio model

In this text however, we will only implement the proportional odds model for our data. Interested readers on the other models can consult the book by Lawal (2003), where all these models are discussed and are implemented in both SAS and SPSS.

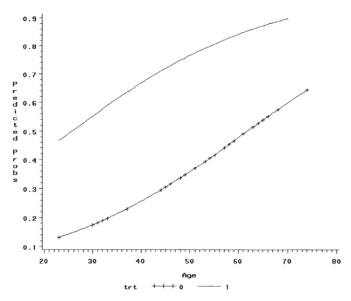

Figure 10.10: Predicted probabilities against age for both treatments when sex = 1

10.6.1 The Cumulative Logit Model

Consider a response variable with J ordered categories assumed to have a multinomial distribution. In our example, $J = 3$, and we have the following structure, (with NI = no improvement, SI = some improvement and MI = marked improvement) and π_j being the outcome probability in response category j.

i	j		
	1	2	3
	NI	SI	MI
	π_1	π_2	π_3

If the response categories are assumed multinomially distributed, then in general,

$$\sum_{j=1}^{J} \pi_j = 1 \quad \text{which implies in our case that} \quad \sum_{j=1}^{3} \pi_j = 1.$$

The cumulative logits are defined in terms of the $\pi's$ as

$$L_j = \ln\left(\sum_{s=1}^{j} \pi_s \Big/ \sum_{s=j+1}^{J} \pi_s\right) = \alpha_j + \beta x_i, \tag{10.27}$$

for $j = 1, 2, \ldots, (J-1)$. Thus for $J = 3$ in our example, we have, the following decomposition of the cumulative model:

$$\ln\left(\frac{\pi_1}{\pi_2 + \pi_3}\right) = \alpha_1 + \beta_1 \text{trt} + \beta_2 \text{sex} + \beta_3 \text{age} = \text{logit}\left(\frac{\text{NI}}{\geq \text{SI}}\right), \tag{10.28a}$$

$$\ln\left(\frac{\pi_1 + \pi_2}{\pi_3}\right) = \alpha_2 + \beta_1^* \text{trt} + \beta_2^* \text{sex} + \beta_3^* \text{age} = \text{logit}\left(\frac{\leq \text{SI}}{\text{MI}}\right). \tag{10.28b}$$

The models in (10.28) have six slopes but two intercepts (cut points): α_1 and α_2 respectively. For the cumulative logit model

$$\beta_j \neq \beta_j^*.$$

The simplest cumulative logit model has

$$L_j = \alpha_j, \quad j = 1, 2, \ldots, (J-1). \tag{10.29}$$

The model implies that the response variable is independent (simultaneously) of the explanatory variables **X**. In this case, the $\{\alpha_j\}$, the cut-point parameters, are nondecreasing in j.

10.6.2   The Proportional Odds Model

The model in (10.28) has also been described as the proportional odds model and has the general form

$$\ln\left[\frac{P(Y \le j \mid x_i)}{P(Y > j \mid x_i)}\right] = \alpha_j + \boldsymbol{\beta}\,\mathbf{X}_i, \quad j = 1, 2, \ldots, (J-1); \; i = 1, 2, \ldots, I. \tag{10.30}$$

We see that there are $(J-1)$ separate proportional odds equations for each possible cut point j, with each having a common slope β and different intercepts α_j. In this formulation, the α_j's are not themselves important, but the slope parameter β's are. The proportional odds model therefore is the cumulative model in (10.28) such that $\beta_j = \beta_j^*$. That is, we have for our data

$$H_0 = \begin{cases} \beta_1 & = \beta_1^* \\ \beta_2 & = \beta_2^* \\ \beta_3 & = \beta_3^* \end{cases} \tag{10.31}$$

versus at least one of the above is not true. Thus the proportional odds model applied to our data has the revised form

$$\ln\left(\frac{\pi_1}{\pi_2 + \pi_3}\right) = \alpha_1 + \beta_1\text{trt} + \beta_2\text{sex} + \beta_3\text{age} \tag{10.32a}$$

$$\ln\left(\frac{\pi_1 + \pi_2}{\pi_3}\right) = \alpha_2 + \beta_1\text{trt} + \beta_2\text{sex} + \beta_3\text{age}. \tag{10.32b}$$

As an example, we employ **proc genmod** and **proc logistic** in what follows to fit the proportional odds model to our data example.

```
data arthrit;
input id treat $ gender $ age improve @@ ;
trt = (treat ='Treat') ;
sex = (gender = 'F');
datalines ;
............
;
run;
proc genmod;
model improve=trt sex age/dist=multinomial
                 link=cumlogit aggregate type1;
run;
                    The SAS System

              The GENMOD Procedure

              Model Information

        Data Set                WORK.ARTHRIT
        Distribution             Multinomial
        Link Function       Cumulative Logit
        Dependent Variable            improve

        Number of Observations Read        84
        Number of Observations Used        84

              Response Profile

          Ordered                Total
          Value    improve    Frequency
            1       0             42
            2       1             14
            3       2             28
```

The **proc genmod** is modeling the probabilities of levels of 'improve' having **lower** ordered values in the response profile table. One way to change this to model the probabilities of **higher** ordered values is to specify the **descending** option in the **proc** statement.

```
        Criteria For Assessing Goodness Of Fit

    Criterion              DF        Value       Value/DF
```

```
--------------------------------------------------------------
Deviance                    121    112.8664      0.9328
Scaled Deviance             121    112.8664      0.9328
Pearson Chi-Square          121    121.0683      1.0006
Scaled Pearson X2           121    121.0683      1.0006
Log Likelihood                     -72.7290
```

Algorithm converged.

Analysis Of Parameter Estimates

Parameter	DF	Estimate	Standard Error	Wald 95% Confidence Limits		Chi-Square	Pr > ChiSq
Intercept1	1	3.7837	1.1438	1.5418	6.0255	10.94	0.0009
Intercept2	1	4.6827	1.1873	2.3556	7.0097	15.55	<.0001
trt	1	-1.7453	0.4759	-2.6780	-0.8126	13.45	0.0002
sex	1	-1.2517	0.5464	-2.3225	-0.1808	5.25	0.0220
age	1	-0.0382	0.0184	-0.0743	-0.0021	4.30	0.0382
Scale	0	1.0000	0.0000	1.0000	1.0000		

NOTE: The scale parameter was held fixed.

LR Statistics For Type 1 Analysis

Source	Deviance	DF	Chi-Square	Pr > ChiSq
Intercepts	137.3244			
trt	124.0146	1	13.31	0.0003
sex	117.4379	1	6.58	0.0103
age	112.8664	1	4.57	0.0325

We also present below the implementation with **proc logistic**. Notice that we are modeling the 0 level as our reference in the above output. Should we wish to reverse this, we can use **descending** in the **proc** statement to effect the change of order of modeling.

```
set arthrit;
proc logistic;
model improve=trt sex age/scale=none aggregate;
units trt=1 age=1 2 3 sex=1;
run;
```

The LOGISTIC Procedure

Model Information

Data Set	WORK.ARTHRIT
Response Variable	improve
Number of Response Levels	3
Model	cumulative logit
Optimization Technique	Fisher's scoring

Number of Observations Read	84
Number of Observations Used	84

Response Profile

Ordered Value	improve	Total Frequency
1	0	42
2	1	14
3	2	28

Probabilities modeled are cumulated over the lower Ordered Values.

Model Convergence Status

Convergence criterion (GCONV=1E-8) satisfied.

Score Test for the Proportional Odds Assumption

Chi-Square	DF	Pr > ChiSq
2.4916	3	0.4768

Deviance and Pearson Goodness-of-Fit Statistics

Criterion	Value	DF	Value/DF	Pr > ChiSq
Deviance	112.8664	121	0.9328	0.6886
Pearson	121.0658	121	1.0005	0.4812

Number of unique profiles: 63

```
                       Model Fit Statistics

                                           Intercept
                              Intercept       and
            Criterion           Only       Covariates
            AIC                173.916       155.458
            SC                 178.778       167.612
            -2 Log L           169.916       145.458

        Testing Global Null Hypothesis: BETA=0

   Test                Chi-Square      DF     Pr > ChiSq

   Likelihood Ratio      24.4580        3        <.0001
   Score                 22.3472        3        <.0001
   Wald                  19.5564        3        0.0002

        Analysis of Maximum Likelihood Estimates

                              Standard      Wald
   Parameter    DF   Estimate    Error   Chi-Square   Pr > ChiSq
   Intercept 0   1    3.7836    1.1530    10.7680       0.0010
   Intercept 1   1    4.6826    1.1949    15.3566       <.0001
   trt           1   -1.7453    0.4772    13.3774       0.0003
   sex           1   -1.2515    0.5321     5.5330       0.0187
   age           1   -0.0382    0.0185     4.2361       0.0396

               Odds Ratio Estimates

                    Point        95% Wald
         Effect   Estimate   Confidence Limits
         trt        0.175      0.069     0.445
         sex        0.286      0.101     0.812
         age        0.963      0.928     0.998

   Association of Predicted Probabilities and Observed Responses

      Percent Concordant    74.6    Somers' D   0.500
      Percent Discordant    24.6    Gamma       0.504
      Percent Tied           0.8    Tau-a       0.309
      Pairs                 2156    c           0.750

             Adjusted Odds Ratios

         Effect     Unit     Estimate
         trt      1.0000      0.175
         sex      1.0000      0.286
         age      1.0000      0.963
         age      2.0000      0.927
         age      3.0000      0.892
```

We notice that both **proc genmod** and **proc logistic** produce identical results. We would therefore restrict our attention to the results from **proc logistic**. The proportional odds model is based on the assumption in (10.31), that the slope parameter are homogeneous (that is equal) as in model (10.30) and (10.28). These hypotheses are tested by the *score test for the proportional odds assumption* in the **proc logistic** output. The score test for this assumption gives a $G^2 = 2.4916$ on 3 degrees of freedom (p-value $= 0.4768$) indicating that the assumption of equal slope is satisfied. Its degree of freedom is $P(J - 2) = 3(3 - 2) = 3$, where P is the number of explanatory variables in the model and J is the number of the response categories. Further, the proportional odds model fits the data well with $G^2 = 112.8664$ on 121 degrees of freedom (p-value $= 0.6886$), and there does not seem to be too much of under-dispersion in the model. This is therefore an acceptable estimated model and since, the π_j's satisfy $\sum_{j=1}^{3} \pi_j = 1$, we have

$$\ln\left(\frac{\hat{\pi}_1}{\hat{\pi}_2 + \hat{\pi}_3}\right) = \ln\left(\frac{\hat{\pi}_1}{1 - \hat{\pi}_1}\right) \qquad = 3.7836 - 1.7453\,\text{trt} - 1.2515\,\text{sex} - 0.0382\,\text{age} \qquad (10.33a)$$

$$\ln\left(\frac{\hat{\pi}_1 + \hat{\pi}_2}{\hat{\pi}_3}\right) = \ln\left(\frac{\hat{\pi}_1 + \hat{\pi}_2}{1 - (\hat{\pi}_1 + \hat{\pi}_2)}\right) = 4.6826 - 1.7453\,\text{trt} - 1.2515\,\text{sex} - 0.0382\,\text{age}. \qquad (10.33b)$$

We next use the estimated proportional odds model to estimate the odds of various combinations of treatment, sex and age. We have selected ages 35, 45, 55 and 65 respectively, with the four combinations of trt and sex. These are contained in the first three columns of the output below, with their corresponding logits, odds, the reciprocal of the estimated odds as well as predicted probabilities and their compliments. We have stored these combinations in SAS data file called **pred** and then use **proc score** to estimate the corresponding logits and odds ratios. The SAS codes and partial output are presented in the following:

```
set arthrit;
proc logistic data=arthrit outest=qq;
model improve=trt sex age/scale=none aggregate;
run;
proc score data=pred score=qq out=dd type=parms;
var trt sex age;
run;
data new1;
set dd;
odds=exp(improve);
recp=1/odds;
p_1=odds/(1+odds);
p_0=1-p_1;
proc print data=new1 noobs;
format p_1 p_0 odds recp improve f8.4;
run;
```

Odds Ratio Estimates

Effect	Point Estimate	95% Wald Confidence Limits	
trt	0.175	0.069	0.445
sex	0.286	0.101	0.812
age	0.963	0.928	0.998

trt	sex	age	improve	odds	recp	p_1	p_0
0	0	35	-1.3357	0.2630	3.8028	0.2082	0.7918
0	0	45	-1.7174	0.1795	5.5699	0.1522	0.8478
0	0	55	-2.0990	0.1226	8.1581	0.1092	0.8908
0	0	65	-2.4807	0.0837	11.9491	0.0772	0.9228
0	1	35	-2.5872	0.0752	13.2931	0.0700	0.9300
0	1	45	-2.9689	0.0514	19.4701	0.0489	0.9511
0	1	55	-3.3505	0.0351	28.5176	0.0339	0.9661
0	1	65	-3.7322	0.0239	41.7692	0.0234	0.9766
1	0	35	-3.0811	0.0459	21.7813	0.0439	0.9561
1	0	45	-3.4627	0.0313	31.9027	0.0304	0.9696
1	0	55	-3.8443	0.0214	46.7274	0.0210	0.9790
1	0	65	-4.2260	0.0146	68.4409	0.0144	0.9856
1	1	35	-4.3326	0.0131	76.1388	0.0130	0.9870
1	1	45	-4.7142	0.0090	111.5193	0.0089	0.9911
1	1	55	-5.0958	0.0061	163.3404	0.0061	0.9939
1	1	65	-5.4775	0.0042	239.2421	0.0042	0.9958

The following conclusions are evident from the results of our analysis:

- At a given age and a given level of sex, those patients receiving treatment are 5.714 (1/0.175) times more likely to have marked improvement over at most some improvement.

- Female patients are 3.497 more likely than their male counterparts to have a marked improvement over at most some improvement or at most some improvement over no improvement for a given age and treatment level.

- A unit increase in age increases the odds of improvement over most improvement by just 1.038.

- For a patient age 35, male and receiving no treatment, the odds of marked improvement over at most some improvement is 3.8028 times higher. This odds increases to 5.5699, 8.1581 and 11.9491 for ages 45, 55 and 65 respectively with corresponding estimated probabilities 0.7918, 0.8478, 0.8908 and 0.9228 respectively for ages 35, 45, 55 and 65 years.

- We see that generally, the odds of improvement increases for patients receiving treatments over those receiving the placebo, for females over males and for increasing age.

10.7   Exercises

10.1 **Domestic Violence.** The data in the file **model_c.txt** is part of the data used by Basu and Famoye (2004) to examine the relation between violence against women and women's economic dependence. The variable separate $= 1$ if the respondent temporarily separated from the spouse and 0 otherwise, conflict index and depend index are variables defined in Basu and Famoye (2004), age_diff is the husband's age minus the wife's age, wife_l $= 1$ if the wife is less educated than the husband and it is zero otherwise, wife_m $= 1$ if the wife is more educated than the husband and it is zero otherwise, years is the total number of years living together and income is the total family income in thousands of dollars. For this problem, use the first 50 cases and the variables separate (y), conflict (x_1) and income (x_2).

    (a) Fit a model involving x_1 and a quadratic term in x_1, and test the significance of x_1^2 being in the model.

    (b) Fit a logistic regression model to the data using the variables y, x_1 and x_2.

    (c) Is this model adequate?

    (d) Find a 95% confidence intervals for the model parameters from the model in (b).

    (e) Graph the estimated logistic regression model.

10.2 Use the domestic violence data for Exercise 10.1 and consider all cases. For this exercise, use the subset where the wife is less educated than the husband and the variables separate (y), conflict (x_1) and income (x_2).

    (a) Fit an appropriate logistic regression model to the data.

    (b) What is the estimated equation? Interpret the estimated parameters of the model.

10.3 The data below are the results of a survey of women, relating frequency of breast self-examination and age. A proportional odds model is employed for the data.

	Freq. of Breast Self-Examination		
Age	Monthly	Occasionally	Never
< 45	91	90	51
45-59	150	200	155
60+	109	198	172

Analyze the above data using cumulative logits, treating frequency of breast self examination as the response. Test the adequacy of the model fit and interpret parameter estimates. What can you deduce from your analysis as being the influence of age on breast self examination?

10.4 **Domestic Violence.** The data in the file **model_d.txt** is part of the data used by Basu and Famoye (2004) to examine the relation between violence against women and women's economic dependence. The variable separate $= 1$ if the respondent temporarily separated from the spouse and 0 otherwise, diverge is an index variable defined in Basu and Famoye (2004), age_diff is the husband's age minus the wife's age, wife_l $= 1$ if the wife is less educated than the husband and it is zero otherwise, wife_m $= 1$ if the wife is more educated than the husband and it is zero otherwise, years is the total number of years living together and income is the total family income in thousands of dollars. For this problem, use the variables separate (y), diverge (x_1) and income (x_2).

    (a) Using the variables y, x_1 and x_2, fit a logistic regression model to the data.

    (b) How would you characterize the model in terms of adequacy?

    (c) Using one independent variable at a time, fit a logistic regression model to the data.

    (d) Which of the two models in (c) fits the data better?

    (d) With your model in (a), predict the expected logit and probability for an individual with diverge $= 1.5$ and income $= 14.2$.

10.5 Use the domestic violence data for Exercise 10.4. For this exercise, use the subset where the wife is less educated than the husband and the variables separate (y), diverge (x_1), and income (x_2). Using the variables y, x_1 and x_2, fit a logistic regression model to the data. Test the adequacy of the model fit and interpret parameter estimates. What can you deduce from your analysis as being the influence of divergence index on separation.

10.6 Use the domestic violence data for Exercise 10.4. For this exercise, use the subset where the age_diff is positive and the variables separate (y), diverge (x_1) and income (x_2).

   (a) Fit a model that includes the explanatory variables diverge and income.

   (b) Use the stepwise procedure to obtain the most parsimonious model.

   (c) With your chosen model, obtain the estimated probabilities for the first 10 cases.

10.7 Use the domestic violence data for Exercise 10.4. For this exercise, use the subset where the age_diff is less than or equal to zero and the variables separate (y), diverge (x_1) and income (x_2).

   (a) Fit a model that includes the explanatory variables diverge, income and their interactions.

   (b) Use the stepwise procedure to obtain the most parsimonious model.

   (c) With your chosen model, obtain the estimated probabilities.

   (d) How good is your model? Obtain a two-way classification of observed and expected counts and test for agreement.

10.8 The data in the table below presents (Hedlund, 1978) the relationship between an ordinal variable, political ideology, and a nominal variable, party affiliation for a sample of voters in the 1976 presidential primary election in Wisconsin.

Party Affiliation	Political Ideology		
	Liberal	Moderate	Conservative
Democrat	143	156	100
Independent	119	210	141
Republican	15	72	127

Analyze the above data using cumulative logits, treating political ideology as the response. Test the adequacy of the model fit and interpret parameter estimates. What can you deduce from your analysis as being the influence of party affiliation on ideology.

10.9 The following data from Guerrero and Johnson (1982) relate to the number of Warsaw girls that have menstruated given 25 groups of ages at menarche of 3918 girls. The total number of girls in each group (n) and the number having experienced menarche (r) are presented in the data **menarche** with the mean age (x) of the group. Fit a logistic model to the data with age as the explanatory variable.

x	n	r	x	n	r	x	n	r	x	n	r
9.21	376	0	11.83	111	17	13.58	105	81	15.33	111	107
10.21	200	0	12.08	100	16	13.83	117	88	15.58	94	92
10.58	93	0	12.33	93	29	14.08	98	79	15.83	114	112
10.83	120	2	12.58	100	39	14.033	97	90	17.58	1049	1049
11.08	90	2	12.83	108	51	14.58	120	113			
11.33	88	5	13.08	99	47	14.83	102	95			
11.58	105	10	13.33	106	67	15.08	122	117			

10.10 The following data relate to the outcome of rate at which blood cells (erythrocytes) settle out of suspension in blood plasma. The response, y, is 1 if erythrocyte sedimentation (ES) exceeds 20mm/h and values below this characterize healthy individuals. Positive response are known to be associated with fibrinogen (x_1) and gamma-globulin (x_2). Fit a parsimonious logistic regression model to these data.

x_1	x_2	y	x_1	x_2	y	x_1	x_2	y	x_1	x_2	y
2.52	38	0	3.15	39	0	3.53	46	1	2.67	39	0
2.56	31	0	2.60	41	0	2.68	34	0	2.29	31	0
2.19	33	0	2.29	36	0	2.60	38	0	2.15	31	0
2.18	31	0	2.35	29	0	2.23	37	0	2.54	28	0
3.41	37	0	5.06	37	1	2.88	30	0	3.93	32	1
2.46	36	0	3.34	32	1	2.65	46	0	3.34	30	0
3.22	38	0	2.38	37	1	2.09	44	1	2.99	36	0
2.21	37	0	3.15	36	0	2.28	36	0	3.32	35	0

10.11 A local health clinic sent fliers to its clients to encourage everyone, but especially older persons at high risk of complications, to get flu shot in time for protection against an expected flu epidemic. In a pilot follow-up study, 50 clients were randomly selected and asked whether they actually received a flu shot. In addition, data were collected on their age (x). A client who received a flu shot was coded $y = 1$, and a client who did not receive a flu shot was coded $y = 0$. A simple logistic regression model is fitted to the following data.

y	Age (in years)
0	38, 41, 43, 34, 31, 54, 63, 38, 28, 42, 36, 45, 47, 53, 42, 42, 48, 46, 44, 46, 35 40, 40, 64, 34, 38, 56, 45, 33
1	52, 46, 41, 57, 49, 53, 39, 53, 49, 49, 46, 54, 63, 56, 64, 52, 46, 57, 56, 46, 47

(a) Find the maximum likelihood estimates of β_0 and β_1. State the fitted response logistic regression function.

(b) Obtain $\exp(\beta_1)$ and interpret the number.

(c) What is the estimated probability that clients aged 55 will receive a flu shot?

(d) Obtain an approximate 95% confidence interval for the regression coefficient β_1. Convert this interval into one for the odds ratio.

(e) What is the estimated age at which 60% of the clients will receive flu shot?

(f) Is the estimated age in (e) reasonable? Explain.

(g) What is the estimated age at which 80% of the clients will receive flu shot?

(h) Is the estimated age in (g) reasonable? Explain.

(i) Based on your result in (c), assess the success of the fliers.

(j) Using the five-step procedure and $\alpha = 0.05$, test the null hypothesis that the regression coefficient β_1 is non-positive. State the p-value of your test.

10.12 In an experiment testing the effect of toxic substance, 1500 experimental insects were divided at random into six groups of 25 each. The insects in each group were exposed to a fixed dose of the toxic substance. A day later, each insect was observed. Death from exposure was scored 1, and survival was scored 0. The results are shown in the following table; x_i denotes the dose level received by the insects in group i and r_i denotes the number of insects that died out of 250 (n_i) in the group.

Group	Time					
x_i	1	2	3	4	5	6
r_i	28	53	93	126	172	197
n_i	250	250	250	250	250	250

(a) Fit a logistic regression response to the data

(b) Obtain $\exp(\beta_1)$ and interpret this number.

(c) What is the estimated probability that an insect dies when the dose level is 3.5?

(d) What is the estimated median lethal dose- that is, the dose for which 50% of the experimental insects are expected to die?

(e) Obtain an approximate 95% confidence interval for β_1. Convert this interval into one for the odds ratio. Interpret this latter interval.

10.13 Consider the data for Example 10.4 in Table 10.4.

(a) Use a statistical software to plot the graph (a half page) of rate against volume for both constriction values on the same page. Are there any outliers or influential observations?

(b) Pregibon (1981) fits a logit response model having logs of volume and rate as explanatory variables. How does this model compare with the one in Example 10.4? Pregibon suggests that the volume for observation 32 should be 0.3 rather than 0.03 as it appears in Table 10.4. Is there any evidence for this suggestion? Identify any lack of fit by carrying out the necessary diagnostics procedures.

10.14 The data below, reported in Woodward *et al.* (1941), examined the relationship between exposure to chloracetic acid and the death of mice. Ten mice were exposed at each dose level and the doses are measured in grams per kilogram of the body weight.

Dose	# Dead	# exposed	Dose	# Dead	# exposed
0.0794	1	10	0.1778	4	10
0.1000	2	10	0.1995	6	10
0.1259	1	10	0.2239	4	10
0.1413	0	10	0.2512	5	10
0.1500	1	10	0.2818	5	10
0.1588	2	10	0.3162	8	10

(a) Fit an appropriate logistic regression model to the data.

(b) What is the estimated equation? Interpret the estimated parameters of the model.

Chapter 11

Count Data Regression Models

11.1 Introduction

In this chapter, our emphasis shifts to data arising as counts. Thus the assumption of discrete distribution error structure will be most appropriate for these type of data. Count data regression analysis is a regression technique in which the dependent variable is a count. The independent variables could take any form (discrete or continuous). In the regression analysis in Chapters 2 through 9, the basic assumption is that the error terms have normal distribution. Thus, the response variables are independent and are distributed as normal random variables. In this chapter, we do not make this assumption since the response (or dependent) variable is a count, and the variability in counts is related to the magnitude of the count so the constant variance assumption required in normal models is violated. In Chapters 2 through 9, the predicted values of the response variables could be negative or non-negative. However, count values are constrained to be non-negative.

The Poisson distribution has been considered in the context of regression analysis for describing count data where the sample mean and sample variance are almost equal [See Frome et al. (1973), Frome (1983), Holford (1983), Cameron and Trivedi (1998), and the references there in.] In many situations, count data are over-dispersed or under-dispersed. Over-dispersion relative to the Poisson is when the sample variance is substantially in excess of the sample mean. Under-dispersion is a situation in which the sample variance is less than the sample mean. Many regression models have been suggested to deal with over-dispersion or under-dispersion. Among these various models are the negative binomial regression model defined and studied by Lawless (1987) and the generalized Poisson regression model defined and studied by Consul and Famoye (1992) and Famoye (1993).

In this chapter, we will discuss the Poisson regression model, the negative binomial regression model, the generalized Poisson regression model and some modified count data regression models.

11.2 Poisson Regression Model

A random variable Y is said to have a Poisson distribution with parameter μ if it takes integer values $0, 1, 2, \ldots,$ with probability mass function (pmf)

$$f(y) = \frac{\mu^y e^{-\mu}}{y!}, \tag{11.1}$$

where $\mu > 0$ is the mean of the Poisson distribution. The mean and variance of the distribution are equal. That is

$$\mathrm{E}(Y) = \mathrm{var}(Y) = \mu.$$

Suppose therefore that $y_1, y_2, \ldots, y_n$ are the responses, which are counts that follow independent Poisson distributions with means μ_i and variances μ_i. If we let $x_1, x_2, \ldots, x_{p-1}$ be a set of explanatory variables, then we can write the model in the form

$$\mu_i = e^{\mathbf{x}_i' \boldsymbol{\beta}}. \tag{11.2}$$

Hence, taking logarithms to base e, we have

$$\log(\mu_i) = \mathbf{x}_i'\boldsymbol{\beta}, \tag{11.3}$$

where the β_j, $j = 1, 2, \ldots, p - 1$ is the parameter which represents the expected change in the log of the mean per unit increase (or decrease) in the explanatory variable x_j. The model in (11.3) is often referred to as a Poisson regression model. It is a generalized linear model with link function $g(x)$, where, $g(x) = \log(x)$. The log link function makes sense because counts are non-negative numbers and this link will guarantee a non-negative estimate of the mean. See Famoye(1995) for examples of other link functions in the literature.

The maximum likelihood estimates of the parameters of the Poisson regression in (11.3) are obtained by using Fisher's scoring algorithm, where the score equation takes the form (see Lawal, 2003)

$$\sum_{i=1}^{n} (y_i - \mu_i) \, \mathbf{x}_i = 0. \tag{11.4}$$

Goodness-of-fit Statistics

The goodness-of-fit test statistics here are the deviance or G^2 and the Pearson's X^2 statistics and have the respective forms

$$\mathrm{D} = 2 \sum_{i=1}^{n} y_i \ln\left(\frac{y_i}{\hat{\mu}_i}\right), \tag{11.5}$$

and

$$X^2 = \sum_{i=1}^{n} \left[\frac{(y_i - \hat{\mu}_i)^2}{\hat{\mu}_i} \right]. \tag{11.6}$$

Both the deviance D and Pearson's X^2 are asymptotically distributed as χ^2 with $n - p$ degrees of freedom.

Example 11.1

The following data (originally from Kutner et al., 2005 pp. 633-634 and reproduced with permission of The McGraw-Hill Companies) relate to a geriatrics study which was designed as a prospective study to investigate the effects of two interventions on the frequency of falls. A random sample of 100 subjects at least 65 years old and in good health was selected and assigned to one of the two interventions: education only ($x_1 = 0$) and education plus aerobic training ($x_1 = 1$). Three variables considered to be important as control variables were gender ($x_2 : 0 =$ female; $1 =$ male), a balance index (x_3), and a strength index (x_4). The higher the balance index, the more stable is the subject; and the higher the strength index, the stronger is the subject. Each subject kept a diary recording the number of falls (y) during the six months of the study. The data in Table 11.1 gives the first four and last four observations.

A model of the form in (11.7) is considered:

$$\ln(\mu_i) = \beta_0 + \beta_1 x_{i1} + \beta_2 x_{i2} + \beta_3 x_{i3} + \beta_4 x_{i4}; \quad i = 1, 2, \ldots, 100. \tag{11.7}$$

The implementation of this model is carried out in SAS using **proc genmod**. The program and partial output for fitting the Poisson regression model to the data in Table 11.1 is presented below.

```
data geriat;
input count x1 x2 x3 x4;
datalines;
   1  1  0  45  70
   1  1  0  62  66
   2  1  1  43  64
   0  1  1  76  48
   . . . . . . . . . . . . . .
   2  0  1  33  55
   4  0  0  69  48
   4  0  1  50  52
   2  0  0  37  56
;
proc print;
```

Subject	Number of Falls	Intervention	Gender	Balance Index	Strength Index
i	y_i	x_1	x_2	x_3	x_4
1	1	1	0	45	70
2	1	1	0	62	66
3	2	1	1	43	64
4	0	1	1	76	48
...	...	...	...	...	...
97	2	0	1	33	55
98	4	0	0	69	48
99	4	0	1	50	52
100	2	0	0	37	56

Table 11.1: Data on frequency of falls

```
run;
proc genmod;
model count=x1 x2 x3 x4/dist=poi link=log type3;
run;
                        The GENMOD Procedure

                        Model Information

            Data Set              WORK.GERIAT
            Distribution          Poisson
            Link Function         Log
            Dependent Variable    count

        Number of Observations Read        100
        Number of Observations Used        100

            Criteria For Assessing Goodness Of Fit

        Criterion             DF        Value        Value/DF
        Deviance              95      108.7899        1.1452
        Scaled Deviance       95      108.7899        1.1452
        Pearson Chi-Square     95      105.5466        1.1110
        Scaled Pearson X2      95      105.5466        1.1110
        Log Likelihood                 79.2067

   Algorithm converged.

                    Analysis Of Parameter Estimates

                        Standard   Wald 95% Confidence    Chi-
   Parameter  DF  Estimate  Error       Limits          Square  Pr > ChiSq
   Intercept   1   0.4895   0.3369   -0.1708    1.1497    2.11     0.1462
   x1          1  -1.0694   0.1332   -1.3304   -0.8084   64.50    <.0001
   x2          1  -0.0466   0.1200   -0.2817    0.1885    0.15     0.6977
   x3          1   0.0095   0.0030    0.0037    0.0153   10.28     0.0013
   x4          1   0.0086   0.0043    0.0001    0.0170    3.95     0.0470
   Scale       0   1.0000   0.0000    1.0000    1.0000

NOTE: The scale parameter was held fixed.

                LR Statistics For Type 3 Analysis

                            Chi-
        Source        DF   Square    Pr > ChiSq
        x1             1   73.52       <.0001
        x2             1    0.15       0.6976
        x3             1   10.28       0.0013
        x4             1    3.98       0.0461
```

Notice that in the model options, we have invoked **dist=poi**, that is the distribution is Poisson, and **link=log** indicating that the link function is **log**. The deviance for this model is 108.7899 on 95 degrees of freedom, p-value = 0.1578 giving a good fit. The estimated parameters of the model are also displayed under 'Analysis of Parameter Estimates', while the Type 3 analysis provides us with the contribution of each explanatory variables to the model as if that variable is entered last. At the 5% level of significance, variable x_2 is not significant, indicating that this variable (gender) can be removed from the model. This therefore leads to the reduced model

$$\ln(\mu_i) = \beta_0 + \beta_1 x_{i1} + \beta_3 x_{i3} + \beta_4 x_{i4}; \quad i = 1, 2, \ldots, 100. \tag{11.8}$$

Again, when the reduced model is implemented in SAS, we have the following output.

```
set geriat;
proc genmod;
model count=x1 x3 x4/dist=poi link=log type3;
run;
                    Criteria For Assessing Goodness Of Fit

          Criterion              DF          Value        Value/DF
          --------------------------------------------------------
          Deviance               96         108.9409        1.1348
          Scaled Deviance        96         108.9409        1.1348
          Pearson Chi-Square      96         105.2878        1.0967
          Scaled Pearson X2       96         105.2878        1.0967
          Log Likelihood                      79.1312

     Algorithm converged.

                    Analysis Of Parameter Estimates

                       Standard   Wald 95% Confidence    Chi-
   Parameter  DF  Estimate   Error        Limits        Square  Pr > ChiSq
   -----------------------------------------------------------------------
   Intercept   1    0.4439   0.3173   -0.1780    1.0658    1.96     0.1618
   x1          1   -1.0778   0.1314   -1.3353   -0.8202   67.26    <.0001
   x3          1    0.0095   0.0030    0.0037    0.0153   10.26     0.0014
   x4          1    0.0090   0.0042    0.0008    0.0172    4.59     0.0321
   Scale       0    1.0000   0.0000    1.0000    1.0000

NOTE: The scale parameter was held fixed.

                    LR Statistics For Type 3 Analysis

                                   Chi-
          Source         DF       Square    Pr > ChiSq
          x1              1        77.16      <.0001
          x3              1        10.25       0.0014
          x4              1         4.60       0.0320
```

Again, a very good fit, with $G^2 = 108.9409$ on 96 degrees of freedom (p-value $= 0.1729$), a better fit than the previous full model. The scale parameter is 1, and hence there is no need to worry about over-dispersion (see Lawal, 2003). The estimated Poisson regression model therefore is

$$\ln(\hat{\mu}_i) = 0.4439 - 1.0778 x_1 + 0.0095 x_3 + 0.0090 x_4. \tag{11.9}$$

We may note here that none of the first-order, second-order or third-order interaction effects is significant and we are therefore left with the estimated reduced model in (11.9) as the final parsimonious estimated Poisson regression equation for the data in Table 11.1.

Interpretation of Estimated Coefficients

For $x_1 = 1$, that is for intervention, the mean number of falls would be reduced by

$$100(1 - e^{\hat{\beta}_1 x_1}) = 100(1 - e^{-1.0778}) = 65.97\%.$$

That is, the mean number of falls will be reduced by about 66% if the individual is on education and aerobic training than, if the individual were just on education only.

Similarly, for a unit change in the balance index, the mean number of falls is expected to increase by

$$100(e^{\hat{\beta}_3 x_3} - 1) = 100(e^{0.0095} - 1) = 1\%.$$

That is, about one-percent change. However, for a 10 unit change in balance index, this change becomes

$$100(e^{\hat{\beta}_3 x_3} - 1) = 100(e^{0.0095 \times 10} - 1) = 10\%.$$

Also for the strength index x_4, a unit increase in strength index, would lead to a mean increase in the number of falls of about 1%.

Example 11.2

The data in Table 11.2 from Breslow and Day (1980, p. 151), relate to the occurrence of esophageal cancer in Frenchmen. Potential risk factors related to the occurrence are age and alcohol consumption where any consumption of wine more than one liter a day is considered high.

Age group (i)	Alcohol consumption (j)	Cancer (k) Yes	Cancer (k) No
25-34	High	1	9
	Low	0	106
35-44	High	4	26
	Low	5	164
45-54	High	25	29
	Low	21	138
55-64	High	42	27
	Low	34	139
65-74	High	19	18
	Low	36	88
75+	High	5	0
	Low	8	31

Table 11.2: Occurrence of esophageal cancer

The Poisson regression model for the above data is of the form

$$\ln(n_{ij}/N_{ij}) = \beta_0 + \beta_1 \text{age}_i + \beta_2 \text{com}_j + \beta_3 \text{agecom}_{ij}. \tag{11.10}$$

That is,

$$\ln(\text{yes}) - \ln(\text{total}) = \beta_0 + \beta_1 \text{age}_i + \beta_2 \text{com}_j + \beta_3 \text{agecom}_{ij}, \tag{11.11}$$

where

- n_{ij} is the number of respondents in age group i, with j alcohol consumption level and whose cancer status is 'yes'.

- N_{ij} is the total number of respondents in a given age group i and alcohol consumption level j.

- agecom_{ij} is the interaction effect between the the i-th level of age and the j-th level of alcohol consumption.

- $\ln(N_{ij})$ or $\ln(\text{total})$, the natural logarithm of the marginal total, is often referred to in Poisson regression as an *offset*.

In many studies, there is unequal sampling effort for some subjects and/or units. If a unit has more sampling effort, it will likely have a larger count associated with it. The offset adjusts the analysis so that this unequal effort does not affect the analysis. Suppose we are interested in the number of cases of a certain type of cancer from cities in a county. In this example, the population size of each city may be used as an offset.

We can also consider a logistic model, actually a *logit* model of the following form to the data in Table 11.2.

$$\ln\left(\frac{\pi_{ij}}{1 - \pi_{ij}}\right) = \beta_0 + \beta_1 \text{age}_i + \beta_2 \text{com}_j + \beta_3 \text{agecom}_{ij}, \tag{11.12}$$

where, π_{ij} is the probability that the response is 'yes' when age group and the alcohol consumption are at levels i and j respectively.

If we consider age group as a categorical variable with five categories, then both models in (11.10) and (11.12) are saturated models, where β_1 and β_2 pertain to main effects of age, alcohol consumption respectively, and β_3 relate to the interaction effects of age and alcohol consumption. The models with age as categorical variable are implemented in SAS as follows: First however, we implement the logit model with the following SAS program and partial output of the Type 3 analysis of effects. The results suggest that the interaction term is not significant, the p-value being computed as 0.6008. Hence for this model, we would be willing to fit a reduced model without the interaction term.

```
data poi3;
input age alchol $ resp $ count@@;
age2=age*age;
datalines;
29.5 high  yes 1 29.5 high no 9
29.5  low yes 0 29.5 low no 106
39.5 high  yes 4 39.5 high no 26
39.5  low yes 5 39.5 low no 164
49.5 high  yes 25 49.5 high no 29
49.5  low yes 21 49.5 low no 138
59.5 high  yes 42 59.5 high no 27
59.5  low yes 34 59.5 low no 139
69.5 high  yes 19 69.5 high no 18
69.5  low yes 36 69.5 low no 88
79.5 high  yes 5 79.5 high no 0
79.5  low yes 8 79.5 low no 31
;
proc print;
run;
proc logistic descending;
class alchol age;
freq count;
model resp=age|alchol/scale=none aggregate=(age alchol);
run;
```

```
                Type 3 Analysis of Effects

                                   Wald
          Effect       DF     Chi-Square    Pr > ChiSq
          age          5        38.0922       <.0001
          alchol       1         0.0011       0.9738
          alchol*age   5         3.6500       0.6008
```

To fit the Poisson regression model, first we performed some data steps in SAS to put the data in an acceptable format for analysis. We computed the offset variable to be $\ln(n)$ and use **proc genmod** in SAS to implement again the full model. The Type 3 analysis indicates that the interaction term with a p-value of 0.0766 is not significant but is a lot better than in the logit model. In both models, we now fit reduced logit and Poisson models to the data.

```
data new1;
set poi3;
if resp='no' then delete;
yes=count;
drop resp count;
output;
run;
data new2;
set poi3;
if resp='yes' then delete;
no=count;
drop resp count;
output;
run;
data comb;
merge new1 new2;
by age;
n=yes+no;
off=log(n);
run;
proc print noobs;
run;
proc genmod;
class alchol age;
model yes=age|alchol/dist=poi link=log offset=off type3;
run;
```

age	alchol	age2	yes	no	n	off
29.5	high	870.25	1	9	10	2.30259
29.5	low	870.25	0	106	106	4.66344
39.5	high	1560.25	4	26	30	3.40120
39.5	low	1560.25	5	164	169	5.12990
49.5	high	2450.25	25	29	54	3.98898
49.5	low	2450.25	21	138	159	5.06890
59.5	high	3540.25	42	27	69	4.23411
59.5	low	3540.25	34	139	173	5.15329
69.5	high	4830.25	19	18	37	3.61092
69.5	low	4830.25	36	88	124	4.82028
79.5	high	6320.25	5	0	5	1.60944
79.5	low	6320.25	8	31	39	3.66356

                    LR Statistics For Type 3 Analysis

```
                                  Chi-
           Source         DF     Square    Pr > ChiSq
           age             5     20.55      0.0010
           alchol          1     17.52     <.0001
           alchol*age      5      9.95      0.0766
```

```
set comb;
proc genmod;
class alchol age;
model yes=age alchol/dist=poi link=log offset=off type3;
run;
```

```
               Criteria For Assessing Goodness Of Fit

        Criterion              DF        Value      Value/DF
        Deviance                5       9.9532       1.9906
        Scaled Deviance         5       9.9532       1.9906
        Pearson Chi-Square      5       7.8175       1.5635
        Scaled Pearson X2       5       7.8175       1.5635
        Log Likelihood               2760.1896
```

```
    Algorithm converged.
```

```
                  Analysis Of Parameter Estimates

                         Standard    Wald 95%        Chi-
    Parameter      DF  Estimate   Error  Confidence Limits  Square  Pr > ChiSq
    Intercept       1   -0.3086  0.1798  -0.6609   0.0438    2.95    0.0861
    age     29.5    1    0.3313  0.2024  -0.0653   0.7280    2.68    0.1016
    age     39.5    1    0.3179  0.1937  -0.0618   0.6976    2.69    0.1008
    age     49.5    1    0.1607  0.1959  -0.2232   0.5446    0.67    0.4120
    age     59.5    1    0.0396  0.1961  -0.3448   0.4240    0.04    0.8401
    age     69.5    1   -0.0233  0.2044  -0.4239   0.3773    0.01    0.9092
    age     79.5    0    0.0000  0.0000   0.0000   0.0000      .       .
    alchol  high    1   -0.4447  0.1045  -0.6496  -0.2398   18.10   <.0001
    alchol  low     0    0.0000  0.0000   0.0000   0.0000      .       .
    Scale           0    1.0000  0.0000   1.0000   1.0000
```

```
NOTE: The scale parameter was held fixed.
```

```
              LR Statistics For Type 3 Analysis

                              Chi-
           Source         DF Square   Pr > ChiSq
           age             5  14.65     0.0120
           alchol          1  19.98    <.0001
```

The corresponding deviance for the logit model is 11.0412 on 5 degrees of freedom (p-value $= 0.0506$), barely fitting the model.

Next, we consider age as having a linear trend effect (rather than categorical). In this case, the basic Poisson and logit models are respectively,

$$\ln (n_{ij}/N_{ij}) = \beta_0 + \beta_1 \text{age}_i + \beta_2 \text{com}_j \qquad (11.13\text{a})$$

$$\text{logit} (\pi_{ij}) = \beta_0 + \beta_1 \text{ age}_i + \beta_2 \text{ com}_j. \qquad (11.13\text{b})$$

These models when implemented in SAS give deviance values of 30.0826 and 31.9315 respectively, both with 9 degrees of freedom. Obviously they give very poor fits. We notice that both give results that are very close. Since the models do not fit, we next consider a quadratic response in age, that is for the Poisson model, a model of the form

$$\ln (n_{ij}/N_{ij}) = \beta_0 + \beta_1 \text{age}_i + \beta_2 \text{ age}_i^2 + \beta_3 \text{com}_j. \qquad (11.14)$$

The model in (11.14) is implemented in SAS with the following program and partial output. The model fits the data with a deviance of 11.2187 on 8 degrees of freedom (p-value $= 0.1896$).

```
set comb;
proc genmod data=comb;
make 'obstats' out=aa;
class alchol;
model yes=age age2 alchol/dist=poi link=log offset=off type3 obstats;
run;
proc print data=aa noobs;
var age alchol yes pred reschi resdev streschi;
format pred reschi resdev streschi 7.3;
run;
```

```
                    Criteria For Assessing Goodness Of Fit

        Criterion              DF         Value        Value/DF
        Deviance                8        11.2187        1.4023
        Scaled Deviance         8        11.2187        1.4023
        Pearson Chi-Square      8        10.6739        1.3342
        Scaled Pearson X2       8        10.6739        1.3342
        Log Likelihood                  438.9020

    Algorithm converged.

                    Analysis Of Parameter Estimates

                            Standard     Wald 95%        Chi-
    Parameter       DF  Estimate  Error  Confidence Limits  Square  Pr > ChiSq
    Intercept        1  -11.0387  1.7973  -14.5613  -7.5161  37.72    <.0001
    age              1    0.2823  0.0622    0.1604   0.4043  20.59    <.0001
    age2             1   -0.0021  0.0005   -0.0031  -0.0010  15.19    <.0001
    alchol    high   1    1.0652  0.1430    0.7849   1.3456  55.46    <.0001
    alchol    low    0    0.0000  0.0000    0.0000   0.0000    .        .
    Scale            0    1.0000  0.0000    1.0000   1.0000

NOTE: The scale parameter was held fixed.

        age    alchol      yes     Pred    Reschi   Resdev   Streschi
        29.5   high          1    0.321    1.199    0.957    1.243
        29.5   low           0    1.172   -1.083   -1.531   -1.239
        39.5   high          4    3.905    0.048    0.048    0.054
        39.5   low           5    7.581   -0.938   -1.000   -1.172
        49.5   high         25   18.884    1.407    1.340    1.686
        49.5   low          21   19.163    0.420    0.413    0.504
        59.5   high         42   42.923   -0.141   -0.141   -0.204
        59.5   low          34   37.090   -0.507   -0.515   -0.722
        69.5   high         19   27.110   -1.558   -1.647   -2.042
        69.5   low          36   31.313    0.838    0.818    1.114
        79.5   high          5    2.857    1.268    1.145    1.408
        79.5   low           8    7.680    0.115    0.115    0.154
```

The corresponding implementation using the logit model is presented in the SAS program and partial output below. Again, the two models are very similar but the Poisson model seems better in all the cases.

```
set poi3;
proc logistic descending;
class alchol(ref=last)/param=ref;
freq count;
model resp=age age2 alchol/scale=none aggregate=(age alchol);
run;
            Deviance and Pearson Goodness-of-Fit Statistics

        Criterion         Value      DF    Value/DF     Pr > ChiSq
        --------------------------------------------------------------
        Deviance         13.0284      8     1.6285        0.1109
        Pearson          11.1345      8     1.3918        0.1942

                Number of unique profiles: 12

            Analysis of Maximum Likelihood Estimates

                            Standard      Wald
        Parameter      DF  Estimate   Error   Chi-Square   Pr > ChiSq
        --------------------------------------------------------------
        Intercept       1  -13.0068   2.0836   38.9695      <.0001
        age             1    0.3438   0.0725   22.4883      <.0001
        age2            1   -0.00247  0.000619 15.8546      <.0001
        alchol   high   1    1.6744   0.1897   77.9075      <.0001

                    Odds Ratio Estimates

                            Point        95% Wald
        Effect             Estimate   Confidence Limits
        age                 1.410     1.223     1.626
        age2                0.998     0.996     0.999
        alchol high vs low  5.335     3.679     7.738
```

The parameter estimates and their standard errors (SE) under these models are displayed in Table 11.3. The estimated Poisson regression model is

$$\ln\left(\hat{\mu}_{ij1}/N_{ij}\right) = -11.0387 + 0.2823\text{age}_i - 0.0021\text{age}_i^2 + 1.0652\text{com}_j. \tag{11.15}$$

Parameters	Poisson model		Logit model	
	Estimates	SE	Estimates	SE
$\hat{\beta}_0$	−11.0387	1.7973	−13.0068	2.0836
$\hat{\beta}_1$	0.2823	0.0622	0.3438	0.0725
$\hat{\beta}_2$	−0.0021	0.0005	−0.0025	0.0006
$\hat{\beta}_3$	1.0652	0.1430	1.6744	0.1897

Table 11.3: Parameter estimates under Poisson and logit models

The plot of estimated log mean of cancer occurrences are presented in Figure 11.1. The odds of occurrence of esophageal cancer is $e^{1.0652} = 2.90$ times higher for high level alcohol consumption subjects than for low alcohol consumption subjects, after adjusting for the effect of age.

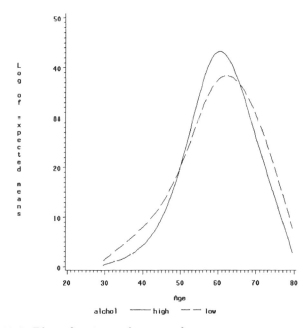

Figure 11.1: Plot of estimated mean of cancer occurrence against age

For a unit increase in age, controlling for level of consumption, the odds increases by $e^{0.2823-0.0021} = 1.32$. However, for two units increase or more, the odds will certainly change considerably.

Similarly, for the logit model, the estimated logit model is

$$\ln\left(\frac{\hat{\pi}_{ij1}}{1 - \hat{\pi}_{ij1}}\right) = -13.0068 + 0.3438\text{age}_i - 0.0025\text{age}_i^2 + 1.6744\text{com}_j, \tag{11.16}$$

with a sketch of the estimated probabilities against age presented in Figure 11.2. The closeness of the results from both the Poisson regression and logit models remind us that the Poisson is the limiting distribution for the binomial.

Example 11.3

A student conducted a project looking at the impact of popping temperature (x_1), amount of oil (x_2), and the popping time (x_3) on the number of inedible kernels of popcorn (y). The data in Table 11.4 was originally presented in Myers et al. (2010, p. 193) and reproduced by permission of John Wiley and Sons, Inc.

In this example, a Poisson regression model of the following form is suggested:

$$\ln(\mu_i) = \beta_0 + \beta_1 x_{i1} + \beta_2 x_{i2} + \beta_3 x_{i3}; \quad i = 1, 2, \dots, 15, \tag{11.17}$$

where μ_i represents the mean number of inedible kernels of popcorn for observation i. The implementation of this in SAS is presented with a partial output.

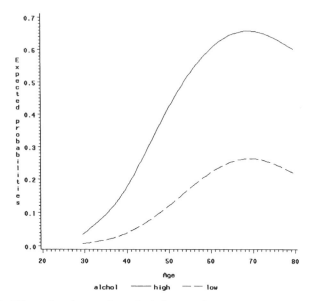

Figure 11.2: Plot of estimated probabilities of cancer occurrence against age

x_1	x_2	x_3	y	x_1	x_2	x_3	y	x_1	x_2	x_3	y
7	4	90	24	6	3	90	16	5	2	90	34
5	3	105	28	5	3	75	126	7	3	75	17
7	3	105	40	6	2	105	34	6	3	90	30
7	2	90	42	5	4	90	32	6	3	90	17
6	4	105	11	6	2	75	32	6	4	75	50

Table 11.4: Number of inedible popcorn kernels

```
data poi1;
input temp oil time count @@;
datalines;
..........
;
proc print;
run;
proc genmod;
model count=temp oil time/dist=poisson link=log type3;
run;
```
```
                    The GENMOD Procedure

                    Model Information

            Data Set            WORK.POI1
            Distribution          Poisson
            Link Function             Log
            Dependent Variable      count

        Number of Observations Read      15
        Number of Observations Used      15

        Criteria For Assessing Goodness Of Fit

    Criterion             DF        Value      Value/DF
    Deviance              11     133.3846      12.1259
    Scaled Deviance       11     133.3846      12.1259
    Pearson Chi-Square    11     147.1083      13.3735
    Scaled Pearson X2     11     147.1083      13.3735
    Log Likelihood              1410.0958

Algorithm converged.

            Analysis Of Parameter Estimates
```

Parameter	DF	Estimate	Standard Error	Wald 95% Confidence Limits		Chi-Square	Pr > ChiSq
Intercept	1	8.2439	0.5310	7.2032	9.2847	241.04	<.0001
temp	1	-0.3462	0.0605	-0.4647	-0.2278	32.80	<.0001
oil	1	-0.0885	0.0595	-0.2052	0.0282	2.21	0.1372
time	1	-0.0267	0.0041	-0.0347	-0.0188	43.52	<.0001
Scale	0	1.0000	0.0000	1.0000	1.0000		

NOTE: The scale parameter was held fixed.

LR Statistics For Type 3 Analysis

Source	DF	Chi-Square	Pr > ChiSq
temp	1	33.39	<.0001
oil	1	2.21	0.1369
time	1	44.57	<.0001

The model in (11.17) from the SAS output does not fit the data at all. It has a deviance of 133.3846 on 11 degrees of freedom. Similarly a model that also includes the three first-order interactions (temp*oil, temp*time, oil*time) gives a deviance of 36.7150 on 8 degrees of freedom- also a very poor fit. We also examine a model that employs the logarithms of temperature, oil and time. The model without interaction terms gives a deviance of 128.2924 on 11 degrees of freedom, while that containing first-order interactions has a deviance of 31.9511 on 8 degrees of freedom. Neither model fits the data. Moreover, in all the four or so models we have examined, the ratio of Deviance/df which measures degree of *over-dispersion* ideally should be close to 1. There is very strong evidence of over-dispersion in our data since for each of the models the ratios are clearly > 1. The effect of over-dispersion is the under estimation of the standard errors of the parameter estimates of the model. Since none of these models fit the data in Table 11.4, we shall re-examine other possible models later in this chapter. However, to control for over-dispersion, let us return to our first model. There, the ratio was computed to be 12.1259. We will re-fit this model, instructing it to scale the parameter to 1 in the estimation procedure. This is accomplished by specifying **dscale** or **scale=deviance** in the model option statement in the SAS program. This will scale the deviance. We can also scale the Pearson's X^2 by specifying **pscale** or **scale=Pearson**. The results of the **dscale** option is presented below.

```
set poi1;
proc genmod;
model count=temp oil time/dist=poisson link=log dscale type3;
run;
```

Criteria For Assessing Goodness Of Fit

Criterion	DF	Value	Value/DF
Deviance	11	133.3846	12.1259
Scaled Deviance	11	11.0000	1.0000
Pearson Chi-Square	11	147.1083	13.3735
Scaled Pearson X2	11	12.1318	1.1029
Log Likelihood		116.2882	

Algorithm converged.

Analysis Of Parameter Estimates

Parameter	DF	Estimate	Standard Error	Wald 95% Confidence Limits		Chi-Square	Pr > ChiSq
Intercept	1	8.2439	1.8490	4.6199	11.8680	19.88	<.0001
temp	1	-0.3462	0.2105	-0.7589	0.0664	2.70	0.1000
oil	1	-0.0885	0.2074	-0.4949	0.3179	0.18	0.6695
time	1	-0.0267	0.0141	-0.0544	0.0009	3.59	0.0582
Scale	0	3.4822	0.0000	3.4822	3.4822		

NOTE: The scale parameter was estimated by the square root of DEVIANCE/DOF.

LR Statistics For Type 3 Analysis

Source	Num DF	Den DF	F Value	Pr > F	Chi-Square	Pr > ChiSq
temp	1	11	2.75	0.1253	2.75	0.0970
oil	1	11	0.18	0.6775	0.18	0.6693
time	1	11	3.68	0.0815	3.68	0.0552

As we can see, the scaled deviance now has a value of 11.000 and a ratio of 1. While the model still does not fit the data well, from the deviance value, the parameter estimates remain the same but the standard errors of the estimated parameters have changed. It is clear from the Type 3 analysis that none of the explanatory variables is significantly important enough to be in the model at the 5% significance level.

11.3   Over-dispersion in Poisson Regression

Over-dispersion in Poisson regression occurs either when the variance of the response variable is much greater than its mean or could also be caused by positive correlation between the count responses or large variations between count responses. When over-dispersion occurs, the standard errors of the parameter estimates are underestimated and this may lead to wrong conclusions with regards to the significance of the predictor variables involved. Over-dispersion therefore often arises when modeling count data. To further illustrate, for example, suppose, the outcome Y_i has a Poisson distribution. That is, $Y_i \sim P(\mu)$. Then,

$$\mathrm{var}(Y_i) = E(Y_i) = \mu.$$

Suppose however, that the variance is proportional to the mean, say,

$$\mathrm{var}(Y_i) \ = \phi E(Y_i) = \phi\mu,$$

where ϕ is to be estimated from our data. There is no dispersion when $\hat{\phi} = 1$. We observe that over-dispersion occurs when $\hat{\phi} > 1$, while under-dispersion similarly occurs when $\hat{\phi} < 1$.

11.3.1   Test for Over-dispersion

A test for over-dispersion proposed by Böhning (1994) is

$$\mathrm{OV} = \sqrt{\frac{n-1}{2}}\left(s^2 - \bar{y}\right)/\bar{y}, \tag{11.18}$$

where OV is approximately normally distributed. For these data, $\bar{y} = 35.5333$ and $s^2 = 741.1238$. Then, the over-dispersion test statistic takes the value 52.5279 (p-value < 0.0001), which indicates strong over-dispersion. Alternatively, over-dispersion can be tested by computing an estimate of ϕ by the residual mean deviation (RMD) for any model applied to the data.

An over-dispersed model manifest itself if the value of deviance/df in the **proc genmod** output is much larger than 1.0. In our example above, we see that an estimate of $\phi = 20.86$ for the data is more than 1.0.

11.3.2   What Causes Over-dispersion?

Over-dispersion in Poisson regression could arise as a result of several reasons or combinations of reasons. These may include (Hilbe, 2007),

- The model omits important explanatory variables.

- The model fails to include a sufficient number of interaction terms.

- A predictor needs to be transformed.

- The link is misspecified.

- The data include outliers.

- There are violations in the distributional assumptions of the count data.

We examine some of these conditions with a simulated data having a Poisson response with parameters $\beta_0 = 2.0$, $\beta_1 = 0.75$, $\beta_2 = 0.30$, and $\beta_3 = -0.70$. We have simulated 10,000 observations and we present the SAS code for generating these along with the first ten simulated values of x_1, x_2, x_3 and y. Thus our data came from

$$Y \sim \mathrm{Poisson}(2.0 + 0.75x_1 + 0.30x_2 - 0.70x_3).$$

```
data new;
seed = 123456;
do i=1 to 10000;
x1=probit(uniform(seed));
x2=probit(uniform(seed));
x3=probit(uniform(seed));
xb1=2.0+0.75*x1+0.30*x2-0.70*x3;
xbeta=exp(xb1);
y=ranpoi(seed,xbeta);
output;
end;
run;
proc print data=new (obs=10);
var x1 x2 x3 y;
run;
```

```
             Obs       x1         x2         x3       y
             -----------------------------------------------
              1      0.64033   -0.60533    0.55202    5
              2     -0.33755   -1.25637   -1.78909   13
              3     -1.74843    0.81845    1.74032    1
              4     -0.41489   -0.41967   -0.31064    6
              5      1.66975   -0.04299    0.55985   23
              6      1.06489    0.59775   -1.70130   69
              7     -1.54969   -0.89676    0.15732    2
              8     -0.68925    1.31332   -0.13864    4
              9      1.58676    0.08698   -0.78473   43
             10      0.37608   -1.15087   -0.35674    6
```

Since we know the data came from a Poisson distribution, let us first fit a Poisson model of the form

$$\ln(\mu) = \beta_0 + \beta_1 x_1 + \beta_2 x_2 + \beta_3 x_3,$$

to the data. This is implemented with the following SAS program and partial output.

```
set new;
proc genmod;
model y=x1 x2 x3/dist=poi link=log type3;
run;
```

```
                   Data Set              WORK.NEW
                   Distribution           Poisson
                   Link Function              Log
                   Dependent Variable           y

              Number of Observations Read       10000
              Number of Observations Used       10000

           Criteria For Assessing Goodness Of Fit
```

Criterion	DF	Value	Value/DF
Deviance	9996	10444.4877	1.0449
Scaled Deviance	9996	10444.4877	1.0449
Pearson Chi-Square	9996	9970.9502	0.9975
Scaled Pearson X2	9996	9970.9502	0.9975
Log Likelihood		281130.9108	

```
              Analysis Of Parameter Estimates
```

Parameter	DF	Estimate	Standard Error	Wald 95% Confidence Limits		Chi-Square	Pr > ChiSq
Intercept	1	1.9972	0.0040	1.9893	2.0051	244276	<.0001
x1	1	0.7501	0.0027	0.7448	0.7555	75599.4	<.0001
x2	1	0.3028	0.0028	0.2974	0.3083	11934.1	<.0001
x3	1	-0.7031	0.0028	-0.7086	-0.6976	63577.6	<.0001
Scale	0	1.0000	0.0000	1.0000	1.0000		

The output gives the dispersion ratio as 1.0449 which is not too far from 1, hence the data as expected is not over-dispersed. The parameter estimates are congruent with the population values and the Wald's 95% confidence intervals contain the true values of each of the parameters. Now, let us imagine that the x_1 predictor variable is missing in the Poisson regression model, then we would have the following, with abbreviated output.

```
proc genmod;
model y=x2 x3/dist=poi link=log;
run;
                 Criteria For Assessing Goodness Of Fit
```

```
        Criterion              DF          Value       Value/DF

        Deviance              9997      85865.4752       8.5891
        Scaled Deviance       9997      85865.4752       8.5891
        Pearson Chi-Square    9997     111295.3035      11.1329
        Scaled Pearson X2     9997     111295.3035      11.1329
        Log Likelihood                 243420.4171

    Algorithm converged.

                    Analysis Of Parameter Estimates

                          Standard  Wald 95% Confidence    Chi-
    Parameter   DF  Estimate  Error        Limits         Square  Pr > ChiSq

    Intercept    1    2.2782  0.0035   2.2714   2.2851    426926     <.0001
    x2           1    0.3026  0.0028   0.2971   0.3080   11974.8     <.0001
    x3           1   -0.7096  0.0028  -0.7151  -0.7041   64133.0     <.0001
    Scale        0    1.0000  0.0000   1.0000   1.0000
```

The dispersion ratio is now 8.5891 which is clearly greater than 1.0, hence there is now a clear evidence of over dispersion in the model employed. The significance of the x_1 predictor can be assessed from the Type 3 results or by noting that $2(-243420.4171 + 281130.9108) = 75420.98$, twice the difference of the log-likelihoods under both models is distributed χ^2 with one degree of freedom. Clearly, the exclusion of x_1 predictor in the model is very significantly felt.

Now suppose that the response variable y is now distributed with the following form (which includes an interaction term $x_{13} = x_1 * x_3$)

$$y \sim \text{Poisson}(2.0 + 0.75x_1 + 0.30x_2 - 0.70x_3 + 0.25x_{13}).$$

Again, we simulate 10,000 samples from such a model with the following SAS program.

```
data new;
seed = 123456;
do i=1 to 10000;
x1=probit(uniform(seed));
x2=probit(uniform(seed));
x3=probit(uniform(seed));
x13=x1*x3;
xb1=2.0+0.75*x1+0.30*x2-0.70*x3+0.25*x13;
xbeta=exp(xb1);
y=ranpoi(seed,xbeta);
output;
end;
proc genmod;
model y=x1 x2 x3 x13/dist=poi link=log type3;
run;
proc genmod;
model y=x1 x2 x3/dist=poi link=log type3;
run;
```

```
                Criteria For Assessing Goodness Of Fit

        Criterion              DF          Value       Value/DF
        -------------------------------------------------------
        Deviance              9995      10276.6401       1.0282
        Scaled Deviance       9995      10276.6401       1.0282
        Pearson Chi-Square    9995       9882.1404       0.9887
        Scaled Pearson X2     9995       9882.1404       0.9887
        Log Likelihood                 235110.8056

                    Analysis Of Parameter Estimates

                          Standard  Wald 95% Confidence    Chi-
    Parameter   DF  Estimate  Error        Limits         Square  Pr > ChiSq
    ------------------------------------------------------------------------
    Intercept    1    1.9944  0.0043   1.9859   2.0029    213114     <.0001
    x1           1    0.7541  0.0032   0.7479   0.7603   56568.3     <.0001
    x2           1    0.3005  0.0028   0.2950   0.3061   11281.0     <.0001
    x3           1   -0.7050  0.0034  -0.7116  -0.6984   44056.6     <.0001
    x13          1    0.2534  0.0026   0.2482   0.2585   9218.13     <.0001
    Scale        0    1.0000  0.0000   1.0000   1.0000

    ------------------------------------------------------------------------
                Criteria For Assessing Goodness Of Fit
```

```
Criterion              DF          Value      Value/DF
---------------------------------------------------------
Deviance              9996     18753.0958      1.8761
Scaled Deviance       9996     18753.0958      1.8761
Pearson Chi-Square    9996     18915.6175      1.8923
Scaled Pearson X2     9996     18915.6175      1.8923
Log Likelihood                230872.5777
```

```
                  Analysis Of Parameter Estimates

                       Standard   Wald 95% Confidence    Chi-
Parameter   DF   Estimate   Error       Limits          Square   Pr > ChiSq
-------------------------------------------------------------------------------
Intercept    1    2.1156   0.0038    2.1081    2.1231   306707     <.0001
x1           1    0.6164   0.0028    0.6109    0.6220   47808.2    <.0001
x2           1    0.3002   0.0028    0.2946    0.3057   11154.7    <.0001
x3           1   -0.5477   0.0029   -0.5534   -0.5420   35208.4    <.0001
Scale        0    1.0000   0.0000    1.0000    1.0000
```

We see that when the interaction term x_{13} is omitted in the second model, the ratio of Value/df increases from 1.0282 to 1.8761, a significant value, indicating slight over dispersion. Further, parameter estimates of x_1 and x_3 are considerably significantly different indicating the strong influence of omission of the x_{13} interaction term on the estimates of these parameters.

Thus, either an omission of an interaction or predictor variable may influence the onset of over-dispersion in Poisson models. However, over-dispersion can be combated either by using the scaling of the test statistics outlined in various texts (see Lawal, 2003) or by employing the negative binomial regression model or generalized Poisson regression model. We explore this approach by using the data in Table 11.4.

11.4   Negative Binomial Regression Model

While the Poisson distribution is often the first to be used for fitting count data, however, if the mean is very much less than the variance of the data, then the data are *over-dispersed* with respect to the Poisson model. Another way we can correct over-dispersion is to fit instead a negative binomial model as an alternative to the over-dispersed Poisson model. The negative binomial has the probability mass function defined as

$$f(y; r, p) = \binom{y + r - 1}{r - 1} p^r (1 - p)^y, \quad y = 0, 1, 2, 3, \ldots \tag{11.19}$$

If we write

$$p = \frac{1}{(1 + \alpha\mu)}, \quad \text{where} \quad \alpha = \frac{1}{r},$$

then the expression in (11.19) can be written as

$$f(y; \mu, \alpha) = \binom{y + 1/\alpha - 1}{1/\alpha - 1} \left(\frac{1}{1 + \alpha\mu}\right)^{1/\alpha} \left(\frac{\alpha\mu}{1 + \alpha\mu}\right)^y. \tag{11.20}$$

The mean and variance of the distribution can be shown to be respectively,

$$E(Y) = \mu, \tag{11.21a}$$

$$V(Y) = \sigma^2 = \mu + \alpha\mu^2. \tag{11.21b}$$

We see that the variance of the negative binomial distribution (NBD) will be equal to that of its mean only if $\alpha = 0$. The parameter α is often called the dispersion parameter.

The expressions in (11.19) and (11.20) can be written more succinctly as (11.22a) and (11.22b) respectively,

$$f(y; r, p) = \frac{\Gamma(y + r)}{y! \, \Gamma(r)} p^r (1 - p)^y \tag{11.22a}$$

$$f(y; \mu, \alpha) = \frac{\Gamma(y + 1/\alpha)}{y! \, \Gamma(1/\alpha)} \left(\frac{1}{1 + \alpha\mu}\right)^{1/\alpha} \left(\frac{\alpha\mu}{1 + \alpha\mu}\right)^y. \tag{11.22b}$$

As we indicated in the previous section, the Poisson distribution applied to the data in Table 11.4 does not provide an adequate fit. The measure of dispersion, the ratio Value/df = 12.1259 $\gg 1$ indicates that there is very strong over-dispersion for the data in Table 11.4. We therefore examine modeling the data with a negative binomial model.

Fitting the Negative Binomial Regression Model

We fit the negative binomial regression (NBR) model to the data in Table 11.4. For comparison, we present below both the SAS codes and output for both the Poisson regression and the negative binomial regression models.

```
data one;
infile 'C:\Research\Book\Data\table164.txt' truncover;
input temp oil time y;
run;
proc genmod;
model y = temp oil time / dist=poisson link=log;
proc genmod;
model y = temp oil time / dist=nb link=log;
run;
```

                          The GENMOD Procedure

                          Model Information

                    Data Set   WORK.ONE
                    Distribution   Poisson
                    Link Function   Log
                    Dependent Variable   y

              Number of Observations Read   15
              Number of Observations Used   15

              Criteria For Assessing Goodness Of Fit

        Criterion   DF   Value   Value/DF
        Deviance   11   133.3846   12.1259
        Scaled Deviance   11   133.3846   12.1259
        Pearson Chi-Square   11   147.1083   13.3735
        Scaled Pearson X2   11   147.1083   13.3735
        Log Likelihood   1410.0958

     Algorithm converged.

                    Analysis Of Parameter Estimates

                              Standard   Wald 95% Confidence   Chi-
    Parameter   DF   Estimate   Error   Limits   Square   Pr > ChiSq
    Intercept   1   8.2439   0.5310   7.2032   9.2847   241.04   <.0001
    temp   1   -0.3462   0.0605   -0.4647   -0.2278   32.80   <.0001
    oil   1   -0.0885   0.0595   -0.2052   0.0282   2.21   0.1372
    time   1   -0.0267   0.0041   -0.0347   -0.0188   43.52   <.0001
    Scale   0   1.0000   0.0000   1.0000   1.0000

    NOTE: The scale parameter was held fixed.

                          The GENMOD Procedure

                          Model Information

                    Data Set   WORK.ONE
                    Distribution   Negative Binomial
                    Link Function   Log
                    Dependent Variable   y

              Number of Observations Read   15
              Number of Observations Used   15

              Criteria For Assessing Goodness Of Fit

        Criterion   DF   Value   Value/DF
        Deviance   11   15.2258   1.3842
        Scaled Deviance   11   15.2258   1.3842
        Pearson Chi-Square   11   17.0352   1.5487
        Scaled Pearson X2   11   17.0352   1.5487
        Log Likelihood   1454.0357

     Algorithm converged.

                    Analysis Of Parameter Estimates

                              Standard   Wald 95% Confidence   Chi-
    Parameter   DF   Estimate   Error   Limits   Square   Pr > ChiSq
    Intercept   1   7.1625   1.3991   4.4204   9.9046   26.21   <.0001
    temp   1   -0.2194   0.1663   -0.5454   0.1065   1.74   0.1870
    oil   1   -0.1561   0.1792   -0.5073   0.1950   0.76   0.3835
    time   1   -0.0207   0.0114   -0.0430   0.0016   3.32   0.0684

```
Dispersion      1      0.2076      0.0840      0.0429      0.3723
```
```
NOTE: The negative binomial dispersion parameter was estimated by maximum likelihood.
```

The NBR model fits the data very well with $D = 15.2258$ on 11 degrees of freedom and the ratio (Value/df) of 1.3842 is very much closer to one in this model compared to the ratio 12.1259 for the Poisson regression model. The parameter estimates for the two models are somewhat close, however, the standard errors of the Poisson regression model are very small compared to the NBR model. This is due to the effect of over-dispersion in the data. Observe that the **proc genmod** does not provide an observed test statistic for the dispersion parameter α.

Using PROC NLMIXED to implement Poisson and negative binomial models

In order for us to be able to use SAS **proc nlmixed**, we need the log-likelihood function of the NBR model. This **proc** can be used for the Poisson regression model without knowing the log-likelihood, since the procedure has Poisson distribution as one of the models that can be specified. The NBR model is not one of the models that can be specified, hence we use the general log-likelihood approach in **proc nlmixed**.

For the Poisson regression model, the component of the log-likelihood function for y_i is given by

$$\ell_i = -\mu_i + y_i \log(\mu_i) - \log(y_i!), \tag{11.23}$$

and of course, the log-likelihood function for the Poisson is the sum of these terms over a random sample of size n, viz:

$$\ell = -\sum_{i=1}^{n} \mu_i + \sum_{i=1}^{n} y_i \log(\mu_i) - \sum_{i=1}^{n} \log(y_i!). \tag{11.24}$$

Restricting ourselves to the single component of the log-likelihood functions therefore, we similarly have the log-likelihood function component for y_i, having the negative binomial distribution as

$$\ell_i = \log \Gamma(y_i + 1/k) - \log \Gamma(y_i + 1) - \log \Gamma(1/k) + y_i \log(k\mu_i) - (y_i + 1/k) \log(1 + k\mu_i). \tag{11.25}$$

In each of the expressions in (11.23) and (11.25), it is assumed that the mean is such that

$$\mu_i = \exp(\mathbf{X}\boldsymbol{\beta}). \tag{11.26}$$

The program and partial output from implementing both the Poisson regression and negative binomial regression models for the data in Table 11.4 using **proc nlmixed** are presented below.

```
data one;
infile 'C:\Research\Book\Data\table164.txt' truncover;
input temp oil time y;
run;
proc nlmixed data=one;
    parms beta0 .1 beta1 .1 beta2 .1 beta3 .1;
    mu=beta0 + beta1*temp + beta2*oil + beta3*time;
    lambda=exp(mu);
    model y~poisson(lambda);

proc nlmixed data=one;
    parms alpha 0.1 beta0 .1 beta1 .1 beta2 .1 beta3 .1;
    mu=exp(beta0 + beta1*temp + beta2*oil + beta3*time);
    de = 1 + alpha*mu;
    cc = lgamma(y+1/alpha)-lgamma(1/alpha)-lgamma(y+1);
    ll = cc + y*log(alpha*mu/de) - log(de)/alpha;
    model y~general(ll);
run;
                            The SAS System

                        The NLMIXED Procedure

                            Specifications
            Data Set                              WORK.ONE
            Dependent Variable                    y
            Distribution for Dependent Variable   Poisson
            Optimization Technique                Dual Quasi-Newton
            Integration Method                    None
                                Dimensions
```

```
            Observations Used              15
            Observations Not Used           0
            Total Observations             15
            Parameters                      4
```

```
                          Parameters
       beta0      beta1       beta2       beta3    NegLogLike
        0.1        0.1         0.1         0.1     568032.032
```

```
                      Iteration History
      Iter   Calls   NegLogLike      Diff     MaxGrad      Slope
        1      8     1142.1711    566889.9    155079.3   -3.27E13
        ..     ..    .........    ........    .......    .......
       14     33     105.978394    1.067E-9    0.00018   -2.12E-9
```

NOTE: GCONV convergence criterion satisfied.

```
                   Fit Statistics

          -2 Log Likelihood              212.0
          AIC (smaller is better)        220.0
          AICC (smaller is better)       224.0
          BIC (smaller is better)        222.8
```

```
                         Parameter Estimates
                 Standard
Parameter Estimate  Error   DF  t Value  Pr > |t|  Alpha   Lower    Upper    Gradient
beta0      8.2439   0.5310  15   15.53   <.0001    0.05   7.1121   9.3757   -2.27E-6
beta1     -0.3462   0.06046 15   -5.73   <.0001    0.05  -0.4751  -0.2174   -0.00002
beta2     -0.08851  0.05955 15   -1.49   0.1579    0.05  -0.2154   0.03841  -5.56E-6
beta3     -0.02674  0.004053 15   -6.60  <.0001    0.05  -0.03538 -0.01810  -0.00018
```

The SAS System

The NLMIXED Procedure
```
                     Specifications
Data Set                             WORK.ONE
Dependent Variable                   y
Distribution for Dependent Variable  General
Optimization Technique               Dual Quasi-Newton
Integration Method                   None
                        Dimensions
```

```
            Observations Used              15
            Observations Not Used           0
            Total Observations             15
            Parameters                      5
```

```
                          Parameters
   alpha     beta0      beta1       beta2       beta3    NegLogLike
    0.1       0.1        0.1         0.1         0.1     866.303723
```

```
                      Iteration History
      Iter   Calls   NegLogLike      Diff     MaxGrad      Slope
        1      6     111.107882    755.1958   7039.976   -2439907
        ..     ..    .........    ........    .......    .......
       25     54     62.0384801    2.885E-9   6.022E-6   -5.77E-9
```

NOTE: GCONV convergence criterion satisfied.

```
                   Fit Statistics

          -2 Log Likelihood              124.1
          AIC (smaller is better)        134.1
```

The SAS System

The NLMIXED Procedure

```
                   Fit Statistics

          AICC (smaller is better)       140.7
          BIC (smaller is better)        137.6
```

```
                         Parameter Estimates
                 Standard
Parameter Estimate  Error   DF  t Value  Pr > |t|  Alpha   Lower    Upper    Gradient
alpha      0.2076   0.08404 15    2.47   0.0260    0.05   0.02849  0.3867   2.402E-7
beta0      7.1625   1.3991  15    5.12   0.0001    0.05   4.1804  10.1445   7.617E-8
```

```
beta1      -0.2194    0.1663    15    -1.32    0.2068    0.05   -0.5739    0.1350    4.95E-7
beta2      -0.1561    0.1792    15    -0.87    0.3972    0.05   -0.5380    0.2257    1.985E-7
beta3     -0.02071   0.01136    15    -1.82    0.0884    0.05   -0.04492   0.003509  6.022E-6
```

The results from both fits agree with our earlier results when **proc genmod** was used. Note that the log-likelihood computed by the **genmod** procedure does not include the constant part (i.e. $\log(y_i!)$). Observe that the **proc nlmixed** provides observed t values for all parameters including the dispersion parameter α. We notice that α in the NBR model is significantly different from zero at 5% level since the p-value of 0.0260 is less than 5%.

11.5   The Generalized Poisson Regression Model

A third count regression model considered in this chapter is the generalized Poisson regression (GPR) model which has the following pmf:

$$f(y_i; \mu_i, \alpha) = \left(\frac{\mu_i}{1 + \alpha\mu_i}\right)^{y_i} \frac{(1 + \alpha y_i)^{y_i - 1}}{y_i!} \exp\left(-\frac{\mu_i(1 + \alpha y_i)}{1 + \alpha\mu_i}\right), \quad y_i = 0, 1, 2, 3, \ldots \tag{11.27}$$

with mean

$$\mathrm{E}(Y_i) = \mu_i \quad \text{and} \quad \mathrm{Var}(Y_i) = \mu_i(1 + \alpha\mu_i)^2. \tag{11.28}$$

The GPR model has been considered for over-dispersed data (see for example Famoye, 1993) and the model has been considered for under-dispersed data (see for example Wang and Famoye, 1997). The GPR model has a dispersion parameter α. The GPR reduces to the Poisson regression when $\alpha = 0$ and the dispersion factor $\mathrm{Var}(Y_i)/\mathrm{E}(Y_i) = (1 + \alpha\mu_i)^2$. If $\alpha > 0$, then $\mathrm{Var}(Y_i) > \mathrm{E}(Y_i)$ and the GPR will model count data with over-dispersion. Similarly, when $\alpha < 0$, then $\mathrm{Var}(Y_i) < \mathrm{E}(Y_i)$ and the GPR will in this case model under-dispersed count data.

Implementing generalized Poisson regression Model

To implement the above model in SAS **proc nlmixed**, we need the log-likelihood function for the model. We consider the component of the log-likelihood function due to a single observation y_i for the count model as this is all we need to specify in **proc nlmixed**. The log-likelihood function for a single observation having the generalized Poisson regression model is given by

$$\ell_i = y_i \log\left(\frac{\mu_i}{1 + \alpha\mu_i}\right) + (y_i - 1)\log(1 + \alpha y_i) - \frac{\mu_i(1 + \alpha y_i)}{1 + \alpha\mu_i} - \log(y_i!). \tag{11.29}$$

In the expression in (11.29), it is assumed that the mean is such that

$$\mu_i = \exp(\mathbf{X}\boldsymbol{\beta}). \tag{11.30}$$

We fit the generalized Poisson regression model to the data in Table 11.4. To do this, we will use SAS **proc nlmixed**. Recall that this **proc** can be used for the Poisson regression model without knowing the log-likelihood, since the procedure has Poisson distribution as one of the models that can be specified. The GPR model is not one of the models that can be specified, hence we use the general log-likelihood approach in **proc nlmixed**. The program and partial output from implementing the GPR model for the data in Table 11.4 using **proc nlmixed** are presented below.

```
data one;
infile 'C:\Research\Book\Data\table164.txt' truncover;
input temp oil time y;
run;

proc nlmixed data=one;
parms alpha=0.1 beta0=.1 beta1=.1 beta2=.1 beta3=.1;
mu=exp(beta0 + beta1*temp + beta2*oil + beta3*time);
de = 1 + alpha*mu; nu = 1 + alpha*y;
ll = y*log(mu/de) + (y-1)*log(nu) - mu*nu/de - lgamma(y+1);
model y~general(ll);
run;
quit;
                        The SAS System
```

```
                            The NLMIXED Procedure
                        Specifications
            Data Set                              WORK.ONE
            Dependent Variable                    y
            Distribution for Dependent Variable   General
            Optimization Technique                Dual Quasi-Newton
            Integration Method                    None

                            Dimensions

                  Observations Used                15
                  Observations Not Used             0
                  Total Observations               15
                  Parameters                        5

                            Parameters

        alpha      beta0      beta1      beta2      beta3   NegLogLike
         0.1        0.1        0.1        0.1        0.1   82.5695175

                        Iteration History

        Iter   Calls   NegLogLike       Diff    MaxGrad       Slope
           1       4   75.2630143   7.306503   17.43142    -1424.47
          ..      ..   ..........   ........   .......     .......
          23      48   61.5400808   1.07E-10    0.00005    -211E-12

        NOTE: GCONV convergence criterion satisfied.

                    Fit Statistics
            -2 Log Likelihood                123.1
            AIC (smaller is better)          133.1
            AICC (smaller is better)         139.7
            BIC (smaller is better)          136.6

                        The SAS System

                    The NLMIXED Procedure

                    Parameter Estimates

                    Standard
Parameter  Estimate     Error   DF  t Value  Pr > |t|  Alpha     Lower     Upper  Gradient
alpha       0.05471   0.01584   15     3.45    0.0035   0.05   0.02094   0.08847  0.000026
beta0        6.9032    1.3978   15     4.94    0.0002   0.05   3.9238    9.8825   3.27E-7
beta1       -0.1787    0.1686   15    -1.06    0.3061   0.05  -0.5380    0.1807   1.506E-6
beta2       -0.1823    0.1852   15    -0.98    0.3405   0.05  -0.5769    0.2124   2.737E-6
beta3      -0.01970   0.01170   15    -1.68    0.1129   0.05  -0.04464  0.005234  0.00005
```

The results from GPR model and NBR model are very similar. The log-likelihood from the NBR model is -62.04 while that of GPR model is -61.54. These results provide a better fit than the log-likelihood of -105.98 from the Poisson regression model. Also, the dispersion parameter in both the GPR model and the NBR model are significantly different from zero.

11.6   Zero-inflated Count Models

Over-dispersion in the Poisson model as observed in the previous sections can also arise as a result of too many occurrences of zeros than would normally be expected from a Poisson model. That is, there could be too many zeros than can be assumed theoretically or expected under such model. If this happens, we would then say that the Poisson model is in this case, **zero-inflated**. In some cases, these zeros can be *structural zeros* in which it is impossible to observe an occurrence. For instance, in a survey to determine the number of bottles of alcohol consumed by respondents per week, there would be individuals in the sample who do not drink alcohol at all. Such people will have the number recorded as zeros but in actual fact, such zeros will be structural as we naturally do not expect them to have a count. On the other hand, if an individual does drink alcohol but he/she did not drink a single bottle of alcohol during the survey period, then such individuals would have a count of zero and the zeros in this case would be referred to as *sampling zeros*.

To illustrate structural and sampling zeros, consider the following example which examines the number of Caretta caretta hatchlings dying from exposure to the sun, Özmen and Famoye (2007). The data originally came from Canbolat (1997). The response y_i $(i = 1, 2, \ldots, 72)$ are the number of C. caretta hatchlings dying from exposure to the sun. The y_i's ranged in size from 0 to 23. There are three qualitative factors: Area (A1-A6), distance from the sea (D1-D4) and year (1991-1993). The data include both structural zeros and sampling zeros. If no C. caretta hatchlings emerge from the nests, the number of C. caretta hatchlings dying from exposure to the sun is zero (structural zero). If C. caretta hatchlings emerge from the nests, the number of hatchlings dying from the exposure to the sun may be zero or greater than zero. The zeros from this second situation are sampling zeros. The data is presented in Table 11.5.

	1991				1992				1993			
Area	D1	D2	D3	D4	D1	D2	D3	D4	D1	D2	D3	D4
A1	0	17	1	7	0	2	5	0*	1	0	0	0
A2	2	7	2	2	0	1	8	0	23	9	7	3
A3	0	0	0	0	0	3	1	0*	3	1	4	0*
A4	0	0	0	0	0	1	0	0	0*	0	0	0*
A5	8	1	0*	0*	4	3	0*	0*	0	0	0	0*
A6	0	0	1	0*	0	1	0	22	0*	0	2	6

Table 11.5: Number of C. caretta hatchlings dying from sun exposure (*: structural zeros)

The frequency distribution reveals that 56.9% of the data are zeros. There are too many zeros in this data. We would therefore expect that there are more zeros in these data than would normally be expected from a Poisson model.

11.6.1 Zero-inflated Poisson Regression Model

Consider the Poisson distribution with parameter μ as defined in (11.1). The zero-inflated Poisson can be modeled by mixed Poisson distributions, comprising the part with zero-inflated and the other part consisting of the zeros that would normally be expected under a Poisson model. A zero modified Poisson distributions have the form in Ridout et al. (1998). Thus, a **zero-inflated Poisson model** has the pmf of the form

$$\Pr(Y_i = y_i) - \begin{cases} \omega_i + (1 - \omega_i)\exp(-\mu_i), & \text{if } y_i = 0 \\ (1 - \omega_i)\dfrac{\exp(-\mu_i)\mu_i^{y_i}}{y_i!}, & \text{if } y_i > 0, \end{cases} \tag{11.31}$$

with $0 \le \omega_i < 1$ is a proportion and

$$E(Y_i) = (1 - \omega_i)\mu_i,$$
$$\text{var}(Y_i) = (1 - \omega_i)\mu_i(1 + \omega_i\mu_i).$$

The zero-inflated Poisson model in (11.31) can also be defined in the form

$$\Pr(Y_i = y_i) = \begin{cases} \omega_i, & \text{for the inflation part } (y_i = 0) \\ (1 - \omega_i)\dfrac{\exp(-\mu_i)\mu_i^{y_i}}{y_i!}, & \text{for } y_i = 0, 1, 2, \dots \end{cases} \tag{11.32}$$

The first part of the expression in (11.32) relates to inflated part where $y_i = 0$, while the second part similarly relates to the Poisson distribution part with $y_i = 0, 1, 2, \dots$ as usual and notice that the variance of the zero-inflated Poisson model is similar to the variance of the negative binomial distribution in (11.21b). Lambert(1992) considers the model of the following form

$$\log(\mu_i) = \mathbf{X}\boldsymbol{\beta} \quad \text{and} \tag{11.33a}$$

$$\log\left(\frac{\omega_i}{1 - \omega_i}\right) = \mathbf{Z}\boldsymbol{\delta}, \tag{11.33b}$$

where $\mathbf{X}$ and $\mathbf{Z}$ are vectors of explanatory variables which may or may not be the same and $\boldsymbol{\beta}$ and $\boldsymbol{\delta}$ are also vectors of parameters. However, when the two sets of covariates are the same, then simpler models are of the form

$$\log(\mu_i) = \mathbf{X}\boldsymbol{\beta} \quad \text{and} \tag{11.34a}$$

$$\log\left(\frac{\omega_i}{1 - \omega_i}\right) = -\tau\mathbf{X}\boldsymbol{\beta}, \tag{11.34b}$$

where τ is a scalar and in this case $\omega_i = (1 + \mu_i^\tau)^{-1}$. When $\tau > 0$, the zero state becomes less likely and when $\tau < 0$, excess zeros become more likely. If ω_i is known, we can fit the general Poisson regression model of the form

$$\log(\mu_i) = \mathbf{X}\boldsymbol{\beta} - \log(1 - \omega_i), \tag{11.35}$$

where $-\log(1 - \omega_i)$ in this case can now be treated as an 'offset' in the usual Poisson regression model.

When ω_i and μ_i are related, the log-likelihood for zero-inflated Poisson (ZIP(τ)) regression model is given by

$$\log(L_\tau) = -\sum_{i=1}^n \log(1 + \mu_i^{-\tau}) + \sum_{y_i=0} \log\left(\mu_i^{-\tau} + \exp(-\mu_i)\right) + \sum_{y_i>0} [y_i \log(\mu_i) - \log(y_i!) - \mu_i]. \tag{11.36}$$

When ω_i and μ_i are not related, the log-likelihood for zero-inflated Poisson (ZIP) regression model is given by

$$\log(L) = -\sum_{i=1}^n \log(1 + \xi_i) + \sum_{y_i=0} \log\left(\xi_i + \exp(-\mu_i)\right) + \sum_{y_i>0} [y_i \log(\mu_i) - \log(y_i!) - \mu_i], \tag{11.37}$$

where $\xi_i = \exp(\mathbf{z}_i\boldsymbol{\delta})$ and $\mu_i = \exp(\mathbf{x}_i\boldsymbol{\beta})$.

In many applications, there is little prior information about how ξ_i is related to μ_i, Lambert(1992). Depending on the data generating process, one can think of a situation in which both ξ_i and μ_i depend on some covariates and a situation in which this is not the case. If we believe that both ξ_i and μ_i depend on different covariates, we use the ZIP model. However, when both ξ_i and μ_i depend on the same covariates, we can use the ZIP(τ) regression model. Alternatively, one can use the ZIP regression model with $\mu_i = \mu_i(x_i)$ and $\xi_i = \xi_i(z_{i1})$, where z_{i1} is a column vector of 1's.

11.6.2   Zero-inflated Negative Binomial Regression Model

The probability mass function for a zero-inflated negative binomial model is given by

$$\Pr(Y_i = y_i) = \begin{cases} \omega_i + (1 - \omega_i)(1 + \alpha\mu_i)^{-\alpha^{-1}}, & y_i = 0 \\ (1 - \omega_i)\dfrac{\Gamma(y_i + \alpha^{-1})}{y_i!\Gamma(\alpha^{-1})} \dfrac{\alpha^{y_i}\mu_i^{y_i}}{(1 + \alpha\mu_i)^{y_i + \alpha^{-1}}}, & y_i > 0, \end{cases} \tag{11.38}$$

with

$$\begin{aligned} \mathrm{E}(Y_i) &= (1 - \omega_i)\mu_i, \quad \text{and} \\ \mathrm{Var}(Y_i) &= (1 - \omega_i)\mu_i\{1 + \omega_i\mu_i + \alpha\mu_i\}, \end{aligned} \tag{11.39}$$

where the parameters μ_i and ω_i depend on the covariates and $\alpha \geq 0$ is a scalar. Thus we have over-dispersion whenever either ω_i or α is greater than 0. Thus the equation in (11.38) reduces to

$$\begin{cases} \text{NB}, & \text{when } \omega_i = 0 \\ \text{ZIP}, & \text{when } \alpha = 0. \end{cases}$$

If ω_i and μ_i are related, the log-likelihood for zero-inflated negative binomial ZINB(τ)) regression model is given by

$$\begin{aligned} \log(L_\tau) = {}&{-}\sum_{i=1}^n \log(1 + \mu_i^{-\tau}) + \sum_{y_i=0} \log\left(\mu_i^{-\tau} + (1 + \alpha\mu_i)^{-1/\alpha}\right) \\ &+ \sum_{y_i>0} \left[y_i \log\left(\frac{\alpha\mu_i}{1 + \alpha\mu_i}\right) - \frac{\log(1 + \alpha\mu_i)}{\alpha} + \log\left(\frac{\Gamma(y_i + \alpha^{-1})}{y_i!\Gamma(\alpha^{-1})}\right)\right]. \end{aligned} \tag{11.40}$$

When ω_i and μ_i are not related, the log-likelihood for zero-inflated negative binomial (ZINB) regression model is given by

$$\begin{aligned} \log(L) = {}&{-}\sum_{i=1}^n \log(1 + \xi_i) + \sum_{y_i=0} \log\left(\xi_i + (1 + \alpha\mu_i)^{-1/\alpha}\right) \\ &+ \sum_{y_i>0} \left[y_i \log\left(\frac{\alpha\mu_i}{1 + \alpha\mu_i}\right) - \frac{\log(1 + \alpha\mu_i)}{\alpha} + \log\left(\frac{\Gamma(y_i + \alpha^{-1})}{y_i!\Gamma(\alpha^{-1})}\right)\right]. \end{aligned} \tag{11.41}$$

When fitting the model to a response variable or a dependent variable with explanatory variables, the ξ_i and μ_i in the log-likelihoods in (11.40) and (11.41), can take the following form:

$$\xi_i = \exp\left(\mathbf{z}_i\boldsymbol{\delta}\right)$$
$$\mu_i = \exp\left(\mathbf{x}_i\boldsymbol{\beta}\right).$$

11.6.3 Zero-inflated Generalized Poisson Regression Model

The probability mass function for a zero-inflated generalized Poisson model is given by

$$\Pr(Y_i = y_i) = \begin{cases} \omega_i + (1-\omega_i)\exp(-\mu_i/(1+\alpha\mu_i), & y_i = 0 \\ (1-\omega_i)\frac{(1+\alpha y_i)^{y_i-1}}{y_i!}\left(\frac{\mu_i}{1+\alpha\mu_i}\right)^{y_i}\exp\left(\frac{-\mu_i(1+\alpha y_i)}{1+\alpha\mu_i}\right), & y_i > 0, \end{cases} \tag{11.42}$$

with

$$\mathrm{E}(Y_i) = (1-\omega_i)\mu_i, \quad \text{and}$$
$$\mathrm{Var}(Y_i) = (1-\omega_i)\mu_i\left\{\omega_i\mu_i + (1+\alpha\mu_i)^2\right\}, \tag{11.43}$$

where the parameters μ_i and ω_i depend on the covariates and α is a scalar. Thus we have over-dispersion whenever either ω_i or α is greater than 0. Thus the equation in (11.42) reduces to

$$\begin{cases} \text{GP}, & \text{when } \omega_i - 0 \\ \text{ZIP}, & \text{when } \alpha = 0. \end{cases}$$

When ω_i and μ_i are related, the log-likelihood for zero-inflated generalized Poisson (ZIGP(τ)) regression model for a sample of size n is given by

$$\begin{aligned} \log(L_\tau) =& \sum_{y_i>0}\left[y_i\log\left(\frac{\mu_i}{1+\alpha\mu_i}\right) + (y_i-1)\log(1+\alpha y_i) - \log(y_i!) - \frac{\mu_i(1+\alpha y_i)}{1+\alpha\mu_i}\right] \\ &\mid \sum_{y_i=0}\log\left(\mu_i^{-\tau} + \exp\left[\frac{-\mu_i}{1+\alpha\mu_i}\right]\right) - \sum_{i=1}^n\log(1+\mu_i^{-\tau}). \end{aligned} \tag{11.44}$$

When ω_i and μ_i are not related, the log-likelihood for zero-inflated generalized Poisson (ZIGP) regression model is given by

$$\begin{aligned} \log(L) =& \sum_{y_i>0}\left[y_i\log\left(\frac{\mu_i}{1+\alpha\mu_i}\right) + (y_i-1)\log(1+\alpha y_i) - \log(y_i!) - \frac{\mu_i(1+\alpha y_i)}{1+\alpha\mu_i}\right] \\ &+ \sum_{y_i=0}\log\left(\xi_i + \exp\left[\frac{-\mu_i}{1+\alpha\mu_i}\right]\right) - \sum_{i=1}^n\log(1+\xi_i), \end{aligned} \tag{11.45}$$

where $\xi_i = \omega_i/(1-\omega_i) = \exp(\mathbf{z}_i\boldsymbol{\delta})$. When the response variable involves explanatory variables, the ξ_i and μ_i in the log-likelihood in (11.45), can take the following form:

$$\xi_i = \exp\left(\mathbf{z}_i\boldsymbol{\delta}\right)$$
$$\mu_i = \exp\left(\mathbf{x}_i\boldsymbol{\beta}\right).$$

11.6.4 Fitting Zero-inflated Regression Models

We consider fitting the zero-inflated regression models in the previous sub-sections to the dataset on domestic violence. The data was used by Famoye and Singh (2006) to illustrate the zero-inflated generalized Poisson regression model. The data is in the file *domestic.txt*. In this section, we will fit the zero-inflated Poisson, zero-inflated negative binomial and zero-inflated generalized Poisson regression models to the data by using the SAS **proc nlmixed**. The response variable, violence, is the number of violent behavior of batterer towards victim. The independent variables used in the analysis are levels of education, employment status, level of income, having family interaction, belonging to a club, and having drug problem. These independent variables are recorded for both victim and batterer. For the descriptive statistics on these variables, see Famoye and Singh (2006). The observed proportion of zeros in the data is 66.4%. The model of interest is

$$\log(\mu_i) = \beta_0 + \beta_1 x_{i1} + \beta_2 x_{i2} + \ldots + \beta_{12} x_{i,12} \quad i = 1, 2, \ldots, 214. \tag{11.46}$$

We use **proc nlmixed** to implement the ZIP(τ), ZINB(τ), and ZIGP(τ) regression models. The programs and the partial results are presented in the following:

```
/* Program for zero-inflated count data regression models */
data one;
infile 'C:\Research\Book\Data\kiya12.txt';
input @1 (educ_v educ_o emp_v emp_o inc_v inc_o fam_v fam_o club_v club_o drugp_v drugp_o)(2.0)
      @26 (minor severe total)(4.0);
y = severe;
run;
proc means data=one;
var educ_v -- severe;
run;

proc iml;
use one;
read all var{y};
read all var{educ_v educ_o emp_v emp_o inc_v inc_o fam_v fam_o club_v club_o drugp_v drugp_o}
     into xx0;
no = nrow(xx0); x0=j(no,1,1.0); xx=x0||xx0;
para1=ginv(t(xx)*xx) * (t(xx)*log(y + 0.5)); para=t(para1);
cnames={b0 b1 b2 b3 b4 b5 b6 b7 b8 b9 b10 b11 b12};
/* creates data set MYDATA with variables b0, b1, ..., b12 */
create mydata from para [colname=cnames];
append from para;
quit;
proc nlmixed data=one tech=trureg;
title "Zero-inflated Poisson regression analysis";
   parms tau 0.1 / data=mydata;
   ave = b0 + b1*educ_v + b2*educ_o + b3*emp_v + b4*emp_o + b5*inc_v + b6*inc_o
       + b7*fam_v + b8*fam_o + b9*club_v + b10*club_o + b11*drugp_v + b12*drugp_o;
   mu=exp(ave);
   d0 = 1 + mu**(-tau); d1=0; d2=0;
   if (y = 0) then d1 = log(mu**(-tau) + exp(-mu));
   if (y > 0) then d2 = y*log(mu) - lgamma(y+1) - mu;
   ll = -log(d0) + d1 + d2;
   model y~general(ll);
run;
 proc nlmixed data=one tech=trureg;
title "Zero-inflated negative binomial regression analysis";
   parms tau 0.1 alpha 0.1 / data=mydata;
   ave = b0 + b1*educ_v + b2*educ_o + b3*emp_v + b4*emp_o + b5*inc_v + b6*inc_o
       + b7*fam_v + b8*fam_o + b9*club_v + b10*club_o + b11*drugp_v + b12*drugp_o;
   mu=exp(ave);
   d0 = 1 + mu**(-tau); dd = alpha*mu; de = 1 + dd; d1=0; d2=0; co=0;
   if (y = 0) then d1 = log(mu**(-tau) + de**(-1/alpha));
   if (y > 0) then co = lgamma(y+1/alpha) - lgamma(1/alpha) - lgamma(y+1);
   if (y > 0) then d2 = y*log(dd/de) - (log(de))/alpha + co;
   ll = -log(d0) + d1 + d2;
   model y~general(ll);
run;
 proc nlmixed data=one tech=trureg;
title "Zero-inflated generalized Poisson regression analysis";
   parms tau 0.1 alpha 0.1 / data=mydata;
   ave = b0 + b1*educ_v + b2*educ_o + b3*emp_v + b4*emp_o + b5*inc_v + b6*inc_o
       + b7*fam_v + b8*fam_o + b9*club_v + b10*club_o + b11*drugp_v + b12*drugp_o;
   mu=exp(ave);
   d0 = 1 + mu**(-tau); de = 1 + alpha*mu; dd = 1 + alpha*y; d1=0; d2=0;
   if (y = 0) then d1 = log(mu**(-tau) + exp(-mu/de));
   if (y > 0) then d2 = y*log(mu/de) + (y-1)*log(dd) - lgamma(y+1) - mu*dd/de;
   ll = -log(d0) + d1 + d2;
   model y~general(ll);
run;
```

The SAS System

The MEANS Procedure

Variable	N	Mean	Std Dev	Minimum	Maximum

educ_v	214	2.2897196	0.7507035	1.0000000	3.0000000
educ_o	214	2.0654206	0.7784784	1.0000000	3.0000000
emp_v	214	0.5046729	0.5011504	0	1.0000000
emp_o	214	0.6588785	0.4751977	0	1.0000000
inc_v	214	2.5654206	1.3083052	1.0000000	5.0000000
inc_o	214	3.0700935	1.4726754	1.0000000	5.0000000
fam_v	214	0.8224299	0.3830465	0	1.0000000
fam_o	214	0.7196262	0.4502350	0	1.0000000
club_v	214	0.2710280	0.4455327	0	1.0000000
club_o	214	0.1915888	0.3944739	0	1.0000000
drugp_v	214	0.1355140	0.3430743	0	1.0000000
drugp_o	214	0.6214953	0.4861515	0	1.0000000
minor	214	5.4906542	7.6051259	0	40.0000000
severe	214	4.2056075	10.6013892	0	65.0000000

Zero-inflated Poisson regression analysis

The NLMIXED Procedure

Specifications

Data Set	WORK.ONE
Dependent Variable	y
Distribution for Dependent Variable	General
Optimization Technique	Trust Region
Integration Method	None

Dimensions

Observations Used	214
Observations Not Used	0
Total Observations	214
Parameters	14

Parameters

tau	b0	b1	b2	b3	b4	b5	b6	b7
0.1	1.187818	-0.07401	-0.16238	0.124456	0.18138	-0.10563	-0.07931	-0.12238

Parameters

b8	b9	b10	b11	b12	NegLogLike
-0.50241	0.436444	-0.34576	0.005086	0.409818	1814.11578

Iteration History

Iter	Calls	NegLogLike	Diff	MaxGrad	Radius
1	33	1478.16041	335.9554	1691.682	-338.02
.	...	....	....	....	....
8	148	641.093631	1.545E-7	5.423E-7	-1.54E-7

NOTE: GCONV convergence criterion satisfied.

Fit Statistics

-2 Log Likelihood	1282.2
AIC (smaller is better)	1310.2
AICC (smaller is better)	1312.3
BIC (smaller is better)	1357.3

Parameter Estimates

Parameter	Estimate	Standard Error	DF	t Value	Pr > |t|	Alpha	Lower	Upper	Gradient
tau	-0.2456	0.06185	214	-3.97	<.0001	0.05	-0.3675	-0.1237	-1.54E-8
b0	3.4206	0.1729	214	19.78	<.0001	0.05	3.0797	3.7615	1.74E-7
b1	-0.3569	0.05496	214	-6.49	<.0001	0.05	-0.4652	-0.2486	3.864E-7
b2	0.03699	0.05270	214	0.70	0.4835	0.05	-0.06689	0.1409	3.815E-7
b3	0.1252	0.08966	214	1.40	0.1640	0.05	-0.05152	0.3019	9.363E-8
b4	0.02111	0.1051	214	0.20	0.8409	0.05	-0.1860	0.2282	1.217E-7
b5	-0.08784	0.03618	214	-2.43	0.0160	0.05	-0.1591	-0.01653	4.06E-7
b6	-0.2012	0.03835	214	-5.25	<.0001	0.05	-0.2768	-0.1256	5.423E-7
b7	0.1245	0.09991	214	1.25	0.2139	0.05	-0.07239	0.3215	1.383E-7
b8	-0.1645	0.06962	214	-2.36	0.0190	0.05	-0.3018	-0.02728	1.147E-7
b9	0.7804	0.1050	214	7.43	<.0001	0.05	0.5734	0.9874	3.587E-8
b10	-0.8548	0.1222	214	-7.00	<.0001	0.05	-1.0956	-0.6139	2.199E-8
b11	-0.7577	0.1275	214	-5.94	<.0001	0.05	-1.0090	-0.5064	1.375E-7
b12	0.6305	0.09289	214	6.79	<.0001	0.05	0.4474	0.8136	1.424E-7

Zero-inflated negative binomial regression analysis

The NLMIXED Procedure

Specifications

Data Set	WORK.ONE
Dependent Variable	y
Distribution for Dependent Variable	General
Optimization Technique	Trust Region
Integration Method	None

Dimensions

Observations Used	214
Observations Not Used	0
Total Observations	214
Parameters	15

Parameters

tau	alpha	b0	b1	b2	b3	b4	b5	b6
0.1	0.1	1.187818	-0.07401	-0.16238	0.124456	0.18138	-0.10563	-0.07931

Parameters

b7	b8	b9	b10	b11	b12	NegLogLike
-0.12238	-0.50241	0.436444	-0.34576	0.005086	0.409818	1259.52797

Iteration History

Iter	Calls	NegLogLike	Diff	MaxGrad	Radius
1	34	657.696761	601.8312	1866.574	-1028.82
..	...	....	....	....	......
13	239	367.666465	1.604E-8	4.068E-8	-1.6E-8

NOTE: GCONV convergence criterion satisfied.

Fit Statistics

-2 Log Likelihood	735.3
AIC (smaller is better)	765.3
AICC (smaller is better)	767.8
BIC (smaller is better)	815.8

Zero-inflated negative binomial regression analysis

The NLMIXED Procedure

Parameter Estimates

Parameter	Estimate	Standard Error	DF	t Value	Pr > |t|	Alpha	Lower	Upper	Gradient
tau	-0.08536	0.09235	214	-0.92	0.3564	0.05	-0.2674	0.09667	-4.07E-8
alpha	1.6298	0.4694	214	3.47	0.0006	0.05	0.7046	2.5550	-3.85E-8
b0	4.0047	0.7770	214	5.15	<.0001	0.05	2.4732	5.5362	-6.63E-9
b1	-0.7394	0.2724	214	-2.71	0.0072	0.05	-1.2763	-0.2025	-2.59E-8
b2	0.2060	0.2675	214	0.77	0.4421	0.05	-0.3212	0.7332	-1.21E-8
b3	0.2549	0.4213	214	0.61	0.5458	0.05	-0.5755	1.0854	-4.58E-9
b4	0.6373	0.5814	214	1.10	0.2743	0.05	-0.5087	1.7833	-6.28E-9
b5	-0.2939	0.1695	214	-1.73	0.0843	0.05	-0.6279	0.04011	-2.94E-8
b6	-0.3198	0.2016	214	-1.59	0.1141	0.05	-0.7171	0.07756	-3.46E-8
b7	0.001488	0.4094	214	0.00	0.9971	0.05	-0.8055	0.8085	-4.59E-9
b8	-0.4466	0.3583	214	-1.25	0.2139	0.05	-1.1529	0.2596	-5.94E-9
b9	1.3569	0.5303	214	2.56	0.0112	0.05	0.3117	2.4022	-707E-12
b10	-1.3514	0.5533	214	-2.44	0.0154	0.05	-2.4419	-0.2608	-4.18E-9
b11	-0.9735	0.4453	214	-2.19	0.0299	0.05	-1.8512	-0.09589	-2.3E-9
b12	1.0584	0.3415	214	3.10	0.0022	0.05	0.3853	1.7314	1.03E-9

Zero-inflated generalized Poisson regression analysis

The NLMIXED Procedure

Specifications

Data Set	WORK.ONE
Dependent Variable	y
Distribution for Dependent Variable	General
Optimization Technique	Trust Region
Integration Method	None

Dimensions

Observations Used	214
Observations Not Used	0
Total Observations	214
Parameters	15

```
                             Parameters
     tau     alpha        b0        b1        b2        b3        b4        b5        b6
     0.1       0.1  1.187818  -0.07401  -0.16238  0.124456   0.18138  -0.10563  -0.07931

                             Parameters
        b7        b8        b9       b10       b11       b12   NegLogLike
  -0.12238  -0.50241  0.436444  -0.34576  0.005086  0.409818   968.009932

                        Iteration History
     Iter   Calls   NegLogLike       Diff    MaxGrad     Radius
        1      34   415.270844   552.7391   548.2601   -609.926
        .     ...   ..........   ........   ........   .......
        9     170   365.844782   1.136E-9   8.192E-9   -1.14E-9

        NOTE: GCONV convergence criterion satisfied.

                      Fit Statistics

        -2 Log Likelihood                731.7
        AIC (smaller is better)          761.7
        AICC (smaller is better)         764.1
        BIC (smaller is better)          812.2
```

Zero-inflated generalized Poisson regression analysis

The NLMIXED Procedure

Parameter Estimates

Parameter	Estimate	Standard Error	DF	t Value	Pr > |t|	Alpha	Lower	Upper	Gradient
tau	-0.1242	0.05701	214	-2.18	0.0304	0.05	-0.2366	0.01105	6.68E 10
alpha	0.3050	0.05553	214	5.49	<.0001	0.05	0.1956	0.4145	-5.73E-9
b0	5.4332	1.2620	214	4.31	<.0001	0.05	2.9457	7.9207	-3.06E-9
b1	-1.5005	0.4967	214	-3.02	0.0028	0.05	-2.4796	-0.5214	-5.39E-9
b2	0.5907	0.3035	214	1.95	0.0529	0.05	-0.00748	1.1889	-4.4E-9
b3	0.3419	0.5027	214	0.68	0.4971	0.05	-0.6489	1.3328	-2.76E-9
b4	1.2458	0.7711	214	1.62	0.1076	0.05	-0.2741	2.7656	-92E-11
b5	-0.4814	0.2154	214	-2.24	0.0264	0.05	-0.9059	-0.05685	-8.19E-9
b6	-0.4183	0.2466	214	-1.70	0.0914	0.05	-0.9044	0.06788	-7.32E-9
b7	0.1804	0.4629	214	0.39	0.6971	0.05	-0.7320	1.0928	-1.5E-9
b8	-0.6656	0.4951	214	-1.34	0.1803	0.05	-1.6415	0.3104	-1.07E-9
b9	1.7158	0.7047	214	2.43	0.0157	0.05	0.3267	3.1050	-483E-12
b10	-1.9866	0.7128	214	-2.79	0.0058	0.05	-3.3915	-0.5817	-335E-12
b11	-1.0645	0.5377	214	-1.98	0.0490	0.05	-2.1243	0.00470	-246E-12
b12	1.5428	0.4019	214	3.84	0.0002	0.05	0.7505	2.3350	-1.44E-9

The log-likelihood statistics for the ZIP(τ), ZINB(τ), and ZIGP(τ) regression models, are respectively -641.09, -367.67, and -365.84. The fit from both the ZINB(τ) and ZIGP(τ) regression models are very similar. However, the parameter τ is significant in both ZIP(τ) and ZIGP(τ) regression models but not in the ZINB(τ) regression model. The dispersion parameter α is significant in both the ZINB(τ) and ZIGP(τ) regression models. The estimated proportions of zeros from ZIP(τ), ZINB(τ) and ZIGP(τ) regression models are, respectively, 63.7%, 66.2% and 65.7%.

We also fitted the ZIP, ZINB, and ZIGP regression models to the data by using $\xi_i = \exp(\delta)$. In this case, we take the covariate **z** to be a vector of 1's. The results are similar to what we have in the previous analysis. The log-likelihood statistics for the ZIP, ZINB, and ZIGP regression models, are respectively -638.60, -367.72, and -365.94. These values are very close to the results from the previous models. The parameter estimates for the covariates are also somewhat close. The parameter estimate for δ in the ZIP, ZINB, and ZIGP regression models, are respectively 0.6595, 0.2144, and 0.3919. This parameter, like τ, is not significant only in the ZIBN regression model. The estimated proportions of zeros from ZIP, ZINB, and ZIGP regression models are, respectively, 66.3%, 66.3%, and 66.4%. These values are remarkably close to the observed proportion of zeros. The following is the SAS program for fitting ZIP, ZINB, and ZIGP regression models.

```
/* Program for zero-inflated count data regression models */
data one;
infile 'C:\Research\Book\Data\kiya12.txt';
input @1 (educ_v educ_o emp_v emp_o inc_v inc_o fam_v fam_o club_v club_o drugp_v drugp_o)(2.0)
    @26 (minor severe total)(4.0);
y = severe;
run;

proc means data=one;
var educ_v -- severe;
run;


proc iml;
```

```
use one;
read all var{y};
read all var{educ_v educ_o emp_v emp_o inc_v inc_o fam_v fam_o club_v club_o drugp_v drugp_o}
    into xx0;
no = nrow(xx0); x0=j(no,1,1.0); xx=x0||xx0;
para1=ginv(t(xx)*xx) * (t(xx)*log(y + 0.5)); para=t(para1);
cnames={b0 b1 b2 b3 b4 b5 b6 b7 b8 b9 b10 b11 b12};
/* creates data set MYDATA with variables b0, b1, ..., b12 */
create mydata from para [colname=cnames];
append from para;
quit;
proc nlmixed data=one tech=trureg;
title "Zero-inflated Poisson regression analysis";
   parms tau 0.1 / data=mydata;
   ave = b0 + b1*educ_v + b2*educ_o + b3*emp_v + b4*emp_o + b5*inc_v + b6*inc_o
       + b7*fam_v + b8*fam_o + b9*club_v + b10*club_o + b11*drugp_v + b12*drugp_o;
   mu=exp(ave);
   d0 = 1 + mu**(-tau); d1=0; d2=0;
   if (y = 0) then d1 = log(mu**(-tau) + exp(-mu));
   if (y > 0) then d2 = y*log(mu) - lgamma(y+1) - mu;
   ll = -log(d0) + d1 + d2;
   model y~general(ll);
   predict (mu**(-tau) + exp(-mu))/(1 + mu**(-tau)) out=zero1(rename=(Pred=f_po));
run;

proc nlmixed data=one tech=trureg;
title "Zero-inflated negative binomial regression analysis";
   parms tau 0.1 alpha 0.1 / data=mydata;
   ave = b0 + b1*educ_v + b2*educ_o + b3*emp_v + b4*emp_o + b5*inc_v + b6*inc_o
       + b7*fam_v + b8*fam_o + b9*club_v + b10*club_o + b11*drugp_v + b12*drugp_o;
   mu=exp(ave);
   d0 = 1 + mu**(-tau); dd = alpha*mu; de = 1 + dd; d1=0; d2=0; co=0;
   if (y = 0) then d1 = log(mu**(-tau) + de**(-1/alpha));
   if (y > 0) then co = lgamma(y+1/alpha) - lgamma(1/alpha) - lgamma(y+1);
   if (y > 0) then d2 = y*log(dd/de) - (log(de))/alpha + co;
   ll = -log(d0) + d1 + d2;
   model y~general(ll);
   predict (mu**(-tau) + de**(-1/alpha))/(1 + mu**(-tau)) out=zero2(rename=(Pred=f_nb));
run;

proc nlmixed data=one tech=trureg;
title "Zero-inflated generalized Poisson regression analysis";
   parms tau 0.1 alpha 0.1 / data=mydata;
   ave = b0 + b1*educ_v + b2*educ_o + b3*emp_v + b4*emp_o + b5*inc_v + b6*inc_o
       + b7*fam_v + b8*fam_o + b9*club_v + b10*club_o + b11*drugp_v + b12*drugp_o;
   mu=exp(ave);
   d0 = 1 + mu**(-tau); de = 1 + alpha*mu; dd = 1 + alpha*y; d1=0; d2=0;
   if (y = 0) then d1 = log(mu**(-tau) + exp(-mu/de));
   if (y > 0) then d2 = y*log(mu/de) + (y-1)*log(dd) - lgamma(y+1) - mu*dd/de;
   ll = -log(d0) + d1 + d2;
   model y~general(ll);
   predict (mu**(-tau) + exp(-mu/de))/(1 + mu**(-tau)) out=zero3(rename=(Pred=f_gp));
run;

data zero;
merge zero1 zero2 zero3;
proc means data=zero;
var f_po f_nb f_gp;
run;
```

11.7   Zero-truncated Count Models

The zero-truncated regression models arise in those situations where there is no zero by nature of the data. An example for instance is the length of stay at an hospital. Once you are admitted, it is deemed that you have spent at least one day in the hospital. Thus the zeros can not be observed in this case for all patients admitted into the hospital. For a random variable Y with a discrete distribution, where the value of $Y = 0$ can not be observed, then the zero-truncated random variable Y_t has the probability mass function

$$\Pr(Y_t = y) = \frac{\Pr(Y = y)}{\Pr(Y > 0)}, \ \ y = 1, 2, 3, \ldots . \tag{11.47}$$

For the zero-truncated Poisson, with parameter μ, $\Pr(Y > 0) = 1 - \Pr(Y = 0) = 1 - \exp(-\mu)$. Hence, the pmf of zero-truncated Poisson random variable Y_t becomes

$$\Pr(Y_t = y) = \frac{\exp(-\mu)\mu^y}{y![1 - \exp(-\mu)]}, \quad y = 1, 2, 3, \ldots, \tag{11.48}$$

Note that the mean of zero-truncated Poisson model is not equal to μ. Suppose the parameter μ in (11.48) is replaced by $\mu_i = \exp(\mathbf{x}_i\boldsymbol{\beta})$, the log-likelihood for the zero-truncated Poisson model in a sample of size n is

$$\log(L) = \sum_{i=1}^{n} \left(-\mu_i + y_i \log \mu_i - \log y_i! - \log[1 - \exp(-\mu_i)] \right).$$

The probability mass function for the zero-truncated negative binomial model is given by

$$\Pr(Y_i = y_i \mid y_i > 0) = \frac{f(y_i; \mu_i, \alpha)}{1 - (1 + \alpha\mu_i)^{-\alpha^{-1}}}, \quad y = 1, 2, 3, \ldots, \tag{11.49}$$

where $f(y_i; \mu_i, \alpha)$ is as defined in (11.20). Hence, the log-likelihood function is

$$
\begin{aligned}
\log(L) = \sum_{i=1}^{n} \Bigg\{ & y_i \left[\log(\alpha\mu_i) - \log(1 + \alpha\mu_i) \right] - \frac{1}{\alpha}\log(1 + \alpha\mu_i) + \log(\Gamma(y_i + 1/\alpha)) \\
& - \log\left(\Gamma\left(\frac{1}{\alpha}\right) \right) - \log(y_i!) - \log\left[1 - (1 + \alpha\mu_i)^{-1/\alpha} \right] \Bigg\},
\end{aligned}
$$

where $\mu_i = \exp(\mathbf{x}_i\boldsymbol{\beta})$.

The probability mass function for the zero-truncated generalized Poisson model is given by

$$\Pr(Y_i = y_i \mid y_i > 0) = \frac{f(y_i; \mu_i, \alpha)}{1 - \exp\left(\frac{-\mu_i}{1 + \alpha\mu_i} \right)}, \quad y = 1, 2, 3, \ldots, \tag{11.50}$$

where $f(y_i; \mu_i, \alpha)$ is as defined in (11.27). Hence, the log-likelihood function is

$$
\begin{aligned}
\log(L) = \sum_{i-1}^{n} \Bigg\{ & y_i \log\left(\frac{\mu_i}{1 + \alpha\mu_i} \right) + (y_i - 1)\log(1 + \alpha y_i) - \frac{\mu_i(1 + \alpha y_i)}{1 + \alpha\mu_i} - \log(y_i!) \\
& - \log\left[1 - \exp\left(\frac{-\mu_i}{1 + \alpha\mu_i} \right) \right] \Bigg\}.
\end{aligned}
$$

where $\mu_i = \exp(\mathbf{x}_i\boldsymbol{\beta})$.

Example 11.4-The shoes

The following data in Table 11.6 was originally analyzed in Simonoff (2003, p. 169) and relates to the number of running shoes purchased by a sample of online registered runners. The explanatory variables are gender (male, female), marital status (1-married, 0 not married), number of runs per week (rpweek), age, income, education, distance ran per week, treadmill (0 or 1), average miles run per week (mpweek) and the dependent variable is the number of running shoes (shoes). Of course here, these data are truncated as we could not have the number zero among these runners.

The following model on selected predictor variables is of interest:

$$\log(\text{shoes}) = \beta_0 + \beta_1\text{distance} + \beta_2\text{rpweek} + \beta_3\text{mpweek} + \beta_4\text{male} + \beta_5\text{age}. \tag{11.51}$$

In this example, since the dependent variable is truncated at zero, we present the results of fitting the truncated and untruncated count models to the data in Table 11.6 based on the predictor variables identified in (11.51). The mean and variance of the response are 2.45 and 2.12. Since the mean is greater than the variance, the data is under-dispersed with respect to the Poisson model. We first fit the Poisson model to the data by using the **proc genmod**. The program and the results are presented below.

male	married	rpweek	age	income	college	dist	treadmill	mpweek	shoes
1	0	6	29.5	57.5	1	1	0	42.5	4
1	1	5	43.5	57.5	1	0	0	37.5	4
1	0	5	29.5	42.5	1	0	0	22.5	2
0	1	6	36.5	57.5	1	1	0	37.5	4
⋮	⋮	⋮	⋮	⋮	⋮	⋮	⋮	⋮	⋮
0	0	4	29.5	42.5	1	0	0	12.5	2
1	1	5	43.5	87.5	1	0	0	12.5	1
1	1	7	29.5	57.5	1	0	0	47.5	1
1	1	5	48.5	42.5	1	1	0	37.5	2

Table 11.6: Number of running shoes purchased from online running logs

```
data one;
infile 'C:\Research\Book\Data\shoes.txt';
input @1 male f2.0 @3 married f2.0 @5 rpweek f2.0 @7 age f5.1 @12 income f6.1
      @18 college f2.0 @20 distance f2.0 @22 treadmill f2.0 @24 mpweek f5.1
      @29 shoes f3.0;
y=shoes;
run;

proc means n mean std var min max data=one;
var shoes;
run;
proc genmod data=one;
title 'Poisson regression model';
model shoes=distance rpweek mpweek male age / dist=poi link=log;
run;
```

```
                        The SAS System

                      The MEANS Procedure

                  Analysis Variable : shoes

   N        Mean        Std Dev       Variance       Minimum       Maximum

  60     2.4500000     1.4546827     2.1161017     1.0000000     8.0000000


                  Poisson regression model

                     The GENMOD Procedure

                      Model Information
              Data Set             WORK.ONE
              Distribution          Poisson
              Link Function             Log
              Dependent Variable      shoes

          Number of Observations Read        60
          Number of Observations Used        60

          Criteria For Assessing Goodness Of Fit

      Criterion              DF        Value       Value/DF
      Deviance               54      28.0216         0.5189
      Scaled Deviance        54      28.0216         0.5189
      Pearson Chi-Square     54      28.4884         0.5276
      Scaled Pearson X2      54      28.4884         0.5276
      Log Likelihood                 -6.1635

   Algorithm converged.

              Analysis Of Parameter Estimates

                          Standard   Wald 95% Confidence    Chi-
   Parameter   DF  Estimate   Error        Limits          Square   Pr > ChiSq
   Intercept    1   -0.3208  0.4931   -1.2872    0.6456      0.42      0.5153
   distance     1    0.4149  0.2114    0.0007    0.8292      3.85      0.0496
   rpweek       1    0.1076  0.0900   -0.0689    0.2840      1.43      0.2322
   mpweek       1    0.0101  0.0081   -0.0058    0.0260      1.54      0.2145
```

male	1	0.0906	0.2002	-0.3018	0.4831	0.20	0.6508
age	1	0.0080	0.0087	-0.0090	0.0251	0.85	0.3570
Scale	0	1.0000	0.0000	1.0000	1.0000		

NOTE: The scale parameter was held fixed.

Based on the results, the ratio of deviance to the degrees of freedom is 0.5189, which shows that the data is under-dispersed. When the data is over-dispersed, one uses either the negative binomial regression model or the generalized Poisson regression model to analyze the data. In this case of under-dispersion, the negative binomial regression model is not appropriate. We will apply the generalized Poisson regression model by using the **proc nlmixed**. The initial values for the parameter estimates are taken to be from the Poisson regression analysis. For the initial estimate of parameter α, one can use $\alpha = 0$. The program and the partial results of **proc nlmixed** for the generalized Poisson regression model are displayed below.

```
data one;
infile 'C:\Research\Book\Data\shoes.txt';
input @1 male f2.0 @3 married f2.0 @5 rpweek f2.0 @7 age f5.1 @12 income f6.1
      @18 college f2.0 @20 distance f2.0 @22 treadmill f2.0 @24 mpweek f5.1
      @29 shoes f3.0;
y=shoes;
run;

proc nlmixed data=one tech=trureg;
title 'Generalized Poisson regression model';
*bounds alpha < 0;
parms b0=-.3208 b1=.4149 b2=.1076 b3=.0101 b4=.0906 b5=.008 alpha=0.0;
mu=exp(b0 + b1*distance + b2*rpweek + b3*mpweek + b4*male + b5*age);
ra=mu/(1+alpha*mu); nra=mu*(1+alpha*y)/(1+alpha*mu);
LL=y*log(ra) + (y-1)*log(1+alpha*y) - nra - lgamma(y+1);
model y~general(LL);
run;
```

```
                   Generalized Poisson regression model

                        The NLMIXED Procedure

                           Specifications

          Data Set                              WORK.ONE
          Dependent Variable                    y
          Distribution for Dependent Variable   General
          Optimization Technique                Trust Region
          Integration Method                    None

                             Dimensions

                  Observations Used            60
                  Observations Not Used         0
                  Total Observations           60
                  Parameters                    7

                             Parameters

     b0        b1        b2        b3        b4        b5     alpha  NegLogLike
 -0.3208    0.4149    0.1076    0.0101    0.0906    0.008        0  93.9340344

                          Iteration History
       Iter    Calls    NegLogLike       Diff     MaxGrad      Radius
          1       19    92.2637125   1.670322    58.84074    -1.69692
          2       28    90.600051    1.663662    14.68264    -2.2766
          3       37    90.5529949   0.047056     1.381714   -0.04465
          4       46    90.5526116   0.000383     0.014711   -0.00038
          5       55    90.5526116   4.237E-8     1.95E-6     -4.24E-8

          NOTE: GCONV convergence criterion satisfied.

                          Fit Statistics

                  -2 Log Likelihood              181.1
                  AIC (smaller is better)        195.1
                  AICC (smaller is better)       197.3
                  BIC (smaller is better)        209.8

                        Parameter Estimates

                      Standard
Parameter  Estimate     Error    DF  t Value  Pr > |t|  Alpha   Lower   Upper  Gradient
b0          -0.3282    0.4235    60   -0.78     0.4414    0.05  -1.1754  0.5190  8.351E-9
```

b1	0.4127	0.1586	60	2.60	0.0116	0.05	0.09553	0.7299	8.75E-9
b2	0.1077	0.07431	60	1.45	0.1524	0.05	-0.04093	0.2564	7.233E-8
b3	0.01023	0.006543	60	1.56	0.1232	0.05	-0.00286	0.02332	1.77E-6
b4	0.1147	0.1650	60	0.70	0.4896	0.05	-0.2154	0.4448	-5.25E-9
b5	0.007518	0.007137	60	1.05	0.2964	0.05	-0.00676	0.02179	1.95E-6
alpha	-0.07248	0.01977	60	-3.67	0.0005	0.05	-0.1120	-0.03294	-1.55E-6

In the above results, the estimate of parameter α is negative and it is significant. Note that the log-likelihood reported in **proc genmod** does not include the constant part. By using **proc nlmixed**, the -2log-likelihood for the Poisson regression model is 187.9. The corresponding result for the generalized Poisson regression model is 181.1. Thus, the generalized Poisson model provides a better fit than the Poisson model. We now fit the zero-truncated models to the data by using **proc nlmixed**. The program and the partial results of **proc nlmixed** to the zero-truncated models are displayed below.

```
data one;
infile 'C:\Research\Book\Data\shoes.txt';
input @1 male f2.0 @3 married f2.0 @5 rpweek f2.0 @7 age f5.1 @12 income f6.1
      @18 college f2.0 @20 distance f2.0 @22 treadmill f2.0 @24 mpweek f5.1
      @29 shoes f3.0;
y=shoes;
run;

proc nlmixed data=one tech=trureg;
title 'Truncated Poisson regression model';
parms b0=.1 b1=.1 b2=.1 b3=.1 b4=.1 b5=.1;
mu=exp(b0 + b1*distance + b2*rpweek + b3*mpweek + b4*male + b5*age);
LL=-mu + y*log(mu) - lgamma(y+1)- log(1-exp(-mu));
model y~general(LL);
run;
```

                    Truncated Poisson regression model

                        The NLMIXED Procedure

                            Specifications

        Data Set   WORK.ONE
        Dependent Variable   y
        Distribution for Dependent Variable   General
        Optimization Technique   Trust Region
        Integration Method   None

                              Dimensions
                    Observations Used   60
                    Observations Not Used   0
                    Total Observations   60
                    Parameters   6

                              Parameters
    b0   b1   b2   b3   b4   b5   NegLogLike
   0.1   0.1   0.1   0.1   0.1   0.1   108227.305

                          Iteration History
        Iter   Calls   NegLogLike   Diff   MaxGrad   Radius
           1   16   39446.9568   68780.35   1541816   -54416.4
          ..   ..
          12   104   84.0795173   2.16E-11   4.517E-8   -216E-13

        NOTE: GCONV convergence criterion satisfied.

                          Fit Statistics
                -2 Log Likelihood   168.2
                AIC (smaller is better)   180.2
                AICC (smaller is better)   181.7
                BIC (smaller is better)   192.7

                        Parameter Estimates

                      Standard
Parameter Estimate   Error   DF t Value Pr > |t| Alpha   Lower   Upper   Gradient
b0   -1.1742   0.6524   60   -1.80   0.0769   0.05   -2.4793   0.1309   9.26E-10
b1   0.5102   0.2337   60   2.18   0.0329   0.05   0.04275   0.9776   2.83E-10
b2   0.1822   0.1146   60   1.59   0.1170   0.05   -0.04699   0.4115   5.262E-9
b3   0.01432   0.009910 60   1.44   0.1538   0.05   -0.00551   0.03414 3.71E-8
b4   0.1315   0.2481   60   0.53   0.5981   0.05   -0.3647   0.6277   7.72E-10
b5   0.01177   0.01027   60   1.15   0.2561   0.05   -0.00876   0.03231 4.517E-8

```
data one;
infile 'C:\Research\Book\Data\shoes.txt';
```

```
input @1 male f2.0 @3 married f2.0 @5 rpweek f2.0 @7 age f5.1 @12 income f6.1
     @18 college f2.0 @20 distance f2.0 @22 treadmill f2.0 @24 mpweek f5.1
     @29 shoes f3.0;
y=shoes;
run;

proc nlmixed data=one tech=trureg;
title 'Truncated generalized Poisson regression model';
parms b0=-.3208 b1=.4149 b2=.1076 b3=.0101 b4=.0906 b5=.008 alpha=0.01;
mu=exp(b0 + b1*distance + b2*rpweek + b3*mpweek + b4*male + b5*age);
ra=mu/(1+alpha*mu); nra=mu*(1+alpha*y)/(1+alpha*mu);
LL=y*log(ra) + (y-1)*log(1+alpha*y) - nra - lgamma(y+1) - log(1-exp(-ra));
model y~general(LL);
run;
```

```
                     Truncated generalized Poisson regression model

                              The NLMIXED Procedure

                                   Specifications

          Data Set                              WORK.ONE
          Dependent Variable                    y
          Distribution for Dependent Variable   General
          Optimization Technique                Trust Region
          Integration Method                    None

                                     Dimensions

                   Observations Used                 60
                   Observations Not Used              0
                   Total Observations                60
                   Parameters                         7

                                     Parameters
         b0         b1        b2        b3        b4        b5     alpha   NegLogLike
    -0.3208     0.4149    0.1076    0.0101    0.0906     0.008      0.01  86.6356493

                                 Iteration History
        Iter   Calls   NegLogLike       Diff    MaxGrad     Radius
           1      19   84.4203427   2.215307   198.9889   -3.67999
           .      ..   ..........   ........   ........    .......
           6      64   83.2052614   5.136E-9   5.965E-7   -5.14E-9

         NOTE: GCONV convergence criterion satisfied.

                                   Fit Statistics
                   -2 Log Likelihood               166.4
                   AIC (smaller is better)         180.4
                   AICC (smaller is better)        182.6
                   BIC (smaller is better)         195.1

                               Parameter Estimates

                      Standard
Parameter  Estimate      Error  DF  t Value  Pr > |t|  Alpha     Lower      Upper    Gradient
b0          -1.0380     0.5866  60    -1.77    0.0819   0.05   -2.2115     0.1354    1.197E-9
b1           0.4829     0.1912  60     2.53    0.0142   0.05    0.1005     0.8653   -2.99E-9
b2           0.1725    0.09903  60     1.74    0.0867   0.05  -0.02562     0.3706   9.305E-9
b3          0.01359   0.008455  60     1.61    0.1132   0.05  -0.00332    0.03050   5.965E-7
b4           0.1441     0.2139  60     0.67    0.5032   0.05   -0.2838     0.5719  -129E-13
b5          0.01088   0.008890  60     1.22    0.2258   0.05  -0.00690    0.02866   5.271E-7
alpha      -0.05108    0.03033  60    -1.68    0.0973   0.05   -0.1117   0.009578   -8.11E-8
```

In the above results, the −2log-likelihood for the zero-truncated Poisson regression model is 168.2, which provides a better fit than the ordinary Poisson regression model. The corresponding result for the zero-truncated generalized Poisson regression model is 166.4. Even though the parameter *alpha* is negative, but it is no longer significant. Furthermore, the fit from both the truncated Poisson and truncated generalized Poisson model are very similar.

11.8 Case Study: The HMO Data

The data in this example was brought to our attention by Joseph Hilbe in his November 2005 workshop on modeling over-dispersed data. The data relate to length of hospital stay from the 1997 MedPar dataset. The data consists of 1495 observations and the following variables: provider's number (provnum), died (1 = died while in hospital; 0 = alive), white (1 = white, 0 = others), patient is an HMO member (hmo), age group (age), emergency admittance (type1), urgent admittance for first available bed (type2) and elective admittance (type3).

We present in Table 11.7 the first five and last five observations from these data. Although, there are 475 patients in the study, there are however 1495 observations in the data, because the observations came from 54 providers (some of which had more than one patient re-admitted), thus the data is clustered. We shall examine if correlation among the clustering subjects (within subjects) affect our parameter estimates. In this example, since the response variable is length of hospital stay, it does not matter how many minutes or hours the subject got admitted, it can not be less than 1 day. Thus the response variable is truncated at los=0. We shall examine the impact of this after our preliminary or exploratory analysis of the data on selected explanatory variables **died, hmo, type2 and type3**.

provnum	died	white	hmo	los	age80	age	type1	type2	type3
30001	0	1	0	4	0	4	1	0	0
30001	0	1	1	9	0	4	1	0	0
30001	1	1	1	3	1	7	1	0	0
30001	0	1	0	9	0	6	1	0	0
30001	1	1	0	1	1	7	1	0	0
$\vdots$	$\vdots$	$\vdots$	$\vdots$	$\vdots$	$\vdots$	$\vdots$	$\vdots$	$\vdots$	$\vdots$
32002	0	1	0	14	0	4	0	0	1
32002	1	1	0	8	0	4	0	0	1
32002	0	1	0	59	0	2	0	0	1
32003	0	1	0	63	0	4	0	1	0
32003	0	1	0	32	0	6	0	1	0

Table 11.7: Hmo data on length of stay

Fitting Various Models to HMO Data

We begin by fitting the Poisson and the negative binomial models to the data using **proc genmod**. Here we shall only report the parameter estimates and the goodness of fit statistics. The generalized Poisson regression model is fitted to the data by using the **proc nlmixed**. The zero-truncated models will also be fitted to the HMO data.

11.8.1   Fitting Poisson and Zero-truncated Poisson Models

We applied the Poisson regression model to the data in Table 11.7 by using the model

$$\ln(\mu) = \beta_0 + \beta_1 \text{hmo} + \beta_2 \text{died} + \beta_3 \text{type2} + \beta_4 \text{type3}. \tag{11.52}$$

Note that one can use **proc nlmixed** to implement the Poisson regression model for the selected explanatory variables. Both **proc genmod** and **proc nlmixed** give the same parameter estimates but different log-likelihood values. In the following program, we used the **proc genmod** for the Poisson model and the **proc nlmixed** for the zero-truncated Poisson model.

```
data one;
infile "C:\Research\Book\Data\hmo.dat";
input provnum died white hmo los age80 age type1 type2 type3;
y=los;
run;
proc genmod data=one;
title 'Poisson regression model';
model los=died hmo type2 type3 / dist=poi link=log;
run;
proc nlmixed data=one tech=trureg;
title 'Truncated Poisson regression model';
parms b0=1 b1=1 b2=1 b3=1 b4=1;
mu=exp(b0 + b1*died + b2*hmo + b3*type2 + b4*type3);
LL=-mu + los*log(mu) - lgamma(los+1)- log(1-exp(-mu));
model los~general(LL);
run;
                        Poisson regression model

                    The GENMOD Procedure
```

```
                       Model Information

              Data Set                 WORK.ONE
              Distribution             Poisson
              Link Function            Log
              Dependent Variable       los

          Number of Observations Read       1495
          Number of Observations Used       1495

          Criteria For Assessing Goodness Of Fit

    Criterion              DF        Value      Value/DF
    Deviance              1490    7978.7475      5.3549
    Scaled Deviance       1490    7978.7475      5.3549
    Pearson Chi-Square    1490    9299.5911      6.2413
    Scaled Pearson X2     1490    9299.5911      6.2413
    Log Likelihood                19434.4741

Algorithm converged.

               Analysis Of Parameter Estimates

                         Standard   Wald 95% Confidence    Chi-
Parameter   DF  Estimate   Error         Limits          Square  Pr > ChiSq
Intercept    1   2.2646   0.0118    2.2414    2.2877     36711.9    <.0001
died         1  -0.2483   0.0181   -0.2838   -0.2129       188.32   <.0001
hmo          1  -0.0754   0.0239   -0.1222   -0.0285         9.94   0.0016
type2        1   0.2499   0.0210    0.2087    0.2910       141.78   <.0001
type3        1   0.7501   0.0262    0.6987    0.8016       817.40   <.0001
Scale        0   1.0000   0.0000    1.0000    1.0000

NOTE: The scale parameter was held fixed.

               Truncated Poisson regression model

                    The NLMIXED Procedure

                        Specifications

    Data Set                              WORK.ONE
    Dependent Variable                    los
    Distribution for Dependent Variable   General
    Optimization Technique                Trust Region
    Integration Method                    None

                         Dimensions

          Observations Used              1495
          Observations Not Used             0
          Total Observations             1495
          Parameters                        5

                         Parameters
     b0        b1        b2        b3        b4     NegLogLike
      1         1         1         1         1     11864.3915

                    Iteration History
   Iter   Calls   NegLogLike      Diff     MaxGrad     Radius
     1     15    8157.08655    3707.305   6947.254   -6277.24
     2     22    6886.40137    1270.685   1031.692   -1150.65
     3     29    6846.7227      39.67867    37.76408   -38.6495
     4     36    6846.65284      0.069863    0.070099   -0.06976
     5     43    6846.65284      3.591E-7    4.141E-7   -3.59E-7

    NOTE: GCONV convergence criterion satisfied.

                    Fit Statistics
          -2 Log Likelihood             13693
          AIC (smaller is better)       13703
          AICC (smaller is better)      13703
          BIC (smaller is better)       13730

                    Parameter Estimates

              Standard
Parameter  Estimate   Error    DF  t Value  Pr > |t|  Alpha    Lower     Upper    Gradient
b0          2.2645   0.01182  1495   191.51   <.0001   0.05    2.2413    2.2877   4.141E-7
b1         -0.2487   0.01812  1495   -13.73   <.0001   0.05   -0.2842   -0.2131   5.075E-8
b2         -0.07551  0.02394  1495    -3.15   0.0016   0.05   -0.1225   -0.02856  2.758E-7
```

b3	0.2501	0.02099	1495	11.91	<.0001	0.05	0.2089	0.2912	8.357E-8
b4	0.7504	0.02624	1495	28.59	<.0001	0.05	0.6989	0.8019	2.038E-8

The results show that all the covariates are significant under the Poisson and zero-truncated Poisson models but there is a very serious over-dispersion, with the over-dispersion parameter being 5.3424 for the Poisson case. The -2log-likelihood for both the Poisson and zero-truncated Poisson models are respectively 13694 and 13693. Because of the clustering of the data, we now fit a generalized estimating Poisson model to the data adjusting for the clusters. Here we have used independent correlation structure. The following are the program and output.

```
data one;
infile "C:\Research\Book\Data\hmo.dat";
input provnum died white hmo los age80 age type1 type2 type3;
y=los;
run;
proc genmod data=one;
class provnum;
model los=hmo died type2 type3/dist=poi link=log;
repeated subject=provnum/type=ind;
run;
```

```
                        The GENMOD Procedure

                     Class Level Information

        Class        Levels   Values

        provnum        54     30001 30002 30003 30006 30007 30008 30009 30010
                              30011 30012 30013 30014 30016 30017 30018 30019
                              30022 30023 30024 30025 30030 30033 30035 30036
                              30037 30038 30043 30044 30055 30059 30060 30061
                              30062 30064 30065 30067 30068 30069 30073 30078
                              30080 30083 ...

                     GEE Model Information

            Correlation Structure              Independent
            Subject Effect             provnum (54 levels)
            Number of Clusters                          54
            Correlation Matrix Dimension                92
            Maximum Cluster Size                         92
            Minimum Cluster Size                          1

        Algorithm converged.

                 Analysis Of GEE Parameter Estimates
                 Empirical Standard Error Estimates

                           Standard   95% Confidence
          Parameter Estimate  Error       Limits        Z Pr > |Z|
          Intercept  2.2646  0.0332   2.1995  2.3297   68.17  <.0001
          hmo       -0.0754  0.0498  -0.1730  0.0222   -1.51  0.1301
          died      -0.2483  0.0628  -0.3713 -0.1253   -3.96  <.0001
          type2      0.2499  0.0641   0.1243  0.3754    3.90  <.0001
          type3      0.7501  0.2165   0.3259  1.1744    3.47  0.0005
```

In the above output, the parameter for the hmo variable is no longer significant. We will now consider the negative binomial regression (NBR) and zero-truncated NBR models.

11.8.2   Fitting NBR and Zero-truncated NBR Models

We fit the negative binomial regression model to the HMO data using both **proc genmod** for the negative binomial model and **proc nlmixed** for the zero-truncated negative binomial model.

```
data one;
infile "C:\Research\Book\Data\hmo.dat";
input provnum died white hmo los age80 age type1 type2 type3;
y=los;
run;

proc genmod data=one;
title 'Negative binomial regression model';
model los=died hmo type2 type3 / dist=negbin link=log;
run;
proc nlmixed data=one tech=trureg;
title 'Truncated negative binomial regression model';
```

```
parms b0=1 b1=1 b2=1 b3=1 b4=1 alpha=0.1;
mu=exp(b0 + b1*died + b2*hmo + b3*type2 + b4*type3);
ra=1+alpha*mu; nra=alpha*mu/ra; ra2=y+1/alpha; cc=ra**(-1/alpha);
LL=y*log(nra)-(log(ra))/alpha+lgamma(ra2)-lgamma(1/alpha)-lgamma(y+1)-log(1-cc);
model y~general(LL);
run;
```

Negative binomial regression model

The GENMOD Procedure

Model Information

Data Set	WORK.ONE
Distribution	Negative Binomial
Link Function	Log
Dependent Variable	los

Number of Observations Read	1495
Number of Observations Used	1495

Criteria For Assessing Goodness Of Fit

Criterion	DF	Value	Value/DF
Deviance	1490	1566.7175	1.0515
Scaled Deviance	1490	1566.7175	1.0515
Pearson Chi-Square	1490	1688.5893	1.1333
Scaled Pearson X2	1490	1688.5893	1.1333
Log Likelihood		21498.8238	

Algorithm converged.

Analysis Of Parameter Estimates

Parameter	DF	Estimate	Standard Error	Wald 95% Confidence Limits		Chi-Square	Pr > ChiSq
Intercept	1	2.2608	0.0270	2.2079	2.3138	7005.19	<.0001
died	1	-0.2370	0.0405	-0.3163	-0.1576	34.25	<.0001
hmo	1	-0.0706	0.0528	-0.1740	0.0328	1.79	0.1809
type2	1	0.2532	0.0499	0.1554	0.3510	25.74	<.0001
type3	1	0.7365	0.0754	0.5887	0.8843	95.41	<.0001
Dispersion	1	0.4352	0.0195	0.3970	0.4734		

NOTE: The negative binomial dispersion parameter was estimated by maximum likelihood.

Truncated negative binomial regression model

The NLMIXED Procedure

Specifications

Data Set	WORK.ONE
Dependent Variable	y
Distribution for Dependent Variable	General
Optimization Technique	Trust Region
Integration Method	None

Dimensions

Observations Used	1495
Observations Not Used	0
Total Observations	1495
Parameters	6

Parameters

b0	b1	b2	b3	b4	alpha	NegLogLike
1	1	1	1	1	0.1	8205.18724

Iteration History

Iter	Calls	NegLogLike	Diff	MaxGrad	Radius
1	16	5606.90273	2598.285	5528.784	-2547.53
.	..	....	....	....	....
9	80	4737.53502	9.53E-10	3.24E-9	-966E-12

NOTE: GCONV convergence criterion satisfied.

Fit Statistics

-2 Log Likelihood	9475.1
AIC (smaller is better)	9487.1
AICC (smaller is better)	9487.1
BIC (smaller is better)	9518.9

```
                        Parameter Estimates

                 Standard
Parameter  Estimate   Error    DF  t Value  Pr > |t|  Alpha    Lower    Upper   Gradient
b0          2.2240   0.03002  1495   74.08   <.0001    0.05    2.1651   2.2829   2.28E-9
b1         -0.2522   0.04471  1495   -5.64   <.0001    0.05   -0.3399  -0.1645   4.06E-10
b2         -0.07542  0.05824  1495   -1.30   0.1955    0.05   -0.1896   0.03881  1.77E-10
b3          0.2685   0.05500  1495    4.88   <.0001    0.05    0.1606   0.3764   2.31E-10
b4          0.7668   0.08304  1495    9.23   <.0001    0.05    0.6039   0.9297   1.06E-10
alpha       0.5325   0.02928  1495   18.19   <.0001    0.05    0.4751   0.5900  -3.24E-9
```

The above results show that the negative binomial model provides an adequate fit to the data as the ratio of deviance to the degrees of freedom is 1.0515, which is very close to 1. The negative binomial model shows lack of dispersion but the hmo parameter now seems insignificant under this model. The -2log-likelihood for the negative binomial model is 9565.2 while that of zero-truncated negative binomial model is 9475.1. The negative binomial model performed better than the Poisson model and the zero-truncated negative binomial model seems to be better than the untruncated negative binomial model. Because of the clustering of the data, we fit a generalized estimating negative binomial model to the data adjusting for the clusters. Here we have used independent correlation structure.

```
data one;
infile "C:\Research\Book\Data\hmo.dat";
input provnum died white hmo los age80 age type1 type2 type3;
y=los;
run;
proc genmod data=one;
class provnum;
model los=hmo died type2 type3/dist=negbin link=log;
repeated subject=provnum/type=ind;
run;
                      The GENMOD Procedure

                     Model Information

            Data Set                WORK.ZTP
            Distribution     Negative Binomial
            Link Function                 Log
            Dependent Variable            los

                 GEE Model Information

       Correlation Structure              Independent
       Subject Effect           provnum (54 levels)
       Number of Clusters                         54
       Correlation Matrix Dimension               92
       Maximum Cluster Size                        92
       Minimum Cluster Size                        1

   Algorithm converged.

           Analysis Of GEE Parameter Estimates
           Empirical Standard Error Estimates

                      Standard   95% Confidence
       Parameter Estimate  Error      Limits         Z Pr > |Z|
       Intercept  2.2608  0.0326  2.1969  2.3247  69.34  <.0001
       hmo       -0.0706  0.0492 -0.1670  0.0259  -1.43   0.1514
       died      -0.2370  0.0581 -0.3508 -0.1231  -4.08  <.0001
       type2      0.2532  0.0628  0.1301  0.3763   4.03  <.0001
       type3      0.7365  0.2113  0.3224  1.1506   3.49   0.0005
```

The results now, after adjusting for the clusters (patients in the same hospitals are more likely to have correlated observations), we see that hmo does not seem to be important under NBR model, but the died, type2 and type3 are still significant under the model. We now consider in the next section fitting the generalized Poisson regression (GPR) and zero-truncated GPR models to the HMO data.

11.8.3 Fitting GPR and Zero-truncated GPR Models

We fit the generalized Poisson and the zero-truncated generalized Poisson models to the HMO data by using the **proc nlmixed**. The program and the partial results are presented as follows:

```
data one;
```

```
infile "C:\Research\Book\Data\hmo.dat";
input provnum died white hmo los age80 age type1 type2 type3;
y=los;
run;
proc nlmixed data=one tech=trureg;
title 'Generalized Poisson regression model';
parms b0=1 b1=1 b2=1 b3=1 b4=1 alpha=0.1;
mu=exp(b0 + b1*died + b2*hmo + b3*type2 + b4*type3);
ra=mu/(1+alpha*mu); nra=mu*(1+alpha*y)/(1+alpha*mu);
LL=y*log(ra) + (y-1)*log(1+alpha*y) - nra - lgamma(y+1);
model y~general(LL);

proc nlmixed data=one tech=trureg;
title 'Truncated generalized Poisson regression model';
parms b0=1 b1=1 b2=1 b3=1 b4=1 alpha=0.1;
mu=exp(b0 + b1*died + b2*hmo + b3*type2 + b4*type3);
ra=mu/(1+alpha*mu); nra=mu*(1+alpha*y)/(1+alpha*mu);
LL=y*log(ra) + (y-1)*log(1+alpha*y) - nra - lgamma(y+1) - log(1-exp(-ra));
model y~general(LL);
run;
```

Generalized Poisson regression model

The NLMIXED Procedure

Specifications

Data Set	WORK.ONE
Dependent Variable	y
Distribution for Dependent Variable	General
Optimization Technique	Trust Region
Integration Method	None

Dimensions

Observations Used	1495
Observations Not Used	0
Total Observations	1495
Parameters	6

Parameters

b0	b1	b2	b3	b4	alpha	NegLogLike
1	1	1	1	1	0.1	6727.42194

Iteration History

Iter	Calls	NegLogLike	Diff	MaxGrad	Radius
1	16	5123.49958	1603.922	4081.83	-1554.73
.	..	....	....	....	....
6	57	4769.90941	1.087E-6	0.000018	-1.09E-6

NOTE: GCONV convergence criterion satisfied.

Fit Statistics

-2 Log Likelihood	9539.8
AIC (smaller is better)	9551.8
AICC (smaller is better)	9551.9
BIC (smaller is better)	9583.7

Parameter Estimates

Parameter	Estimate	Standard Error	DF	t Value	Pr > |t|	Alpha	Lower	Upper	Gradient
b0	2.2595	0.02785	1495	81.13	<.0001	0.05	2.2048	2.3141	-1.53E-6
b1	-0.2334	0.04146	1495	-5.63	<.0001	0.05	-0.3148	-0.1521	-2.01E-8
b2	-0.06922	0.05368	1495	-1.29	0.1974	0.05	-0.1745	0.03607	-1.96E-8
b3	0.2548	0.05255	1495	4.85	<.0001	0.05	0.1517	0.3578	-8.33E-8
b4	0.7286	0.08774	1495	8.30	<.0001	0.05	0.5565	0.9007	-3.51E-8
alpha	0.1400	0.005024	1495	27.87	<.0001	0.05	0.1302	0.1499	-0.00002

Truncated generalized Poisson regression model

The NLMIXED Procedure

Specifications

Data Set	WORK.ONE
Dependent Variable	y
Distribution for Dependent Variable	General
Optimization Technique	Trust Region
Integration Method	None

Dimensions

```
              Observations Used              1495
              Observations Not Used             0
              Total Observations             1495
              Parameters                        6
```

```
                        Parameters

  b0       b1       b2       b3       b4    alpha    NegLogLike
   1        1        1        1        1      0.1    6638.51777
```

```
                  Iteration History

     Iter   Calls   NegLogLike      Diff    MaxGrad     Radius
       1      16    5088.07893   1550.439   4288.466   -1503.67
       2      25    4793.28108   294.7979   1333.775   -289.532
       3      33    4742.73441   50.54666   422.9061   -46.3058
       4      41    4739.4968    3.237613   51.87784    -2.9993
       5      49    4739.44227   0.054536   1.068159   -0.05382
       6      57    4739.44224   0.000023   0.000469   -0.00002
```

```
     NOTE: GCONV convergence criterion satisfied.
```

```
                  Fit Statistics
         -2 Log Likelihood            9478.9
         AIC (smaller is better)      9490.9
         AICC (smaller is better)     9490.9
         BIC (smaller is better)      9522.7
```

```
                  Parameter Estimates

                  Standard
Parameter Estimate  Error   DF  t Value  Pr > |t|  Alpha    Lower   Upper   Gradient
b0         2.2370  0.02972  1495  75.28   <.0001    0.05    2.1787  2.2953  -0.00002
b1        -0.2426  0.04415  1495  -5.50   <.0001    0.05   -0.3293 -0.1560   2.64E-6
b2        -0.07213 0.05713  1495  -1.26    0.2069   0.05   -0.1842  0.03993  3.285E-7
b3         0.2642  0.05602  1495   4.72   <.0001    0.05    0.1543  0.3741   -8.94E-8
b4         0.7453  0.09407  1495   7.92   <.0001    0.05    0.5608  0.9298   -3.33E-7
alpha      0.1545  0.006207 1495  24.89   <.0001    0.05    0.1423  0.1667  -0.00047
```

The -2log-likelihood under the generalized Poisson model and the zero-truncated generalized Poisson model are respectively 9539.8 and 9478.9. Again, the zero-truncated model appears to have done better in fitting the data. In comparing the negative binomial model with the generalized Poisson model, the generalized Poisson model seems to have done better. However, there is no difference between the zero-truncated negative binomial model and zero-truncated generalized Poisson model. In all these models, the parameter for the variable hmo is insignificant.

11.9 Exercises

11.1 Derive the mean and variance of the zero-truncated Poisson distribution in (11.48).

11.2 Derive the mean and variance of the zero-truncated negative binomial distribution.

11.3 Derive the mean and variance of the zero-truncated generalized Poisson distribution.

11.4 Cameron et al. (1988) analyzed various measures of health-care utilization by using a sample of 5190 single-person households from the 1977-78 Australian Health Survey. The data are obtained from the Journal of Applied Econometrics 1997 Data Archive. The first 1000 cases of the data are summarized in Table 11.8 for the following variables: the number of consultations with a doctor or specialist in the past 2 weeks (y_1), the number of consultations with non-doctor health professionals in the past 2 weeks (y_2), and the number of admissions to a hospital, psychiatric hospital, nursing or convalescent home in the past 12 months (y_3). The number and the corresponding frequencies are given in Table 11.8.

   (a) Fit Poisson distribution to each of the three sets of data in Table 11.8. State the estimate of the Poisson parameter μ for each set.

   (b) Conduct the goodness of fit test for the estimated Poisson distributions in (a).

   (c) Fit the zero-inflated Poisson to each set of data in Table 11.8 and comment on your results.

Number (y_i)	0	1	2	3	4	5	6	7	8	9	10	11
Frequency for y_1	743	165	30	23	9	12	12	5	1			
Frequency for y_2	850	77	26	4	8	4	7	15	2	3	1	3
Frequency for y_3	769	171	39	13	3	5						

Table 11.8: Number of doctor and non-doctor consultations, and admissions

11.5 For each of the data sets in Table 11.8:

(a) Fit the negative binomial distribution.

(b) Fit the generalized Poisson distribution.

(c) Conduct the goodness of fit test for the estimated negative binomial and generalized Poisson distributions in (a) and (b).

(d) Compare your results in (a)-(c) and comment on your findings.

(e) Compare your results in (c) with the Poisson fit in Problem 11.4.

11.6 For each of the data sets in Table 11.8:

(a) Fit the zero-inflated Poisson distribution.

(b) Fit the zero-inflated negative binomial distribution.

(c) Fit the zero-inflated generalized Poisson distribution.

(d) Conduct the goodness of fit test for the estimated distributions in (a)-(c).

(e) Compare your results in (d) and comment on your findings.

(e) Compare your results in (c) with the fits in Problems 11.4 and 11.5.

11.7 Cameron et al. (1988) analyzed various measures of health-care utilization by using a sample of 5190 single-person households from the 1977-78 Australian Health Survey. The data are obtained from the Journal of Applied Econometrics 1997 Data Archive. The first 1000 cases of the data are summarized in Table 11.9 for the following variables: the total number of prescribed medications used in past 2 days (y_1) and the total number of non-prescribed medications used in past 2 days (y_2).

Number (y_i)	0	1	2	3	4	5	6	7	8
Frequency for y_1	281	257	190	102	77	45	23	8	17
Frequency for y_2	750	185	46	9	5	3	2		

Table 11.9: Number of prescribed and non-prescribed medications

(a) Fit Poisson distribution to both data sets.

(b) Fit negative binomial distribution to both data sets.

(c) Fit generalized Poisson distribution to both data sets.

11.8 For the data in Table 11.9, the frequency (f) for the total number of prescribed and non-prescribed medications (y) is given in Table 11.10.

Number (y)	0	1	2	3	4	5	6	7	8
Frequency (f)	207	233	218	129	90	54	33	13	23

Table 11.10: Total number of prescribed and non-prescribed medications

   (a) Fit Poisson distribution to the data in Table 11.10.

   (b) Fit negative binomial distribution to the data in Table 11.10.

   (c) Fit generalized Poisson distribution to the data in Table 11.10.

11.9 For the data in Table 11.8, the frequency (f) for the total number of consultations with doctors and non-doctors (y) is given in Table 11.11.

y	1	2	3	4	5	6	7	8	9	10	11	12	13	14	17
f	638	203	52	28	15	15	11	15	8	2	4	2	4	2	1

Table 11.11: Total number of doctor and non-doctor consultations

   (a) Fit the zero-truncated Poisson distribution to the data in Table 11.11.

   (b) Fit the zero-truncated negative binomial distribution to the data in Table 11.11.

   (c) Fit the zero-truncated generalized Poisson distribution to the data in Table 11.11.

11.10 The following data from Böhning (1998) gives the distribution of the number of of criminal acts of 4301 people with deviating behavior. The data is presented in Table 11.12.

Number of Criminal acts	Number of people	Number of Criminal acts	Number of people
0	4037	3	9
1	219	4	5
2	29	5	2

Table 11.12: Data originally presented in Dieckman (1981)

   (a) Fit a Poisson and a ZIP distribution to these data. Comment on your results. Does either fit well?

   (b) Fit a ZINB distribution to the data in Table 11.12. Compare this model to the ones in (a).

   (c) How can we justify the use of ZINB here?

   (d) Fit a ZIGP distribution to the data in Table 11.12. Compare this model to the ones in (a) and (b).

11.11 The data in Table 11.13 is from Lindsey and Merch (1992) (reprinted with permission from Elsevier) and relate to the number of occupants in homes, based on a postal survey.

Number of occupants	Number of houses	Number of occupants	Number of houses
1	436	5	1
2	133	6	0
3	19	7	1
4	2		

Table 11.13: Distribution of occupants in houses

   (a) Fit a zero-truncated Poisson distribution and justify why you would use this distribution.

   (b) Fit a zero-truncated generalized Poisson distribution and compare your result with the fit in (a).

11.12 Number of injuries for 16 year olds. The US Department of Health and Human Services (2003, ICPSR 4372) undertook 2001-2002 Health Behavior in School-Aged Children (HBSC) Survey. The HBSC survey collected data on a wide range of health issues and factors that may influence them. For this problem, we used the student survey and selected high school students who are 16 years old. The response variable y is the number of injuries during the past twelve months. The covariates are x_1(gender, $1 = $ female and $0 = $ male), x_2(race, $1 = $ white and $0 = $ others), and x_3(frequency of going to school or to bed hungry, $0 = $ never to $3 = $ always). After excluding cases having missing information we have 1137 cases. Assume that the marginal mean of y has a log-linear relationship with the covariates through

$$\log(\mu_i) = \beta_0 + \beta_1 x_{i1} + \beta_2 x_{i2} + \beta_3 x_{i3}, \quad i = 1, 2, \ldots, 1137.$$

The above regression function relates the logarithm of the marginal mean to the explanatory variables. The data for the exercise is in the file **injury16.txt**.

(a) Obtain the descriptive statistics for the response and the independent variables.

(b) Fit a Poisson regression model to the data.

(c) Fit a negative binomial regression model to the data.

(d) Fit a generalized Poisson regression model to the data.

(e) Compare the fitted models in (b)-(d). Compare the parameter estimates from the three models. Comment on your findings.

(f) Use each of the three models to test the hypothesis that the frequency of going to school or bed hungry influences the number of injuries. State your null and alternative hypotheses, your p-value and an appropriate conclusion.

(g) Comment on the statement "There is no significance difference between the number of injuries sustained by male and female students".

11.13 Number of injuries for 16 year olds. Use the data for Problem 11.12 to answer the following questions.

(a) Fit a zero-inflated Poisson regression model to the data.

(b) Fit a zero-inflated negative binomial regression model to the data.

(c) Fit a zero-inflated generalized Poisson regression model to the data.

(d) Compare the fitted models in (a)-(c). Compare the parameter estimates from the three models. Comment on your findings.

(e) Compare your results with corresponding results in Problem 11.12.

11.14 Number of injuries for 15 year olds. The US Department of Health and Human Services (2003, ICPSR 4372) undertook 2001-2002 Health Behavior in School-Aged Children (HBSC) Survey. The HBSC survey collected data on a wide range of health issues and factors that may influence them. For this problem, we used the student survey and selected high school students who are 15 years old. The response variable y is the number of injuries during the past twelve months. The covariates are x_1(gender, $1 = $ female and $0 = $ male), x_2(race, $1 = $ white and $0 = $ others), and x_3(frequency of going to school or to bed hungry, $0 = $ never to $3 = $ always). After excluding cases having missing information we have 2591 cases. Assume that the marginal mean of y has a log-linear relationship with the covariates through

$$\log(\mu_i) = \beta_0 + \beta_1 x_{i1} + \beta_2 x_{i2} + \beta_3 x_{i3}, \quad i = 1, 2, \ldots, 2591.$$

The above regression function relates the logarithm of the marginal mean to the explanatory variables. The data for the exercise is in the file **injury15.txt**.

(a) Obtain the descriptive statistics for the count response and the independent variables.

(b) Fit a Poisson regression model to the data.

(c) Fit a negative binomial regression model to the data.

(d) Fit a generalized Poisson regression model to the data.

(e) Compare the fitted models in (b)-(d). Compare the parameter estimates from the three models. Comment on your findings.

(f) Use each of the three models to test the hypothesis that the frequency of going to school or bed hungry influences the number of injuries. State your null and alternative hypotheses, your p-value and an appropriate conclusion.

(g) Comment on the statement "There is no significance difference between the number of injuries sustained by male and female students".

11.15 Number of injuries for 15 year olds. Use the data for Problem 11.14 to answer the following questions.

(a) Fit a zero-inflated Poisson regression model to the data.

(b) Fit a zero-inflated negative binomial regression model to the data.

(c) Fit a zero-inflated generalized Poisson regression model to the data.

(d) Compare the fitted models in (a)-(c). Compare the parameter estimates from the three models. Comment on your findings.

(e) Compare your results with corresponding results in Problem 11.14.

11.16 Number of things to control weights for 16 year olds. The US Department of Health and Human Services (2003, ICPSR 4372) undertook 2001-2002 Health Behavior in School-Aged Children (HBSC) Survey. The HBSC survey collected data on a wide range of health issues and factors that may influence them. For this problem, we used the student survey and selected high school students who are 16 years old. The response variable y is the number of things students did to control their weights. These things range from exercise, skip meals to things like smoke more. The covariates are x_1(gender, $1 =$ female and $0 =$ male), x_2(race, $1 =$ white and $0 =$ others), and x_3(frequency of going to school or to bed hungry, $0 =$ never to $3 =$ always). After excluding cases having missing information we have 985 cases. The marginal mean of y has a log-linear relationship with the covariates through

$$\log(\mu_i) = \beta_0 + \beta_1 x_{i1} + \beta_2 x_{i2} + \beta_3 x_{i3}, \quad i = 1, 2, \ldots, 985.$$

The above regression function relates the logarithm of the marginal mean to the explanatory variables. The data for the exercise is in the file **weight16.txt**.

(a) Obtain the descriptive statistics for the response variable.

(b) Fit a Poisson regression model to the data.

(c) Fit a negative binomial regression model to the data.

(d) Fit a generalized Poisson regression model to the data.

(e) Compare the fitted models in (b)-(d). Compare the parameter estimates from the three models. Comment on your findings.

(f) Use each of the three models to test the hypothesis that the frequency of going to school or bed hungry influences the number of things the students did to control their weights. State your null and alternative hypotheses, your p-value and an appropriate conclusion.

(g) Comment on the statement "There is no significance difference between the number of things done by male and female students to control their weights".

11.17 Number of things to control weights for 16 year olds. Use the data for Problem 11.16 to answer the following questions.

(a) Fit a zero-truncated Poisson regression model to the data.

(b) Fit a zero-truncated negative binomial regression model to the data.

(c) Fit a zero-truncated generalized Poisson regression model to the data.

(d) Compare the fitted models in (a)-(c). Compare the parameter estimates from the three models. Comment on your findings.

(e) Compare your results with corresponding results in Problem 11.16.

(f) Based on your findings in (e), which model will you recommend for analyzing the data?

11.18 **Number of things to control weights for 15 year olds.** The US Department of Health and Human Services (2003, ICPSR 4372) undertook 2001-2002 Health Behavior in School-Aged Children (HBSC) Survey. The HBSC survey collected data on a wide range of health issues and factors that may influence them. For this problem, we used the student survey and selected high school students who are 15 years old. The response variable y is the number of things students did to control their weights. These things range from exercise, skip meals to things like smoke more. The covariates are x_1(gender, 1 = female and 0 = male), x_2(race, 1 = white and 0 = others), and x_3(frequency of going to school or to bed hungry, 0 = never to 3 = always). After excluding cases having missing information we have 2171 cases. The marginal mean of y has a log-linear relationship with the covariates through

$$\log(\mu_i) = \beta_0 + \beta_1 x_{i1} + \beta_2 x_{i2} + \beta_3 x_{i3}, \quad i = 1, 2, \ldots, 2171.$$

The above regression function relates the logarithm of the marginal mean to the explanatory variables. The data for the exercise is in the file **weight15.txt**.

(a) Obtain the descriptive statistics for the response variable.

(b) Fit a Poisson regression model to the data.

(c) Fit a negative binomial regression model to the data.

(d) Fit a generalized Poisson regression model to the data.

(e) Compare the fitted models in (b)-(d). Compare the parameter estimates from the three models. Comment on your findings.

(f) Use each of the three models to test the hypothesis that the frequency of going to school or bed hungry influences the number of things the students did to control their weights. State your null and alternative hypotheses, your p-value and an appropriate conclusion.

(g) Comment on the statement "There is no significance difference between the number of things done by male and female students to control their weights".

11.19 **Number of things to control weights for 15 year olds.** Use the data for Problem 11.18 to answer the following questions.

(a) Fit a zero-truncated Poisson regression model to the data.

(b) Fit a zero-truncated negative binomial regression model to the data.

(c) Fit a zero-truncated generalized Poisson regression model to the data.

(d) Compare the fitted models in (a)-(c). Compare the parameter estimates from the three models. Comment on your findings.

(e) Compare your results with corresponding results in Problems 11.18.

(f) Based on your findings in (e), which model will you recommend for analyzing the data?

Chapter 12

Regression with Censored or Truncated Data

12.1   Survival Analysis

In the health sciences, survival analysis is often used to model the time duration until the occurrence of an event-usually death. These survival time durations arise as a result of subjects being followed over time until they reached a specified endpoint or the event of interest occurs. An example of this is time to death of females who are diagnosed with breast cancer. Here, the event is death. Another example is the length of time a particular disease in humans remain in remission. Other applications of survival analysis can be found in engineering-usually referred to as time-to-failure or accelerated failure time model, where for example, the time it takes for an electric bulb to burn out in hours clearly falls into this category. Another example in the behavioral sciences is the recidivism study involving the duration in months or weeks released inmates are re-arrested.

A full understanding of the survival analysis techniques requires that we clearly spell out the following terms:

- Proper determination of the time to an event, which necessitates the specification of *starting time* and *end time*.

- Complete and Censored data

- Left and Right censoring

We now discuss each of the above terms in turn.

12.2   Start and End Times

Suppose the study starts at a particular time, Jan 1, 2004 and is to run for say, two years (December 2005). Thus, patients are enrolled from Jan 1, 2004. At this time the investigator collected all other baseline information on each subject. The treatments are then administered and the patients are then followed through time, with follow ups at regular intervals and on each follow-up, data are collected on each subject including information on the status of the subject. Because investigators do not often have large enough number of subjects at the beginning of the study, subjects are therefore often enrolled (or enter the study) between January 1, 2004 and December 2005. The starting time for each subject being the time the disease condition is first diagnosed.

However, in this kind of study, subjects often leave the study either through death or lost through follow-up or willingly leaving the study. In other situations, some patients are not followed until death because of the expiration of the study in 2005. We present in Table 1, typical scenarios for ten subjects to illustrate these possibilities.
In Table 12.1, we use the following notations:

x means the subject is still alive at the specified date at the end of the study.

o means that the subject was lost or dropped out before the end of the study

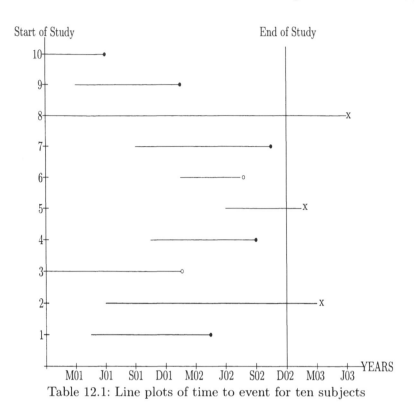

Table 12.1: Line plots of time to event for ten subjects

• means that the subject died at the given date

M01, M02, M03 refer respectively to March 2001, 2002 and 2003. Similarly, J01 for June 2001, S01 for September 2001 and D01 for December 2001.

- The starting date for the study is January 1, 2001, while

- The end of the study is on December 31, 2002.

We present the actual data for these ten subjects in Table 12.2, where for instance, subject 1 was diagnosed on April 15, 2001 and died on May 12, 2002. On the other hand, subject 5 was diagnosed on June 3, 2002 and was still alive on February 12, 2003. Subject 6, was diagnosed on February 18, 2002 and was lost or dropped from the study on August 10, 2002. In terms of duration days or weeks, we present in columns 5 and 6 of Table 12.2, the survival times for these ten subjects in days and weeks after enrollment (or diagnosis) into the study.

```
data surv1;
input subj $ stdate lsdate;
informat stdate lsdate mmddyy8.;
format stdate lsdate mmddyy8.;
datalines;
1 04/15/01 05/12/02
2 06/01/01 03/31/03
3 01/01/01 02/10/02
4 07/18/01 09/02/02
5 06/03/02 02/12/03
6 02/18/02 08/10/02
7 09/01/01 11/26/02
8 01/01/01 05/26/03
9 03/04/01 02/20/02
10 01/01/01 05/26/01
;
data new;
set surv1;
actual=datdif(stdate, lsdate, 'act/act');
format stdate lsdate mmddyy8.;
proc print data=new;
run;
data key;
```

```
set surv1;
days=-intck('day', lsdate, stdate);
weeks=-intck('week', lsdate, stdate);
month=-intck('month', lsdate, stdate);
year=intck('year', lsdate, stdate);
dur=lsdate-stdate;
format stdate lsdate mmddyy8.;
title 'Days Between Project Start and Project End';
run;
```

```
          Days Between Project Start and Project End

Obs  subj    stdate    lsdate   days  weeks  month  year   dur
 1     1   04/15/01  05/12/02   392    56     13    -1    392
 2     2   06/01/01  03/31/03   668    96     21    -2    668
 3     3   01/01/01  02/10/02   405    58     13    -1    405
 4     4   07/18/01  09/02/02   411    59     14    -1    411
 5     5   06/03/02  02/12/03   254    36      8    -1    254
 6     6   02/18/02  08/10/02   173    24      6     0    173
 7     7   09/01/01  11/26/02   451    65     14    -1    451
 8     8   01/01/01  05/26/03   875   125     28    -2    875
 9     9   03/04/01  02/20/02   353    50     11    -1    353
10    10   01/01/01  05/26/01   145    20      4     0    145
```

Subject	Starting Date	Date of Last contact	Date of Death	Days	weeks	censoring
1	04/15/01	.	05/12/02	392	56	
2	06/01/01	03/31/03	.	668	96	Yes
3	01/01/01	02/10/02		405	58	Yes
4	07/18/01	.	09/02/02	411	59	
5	06/03/02	02/12/03		254	36	Yes
6	02/18/02	08/10/02		173	24	Yes
7	09/01/01	.	11/26/02	451	65	
8	01/01/01	05/26/03	.	875	125	Yes
9	03/04/01	.	02/20/02	353	50	
10	01/01/01	.	05/26/01	145	20	

Table 12.2: Survival times for ten subjects

12.3   Describing Event Times

If we use T to denote the survival time, then, the survival function, designated as $S(t)$ is defined as the probability that an individual survives past time t. That is,

$$S(t) = \Pr(T > t) = 1 - F(t), \qquad (12.1)$$

where

$$F(t) = \Pr(T \le t),$$

is the cumulative distribution function and T is a random variable with the probability density function $f(t)$ defined by

$$f(t) = \frac{dF(t)}{dt}.$$

12.4   Estimating the Survival Function $S(t)$

From (12.1), we have $S(t) = 1 - F(t)$, hence, $0 < S(t) < 1$. That is, $S(t)$ is a decreasing function of t. For situations involving censored data, survival functions or curves can be estimated by the method of *product-limit* or the **Kaplan-Meier estimator** and the **Life-Table method**.

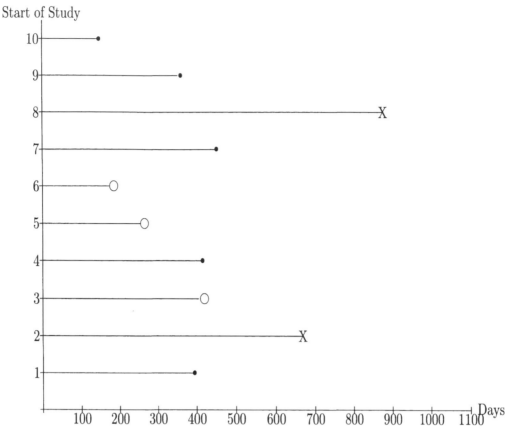

Table 12.3: Time to death(event) for all the ten subjects, starting from 0

12.4.1   The Kaplan-Meier Method

We illustrate the Kaplan-Meier method with the following data in Table 12.4 from Dunn and Clark (2009, p. 210) reproduced by permission of John Wiley and Sons, Inc. The data consist of 40 patients, listed in columns 1, 4 and 7. In columns 2, 5, and 8 are listed the known days the patients lived after entering the study. In columns 3, 6 and 9 are also listed their respective status: 1 if the patient died, 2 if the patient was lost to follow-up, and 3 if the patient was withdrawn alive. Thus, status categories 2 and 3 correspond to censoring as these patients did not experience the event 'death'. Thus for instance, patient 2 died at 39 days, while patients 20 and 37 were withdrawn alive. The Kaplan-Meier method can be implemented in SAS with **proc lifetest**.

```
data survfn1;
input patnt days status @@;
if status=2 or status=3 then censor=1;
else censor=0;
datalines;
1 21 1 2 39 1 3 77 1 4 133 1 5 141 2 6 152 1 7 153 1
8 161 1 9 179 1 10 184 1 11 197 1 12 199 1 13 214 1
14 228 1 15 256 2 16 260 1 17 261 1 18 266 1 19 269 1
20 287 3 21 295 1 22 308 1 23 311 1 24 321 2 25 326 1
26 355 1 27 361 1 28 374 1 29 398 1 30 414 1 31 420 1
32 468 2 33 483 1 34 489 1 35 505 1 36 539 1 37 565 3
38 618 1 39 793 1 40 794 1
;
proc print noobs;
run;
proc lifetest data=survfn1 plots=(s,ls,lls) graphics;
time days*censor(1);
run;
```

In the above SAS program, we have specified that Status 2 or 3 be coded as category 1 of the censoring variable. The **proc** statement requests that plots be made of

Patient	Days	Status	Patient	Days	Status	Patient	Days	Status
1	21	1	15	256	2	29	398	1
2	39	1	16	260	1	30	414	1
3	77	1	17	261	1	31	420	1
4	133	1	18	266	1	32	468	2
5	141	2	19	269	1	33	483	1
6	152	1	20	287	3	34	489	1
7	153	1	21	295	1	35	505	1
8	161	1	22	308	1	36	539	1
9	179	1	23	311	1	37	565	3
10	184	1	24	321	2	38	618	1
11	197	1	25	326	1	39	793	1
12	199	1	26	355	1	40	794	1
13	214	1	27	361	1			
14	228	1	28	374	1			

Table 12.4: Patient data from Dunn and Clark (2009)

1. **S**, the estimated survival function against time in Figure 12.1

2. **LS**, the negative log of the estimated survival function against time, and,

3. **LLS**, the log of the negative log of the estimated survival function against the log of time.

Both the LS and LLS plots in Figure 12.2 give us ideas as to whether the distribution of days follows an exponential or a Weibull distribution. We present the output from the above analysis as follows:

```
          patnt    days   status   censor
            1       21       1        0
            2       39       1        0
            3       77       1        0
            4      133       1        0
            5      141       2        1
            6      152       1        0
            7      153       1        0
            8      161       1        0
            9      179       1        0
           10      184       1        0
         ...........................
           37      565       3        1
           38      618       1        0
           39      793       1        0
           40      794       1        0
```

```
              The LIFETEST Procedure

          Product-Limit Survival Estimates

                            Survival
                            Standard   Number    Number
   days    Survival  Failure  Error    Failed     Left
  0.000    1.0000      0        0         0         40
 21.000    0.9750    0.0250   0.0247      1         39
 39.000    0.9500    0.0500   0.0345      2         38
 77.000    0.9250    0.0750   0.0416      3         37
133.000    0.9000    0.1000   0.0474      4         36
141.000*     .          .        .        4         35
152.000    0.8743    0.1257   0.0526      5         34
153.000    0.8486    0.1514   0.0570      6         33
161.000    0.8229    0.1771   0.0608      7         32
179.000    0.7971    0.2029   0.0641      8         31
184.000    0.7714    0.2286   0.0670      9         30
197.000    0.7457    0.2543   0.0695     10         29
199.000    0.7200    0.2800   0.0717     11         28
214.000    0.6943    0.3057   0.0736     12         27
228.000    0.6686    0.3314   0.0752     13         26
256.000*     .          .        .       13         25
```

260.000	0.6418	0.3582	0.0768	14	24
261.000	0.6151	0.3849	0.0782	15	23
266.000	0.5883	0.4117	0.0792	16	22
269.000	0.5616	0.4384	0.0800	17	21
287.000*	.	.	.	17	20
295.000	0.5335	0.4665	0.0808	18	19
308.000	0.5054	0.4946	0.0813	19	18
311.000	0.4774	0.5226	0.0814	20	17
321.000*	.	.	.	20	16
326.000	0.4475	0.5525	0.0816	21	15
355.000	0.4177	0.5823	0.0815	22	14
361.000	0.3879	0.6121	0.0809	23	13
374.000	0.3580	0.6420	0.0800	24	12
398.000	0.3282	0.6718	0.0787	25	11
414.000	0.2984	0.7017	0.0770	26	10
420.000	0.2685	0.7315	0.0749	27	9
468.000*	.	.	.	27	8
483.000	0.2350	0.7650	0.0726	28	7
489.000	0.2014	0.7986	0.0696	29	6
505.000	0.1678	0.8322	0.0656	30	5
539.000	0.1343	0.8657	0.0604	31	4
565.000*	.	.	.	31	3
618.000	0.0895	0.9105	0.0544	32	2
793.000	0.0448	0.9552	0.0417	33	1
794.000	0	1.0000	0	34	0

NOTE: The marked survival times are censored observations.

Summary Statistics for Time Variable days

Quartile Estimates

	Point	95% Confidence Interval	
Percent	Estimate	[Lower	Upper)
75	483.000	361.000	618.000
50	311.000	260.000	398.000
25	197.000	153.000	266.000

The LIFETEST Procedure

| Mean | Standard Error |
| 348.869 | 33.808 |

Summary of the Number of Censored and Uncensored Values

			Percent
Total	Failed	Censored	Censored
40	34	6	15.00

From the output, we notice that the median survival time of patients in this study is 311 days with a 95% confidence interval (260, 398).

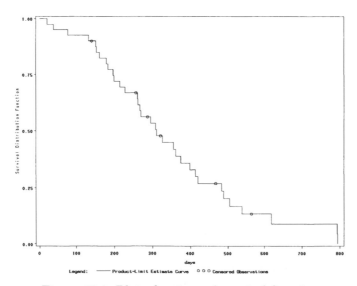

Figure 12.1: Plot of estimated survival function

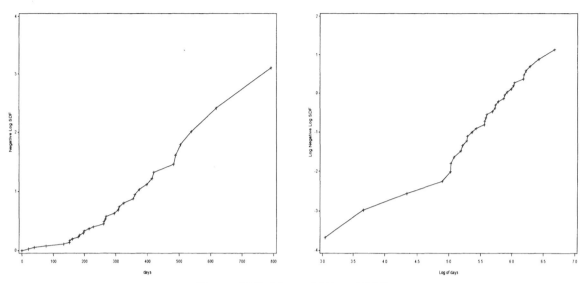

Figure 12.2: Plots of LS and LLS

12.4.2 The Life-Table Method

The Life-Table method again can be implemented in SAS by specifying **method=LT**, where LT stands for Life-Table.

```
set survfn1;
proc lifetest data=survfn1 method=lt plots=(s,ls,lls) graphics;
time days*censor(1);
symbol1 v=dot height=.03in;
run;
```

The LIFETEST Procedure

Life Table Survival Estimates

Interval [Lower,	Upper)	Number Failed	Number Censored	Effective Sample Size	Conditional Probability of Failure	Conditional Probability Standard Error	Survival
0	100	3	0	40.0	0.0750	0.0416	1.0000
100	200	8	1	36.5	0.2192	0.0685	0.9250
200	300	7	2	27.0	0.2593	0.0843	0.7223
300	400	7	1	18.5	0.3784	0.1128	0.5350
400	500	4	1	10.5	0.3810	0.1499	0.3326
500	600	2	1	5.5	0.3636	0.2051	0.2059
600	700	1	0	3.0	0.3333	0.2722	0.1310
700	800	2	0	2.0	1.0000	0	0.0873

Interval [Lower,	Upper)	Failure	Survival Standard Error	Median Residual Lifetime	Median Standard Error
0	100	0	0	317.3	39.0529
100	200	0.0750	0.0416	235.8	37.8163
200	300	0.2777	0.0712	185.9	34.3317
300	400	0.4650	0.0806	151.4	49.0894
400	500	0.6674	0.0784	152.9	68.5463
500	600	0.7941	0.0696	164.3	100.5
600	700	0.8690	0.0612	.	.
700	800	0.9127	0.0542	.	.

Evaluated at the Midpoint of the Interval

Interval [Lower,	Upper)	PDF	PDF Standard Error	Hazard	Hazard Standard Error
0	100	0.000750	0.000416	0.000779	0.00045
100	200	0.00203	0.000640	0.002462	0.000864
200	300	0.00187	0.000636	0.002979	0.001113
300	400	0.00202	0.000676	0.004667	0.001715
400	500	0.00127	0.000581	0.004706	0.002287
500	600	0.000749	0.000492	0.004444	0.003064
600	700	0.000437	0.000411	0.004	0.003919
700	800	0.000873	0.000542	0.02	0

Summary of the Number of Censored and Uncensored Values

			Percent
Total	Failed	Censored	Censored
40	34	6	15.00

The effective sample size is computed as follows: For the first interval, it is computed as

$$(n - a_1 - x) - \frac{x}{2} = 40 - 0 = 40.0,$$

where n is the sample size, and a_1 is the lagged number of failures in the previous intervals and x is the number of censored cases for that interval. For the second interval, we have

$$(40 - 3) - \frac{1}{2} = 37 - 0.5 = 36.5.$$

Similarly for the third interval, we have,

$$(40 - 3 - 8 - 1) - \frac{2}{2} = 28 - 1 = 27.0.$$

As it can be seen from this method, the time is first categorized into classes with equal intervals. In this case, SAS creates 8 categories (0, 100) up to (700, 800) with 100 day intervals. For instance, the 400-day survival rate is 0.3326 with a standard error of 0.0784. On the other hand, the estimated median residual lifetime initially was 317.3 days and has been dropping with it being 152.9 days at the beginning of the 400 days. The survival probability is the probability that the subject survives past the lower limit of that interval. We present the corresponding S plot under this method in Figure 12.3. Similarly, the negative log survival function LS and the log negative log survival function LLS are presented in Figure 12.4.

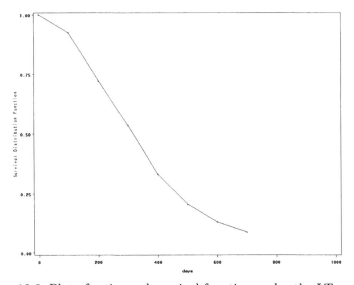

Figure 12.3: Plot of estimated survival function under the LT method

12.4.3   What if we Have Treatment Groups?

We can illustrate this case from the following example taken from Sedmak et al. (1989) which relates to female breast cancer patients originally classified as lymph node negative by standard light microscopy (SLM). The data in Table 12.5 give the times to death in months of a random sample of 45 female breast cancer patients with a minimum of 10-year follow up from the Ohio State University Hospitals Cancer Registry. Of these 45 patients, 36 were immunoperoxidase negative while the remaining 9 were negative. A status of 2 denotes a censored observation. We present below the SAS program and a partial output for the analysis of the data in Table 12.5.

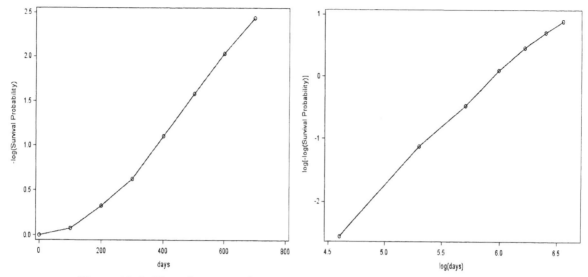

Figure 12.4: Plot of estimated LS and LLS functions under the LT method

	Immunoperoxidase Neg.								Immunoperoxidase Pos.		
Sub	Time	Status	Sub	Time	Status	Sub	Time	Status	Sub	Time	Status
1	19	1	13	67	1	25	143	2	37	22	1
2	25	1	14	74	1	26	148	2	38	23	1
3	30	1	15	78	1	27	151	2	39	38	1
4	34	1	16	86	1	28	152	2	40	42	1
5	37	1	17	122	2	29	153	2	41	73	1
6	46	1	18	123	2	30	154	2	42	77	1
7	47	1	19	130	2	31	156	2	43	89	1
8	51	1	20	130	2	32	162	2	44	115	1
9	56	1	21	133	2	33	164	2	45	144	2
10	57	1	22	134	2	34	165	2			
11	61	1	23	136	2	35	182	2			
12	66	1	24	141	2	36	189	2			

Table 12.5: Times to death for 45 breast cancer patients

```
data breast;
input tsurv censor treat @@;
datalines;
 19 1 1 25 1 1 30 1 1 34 1 1 37 1 1 46 1 1 47 1 1 51 1 1
 56 1 1 57 1 1 61 1 1 66 1 1 67 1 1 74 1 1 78 1 1 86 1 1
 122 0 1 123 0 1 130 0 1 130 0 1 133 0 1 134 0 1 136 0 1
 141 0 1 143 0 1 148 0 1 151 0 1 152 0 1 153 0 1 154 0 1
 156 0 1 162 0 1 164 0 1 165 0 1 182 0 1 189 0 1 22 1 2
 23 1 2 38 1 2 42 1 2 73 1 2 77 1 2 89 1 2 115 1 2 144 0 2
 ;
run;
proc lifetest plots=(s);
time tsurv*censor(0);
strata treat;
run;
```

```
                    The LIFETEST Procedure

                   Stratum 1: treat = 1

                Product-Limit Survival Estimates

                              Survival
                              Standard   Number    Number
         tsurv   Survival   Failure   Error    Failed    Left
         0.000   1.0000        0        0         0       36
```

tsurv	Survival	Failure	Survival Standard Error	Number Failed	Number Left
19.000	0.9722	0.0278	0.0274	1	35
25.000	0.9444	0.0556	0.0382	2	34
30.000	0.9167	0.0833	0.0461	3	33
34.000	0.8889	0.1111	0.0524	4	32
37.000	0.8611	0.1389	0.0576	5	31
46.000	0.8333	0.1667	0.0621	6	30
47.000	0.8056	0.1944	0.0660	7	29
51.000	0.7778	0.2222	0.0693	8	28
56.000	0.7500	0.2500	0.0722	9	27
57.000	0.7222	0.2778	0.0747	10	26
61.000	0.6944	0.3056	0.0768	11	25
66.000	0.6667	0.3333	0.0786	12	24
67.000	0.6389	0.3611	0.0801	13	23
74.000	0.6111	0.3889	0.0812	14	22
78.000	0.5833	0.4167	0.0822	15	21
86.000	0.5556	0.4444	0.0828	16	20
122.000*	.	.	.	16	19
123.000*	.	.	.	16	18
130.000*	.	.	.	16	17
130.000*	.	.	.	16	16
133.000*	.	.	.	16	15
134.000*	.	.	.	16	14
136.000*	.	.	.	16	13
141.000*	.	.	.	16	12
143.000*	.	.	.	16	11
148.000*	.	.	.	16	10
151.000*	.	.	.	16	9
152.000*	.	.	.	16	8
153.000*	.	.	.	16	7
154.000*	.	.	.	16	6
156.000*	.	.	.	16	5
162.000*	.	.	.	16	4
164.000*	.	.	.	16	3
165.000*	.	.	.	16	2
182.000*	.	.	.	16	1
189.000*	.	.	.	16	0

NOTE: The marked survival times are censored observations.

Summary Statistics for Time Variable tsurv

Quartile Estimates

Percent	Point Estimate	95% Confidence Interval [Lower	Upper)
75	.	.	.
50	.	67.000	.
25	56.500	37.000	78.000

The LIFETEST Procedure

Mean	Standard Error
70.944	3.626

NOTE: The mean survival time and its standard error were underestimated because the largest observation was censored and the estimation was restricted to the largest event time.

The LIFETEST Procedure

Stratum 2: treat = 2

Product-Limit Survival Estimates

tsurv	Survival	Failure	Survival Standard Error	Number Failed	Number Left
0.000	1.0000	0	0	0	9
22.000	0.8889	0.1111	0.1048	1	8
23.000	0.7778	0.2222	0.1386	2	7
38.000	0.6667	0.3333	0.1571	3	6
42.000	0.5556	0.4444	0.1656	4	5
73.000	0.4444	0.5556	0.1656	5	4
77.000	0.3333	0.6667	0.1571	6	3
89.000	0.2222	0.7778	0.1386	7	2
115.000	0.1111	0.8889	0.1048	8	1
144.000*	.	.	.	8	0

NOTE: The marked survival times are censored observations.

Summary Statistics for Time Variable tsurv

Quartile Estimates

Percent	Point Estimate	95% Confidence Interval [Lower	Upper)
75	89.000	42.000	.
50	73.000	38.000	89.000
25	38.000	22.000	77.000

Mean	Standard Error
66.000	12.256

NOTE: The mean survival time and its standard error were underestimated because the largest observation was censored and the estimation was restricted to the largest event time.

Summary of the Number of Censored and Uncensored Values

Stratum	treat	Total	Failed	Censored	Percent Censored
1	1	36	16	20	55.56
2	2	9	8	1	11.11
Total		45	24	21	46.67

The LIFETEST Procedure

Testing Homogeneity of Survival Curves for TSURV over Strata

Rank Statistics

treat	Log-Rank	Wilcoxon
1	-4.1873	-130.00
2	4.1873	130.00

Covariance Matrix for the Log-Rank Statistics

treat	1	2
1	3.19123	-3.19123
2	-3.19123	3.19123

Covariance Matrix for the Wilcoxon Statistics

treat	1	2
1	3884.00	-3884.00
2	-3884.00	3884.00

Test of Equality over Strata

Test	Chi-Square	DF	Pr > Chi-Square
Log-Rank	5.4943	1	0.0191
Wilcoxon	4.3512	1	0.0370
-2Log(LR)	5.6708	1	0.0172

```
proc lifetest method=lt plots=(s);
time tsurv*censor(0);
strata treat;
run;
The LIFETEST Procedure
```

Stratum 1: treat = 1

Life Table Survival Estimates

Interval [Lower,	Upper)	Number Failed	Number Censored	Effective Sample Size	Conditional Probability of Failure	Conditional Probability Standard Error	Survival
0	20	1	0	36.0	0.0278	0.0274	1.0000
20	40	4	0	35.0	0.1143	0.0538	0.9722
40	60	5	0	31.0	0.1613	0.0661	0.8611
60	80	5	0	26.0	0.1923	0.0773	0.7222
80	100	1	0	21.0	0.0476	0.0465	0.5833
100	120	0	0	20.0	0	0	0.5556
120	140	0	7	16.5	0	0	0.5556
140	160	0	8	9.0	0	0	0.5556
160	180	0	3	3.5	0	0	0.5556

```
180        .      0      2     1.0         0            0      0.5556
```

Interval [Lower,	Upper)	Failure	Survival Standard Error	Median Residual Lifetime	Median Standard Error
0	20	0	0	.	.
20	40	0.0278	0.0274	.	.
40	60	0.1389	0.0576	.	.
60	80	0.2778	0.0747	.	.
80	100	0.4167	0.0822	.	.
100	120	0.4444	0.0828	.	.
120	140	0.4444	0.0828	.	.
140	160	0.4444	0.0828	.	.
160	180	0.4444	0.0828	.	.
180	.	0.4444	0.0828	.	.

Evaluated at the Midpoint of the Interval

Interval [Lower,	Upper)	PDF	PDF Standard Error	Hazard	Hazard Standard Error
0	20	0.00139	0.00137	0.001408	0.001408
20	40	0.00556	0.00262	0.006061	0.003025
40	60	0.00694	0.00288	0.008772	0.003908
60	80	0.00694	0.00288	0.010638	0.004731
80	100	0.00139	0.00137	0.002439	0.002438
100	120	0	.	0	.
120	140	0	.	0	.
140	160	0	.	0	.
160	180	0	.	0	.
180	.	.	.	.	.

The LIFETEST Procedure

Stratum 2: treat = 2

Life Table Survival Estimates

Interval [Lower,	Upper)	Number Failed	Number Censored	Effective Sample Size	Conditional Probability of Failure	Conditional Probability Standard Error	Survival
0	20	0	0	9.0	0	0	1.0000
20	40	3	0	9.0	0.3333	0.1571	1.0000
40	60	1	0	6.0	0.1667	0.1521	0.6667
60	80	2	0	5.0	0.4000	0.2191	0.5556
80	100	1	0	3.0	0.3333	0.2722	0.3333
100	120	1	0	2.0	0.5000	0.3536	0.2222
120	140	0	0	1.0	0	0	0.1111
140	160	0	1	0.5	0	0	0.1111

Interval [Lower,	Upper)	Failure	Survival Standard Error	Median Residual Lifetime	Median Standard Error
0	20	0	0	65.0000	15.0000
20	40	0	0	45.0000	15.0000
40	60	0.3333	0.1571	40.0000	12.2474
60	80	0.4444	0.1656	30.0000	22.3607
80	100	0.6667	0.1571	30.0000	17.3205
100	120	0.7778	0.1386	20.0000	14.1421
120	140	0.8889	0.1048	.	.
140	160	0.8889	0.1048	.	.

Evaluated at the Midpoint of the Interval

Interval [Lower,	Upper)	PDF	PDF Standard Error	Hazard	Hazard Standard Error
0	20	0	.	0	.
20	40	0.0167	0.00786	0.02	0.011314
40	60	0.00556	0.00524	0.009091	0.009053
60	80	0.0111	0.00693	0.025	0.017116
80	100	0.00556	0.00524	0.02	0.019596
100	120	0.00556	0.00524	0.033333	0.031427
120	140	0	.	0	.
140	160	0	.	0	.

Summary of the Number of Censored and Uncensored Values

Stratum	TREAT	Total	Failed	Censored	Percent Censored
1	1	36	16	20	55.56

```
       2              2              9        8        1        11.11
-------------------------------------------------------------------------
   Total                           45       24       21        46.67
```

The LIFETEST Procedure

Testing Homogeneity of Survival Curves for tsurv over Strata

Rank Statistics

```
          treat        Log-Rank     Wilcoxon
          1             -4.1873      -130.00
          2              4.1873       130.00
```

Covariance Matrix for the Log-Rank Statistics

```
          treat             1               2
          1            3.19123        -3.19123
          2           -3.19123         3.19123
```

Covariance Matrix for the Wilcoxon Statistics

```
          treat             1               2
          1            3884.00        -3884.00
          2           -3884.00         3884.00
```

Test of Equality over Strata

```
                                            Pr >
        Test        Chi-Square     DF     Chi-Square
        Log-Rank      5.4943        1       0.0191
        Wilcoxon      4.3512        1       0.0370
        -2Log(LR)     5.6708        1       0.0172
```

The product limit method and the life table method give the same results for the test of the homogeneity of survival curves for the two groups. All the three tests indicate that the null hypothesis of homogeneity curves for the two groups is not tenable (p-value $< .05$). Hence, we can say that the survival curves of the two groups significantly differ, with group 1 more likely to survive longer than group 2, with mean survival times being 70.94 and 66 days respectively (note that these estimates are underestimated because the largest observations in both groups are censored). We present in Figure 12.5 the survival curves for both groups under both methods of estimation.

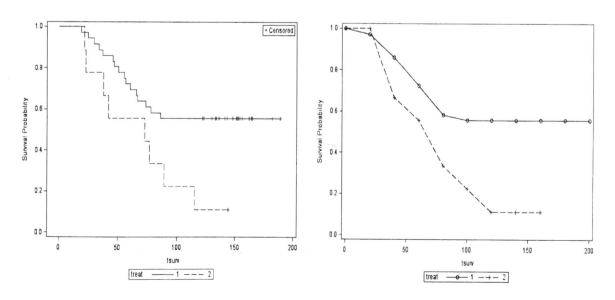

Figure 12.5: Plot of estimated survival curves under both K-M and LT methods

12.5   Hazard Function

If we let T be a random variable denoting the time of event occurrence, then, the hazard function, denoted by $h(t)$ is the probability that a subject experiences the event of interest in a small interval Δt, given that the subject has survived to the beginning of this interval. In other words, the hazard function is the conditional probability of experiencing the event between time t and $t + \Delta t$, given that the subject survived to at least time t. We can express this mathematically as

$$h(t) = \lim_{\Delta t \to 0} \frac{W(t, t + \Delta t)}{\Delta t}, \tag{12.2}$$

where

$$W(t, t + \Delta t) = \Pr(t < T < t + \Delta t \mid T \geq t).$$

The hazard function is sometimes referred to as the *conditional failure rate* in reliability, the instantaneous failure rate, or the age-specific failure rate in epidemiology, the force of mortality in demography or simply as the hazard rate function.

Some Basic Properties of $h(t)$

- $h(t) \geq 0$ and can be greater than 1. Thus, $h(t)$ is not a probability.

- If $h(t)$ is constant over time t, then $E(T) = 1/h(t)$. For instance, if $h(t) = 0.32$ for all t, and time is measured in months, then $1/0.32 = 3.125$ days is the expected length of time until the event occurs.

- $h(t) = $ (No of occurrence of events)/(Unit of time), where unit of time could be days, weeks, months, or years.

12.5.1   Types of Hazard Functions

The hazard function is related to the survival function with the following expression

$$S(t) = \exp\left(-\int_0^t h(u)\, du\right). \tag{12.3}$$

With $S(t)$ defined above, we have,

$$h(t) = \frac{f(t)}{S(t)} = -\frac{d}{dt} \ln[S(t)].$$

Thus,

$$f(t) = h(t) \exp\left(-\int_0^t h(u)\, du\right).$$

Similarly, the cumulative hazard function $H(t)$ is defined as

$$H(t) = \int_0^t h(u)\, du = -\ln[S(t)].$$

Thus,

$$S(t) = \exp[-H(t)].$$

The integral term in (12.3) is referred to as the **integrated hazard**. We now consider briefly, the following hazard functions: the exponential, the Gompertz and the Weibull models.

(a) The Exponential Model

The Exponential or **constant hazard** model has the random variable T being distributed exponentially. Thus, if $h(t) = \lambda$ for all t, then, the probability distribution for T is given by

$$f(t) = \lambda \exp(-\lambda t). \tag{12.4}$$

(b) **The Gompertz Model**

The Gompertz model has

$$\ln h(t) = \mu + \alpha\, t,$$

where α can be positive or negative. Thus,

$$h(t) = \exp(\mu + \alpha\, t) = \lambda_0 \exp(\alpha\, t),$$

with $\lambda_0 = e^\mu$. That is, when $t = 0$, then $h(t) = e^\mu$.

(c) **The Weibull Model**

Here, the model becomes

$$\ln h(t) = \mu + \alpha \ln t, \quad \text{where } \alpha > -1,$$

and therefore

$$h(t) = \lambda_0 t^\alpha, \quad \text{with } \lambda_0 = e^\mu.$$

Suppose we now have explanatory variables $x_1, x_2, \ldots, x_p$, then, including the explanatory variables in each of the above models we would have

$$\ln h(t) = \beta_0 + \beta_1 x_1 + \beta_2 x_2 + \cdots + \beta_p x_p, \qquad \text{(Exponential)} \qquad (12.5a)$$
$$\ln h(t) = \beta_0 + \beta_1 x_1 + \beta_2 x_2 + \cdots + \beta_p x_p + \alpha\, t, \qquad \text{(Gompertz)} \qquad (12.5b)$$
$$\ln h(t) = \beta_0 + \beta_1 x_1 + \beta_2 x_2 + \cdots + \beta_p x_p + \alpha \ln t. \qquad \text{(Weibull)} \qquad (12.5c)$$

The expressions in (12.5) are often described as *accelerated failure time* (AFT) models and are members of the family of models known as the *proportional hazard* models. In general, an accelerated failure time model with covariates can be written in the form

$$\ln T = \beta_0 + \underbrace{\beta_1 x_1 + \beta_2 x_2 + \cdots + \beta_p x_p}_{\text{covariate parameters}} + \sigma\varepsilon \qquad (12.6a)$$

$$= \beta \mathbf{X} + \sigma\varepsilon, \qquad (12.6b)$$

where σ is the shape parameter and ε is the error distribution. Other AFT models are the following:

The log-normal

It has $h(t) = 0$ when $t = 0$. Examples of data best described for this model are (i) change of jobs, (ii) change of residence or (iii) change of promotion within a job. The log-normal hazard function increases to a maximum, then decreases asymptotically to zero as t tends to infinity. For the log-normal, ε has a standard normal distribution. That is,

$$f(\varepsilon) = \frac{e^{-\frac{\varepsilon^2}{2}}}{\sqrt{2\pi}}.$$

The log-logistic

This has

$$S(t) = \frac{1}{1 + \lambda t^\alpha}, \quad \text{where} \quad \alpha = 1/\sigma \quad \text{and} \quad \lambda = \exp\left(-\beta_0/\sigma\right).$$

Further, if $\sigma = 1$, the log-logistic behaves like the log-normal. If $\sigma > 1$, then $h(t)$ is a monotone decreasing function from infinity and if $\sigma > 1$, $h(t)$ is a monotone decreasing function from a constant. The ε in this model has a logistic distribution. That is,

$$f(\varepsilon) = \frac{e^\varepsilon}{(1 + e^\varepsilon)^2}.$$

The Generalized Gamma

This distribution is very useful because the exponential, the Weibull and the log-normal are special cases of this distribution. Thus, it is very appropriate for finding alternative models. For this distribution, ε has the following probability density function

$$f(\varepsilon) = \frac{\mid \theta \mid [\exp(\theta\varepsilon)/\theta^2]^{(1/(\theta^2))} \exp[-\exp(\theta\varepsilon)/\theta^2]}{\Gamma(1/\theta^2)}, \quad -\infty < \varepsilon < \infty. \tag{12.7}$$

Thus,

1. if $\theta = 1$, then (12.7) becomes

$$f(\varepsilon) = \frac{[\exp(\varepsilon)] \exp[-\exp(\varepsilon)]}{\Gamma(1)}, \quad -\infty < \varepsilon < \infty$$
$$= \exp(\varepsilon - e^\varepsilon), \quad \text{the extreme value distribution.}$$

The model in this case reduces to the Weibull distribution.

2. If $\theta = 0$, then the model reduces to the log-normal distribution.

To implement any of the above AFT models, SAS **proc lifereg** will be used.

12.5.2   Example: The HMO data

The following data from Hosmer and Lemeshow (1999, p. 4) was reproduced by permission of John Wiley and Sons, Inc. The data relates to a large HMO wishing to evaluate the survival time of HIV+ members using a follow up study. The study started enrolling subjects from January 1, 1989 to December 31, 1991. The study ended on December 31, 1995. The data consist of 100 subjects. The variables are

- time: the follow-up time in months between the entry date and the end date

- age: the age of the subject at the start of the follow-up in years

- drug: history of prior intravenous (IV) drug usage (1 = yes, 0 = no)

- censor: status of subject at the end of study (1 = death due to AIDS, 0 = lost to follow-up or still alive)

We present in Table 12.6 the data for the first five and the last five subjects in the study. The original data has the actual date of entry into the study and the end date. The full data is presented in the accompanying CD as hiv.dat.

ID	Time	Age	Drug	Censor
1	5	46	0	1
2	6	35	1	0
3	8	30	1	1
4	3	30	1	1
5	22	36	0	1
⋮	⋮	⋮	⋮	⋮
96	1	34	1	1
97	5	28	0	1
98	60	29	0	0
99	2	35	1	0
100	1	34	1	1

Table 12.6: Data for the HMO HIV+ study

The original data carries the entry date and the end date in the Julian format (that is, day, month, year). We have entered these dates in SAS and again use SAS functions **intck** and **round** to compute the months. The latter function seems to give the same number of months as contained in the data presented in Hosmer and Lemeshow (1999, p. 4). We present the SAS program and output for the first five and last five subjects in the study.

```
data surv1;
input subj $ stdate lsdate;
informat stdate lsdate ddmmyy8.;
format stdate lsdate ddmmyy8.;
datalines;
1 15/05/90 14/10/90
2 19/09/89 20/03/90
3 21/04/91 20/12/91
4 03/01/91 04/04/91
5 18/09/89 19/07/91
96 02/08/91 01/09/91
97 22/05/91  21/10/91
98 02/04/90 01/04/95
99 01/05/91 30/06/91
100 11/05/89 10/06/89
;
data new;
set surv1;
month=-intck('month', lsdate, stdate);
dur=lsdate-stdate;
month2=round(dur/30.25);
format stdate lsdate ddmmyy8.;
title 'Days Between Project Start and Project End';
run;
proc print noobs;
run;
        Days Between Project Start and Project End
```

subj	stdate	lsdate	month	dur	month2
1	15/05/90	14/10/90	5	152	5
2	19/09/89	20/03/90	6	182	6
3	21/04/91	20/12/91	8	243	8
4	03/01/91	04/04/91	3	91	3
5	18/09/89	19/07/91	22	669	22
96	02/08/91	01/09/91	1	30	1
97	22/05/91	21/10/91	5	152	5
98	02/04/90	01/04/95	60	1825	60
99	01/05/91	30/06/91	1	60	2
100	11/05/89	10/06/89	1	30	1

To implement the accelerated failure time models for these data, we use SAS **proc lifereg**. The model we wish to fit is

$$\ln T = \beta_0 + \beta_1 \text{age}_i + \beta_2 \text{drug}_i + \sigma \varepsilon. \tag{12.8}$$

The SAS program and the partial output for implementing the model in (12.8) for the case when ε has the exponential distribution is displayed below.

```
data hmo;
input time age drug censor @@;
datalines;
5 46 0 1  6 35 1 0 8 30 1 1 3 30 1 1 22 36 0 1
1 32 1 0 7 36 1 1 9 31 1 1 3 48 0 1 12 47 0 1
2 28 1 0 12 34 0 1 1 44 1 1 15 32 1 1 34 36 0 1
1 36 0 1 4 54 0 1 19 35 0 0 3 44 1 0 2 38 0 1
2 40 0 0 6 34 1 1 60 25 0 0 11 32 0 1 2 42 1 0
5 47 0 1 4 30 0 0 1 47 1 1 13 41 0 1 3 40 1 1
2 43 0 1 1 41 0 1 30 30 0 1 7 37 0 1 4 42 1 1
8 31 1 1 5 39 1 1 10 32 0 1 2 51 0 1 9 36 0 1
36 43 0 1 3 39 0 1 9 33 0 1 3 45 1 1 35 33 0 1
8 28 0 1 11 31 0 1 56 20 1 0 2 44 0 0 3 39 1 1
15 33 0 1 1 31 0 1 10 33 0 1 1 50 1 1 7 36 1 1
3 30 1 1 3 42 1 1 2 32 1 1 32 34 0 1 3 38 1 1
10 33 0 0 11 39 1 1 3 39 1 1 7 33 1 1 5 34 1 1
31 34 0 1 5 46 1 1 58 22 0 1 1 44 1 1 3 37 0 0
43 25 0 1 1 38 0 1 6 32 0 1 53 34 0 1 14 29 0 1
4 36 1 1 54 21 0 1 1 26 1 1 1 32 1 1 8 42 0 1
5 40 1 1 1 37 1 1 1 47 0 1 2 32 1 1 7 41 1 0
1 46 1 0 10 26 1 1 24 30 0 0 7 32 1 1 12 31 1 0
4 35 0 1 57 36 0 1 1 41 1 1 12 36 1 0 7 35 1 1
1 34 1 1 5 28 0 1 60 29 0 0 2 35 1 0 1 34 1 1
```

```
;
proc print;
run;
proc lifereg data=hmo;
model time*censor(0)=age drug/dist=exponential;
run;
```

The LIFEREG Procedure

Model Information

```
              Data Set                    WORK.HMO
              Dependent Variable          Log(time)
              Censoring Variable            censor
              Censoring Value(s)                 0
              Number of Observations           100
              Noncensored Values                80
              Right Censored Values             20
              Left Censored Values               0
              Interval Censored Values           0
              Name of Distribution     Exponential
              Log Likelihood          -130.3970822

              Number of Observations Read      100
              Number of Observations Used      100
```

Algorithm converged.

Type III Analysis of Effects

		Wald	
Effect	DF	Chi-Square	Pr > ChiSq
age	1	32.3956	<.0001
drug	1	20.3276	<.0001

Analysis of Parameter Estimates

Parameter	DF	Estimate	Standard Error	95% Confidence Limits		Chi-Square	Pr > ChiSq
Intercept	1	6.1516	0.6062	4.9635	7.3398	102.98	<.0001
age	1	-0.0921	0.0162	-0.1238	-0.0604	32.40	<.0001
drug	1	-1.0099	0.2240	-1.4489	-0.5709	20.33	<.0001
Scale	0	1.0000	0.0000	1.0000	1.0000		
Weibull Shape	0	1.0000	0.0000	1.0000	1.0000		

Lagrange Multiplier Statistics

Parameter	Chi-Square	Pr > ChiSq
Scale	5.2252	0.0223

In the program, we specify in the model statement the dependent time variable, the censoring variable with the corresponding level, in this case, the zeros. On the right hand side of the model statement, we have specified the covariates with the relevant distribution. In this case, we are employing the exponential model. Similar results can be obtained for fitting the other AFT models by simply specifying the distribution in the model statement. For instance, the following will fit the Weibull, log-normal, log-logistic and the gamma AFT models to the data. We present in Table 12.7 the log-likelihood as well as the Akaike information criterion for each of these models.

```
model time*censor(0)=age drug/dist=weibull;
model time*censor(0)=age drug/dist=lognormal;
model time*censor(0)=age drug/dist=loglogistic;
model time*censor(0)=age drug/dist=gamma;
```

Model	Log likelihood	AIC
Exponential	−130.3971	266.7942
Weibull	−128.5023	265.0046
Log normal	−128.8328	265.6656
Log logistic	−129.1061	266.2122
Gamma	−127.5540	265.1080

Table 12.7: Results of fitting the AFT models to our data

To compute the AIC, since the AFT model implementations do not produce goodness-of-fit test statistics for the overall model fit, it has been suggested that we can compute the minimum Akaike information criterion (AIC) defined in this case as

$$\text{AIC} = -2 * \ln(\text{likelihood}) + 2(p + 1 + k), \tag{12.9}$$

where,

(i) p is the number of covariates in the model (in our case, $p = 2$)

(ii) k is defined as

$$k = \begin{cases} 0, & \text{if exponential} \\ 1, & \text{if Weibull, log-logistic or log-normal} \\ 2, & \text{if gamma.} \end{cases} \tag{12.10}$$

Thus, for the exponential, the AIC is computed as $-2(-130.3971) + 2(2 + 1 + 0) = 266.7942$. Similar calculations lead to the AIC values presented in Table 12.7. From Table 12.7, we notice that the Weibull model provides the best fit in terms of smallest AIC. The gamma model, though provides the largest log-likelihood, has an AIC that is higher than that of the Weibull model. Further, the Weibull model is a much simpler model than the gamma. A test of the hypothesis

$$H_0 : \sigma = 1$$
$$H_a : \sigma \neq 1,$$

which tests whether the Weibull scale equals 1 or not for the exponential can be tested by calculating

$$-2(\text{log-likelihood}_{\text{Weibull}} - \text{log-likelihood}_{\text{expo}}) = 2(-128.5023 + 130.3971) = 3.7892.$$

This statistic is distributed as χ^2 with 1 degree of freedom. The null hypothesis is rejected if $3.7892 > \chi^2_{\alpha, 1}$, the tabulated $\chi^2_{\alpha, 1}$ value in Table 3 of the Appendix. Since $\chi^2_{0.05, 1} = 3.841$, H_0 cannot be rejected. Alternatively, the test has a p-value of 0.0516 which is clearly not significant at $\alpha = 0.05$ level. Hence, it can be concluded that the exponential model provides as good a fit to this model as the slightly more complicated Weibull model. Thus, we would adopt the exponential model for this data. Based on the parameter estimates, the estimated AFT model under the exponential model is

$$\widehat{\log T} = 6.1516 - 0.0921\text{age} - 1.0099\text{drug}. \tag{12.11}$$

The percent change in a unit increase in the covariates are often computed as $100(e^{\hat{\beta}} - 1)$. Thus in our example, the percentage changes are:

$$100(e^{-0.0921} - 1) = -8.8\%, \quad 100(e^{-1.0099} - 1) = -63.6\%.$$

In other words,

- each additional age of a subject (controlling for intravenous drug) at entering the study decreases the survival time by 8.8%.

- subjects on intravenous drug can expect their survival times to be decreased by about 63.6% or put in another way, subjects on intravenous drug are $e^{-1.0099} = 0.36$ times more likely than those not on intravenous drug to survive longer.

12.6 Proportional Hazards Model

Cox (1972) proposed a general method for modeling the hazard function $h(t)$ which unlike the AFT models discussed earlier, allows time dependent covariates. Cox general method can be written as

$$\log h(t) = \log \alpha(t) + \beta_1 x_1 + \beta_2 x_2 + \cdots + \beta_p x_p, \tag{12.12}$$

where $\alpha(t)$ is any function of t. The result in (12.12) can also be written in the form

$$h(t) = \alpha(t) \exp\left(\beta_1 x_1 + \beta_2 x_2 + \cdots + \beta_p x_p\right). \tag{12.13}$$

Clearly, $\alpha(t)$ can be assumed as the baseline hazard function when all the explanatory variables are zero. The expression in (12.12) is called the proportional odds model for the following reason. Suppose we take two individuals with hazards $h_1(t)$ and $h_2(t)$ respectively. Then using the form in (12.13), we have

$$\frac{h_1(t)}{h_2(t)} = \frac{\alpha(t)\exp\left(\beta_1 x_{11} + \beta_2 x_{12} + \cdots + \beta_p x_{1p}\right)}{\alpha(t)\exp\left(\beta_1 x_{21} + \beta_2 x_{22} + \cdots + \beta_p x_{2p}\right)} = \frac{\exp\left(\beta_1 x_{11} + \beta_2 x_{12} + \cdots + \beta_p x_{1p}\right)}{\exp\left(\beta_1 x_{21} + \beta_2 x_{22} + \cdots + \beta_p x_{2p}\right)}$$

$$= e^{\beta_1(x_{11}-x_{21})}e^{\beta_2(x_{12}-x_{22})}\cdots e^{\beta_p(x_{1p}-x_{2p})},$$

which as we can see does not depend on time and is therefore constant. The parameters of the Cox's proportional hazard model is estimated by the method of partial likelihood which are not presented here. SAS **proc phreg** can be used to implement the Cox proportional hazard model. We now apply this model to the data in Table 12.6

```
set hmo;
proc phreg data=hmo;
model time*censor(0)=age drug;
run;
```

                         The PHREG Procedure

                         Model Information

                Data Set   WORK.HMO
                Dependent Variable   time
                Censoring Variable   censor
                Censoring Value(s)   0
                Ties Handling   BRESLOW

            Number of Observations Read   100
            Number of Observations Used   100

        Summary of the Number of Event and Censored Values

                                              Percent
            Total   Event   Censored   Censored
            100   80   20   20.00

                         Convergence Status

        Convergence criterion (GCONV=1E-8) satisfied.

                    Model Fit Statistics

                         Without   With
            Criterion   Covariates   Covariates
            -2 LOG L   598.390   563.408
            AIC   598.390   567.408
            SBC   598.390   572.172

            Testing Global Null Hypothesis: BETA=0

        Test   Chi-Square   DF   Pr > ChiSq
        Likelihood Ratio   34.9819   2   <.0001
        Score   34.3038   2   <.0001
        Wald   32.4859   2   <.0001

            Analysis of Maximum Likelihood Estimates

                     Parameter   Standard   Hazard
Variable   DF   Estimate   Error   Chi-Square   Pr > ChiSq   Ratio
age   1   0.09151   0.01849   24.5009   <.0001   1.096
drug   1   0.94108   0.25550   13.5662   0.0002   2.563
```

The parameter estimates for age and drug are respectively 0.09151 and 0.94108.

- Thus, for age, the hazard ratio for a unit increase in age is $e^{0.09151} = 1.1$. That is, keeping drug level constant, the hazard ratio of 1.1 indicates that an individual is 1.1 times more likely to die during the study for a unit increase in age.

- On the other hand, for a given age, an individual with a history of drug use would die at about $e^{0.94108} = 2.56$ the rate of an individual without the drug history.

- Also for a five year increase in age, an individual with $x + 5$ years of age would die at about $e^{5(0.09151)} = 1.8$ the rate of an individual with $x$ number of years in age after adjusting for the effect of drug or for a given drug level.
```

For two given ages, say, 35 and 45, we can compute the survival functions for the two drug categories using the following SAS program. In the program (just for age 35), we first create an input covariate file called **new**, which will appear in the baseline command as an input. Lower and upper confidence limits for the survival function are generated with options **lower** and **upper** respectively. The **method=ch** requests the empirical cumulative hazard function estimates to be computed. The **cltype=loglog** specifies that normal theory be employed to compute the confidence intervals for $\log(-\log(S(t, z)))$. A partial output up to time $= 10$ for both drug levels at age 35 is presented. There are 28 such observations for each drug level.

```
set hmo;
data new;
input age drug;
datalines;
35 1
35 0
;
run;
proc phreg data=hmo;
model time*censor(0)=age drug ;
baseline  out=aa survival=s lower=lower_C upper=upper_C
        covariates=new/method = ch nomean cltype = loglog;
run;
proc print data=aa;
run;
proc sort data= aa;
  by time;
run;
symbol1 color=b i=stepj1 line=1 height=.03in;
symbol2 color=b i=stepj1 line=4 height-.03in,
axis1 order = (0 to 1 by .2) label=(a= 90 'Estimated Survival Function, S(t)') minor = none;
axis2 order = (0 to 60 by 10) label = ('Months') minor = none;
proc gplot data =aa;
  plot s*time=drug/vaxis = axis1 haxis = axis2 noframe;
run;
```

Obs	age	drug	time	s	lower_C	upper_C
1	35	1	0	1.00000	.	.
2	35	1	1	0.84875	0.75278	0.90965
3	35	1	2	0.78812	0.67791	0.86429
4	35	1	3	0.66329	0.53572	0.76335
5	35	1	4	0.60299	0.46843	0.71360
6	35	1	5	0.49673	0.35890	0.62017
7	35	1	6	0.46127	0.32320	0.58855
8	35	1	7	0.36210	0.23004	0.49549
9	35	1	8	0.28924	0.16375	0.42722
10	35	1	9	0.23844	0.12192	0.37655
11	35	1	10	0.19270	0.08803	0.32767
29	35	0	0	1.00000	.	.
30	35	0	1	0.93801	0.88340	0.96751
31	35	0	2	0.91127	0.84369	0.95048
32	35	0	3	0.85198	0.75939	0.91098
33	35	0	4	0.82087	0.71719	0.88940
34	35	0	5	0.76107	0.63999	0.84618
35	35	0	6	0.73939	0.61314	0.82997
36	35	0	7	0.67275	0.53464	0.77808
37	35	0	8	0.61628	0.47283	0.73138
38	35	0	9	0.57154	0.42593	0.69304
39	35	0	10	0.52596	0.37990	0.65276

Figure 12.6 displays the plot of the survival functions for both drug levels given that the age is 35. Similarly, Figure 12.7 displays a similar graph for a given age at 44 years.

12.7 Truncated Regression Model

Before we fully discuss truncated regression models, first let us examine the effect of truncation on the distributional assumption of the linear model. We say that a random variable X with probability density function $f(x)$ is truncated if $f(x)$ is only defined for $x > a$ or it could be for $x < a$. In this case, we would have the following:

$$f(x \mid x > a) = \frac{f(x)}{\Pr(X > a)}.$$

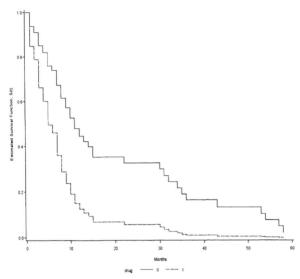

Figure 12.6: Plot of estimated survival function against time (age=35)

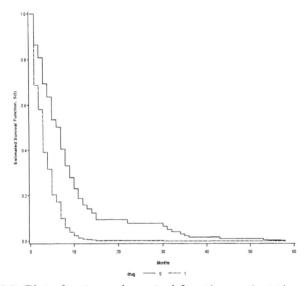

Figure 12.7: Plot of estimated survival function against time (age=44)

In this case,

$$E(X \mid x > a) > E(X),$$
$$\mathrm{Var}(X \mid x > a) < \mathrm{Var}(X).$$

If X has the normal distribution, then, $X \sim N(\mu, \sigma^2)$ and the expressions for the mean and variance become

$$E(X \mid x > a) = \mu + \sigma\lambda(\alpha), \tag{12.14}$$

and

$$\mathrm{Var}(X \mid x > a) = \sigma^2[1 - \delta(\alpha)], \quad 0 < \delta(\alpha) < 1, \tag{12.15}$$

where, for expressions in (12.14) and (12.15), we have,

$$\alpha = \frac{a - \mu}{\sigma}, \quad \lambda(\alpha) = \frac{\phi(\alpha)}{1 - \Phi(\alpha)} \quad \text{and} \quad \delta(\alpha) = \lambda(\alpha)\left[\lambda(\alpha) - 1\right].$$

For truncated data, the sample size is reduced because part of the observation is assumed to be missing. Thus an OLS estimation conducted on the truncated data would only be valid for those subjects satisfying the condition $x > a$. However, if we were to apply our results to the general population then the OLS will give us misleading results and in this case, we would need to employ the techniques developed for estimating such models which are often referred to as **tobit models**. Generally, if interest is in the sub-population having observations y_i such that $x_i > a$, then we would employ the OLS method of estimation. However, if interest is on the entire population, then the method of maximum likelihood estimation will be employed.

Example: Job Prestige

This example employs the data mentioned by Long (1997, p. 19) and reproduced by permission of the author. The data relate to the prestige of a scientist's first academic job for a sample of biochemists. The dependent variable is the prestige of the job (job) which is rated on a continuous scale of 1.00 to 5.00, where (1.00-1.99) represents adequate school, (2.00-2.99) represents good school, (3.00-3.98) represents strong school, and (3.99+) represents distinguished school. For cases where prestige job values were not available for departments without graduate programs or for graduate programs, where programs are rated below adequate, the prestige values are coded 1.0 in these cases. The response variable is job and the predictor variables in the data are: gender (Male or Female), phd (Prestige of Ph.D. department), ment (citations received by mentors), fel (fellowship status, NFL-non Fellows and FL-Fellows), art (number of articles published) and cit (number of citations received).

The data has a total of 408 observations. For those cases where the prestige values were not available, these cases were coded ones, with the ones representing some 'true' ones, and others representing censored values that are less than one but whose 'true' values are unknown. A model of the following form is of interest:

$$y_i = \beta_0 + \beta_1 \text{gender}_i + \beta_2 \text{phd}_i + \beta_3 \text{ment}_i + \beta_4 \text{fel}_i + \beta_5 \text{art}_i + \beta_6 \text{cit}_i + \varepsilon_i. \tag{12.16}$$

Here,

$$\text{fem} = \begin{cases} 1, & \text{if Female} \\ 0, & \text{if Male,} \end{cases} \qquad \text{fel} = \begin{cases} 1, & \text{if FL} \\ 0, & \text{if NFL.} \end{cases}$$

Our initial analysis ignored the censoring values and fits a multiple regression model using the ordinary least squares approach to the entire data resulting in the following output. We have reproduced the first 10 observations in the data.

```
data new;
infile "c:\reg\poisson\tobit2.txt";
input job gender$ phd ment felw$ art cit;
if gender='Female' then fem=1;
else fem=0;
if felw='NFL' then fel=0;
else fel=1;
proc print data=new (obs=10) noobs;
run;
proc means data=new;
title 'descriptive statistics for data';
var job;
run;
proc reg;
model job=fem phd ment fel art cit/stb clb;
run;
```

job	gender	phd	ment	felw	art	cit	fem	fel
1.89	Male	4.29	89	FL	4	43	0	1
3.52	Male	3.52	532	FL	1	1	0	1
4.64	Male	4.64	146	FL	1	5	0	1
1.94	Male	1.81	0	NFL	1	4	0	0
2.52	Male	3.59	37	FL	4	57	0	1
1.00	Male	4.29	1	NFL	0	0	0	0
1.29	Male	1.80	4	FL	6	60	0	1
3.84	Male	4.29	0	NFL	1	1	0	0
2.11	Male	2.11	11	FL	7	22	0	1
1.83	Male	4.34	52	NFL	2	37	0	0

descriptive statistics for data

The MEANS Procedure

Variable	N	Mean	Std Dev	Minimum	Maximum
job	408	2.2334314	0.9736029	1.0000000	4.8000000

The REG Procedure
Model: MODEL1
Dependent Variable: job

Number of Observations Read   408
Number of Observations Used   408

Analysis of Variance

Source	DF	Sum of Squares	Mean Square	F Value	Pr > F
Model	6	81.05848	13.50975	17.78	<.0001
Error	401	304.73792	0.75994		
Corrected Total	407	385.79640			

Root MSE	0.87175	R-Square	0.2101
Dependent Mean	2.23343	Adj R-Sq	0.1983
Coeff Var	39.03179		

Parameter Estimates

Variable	DF	Parameter Estimate	Standard Error	t Value	Pr > |t|	Standardized Estimate
Intercept	1	1.06718	0.16614	6.42	<.0001	0
fem	1	-0.13919	0.09023	-1.54	0.1237	-0.06981
phd	1	0.27268	0.04932	5.53	<.0001	0.26712
ment	1	0.00119	0.00070116	1.69	0.0913	0.07987
fel	1	0.23414	0.09482	2.47	0.0140	0.11701
art	1	0.02280	0.02888	0.79	0.4303	0.05284
cit	1	0.00448	0.00197	2.28	0.0234	0.15208

Parameter Estimates

Variable	DF	95% Confidence Limits	
Intercept	1	0.74058	1.39379
fem	1	-0.31659	0.03820
phd	1	0.17573	0.36964
ment	1	-0.00019171	0.00257
fel	1	0.04773	0.42055
art	1	-0.03398	0.07958
cit	1	0.00060865	0.00835

Next, we fit the same regression model, again using OLS, but on the reduced sample after dropping out all the cases that have been censored to one. Thus we have truncated data by dropping all prestige ratings less than 1. The resulting analysis is again presented in the following output. This is accomplished in SAS with the **where job ne 1** in the SAS code. The resulting analysis indicates that there are now 309 observations in the sample data and the analysis is based on these 309 observations in this case.

```
set new;
proc reg;
where job ne 1;
model job=fem phd ment fel art cit/stb clb;
run;
```

The REG Procedure
Model: MODEL1
Dependent Variable: job

Number of Observations Read   309
Number of Observations Used   309

Analysis of Variance

Source	DF	Sum of Squares	Mean Square	F Value	Pr > F
Model	6	37.63651	6.27275	12.69	<.0001
Error	302	149.29099	0.49434		
Corrected Total	308	186.92750			

Root MSE	0.70309	R-Square	0.2013
Dependent Mean	2.62861	Adj R-Sq	0.1855

```
        Coeff Var           26.74776
```

Parameter Estimates

Variable	DF	Parameter Estimate	Standard Error	t Value	Pr > |t|	Standardized Estimate
Intercept	1	1.41278	0.16214	8.71	<.0001	0
fem	1	0.10145	0.08548	1.19	0.2362	0.06178
phd	1	0.29738	0.04675	6.36	<.0001	0.35402
ment	1	0.00077836	0.00061135	1.27	0.2039	0.06949
fel	1	0.14053	0.08979	1.57	0.1186	0.08496
art	1	0.00590	0.02483	0.24	0.8124	0.01817
cit	1	0.00210	0.00166	1.27	0.2049	0.09759

Parameter Estimates

Variable	DF	95% Confidence Limits	
Intercept	1	1.09372	1.73185
fem	1	-0.06677	0.26967
phd	1	0.20539	0.38937
ment	1	-0.00042468	0.00198
fel	1	-0.03617	0.31723
art	1	-0.04296	0.05476
cit	1	-0.00115	0.00536

In our next analysis, we would now employ Tobit regression on the data. The advantage of Tobit regression over the former analysis is that the method utilizes all the observations in the data, including the censored observations. We shall employ **proc lifereg** to accomplish this after some suitable transformation. To fit the Tobit model in SAS, we need to generate two dependent variables 'tobit' and 'job'. The latter is our original dependent variable. If the variable is censored, that is job = 1, then tobit = ., otherwise, tobit=job.

```
data new1;
set new;
if job = 1 then tobit=.;
else tobit=job;
run;
proc lifereg data=new1;
model (tobit, job)=fem phd ment fel art cit/ d=normal;
run;
```

The LIFEREG Procedure

Model Information

Data Set	WORK.NEW1
Dependent Variable	tobit
Dependent Variable	job
Number of Observations	408
Noncensored Values	309
Right Censored Values	0
Left Censored Values	99
Interval Censored Values	0
Name of Distribution	Normal
Log Likelihood	-560.2520876

```
        Number of Observations Read        408
        Number of Observations Used        408
```

Algorithm converged.

Type III Analysis of Effects

Effect	DF	Wald Chi-Square	Pr > ChiSq
fem	1	4.1272	0.0422
phd	1	25.4668	<.0001
ment	1	2.2917	0.1301
fel	1	7.0551	0.0079
art	1	0.8628	0.3530
cit	1	4.2289	0.0397

Analysis of Parameter Estimates

Parameter	DF	Estimate	Standard Error	95% Confidence Limits		Chi-Square	Pr > ChiSq
Intercept	1	0.6854	0.2183	0.2576	1.1132	9.86	0.0017
fem	1	-0.2368	0.1166	-0.4654	-0.0083	4.13	0.0422
phd	1	0.3226	0.0639	0.1973	0.4479	25.47	<.0001
ment	1	0.0013	0.0009	-0.0004	0.0031	2.29	0.1301

```
fel      1    0.3253   0.1225    0.0853   0.5653   7.06    0.0079
art      1    0.0339   0.0365   -0.0376   0.1054   0.86    0.3530
cit      1    0.0051   0.0025    0.0002   0.0099   4.23    0.0397
Scale    1    1.0872   0.0465    0.9997   1.1824
```

The output tells us that

- Our results agree with those presented in Long (1997) using STATA.

- We have 408 observations, with 99 observations left censored.

- The log-likelihood is -560.252088.

- The scale parameter is 1.0872 and corresponds to the standard deviation of the Tobit model

- The variables gender, phd, fellowship and citation numbers are statistically significant at the 5% level.

The model employed above fits the Tobit model to only the left censored data. Some times we wish to extend the model to fit Tobit model to both left censored and right censored data. The following SAS codes and relevant output provides this fit to the data.

```
if job=1 then tobit1=.;
else if job=4.8 then tobit1=4.8;
else tobit1=job;
if job=1 then tobit2=0;
else if job=4.8 then tobit2=.;
else tobit2=job;

proc lifereg data=new1;
model (tobit1, tobit2)=fem phd ment fel art cit/ d=normal;
run;
```

```
                    The LIFEREG Procedure

                    Model Information

        Data Set                        WORK.NEW1
        Dependent Variable                 tobit1
        Dependent Variable                 tobit2
        Number of Observations                408
        Noncensored Values                    308
        Right Censored Values                   1
        Left Censored Values                   99
        Interval Censored Values                0
        Name of Distribution               Normal
        Log Likelihood              -668.9393552

        Number of Observations Read           408
        Number of Observations Used           408
```

```
Algorithm converged.
```

```
                Type III Analysis of Effects

                              Wald
        Effect    DF    Chi-Square    Pr > ChiSq
        fem        1        6.0026        0.0143
        phd        1       17.9147        <.0001
        ment       1        1.8807        0.1703
        fel        1        6.9722        0.0083
        art        1        1.1480        0.2840
        cit        1        3.4964        0.0615
```

```
              Analysis of Parameter Estimates

                        Standard   95% Confidence    Chi-
    Parameter DF Estimate   Error       Limits      Square Pr > ChiSq
    Intercept  1   0.1304   0.3015   -0.4606  0.7214   0.19    0.6654
    fem        1  -0.3957   0.1615   -0.7122 -0.0791   6.00    0.0143
    phd        1   0.3743   0.0884    0.2010  0.5476  17.91    <.0001
    ment       1   0.0017   0.0012   -0.0007  0.0041   1.88    0.1703
    fel        1   0.4476   0.1695    0.1154  0.7798   6.97    0.0083
    art        1   0.0544   0.0507   -0.0451  0.1538   1.15    0.2840
    cit        1   0.0064   0.0034   -0.0003  0.0132   3.50    0.0615
    Scale      1   1.5113   0.0656    1.3881  1.6456
```

In this analysis, we define

$$\text{tobit1} = \begin{cases} ., & \text{if job=1} \\ 4.8, & \text{if job=4.8} \;; \\ \text{job}, & \text{elsewhere}, \end{cases} \qquad \text{tobit2} = \begin{cases} 0, & \text{if job=1} \\ ., & \text{if job=4.8} \\ \text{job}, & \text{elsewhere}. \end{cases}$$

The (.) implies that the dependent variable $\text{tobit}_i, i = 1, 2$ is considered missing. Thus tobit1 assumes (.) when it is left censored while tobit2 assumes (.) when it is right censored. The output in this case indicates that there are 99 left censored observations and 1 right censored observation. The effect of gender is more pronounced in this case, while the effect of cit (citation) does not seem to be important any more.

Example 12.2

We give below a simulated example involving 200 random samples from a normal population satisfying

$$y_i = 20 + 2x_i + \varepsilon_i, \quad i = 1, 2, 3, \dots, 200,$$

where $\varepsilon_i \sim N(0, 1)$. We present the first ten observations from this simulation study in the corresponding SAS output that follows. Here, from the **proc means** result, $0.325 < x < 99.814$, while $19.275 < y < 218.808$.

```
data simul;
seed = 123456;
do i=1 to 200;
x=100*(uniform(seed));
y=20+2*x+probit(uniform(seed));
output;
drop seed i;
end;
run;
proc means;
var x y;
run;
                    The MEANS Procedure

  Variable    N        Mean        Std Dev       Minimum        Maximum
  -------------------------------------------------------------------------
  x          200    49.9432718    29.0477951     0.3249454     99.8144010
  y          200   119.8030920    57.9438998    19.2754519    218.8075789
  -------------------------------------------------------------------------
```

When the regression model

$$y_i = \beta_0 + \beta_1 x_i + \varepsilon_i,$$

is applied to the simulated data, we have the following results with corresponding 95% confidence intervals.

```
set simul;
proc reg;
model y=x/clb;
run;
                       The REG Procedure
                        Model: MODEL1
                    Dependent Variable: y

              Number of Observations Read        200
              Number of Observations Used        200

                    Analysis of Variance

                              Sum of        Mean
  Source              DF     Squares      Square    F Value    Pr > F
  Model                1      667922      667922     602486    <.0001
  Error              198   219.50493     1.10861
  Corrected Total    199      668142

          Root MSE             1.05291    R-Square    0.9997
          Dependent Mean     119.80309    Adj R-Sq    0.9997
          Coeff Var            0.87886

                    Parameter Estimates
```

Variable	DF	Parameter Estimate	Standard Error	t Value	Pr > |t|	95% Confidence Limits	
Intercept	1	20.19372	0.14836	136.11	<.0001	19.90115	20.48630
x	1	1.99445	0.00257	776.20	<.0001	1.98938	1.99952

The models fits very well and the estimated 95% confidence intervals contain the population parameters of the model. Now suppose that some of the dependent observations y_i are constrained such that they have to be greater than 80. Then this implies that we have to fit our regression model to all values of $y_i > 80$. That is, the data is now been segmented into two groups and clearly, there would be fewer observations in this reduced sample than the original data. This truncated regression model is implemented with the following SAS program and partial output. There are now 142 observations in the truncated data satisfying $y_i > 80$.

```
set simul;
proc reg;
where y >80;
model y=x/clb;
run;
```

```
                       Model: MODEL1
                   Dependent Variable: y

           Number of Observations Read        142
           Number of Observations Used        142

                    Analysis of Variance
```

Source	DF	Sum of Squares	Mean Square	F Value	Pr > F
Model	1	229878	229878	210365	<.0001
Error	140	152.98574	1.09276		
Corrected Total	141	230031			

Root MSE	1.04535	R-Square	0.9993
Dependent Mean	149.16717	Adj R-Sq	0.9993
Coeff Var	0.70079		

```
                    Parameter Estimates
```

Variable	DF	Parameter Estimate	Standard Error	t Value	Pr > |t|	95% Confidence Limits	
Intercept	1	20.18595	0.29458	68.52	<.0001	19.60355	20.76835
x	1	1.99442	0.00435	458.66	<.0001	1.98583	2.00302

12.8   Exercises

12.1 Suppose that one is interested in examining the survival times of individuals with leukemia. The following data are the times, in months, to remission of 20 such patients.

> 1.50, 1.50, 1.50, 1.50, 1.75,
> 2.25, 2.50, 2.50, 2.75, 3.25,
> 4.00, 4.25, 4.75, 5.00, 5.50,
> 5.75, 6.25, 8.00, 8.00, 8.50

(a) What is the median survival time?

(b) For fixed intervals of length two months, use the life table method to estimate the survival function $s(t)$.

(c) Is the life table cross-sectional or longitudinal? Explain.

(d) Construct a survival curve for this sample of patients.

12.2 The following data represent survival times, in months, for 11 lymphoma patients. Values with asterisks (*) denote censored observations: 1*, 3, 4*, 5, 5, 6*, 7, 7, 7*, 8*, 8

(a) What is the modal survival time?

(b) Use the Kaplan-Meier method to estimate the survival function $s(t)$.

(c) Construct a graph for the product-limit curve.

(d) Use the life table method to estimate the survival function $s(t)$.

12.3 Consider a clinical trial in which 10 patients are observed to have the following survival pattern (in months). The plus (+) values are patients who are lost to follow-up. The values are as follows: 1, 2, 3, 3+, 4, 4+, 5, 5+, 8, 9+

    (a) What is the median survival time?

    (b) Use the product-limit method to estimate the survival function $s(t)$.

    (c) Construct a survival curve for this sample of patients.

    (d) Use the life table method to estimate the survival function $s(t)$.

12.4 Two groups of rats with different pretreatment regimes were exposed to a certain type of carcinogen. The time to mortality from cancer in the two groups was recorded and asterisk (*) denotes censored observation.

Group	Time
1	143, 164, 188, 188, 190, 192, 206, 209, 213, 216, 216*, 220, 227, 230, 234, 244*, 246, 265, 304
2	142, 156, 163, 198, 204*, 205, 232, 232, 233, 233, 233, 233, 240, 261, 280, 280, 296, 296, 323, 344*

    (a) For each group, use the product-limit method to estimate the survival function $s(t)$.

    (b) For each group, construct a graph for the product-limit curve.

    (c) Carry out a test to compare the distributions of survival times for the two groups.

    (d) For each group, use the life table method to estimate the survival function $s(t)$.

12.5 The data in the following table are on two samples of 21 patients each, sample 1 was given an experimental drug and sample 2 was given a placebo. The times to remission of leukemia patients are given in weeks and values with asterisks (*) denote censored observations.

Sample	Time
1	6*, 6, 6, 6, 7, 9*, 10*, 10, 11*, 13, 16, 17*, 19*, 20*, 22, 23, 25*, 32*, 32*, 34*, 35*
2	1, 1, 2, 2, 3, 4, 4, 5, 5, 8, 8, 8, 8, 11, 11, 12, 12, 15, 17, 22, 23

    (a) What is the median survival time in each sample?

    (b) Use the product-limit method to estimate the survival function $s(t)$ for the two sets of patients.

    (c) Use the log-rank test to evaluate the null hypothesis that the distributions of survival times are identical in the two groups.

    (d) For each set of patients, use the life table method to estimate the survival function $s(t)$.

Chapter 13

Nonlinear Regression

13.1 Introduction

In the introductory chapter to this text, we briefly describe linear and nonlinear regression models. In this chapter, we will focus on nonlinear regression models, their estimation and other characteristics. Most biological or physical processes or phenomena often give rise to nonlinear models. Examples of such processes are growth responses, radioactive decay, etc.

In the above examples, the parameters of the nonlinear models have direct interpretations in terms of the processes under consideration. First let us distinguish between linear and nonlinear models. As indicated in the introductory chapter, a model is said to be linear in its parameters ($\boldsymbol{\beta}$) if it is of the form

$$y_i = \sum_{j=1}^{p} \beta_j x_{ij} + \varepsilon_i. \tag{13.1}$$

A very good distinction between linear and nonlinear models as a consequence of the linearization in parameters is that, a model will be said to be nonlinear, if the first-order derivatives of the model with respect to the model parameters depends on one or more parameters. For example, the quadratic model $y = \beta_0 + \beta_1 x + \beta_2 x^2 + \varepsilon$ is parabolic in shape, however, it is a linear model because its derivatives are

$$\frac{\partial y}{\partial \beta_0} = 1, \quad \frac{\partial y}{\partial \beta_1} = x \quad \text{and} \quad \frac{\partial y}{\partial \beta_2} = x^2. \tag{13.2}$$

None of the partial derivatives in (13.2) depends on the model parameter, and therefore the model is linear. On the other hand, consider the model

$$y = \beta_0 + \beta_2 e^{\beta_1 x} + \varepsilon. \tag{13.3}$$

The partial derivatives with respect to the parameters of the model are

$$\frac{\partial y}{\partial \beta_0} = 1, \quad \frac{\partial y}{\partial \beta_1} = \beta_2(x e^{\beta_1 x}) \quad \text{and} \quad \frac{\partial y}{\partial \beta_2} = e^{\beta_1 x}. \tag{13.4}$$

Clearly, some of these derivatives in (13.4) involve other model parameters, therefore the model in (13.3) is nonlinear. We can show that while the models in (13.5) are all linear, those in (13.6) are nonlinear.

$$\left.\begin{aligned}
y &= \beta_0 + \beta_1 x_1 + \varepsilon \\
y &= \beta_0 + \beta_1 x + \beta_2 x^2 + \varepsilon \\
y &= \beta_0 + \beta_1 x_1 + \beta_2 x_2 + \beta_3 x_2^2 + \beta_4 x_1 x_3 + \varepsilon \\
y &= \beta_0 + \beta_1 x_1^{5/2} + \beta_2 x_2 + \beta_3 e^{x_1 x_2} + \varepsilon
\end{aligned}\right\} \tag{13.5}$$

$$\left.\begin{aligned} y &= \beta_0 + \beta_2 e^{\beta_1 x} + \varepsilon \\ y &= \frac{\alpha}{1 + \beta e^{(-kx)}} + \varepsilon \\ y &= \alpha e^{[-\beta e^{-kx}]} + \varepsilon \\ y &= \frac{\alpha}{[1 + \beta e^{-kx}]^{1/\delta}} + \varepsilon \end{aligned}\right\} \tag{13.6}$$

The last three models in (13.6) are respectively, the logistic, the Gompertz and the Richards growth models. As an example of growth curves which are often nonlinear, most of these are of the exponential form and are often described as *exponential growth curves*. We present below plots of the one-term exponential models

$$y_i = \alpha e^{\beta x_i} + \varepsilon_i. \tag{13.7}$$

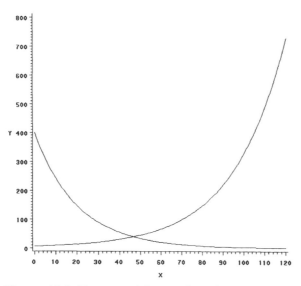

Figure 13.1: Exponential growth and decay curves

In the growth curve model in (13.7), if $\beta > 0$ then we have exponential growth that starts slowly and then accelerates. The plot in Figure 13.1 gives an exponential growth curve having

$$\hat{y}_i = 6e^{0.04\,x_i}. \tag{13.8}$$

On the other hand, if $\beta < 0$, then we have exponential decay, and again the plot in Figure 13.1 gives an exponential decay curve from top left to right down curve and having

$$\hat{y}_i = 400e^{-0.05\,x_i}. \tag{13.9}$$

The problem however with exponential models is their rapid and unlimited increase as x increases and this sometimes make them unrealistic for real life phenomena. An exponential model that includes limitation on growth is of the following form

$$y_i = \alpha(1 - e^{-\beta x_i}) + \varepsilon_i. \tag{13.10}$$

The model in (13.10) is often referred to as the *negative exponential* model and we present in Figure 13.2 the plots for $\alpha = 20$ and two β values of -0.06 and -0.03, with the $\beta = -0.06$ curve being steeper.

Other exponential curves include the *two-term exponential* curves which rise steeply from zero and then fall slowly to zero asymptotically. These curves are of the form

$$y_i = \frac{\gamma_1}{\gamma_1 - \gamma_2}(e^{-\gamma_2 x_i - \gamma_1 x_i}) + \varepsilon_i. \tag{13.11}$$

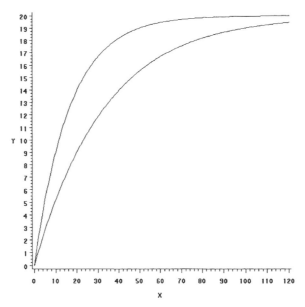

Figure 13.2: Negative exponential growth curves

By using equation (13.11), Figure 13.3 gives the two-term exponential curves with the higher curve having $\gamma_1 = 0.05$ and $\gamma_2 = 0.04$.

$$y_i = \frac{0.05}{0.05 - 0.04}(e^{-0.04x_i - 0.05x_i}) + \varepsilon_i,$$
$$y_i = \frac{0.05}{0.05 - 0.11}(e^{-0.11x_i - 0.05x_i}) + \varepsilon_i.$$

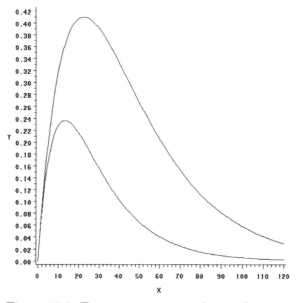

Figure 13.3: Two-term exponential growth curves

Other nonlinear growth curves include the logistic curve that we have discussed in Chapter 10. Others are:

- Drug Responsiveness Model: These are often used in pharmacological settings and the model describes the effects of varying dose levels of drugs. The model has the form

$$y_i = \beta_0 - \frac{\beta_0}{1 + \left(\dfrac{x_i}{\beta_2}\right)^{\beta_1}} + \varepsilon_i. \tag{13.12}$$

In the above model, x_i is the dosage at level i and y is the response variable as a percentage of maximum possible responsiveness. β_0 describes the expected response at the dose saturation, β_2 is the concentration that produces a half-maxima response; and the β_1 parameter determines the slope of the function.

- Piece-wise linear regression Models: We have seen in Chapter 8 the concept of piece-wise regression in the context of the use of dummy variables in regression models. This occurs when the relationship between a response variable and one or more independent variables changes over the range of the independent variables. The general result for a single independent variable is of the form

$$y_i = \beta_0 + \beta_1 x_i (x_i \le q) + \beta_2 x_i (x_i > q).$$

The above model has a common intercept β_0 and a slope given by

$$\beta_1' = \begin{cases} \beta_1, & \text{if } x_i \le q \\ \beta_2, & \text{if } x_i > q. \end{cases}$$

In the above setting, q is the knot-point and it is often unknown. As we have in this situation, we can replace its value with an additional unknown parameter β_3.

- Breakpoint regression: One could also adjust the equation above to reflect a "jump" in the regression line. For example, imagine that after the older machines are put on-line, the per-unit-cost jumps to a higher level, and then slowly goes down as volume continues to increase. In that case, simply specify an additional intercept (b_3), so that

$$y = (b_0 + b_1 x)(x \le 500) + (b_3 + b_2 x)(x > 500).$$

13.2   Estimating Nonlinear Models

Some times the method of ordinary least squares can be applied to our nonlinear model data after a suitable transformation, usually, logarithmic. For instance, the nonlinear model $y = \theta_0 x^{\theta_1} \varepsilon$ can be transformed into the model $y' = \beta_0 + \beta_1 x' + \varepsilon'$, where, $y' = \ln(y)$, $\beta_0 = \ln(\theta_0)$, $\beta_1 = \theta_1$, $x' = \ln x$, and $\varepsilon' = \ln \varepsilon$. However, if the model is such that we can not find a suitable transformation that would make the model linear, then the parameters of this untransformable nonlinear regression model would have to be estimated by iterative procedures. For example the model

$$y_i = \beta_0 x_i^{\beta_1} + \varepsilon_i,$$

can not be linearly transformed.

Generally, we have a nonlinear model

$$y_i = f(\mathbf{x}_i, \boldsymbol{\beta}) + \varepsilon_i, \tag{13.13}$$

where $f(\mathbf{x}_i, \boldsymbol{\beta})$ is a nonlinear function, $\mathbf{x}_i$ is a vector of explanatory variables on observation i, $\boldsymbol{\beta}$ are the parameters of the model and ε_i are the random error terms having the properties of independence and identically and normally distributed with mean zero and variance σ^2. Now let us consider for example the following nonlinear function

$$f(\mathbf{x}, \boldsymbol{\beta}) = \frac{\beta_1 x}{\beta_2 + x}. \tag{13.14}$$

The model in (13.14) is the **Michaelis-Menten** (1913) model which has received wide usage for enzymatic reactions amongst others. Here as x increases, the function approaches an asymptote, β_1, and β_2 can be assumed to be the value of x at which the function has reached half its asymptotic value. We give a practical example in what follows.

Example 13.1 (The Puromycin Data)

The following example from Bates and Watts (1988, p. 269) gives the relationship between the velocity of an enzymatic reaction (chemical kinetics), y, and the substrate concentration, x. The data for this experiment is presented in Table 13.1.

x	.02	.02	.06	.06	.11	.11	.22	.22	.56	.56	1.10	1.10
y	76	47	97	107	123	139	159	152	191	201	207	200

Table 13.1: Reaction velocity and substrate concentration data

We present in the following figure the plot of reaction velocity (y) against substrate concentration (x).

Estimation Procedure

Given a nonlinear regression function $f(x, \beta)$, let us assume for now that we have one regressor variable, then we wish to solve

$$\hat{\beta} = \sum_{i=1}^{n} [y_i - f(x_i, \beta)]^2, \tag{13.15}$$

where x_i is the i-th observation on the regressor

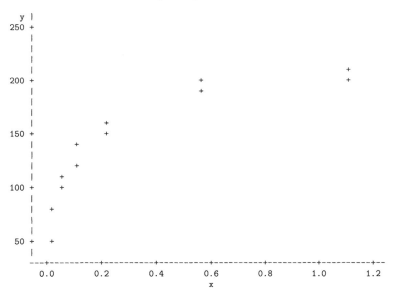

Solving (13.15) is often accomplished either by the method of steepest descent or the by the use of the Gauss-Newton or the Newton-Raphson iterative procedure. Thus given an initial value $\hat{\beta}^0$, we obtain an approximate vector of solution $\eta_i^0(\beta)$ such that

$$f(x_i, \beta) \simeq f(x_i, \hat{\beta}^0) + \sum_{j=1}^{p} \left. \frac{\partial f}{\partial \beta_j} \right|_{(x_i, \beta) = (x_i, \hat{\beta}^0)}, (\beta_i - \beta_i^0) = \eta_i^0(\beta).$$

The solution is thus obtained by successive iteration until convergence is attained. However, we do not know the initial estimates of the parameters $\hat{\beta}^0$. SAS has **proc nlin** for implementing nonlinear models. However, we need initial parameter estimates. There are two approaches we can adopt in this example to determine the initial estimates of β_1 and β_2. A cursory look at the plot of observed velocity versus concentration indicates that the function approaches the asymptotic value, which is, β_1 as x increases. Thus, we can assume initially that $\hat{\beta}_1^0$ is between 200 and 207. We will choose 200 for $\hat{\beta}_1^0$ here. Also, by definition, β_2 is the value of x at which the function would reach half its asymptotic value, say, 100. Thus this value would be between 0.1 and 0.2. Again, let us assume an initial value $\hat{\beta}_2^0 = 0.1$ for this example. The second approach is to consider the model in (13.14). We have

$$\lim_{x \to \infty} \left(\frac{\beta_1 x}{\beta_2 + x} \right) = \beta_1.$$

Since the largest value of of y in the data is 207, we can take this value as the initial value for β_1. For β_2, we first arrange the expression in (13.14) as a function of β_2 giving

$$\beta_2 = \frac{\beta_1 x}{y} - x.$$

Substituting, $\beta_1 = 207$ for various values of observed pairs (x_i, y_i), we see that the initial values of β_2 can range between 0 and 0.068. This is described as the grid method in **proc nlin** in SAS, but first, let us implement in SAS the Michaelis-Menten nonlinear model to our data in Table 13.1 with the following SAS codes and output with our assumed initial parameter estimates in the first case.

```
options formchar="|----|+|---+=|-/\<>*";
data funt;
input x y @@;
datalines;
0.02 76 0.02 47 0.06 97 0.06 107 .11 123 .11 139 .22 159 .22 152 .56 191 .56 201 1.1 207 1.1 200
;
run;
proc plot data=funt vpct=50;
plot y*x='+';
run;
proc nlin data=funt method=gauss hougaard;
parameters beta1=200 beta2=0.1;
model y = (beta1*x)/(beta2+x);
run;
```

```
                         The NLIN Procedure
                       Dependent Variable y
                       Method: Gauss-Newton

                          Iterative Phase
                                                   Sum of
            Iter       beta1       beta2          Squares
               0       200.0      0.1000           7964.2
               1       212.0      0.0543           1593.2
               2       211.8      0.0623           1201.0
               3       212.6      0.0639           1195.5
               4       212.7      0.0641           1195.4
               5       212.7      0.0641           1195.4
               6       212.7      0.0641           1195.4

      NOTE: Convergence criterion met.


                        Estimation Summary

            Method                     Gauss-Newton
            Iterations                            6
            R                           6.096E-6
            PPC(beta2)                  2.489E-6
            RPC(beta2)                  0.000026
            Object                      4.387E-9
            Objective                   1195.449
            Observations Read                  12
            Observations Used                  12
            Observations Missing                0

      NOTE: An intercept was not specified for this model.

                         Sum of       Mean              Approx
      Source         DF  Squares      Square   F Value   Pr > F

      Model           2  270214       135107   1130.18   <.0001
      Error          10  1195.4       119.5
      Uncorrected Total 12 271409

                          Approx     Approximate 95%
      Parameter  Estimate Std Error  Confidence Limits     Skewness
```

| beta1 | 212.7 | 6.9471 | 197.2 | 228.2 | 0.0961 |
| beta2 | 0.0641 | 0.00828 | 0.0457 | 0.0826 | 0.3207 |

```
            Approximate Correlation Matrix
                        beta1           beta2
            beta1    1.0000000       0.7650835
            beta2    0.7650835       1.0000000
```

In the above program, we have asked SAS to employ the Gauss-Newton Method (which is the default in SAS). We also asked that the skewness measures be calculated by the use of the **hougaard** option. The initial parameter estimates are specified in the **parms** statement and the **model** statement contains the mathematical expression of the model, excluding the error term.

The first part of the output contains the iteration history of the model fitting. The first iteration contains the starting values and in this case we have a total of six iterations before convergence is achieved. Convergence is based on the Bates and Watts (1981) relative offset measure. When this measure is less than 10^{-5}, then convergence is assumed achieved. We notice that with each iteration cycle, the residual SS is being gradually reduced until no further drop is significant. The residual SS here is 1195.449 (also the objective) with the error mean square (MSE) being 119.5.

The parameter estimates of the model are next provided after the usual ANOVA table. These estimates also include their standard errors, approximate 95% confidence intervals (all asymptotic) and upon request, the skewness measures. The Hougaard's measure (g_i) of skewness assesses whether the parameter is close to linearity or whether it still contains considerable nonlinearity. The measure is interpreted as follows:

$$\text{If} \begin{cases} |g_i| < 0.1, & \text{estimates of } \beta_i \text{ is close to linear} \\ 0.1 < |g_i| < 0.25, & \text{the estimates are reasonably close to linear} \\ |g_i| > 0.25, & \text{skewness is apparent} \\ |g_i| > 1.0, & \text{estimator exhibits strong nonlinear behavior.} \end{cases}$$

We see that while $\hat{\beta}_1$ can be considered linear, that of $\hat{\beta}_2$ is suspect and we should be very careful with the inference we draw from this analysis. The output also includes correlation matrix among the parameter estimates indicating that changing one parameter value does affect the other parameter values. Thus, it is important to note the strength of the correlations among the parameters.

The Grid Method in SAS for Starting Values

Sometimes, if we are not sure of what the starting values are, we can use the grid approach to accomplish this. For each combination of grid values, residual SS are calculated until the lowest one found and these values will be used as the starting values. We illustrate this with the following SAS program and partial output.

```
set funt;
proc nlin data=funt method=Gauss hougaard;
parameters  beta1=200 to 207 by 1.0
            beta2= 0 to 0.7 by 0.1;
model y = (beta1*x) / (beta2+x);
output out=aa r=resid p=pred student=stud h=lev;
run;
                    The NLIN Procedure
                    Dependent Variable y


                    Grid Search
                                       Sum of
                beta1        beta2     Squares
                200.0            0     71809.0
                201.0            0     73223.0
                202.0            0     74661.0

                .....       ......     .......
                205.0       0.7000     99854.9
                206.0       0.7000     99287.4
                207.0       0.7000     98722.4


                    The NLIN Procedure
                    Dependent Variable y
                    Method: Gauss-Newton


                    Iterative Phase
```

```
                            Sum of
Iter      beta1      beta2   Squares
   0      207.0     0.1000    5918.0
   1      212.0     0.0558    1464.8
   2      212.0     0.0627    1198.8
   3      212.6     0.0640    1195.5
   4      212.7     0.0641    1195.4
   5      212.7     0.0641    1195.4
   6      212.7     0.0641    1195.4
```

NOTE: Convergence criterion met.

```
              Estimation Summary

     Method                Gauss-Newton
     Iterations                       6
     R                        4.596E-6
     PPC(beta2)               1.877E-6
     RPC(beta2)               0.000019
     Object                   2.495E-9
     Objective                1195.449
     Observations Read              12
     Observations Used              12
     Observations Missing            0
```

NOTE: An intercept was not specified for this model.

```
                      Sum of       Mean                   Approx
Source            DF  Squares      Square     F Value     Pr > F
Model              2   270214      135107     1130.18     <.0001
Error             10   1195.4       119.5
Uncorrected Total 12   271409
```

```
                        Approx     Approximate 95%
Parameter  Estimate   Std Error   Confidence Limits   Skewness
beta1         212.7      6.9472      197.2    228.2     0.0961
beta2        0.0641     0.00828     0.0457   0.0826     0.3207
```

```
       Approximate Correlation Matrix
                  beta1          beta2
       beta1   1.0000000      0.7650836
       beta2   0.7650836      1.0000000
```

The grid search procedure utilizes a search within the grid $200 \leq \beta_1 \leq 207$ and $0 \leq \beta_2 \leq 0.7$. We have asked SAS to search in increments of 1 for β_1 and 0.1 for β_2. Of the 64 combination of paired values (β_1, β_2), the pair $(207, 0.1)$ produces the lowest residual sum of squares and is therefore adopted as the starting values in the Gauss-Newton computational procedure. The results obtained are identical to our previous result, namely that $\hat{\beta}_1 = 212.7$ and $\hat{\beta}_2 = 0.0641$ and hence the estimated velocity function is

$$\hat{y}_i = \frac{212.7 x_i}{0.0641 + x_i}.$$

We also present results from the use of the Newton-Raphson and Marquardt method using our original starting values. Both results are identical to the results from the Gauss-Newton method adopted earlier. Note the different skewness measures with the Newton method and the Marquardt method, and that the Newton method converges at the 7th iteration.

```
proc nlin data=funt method=newton hougaard;
parameters  beta1=200 beta2=0.1;
model y = (beta1*x) / (beta2+x);
output out=aa r=resid p=pred student=stud h=lev;
run;

/* Use the Marquardt Method8 */
proc nlin data=funt method=marquardt hougaard;
parameters  beta1=200 beta2=0.1;
model y = (beta1*x) / (beta2+x);
output out=aa r=resid p=pred student=stud h=lev;
run;
                    The SAS System

                  The NLIN Procedure
                  Dependent Variable y
                     Method: Newton
```

```
                    Iterative Phase
                                          Sum of
        Iter     beta1       beta2      Squares
          0      200.0      0.1000       7964.2
          1      176.1      0.0271       5589.9
          2      183.8      0.0328       3917.0
          3      199.3      0.0459       1871.6
          4      209.4      0.0583       1254.1
          5      212.3      0.0634       1196.3
          6      212.7      0.0641       1195.4
          7      212.7      0.0641       1195.4
```

NOTE: Convergence criterion met.

```
                 Estimation Summary

        Method                      Newton
        Iterations                       7
        Subiterations                    4
        Average Subiterations     0.571429
        R                          1.615E-7
        PPC(beta2)                 6.629E-8
        RPC(beta2)                  0.00017
        Object                     1.712E-7
        Objective                 1195.449
        Observations Read               12
        Observations Used               12
        Observations Missing             0
```

NOTE: An intercept was not specified for this model.

Source	DF	Sum of Squares	Mean Square	F Value	Approx Pr > F
Model	2	270214	135107	1130.18	<.0001
Error	10	1195.4	119.5		
Uncorrected Total	12	271409			

Parameter	Estimate	Approx Std Error	Approximate 95% Confidence Limits		Skewness
beta1	212.7	7.1607	196.7	228.6	0.0877
beta2	0.0641	0.00871	0.0447	0.0835	0.2755

```
           Approximate Correlation Matrix
                        beta1          beta2
        beta1       1.0000000      0.7808405
        beta2       0.7808405      1.0000000
```

```
                 The NLIN Procedure
               Dependent Variable y
                 Method: Marquardt

                    Iterative Phase
                                          Sum of
        Iter     beta1       beta2      Squares
          0      200.0      0.1000       7964.2
          1      212.0      0.0543       1593.2
          2      211.8      0.0623       1201.0
          3      212.6      0.0639       1195.5
          4      212.7      0.0641       1195.4
          5      212.7      0.0641       1195.4
          6      212.7      0.0641       1195.4
```

NOTE: Convergence criterion met.

```
                 Estimation Summary

        Method                   Marquardt
        Iterations                       6
        R                          6.096E-6
        PPC(beta2)                 2.489E-6
        RPC(beta2)                0.000026
        Object                     4.387E-9
        Objective                 1195.449
        Observations Read               12
        Observations Used               12
        Observations Missing             0
```

NOTE: An intercept was not specified for this model.

Source	DF	Sum of Squares	Mean Square	F Value	Approx Pr > F
Model	2	270214	135107	1130.18	<.0001
Error	10	1195.4	119.5		
Uncorrected Total	12	271409			

Parameter	Estimate	Approx Std Error	Approximate 95% Confidence Limits		Skewness
beta1	212.7	6.9471	197.2	228.2	0.0961
beta2	0.0641	0.00828	0.0457	0.0826	0.3207

Approximate Correlation Matrix

	beta1	beta2
beta1	1.0000000	0.7650835
beta2	0.7650835	1.0000000

The predicted, residual and studentized residuals as well as the leverages are presented in the output below.

```
proc print data=aa;
var x y pred resid stud lev;
format pred resid stud lev 8.3;
run;
```

Obs	x	y	pred	resid	stud	lev
1	0.02	76	50.566	25.434	2.487	0.125
2	0.02	47	50.566	-3.566	-0.349	0.125
3	0.06	97	102.811	-5.811	-0.592	0.193
4	0.06	107	102.811	4.189	0.427	0.193
5	0.11	123	134.362	-11.362	-1.123	0.144
6	0.11	139	134.362	4.638	0.458	0.144
7	0.22	159	164.685	-5.685	-0.549	0.104
8	0.22	152	164.685	-12.685	-1.226	0.104
9	0.56	191	190.833	0.167	0.017	0.177
10	0.56	201	190.833	10.167	1.025	0.177
11	1.10	207	200.969	6.031	0.640	0.257
12	1.10	200	200.969	-0.969	-0.103	0.257

We also present both the residual and studentized residual plots against the predicted values below. It looks like the first observation has an unusual high value of studentized residual and might be considered an outlier in this case.

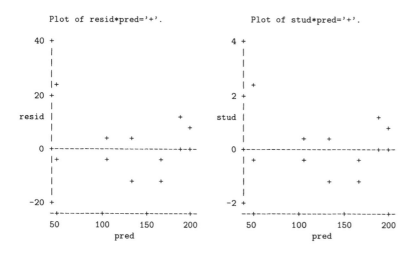

For nonlinear models, the coefficient of multiple determination R^2 is not readily defined as most models do not have the intercepts. However, a pseudo R^2 can be defined as

$$R^2 = 1 - \frac{\text{Residual SS}}{\text{Total SS}}.$$

For our model, this is calculated as $1 - 1195.449/30858.917 = 0.9613$ and the predicted model and the observed are overlayed in a plot in Figure 13.4.

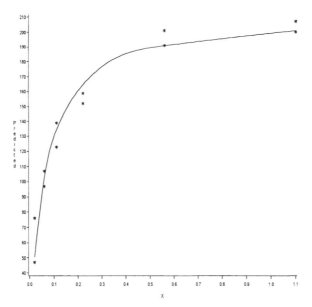

Figure 13.4: Predicted M-M model superimposed over observed values

Example 13.2 (Drug Responsiveness)

The following example is taken from Kutner et al. (2005, p. 550) (reproduced with permission of The McGraw-Hill Companies) and relates to a pharmacologist modeling the responsiveness to a drug using the nonlinear regression model discussed in (13.12), which is reproduced below.

$$y_i = \beta_0 - \frac{\beta_0}{1 + \left(\dfrac{x_i}{\beta_2}\right)^{\beta_1}} + \varepsilon_i.$$

The data for 19 cases at 13 dose levels are presented in Table 13.2. We see that the observations are replicated at $x = 3.5, 4.0, 4.5, 5.0, 5.5$ and 6.0. The plot of the observed responsiveness (y) against dose levels is presented in the SAS program with partial output.

dose	Responsiveness	
x	y	
1.0	0.5	
2.0	2.3	
3.0	3.4	
3.5	11.5	10.9
4.0	24.0	25.3
4.5	39.6	37.9
5.0	54.7	56.8
5.5	70.8	68.4
6.0	82.1	80.6
6.5	89.2	
7.0	94.8	
8.0	96.2	
9.0	96.4	

Table 13.2: Drug responsiveness data

To fit the model in (13.12), we realize that we do not have initial values. Therefore, we use starting values

$\beta_0 = 0, \beta_1 = 0$ and $\beta_2 = 0.1$. The choice of $\beta_2 = 0.1$ is motivated by the fact that we do not want the denominator in the expression involving x to be zero. The results of this analysis together with computed predicted values and residuals are in the following SAS output.

```
data drug;
input y x;
datalines;
..........
;
proc print;
run;
proc plot vpct=50;
plot y*x='+';
run;
proc nlin data=drug method=Marquardt hougaard;
parms beta0=0 beta1=0 beta2=0.1;
u=(x/beta2)**(beta1);
model y=beta0-beta0/(1+u);
output out=aa r=resid p=pred student=stud h=lev;
run;
goptions vsize=6in hsize=6in;
symbol1 c=black i=spline value=none line=1  height=.75;
symbol2 c=r    value=plus  line=2 height=.75;
axis1 label=(angle=-90 rotate=90 'Predicted');
axis2 label=('x');
proc gplot data=aa;
plot pred*x y*x/overlay vaxis=axis1 haxis=axis2 noframe;
run;
```

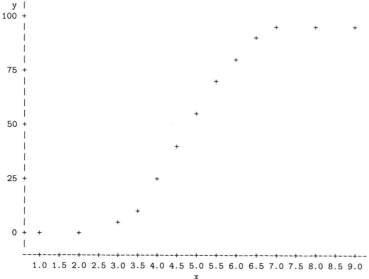

Plot of y*x. Symbol used is '+'.

```
The NLIN Procedure
Dependent Variable y
Method: Marquardt

Iterative Phase
                                        Sum of
      Iter      beta0      beta1      beta2     Squares
        0          0          0     0.1000     69126.2
        1    99.5158          0     0.1000     22085.1
        .      .....      .....      .....     .......
        9      100.4     6.4687     4.8170     35.7344
       10      100.3     6.4810     4.8154     35.7149
       11      100.3     6.4802     4.8155     35.7149
       12      100.3     6.4802     4.8155     35.7149

NOTE: Convergence criterion met.

Estimation Summary

Method                    Marquardt
Iterations                      12
Subiterations                   12
Average Subiterations            1
```

```
R                          4.169E-6
PPC(gamma1)                4.664E-7
RPC(gamma1)                7.536E-6
Object                     4.218E-9
Objective                 35.71489
Observations Read               19
Observations Used               19
Observations Missing             0
```

NOTE: An intercept was not specified for this model.

Source	DF	Sum of Squares	Mean Square	F Value	Approx Pr > F
Model	3	69090.5	23030.2	10317.3	<.0001
Error	16	35.7149	2.2322		
Uncorrected Total	19	69126.2			

Parameter	Estimate	Approx Std Error	Approximate 95% Confidence Limits		Skewness
beta0	100.3	1.1741	97.8512	102.8	0.0863
beta1	6.4802	0.1943	6.0683	6.8922	0.0919
beta2	4.8155	0.0280	4.7561	4.8749	0.0726

Approximate Correlation Matrix

	beta0	beta1	beta2
beta0	1.0000000	-0.6960587	0.8177676
beta1	-0.6960587	1.0000000	-0.5314295
beta2	0.8177676	-0.5314295	1.0000000

The estimated nonlinear regression equation is therefore

$$\hat{y}_i = 100.3 - \frac{100.3}{1 + \left(\dfrac{x_i}{4.8155}\right)^{6.4802}}.$$

The estimated curve is plotted in Figure 13.5 with the observed values superimposed over it. Clearly this model fits the data very well.

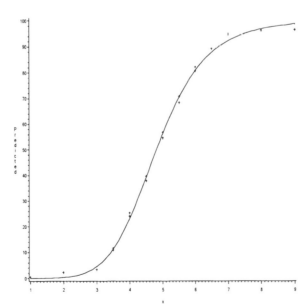

Figure 13.5: Estimated drug responsiveness curve

The skewness measures of the parameters satisfy the linearity assumption of the parameter estimates. Hence, we can assume that the large sample inferences will be appropriate here. Consequently, we can conduct the lack of fit test for this model since observations are replicated. The pure error sum of squares are computed at $x = 3.5, 4.0, 4.5, 5.0, 5.5, 6.0$ respectively to give 8.66, where for instance, the pure error at say, $x = 5.0$ is computed as $(54.7 - 56.8)^2/2 = 2.205$. Hence, we can modify the ANOVA Table as follows:

Source	d.f.	SS	MS	F	p-value
Model	3	69090.5	23030.2	10317.3	
Error	16	35.7149	2.2322		
Lack of Fit	10	27.0349	2.7035	1.87	0.2290
Pure Error	6	8.66	1.4467		
Uncorrected Total	19	69126.2			

The corresponding p-value for the computed lack of fit F test is 0.2290 which clearly indicates that the null hypothesis of the adequacy of the model holds. Also the following predicted values and residuals indicate a very good fit of this model to the data.

```
Obs    x      y      pred     resid      stud      lev
--------------------------------------------------------
 1    1.0    0.5    0.004    0.496     0.332    0.000
 2    2.0    2.3    0.337    1.963     1.315    0.001
 3    3.0    3.4    4.465   -1.065    -0.735    0.058
 4    3.5   11.5   11.265    0.235     0.170    0.147
 5    3.5   10.9   11.265   -0.365    -0.265    0.147
 6    4.0   24.0   23.183    0.817     0.608    0.190
 7    4.0   25.3   23.183    2.117     1.574    0.190
 8    4.5   39.6   39.327    0.273     0.199    0.155
 9    4.5   37.9   39.327   -1.427    -1.039    0.155
10    5.0   54.7   56.251   -1.551    -1.129    0.154
11    5.0   56.8   56.251    0.549     0.400    0.154
12    5.5   70.8   70.531    0.269     0.197    0.163
13    5.5   68.4   70.531   -2.131    -1.559    0.163
14    6.0   82.1   80.888    1.212     0.876    0.142
15    6.0   80.6   80.888   -0.288    -0.208    0.142
16    6.5   89.2   87.774    1.426     1.028    0.138
17    7.0   94.8   92.176    2.624     1.931    0.173
18    8.0   96.2   96.734   -0.534    -0.429    0.305
19    9.0   96.4   98.626   -2.226    -1.961    0.422
```

13.3   Problems with the Newton Procedure

The next example illustrates the need to choose carefully the computational procedure for estimating the parameters of a nonlinear function. We shall illustrate this with the following data taken from Myers (1990, p. 446) and reproduced by permission of Brooks/Cole, a part of Cengage Learning, Inc. The data relate to the study of age and growth characteristics in selected freshwater mussel species in Southwest Virginia. For a particular mussel, age (x) in years and length (y) in inches are related by the regression function

$$y_i = \alpha - \exp[-(\beta + \gamma x_i)] + \varepsilon_i, \quad i = 1, 2, \ldots, n. \tag{13.16}$$

The following data in Table 13.3 are measurements on 20 female mussels and we wish to fit the above model to the data.

Subjects	length (y)	Age (x)	Subjects	Length (y)	Age (x)
1	29.2	4	11	36.9	8
2	28.6	4	12	40.2	8
3	29.4	5	13	39.2	9
4	33.0	5	14	40.6	9
5	28.2	6	15	35.2	10
6	33.9	6	16	43.3	10
7	33.1	6	17	42.3	13
8	33.2	7	18	41.4	13
9	31.4	7	19	45.2	14
10	37.8	7	20	47.1	18

Table 13.3: Length and age of 20 female mussels from Virginia

The following is the SAS program and a partial output from fitting the model in (13.16) to the data in Table 13.3.

```
data vg;
input y x @@;
datalines;
29.2 4 28.6 4 29.4 5 33 5 28.2 6 33.9 6
33.1 6 33.2 7 31.4 7 37.8 7 36.9 8 40.2 8
39.2 9 40.6 9 35.2 10 43.3 10 42.3 13
41.4 13 45.2 14 47.1 18
;
proc print;
run;
proc plot vpct=50;
plot y*x='+';
run;
proc nlin data=vg method=gauss hougaard;
parms alpha=0 beta=0 gamma=0;
u=beta+gamma*x;
model y=alpha-exp(-u);
run;
proc nlin data=vg method=marquardt hougaard;
parms alpha=0 beta=0 gamma=0;
u=beta+gamma*x;
model y=alpha-exp(-u);
run;
proc nlin data=vg method=newton hougaard;
parms alpha=0 beta=0 gamma=0;
u=beta+gamma*x;
model y=alpha-exp(-u);
run;
```

                    Plot of y*x. Symbol used is '+'.

```
  y |
 50 +                                                          +
    |
    |
    |
 45 +                                                +
    |                         +
    |                                   +
    |                                   +
 40 +              +    +
    |                   +
    |         +    +
    |
 35 +                   +
    |      +    +
    |   +  +
    |         +
 30 +   +
    | +         +
    |
    |
 25 +
    |
    ---+----+----+----+----+----+----+----+----+----+----+----+----+----+--
       4    5    6    7    8    9   10   11   12   13   14   15   16   17   18
                                   x
```

                        The NLIN Procedure
                      Dependent Variable y
                      Method: Gauss-Newton

                          Iterative Phase

Iter	alpha	beta	gamma	Sum of Squares
0	0	0	0	28700.7
1	25.6816	0	1.3939	2959.3
. . . .
14	52.0536	-3.6021	0.1089	111.8
15	52.0815	-3.6017	0.1089	111.8

   NOTE: Convergence criterion met.

                      Estimation Summary

Method	Gauss-Newton
Iterations	15
Subiterations	24
Average Subiterations	1.6
R	9.465E-6
PPC(alpha)	3.344E-7
RPC(alpha)	0.000536

```
            Object                   0.000206
            Objective                111.7914
            Observations Read             20
            Observations Used             20
            Observations Missing           0
```

		Sum of	Mean		Approx
Source	DF	Squares	Square	F Value	Pr > F
Model	2	523.9	262.0	39.84	<.0001
Error	17	111.8	6.5760		
Corrected Total	19	635.7			

		Approx	Approximate 95%		
Parameter	Estimate	Std Error	Confidence Limits		Skewness
alpha	52.0815	7.5807	36.0877	68.0753	2.7996
beta	-3.6017	0.1159	-3.8462	-3.3573	-0.1125
gamma	0.1089	0.0604	-0.0186	0.2363	0.2973

Approximate Correlation Matrix

	alpha	beta	gamma
alpha	1.0000000	-0.4705537	-0.9700407
beta	-0.4705537	1.0000000	0.2549546
gamma	-0.9700407	0.2549546	1.0000000

```
                         The NLIN Procedure
                       Dependent Variable y
                          Method: Newton
```

Iterative Phase

				Sum of
Iter	alpha	beta	gamma	Squares
0	0	0	0	28700.7
1	37.4761	-0.2998	0.0336	612.5
2	37.1283	-0.2998	0.0767	600.3
3	37.0391	-0.2998	0.1040	598.0
4	37.0100	-0.2998	0.1145	597.8
5	37.0069	-0.2998	0.1158	597.8
6	37.0069	-0.2998	0.1158	597.8

NOTE: Convergence criterion met but a note in the log indicates
a possible problem with the model.

Estimation Summary

```
            Method                    Newton
            Iterations                     6
            R                       4.579E-9
            PPC(gamma)              2.704E-8
            RPC(gamma)              0.000149
            Object                   6.4E-10
            Objective               597.7832
            Observations Read             20
            Observations Used             20
            Observations Missing           0
```

		Sum of	Mean		Approx
Source	DF	Squares	Square	F Value	Pr > F
Model	1	37.9248	37.9248	1.14	0.2994
Error	18	597.8	33.2102		
Corrected Total	19	635.7			

NOTE: The (approximate) Hessian is singular.

		Approx	Approximate 95%		
Parameter	Estimate	Std Error	Confidence Limits		Skewness
alpha	37.0069	1.4439	33.9734	40.0403	.
beta	-0.2998	.	.	.	.
gamma	0.1158	0.1644	-0.2296	0.4612	.

Approximate Correlation Matrix

	alpha	beta	gamma
alpha	1.0000000	.	-0.4511318
beta	.	.	.
gamma	-0.4511318	.	1.0000000

It is observed that both the Gauss-Newton and Marquardt methods converged successfully giving a minimum sum of square as 111.8 (output from Marquardt method are not shown). However, the Newton method failed to successfully converge. As indicated in the output, the convergence criterion is met with a possible problem. The iteration stops after the sixth step and the sum of squares of 597.8 is no where near the result from the Gauss-Newton and the Marquardt methods. For this data, the Newton method failed to yield an acceptable results.

13.4 Exercises

13.1 For each of the following functions, state whether it is linear or nonlinear. In the case of a nonlinear function, state whether or not it can be made linear through a transformation. If it can be made linear, obtain the linear form and explain your steps.

(a) $y = \exp\left(\log \beta_0 + \beta_1 x\right) + \varepsilon$

(b) $\exp(y) = \exp(\beta_0 + \beta_1 x) + \varepsilon$

(c) $y = \beta_0 x_1^{\beta_1} x_2^{\beta_2 + \beta_3} + \varepsilon$

(d) $y = \beta_1 x^{-\beta_2} + \varepsilon$

(e) $y = \beta_1 \left(1 + \beta_2 \exp(-\beta_3 x)\right) + \varepsilon$

(f) $y = \theta x_1^{\beta_1} x_2^{\beta_2} x_3^{\beta_3} + \varepsilon$

(g) $y = \beta_0 + \theta \beta_1^{x_1} \beta_2^{x_2} + \varepsilon$

(h) $y = \beta_0 + \beta_1 \ln\left(1 + \beta_2 \exp(-\beta_3 x)\right) + \varepsilon$

(i) $y = \beta_1 x^{\beta_2} / \left(\beta_3 + x^{\beta_2}\right) + \varepsilon$

13.2 **Diamond.** The data in the file **diamond.txt** is taken from the diamond data submitted by Chu, S. to the Journal of Statistics Education data archive (see also Chu, S., 1996). The data contains the size of diamonds (in carats) and the price of ring in Singapore dollars. To model the price (y) as a function of the size, the following regression model is suggested: $y = \beta_1 x^{\beta_2} \varepsilon$.

(a) Linearize the model and fit a linear regression model to the data.

(b) Construct an approximate 95% confidence interval for the parameters of the linear regression model.

(c) Fit a nonlinear regression model to the data.

(d) Construct an approximate 95% confidence interval for the parameters of the nonlinear regression model.

(e) Compare the parameter estimates from the linear and nonlinear regression models. Comment on your findings.

(f) Compare the fitted values from the linear and nonlinear regression models. Comment on your findings.

13.3 **Diamond.** The data in the file **diamond.txt** is taken from the diamond data submitted by Chu, S. to the Journal of Statistics Education data archive (see also Chu, S., 1996). The data contains the size of diamonds (in carats) and the price of ring in Singapore dollars. To model the price (y) as a function of the size, the following regression model is suggested: $y = \exp(\beta_0 + \beta_1 x) + \varepsilon$.

(a) Fit the nonlinear regression model to the data.

(b) Construct an approximate 95% confidence interval for the parameters.

(c) Obtain the fitted values and the residuals from the nonlinear regression models. Do the residuals follow a normal distribution? Comment on your findings.

13.4 **Galileo's Experiment.** The following is part of the data submitted by Dickey, D.A. to the Journal of Statistics Education data archive (see also Dickey and Arnold, 1995). The data is from an experiment in which Galileo rolled a ball down an inclined plane above a floor. The data contains the height (y) of the ball rolled down and the horizontal distance (x) traveled before landing.

x	573	534	495	451	395	337	253
y	1000	800	600	450	300	200	100

To model the height (y) as a function of the distance (x), the following regression model is suggested: $y = \left(\beta_0 x^2\right)/(1 + \beta_1 x) + \varepsilon$.

(a) Fit the nonlinear regression model to the data.

(b) Construct an approximate 95% confidence interval for the parameters.

(c) Obtain the fitted values and the residuals from the nonlinear regression models. Do the residuals follow a normal distribution? Comment on your findings.

13.5 **Galileo's Experiment.** The following is part of the data submitted by Dickey, D.A. to the Journal of Statistics Education data archive (see also Dickey and Arnold, 1995). The data is from an experiment in which Galileo rolled a ball down an inclined plane above a floor and the ball crossed a horizontal shelf built into the end of the ramp. The data contains the height (y) of the ball rolled down and the horizontal distance (x) traveled before landing.

x	1500	1340	1328	1172	800
y	1000	828	800	600	300

To model the height (y) as a function of the distance (x), the following regression model is suggested: $y = \left(\beta_0 x^2\right)/(1 + \beta_1 x) + \varepsilon$.

(a) Fit the nonlinear regression model to the data.

(b) Construct an approximate 95% confidence interval for the parameters.

(c) Obtain the fitted values and the residuals from the nonlinear regression models. Do the residuals follow a normal distribution? Comment on your findings.

Chapter 14

One-Way Analysis of Variance

14.1 Introduction

Consider a single factor experiment having k levels. For instance, the factor could be diets having say, four levels corresponding to three new diets A, B, C and a control or currently available diet in the market labeled D. Suppose we wish to conduct an experiment involving animals. We shall follow the following principles of experimentation, namely,

1. Randomization,

2. Replication, and,

3. Local control.

To illustrate the above three principles, suppose that an animal nutritionist has four diets A, B, C, D as described above and wants to allocate five rats to each diet. The response is the weight gained after 6 weeks. The diets will then be randomly allocated to the 20 rats. The 20 rats are treated alike in all other respects except for type of diet. The rats are in different pens in the same area and fed individually. The 20 rats are being subjected to the same environmental conditions. The diets (treatments) are compared in as nearly equitable manner as possible. One possible random assignments is to do the following:

STEP 1: Label the animals from 01 to 20.

STEP 2: Randomly select five animals using random numbers table. Suppose these are {11, 04, 01, 07, 12}.

STEP 3: Remove the five digits in step 2 from the pool and again perform step 2, yielding the digits {02, 10, 19, 18, 14}.

STEP 4: Repeat step 3 by removing the digits generated in steps 2 and 3 and again randomly select five digits from the remaining 10 digits yielding once again {06, 17, 05, 09, 03}. The remaining five digits are therefore {08, 13, 15, 16, 20}.

STEP 5: If we again randomize the diets, say, B, D, A, C, then, diet B will be the first group and is assigned to the animals selected from step 2. The last group, which is diet C received rats numbered {08, 13, 15, 16, 20}.

Table 14.1 gives the final random allocation for this experiment.

Since each diet receives 5 animals, the five animals for each diet therefore constitute what we call *replications*. Thus each diet in the experiment is equally replicated five times, leading to a total of $4 \times 5 = 20$ experimental units or animals in the study. A random allocation of animals to treatments can readily be conducted in SAS as follows with the corresponding output.

Rats	Diet	Rats	Diet	Rats	Diet	Rats	Diet
1	B	6	A	11	B	16	C
2	D	7	B	12	B	17	A
3	A	8	C	13	C	18	D
4	B	9	A	14	D	19	D
5	A	10	D	15	C	20	C

Table 14.1: A layout with 4 treatments and 5 replications

```
data aa;
do subj=1 to 20;
do j=1 to 4;
if (5*j-4 le subj le 5*j) then dg=j;
end;
if dg=1 then diet='A';
if dg=2 then diet='B';
if dg=3 then diet='C';
if dg=4 then diet='D';
output;
end;
drop j;
proc print data=aa;
title 'Unrandomized assignment of subjects to diets';
run;
proc plan seed=82844;
factors subj=20;
output data=aa out=bb;
run;
proc print data=bb;
title1 'Random assignments of subjects to diets';
run;
```

```
                  Unrandomized assignment of subjects to diets

                    Obs     subj     dg      diet
                     1        1       1        A
                     2        2       1        A
                     3        3       1        A
                     4        4       1        A
                     5        5       1        A
                     6        6       2        B
                     7        7       2        B
                     8        8       2        B
                     9        9       2        B
                    10       10       2        B
                    11       11       3        C
                    12       12       3        C
                    13       13       3        C
                    14       14       3        C
                    15       15       3        C
                    16       16       4        D
                    17       17       4        D
                    18       18       4        D
                    19       19       4        D
                    20       20       4        D

                  Unrandomized assignment of subjects to diets

                     The PLAN Procedure

              Factor      Select      Levels      Order
              subj          20          20        Random

        ---------------------------subj---------------------------
         1  2  3 14  8 18  4  7 10 17  9 16 11 19  5 15  6 12 13 20

                  Random assignments of subjects to diets

                    Obs     subj     dg      diet
                     1        1       1        A
                     2        2       1        A
                     3        3       1        A
                     4       14       1        A
                     5        8       1        A
                     6       18       2        B
```

7	4	2	B
8	7	2	B
9	10	2	B
10	17	2	B
11	9	3	C
12	16	3	C
13	11	3	C
14	19	3	C
15	5	3	C
16	15	4	D
17	6	4	D
18	12	4	D
19	13	4	D
20	20	4	D

The above layout is referred to as the *completely randomized design (CRD)*, and it is the simplest of all experimental designs, both in terms of analysis and experimental layout. Here, treatments are randomly allocated to the experimental units entirely at random. Thus, if a treatment is to be applied to 5 experimental units, then each unit is deemed to have the same chance of receiving the treatment as any other unit. The CRD is often used if we believe that the experimental material is homogeneous or uniform. In this case, the experimental units are regarded as a group and the investigator believes that the experimental material available contains only non-assignable variation, and that it would be impossible to try to group the material into blocks or some other subgroups such that the variation among sub-groups is larger than among units within sub-groups as far as the response variable under investigation is concerned. Usually, though not necessarily, the random assignment is restricted in such a manner as to have an equal number of experimental units assigned to each treatment. The CRD should therefore be used where extraneous factors can easily be controlled, such as in laboratories or green houses. The CRD is usually the design choice in pilot studies where experimental units and conditions are homogeneous.

The above layout for an experiment in a completely randomized design might be appropriate for 20 pots on a greenhouse bench or a series of soil analysis involving five treatments. It involves zero-way or no elimination of heterogeneity in the experimental material. The total variation in the experiment may be written as the sum of 'variation among treatment means' and 'error variation'.

Example 14.1

Carbon dioxide (CO_2) is known to have a critical effect on microbiological growth. Small amounts of CO_2 stimulate the growth of many organisms, while high concentrations inhibit the growth of most. The latter effect is used commercially when perishable food products are stored. A study is conducted to investigate the effect of CO_2 on the growth rate of *Pseudomonas fragi*, a food spoiler. Carbon dioxide is administered at five different atmospheric pressures. The response noted was the percentage change in cell mass after a 1-hour growing time. Ten cultures are used at each level. The following is the data obtained.

Factor (CO_2 pressure in atmospheres) level

0.0	0.083	0.29	0.50	0.86
62.6	50.9	45.5	29.5	24.9
59.6	44.3	41.1	22.8	17.2
64.5	47.5	29.8	19.2	7.8
59.3	49.5	38.3	20.6	10.5
58.6	48.5	40.2	29.2	17.8
64.6	50.4	38.5	24.1	22.1
50.9	35.2	30.2	22.6	22.6
56.2	49.9	27.0	32.7	16.8
52.3	42.6	40.0	24.4	15.9
62.8	41.6	33.9	29.6	8.8

Here, we are interested in testing for the overall significant differences between the five pressure levels. Denoting the population means by $\mu_1, \mu_2, \mu_3, \mu_4$ and μ_5 respectively, that is, we wish to test

$$H_0 : \mu_1 = \mu_2 = \mu_3 = \mu_4 = \mu_5$$
$$H_a : \text{at least two of the } \mu\text{'s are not equal.}$$

Further, we wish to conduct the following tests or comparisons, using both reference cell and effect coding schemes.

(a) level 1 mean = level 2 mean

(b) level 1 mean = level 3 mean

(c) level 1 mean = level 5 mean

(d) level 3 mean = level 5 mean

14.2   Regression Approach Methods

Multiple regression methods can be used to analyze the one-way ANOVA data using two coding schemes, namely,

(a) Reference Cell Coding scheme

(b) Effect Coding scheme.

14.2.1   Cell Reference Coding Scheme

As discussed in Chapter 8, one of the categories of the categorical variable, diet, is adopted as the reference cell. This is usually the last category of the variable (it could also be the first category of the variable). We adopt here the last category as the reference cell, that is pressure 0.86. In this case, since there are five levels, then we would create $5 - 1 = 4$ dummy variables, x_1, x_2, x_3 and x_4, where,

$$x_1 = \begin{cases} 1, & \text{if apl is } 0.000 \\ 0, & \text{elsewhere,} \end{cases} \qquad x_2 = \begin{cases} 1, & \text{if apl is } 0.083 \\ 0, & \text{elsewhere,} \end{cases}$$

$$x_3 = \begin{cases} 1, & \text{if apl is } 0.290 \\ 0, & \text{elsewhere,} \end{cases} \qquad x_4 = \begin{cases} 1, & \text{if apl is } 0.500 \\ 0, & \text{elsewhere,} \end{cases}$$

and where apl refers to atmospheric pressure level. This results in the following:

apl	x_1	x_2	x_3	x_4
0.000	1	0	0	0
0.083	0	1	0	0
0.290	0	0	1	0
0.500	0	0	0	1
0.860	0	0	0	0

Since we have assumed that the variances from the five populations are homogeneous, our regression equation for the above data becomes

$$y = \mu + \tau_1 x_1 + \tau_2 x_2 + \tau_3 x_3 + \tau_4 x_4 + \varepsilon. \tag{14.1}$$

A test of the overall significance of the apl means is obtained from the test of the following hypotheses,

$$H_0 : \tau_1 = \tau_2 = \tau_3 = \tau_4 = 0$$
$$H_a : \text{at least one of these parameters} \neq 0. \tag{14.2}$$

For this model therefore, if we let $\bar{y}_{i\cdot}$ be the apl means from samples $i = 1, 2, 3, 4, 5$ and $\bar{y}_{\cdot\cdot}$ be the overall mean, then, under this model we have

$$\hat{\mu} = \bar{y}_{5\cdot}$$

$$\hat{\tau}_i = \bar{y}_{i\cdot} - \bar{y}_{5\cdot}, \quad i = 1, 2, 3, 4.$$

The observed means are

$$\bar{y}_{1\cdot} = 59.14, \ \bar{y}_{2\cdot} = 46.04, \ \bar{y}_{3\cdot} = 36.45, \ \bar{y}_{4\cdot} = 25.47, \ \bar{y}_{5\cdot} = 16.44, \ \text{and} \ \bar{y}_{\cdot\cdot} = 36.708.$$

The regression analysis is implemented with the following SAS program and its corresponding output.

```
data noval;
do rep=1 to 10;
do level=1 to 5;
input y @@;
output;
end; end;
datalines;
 62.6 50.9 45.5 29.5 24.9 59.6 44.3 41.1 22.8 17.2
 64.5 47.5 29.8 19.2 7.8 59.3 49.5 38.3 20.6 10.5
 58.6 48.5 40.2 29.2 17.8 64.6 50.4 38.5 24.1 22.1
 50.9 35.2 30.2 22.6 22.6 56.2 49.9 27.0 32.7 16.8
 52.3 42.6 40.0 24.4 15.9 62.8 41.6 33.9 29.6 8.8
;
data new;
set noval;
if level=1 then x1=1; else x1=0;
if level=2 then x2=1; else x2=0;
if level=3 then x3=1; else x3=0;
if level=4 then x4=1; else x4=0;
proc print;
run;
proc reg data=new;
model y=x1 x2 x3 x4;
test1: test x1-x2;
test2: test x1-x3;
test3: test x1;
test4: test x3;
run;
```

```
                     The SAS System

           Obs   rep   level     y     x1   x2   x3   x4
            1     1      1      62.6    1    0    0    0
            2     1      2      50.9    0    1    0    0
            3     1      3      45.5    0    0    1    0
            4     1      4      29.5    0    0    0    1
            5     1      5      24.9    0    0    0    0
           .............................................
           45     9      5      15.9    0    0    0    0
           46    10      1      62.8    1    0    0    0
           47    10      2      41.6    0    1    0    0
           48    10      3      33.9    0    0    1    0
           49    10      4      29.6    0    0    0    1
           50    10      5       8.8    0    0    0    0
```

```
                   The REG Procedure
                    Model: MODEL1
                Dependent Variable: y

                 Analysis of Variance
```

Source	DF	Sum of Squares	Mean Square	F Value	Pr > F
Model	4	11274	2818.57970	101.63	<.0001
Error	45	1248.03800	27.73418		
Corrected Total	49	12522			

Root MSE	5.26632	R-Square	0.9003	
Dependent Mean	36.70800	Adj R-Sq	0.8915	
Coeff Var	14.34653			

```
                  Parameter Estimates
```

Variable	DF	Parameter Estimate	Standard Error	t Value	Pr > |t|

```
Intercept   1      16.44000      1.66536      9.87    <.0001
x1          1      42.70000      2.35517     18.13    <.0001
x2          1      29.60000      2.35517     12.57    <.0001
x3          1      20.01000      2.35517      8.50    <.0001
x4          1       9.03000      2.35517      3.83    0.0004
```

             Test test1 Results for Dependent Variable y

| | | Mean | | |
Source	DF	Square	F Value	Pr > F
Numerator	1	858.05000	30.94	<.0001
Denominator	45	27.73418		

             Test test2 Results for Dependent Variable y

| | | Mean | | |
Source	DF	Square	F Value	Pr > F
Numerator	1	2574.18050	92.82	<.0001
Denominator	45	27.73418		

             Test test3 Results for Dependent Variable y

| | | Mean | | |
Source	DF	Square	F Value	Pr > F
Numerator	1	9116.45000	328.71	<.0001
Denominator	45	27.73418		

             Test test4 Results for Dependent Variable y

| | | Mean | | |
Source	DF	Square	F Value	Pr > F
Numerator	1	2002.00050	72.19	<.0001
Denominator	45	27.73418		

The estimates of the regression parameters here are, $\hat{\mu} = 16.44 = \bar{y}_{5.}$, while

$$\hat{\tau}_1 = 42.70 = \bar{y}_{1.} - \bar{y}_{5.} = 59.14 - 16.44 = 42.70$$
$$\hat{\tau}_2 = 29.60 = \bar{y}_{2.} - \bar{y}_{5.} = 46.04 - 16.44 = 29.60$$
$$\hat{\tau}_3 = 20.01 = \bar{y}_{3.} - \bar{y}_{5.} = 36.45 - 16.44 = 20.01$$
$$\hat{\tau}_4 = 9.03 = \bar{y}_{4.} - \bar{y}_{5.} = 25.47 - 16.44 = 9.03.$$

The global hypotheses in (14.2) is tested with the overall F value in the regression ANOVA table. The corresponding F value is $F^* = 101.63$ with 4 and 45 degrees of freedom, with a corresponding p-value of $< .0001$. Hence, there are significant differences in the five means of the populations.

Under the cell reference coding scheme, the selected hypotheses above reduce to the following:

$$\mu_1 = \mu_2 \quad \text{becomes} \quad \tau_1 - \tau_2 = 0$$
$$\mu_1 = \mu_3 \quad \text{becomes} \quad \tau_1 - \tau_3 = 0$$
$$\mu_1 = \mu_5 \quad \text{becomes} \quad \tau_1 = 0$$
$$\mu_3 = \mu_5 \quad \text{becomes} \quad \tau_3 = 0,$$

since $\tau_5 = 0$ under this coding scheme. The above are accordingly implemented in the previous SAS program as tests 1 to 4 respectively. The comparisons are all significant. We caution however that this kind of comparison has complications that are examined at a later section in this chapter.

14.2.2   Effect Coding Scheme

Here, the four dummy variables x_1, x_2, x_3 and x_4 are defined as

$$x_1 = \begin{cases} 1, & \text{if apl is } 0.000 \\ -1, & \text{if apl is } 0.860 \\ 0, & \text{elsewhere,} \end{cases} \qquad x_2 = \begin{cases} 1, & \text{if apl is } 0.083 \\ -1, & \text{if apl is } 0.860 \\ 0, & \text{elsewhere,} \end{cases}$$

$$x_3 = \begin{cases} 1, & \text{if apl is } 0.290 \\ -1, & \text{if apl is } 0.860 \\ 0, & \text{elsewhere,} \end{cases} \qquad x_4 = \begin{cases} 1 & \text{if apl is } 0.500 \\ -1 & \text{if apl is } 0.860 \\ 0 & \text{elsewhere,} \end{cases}$$

which results in the following:

apl	x_1	x_2	x_3	x_4
0.000	1	0	0	0
0.083	0	1	0	0
0.290	0	0	1	0
0.500	0	0	0	1
0.860	-1	-1	-1	-1

Again, based on our assumption of homogeneity of populations, our regression equation becomes

$$y = \mu + \tau_1 x_1 + \tau_2 x_2 + \tau_3 x_3 + \tau_4 x_4 + \varepsilon. \tag{14.3}$$

A test of the overall significance of the apl means is obtained from the test of the following hypotheses:

$$H_0 : \tau_1 = \tau_2 = \tau_3 = \tau_4 = 0$$
$$H_a : \text{at least one of these parameters} \neq 0. \tag{14.4}$$

For this model therefore, if we let $\bar{y}_{i.}$ be the apl means from samples $i = 1, 2, 3, 4, 5$ and $\bar{y}_{..}$ be the overall mean, then, under this model we have

$$\hat{\mu} = \bar{y}_{..}$$
$$\hat{\tau}_i = \bar{y}_{i.} - \bar{y}_{..}, \quad i = 1, 2, 3, 4.$$

The observed means are

$$\bar{y}_{1.} = 59.14, \ \bar{y}_{2.} = 46.04, \ \bar{y}_{3.} = 36.45, \ \bar{y}_{4.} = 25.47, \ \bar{y}_{5.} = 16.44, \text{ and } \bar{y}_{..} = 36.708.$$

The regression analysis is implemented with the following SAS program and its corresponding output.

```
data nova2;
set noval;
if level=1 then x1=1; else if level=5 then x1=-1; else x1=0;
if level=2 then x2=1; else if level=5 then x2=-1; else x2=0;
if level=3 then x3=1; else if level=5 then x3=-1; else x3=0;
if level=4 then x4=1; else if level=5 then x4=-1; else x4=0;
proc print; run;
proc reg data=nova2;
model y=x1 x2 x3 x4;
test1: test x1-x2;
test2: test x1-x3;
test3: test 2*x1+x2+x3+x4;
test4: test x1+x2+2*x3+x4;
run;
```

Obs	rep	level	y	x1	x2	x3	x4
1	1	1	62.6	1	0	0	0
2	1	2	50.9	0	1	0	0
3	1	3	45.5	0	0	1	0
4	1	4	29.5	0	0	0	1
5	1	5	24.9	-1	-1	-1	-1

```
..............................................................
       46    10    1      62.8     1     0     0     0
       47    10    2      41.6     0     1     0     0
       48    10    3      33.9     0     0     1     0
       49    10    4      29.6     0     0     0     1
       50    10    5       8.8    -1    -1    -1    -1
```

```
                        The REG Procedure
                         Model: MODEL1
                      Dependent Variable: y

                      Analysis of Variance

                                   Sum of        Mean
Source                  DF         Squares      Square    F Value   Pr > F
Model                    4           11274   2818.57970    101.63   <.0001
Error                   45      1248.03800     27.73418
Corrected Total         49           12522


            Root MSE             5.26632     R-Square    0.9003
            Dependent Mean      36.70800     Adj R-Sq    0.8915
            Coeff Var           14.34653

                       Parameter Estimates

                      Parameter     Standard
Variable     DF       Estimate        Error    t Value   Pr > |t|
Intercept     1        36.70800      0.74477     49.29    <.0001
x1            1        22.43200      1.48954     15.06    <.0001
x2            1         9.33200      1.48954      6.27    <.0001
x3            1        -0.25800      1.48954     -0.17    0.8633
x4            1       -11.23800      1.48954     -7.54    <.0001

          Test test1 Results for Dependent Variable y

                                 Mean
Source               DF         Square    F Value   Pr > F
Numerator             1      858.05000      30.94   <.0001
Denominator          45       27.73418


          Test test2 Results for Dependent Variable y

                                 Mean
Source               DF         Square    F Value   Pr > F
Numerator             1     2574.18050      92.82   <.0001
Denominator          45       27.73418


          Test test3 Results for Dependent Variable y

                                 Mean
Source               DF         Square    F Value   Pr > F
Numerator             1     9116.45000     328.71   <.0001
Denominator          45       27.73418


          Test test4 Results for Dependent Variable y

                                 Mean
Source               DF         Square    F Value   Pr > F
Numerator             1     2002.00050      72.19   <.0001
Denominator          45       27.73418
```

The estimates of the regression parameters are, $\hat{\mu} = 36.708 = \bar{y}_{..}$, while

$$\hat{\tau}_1 = 22.432 \quad = \bar{y}_{1.} - \bar{y}_{..} = 59.14 - 36.708 \quad = 22.432$$
$$\hat{\tau}_2 = 9.332 \quad = \bar{y}_{2.} - \bar{y}_{..} = 46.04 - 36.708 \quad = 9.332$$
$$\hat{\tau}_3 = -0.258 = \bar{y}_{3.} - \bar{y}_{..} = 36.45 - 66.708 \quad = -0.258$$
$$\hat{\tau}_4 = -11.238 = \bar{y}_{4.} - \bar{y}_{..} = 25.47 - 36.708 = -11.238.$$

The global hypotheses in (14.2) is tested with the overall F value in the regression ANOVA table. The corresponding F value is $F^* = 101.63$ with 4 and 45 degrees of freedom, with a corresponding p-value of $< .0001$. Hence there are significant differences in the five means of the populations. We may note here that while parameter estimates under both coding schemes are different, the analysis of variance table are the same and the same conclusions are drawn as in the previous coding scheme.

Under the effect coding scheme, the selected hypotheses above reduce to the following:

$$\mu_1 = \mu_2 \quad \text{becomes} \quad \tau_1 - \tau_2 = 0$$
$$\mu_1 = \mu_3 \quad \text{becomes} \quad \tau_1 - \tau_3 = 0$$
$$\mu_1 = \mu_5 \quad \text{becomes} \quad 2\tau_1 + \tau_2 + \tau_3 + \tau_4 = 0$$
$$\mu_3 = \mu_5 \quad \text{becomes} \quad \tau_1 + \tau_2 + 2\tau_3 + \tau_4 = 0,$$

since $\tau_5 = -(\tau_1 + \tau_2 + \tau_3 + \tau_4)$ under this coding scheme. The above are accordingly implemented in the previous SAS program as tests 1 to 4 respectively. The comparisons are all significant. We caution again, however that this kind of comparison has complications that are examined at a later section in this chapter. Alternatively, the transformation in the previous SAS program could have been accomplished with the following array statement in SAS.

```
data nova2;
set nova1;
array xll[4] x1-x4;
do i=1 to 4;
if level=i then xll[i]=1; else if level=5 then xll[i]=-1; else xll[i]=0;
end;
output;
drop i;
run;
```

14.3   ANOVA Model Approach

The regression approach model under the effect coding scheme in (14.3) can be written more succinctly as

$$y_{ij} = \mu + \tau_i + \varepsilon_{ij}, \quad i = 1, 2, \ldots, 5 \quad \text{and} \quad j = 1, 2, \ldots, n, \tag{14.5}$$

with $\sum \tau_i = 0$, where

- y_{ij} is the observed response for treatment i in replicate j

- μ is the overall mean

- τ_i is the effect of treatment i, and

- ε_{ij} is the random error term for treatment i in replicate j.

The following assumptions are needed for the analysis.

1. The ε_{ij} are assumed to be normally and independently distributed with mean 0 and constant variance σ^2. That is, we assume that $\varepsilon_{ij} \sim \text{NID}(0, \sigma^2)$.

2. The populations are homogeneous. In other words, the populations have a common variance, σ^2, and we would therefore need to test the homogeneity hypotheses:

$$H_0 : \sigma_1^2 = \sigma_2^2 = \cdots = \sigma_5^2$$
$$H_a : \text{at least two of the variances are unequal.} \tag{14.6}$$

3. The μ and τ_i are assumed to be *fixed* and that $\sum_{i=1}^{5} \tau_i = 0$.

The hypotheses of interest in the study is given by

$$H_0 : \tau_1 = \tau_2 = \tau_3 = \tau_4 = \tau_5$$
$$H_a : \text{at least two of the } \tau_i\text{'s are unequal.} \tag{14.7}$$

The model in (14.5) is called the factor effects model, where τ_i is the effect of treatment or factor i. An equivalent formulation of one-way ANOVA model is

$$y_{ij} = \mu_i + \varepsilon_{ij}, \tag{14.8}$$

where y_{ij} is the observed response for treatment or factor i in replicate j, μ_i are parameters and ε_{ij} is the random error term for factor i in replicate j. The model in (14.8) is called the cell means model. By writing the μ_i in cell mean model (14.8) as

$$\mu_i = \mu + (\mu_i - \mu) = \mu + \tau_i,$$

the cell means model can be re-expressed as the factor effects model in (14.5). An equivalent hypotheses to (14.7) is given by

$$H_0 : \mu_1 = \mu_2 = \mu_3 = \mu_4 = \mu_5$$

$$H_a : \text{at least two of the means are unequal.}$$

Data Presentation

Suppose we present the data layout for such an experiment resulting from the expression in equation (14.5), where y_{ij} is the observed response for treatment i in replicate j, $y_{i.}$ is the total response yield for treatment i, and $y_{..}$ is the overall total or grand total, which may sometimes be denoted by G. The display below gives this representation for a one-way ANOVA layout.

Treatments	Replicates (j)					Total
i	1	2	3	$\cdots$	b_i	
1	y_{11}	y_{12}	y_{13}	$\cdots$	y_{1b_1}	$y_{1.}$
2	y_{21}	y_{22}	y_{23}	$\cdots$	y_{2b_2}	$y_{2.}$
3	y_{31}	y_{32}	y_{33}	$\cdots$	y_{3b_3}	$y_{3.}$
$\vdots$	$\vdots$	$\vdots$	$\vdots$	$\vdots$	$\vdots$	$\vdots$
i	y_{i1}	y_{i2}	y_{i3}	$\cdots$	y_{ib_i}	$y_{i.}$
$\vdots$	$\vdots$	$\vdots$	$\vdots$	$\vdots$	$\vdots$	$\vdots$
k	y_{k1}	y_{k2}	y_{k3}	$\cdots$	y_{kb_k}	$y_{k.}$
Total						$y_{..}$

14.3.1   Calculations for Analysis of Variance Table

The total sum of squares is computed as

$$\text{SS Tot} = y_{11}^2 + y_{12}^2 + \cdots + y_{kb_k}^2 - \frac{y_{..}^2}{n}, \tag{14.9}$$

where $n = b_1 + b_2 + \cdots + b_k$ is the total number of observations. It is assumed that treatments $1, 2, \ldots, k$ are each replicated $b_1, b_2, \ldots, b_k$ times. If treatments are equally replicated (i.e. $b_1 = b_2 = \ldots = b_k = b$, then $n = kb$. Alternatively, the total SS in (14.9) can also be expressed as,

$$\text{SS Tot} = \sum_{i=1}^{k} \sum_{j=1}^{b_i} (y_{ij} - \bar{y})^2, \tag{14.10}$$

where $\bar{y}_{..} = \dfrac{y_{..}}{n}$. The total sum of squares is based on $n - 1$ degrees of freedom. The treatment SS is also computed as

$$\text{SS Trt} = \frac{y_{1.}^2}{b_1} + \frac{y_{2.}^2}{b_2} + \cdots + \frac{y_{k.}^2}{b_k} - \frac{y_{..}^2}{n}. \tag{14.11}$$

An alternative way to compute (14.11) is

$$\text{SS Trt} = \sum_{i=1}^{k}\sum_{j=1}^{b_i}(\bar{y}_{i.} - \bar{y})^2 = \sum_{i=1}^{k} b_i(\bar{y}_{i.} - \bar{y})^2, \tag{14.12}$$

where $\bar{y}_{i.}$ is the average for treatment level i. Thus, $\bar{y}_{i.} = y_{i.}/b_i$. The treatment SS is based on $k-1$ degrees of freedom and the corresponding Error SS is obtained by subtraction, viz:

$$\text{Error SS} = \sum_{i=1}^{k}\sum_{j=1}^{b_i}(y_{ij} - \bar{y})^2 - \sum_{i=1}^{k}\sum_{j=1}^{b_i}(\bar{y}_{i.} - \bar{y})^2 = \sum_{i=1}^{k}\sum_{j=1}^{b_i}(y_{ij} - \bar{y}_{i.})^2, \tag{14.13}$$

leading to the following analysis of variance table.

Source	d.f.	SS	MS	F
Treatments	$k-1$	SS Trt	$\text{TMS} = \dfrac{\text{SS Trt}}{k-1}$	$F^* = \dfrac{\text{TMS}}{\text{EMS}}$
Error	$n-k$	SSE	$\text{MSE} = \dfrac{\text{SSE}}{n-k} = S^2$	
Total	$n-1$	SS Tot		

For our data, first let us display the summary statistics:

	Levels of CO_2 (apl)				
	0.0	0.083	0.29	0.50	0.86
$\sum y_{i.}$	591.4	460.4	364.5	254.7	164.4
$\bar{y}_{i.}$	59.14	46.04	36.45	25.47	16.44
S_i^2	23.0849	25.5293	35.2117	20.1001	34.7449

Hence,

$$\text{Total SS} = 62.6^2 + 59.6^2 + \cdots + 8.8^2 - \frac{1835.4^2}{50} = 12522.3568$$

$$\text{Levels SS} = \frac{591.4^2}{10} + \frac{460.4^2}{10} + \cdots + \frac{164.4^2}{10} - \frac{1835.4^2}{50} = 11274.3188$$

The error SS is obtained by subtraction. That is, error SS = Total SS − Levels SS, which equals

$$12522.3568 - 11274.3168 = 1248.0380.$$

The analysis of variance table is therefore obtained as follows:

Source	d.f.	SS	MS	F
Levels	4	11274.3188	2818.5797	101.63
Error	45	1248.0380	27.7342	
Total	49	12522.3568		

Before commencing to conduct tests of significance, etc., it is important to test for the underlying assumptions.

(a) The assumption that the error terms ε_{ij} are normally distributed randomly and independently will be tested by conducting the *normality test* and we shall adopt the Anderson-Darling test for normality in this text. Here, the hypotheses of interest can be written as

$$H_0 : \varepsilon_{ij} \text{ is normal}$$
$$H_a : \varepsilon_{ij} \text{ is not normal.}$$

The normality test using **proc univariate** in SAS produces the following goodness-of-fit test statistics. The p-value from the Anderson-Darling is 0.211 which indicates that we would fail to reject H_0. That is, the data could be assumed to have come from normally distributed populations.

```
                          The UNIVARIATE Procedure

            Goodness-of-Fit Tests for Normal Distribution

       Test                ---Statistic----    -----p Value-----
       Kolmogorov-Smirnov   D    0.09803323    Pr > D      >0.150
       Cramer-von Mises     W-Sq 0.07808004    Pr > W-Sq    0.221
       Anderson-Darling     A-Sq 0.49802112    Pr > A-Sq    0.211
```

(b) The assumption that the samples were drawn from homogeneous populations, that is, the assumption that

$$\sigma_1^2 = \sigma_2^2 = \sigma_3^2 = \sigma_4^2 = \sigma_1^5 = \sigma^2 (\text{Unknown}),$$

will be tested by the use of "Bartlett's test of Homogeneity of Variance" or by "Levene's test". We shall adopt Bartlett's test in this text. The results of conducting this test in SAS is displayed below.

```
        Bartlett's Test for Homogeneity of  y Variance

        Source       DF     Chi-Square    Pr > ChiSq
        ------------------------------------------------
        Level         4       1.0701        0.8990
```

The resulting p-value of 0.8990 indicates that we would fail to reject the null hypothesis, H_0, of equality of variances.

$$H_0 : \sigma_1^2 = \sigma_2^2 = \sigma_3^2 = \sigma_4^2 = \sigma_5^2$$
$$H_a : \text{at least two of the } \sigma\text{'s are not equal.}$$

14.3.2   Tests of Significance

The primary interest is to test the following hypotheses concerning the population means from which the sample were drawn in this study. That is,

$$H_0 : \mu_1 = \mu_2 = \mu_3 = \mu_4 = \mu_5$$
$$H_a : \text{at least two of the } \mu\text{'s are not equal.}$$

From the analysis of variance table presented earlier, the above hypotheses are tested by the computed F value in the ANOVA table. Our decision rule would be to reject H_0 if $F^* \geq f_{\alpha,4,45}$.

Since the computed F value is $101.63 \ggg f_{0.05,4,45} = 5.19$, we would therefore strongly reject H_0 and conclude that there are significant differences in the population means of the five levels of CO_2 pressures. Alternatively, from the SAS output, the corresponding p-value is $< .0001 \lll 0.05$, which again indicates that we would strongly reject H_0. We present the SAS program and partial output for this analysis, as well as the options required to test the two assumptions described above.

```
data nova2;
set nova1;
proc glm data=nova2 order=data;
class level;
model y=level;
means level/hovtest=bartlett lsd tukey;
```

```
output out=aa r=resid;
contrast '1 vs 2' level 1 -1 0 0 0;
contrast '1 vs 3' level 1 0 -1 0 0;
contrast '1 vs 5' level 1 0 0 0 -1;
contrast '3 vs 5' level 0 0 1 0 -1;
run;
proc univariate data=aa normal;
var resid;
run;
```

The GLM Procedure

Class Level Information

Class	Levels	Values
level	5	1 2 3 4 5

Number of observations   50

The GLM Procedure

Dependent Variable: y

Source	DF	Sum of Squares	Mean Square	F Value	Pr > F
Model	4	11274.31880	2818.57970	101.63	<.0001
Error	45	1248.03800	27.73418		
Corrected Total	49	12522.35680			

R-Square	Coeff Var	Root MSE	y Mean
0.900335	14.34653	5.266325	36.70800

Source	DF	Type I SS	Mean Square	F Value	Pr > F
level	4	11274.31880	2818.57970	101.63	<.0001

Source	DF	Type III SS	Mean Square	F Value	Pr > F
level	4	11274.31880	2818.57970	101.63	<.0001

The GLM Procedure

Bartlett's Test for Homogeneity of y Variance

Source	DF	Chi-Square	Pr > ChiSq
level	4	1.0701	0.8990

The SAS System

The GLM Procedure

t Tests (LSD) for y

NOTE: This test controls the Type I comparisonwise error rate, not the experimentwise error rate.

Alpha	0.05
Error Degrees of Freedom	45
Error Mean Square	27.73418
Critical Value of t	2.01410
Least Significant Difference	4.7436

Means with the same letter are not significantly different.

t Grouping	Mean	N	level
A	59.140	10	1
B	46.040	10	2
C	36.450	10	3
D	25.470	10	4
E	16.440	10	5

The GLM Procedure

Tukey's Studentized Range (HSD) Test for y

```
NOTE: This test controls the Type I experimentwise error rate, but it generally has
                 a higher Type II error rate than REGWQ.

              Alpha                                      0.05
              Error Degrees of Freedom                     45
              Error Mean Square                      27.73418
              Critical Value of Studentized Range    4.01842
              Minimum Significant Difference          6.6921

        Means with the same letter are not significantly different.

           Tukey Grouping           Mean        N     level
                      A            59.140       10     1

                      B            46.040       10     2

                      C            36.450       10     3

                      D            25.470       10     4

                      E            16.440       10     5
```

```
                          The GLM Procedure
Dependent Variable: y

Contrast                 DF     Contrast SS      Mean Square    F Value    Pr > F
----------------------------------------------------------------------------------
1 vs 2                    1      858.050000       858.050000      30.94    <.0001
1 vs 3                    1     2574.180500      2574.180500      92.82    <.0001
1 vs 5                    1     9116.450000      9116.450000     328.71    <.0001
3 vs 5                    1     2002.000500      2002.000500      72.19    <.0001
```

```
                       The UNIVARIATE Procedure
                          Variable:  resid

                         Tests for Normality

        Test                  --Statistic---     -----p Value------
        Shapiro-Wilk          W   0.962699       Pr < W      0.1153
        Kolmogorov-Smirnov    D   0.133512       Pr > D      0.0243
        Cramer-von Mises      W-Sq  0.1142       Pr > W-Sq   0.0742
        Anderson-Darling      A-Sq  0.682836     Pr > A-Sq   0.0741
```

14.3.3   Standard Errors

The estimated standard error of a treatment mean is

$$\sqrt{\frac{S^2}{b_i}} = \sqrt{\frac{27.7342}{10}} = 1.6654,$$

where b_i is the number of observations for treatment level i. The estimated standard error of a difference between two treatment means based on b_i and b_j observations $(\bar{y}_{i.} - \bar{y}_{j.})$, $i \neq j$, is

$$\sqrt{\frac{S^2}{b_i} + \frac{S^2}{b_j}} \quad \text{or} \quad \sqrt{\frac{2S^2}{b}} \quad \text{if } b_i = b_j = b.$$

Hence, in our example where all treatments are equally replicated, this becomes

$$\sqrt{\frac{2S^2}{b}} = \sqrt{\frac{2 \times 27.7342}{10}} = 2.3552.$$

We can therefore present the results as follows:

```
        Levels           Means
           1             59.14
           2             46.04     S.E. of difference between
           3             36.45     any two means 2.3552 (45 d.f.)
           4             25.47
           5             16.44
```

Without further knowledge about the five levels, the making of particular comparisons between pairs of levels is rather dangerous. However, it is clear that level 5 is the best atmospheric pressure as it has the lowest percentage change.

14.4 Multiple Comparisons Procedures

We have seen that although the F test above indicates that there are significant differences among the level (treatment) means, however, this does not tell or indicate which specific means are significantly different. To do this we have to conduct paired comparisons. In this example, since we have five population means, we would therefore need to make $\binom{5}{2} = 10$ such paired comparisons, leading to what is often known as *multiple comparison tests*-because we are making more comparisons than allowed by the treatment degrees of freedom (Here, $10 > 4$). We consider some of the procedures for conducting multiple comparisons in what follows.

14.4.1 Fisher's Least Significant Difference-LSD

For any two means, say, μ_i and μ_j, with $i \neq j$, the hypothesis of interest is

$$H_0 : \mu_i = \mu_j$$
$$H_a : \mu_i \neq \mu_j, \quad i \neq j.$$

To conduct the test, the standard error for the difference between the two sample means is $\sqrt{\frac{S^2}{b_1} + \frac{S^2}{b_2}}$. In our example, this equals $\sqrt{\frac{2S^2}{10}}$ since $b_1 = b_2 = 10$. The least significance difference is

$$\text{LSD} = \sqrt{\frac{2S^2}{10}} \times t_{0.025,45} = 2.3552 \times 2.0141 = 4.7436,$$

where $t_{0.025,45} = 2.0141$ is the Students' t-distribution percentile based on the error degrees of freedom of 45 at the $\alpha = 0.05$ level of significance. In other words, the two means μ_i and μ_j will be significantly different at $\alpha = 0.05$ level of significance if

$$|\bar{y}_{i.} - \bar{y}_{j.}| \geq 4.7436, \quad \text{for } i \neq j. \tag{14.14}$$

The implementation of this in SAS is presented in a summarized result below and is accomplished in SAS by specifying after the model statement **means level/lsd**, where **level** is the factor name. The result is given in the previous section.

Alternatively, we could use confidence interval approach to implement the same. Here, 95% confidence intervals for the ten pairs of comparisons can be obtained as follows:

$$(\bar{y}_{i.} - \bar{y}_{j.}) \pm t_{0.025,45} \times \sqrt{\frac{2S^2}{10}} = (\bar{y}_{i.} - \bar{y}_{j.}) \pm 4.7436.$$

For levels 1 and 2 for instance, this becomes

$$(59.14 - 46.04) \pm 4.7436 = 13.10 \pm 4.7436 = (8.3564, 17.8436).$$

Since this interval does not include zero, therefore, we can conclude that there is a significant difference between level means 1 and 2. The procedure is then repeated for the other 9 paired comparisons. This can readily be accomplished in SAS by specifying after the model statement the following **means level/lsd cldiff**.

```
                    The GLM Procedure

                   t Tests (LSD) for y

      NOTE: This test controls the Type I comparisonwise error rate, not the
                    experimentwise error rate.

            Alpha                          0.05
            Error Degrees of Freedom         45
            Error Mean Square          27.73418
            Critical Value of t         2.01410
            Least Significant Difference  4.7436
```

Comparisons significant at the 0.05 level are indicated by ***.

level Comparison	Difference Between Means	95% Confidence Limits		
1 - 2	13.100	8.356	17.844	***
1 - 3	22.690	17.946	27.434	***
1 - 4	33.670	28.926	38.414	***
1 - 5	42.700	37.956	47.444	***
2 - 1	-13.100	-17.844	-8.356	***
2 - 3	9.590	4.846	14.334	***
2 - 4	20.570	15.826	25.314	***
2 - 5	29.600	24.856	34.344	***
3 - 1	-22.690	-27.434	-17.946	***
3 - 2	-9.590	-14.334	-4.846	***
3 - 4	10.980	6.236	15.724	***
3 - 5	20.010	15.266	24.754	***
4 - 1	-33.670	-38.414	-28.926	***
4 - 2	-20.570	-25.314	-15.826	***
4 - 3	-10.980	-15.724	-6.236	***
4 - 5	9.030	4.286	13.774	***
5 - 1	-42.700	-47.444	-37.956	***
5 - 2	-29.600	-34.344	-24.856	***
5 - 3	-20.010	-24.754	-15.266	***
5 - 4	-9.030	-13.774	-4.286	***

Notice that SAS produced 20 rather than 10 comparisons for this example. SAS not only computes confidence intervals for $\mu_i - \mu_j$ but also for $\mu_j - \mu_i$, since SAS is not sure what the reader prefers. The result obtained for μ_1 and μ_2 agree with our computed results. Before we discuss the results from both approaches, we notice that both approaches warns us with *NOTE: This test controls the Type I comparison-wise error rate, not the experiment-wise error rate*. It is therefore important that we discuss briefly experiment-wise error rate.

14.4.2   Experiment-wise Error Rate (EER)

With several comparisons on the means, the experiment-wise error rate (EER) is the probability that one or more of the comparison tests results in a Type I error (that is, the probability of rejecting at least one correct null hypothesis under several other competing null hypotheses-which may be true or false). If the comparisons are independent, then the experiment-wise error rate is

$$\alpha^* = 1 - (1 - \alpha)^h,$$

where α^* is the experiment-wise error rate, α is the per-comparison error rate or specified level of significance, and h is the total number of comparisons. In our example for instance where there are 10 independent comparisons to be made at the 0.05 level of significance each, then the probability that at least one of them would result in a Type I error is

$$1 - (1 - 0.05)^{10} = 0.4013.$$

Clearly, a Type I error rate of 0.4013 is unacceptable. However, if the comparisons are not independent then the experiment-wise error rate is less than $1 - (1 - \alpha)^h$ and regardless of whether the comparisons are independent, $\alpha^* \leq h\alpha$. In our example for instance, $0.4013 < 10(0.05) = 0.50$.

Although Fisher's LSD does not control the experiment-wise error rate, the results obtained indicate that there are significant differences in the ten pairs. Thus Level 5 gives the lowest mean, while level 1 gives the highest rate of change of bacteria. Since lowest is best here, we would recommend level 5 as the best.

We now consider multiple comparisons procedures that endeavor to control the experiment-wise error rates in the following sections.

14.4.3   The Tukey Test

The Tukey test procedure uses the critical value $q_\alpha(k, \nu)$ which is obtained from tables of significant studentized ranges (two-tailed Table 7 in the Appendix) having k treatments, in the expression in (14.15). Here, α is the upper tail of the q distribution and ν is the number of degrees of freedom on which the mean square error (MSE) is based and b_i is the number of observations on which the means are based. That is, we need to obtain

$$\text{Significance Difference (SD)} = q_\alpha(k, \nu) \times \sqrt{\frac{MSE}{b_i}}. \tag{14.15}$$

In our example, $b_i = 10$, and if $\alpha = 0.05$ and $\sqrt{MSE/10} = \sqrt{27.7342/10} = 1.6654$, hence $q_{0.05}(5, 45) = 4.025$ where $k = 5$ is the number of means to be compared. Now SD $= 1.6654 \times 4.025 = 6.7032$, approximately. When testing differences between the various means, if the difference between any two means is larger than the SD $= 6.7032$, then the means are assumed to be significantly different. That is, if

$$|\bar{y}_{i.} - \bar{y}_{j.}| \geq 6.7032, \quad \text{for } i \neq j. \tag{14.16}$$

The implementation of this in SAS is presented in a summarized result below and is accomplished in SAS by specifying after the model statement **means level/tukey lines**.
In general, to conduct Tukey's test, we would do the following:

(a) Calculate the SD for a specified α level as in (14.15).

(b) Rank the treatment means from smallest to largest.

(c) For those treatment means not indicating significance, place a bar under those pairs. Any pair not connected by an underbar implies significant difference in the population means.

We may note that Tukey's procedure ensures that all comparisons are made at the specified α significance value. For our example, the results indicate significance differences between all means.

Results of Tukey's Test

$\bar{y}_{5.}$	$\bar{y}_{4.}$	$\bar{y}_{3.}$	$\bar{y}_{2.}$	$\bar{y}_{1.}$
16.44	25.47	36.45	46.04	59.14

14.4.4 Other Multiple Comparisons Procedures

Before we discuss some other multiple comparisons procedures, let us first define *contrasts* and the concept of orthogonal contrasts.

Definition of Contrast

For k treatments having population means $\mu_1, \mu_2, \ldots, \mu_k$, a contrast, L, is a linear combination of the k treatment means, that is

$$L = c_1\mu_1 + c_2\mu_2 + \cdots + c_k\mu_k = \sum_{i=1}^{k} c_i\mu_i, \tag{14.17}$$

where $c_1, c_2, \ldots, c_k$ are constants such that $\sum_{i=1}^{k} c_i = 0$.

Consider for example four treatments A, B, C, and D. One might be interested in the following comparisons (contrasts)

$$L_1 = \mu_A - \frac{\mu_B + \mu_C + \mu_D}{3},$$
$$L_2 = \mu_B - \frac{\mu_C + \mu_D}{2},$$
$$L_3 = \mu_C - \mu_D.$$

(a) In L_1, $c_A = 1, c_B = c_C = c_D = -\frac{1}{3}$. Hence, L_1 is a linear combination of the four means and $\sum c_i = 0$. Thus, L_1 is a contrast.

(b) In L_2, $c_A = 0, c_B = 1, c_C = c_D = -\frac{1}{2}$. Hence, L_2 is a linear combination of the four means and again, $\sum c_i = 0$. Thus, L_2 is also a contrast.

(c) Similarly, in L_3, $c_A = 0, c_B = 0, c_C = 1, c_D = -1$. Again, this indicates that L_3 is a linear combination of the means and that $\sum c_i = 0$ in this case.

Alternatively, we may decide not to work with fractions and re-write the contrasts, say, L_1 as $L_1 : 3\mu_A - \mu_B - \mu_C - \mu_D$. The results of removing fractions are displayed in the following table.

Contrasts	μ_A	μ_B	μ_C	μ_D
L_1	3	-1	-1	-1
L_2	0	2	-1	-1
L_3	0	0	1	-1

Definition of Orthogonal Contrasts

Two contrasts, say,

$$L_1 = \sum_{i=1}^{k} c_i \mu_i \quad \text{and} \quad L_2 = \sum_{i=1}^{k} d_i \mu_i,$$

where $c_1, c_2, \ldots, c_k$ and $d_1, d_2, \ldots, d_k$ are constants with $\sum_{i=1}^{k} c_i = 0$ and $\sum_{i=1}^{k} d_i = 0$ are said to be orthogonal if and only if

$$\sum_{i=1}^{k} c_i \, d_i = 0.$$

For instance, in the four treatments example above, the pairs of contrasts (L_1, L_2), (L_1, L_3) and (L_2, L_3) are orthogonal. That is, L_1, L_2 and L_3 are pairwise orthogonal. This has implications for partitioning the treatment SS into the three components represented by the contrasts. We shall examine an example of this in a later section in this chapter.

14.4.5   Scheffé's Test

To implement Scheffé's multiple comparison procedure, we do the following:

(i) For each of the contrasts, say, $L = \sum_{i=1}^{k} c_i \mu_i$, calculate

$$\hat{L} = \sum_{i=1}^{k} c_i \bar{y}_i. \quad \text{and} \quad U = \sqrt{(k-1)f_{\alpha, \, k-1, \, \nu} MSE \sum_{i=1}^{k} \left(\frac{c_i^2}{b_i} \right)},$$

where

- k is the number of treatment means.
- b_i is the number of observations in which treatment mean i is based.
- $\bar{y}_i$. is the sample mean for treatment i.
- MSE is the error mean square from the analysis of variance table.
- $f_{\alpha, k-1, \nu}$ is the tabulated value from the F distribution table with $\nu_1 = k - 1$, $\nu_2 = \nu$, with ν being the degrees of freedom associated with mean square error (MSE).

(ii) Calculate the confidence interval $\hat{L} \pm U$ for each contrast.

For example, for each of the ten pairwise comparisons for our data, $b_i = b_j = 10$. Hence the comparison, say, μ_i versus μ_j has a linear contrast $L = \mu_i - \mu_j$, with $c_i = 1$, $c_j = -1$ and $c_l = 0$ for $l \neq i$ or j. Hence, for this pair, we have

$$U_{ij} = \sqrt{(k-1)f_{.05, 4, 45}MSE\left(\frac{1}{b_i} + \frac{1}{b_j}\right)} = \sqrt{4(2.59)(27.7342)\left(\frac{1}{10} + \frac{1}{10}\right)} = 7.5806.$$

Here, $f_{.05,4,45}$ is obtained to be 2.59 approximately by interpolation. Thus, if the difference between any two means is larger than $U_{ij} = 7.5806$, then the means are assumed to be significantly different. That is, if

$$|\bar{y}_{i.} - \bar{y}_{j.}| \geq 7.5808, \quad \text{for } i \neq j. \tag{14.18}$$

The implementation of this in SAS is presented in a summarized result below and is accomplished in SAS by specifying after the model statement **means level/scheffe lines**.

```
proc glm data=nova2 order=data;
class level;
model y=level;
means level/scheffe lines;
run;
                      The GLM Procedure

                  Scheffe's Test for y

     NOTE: This test controls the Type I experimentwise error rate.

             Alpha                          0.05
             Error Degrees of Freedom         45
             Error Mean Square           27.73418
             Critical Value of F          2.57874
             Minimum Significant Difference 7.5641

       Means with the same letter are not significantly different.

     Scheffe Grouping        Mean     N    level
                    A       59.140    10    1

                    B       46.040    10    2

                    C       36.450    10    3

                    D       25.470    10    4

                    E       16.440    10    5
```

The results obtained here are consistent with those obtained earlier with **lsd** and **tukey** tests.

14.4.6   The Bonferroni Test

To implement Bonferroni multiple comparison procedure, we do the following:

(i) For each of the contrasts, say, $L = \sum_{i=1}^{k} c_i \mu_i$, calculate

$$\hat{L} = \sum_{i=1}^{k} c_i \bar{y}_{i.} \quad \text{and} \quad B = t_{\alpha/(2g), \nu} S \sqrt{\sum_{i=1}^{k} \left(\frac{c_i^2}{b_i}\right)},$$

where

- k is the number of treatment means.
- b_i is the number of observations in which treatment mean i is based.
- g is the number of contrasts. For pairwise comparisons, then, $g = \binom{k}{2}$.

- $S = \sqrt{MSE}$.

- $t_{\alpha/(2g),\nu}$ is the tabulated value from the t-distribution table with ν degrees of freedom and upper tail area of $\alpha/(2g)$.

(ii) Calculate the confidence interval $\hat{L} \pm B$ for each contrast.

For example, for each of the ten pairwise comparisons for our data, $b_i = b_j = 10$. Hence the comparison, say, μ_i versus μ_j has a linear contrast $L = \mu_i - \mu_j$, with $c_i = 1$, $c_j = -1$ and $c_l = 0$ for $l \neq i$ or j. Therefore $t_{.05/(2g),\nu} = t_{0.0025,45} = 2.95$ approximately, that is, $(0.05)/(2 \times 10) = 0.0025$.

$$B_{ij} = t_{.0025}S\sqrt{\left(\frac{1}{b_i} + \frac{1}{b_j}\right)} = (2.95)(2.3552) = 6.9478.$$

Again, if the difference between any two means is larger than $B_{ij} = 6.9478$, then the means are assumed to be significantly different. That is, if

$$|\bar{y}_{i.} - \bar{y}_{j.}| \geq 6.9478, \quad \text{for } i \neq j. \tag{14.19}$$

The implementation of this in SAS is presented in a summarized result below and is accomplished in SAS by specifying after the model statement **means level/bon lines**.

```
                    The GLM Procedure

              Bonferroni (Dunn) t Tests for y

NOTE: This test controls the Type I experimentwise error rate, but it generally has
        a higher Type II error rate than REGWQ.

        Alpha                              0.05
        Error Degrees of Freedom             45
        Error Mean Square              27.73418
        Critical Value of t             2.95208
        Minimum Significant Difference   6.9527


    Means with the same letter are not significantly different.

      Bon Grouping         Mean      N    level
                 A       59.140     10    1

                 B       46.040     10    2

                 C       36.450     10    3

                 D       25.470     10    4

                 E       16.440     10    5
```

The results are consistent with earlier results. Thus the fifth level gives the best result in this study.

14.5   Partitioning the Treatments SS

For our example, the treatment degrees of freedom is 4, hence we can form at most 4 contrasts each based on 1 degree of freedom, in this study. Suppose the contrasts so formed are presented in the following table.

Contrasts	μ_1	μ_2	μ_3	μ_4	μ_5	Divisor
L_1	4	-1	-1	-1	-1	4
L_2	0	3	-1	-1	-1	3
L_3	0	0	2	1	-1	2
L_4	0	0	0	1	-1	1

The above contrasts correspond respectively to the following null hypotheses:

$$H_0 : L_1 = \mu_1 - \frac{\mu_2 + \mu_3 + \mu_4 + \mu_5}{4}$$

$$H_0 : L_2 = \mu_2 - \frac{\mu_3 + \mu_4 + \mu_5}{3}$$

$$H_0 : L_3 = \mu_3 - \frac{\mu_4 + \mu_5}{2}$$

$$H_0 : L_4 = \mu_4 - \mu_5.$$

These contrasts can be estimated and tested in SAS with the following SAS statements and corresponding partial output.

```
data nova2;
set nova1;
proc glm data=nova2 order=data;
class level;
model y=level;
contrast '1 vs others' level 4 -1 -1 -1 -1;
contrast '2 vs 3, 4 & 5' level 0 3 -1 -1 -1;
contrast '3 vs 4 & 5' level 0 0 2 -1 -1;
contrast '4 vs 5' level 0 0 0 1 -1;
estimate '1 vs others' level 4 -1 -1 -1 -1/divisor=4;
estimate '2 vs 3, 4 & 5' level 0 3 -1 -1 -1/divisor=3;
estimate '3 vs 4 & 5' level 0 0 2 -1 -1/divisor=2;
estimate '4 vs 5' level 0 0 0 1 -1;
run;
```

Source	DF	Type I SS	Mean Square	F Value	Pr > F
level	4	11274.31880	2818.57970	101.63	<.0001

Source	DF	Type III SS	Mean Square	F Value	Pr > F
level	4	11274.31880	2818.57970	101.63	<.0001

Contrast	DF	Contrast SS	Mean Square	F Value	Pr > F
1 vs others	1	6289.932800	6289.932800	226.79	<.0001
2 vs 3, 4 & 5	1	2976.048000	2976.048000	107.31	<.0001
3 vs 4 & 5	1	1600.633500	1600.633500	57.71	<.0001
4 vs 5	1	407.704500	407.704500	14.70	0.0004

Parameter	Estimate	Standard Error	t Value	Pr > |t|
1 vs others	28.0400000	1.86192702	15.06	<.0001
2 vs 3, 4 & 5	19.9200000	1.92298996	10.36	<.0001
3 vs 4 & 5	15.4950000	2.03963886	7.60	<.0001
4 vs 5	9.0300000	2.35517209	3.83	0.0004

We note the following from the above partial SAS output for this example:

1. First, because the design is balanced (each treatment being equally replicated), the Type I and Type III SS are the same. This is often not the case for unbalanced designs as we would see later in this text.

2. That the total of all contrasts SS equals the original Treatment (level) SS of 11274.3188. That is,

$$6289.9328 + 2976.0480 + 1600.6335 + 407.7045 = 11274.3188$$

The Sum of squares for the contrasts add up to the original treatment SS in this case because the contrasts are pairwise orthogonal. That is, for constants c_i and d_j, we have

$$\sum_{i=1}^{5} \sum_{j=1}^{5} c_i\, d_j = 0, \quad \text{for all pairs of contrasts.}$$

Calculating Contrasts SS

For any contrast $L = \sum_{i=1}^{k} c_i \mu_i$, the SS is obtained as

$$SS(L) = \frac{\left(\sum_{i=1}^{k} c_i y_{i.}\right)^2}{b_i \sum c_i^2}, \tag{14.20}$$

where $y_{i.}$ is the total for treatment i. For example for the L_1 contrast, $b_i = 10$, $\sum c_i^2 = 4^2 + 4 = 20$, hence, $SS(L_1)$ is

$$\frac{[(591.4)(4) + (460.4)(-1) + (364.5)(-1) + (254.7)(-1) + (164.4)(-1)]^2}{(20)(10)} = \frac{1121.6^2}{200}$$
$$= 6289.9328.$$

Similarly, $SS(L_2)$ is

$$\frac{[(460.4)(3) + (364.5)(-1) + (254.7)(-1) + (164.4)(-1)]^2}{(12)(10)} = \frac{597.6^2}{120} = 2976.0480.$$

$SS(L_3)$ is computed as

$$\frac{[(364.5)(2) + (254.7)(-1) + (164.4)(-1)]^2}{(6)(10)} = \frac{309.9^2}{60} = 1600.6335.$$

Finally, $SS(L_4)$ is similarly calculated as

$$\frac{[(254.7)(1) + (164.4)(-1)]^2}{(2)(10)} = \frac{90.3^2}{20} = 407.7045.$$

Each of the above calculated contrasts SS is based on 1 degree of freedom. To obtain corresponding estimates of means, we note that for contrast L_1 for instance, this equals

$$\bar{y}_{1.} - \frac{\bar{y}_{2.} + \bar{y}_{3.} + \bar{y}_{4.} + \bar{y}_{5.}}{4} = 59.14 - \frac{46.04 + 36.45 + 25.47 + 16.44}{4} = 59.14 - 31.10 = 28.04,$$

with corresponding standard error calculated as

$$\sqrt{\frac{S^2}{10} + \frac{4S^2}{160}} = \sqrt{\frac{27.7342}{10} + \frac{4(27.7342)}{160}} = 1.8619.$$

Similar calculations lead to the standard errors of the estimates presented in the SAS output. Of course here, we have only four contrasts that are all significant, indicating again that level 5 gives the lowest growth rate of the bacteria and hence the best in this example.

14.6   One-Way ANOVA with a Quantitative Treatment

The single treatment factor investigated in the one way classification analysis of variance can be either quantitative or qualitative. A quantitative factor is one whose level can be associated with points on a numerical scale, such as temperature, pressure, time, dosage levels. Qualitative factors, on the other hand, are factors in which the levels cannot be arranged in order of magnitude. In example 14.1, the CO_2 levels are quantitative, but the levels are not equally spaced. Note that the analysis in the example focused on the differences in the treatment means.

In so far as the initial design and analysis of the experiment are concerned, both types of factors are treated identically. However, in experiments relating to quantitative factors, we are interested not only in the differences in the treatment means, but also in determining if the treatment means are functionally related to the ordered values of the factor. In general, we are interested in finding a mathematical relationship between the factor and the response. The general procedure for this problem, we will recall, is called regression analysis. However, if the levels of the factor are equally spaced, a simple procedure using orthogonal polynomial coefficients may be readily employed.

The procedure consists of computing a linear, quadratic, cubic, quartic, quintic, etc. effect and sum of squares for the factor. Each effect (which forms a contrast) has a single degree of freedom and they are computed from the treatment totals at the a factor levels as in the previous example, and the corresponding sum of squares is found from equation (14.20). It is possible to extract polynomial effects up through order $a - 1$ if there are a factor levels used in the experiment.

Example 14.2

The data with slight modifications is from Montgomery (2009, p. 115) and reproduced by permission of John Wiley and Sons, Inc. The tensile strength of synthetic fiber used to make cloth for men's shirts is of interest to a manufacturer. It is suspected that strength is affected by the percentage of cotton in the fiber. Five levels of cotton percentage are of interest, 15%, 20%, 25%, 30% and 35%. Five observations are taken at each level of cotton percentage. Table 14.2 gives the data for this experiment.

Percentage of Cotton	\multicolumn{5}{c}{Observations}	Total $y_{i.}$

Percentage of Cotton	1	2	3	4	5	Total $y_{i.}$
15	7	7	15	11	9	49
20	12	17	12	18	18	77
25	14	18	18	19	19	88
30	19	25	22	19	23	108
35	7	10	11	15	11	54
Total						376

Table 14.2: Tensile strength as functions of percentage of cotton

Analysis of Variance for the Experiment:

$$\text{Correction Factor (CF)} = 376^2/25$$

$$\text{Total SS} = 7^2 + 12^2 + 14^2 + \cdots + 11^2 - CF = 636.96$$

$$\text{Treatment SS} = \frac{49^2}{5} + \frac{77^2}{5} + \frac{88^2}{5} + \frac{108^2}{5} + \frac{54^2}{5} - \text{CF} = 475.76$$

Source	d.f.	SS	MS	F
Treatments	4	475.75	118.94	14.76
Error	20	161.20	8.06	
Total	24	636.96		

Table 14.3: ANOVA table for the data in Table 14.2

The tabulated $f_{.05,\,4,\,20}$ is 2.87. Thus, we would reject H_0 and conclude that the percentage of cotton in the fiber significantly affects the strength. In order to partition the treatment SS into four components (corresponding to the treatments degree of freedom), we may use the fact that the percentage of cotton are equally spaced. This enables us to make use of table of coefficients of orthogonal polynomials. From Table 6 in the Appendix, we have $(5-1) = 4$ as the degrees of freedom. The coefficients are presented in Table 14.4.

Like in the previous section, we note here too that each component forms a contrast and the four contrasts are mutually orthogonal. Since they are orthogonal, the addition of the sum of squares for the four components will equal the original treatment SS. The treatment totals $y_{i.}$ are given in row 1 of Table 14.4.

Treatment Totals	49	77	88	108	54
Linear	−2	−1	0	1	2
Quadratic	2	−1	−2	−1	2
Cubic	−1	2	0	−2	1
Quartic	1	−4	6	−4	1

Table 14.4: Calculation of components SS

$$\text{Linear SS} = \frac{\{(-2) \times 49 + (-1) \times 77 + (0) \times 88 + (+1) \times 108 + (+2) \times 54\}^2}{5 \times 10} = 33.62.$$

Similarly, the quadratic SS = 343.21 and the cubic SS = 64.98, while the quartic SS = 33.95. The revised analysis of variance is shown in Table 14.5.

Source	d.f.	SS	MS	F
Percentage Cotton	4	475.76	118.94	14.76
Linear	1	33.62	33.62	4.17
Quadratic	1	343.21	343.21	42.58
Cubic	1	64.98	64.98	8.06
Quartic	1	33.95	33.95	4.21
Error	20	161.20	8.06	
Total	24	636.96		

Table 14.5: Revised analysis of variance table

At the 5% significance level, $f_{.05, 1, 20} = 4.35$, since for the quartic, $4.21 < 4.35$, hence, we would conclude that the model does not need a fourth-degree polynomial term. Both the quadratic and cubic terms are significant, indicating that a cubic polynomial will fully describe the response of tensile strength to increasing cotton percentage in the fabric.

The analysis of variance and accompanying contrast comparisons are carried out in SAS with the following statements and partial output.

```
data quant;
do pct=15 to 35 by 5;
input y @@;
output;
end;
datalines;
7 12 14 19 7
7 17 18 25 10
15 12 18 22 11
11 18 19 19 15
9 18 19 23 11
;
run;
proc print; run;
proc glm;
class pct;
model y=pct;
means pct;
output out=aa r=resid;
contrast 'linear' pct -2 -1 0 1 2;
contrast 'quadratic' pct 2 -1 -2 -1 2;
contrast 'cubic' pct -1 2 0 -2 1;
contrast 'quartic' pct 1 -4 6 -4 1;
run;
data new;
set quant;
x1=pct; x2=x1*x1; x3=x2*x1; x4=x3*x1;
proc reg;
model y=x1-x4/ss1 ss2;
test1: test x4;
test2: test x3,x4;
run;
proc reg;
model y=x1-x3;
output out=aa p=pred;
run;
proc sort data=aa;
by x1;
run;

goptions cback=white colors=(black( vsize=6in hsize=6in;
symbol1 i=spline value=none height=.75 c=red;
axis1 label=(angle=-90 rotate=90 'PREDICTED');
axis2 label=('% OF COTTON');
proc gplot data=aa;
```

```
plot pred*x1/vaxis=axis1 haxis=axis2;
run;
```
The GLM Procedure

Class Level Information

Class	Levels	Values
pct	5	15 20 25 30 35

Number of observations   25

The GLM Procedure

Dependent Variable: y

Source	DF	Sum of Squares	Mean Square	F Value	Pr > F
Model	4	475.7600000	118.9400000	14.76	<.0001
Error	20	161.2000000	8.0600000		
Corrected Total	24	636.9600000			

R-Square	Coeff Var	Root MSE	y Mean
0.746923	18.87642	2.839014	15.04000

Source	DF	Type I SS	Mean Square	F Value	Pr > F
pct	4	475.7600000	118.9400000	14.76	<.0001

Source	DF	Type III SS	Mean Square	F Value	Pr > F
pct	4	475.7600000	118.9400000	14.76	<.0001

Dependent Variable: y

Contrast	DF	Contrast SS	Mean Square	F Value	Pr > F
linear	1	33.6200000	33.6200000	4.17	0.0545
quadratic	1	343.2142857	343.2142857	42.58	<.0001
cubic	1	64.9800000	64.9800000	8.06	0.0101
quartic	1	33.9457143	33.9457143	4.21	0.0535

The REG Procedure
Model: MODEL1
Dependent Variable: y

Analysis of Variance

Source	DF	Sum of Squares	Mean Square	F Value	Pr > F
Model	4	475.76000	118.94000	14.76	<.0001
Error	20	161.20000	8.06000		
Corrected Total	24	636.96000			

Root MSE	2.83901	R-Square	0.7469
Dependent Mean	15.04000	Adj R-Sq	0.6963
Coeff Var	18.87642		

Parameter Estimates

Variable	DF	Parameter Estimate	Standard Error	t Value	Pr > |t|	Type I SS	Type II SS
Intercept	1	-406.40000	231.51806	-1.76	0.0945	5655.04000	24.83549
x1	1	73.77667	40.62993	1.82	0.0844	33.62000	26.57547
x2	1	-4.80767	2.58509	-1.86	0.0777	343.21429	27.87738
x3	1	0.13773	0.07087	1.94	0.0662	64.98000	30.44476
x4	1	-0.00145	0.00070817	-2.05	0.0535	33.94571	33.94571

The REG Procedure
Model: MODEL1

Test test1 Results for Dependent Variable y

Source	DF	Mean Square	F Value	Pr > F
Numerator	1	33.94571	4.21	0.0535
Denominator	20	8.06000		

The REG Procedure
Model: MODEL1

Test test2 Results for Dependent Variable y

Mean

```
Source              DF       Square   F Value    Pr > F
Numerator            2     49.46286      6.14    0.0084
Denominator         20      8.06000
```

```
                    The REG Procedure
                      Model: MODEL1
                  Dependent Variable: y

                  Analysis of Variance

                           Sum of        Mean
Source              DF     Squares      Square   F Value    Pr > F
Model                3   441.81429   147.27143     15.85    <.0001
Error               21   195.14571     9.29265
Corrected Total     24   636.96000


      Root MSE             3.04839   R-Square    0.6936
      Dependent Mean      15.04000   Adj R-Sq    0.6499
      Coeff Var           20.26852

                  Parameter Estimates

                    Parameter    Standard
Variable     DF      Estimate       Error   t Value    Pr > |t|
Intercept     1      62.61143    39.75744      1.57      0.1302
x1            1      -9.01143     5.19661     -1.73      0.0976
x2            1       0.48143     0.21605      2.23      0.0369
x3            1      -0.00760     0.00287     -2.64      0.0152
```

Based on the results, a cubic polynomial of the form

$$y_i = \beta_0 + \beta_1 x_i + \beta_2 x_i^2 + \beta_3 x_i^3 + \varepsilon_i, \tag{14.21}$$

is fitted to the data, resulting in the estimated regression model

$$\hat{y}_i = 62.6114 - 9.0114x_i + 0.4814x_i^2 - 0.0076x_i^3. \tag{14.22}$$

The plot of the estimated regression response profile is presented in Figure 14.1.

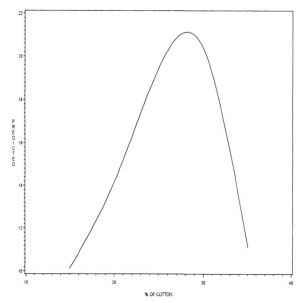

Figure 14.1: Plot of estimated regression function

14.7 Correspondence Between Regression and ANOVA

In this section, we present the correspondence between the coding schemes employed in regression analysis with **proc reg** and contrast specifications in general linear model (GLM) as employed in **proc glm** in SAS.

14.7.1   Cell Reference Coding in Regression Analysis

Consider the data for Example 14.1. For this coding scheme, we give the corresponding coding implementation in **proc reg** and equivalent contrast statements in **proc glm**, where level of apl refers to atmospheric pressure level and the response variable ($i = 0.000, 0.083, 0.290, 0.500, 0.860$).

<table>
<tr><th colspan="5">proc reg</th><th colspan="5">proc glm</th></tr>
<tr><th>apl</th><th>x_1</th><th>x_2</th><th>x_3</th><th>x_4</th><th>apl</th><th>C_1</th><th>C_2</th><th>C_3</th><th>C_4</th></tr>
<tr><td>0.000</td><td>1</td><td>0</td><td>0</td><td>0</td><td>0.000</td><td>1</td><td>0</td><td>0</td><td>0</td></tr>
<tr><td>0.083</td><td>0</td><td>1</td><td>0</td><td>0</td><td>0.083</td><td>0</td><td>1</td><td>0</td><td>0</td></tr>
<tr><td>0.290</td><td>0</td><td>0</td><td>1</td><td>0</td><td>0.290</td><td>0</td><td>0</td><td>1</td><td>0</td></tr>
<tr><td>0.500</td><td>0</td><td>0</td><td>0</td><td>1</td><td>0.500</td><td>0</td><td>0</td><td>0</td><td>1</td></tr>
<tr><td>0.860</td><td>0</td><td>0</td><td>0</td><td>0</td><td>0.860</td><td>-1</td><td>-1</td><td>-1</td><td>-1</td></tr>
</table>

In general for this coding scheme and k factor levels, we have under regression analysis,

$$x_i = \begin{cases} 1, & \text{for factor level } i \\ 0, & \text{elsewhere} \end{cases} \qquad i = 1, 2, \ldots, (k-1), \tag{14.23}$$

and the corresponding contrasts C_i are generated as

$$C_i = \begin{cases} 1, & \text{for factor level } i \\ -1, & \text{if factor level is last} \\ 0, & \text{elsewhere} \end{cases} \qquad i = 1, 2, \ldots, (k-1). \tag{14.24}$$

The results of implementing the two approaches to the data for Example 14.1 in SAS is presented below.

```
data nova2;
set nova1;
array xll[4] x1-x4;
do i=1 to 4;
if level=i then xll[i]=1; else xll[i]=0;
end;
output;
drop i;
run;
proc reg data=nova2;
model y=x1 x2 x3 x4;
run;
proc glm data=nova1;
class level;
model y=level;
estimate 'level 1 vs level 5' level 1 0 0 0 -1;
estimate 'level 2 vs level 5' level 0 1 0 0 -1;
estimate 'level 3 vs level 5' level 0 0 1 0 -1;
estimate 'level 4 vs level 5' level 0 0 0 1 -1;
run;
```

```
                       The REG Procedure
                        Model: MODEL1
                     Dependent Variable: y


                     Parameter Estimates

                       Parameter      Standard
    Variable    DF      Estimate         Error    t Value    Pr > |t|
    ----------------------------------------------------------------
    Intercept    1      16.44000       1.66536       9.87      <.0001
    x1           1      42.70000       2.35517      18.13      <.0001
    x2           1      29.60000       2.35517      12.57      <.0001
    x3           1      20.01000       2.35517       8.50      <.0001
    x4           1       9.03000       2.35517       3.83      0.0004


                     The GLM Procedure

                  Class Level Information

              Class        Levels     Values
              level             5      1 2 3 4 5

                          Standard
```

```
Parameter                Estimate      Error   t Value   Pr > |t|
------------------------------------------------------------------
level 1 vs level 5     42.7000000   2.35517209    18.13    <.0001
level 2 vs level 5     29.6000000   2.35517209    12.57    <.0001
level 3 vs level 5     20.0100000   2.35517209     8.50    <.0001
level 4 vs level 5      9.0300000   2.35517209     3.83    0.0004
```

14.7.2   Effect Coding Scheme

Here, we give the equivalent coding scheme in regression analysis and the corresponding contrasts (C_i) statements in the general linear model approach.

proc reg						**proc glm**				
apl	x_1	x_2	x_3	x_4		apl	C_1	C_2	C_3	C_4
0.000	1	0	0	0		0.000	4/5	−1/5	−1/5	−1/5
0.083	0	1	0	0		0.083	−1/5	4/5	−1/5	−1/5
0.290	0	0	1	0		0.290	−1/5	−1/5	4/5	−1/5
0.500	0	0	0	1		0.500	−1/5	−1/5	−1/5	4/5
0.860	−1	−1	−1	−1		0.860	−1/5	−1/5	−1/5	−1/5

In general for this coding scheme and k factor levels, we have under regression analysis,

$$x_i = \begin{cases} 1, & \text{for factor level } i \\ -1, & \text{if factor level is last} \\ 0, & \text{elsewhere} \end{cases} \quad i = 1, 2, \ldots, (k-1), \qquad (14.25)$$

with the corresponding contrast (C_i) in GLM being in general:

$$C_i = \begin{cases} \dfrac{(k-1)}{k}, & \text{for factor level } i \\ \dfrac{-1}{k}, & \text{elsewhere} \end{cases} \quad i = 1, 2, \ldots, (k-1). \qquad (14.26)$$

Again, the implementation of these in SAS with a summary parameter output are presented in the following:

```
data nova2;
set nova1;
array xll[4] x1-x4;
do i=1 to 4;
if level=i then xll[i]=1; else if level=5 then xll[i]=-1; else xll[i]=0;
end;
output;
drop i;
run;
proc reg data=nova2;
model y=x1 x2 x3 x4;
run;
proc glm data=nova1;
class level;
model y=level;
estimate 'level 1 vs level A' level 4 -1 -1 -1 -1/divisor=5;
estimate 'level 2 vs level B' level -1 4 -1 -1 -1/divisor=5;
estimate 'level 3 vs level C' level -1 -1 4 -1 -1/divisor=5;
estimate 'level 4 vs level D' level -1 -1 -1 4 -1/divisor=5;
run;
```

The REG Procedure

Parameter Estimates

```
                   Parameter    Standard
Variable    DF     Estimate       Error    t Value   Pr > |t|
Intercept    1     36.70800      0.74477     49.29    <.0001
X1           1     22.43200      1.48954     15.06    <.0001
X2           1      9.33200      1.48954      6.27    <.0001
X3           1     -0.25800      1.48954     -0.17    0.8633
X4           1    -11.23800      1.48954     -7.54    <.0001
```

```
The GLM Procedure

Class Level Information

         Class        Levels      Values
         level          5         1 2 3 4 5

                            Standard
Parameter                Estimate       Error    t Value   Pr > |t|
level 1 vs level A     22.4320000    1.48954161    15.06    <.0001
level 2 vs level B      9.3320000    1.48954161     6.27    <.0001
level 3 vs level C     -0.2580000    1.48954161    -0.17    0.8633
level 4 vs level D    -11.2380000    1.48954161    -7.54    <.0001
```

The contrast statements in **proc glm** could have been written equivalently as

```
estimate 'level 1 vs level A' level 0.8 -0.2 -0.2 -0.2 -0.2;
estimate 'level 2 vs level B' level -0.2 0.8 -0.2 -0.2 -0.2;
estimate 'level 3 vs level C' level -0.2 -0.2 0.8 -0.2 -0.2;
estimate 'level 4 vs level D' level -0.2 -0.2 -0.2 0.8 -0.2;
```

and the results would have been the same. We may also note here that SAS will not often accept putting the contrasts as, for example: **estimate 'level 1 vs level A' level** $4/5$ $-1/5$ $-1/5$ $-1/5$ $-1/5$. We see here that the 'effect coding schemes' compare the mean of a given level with the overall mean of all the levels. Thus for instance, the four contrast statements above test the following hypotheses:

$$(C_1)\ H_0 : \mu_1 = \frac{\mu_2 + \mu_3 + \mu_4 + \mu_5}{4}$$

$$(C_2)\ H_0 : \mu_2 = \frac{\mu_1 + \mu_3 + \mu_4 + \mu_5}{4}$$

$$(C_3)\ H_0 : \mu_3 = \frac{\mu_1 + \mu_2 + \mu_4 + \mu_5}{4}$$

$$(C_4)\ H_0 : \mu_4 = \frac{\mu_1 + \mu_2 + \mu_3 + \mu_5}{4}.$$

The above will give the appropriate tests of significance but the estimates will be different from the correct values. We can change this by multiplying both sides above by $(k-1)/k$ which in our case would be $4/5$. These as can be seen will lead to the correct contrast coefficients in GLM.

Other forms of coding that are suitable for factor variables whose levels are ordinal in nature are the Helmert and orthogonal coding schemes. The contrast in section 14.5 is the Helmert scheme. In our example here, the corresponding hypotheses will be:

$$(C_1)\ H_0 : \mu_1 = \frac{\mu_2 + \mu_3 + \mu_4 + \mu_5}{4}$$

$$(C_2)\ H_0 : \mu_2 = \frac{\mu_3 + \mu_4 + \mu_5}{3}$$

$$(C_3)\ H_0 : \mu_3 = \frac{\mu_4 + \mu_5}{2}$$

$$(C_4)\ H_0 : \mu_4 = \mu_5.$$

proc reg

apl	x_1	x_2	x_3	x_4
0.000	4/5	0	0	0
0.083	−1/5	3/4	0	0
0.290	−1/5	−1/4	2/3	0
0.500	−1/5	−1/4	−1/3	1/2
0.860	−1/5	−1/4	−1/3	−1/2

proc glm

apl	C_1	C_2	C_3	C_4
0.000	4	0	0	0
0.083	−1	3	0	0
0.290	−1	−1	2	0
0.500	−1	−1	−1	1
0.860	−1	−1	−1	−1

```
data nova2;
set nova1;
if level=1 then x1=4/5; else x1=-1/5;
if level=1 then x2=0; else if level=2 then x2=3/4; else x2=-1/4;
if level=1 or level=2 then x3=0; else if level=3 then x3=2/3; else x3=-1/3;
if level=1 or level=2 or level=3 then x4=0; else if level=4 then x4=1/2; else x4=-1/2;
```

```
proc print;
run;
proc reg data=nova2;
model y=x1 x2 x3 x4;
run;

proc glm data=nova1;
class level;
model y=level;
estimate 'level 1 vs level2 3 ,3,4, & 5' level 4 -1 -1 -1 -1/divisor=4;
estimate 'level 2 vs levels 3, 4 & 5' level 0 3 -1 -1 -1/divisor=3;
estimate 'level 3 vs levels 4 & 5' level 0 0 2 -1 -1/divisor=2;
estimate 'level 4 vs level 5' level 0 0 0 1 -1;
run;
```

The REG Procedure

Parameter Estimates

Variable	DF	Parameter Estimate	Standard Error	t Value	Pr > |t|
Intercept	1	36.70800	0.74477	49.29	<.0001
x1	1	28.04000	1.86193	15.06	<.0001
x2	1	19.92000	1.92299	10.36	<.0001
x3	1	15.49500	2.03964	7.60	<.0001
x4	1	9.03000	2.35517	3.83	0.0004

The GLM Procedure

Class Level Information

Class	Levels	Values
level	5	1 2 3 4 5

Parameter	Estimate	Standard Error	t Value	Pr > |t|
level 1 vs level2 3 ,3,4, & 5	28.0400000	1.86192702	15.06	<.0001
level 2 vs levels 3, 4 & 5	19.9200000	1.92298996	10.36	<.0001
level 3 vs levels 4 & 5	15.4950000	2.03963886	7.60	<.0001
level 4 vs level 5	9.0300000	2.35517209	3.83	0.0004

14.8   Exercises

14.1 The table below relates to the outcome of an experiment in which five subjects were assigned at random to each of the four dosages of a drug. The dependent variable y is a physiological measure that presumably is influenced by the amount of the drug administered.

Dosage Levels			
0	5	10	15
10	9	14	17
8	13	13	15
12	12	11	14
11	10	12	18
9	11	15	16
50	55	65	80

The experiment was designed to see if there is a linear trend in the response. The *one-factor effect model*

$$y_{ij} = \mu + \tau_i + \varepsilon_{ij}, \quad i = 1, 2, 3, 4; \quad j = 1, 2, \ldots, 5,$$

where $\tau_i = \mu_i - \mu$, is suggested.

(a) Complete the following analysis of variance table, explaining how the treatment SS of 105 was obtained.

Source	df	SS	MS	F
Treatments	---	105	—	—
Error	—	—	—	
Total	—	145		

(b) Test whether the four dosage levels have significantly different effects on the physiological measure y (Use $\alpha = 0.05$).

(c) Are there any assumptions that must be satisfied? State them and conduct the appropriate tests for the validity of these assumptions.

(d) The treatments SS has been partitioned into linear, quadratic and cubic components using tables of orthogonal polynomial coefficients displayed below.

Linear	-3	-1	1	3
Quadratic	1	-1	-1	1
Cubic	-1	3	-3	1

How was the linear SS computed? Test your results and suggest the most suitable regression model for the data.

(e) What relationship, if any, is there between the single observed F value of part (a) and the three observed F values of part (d)?

(f) Suppose a regression approach to modeling the data which utilizes indicator variables. (i.e. treating dosage levels as a categorical variable) is employed with an equivalent regression model, utilizing effect coding scheme,

$$y = \mu + \alpha_1 x_1 + \alpha_2 x_2 + \alpha_3 x_3 + \varepsilon.$$

How are x_1, x_2 and x_3 defined in this case?

(g) Set up hypotheses in terms of the parameters in (f) to test whether the four groups have significantly different means.

(h) In terms of the parameters in the regression approach, write down the corresponding equivalents for the hypotheses of linear effects in (d).

14.2 A graduate student has conducted a four-group study in which he tested the following three planned comparisons:

	μ_1	μ_2	μ_3	μ_4
L_1	1	-1	0	0
L_2	1	1	-2	0
L_3	1	1	1	-3

The sum of squares (SS) for the three comparisons are 75, 175, and 125, respectively. The value of MSE equals 25, and there were 11 subjects in each group. The student's adviser wonders whether the omnibus F test of $H_0 : \mu_1 = \mu_2 = \mu_3 = \mu_4$ would be statistically significant for these data. Can you help?

(a) Is it possible to perform the test of the omnibus null hypothesis from the available information? If so, is the test significant? If it is not possible, explain why not.

(b) Find the observed F value for each of the planned comparisons tested by the student. Which, if any, are statistically significant at $\alpha = 0.05$?

(c) What relationship, if any, is there between the single observed F value of part (a) and the three observed F values of part (b)?

(d) For each of the L_i, indicate the null hypothesis that is being tested. Find the observed F value for each of the planned comparisons tested by the student. Which, if any, are statistically significant at $\alpha = 0.05$?

(e) For the equivalent regression model

$$y = \mu + \alpha_1 x_1 + \alpha_2 x_2 + \alpha_3 x_3 + \varepsilon,$$

what are the explanatory variables in the model? Define them.

(f) Test whether the four groups have significantly different means (Use $\alpha = .05$). State the null and the alternative hypotheses in this case.

(g) In terms of the parameters in your regression approach, write down the corresponding equivalents for the hypotheses employed in (b) by the student.

(h) Test for the validity of your analysis under the ANOVA or regression procedures.

14.3 For a one-factor experiment having four levels, let μ_1, μ_2, μ_3 and μ_4 denote their respective population means. A regression model has for the effect coding case, the following form:

$$y = \mu + \alpha_1 x_1 + \alpha_2 x_2 + \alpha_3 x_3 + \varepsilon,$$

where for $i = 1, 2, 3$,

$$x_i = \begin{cases} 1, & \text{if level} = i \\ -1, & \text{if level} = 4 \\ 0, & \text{otherwise.} \end{cases}$$

Write down the corresponding expressions from the above model to test the following hypotheses:

(a) $H_0 : \mu_1 = \mu_2 = \mu_3 = \mu_4$ versus H_a : at least two of these means are unequal.

(b) $H_0 : \mu_1 = \mu_2$ versus $H_a : \mu_1 \neq \mu_2$.

(c) $H_0 : \mu_1 = \mu_4$ versus $H_a : \mu_1 \neq \mu_4$.

(d) $H_0 : \mu_4 = (\mu_1 + \mu_2 + \mu_3)/3$ versus $H_a : \mu_4 \neq (\mu_1 + \mu_2 + \mu_3)/3$.

14.4 In a study of the effects of television commercials on 7-year-old children, the attention span of children watching commercials for clothing(C), food products (F), and toys (T) is measured. To reduce the effects of outliers, only the median attention span in seconds for each commercial is used.

	C	21	30	23	37
Commercial	F	32	51	46	30
	T	48	59	56	60

(a) Obtain the point estimates for the population mean of each commercial.

(b) Obtain SS(commercial), SS(total), and SS(error).

(c) Obtain an appropriate analysis of variance table and test the null hypothesis of equal means by using $\alpha = 0.01$.

(d) Should a post-hoc analysis be performed on commercial means? If yes, carry out the test.

14.5 Three car models A, B, and C are studied in tests that involve several cars of each model. In each case, the car is run on exactly 1 gallon of gas until the fuel supply is exhausted. The distances traveled in miles are shown in the following table.

A	16	20	18	18				
B	14	16	16	17	15	18		
C	20	21	19	22	18	24	18	20

(a) Name the appropriate design for the study.

(b) Is there a sufficient evidence to indicate that the mean distances differ among the three car models? Test using $\alpha = 0.01$.

(c) Should post-hoc analysis be performed? If yes, state which pairs of means are different at the significance level 0.01.

14.6 A consumer agency wanted to find out if the mean time it takes for each of three brands of medicines to provide relief from headache is the same. The first drug was administered to six randomly selected patients, the second to four randomly selected patients, and the third to five randomly selected patients. The following table gives the time (in minutes) taken by each patient to get relief from headache after taking the medicine.

Drug							b_i	$y_{i.}$
A	25	38	40	65	47	55	6	270
B	15	21	19	25			4	80
C	44	39	52	58	72		5	265

(a) Set up the analysis of variance table for the problem.

(b) Do the data provide sufficient evidence to indicate that the mean time taken to provide relief from a headache is the same for all three drugs?

(c) Should a post-hoc analysis be performed on drug means? Why or why not?

(d) Find a 95% confidence interval for $\mu_A - \mu_C$. Interpret the interval.

(e) Name the design used in this problem (or study).

14.7 For the one-way analysis of variance model

$$y_{ij} = \mu + \alpha_i + \varepsilon_{ij}, \quad i = 1, 2, \ldots, k; \quad j = 1, 2, \ldots, n,$$

where the ε_{ij} are the values of nk independent random variables having normal distributions with zero means and the common variance σ^2, show that

$$E\left\{ n \sum_{i=1}^{k} \left(\bar{Y}_{i.} - \bar{Y}_{..}\right)^2 / (k-1) \right\} = \sigma^2 + n \sum_{i=1}^{k} \alpha_i^2 / (k-1).$$

Use the fact that $\sum_{i=1}^{k} \alpha_i = 0$.

Chapter 15

Two-Factor Analysis of Variance

15.1 The Mean Model

A two factor experiment consists say, two factors A and B each having a and b levels respectively. The resulting ab treatment combinations can be laid out in a completely randomized design. If all treatment combinations are equally replicated, then we say that the design is balanced. Otherwise, the design is unbalanced or non-orthogonal. We will however, start our discussion of two-factor experiments by considering a balanced case with interaction present. When interaction is present, the general model can be written as

$$\mu_{ij} = \mu + \alpha_i + \beta_j + (\alpha\beta)_{ij}, \quad i = 1, 2, \ldots, a, \quad j = 1, 2, \ldots, b, \tag{15.1}$$

with $\sum_i \alpha_i = 0 = \sum_j \beta_j = \sum_i (\alpha\beta)_{ij} = \sum_j (\alpha\beta)_{ij}$.

Thus,

$$\alpha_i = \mu_{i.} - \mu_{..}, \quad \beta_j = \mu_{.j} - \mu_{..}, \quad \text{and} \quad (\alpha\beta)_{ij} = \mu_{ij} - \mu_{i.} - \mu_{.j} + \mu_{..},$$

where in (15.1), μ is the overall mean, α_i is the main effect of the i-th level of factor A, β_j is the main effect of the j-th level of factor B, and $(\alpha\beta)_{ij}$ is the interaction effect between level i of factor A and level j of factor B.

Example 15.1

Consider a simple two-factor study in which effects of cost engineer and job type on the total cost (in thousands of dollars) required to complete the job are of interest. The corresponding mean scores are presented in Table 15.1.

| Factor A | Factor B | | | Row |
	Job 1 $j = 1$	Job 2 $j = 2$	Job 3 $j = 3$	Average
Engineer 1 $i = 1$	8 μ_{11}	11 μ_{12}	17 μ_{13}	12 $\mu_{1.}$
Engineer 2 $i = 2$	8 μ_{21}	11 μ_{22}	17 μ_{23}	12 $\mu_{2.}$
Column Average	8 $\mu_{.1}$	11 $\mu_{.2}$	17 $\mu_{.3}$	12 $\mu_{..}$

Table 15.1: Observed mean responses for a 2×3 table

For the data in Table 15.1,

$$\alpha_1 = \mu_{1.} - \mu_{..} = 12 - 12 = 0,$$
$$\alpha_2 = \mu_{2.} - \mu_{..} = 12 - 12 = 0,$$
$$\beta_1 = \mu_{.1} - \mu_{..} = 8 - 12 \ = -4,$$
$$\beta_2 = \mu_{.2} - \mu_{..} = 11 - 12 = -1,$$
$$\beta_3 = \mu_{.3} - \mu_{..} = 17 - 12 = 5.$$

Similarly,

$$(\alpha\beta)_{11} = \mu_{11} - \mu_{1.} - \mu_{.1} + \mu_{..} = 8 - 12 - 8 + 12 = 0.$$

From the above,

$$(\alpha\beta)_{11} + (\alpha\beta)_{21} = 0,$$
$$(\alpha\beta)_{12} + (\alpha\beta)_{22} = 0,$$
$$(\alpha\beta)_{13} + (\alpha\beta)_{23} = 0,$$
$$(\alpha\beta)_{11} + (\alpha\beta)_{12} + (\alpha\beta)_{13} = 0,$$
$$(\alpha\beta)_{21} + (\alpha\beta)_{22} + (\alpha\beta)_{23} = 0.$$

For these data, the interaction is zero.

15.2   Additive Factor Effects

Each mean response μ_{ij} can be obtained by adding the respective factor main effects and the overall mean $\mu_{..}$. That is,

$$\mu_{ij} = \mu_{i.} + \mu_{.j} - \mu_{..}.$$

The above can be written alternatively as

$$\mu_{ij} = \mu_{..} + \alpha_i + \beta_j,$$

since

$$\alpha_i = \mu_{i.} - \mu_{..},$$
$$\beta_j = \mu_{.j} - \mu_{..},$$

with $\sum_i \alpha_i = 0 = \sum_j \beta_j$. The above is referred to as the *Additive Factor Effect*, which implies that the interactions are absent. In general, for an additive factor effects model, we have

$$\mu_{ij} = \mu_{ij'} + \mu_{i'j} - \mu_{i'j'},$$

where $i \neq i'$ and $j \neq j'$.

Example 15.2

Consider a simple two-factor study, shown in Table 15.2, in which effects of cost engineer and job type on the total cost (in thousands of dollars) are required to complete the job are of interest.

Factor A	Factor B			Row
	Job 1	Job 2	Job 3	Average
	$j = 1$	$j = 2$	$j = 3$	
Engineer 1	8	12	19	13
$i = 1$	μ_{11}	μ_{12}	μ_{13}	$\mu_{1.}$
Engineer 2	10	10	13	11
$i = 2$	μ_{21}	μ_{22}	μ_{23}	$\mu_{2.}$
Column	9	11	16	12
Average	$\mu_{.1}$	$\mu_{.2}$	$\mu_{.3}$	$\mu_{..}$

Table 15.2: A two-factor example

For these data,

$$\alpha_1 = \mu_{1.} - \mu_{..} = 13 - 12 = 1,$$
$$\alpha_2 = \mu_{2.} - \mu_{..} = 11 - 12 = -1,$$
$$\beta_1 = \mu_{.1} - \mu_{..} = 9 - 12 = -3,$$
$$\beta_2 = \mu_{.2} - \mu_{..} = 11 - 12 = -1,$$
$$\beta_3 = \mu_{.3} - \mu_{..} = 16 - 12 = 4.$$

Similarly,

$$(\alpha\beta)_{11} = \mu_{11} - \mu_{1.} - \mu_{.1} + \mu_{..} = 8 - 13 - 9 + 12 = -2.$$

From the above, we have the following table:

i	j			Row
	1	2	3	Average
1	-2	0	2	0
2	2	0	-2	0
Col	0	0	0	0

For the data in Table 15.1, let $i = 1$, $i' = 2$, $j = 1$, $j' = 2$, then,

$$\mu_{11} = \mu_{12} + \mu_{21} - \mu_{22} = 11 + 8 - 11 = 8,$$
$$\mu_{11} = \mu_{13} + \mu_{21} - \mu_{23} = 17 + 8 - 17 = 8.$$

In the latter case, $i = 1$, $i' = 2$, $j = 1$, $j' = 3$. Corresponding results for the data in Table 15.2, are:

$$\mu_{11} = \mu_{12} + \mu_{21} - \mu_{22} = 12 + 10 - 10 = 12,$$
$$\mu_{11} = \mu_{13} + \mu_{21} - \mu_{23} = 19 + 10 - 13 = 16.$$

In the first case, the results match μ_{11} while in the latter case, both 12 and 16 are not equal to $\mu_{11} = 8$. Hence, additivity is not present in the second table and the factors therefore have interaction.

15.3   Two-Level Factor Factorial Experiments: The 2^n Series

A 2^n factorial experiment is an experiment involving n factors each at 2 levels designated 0, 1. The simplest of the design is when $n = 2$, i.e. two factors each at two levels- 2^2. For this situation, suppose the two factors are A and B, with the levels designated 0, 1 respectively. Then, there will be four (2×2) treatment combinations (0 0), (1 0), (0 1) and (1 1).

If a_0 and b_0 denote the zero level for both factors and a_1 and b_1 also denote the upper levels of the two factors, then the four treatment combinations can be put in a table as follows:

Factor	Factor B	
A	b_0	b_1
a_0	$a_0 b_0$	$a_0 b_1$
a_1	$a_1 b_0$	$a_1 b_1$

These are sometimes written as

$$(1) = a_0 b_0; \quad \text{a} = a_1 b_0; \quad \text{b} = a_0 b_1; \quad \text{ab} = a_1 b_1$$

Since there are 4 treatment combinations, it follows that there are 3 degrees of freedom for treatments made up as follows in Table 15.3.

Source	d.f.	
A	1	Main effect of A
B	1	Main effect of B
AB	1	Interaction effect of A and B.

Table 15.3: Structure of ANOVA table

(a) The **main effect** of a factor is a measure of the change in the level of the factor averaged over all levels of the other factors.

(b) The **interaction** is the differential response to one factor in combination with varying levels of a second factor applied simultaneously. That is, interaction is an additional effect due to the combined influence of two (or more) factors.

Effects in the 2^2 Factorial Experiments

Consider the 2×2 table of means in Table 15.4. If we let $\mu_{ij}, i = 0, 1; j = 0, 1$ be the expected response from treatment combination ij, then $\mu_{ij} = \{\mu_{00}, \mu_{01}, \mu_{10}, \mu_{11}\}$.

Level of A	level of B		Simple effect of B $\mu[A_i B]$
	b_0	b_1	
a_0	μ_{00}	μ_{01}	$\mu[A_0 B] = \mu_{01} - \mu_{00}$
a_1	μ_{10}	μ_{11}	$\mu[A_1 B] = \mu_{11} - \mu_{10}$
Simple effect of A $\mu[AB_j]$	$\mu[AB_0] = \mu_{10} - \mu_{00}$	$\mu[AB_1] = \mu_{11} - \mu_{01}$	

Table 15.4: Population means and simple effects in a 2^2 factorial

Simple Effects

The *Simple effect* of A at level b_0 of B is defined as

$$\mu[AB_0] = \mu_{10} - \mu_{00}. \tag{15.2}$$

That is, the simple effect of A at level b_0 of B is the amount of change in the expected response when the level of A is changed from a_1 to a_0, with the level of B held constant at b_0. Similarly, the simple effect of A at level b_1 of B is defined as

$$\mu[AB_1] = \mu_{11} - \mu_{01}, \tag{15.3}$$

which again can be interpreted as the amount of change in the response when level of A is changed from a_1 to a_0, while the level of B is kept constant at b_1.

The main effect of A therefore, denoted as $\mu[A]$ is defined as the average of the simple effects $\mu[AB_0]$ and $\mu[AB_1]$ in (15.2) and (15.3) respectively.

$$\mu[A] = \{\mu[AB_1] + \mu[AB_0]\}/2$$
$$= (\mu_{11} - \mu_{01} + \mu_{10} - \mu_{00})/2. \tag{15.4}$$

Similarly, the main effect of B is defined as the average of the simple effects of B at a_0 and a_1 respectively. That is,

$$\mu[B] = \{\mu[A_1B] + \mu[A_0B]\}/2$$
$$= (\mu_{11} - \mu_{10} + \mu_{01} - \mu_{00})/2. \tag{15.5}$$

Both main effects can be estimated from the table of observed means $\bar{y}_{ij}$ as

$$\hat{\mu}[A] = (\bar{y}_{11} - \bar{y}_{01} + \bar{y}_{10} - \bar{y}_{00})/2, \tag{15.6a}$$
$$\hat{\mu}[B] = (\bar{y}_{11} - \bar{y}_{10} + \bar{y}_{01} - \bar{y}_{00})/2. \tag{15.6b}$$

Example 15.2a

Consider the following table of means from a 2×2 factorial experiment with factors A and B.

	B	
A	b_0	b_1
a_0	33	63
a_1	22	52

The simple effects of A at b_0 and b_1 are respectively, $22 - 33 = -11$ and $52 - 63 = -11$. Hence Main effect of A is $(-11 - 11)/2 = -11$ or it could have been computed as $(52 - 63 + 22 - 33)/2 = -11$. Similarly the main effect of B is $(52 - 22 + 63 - 33)/2 = 30$.

Interaction Effects

Refer again to 2×2 table of population means in Table 15.4. Factors A and B are said to have interaction or to *interact* if the simple effect of A changes with the level of B. Thus the interaction term $\mu[AB]$ is defined as

$$\mu[AB] = (\mu[AB_1] - \mu[AB_0])/2 = (\mu[A_1B] - \mu[A_0B])/2. \tag{15.7}$$

If there is no interaction, then $\mu[AB] = 0$. Thus a non-zero value for $\mu[AB]$ is an indication of the presence of interaction between factors A and B.

Example 15.2b

Consider again the table of means from a 2×2 factorial experiment with factors A and B used in the previous example. The simple effects from four different table of means (a)-(d) are presented in Table 15.5.

A	b_0	b_1	$\mu[A_iB]$
a_0	33	63	30
a_1	22	52	30
$\mu[AB_j]$	-11	-11	

(a)

A	b_0	b_1	$\mu[A_iB]$
a_0	12	32	20
a_1	4	10	6
$\mu[AB_j]$	-8	-22	

(b)

A	b_0	b_1	$\mu[A_iB]$
a_0	33	13	-20
a_1	22	42	20
$\mu[AB_j]$	-11	29	

(c)

A	b_0	b_1	$\mu[A_iB]$
a_0	16	23	7
a_1	34	23	-11
$\mu[AB_j]$	18	0	

(d)

Table 15.5: Simple and interaction effects for four different tables of means

In Table 15.5, (a) gives identical simple effects with $\mu[AB] = 0$, (b) gives unequal simple effects with the same signs with $\mu[AB] = -7$, (c) gives unequal simple effects with opposite signs with $\mu[AB] = 20$, and (d) gives unequal simple effects with the same signs and has $\mu[AB] = -9$. In Figure 15.1, which corresponds to Table 15.5(a), $\mu[AB] = 0$ because the simple effects of B are the same at both levels of A. Thus the figure depicts the case when no interaction is present, which leads to parallel lines.

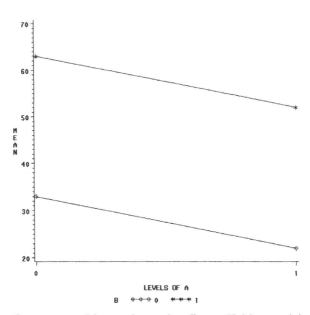

Figure 15.1: Identical simple effects- Table 15.5(a)

In Figure 15.2 which corresponds to the table of means in Table 15.5(b), we have the simple effects of B changing with the level of A (and vice versa), indicating the presence of interaction. The interaction effect here is $\mu[AB] = -7$. The simple effects of A are both negative, while those of B are both positive. For A, the expected response decreases from a_0 to a_1 at both levels of B. The interaction presented in Figure 15.2 therefore represents a *quantitative interaction*, because changing the levels of any one factor results in a change in the magnitude of the simple effects (but not the direction) of the other factor. Further, both lines in Figure 15.2 have downward slopes and quantitative interaction has a pattern of having lines not being parallel but have the same direction for their slopes.

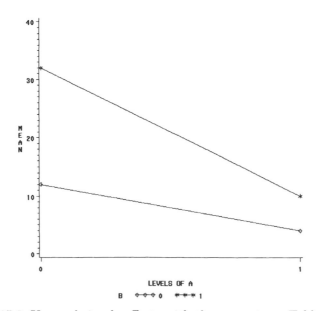

Figure 15.2: Unequal simple effects with the same signs- Table 15.5(b)

In Figure 15.3 which also corresponds to the table of means in Table 15.5(c), we have the simple effect of B at a_0 is negative, while its simple effect at a_1 is positive. That is, the expected responses of factor B decreases at a_0 and increases at a_1. The interaction therefore, is due to the difference in the signs of the simple effects. The interaction plotted in Figure 15.3 represents therefore a *qualitative interaction* because changing the level of any one factor results in a change in the direction (sign, $-$ to $+$) of the simple effect of the other factor. Again, the pattern of the plot in this figure is non-parallel but the slopes have different directions.

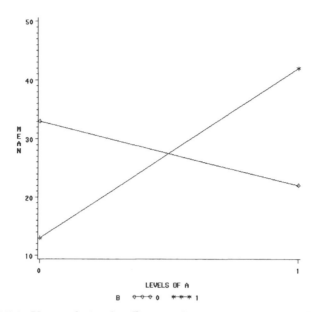

Figure 15.3: Unequal simple effects with opposite signs- Table 15.5(c)

The interaction plot in Figure 15.4 is similar to that in Figure 15.3 except that it is a variation of the former. Here, the pattern is that the lines are not parallel and have different slopes except that there are no increase in the simple effect of A at b_1 and therefore is flat.

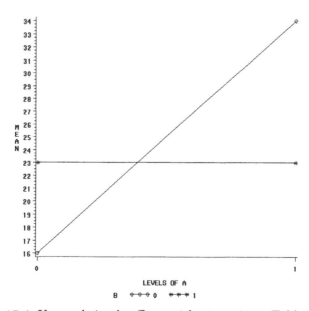

Figure 15.4: Unequal simple effects with same signs- Table 15.5(d)

15.4   Structure of Observations in a Two-Way ANOVA

We present in the following table the layout of the general structure of observations in a balanced two-way ANOVA.

Factor A Levels	Factor B Levels						Total
	1	2	$\cdots$	j	$\cdots$	b	
1	$y_{111}, \ldots, y_{11n}$	$y_{121}, \ldots, y_{12n}$	$\cdots$	$y_{1j1}, \ldots, y_{1jn}$	$\cdots$	$y_{1b1}, \ldots, y_{1bn}$	$y_{1..}$
2	$y_{211}, \ldots, y_{21n}$	$y_{221}, \ldots, y_{22n}$	$\cdots$	$y_{2j1}, \ldots, y_{2jn}$	$\cdots$	$y_{2b1}, \ldots, y_{2bn}$	$y_{2..}$
$\vdots$	$\vdots$	$\vdots$	$\cdots$	$\vdots$	$\cdots$	$\vdots$	$\vdots$
i	$y_{i11}, \ldots, y_{i1n}$	$y_{i21}, \ldots, y_{i2n}$	$\cdots$	$y_{ij1}, \ldots, y_{ijn}$	$\cdots$	$y_{ib1}, \ldots, y_{ibn}$	$y_{i..}$
$\vdots$	$\vdots$	$\vdots$	$\vdots$	$\vdots$	$\vdots$	$\vdots$	$\vdots$
a	$y_{a11}, \ldots, y_{a1n}$	$y_{a21}, \ldots, y_{a2n}$	$\cdots$	$y_{aj1}, \ldots, y_{ajn}$	$\cdots$	$y_{ab1}, \ldots, y_{abn}$	$y_{a..}$
Total	$y_{.1.}$	$y_{.2.}$	$\cdots$	$y_{.j.}$	$\cdots$	$y_{.b.}$	$y_{...}$

Let y_{ijk} denote the kth observation on the (i, j) treatment combination, where $i = 1, 2, \ldots, a$, $j = 1, 2, \ldots, b$, and $k = 1, 2, \ldots, n$. Thus we have a total of ab treatment combinations each having n replicates. Hence, a total of abn observations in the experiment. The two-way balanced ANOVA model with interactions have the following general linear model

$$y_{ijk} = \mu + \alpha_i + \beta_j + (\alpha\beta)_{ij} + \varepsilon_{ijk}, \tag{15.8}$$

with

$$\sum_i \alpha_i = \sum_j \beta_j = \sum_i (\alpha\beta)_{ij} = \sum_j (\alpha\beta)_{ij} = 0, \tag{15.9}$$

and the ε_{ijk} are independent and identically random error terms distributed as $N(0, \sigma^2)$. In the above set up, μ is the overall (or grand) mean, α_i is the effect of factor A at level i, β_j is similarly the effect of factor B at level j and $(\alpha\beta)_{ij}$ is the interaction effect of the two factors at levels i and j respectively. The *least squares* estimates of the parameters of the above model in (15.8) with the constraints in (15.9) give the following parameter estimates:

$$\hat{\mu} = \bar{y}...$$
$$\hat{\alpha}_i = \bar{y}_{i..} - \bar{y}...$$
$$\hat{\beta}_j = \bar{y}_{.j.} - \bar{y}...$$
$$\widehat{(\alpha\beta)}_{ij} = \bar{y}_{ij.} - \bar{y}_{i..} - \bar{y}_{.j.} + \bar{y}....$$

We illustrate in the following example, the analysis of a 3×4 two-way balanced ANOVA data.

Example 15.3

The following data are from Kleinbaum et al. (2008, p. 549) and reproduced by permission of Brooks/Cole, a part of Cengage Learning, Inc. The data relate to the diameter (y) of three species of pine trees which were compared at each of four locations, using samples of five trees per species at each location. Thus each of the 12 treatment combinations were replicated 5 times, resulting in a total of 60 observations in the study. We present the SAS codes and partial output for the analysis of these data.

Species	Location			
	1	2	3	4
	23	25	21	14
	15	20	17	17
A	26	21	16	19
	13	16	24	20
	21	18	27	24
	28	30	19	17
	22	26	24	21
B	25	26	19	18
	19	20	25	26
	26	28	29	23
	18	15	23	18
	10	21	25	12
C	12	22	19	23
	22	14	13	22
	13	12	22	19

```
data anova2;
do S='A', 'B','C';
do rep=1 to 5;
do L=1 to 4;
input y @@;
output;
end; end; end;
datalines;
23 25 21 14 15 20 17 17 26 21 16 19 13 16 24 20 21 18 27 24 28 30 19 17
22 26 24 21 25 26 19 18 19 20 25 26 26 28 29 23 18 15 23 18 10 21 25 12
12 22 19 23 22 14 13 22 13 12 22 19
;
proc glm;
class S L;
model y=S|L;
means S L/tukey;
lsmeans S*L/slice=(S L);
run;
```

```
                     The GLM Procedure

                   Class Level Information

            Class        Levels   Values
            S                 3   A B C
            L                 4   1 2 3 4

            Number of Observations Read        60
            Number of Observations Used        60
```

```
Dependent Variable: y

                          Sum of
Source               DF        Squares     Mean Square   F Value   Pr > F
Model                11    504.583333       45.871212      2.51   0.0137
Error                48    875.600000       18.241667
Corrected Total      59   1380.183333

          R-Square     Coeff Var     Root MSE       y Mean

          0.365592      20.95352     4.271026     20.38333

Source               DF     Type I SS     Mean Square   F Value   Pr > F
S                     2   344.9333333    172.4666667      9.45   0.0003
L                     3    46.0500000     15.3500000      0.84   0.4779
S*L                   6   113.6000000     18.9333333      1.04   0.4128

Source               DF   Type III SS     Mean Square   F Value   Pr > F
S                     2   344.9333333    172.4666667      9.45   0.0003
L                     3    46.0500000     15.3500000      0.84   0.4779
S*L                   6   113.6000000     18.9333333      1.04   0.4128
```

```
                    Tukey's Studentized Range (HSD) Test for y

NOTE: This test controls the Type I experimentwise error rate, but it generally has
               a higher Type II error rate than REGWQ.

            Alpha                                    0.05
            Error Degrees of Freedom                   48
            Error Mean Square                    18.24167
            Critical Value of Studentized Range   3.42026
            Minimum Significant Difference         3.2665


       Means with the same letter are not significantly different.

            Tukey Grouping           Mean      N    S
                        A          23.550     20    B

                        B          19.850     20    A
                        B
                        B          17.750     20    C

                         The GLM Procedure

                 Tukey's Studentized Range (HSD) Test for y

NOTE: This test controls the Type I experimentwise error rate, but it generally has
               a higher Type II error rate than REGWQ.

            Alpha                                    0.05
            Error Degrees of Freedom                   48
            Error Mean Square                    18.24167
            Critical Value of Studentized Range   3.76375
            Minimum Significant Difference         4.1506


       Means with the same letter are not significantly different.

            Tukey Grouping           Mean      N    L
                        A          21.533     15    3
                        A
                        A          20.933     15    2
                        A
                        A          19.533     15    1
                        A
                        A          19.533     15    4

                         The GLM Procedure
                       Least Squares Means

                 S     L        y LSMEAN
                 A     1       19.6000000
                 A     2       20.0000000
                 A     3       21.0000000
                 A     4       18.8000000
                 B     1       24.0000000
                 B     2       26.0000000
                 B     3       23.2000000
                 B     4       21.0000000
                 C     1       15.0000000
                 C     2       16.8000000
                 C     3       20.4000000
                 C     4       18.8000000

                         The GLM Procedure
                       Least Squares Means

                 S*L Effect Sliced by S for y

                      Sum of
       S     DF      Squares     Mean Square    F Value   Pr > F
       A      3    12.550000       4.183333       0.23    0.8755
       B      3    64.150000      21.383333       1.17    0.3301
       C      3    82.950000      27.650000       1.52    0.2224

                         The GLM Procedure
                       Least Squares Means

                 S*L Effect Sliced by L for y

                      Sum of
       L     DF      Squares     Mean Square    F Value   Pr > F
       1      2   202.533333     101.266667       5.55    0.0068
```

2	2	218.133333	109.066667	5.98	0.0048
3	2	21.733333	10.866667	0.60	0.5552
4	2	16.133333	8.066667	0.44	0.6452

The analysis indicates that the interaction term $S * L$ is not significant, nor is the main effect of factor L significant. However, there is strong significance difference between the means of factor S (species). We also request SAS to conduct Tukey's pairwise comparison tests. The results of this analysis indicate that while species A and C means are not significantly different at the 5% significance level, however, the mean for species B is significantly different from both the means of A and C. That is, we can display this result succinctly as:

$$\begin{array}{ccc} \mu_B & \mu_A & \mu_C \\ 23.55 & 19.85 & 17.75 \end{array}$$

Further we also use the **slice** command in **proc glm** to test for simple effects of S at L and those of L at various levels of S. For instance the test for the simple effects of S at levels L = 1 gives a SS = 202.5333 with a p-value of 0.0068. Similar results indicate that the simple effect of S at location 2 is also significant while those for locations 3 and 4 are not significant. The simple effects of locations at the three levels, A, B and C of factor S are not all significant.

15.5 Analysis with Quantitative Factor Levels

Example 15.4

The data in Table 15.6 refer to an experiment involving two factors A and B. A has 4 levels and B has 3 levels. The experiment was replicated twice ($r = 2$), and for illustrative purposes, we are assuming that the levels of the two factors are equally spaced.

Replicate	Factor B	Factor A a₁	a₂	a₃	a₄
1	b_1	7	8	9	7
	b_2	5	6	11	10
	b_3	4	6	10	12
2	b_1	7	9	9	8
	b_2	6	6	10	11
	b_3	6	7	10	12

Table 15.6: Coded data for this example

Since the number of levels for factors A and B are 4 and 3 respectively, it is possible to evaluate the linear, quadratic and cubic effects of treatment A as well as the linear and quadratic effects of B. The joint effects (interaction) are measured by subdividing the interaction sum of squares into $(A_L B_L), \ldots, (A_C B_Q)$. The treatment totals formed from the data in this example are presented in Table 15.7.

	a₁	a₂	a₃	a₄	Total
b_1	14	17	18	15	64
b_2	11	12	21	21	65
b_3	10	13	20	24	67
Total	35	42	59	60	196

Table 15.7: Treatment sums formed from data in Table 15.6

Replicates Total for Rep1 and Rep2 are respectively, Rep1 = 95 and Rep2 = 101.

Analysis

Here, $r = 2, a = 4, b = 3$, therefore, we have a total of 24 observations. The relevant sums of squares are computed as follows:

$$\text{Total SS} = 7^2 + 5^2 + \cdots + 12^2 - \frac{196^2}{24} \qquad = 117.33$$

$$\text{Replicate SS} = \frac{95^2}{12} + \frac{101^2}{12} - \frac{196^2}{24} \qquad = 1.50$$

$$\text{SS(A)} = \frac{35^2 + 42^2 + \cdots + 60^2}{6} - \frac{196^2}{24} \qquad = 77.67$$

$$\text{SS(B)} = \frac{64^2 + 65^2 + 67^2}{8} - \frac{196^2}{24} \qquad = 0.583$$

$$\text{SS(AB)} = \frac{14^2 + 17^2 + \cdots + 24^2}{2} - \text{CF} - \text{SS(A)} - \text{SS(B)}$$
$$= 112.33 - \text{SS(A)} - \text{SS(B)} = 34.08.$$

We present in Table 15.8 the initial analysis of variance for the data in Table 15.6.

Source	d.f.	SS	MS	F
Replicates	1	1.5	1.5	
A	3	77.67	25.89	80.91*
B	2	0.58	0.29	0.91
AB	6	34.08	5.68	17.75*
Error	11	3.50	0.32	
Total	23	117.33		

Table 15.8: Initial analysis of variance

We see that both main effect of A and the interaction terms AB are highly significant at the 5% level. The above analysis is carried out in SAS and the following is the program and partial output.

```
data new;
do rep=1 to 2;
do B=1 to 3;
do A=1 to 4;
input y @@;
output;
end; end; end;
datalines;
7 8 9 7 5 6 11 10 4 6 10 12
7 9 9 8 6 6 10 11 6 7 10 12
;
proc glm;
class rep A B;
model y= rep A|B/solution;
run;
```

```
                    The GLM Procedure

                Class Level Information

        Class        Levels    Values
        rep               2    1 2
        A                 4    1 2 3 4
        B                 3    1 2 3

        Number of observations    24
```

Dependent Variable: y

Source	DF	Sum of Squares	Mean Square	F Value	Pr > F
Model	12	113.8333333	9.4861111	29.81	<.0001
Error	11	3.5000000	0.3181818		
Corrected Total	23	117.3333333			

R-Square	Coeff Var	Root MSE	y Mean
0.970170	6.907054	0.564076	8.166667

Source	DF	Type I SS	Mean Square	F Value	Pr > F
rep	1	1.50000000	1.50000000	4.71	0.0527
A	3	77.66666667	25.88888889	81.37	<.0001
B	2	0.58333333	0.29166667	0.92	0.4283
A*B	6	34.08333333	5.68055556	17.85	<.0001

Source	DF	Type III SS	Mean Square	F Value	Pr > F
rep	1	1.50000000	1.50000000	4.71	0.0527
A	3	77.66666667	25.88888889	81.37	<.0001
B	2	0.58333333	0.29166667	0.92	0.4283
A*B	6	34.08333333	5.68055556	17.85	<.0001

We shall now partition the three sum of squares into their various components by making use of coefficients of orthogonal polynomials. For factor A where $k = 4$, there are linear, quadratic and cubic components. We give below their orthogonal coefficients with the treatment totals.

Linear (L)	-3	-1	1	3
Quadratic (Q)	1	-1	-1	1
Cubic (C)	-1	3	-3	1
Totals	35	42	59	60

For factor A, we calculate below, the linear, quadratic and cubic SS, which we have denoted here respectively as, A_L, A_Q, and A_C.

$$A_L = \frac{[35(-3) + 42(-1) + 59(1) + 60(3)]^2}{6\{(-3)^2 + (-1)^2 + 1^2 + 3^2\}} = 70.53$$

$$A_Q = \frac{[35(1) + 42(-1) + 59(1) + 60(1)]^2}{6\{4\}} = 1.50$$

$$A_C = \frac{[35(-1) + 42(3) + 59(-3) + 60(1)]^2}{6\{20\}} = 5.63.$$

Similarly for factor B where $k = 3$, both linear and quadratic components are again calculated from the totals for factor B levels.

Linear	-1	0	1
Quadratic	1	-2	1
Totals	64	65	67

Hence,

$$B_L = \frac{[64(-1) + 65(0) + 67(1)]^2}{8\{2\}} = 0.56$$

$$B_Q = \frac{[64(1) + 65(-2) + 67(1)]^2}{8\{6\}} = 0.02.$$

To obtain the interaction contrasts; we first obtain the A contrasts of each level of B by using

$$L_A^1 = (-3, -1, 1, 3), \quad Q_A^1 = (1, -1, -1, 1), \quad C_A^1 = (-1, 3, -3, 1).$$

For the first level of B and using the sums in Table 15.7, we have

$$b = 1 : -3(14) - 1(17) + 1(18) + 3(15) = 4$$
$$b = 2 : -3(11) - 1(12) + 1(21) + 3(21) = 39$$
$$b = 3 : -3(10) - 1(13) + 1(20) + 3(24) = 49.$$

For the second level of B and using the sums in Table 15.7, we have

$$b = 1 : 1(14) - 1(17) - 1(18) + 1(15) = -6$$
$$b = 2 : 1(11) - 1(12) - 1(21) + 1(21) = -1$$
$$b = 3 : 1(10) - 1(13) - 1(20) + 1(24) = 1.$$

For the third level of B and using the sums in Table 15.7, we also have

$$b = 1 : -1(14) + 3(17) - 3(18) + 1(15) = -2$$
$$b = 2 : -1(11) + 3(12) - 3(21) + 1(21) = -17$$
$$b = 3 : -1(10) + 3(13) - 3(20) + 1(24) = -7.$$

The A contrasts are given in Table 15.9. To get the linear B and cubic A entry in Table 15.9, use $L_B^1 = (-1, 0, 1)$ to obtain $-1(-2) + 0(-17) + 1(-7) = -5$.

	Linear A	Quadratic A	Cubic A	Factor B Divisors
B_1	4	−6	−2	
B_2	39	−1	−17	
B_3	49	1	−7	
Linear B	45	7	−5	2
Quadratic B	−25	−3	25	6
A divisors	20	4	20	

Table 15.9: Factor A contrasts

Then we multiply $L_B^1 = (-1, 0, 1)$, $Q_B^1 = (1, -2, 1)$, thus we have

$$L_A^1 \times L_B^1 = 4(-1) + 39(0) + 49(1) = 45,$$

Hence, the sum of squares (SS) for $L_A^1 \times L_B^1$ is

$$\mathrm{SS}(L_A^1 \times L_B^1) = \mathrm{SS}(\mathrm{A}_L\mathrm{B}_L) = \frac{45^2}{2 \times 20 \times 2} = 25.31.$$

Similarly,

$$\mathrm{SS}(\mathrm{A}_L\mathrm{B}_Q) = \frac{(-25)^2}{2 \times 20 \times 6} = 2.60$$

$$\mathrm{SS}(\mathrm{A}_Q\mathrm{B}_L) = \frac{7^2}{2 \times 4 \times 2} = 3.60$$

$$\mathrm{SS}(\mathrm{A}_Q\mathrm{B}_Q) = \frac{(-3)^2}{2 \times 4 \times 6} = 0.19$$

$$\mathrm{SS}(\mathrm{A}_C\mathrm{B}_L) = \frac{(-5)^2}{2 \times 20 \times 2} = 0.31$$

$$\mathrm{SS}(\mathrm{A}_C\mathrm{B}_Q) = \frac{(25)^2}{2 \times 20 \times 6} = 2.60.$$

Table 15.10 gives the full analysis of variance for the data in Table 15.7.

From the F tables in Table 4 of the Appendix, $f_{0.05,1,11} = 4.84$, hence, $A_L, A_C, A_LB_L, A_LB_Q, A_QB_L$ and A_CB_Q are therefore found to be significant at $\alpha = 0.05$. Obviously, the response of factor A can be least described by a third degree polynomial. From Table 15.7, the averages of the entries are presented in the following table:

Source	d.f.	SS	MS	F
Replicates	1	1.5	1.5	
Treatments				
A_L	1	70.53	70.53	240.4 *
A_Q	1	1.50	1.50	4.69
A_C	1	5.63	5.63	17.59*
B_L	1	0.56	0.56	1.75
B_Q	1	0.02	0.02	0.06
$A_L B_L$	1	25.31	25.31	79.09*
$A_L B_Q$	1	2.60	2.60	8.12
$A_Q B_L$	1	3.06	3.06	9.56*
$A_Q B_Q$	1	0.19	0.19	0.59
$A_C B_L$	1	0.31	0.31	0.97
$A_C B_Q$	1	2.60	2.60	8.12*
Error	11	3.50	0.32	
Total	23	117.33		

Table 15.10: Full analysis of variance table

		A			B
B	1	2	3	4	mean
1	7.0	8.5	9.0	7.5	8.0
2	5.5	6.0	10.5	10.5	8.125
3	5.0	6.5	10.0	12.0	8.375
A mean	5.833	7.000	9.833	10.000	8.167

		A			B
B	1	2	3	4	effect
1	1.3333	1.6667	-0.6667	-2.3333	-0.1667
2	-0.2917	-0.9583	0.7083	0.5417	-0.0417
3	-1.0416	-0.7084	-0.7500	2.5000	0.2083
A effect	-2.3333	-1.1667	1.6667	1.8333	$\hat{\mu} = 0$

Table 15.11: Table of parameter estimates

We have

$$\hat{\alpha}_1 = -2.3333, \ \hat{\alpha}_2 = -1.1667, \ \hat{\alpha}_3 = 1.6667, \text{ and } \hat{\alpha}_4 = 1.8333.$$

Similarly,

$$\hat{\beta}_1 = -0.1667, \ \hat{\beta}_2 = -0.0417, \text{ and } \hat{\beta}_3 = 0.2083.$$

Further,

$$\widehat{\alpha\beta}_{11} = 1.3333, \ \widehat{\alpha\beta}_{21} = 1.1667, \ \ldots, \ \widehat{\alpha\beta}_{43} = 2.5000.$$

Obtaining Interaction Components in SAS

$$A_L = \{-3, -1, 1, 3\}, \quad A_Q = \{1, -1, -1, 1\}, \quad \text{and} \quad A_C = \{-1, 3, -3, 1\}$$

Similarly,

$$B_L = \{-1, 0, 1\}, \quad \text{and} \quad B_Q = \{1, -2, 1\}$$

Hence the coefficients for the interaction component $A_L B_L$ and $A_L B_Q$ that will be used in the contrast statements are presented respectively for the 4×3 table of interaction means as:

3	1	−1	−3
0	0	0	0
−3	−1	1	3

−3	−1	1	3
6	2	−2	−6
−3	−1	1	3

For example, the first column of the contrast table for the $A_L B_L$ component is obtained as

$$-3 \times -1 = 3, \quad -3 \times 0 = 0, \quad -3 \times 1 = 3,$$

and the first column for the $A_L B_Q$ are similarly obtained as

$$-3 \times 1 = -3, \quad -3 \times -2 = 6, \quad -3 \times 1 = -3.$$

The coefficients for the $A_Q B_L$ and $A_Q B_Q$ are presented respectively in the following:

−1	1	1	−1
0	0	0	0
1	−1	−1	1

1	−1	−1	1
−2	2	2	−2
1	−1	−1	1

Also, the coefficients for the $A_C B_L$ and $A_C B_Q$ are presented respectively in the following:

−1	−3	3	−1
0	0	0	0
−1	3	−3	1

−1	3	−3	1
2	−6	6	−2
−1	3	−3	1

We implement these contrasts in SAS with the following contrast statements derived from our earlier calculations. We present in the following the contrast results along with the SAS program.

```
set new;
proc glm;
class rep A B;
model y = rep A|B;
contrast 'LL' A*B 3  0 -3  1 0 -1 -1 0 1 -3 0 3;
contrast 'LQ' A*B -3 6 -3 -1 2 -1 1 -2 1 3 -6 3;
contrast 'QL' A*B -1 0 1 1 0 -1 1 0 -1 -1 0 1;
contrast 'QQ' A*B 1 -2 1 -1 2 -1 -1 2 -1 1 -2 1;
contrast 'CL' A*B 1 0 -1 -3 0 3 3 0 -3 -1 0 1;
contrast 'CQ' A*B -1 2 -1 3 -6 3 -3 6 -3 1 -2 1;
run;
        Contrast             DF    Contrast SS    Mean Square    F Value    Pr > F
        ------------------------------------------------------------------------
        LL                   1     25.31250000    25.31250000    79.55     <.0001
        LQ                   1      2.60416667     2.60416667     8.18     0.0155
        QL                   1      3.06250000     3.06250000     9.62     0.0101
        QQ                   1      0.18750000     0.18750000     0.59     0.4589
        CL                   1      0.31250000     0.31250000     0.98     0.3430
        CQ                   1      2.60416667     2.60416667     8.18     0.0155
```

We observe that the sums of squares computed for each of the six components in SAS agree with our calculated values in Table 15.10. The plot of the significant AB interaction is presented in Figure 15.5.

15.6 Two-Way ANOVA in Randomized Block Designs

For two factors A and B with ab treatment combinations arranged in a randomized complete block design, the appropriate model is

$$y_{ijk} = \mu + \gamma_i + \alpha_j + \beta_k + (\alpha\beta)_{jk} + \varepsilon_{ijk}, \tag{15.10}$$

with $\sum_i \gamma_i = \sum_j \alpha_j = \sum_k \beta_k = \sum_j (\alpha\beta)_{jk} = \sum_k (\alpha\beta)_{jk} = 0$. The block effect, factor B effect and factor A effect are respectively measured by γ_i, α_j and β_k. The model in (15.10) assumes that there is no treatment-block interactions which means that the model is additive with respect to blocks. We give a practical example in the following data.

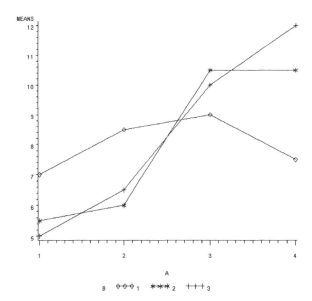

Figure 15.5: Plot of the significant AB interaction term

| Time of | | Blocks | | | | |
Covering	Variety	I	II	III	IV	Total
February	V	10.2	10.1	12.1	12.3	44.7
	R	11.1	9.8	8.6	9.4	38.9
	F	6.8	9.5	9.5	10.3	36.1
	G	5.3	7.5	4.6	7.3	24.7
March	V	8.0	9.7	12.0	7.8	37.5
	R	9.7	7.9	10.3	11.2	39.1
	F	8.6	9.6	9.5	10.0	37.7
	G	3.4	4.2	7.3	7.6	22.5
April	V	2.0	6.1	4.8	6.7	19.6
	R	10.9	8.4	6.5	9.2	35.0
	F	2.2	4.9	4.4	3.6	15.1
	G	2.1	0.9	3.4	2.3	8.7
Block Totals		80.3	88.6	93.0	97.7	359.6

Table 15.12: Data for the 4×3 factorial experiment in this example

Example 15.5

An experiment was conducted on strawberries under cloches to investigate the response of four varieties to three times of covering. A randomized block design was used, with four blocks and twelve treatment combinations. Table 15.12 gives the results from this experiment.

Here, there are two factors: variety and times of covering. Variety is at 4 levels (V, R, F, G) while time of covering has 3 levels (Feb, Mar, Apr). Thus we have a total of $4 \times 3 = 12$ treatment combinations. In this experiment, each block must have 12 plots and each treatment combination must be present in each block.

The initial analysis of variance (ignoring the factorial structure of treatments, that is, treating experiment as 4 blocks of 12 treatments each) gives the following ANOVA table:

Source	d.f.	SS	MS	F
Blocks	3	13.70	4.57	
Treatments	11	356.02	32.37	15.40 ***
Error	33	69.38	2.102	
Total	47	439.16		

The Treatments F value of 15.40 is highly significant at the 0.01% level. We present in Table 15.13 the two-way interaction table for times and varieties.

Covering	Variety				
Time	V	R	F	G	Totals
February	44.7	38.9	36.1	24.7	144.4
March	37.5	39.1	37.7	22.5	136.8
April	19.6	35.0	15.1	8.7	78.4
Variety Totals	101.8	113.0	88.9	55.9	359.6

Table 15.13: Two-way interaction table for times and varieties

The relevant SS are calculated as follows:

$$\text{SS Main effect of Varieties} = \frac{101.8^2}{12} + \frac{113^2}{12} + \frac{88.9^2}{12} + \frac{55.9^2}{12} - \text{CF} = 152.69,$$

since each of $101.8, \ldots, 55.9$ comes from 12 observations.

Source	d.f.	SS	MS	F
Blocks	3	13.70	4.57	
Varieties	3	152.69	50.90	24.2 ***
Times	2	163.01	81.50	38.8 ***
Varieties x Times	6	40.32	6.72	3.2*
Error	33	69.38	2.102	
Total	47	439.16		

Table 15.14: Full analysis of variance table for the data in Table 15.12

Similarly,

$$\text{SS Main effect of Times} = \frac{144.4^2}{16} + \frac{136.8^2}{16} + \frac{78.4^2}{16} - \text{CF} = 163.01.$$

$$\text{Interaction SS} = \frac{44.7^2}{4} + \frac{38.9^2}{4} + \cdots + \frac{8.7^2}{4} - \text{CF} - \text{SS(Varieties)} - \text{SS (Times)} = 40.32.$$

This could have been obtained as Treatment SS − Varieties SS − Times SS. The full analysis of variance is presented in Table 15.14. As the interaction SS is significant the results are presented in a two-way table of treatment means. The plot of these means is presented in Figure 15.6.

Covering	Variety				Time
Time	V	R	F	G	mean
February	11.2	9.7	9.0	6.2	9.0
March	9.4	9.8	9.4	5.6	8.5
April	4.9	8.8	3.8	2.2	4.9
Variety Mean	8.5	9.4	7.4	4.7	7.5

Table 15.15: Table of treatment means

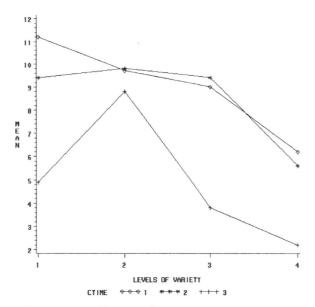

Figure 15.6: Time and variety interaction plot

The standard error of difference between two values in Table 15.15 is $-\sqrt{\frac{2S^2}{4}} = 1.03$. The standard error of difference between two variety means $= \sqrt{\frac{2S^2}{16}} = 0.59$.

Summary of Results

For each variety, the difference between the means for the first two covering times was not significant; for all varieties except variety R the third covering times gave a significantly lower yield than the other times. For the first two covering times, variety G gave a significantly lower yield than the other three varieties; for the third covering time variety R gave a significantly higher yield than the other three varieties. (All significance statements refer to the 5% significance levels). The SAS implementation of the analysis of the data in Table 15.12 is presented in the following.

```
data ex102;
title 'Two way ANOVA with interaction';
do time='FEB', 'MARCH', 'APRIL';
do variety='V','R','F','G';
do blocks=1 TO 4;
input y @@;
output;
end; end; end;
datalines;
10.2 10.1 12.1 12.3 11.1 9.8 8.6 9.4 6.8 9.5 9.5 10.3 5.3 7.5 4.6 7.3
8.0 9.7 12 7.8 9.7 7.9 10.3 11.2 8.6 9.6 9.5 10 3.4 4.2 7.3 7.6
2 6.1 4.8 6.7 10.9 8.4 6.5 9.2 2.2 4.9 4.4 3.6 2.1 0.9 3.4 2.3
;
run;
proc glm;
class blocks time variety;
model y=blocks time|variety;
run;
```

```
                  The GLM Procedure

             Class Level Information

        Class       Levels    Values
        blocks           4    1 2 3 4
        time             3    APR FEB MAR
        variety          4    F G R V

          Number of observations     48

                  The GLM Procedure
```

Dependent Variable: y

Source	DF	Sum of Squares	Mean Square	F Value	Pr > F
Model	14	369.7033333	26.4073810	12.55	<.0001
Error	33	69.4533333	2.1046465		
Corrected Total	47	439.1566667			

R-Square	Coeff Var	Root MSE	y Mean
0.841848	19.36472	1.450740	7.491667

Source	DF	Type I SS	Mean Square	F Value	Pr > F
blocks	3	13.6916667	4.5638889	2.17	0.1104
time	2	163.0066667	81.5033333	38.73	<.0001
variety	3	152.6850000	50.8950000	24.18	<.0001
time*variety	6	40.3200000	6.7200000	3.19	0.0139

Source	DF	Type III SS	Mean Square	F Value	Pr > F
blocks	3	13.6916667	4.5638889	2.17	0.1104
time	2	163.0066667	81.5033333	38.73	<.0001
variety	3	152.6850000	50.8950000	24.18	<.0001
time*variety	6	40.3200000	6.7200000	3.19	0.0139

15.7   Constructing Possible Contrasts for Interaction

Since the interaction term is significant, further analysis of the data would now focus on this significant interaction term as main effects now become secondary. We present in the next section possible comparison contrasts for the 6 degrees of freedom interaction terms. It should be noted here that other possible constructs could also be so construed. This is to understand the structure of the data.

1. The average y of covering time in April for variety V to the average y in February and March for Variety V.

$$H_0 : \mu_{31} = \frac{1}{2}(\mu_{11} + \mu_{21})$$

2. The average y of covering time in April for variety R to the average y in February and March for Variety R.

$$H_0 : \mu_{32} = \frac{1}{2}(\mu_{12} + \mu_{22})$$

3. The average y of covering time in April for variety F to the average y in February and March for Variety F.

$$H_0 : \mu_{33} = \frac{1}{2}(\mu_{13} + \mu_{23})$$

4. The average y of covering time in April for variety G to the average y in February and March for Variety G.

$$H_0 : \mu_{34} = \frac{1}{2}(\mu_{14} + \mu_{24})$$

5. The average y of covering time in February for variety V to the average y at covering time in March for variety V. That is,

$$H_0 : \mu_{11} = \mu_{21}$$

6. The average y of covering time in April for variety V to the average of y in April of varieties F and G.

$$H_0 : \mu_{31} = \frac{1}{2}(\mu_{33} + \mu_{34}) \tag{15.11}$$

Table 15.16 contains the population means μ_{11} to μ_{34}.
We recall that the interaction can be written as

$$\mu_{ij} = \mu + \alpha_i + \beta_j + (\alpha\beta)_{ij}.$$

Hence, the first contrast can be written as

$$\mu_{31} = \frac{1}{2}(\mu_{11} + \mu_{21}) \implies 2\mu_{31} - (\mu_{11} + \mu_{21}) = 0.$$

	Variety			
Covering	V	R	F	G
Time	1	2	3	4
1	μ_{11}	μ_{12}	μ_{13}	μ_{14}
2	μ_{21}	μ_{22}	μ_{23}	μ_{24}
3	μ_{31}	μ_{32}	μ_{33}	μ_{34}

Table 15.16: Table of population means

Therefore,

$$0 = 2\mu + 2\alpha_3 + 2\beta_1 + 2(\alpha\beta)_{31} - [\mu + \alpha_1 + \beta_1 + (\alpha\beta)_{11} + \mu + \alpha_2 + \beta_1 + (\alpha\beta)_{21}]$$
$$= -\alpha_1 - \alpha_2 + 2\alpha_3 - (\alpha\beta)_{11} - (\alpha\beta)_{21} + 2(\alpha\beta)_{31}.$$

The contrast in SAS will therefore be written as

```
contrast 'one' time -1 -1 2
         time*variety -1 0 0 0 -1 0 0 0 2 0 0 0;

estimate 'one' time -1 -1 2
         time*variety -1 0 0 0 -1 0 0 0 2 0 0 0/divisor=2;
```

Similarly, we can show that the second contrast can be finally written as

$$0 = -\alpha_1 - \alpha_2 + 2\alpha_3 - (\alpha\beta)_{12} - (\alpha\beta)_{22} + 2(\alpha\beta)_{32}.$$

Similar results are also obtained for the third and fourth contrasts. For the fifth contrast, we have,

$$\mu_{11} = \mu_{21} \Longrightarrow \mu + \alpha_1 + (\alpha\beta)_{11} - \mu - \alpha_2 - (\alpha\beta)_{21} = 0.$$

That is,

$$\alpha_1 - \alpha_2 + (\alpha\beta)_{11} - (\alpha\beta)_{21} = 0.$$

Therefore, the SAS contrast statement becomes

```
contrast 'five' time 1 -1 0
         time*variety 1 0 0 0 -1 0 0 0 0 0 0 0;
```

The sixth and final contrast can be written as

$$\mu_{31} = \frac{1}{2}(\mu_{33} + \mu_{34}) \Longrightarrow 2\mu_{31} - (\mu_{33} + \mu_{34}) = 0.$$

Hence,

$$0 = 2\mu + 2\alpha_3 + 2\beta_1 + 2(\alpha\beta)_{31} - [\mu + \alpha_3 + \beta_3 + (\alpha\beta)_{33} + \mu + \alpha_3 + \beta_4 + (\alpha\beta)_{34}]$$
$$= 2\beta_1 - \beta_3 - \beta_4 + 2(\alpha\beta)_{31} - (\alpha\beta)_{33} - (\alpha\beta)_{34}.$$

The contrast in SAS will therefore be written as

```
contrast 'six' variety 2 0 -1 -1
         time*variety  0 0 0 0 0 0 0 0 2 0 -1 -1;

estimate 'six' variety 2 0 -1 -1
         time*variety 0 0 0 0 0 0 0 0 2 0 -1 -1/divisor=2;
```

The above contrasts are implemented in SAS with the following program and a partial output is presented. Note the **order=data** in the PROC statement. This is necessary so as to not change the order of the levels of the factor variables into alphabetical ordering.

```
set ex102;
proc glm order=data;
class blocks time variety;
model y=blocks time|variety;
lsmeans time variety time*variety;
contrast 'one' time -1 -1 2
      time*variety -1 0 0 0 -1 0 0 0 2 0 0 0;
estimate 'one' time -1 -1 2
      time*variety -1 0 0 0 -1 0 0 0 2 0 0 0/divisor=2;
contrast 'two' time -1 -1 2
      time*variety  0 -1 0 0 0 -1 0 0 0 2 0 0;
estimate 'two' time -1 -1 2
      time*variety  0 -1 0 0 0 -1 0 0 0 2 0 0/divisor=2;
contrast 'three' time -1 -1 2
        time*variety  0 0 -1 0 0 0 -1 0 0 0 2 0;
estimate 'three' time -1 -1 2
        time*variety  0 0 -1 0 0 0 -1 0 0 0 2 0/divisor=2;
contrast 'four' time -1 -1 2
      time*variety  0 0 0 -1 0 0 0 -1 0 0 0 2;
estimate 'four' time -1 -1 2
      time*variety  0 0 0 -1 0 0 0 -1 0 0 0 2/divisor=2;
contrast 'five' time 1 -1 0
      time*variety 1 0 0 0 -1 0 0 0 0 0 0 0;
estimate 'five' time 1 -1 0
      time*variety 1 0 0 0 -1 0 0 0 0 0 0 0;
contrast 'six' variety 2 0 -1 -1
      time*variety 0 0 0 0 0 0 0 0 2 0 -1 -1;
estimate 'six' variety 2 0 -1 -1
      time*variety 0 0 0 0 0 0 0 0 2 0 -1 -1/divisor=2;
run;
```

                              The GLM Procedure
                            Least Squares Means

                        time   y LSMEAN
                        FEB   9.02500000
                        MAR   8.55000000
                        APR   4.90000000

                        variety   y LSMEAN
                        V   8.48333333
                        R   9.41666667
                        F   7.40833333
                        G   4.65833333

                  time   variety   y LSMEAN
                  FEB   V   11.1750000
                  FEB   R   9.7250000
                  FEB   F   9.0250000
                  FEB   G   6.1750000
                  MAR   V   9.3750000
                  MAR   R   9.7750000
                  MAR   F   9.4250000
                  MAR   G   5.6250000
                  APR   V   4.9000000
                  APR   R   8.7500000
                  APR   F   3.7750000
                  APR   G   2.1750000

                            The GLM Procedure

Dependent Variable: y

Contrast	DF	Contrast SS	Mean Square	F Value	Pr > F
one	1	77.04166667	77.04166667	36.61	<.0001
two	1	2.66666667	2.66666667	1.27	0.2684
three	1	79.20666667	79.20666667	37.63	<.0001
four	1	37.00166667	37.00166667	17.58	0.0002
five	1	6.48000000	6.48000000	3.08	0.0886
six	1	9.88166667	9.88166667	4.70	0.0376

Parameter	Estimate	Standard Error	t Value	Pr > |t|
one	-5.37500000	0.88839317	-6.05	<.0001
two	-1.00000000	0.88839317	-1.13	0.2684
three	-5.45000000	0.88839317	-6.13	<.0001
four	-3.72500000	0.88839317	-4.19	0.0002
five	1.80000000	1.02582807	1.75	0.0886
six	1.92500000	0.88839317	2.17	0.0376

15.8 Random Effects Models in Two-Factor Studies

In the previous analysis in this chapter, we have assumed in Example 15.3 for instance that all factor levels of interest, that is, locations (1, 2, 3, 4) and species (A, B, C) are included in the experiment. That is, there are only 4 locations and three known species. Therefore, the resulting model is a *fixed effects model*. Similarly, in Example 15.5, we have also assumed that all covering times of interest (February, March, April) and all varieties of interest (V, R, F, G) are included in the study, again resulting in a *fixed effects model*. Fixed effects models are often referred to as two-factor ANOVA Model I. Random effects models in this case can be of three types, namely,

a. Both factors are random. This would result in two-factor ANOVA *random effects model*. This will be called *random effects model* IIa.

b. One factor (say row factor) is random and the other factor (say column factor) is fixed. This will result in a two-factor ANOVA *mixed effects model*. This will be called *random effects model* IIb.

c. One factor (say column factor) is random and the other factor (the row factor) is fixed. This will result in a two-factor ANOVA *mixed effects model*. This will be called *random effects model* IIc.

We now discuss each of the above random effects models in turn.

15.8.1 Random Effects Model IIa

For example, if in Example 15.5 for instance, it could be that of say eight covering dates, only three times, February, March and April were randomly selected (the actual chance of this sequence occurring is very small, but let us suppose this is a realized outcome) from a possible 8 covering times (January-August). Similarly, let us also assume that there are more than four varieties under study, but we have just randomly selected varieties V, R, F and G. The resulting ANOVA model has both factors being random, and we would have the following model

$$y_{ijk} = \mu + A_i + B_j + C_{ij} + \varepsilon_{ijk}, \tag{15.12}$$

where,

$$A_i \sim N(0, \sigma_A^2), \quad i = 1, 2, \ldots, a$$
$$B_j \sim N(0, \sigma_B^2), \quad j = 1, 2, \ldots, b$$
$$C_{ij} \sim N(0, \sigma_{AB}^2),$$
$$\varepsilon_{ijk} \sim N(0, \sigma^2), \quad k = 1, 2, \ldots, r.$$

Further, A_i, B_j, C_{ij} and ε_{ijk} are independently distributed with the above properties.

The fixed model hypotheses in (15.13a) to (15.13c) become respectively, hypotheses of interest in (15.14a) to (15.14c) under the random effects model in (15.12).

$$H_0 : \alpha_1 = \alpha_2 = \cdots = \alpha_a = 0 \tag{15.13a}$$
$$H_0 : \beta_1 = \beta_2 = \cdots = \beta_b = 0 \tag{15.13b}$$
$$H_0 : \alpha\beta_{11} = \alpha\beta_{12} = \cdots = \alpha\beta_{ab} = 0 \tag{15.13c}$$

$$H_0 : \sigma_A^2 = 0 \tag{15.14a}$$
$$H_0 : \sigma_B^2 = 0 \tag{15.14b}$$
$$H_0 : \sigma_{AB}^2 = 0. \tag{15.14c}$$

The expected means under model IIa are given in Table 15.17.

Source	d.f.	MS	E(MS)	F
A	$a-1$	MS_a	$\sigma^2 + r\sigma^2_{AB} + rb\sigma^2_A$	MS_a/MS_{ab}
B	$b-1$	MS_b	$\sigma^2 + r\sigma^2_{AB} + ra\sigma^2_B$	MS_b/MS_{ab}
AB	$(a-1)(b-1)$	MS_{ab}	$\sigma^2 + r\sigma^2_{AB}$	MS_{ab}/MS_e
Error	$ab(r-1)$	MS_e	σ^2	
Total	$abr-1$			

Table 15.17: Table of expected mean squares under model IIa

To test the hypothesis in (15.14c), we compute

$$F^* = \frac{MS_{ab}}{MS_e}.$$

Under $H_0 : \sigma^2_{AB} = 0$, we would expect F equals 1. Large computed values of F therefore would indicate that the null hypothesis would be rejected. F^* will be distributed as an F with $(a-1)(b-1)$ and $ab(r-1)$ degrees of freedom. Similarly, to test hypothesis in (15.14b), we compute

$$F^* = \frac{MS_b}{MS_{ab}}. \tag{15.15}$$

The computed F in (15.15) is under H_0 distributed as an F with $(b-1)$ and $(a-1)(b-1)$ degrees of freedom. Finally, to also test the hypothesis in (15.14a), we compute

$$F^* = \frac{MS_a}{MS_{ab}}. \tag{15.16}$$

The computed F^* in (15.16) is under H_0 distributed as an F with $(a-1)$ and $(a-1)(b-1)$ degrees of freedom.

Example 15.5 continued

For the data in Table 15.12, suppose both factor variable levels are assumed to be random selections. Then both factors will be assumed to be random and the model in (15.12) will apply in this case, leading to the following analysis from SAS.

```
set ex102;
proc glm order=data;
class blocks time variety;
model y=blocks time|variety;
random time variety time*variety/test;
run;
                    The GLM Procedure

Dependent Variable: y

                        Sum of
Source              DF     Squares   Mean Square   F Value   Pr > F
Model               14   369.7033333   26.4073810    12.55   <.0001
Error               33    69.4533333    2.1046465
Corrected Total     47   439.1566667

          R-Square   Coeff Var    Root MSE     y Mean
          0.841848    19.36472    1.450740    7.491667

Source              DF    Type I SS   Mean Square   F Value   Pr > F
blocks               3    13.6916667    4.5638889     2.17   0.1104
time                 2   163.0066667   81.5033333    38.73   <.0001
variety              3   152.6850000   50.8950000    24.18   <.0001
time*variety         6    40.3200000    6.7200000     3.19   0.0139

Source              DF   Type III SS   Mean Square   F Value   Pr > F
blocks               3    13.6916667    4.5638889     2.17   0.1104
time                 2   163.0066667   81.5033333    38.73   <.0001
variety              3   152.6850000   50.8950000    24.18   <.0001
time*variety         6    40.3200000    6.7200000     3.19   0.0139
```

```
Source              Type III Expected Mean Square
blocks              Var(Error) + Q(blocks)
time                Var(Error) + 4 Var(time*variety) + 16 Var(time)
variety             Var(Error) + 4 Var(time*variety) + 12 Var(variety)
time*variety        Var(Error) + 4 Var(time*variety)

                         The GLM Procedure
          Tests of Hypotheses for Mixed Model Analysis of Variance

Dependent Variable: y
Source               DF     Type III SS   Mean Square   F Value   Pr > F
blocks                3       13.691667      4.563889      2.17   0.1104
time*variety          6       40.320000      6.720000      3.19   0.0139
Error: MS(Error)     33       69.453333      2.104646

Source               DF     Type III SS   Mean Square   F Value   Pr > F
time                  2      163.006667     81.503333     12.13   0.0078
variety               3      152.685000     50.895000      7.57   0.0183
Error                 6       40.320000      6.720000
Error: MS(time*variety)
```

The computed F values as expected for hypotheses (15.14c), (15.14b), and (15.14a) are obtained respectively as

$$F^* = \frac{6.72}{2.1046} = 3.19; \quad F^* = \frac{50.8950}{6.720} = 7.57; \quad F^* = \frac{81.5033}{6.720} = 12.13.$$

These tests are accomplished in SAS with the option **test** in the random line statements. SAS will use the relevant denominator mean squares to compute the F values and subsequently compute the relevant p-values. We also note that SAS generates the expected mean squares under this model automatically.

15.8.2   Other Tests of Significance

Assuming that the interaction were not significant in this case and we would like to concentrate on the main effects of time and variety. We can conduct multiple comparison tests as usual except that we need to specify the denominator mean square in this case. For instance, the SAS program below will conduct Tukey's test on the means of time and variety, using the appropriate denominator mean square specified by the **E**, error option. Also, a test of the hypothesis on the means of the factor variable time, $H_0 : (\mu_1 + \mu_2)/2 = \mu_3$, can also be conducted by specifying the appropriate error mean square with the use of the **E** option. A modified sample output is presented below.

```
set ex102;
proc glm order=data;
class blocks time variety;
model y=blocks time|variety;
random time|variety;
means time variety/tukey E=time*variety;
contrast '1 & 2 vs 3' time .5 .5 -1/E=time*variety;
run;
                         The GLM Procedure

Source              Type III Expected Mean Square
blocks              Var(Error) + Q(blocks)
time                Var(Error) + 4 Var(time*variety) + 16 Var(time)
variety             Var(Error) + 4 Var(time*variety) + 12 Var(variety)
time*variety        Var(Error) + 4 Var(time*variety)

              Tukey's Studentized Range (HSD) Test for y

NOTE: This test controls the Type I experimentwise error rate, but it generally has
            a higher Type II error rate than REGWQ.

          Alpha                                   0.05
          Error Degrees of Freedom                   6
          Error Mean Square                       6.72
          Critical Value of Studentized Range  4.33902
          Minimum Significant Difference         2.812

        Means with the same letter are not significantly different.

          Tukey Grouping        Mean      N    time
                     A         9.0250     16    FEB
                     A
                     A         8.5500     16    MAR
```

```
         B        4.9000    16    APR
```
```
           Tukey's Studentized Range (HSD) Test for y
```
```
NOTE: This test controls the Type I experimentwise error rate, but it generally has
                a higher Type II error rate than REGWQ.
```
```
         Alpha                                    0.05
         Error Degrees of Freedom                    6
         Error Mean Square                        6.72
         Critical Value of Studentized Range   4.89559
         Minimum Significant Difference         3.6635
```
```
       Means with the same letter are not significantly different.
```
```
       Tukey Grouping          Mean     N    variety
                      A         9.417    12   R
                      A
                      A         8.483    12   V
                      A
                  B   A         7.408    12   F
                  B
                  B             4.658    12   G
```
```
Dependent Variable: y
```
```
     Tests of Hypotheses Using the Type III MS for time*variety as an Error Term
```
```
Contrast                    DF    Contrast SS    Mean Square   F Value   Pr > F
1 & 2 vs 3                   1    161.2016667    161.2016667    23.99    0.0027
```

15.8.3   Estimating the Variance Components

For this model, we would like to estimate the components of variance, namely, σ^2, σ^2_{AB}, σ^2_B and σ^2_A. In this example, $a = 3, b = 4$ and $r = 4$. From the expected mean squares in Table 15.17, we compute the estimate

$$\hat{\sigma}^2 = \text{MS}_e = 2.104646.$$

Similarly,

$$\hat{\sigma}^2_{AB} = \frac{\text{MS}_{ab} - \text{MS}_e}{r} = \frac{6.720 - 2.1046}{4} = 1.1539.$$

Also,

$$\hat{\sigma}^2_B = \frac{\text{MS}_b - \text{MS}_{ab}}{ra} = \frac{50.8950 - 6.7200}{12} = 3.6813.$$

$$\hat{\sigma}^2_A = \frac{\text{MS}_a - \text{MS}_{ab}}{rb} = \frac{81.5033 - 6.7200}{16} = 4.6740.$$

The above variance components can also be estimated with SAS by employing **proc varcomp**. The option **fixed = 1** specifies that the first factor in the model statement, namely blocks should be assumed fixed, since we do not want to compute variance components for it. The results from the SAS output agree with those computed from our table of expected mean squares.

```
proc varcomp method=type1;
class blocks time variety;
model y=blocks time|variety/fixed=1;
run;
            Variance Components Estimation Procedure
```
```
                Class Level Information
```
```
         Class      Levels    Values
         blocks         4     1 2 3 4
         time           3     APR FEB MAR
         variety        4     F G R V
```
```
            Number of observations    48
```
```
            Dependent Variable:    y
```

```
                          Type 1 Analysis of Variance

                                  Sum of
      Source               DF     Squares      Mean Square
      blocks                3     13.691667      4.563889
      time                  2    163.006667     81.503333
      variety               3    152.685000     50.895000
      time*variety          6     40.320000      6.720000
      Error                33     69.453333      2.104646
      Corrected Total      47    439.156667          .

                          Type 1 Analysis of Variance

      Source          Expected Mean Square
      blocks          Var(Error) + Q(blocks)
      time            Var(Error) + 4 Var(time*variety) + 16 Var(time)
      variety         Var(Error) + 4 Var(time*variety) + 12 Var(variety)
      time*variety    Var(Error) + 4 Var(time*variety)
      Error           Var(Error)
      Corrected Total     .

                          Type 1 Estimates

            Variance Component         Estimate

            Var(time)                   4.67396
            Var(variety)                3.68125
            Var(time*variety)           1.15384
            Var(Error)                  2.10465
```

The overall total variance therefore is

$$\hat{\sigma}^2_{all} = \hat{\sigma}^2 + \hat{\sigma}^2_{AB} + \hat{\sigma}^2_B + \hat{\sigma}^2_A = 2.1047 + 1.1538 + 3.6813 + 4.6740 = 11.6138.$$

15.8.4   Random Effects Model IIb

Let us suppose for the data in Table 15.12, only the covering times are random and that the variety levels are fixed. This will lead to what we shall describe as a *random effects mixed model* of type IIb. The resulting ANOVA model for this mixed model would be

$$y_{ijk} = \mu + A_i + \beta_j + C_{ij} + \varepsilon_{ijk}, \tag{15.17}$$

where

$$A_i \sim N(0, \sigma^2_A) \qquad i = 1, 2, \ldots, a$$
$$C_{ij} \sim N(0, \sigma^2_{AB}) \quad j = 1, 2, \ldots, b$$
$$\varepsilon_{ijk} \sim N(0, \sigma^2) \qquad k = 1, 2, \ldots, r.$$

Further, A_i, C_{ij} and ε_{ijk} are independently distributed with the above properties. The expected mean squares are given in Table 15.18.

Source	d.f.	MS	E(MS)	F
A	$a-1$	MS_a	$\sigma^2 + r\sigma^2_{AB} + rb\sigma^2_A$	MS_a/MS_{ab}
B	$b-1$	MS_b	$\sigma^2 + r\sigma^2_{AB} + ra\sum_{j=1}^b \frac{\beta_j^2}{b-1}$	MS_b/MS_{ab}
AB	$(a-1)(b-1)$	MS_{ab}	$\sigma^2 + r\sigma^2_{AB}$	MS_{ab}/MS_e
Error	$ab(r-1)$	MS_e	σ^2	
Total	$abr-1$			

Table 15.18: Table of expected mean squares under model IIb

As in the previous case, the fixed effects model hypotheses in (15.13a) to (15.13c) become respectively under model IIb,

$$H_0 : \sigma_A^2 = 0 \tag{15.18a}$$

$$H_0 : \beta_1 = \beta_2 = \cdots = \beta_b = 0 \tag{15.18b}$$

$$H_0 : \sigma_{AB}^2 = 0. \tag{15.18c}$$

Note that the tests in (15.18a) to (15.18c) are similar to those in (15.13a) to (15.13c). The following is the SAS program and a partial output to implement the model in (15.17).

```
set ex102;
proc glm order=data;
class blocks time variety;
model y=blocks time|variety;
random time time*variety/test;
run;
```

```
                    The GLM Procedure
```

Dependent Variable: y

Source	DF	Sum of Squares	Mean Square	F Value	Pr > F
Model	14	369.7033333	26.4073810	12.55	<.0001
Error	33	69.4533333	2.1046465		
Corrected Total	47	439.1566667			

R-Square	Coeff Var	Root MSE	y Mean
0.841848	19.36472	1.450740	7.491667

Source	DF	Type I SS	Mean Square	F Value	Pr > F
blocks	3	13.6916667	4.5638889	2.17	0.1104
time	2	163.0066667	81.5033333	38.73	<.0001
variety	3	152.6850000	50.8950000	24.18	<.0001
time*variety	6	40.3200000	6.7200000	3.19	0.0139

Source	DF	Type III SS	Mean Square	F Value	Pr > F
blocks	3	13.6916667	4.5638889	2.17	0.1104
time	2	163.0066667	81.5033333	38.73	<.0001
variety	3	152.6850000	50.8950000	24.18	<.0001
time*variety	6	40.3200000	6.7200000	3.19	0.0139

```
                    The GLM Procedure
```

Source	Type III Expected Mean Square
blocks	Var(Error) + Q(blocks)
time	Var(Error) + 4 Var(time*variety) + 16 Var(time)
variety	Var(Error) + 4 Var(time*variety) + Q(variety)
time*variety	Var(Error) + 4 Var(time*variety)

```
                    The GLM Procedure
       Tests of Hypotheses for Mixed Model Analysis of Variance
```

Dependent Variable: y

Source	DF	Type III SS	Mean Square	F Value	Pr > F
blocks	3	13.691667	4.563889	2.17	0.1104
time*variety	6	40.320000	6.720000	3.19	0.0139
Error: MS(Error)	33	69.453333	2.104646		

Source	DF	Type III SS	Mean Square	F Value	Pr > F
time	2	163.006667	81.503333	12.13	0.0078
variety	3	152.685000	50.895000	7.57	0.0183
Error	6	40.320000	6.720000		

Error: MS(time*variety)

We present again the variance components for this model using **proc varcomp**.

```
set ex102;
proc varcomp method=type1;
class blocks time variety;
model y=blocks variety|time/fixed=2;
run;
```

```
                Variance Components Estimation Procedure

                        Class Level Information

              Class         Levels      Values

              blocks            4       1 2 3 4

              time              3       APR FEB MAR

              variety           4       F G R V

                  Number of observations    48

                  Dependent Variable:    y

                  Type 1 Analysis of Variance

                                    Sum of
          Source            DF      Squares      Mean Square
          blocks             3      13.691667      4.563889
          variety            3     152.685000     50.895000
          time               2     163.006667     81.503333
          time*variety       6      40.320000      6.720000
          Error             33      69.453333      2.104646
          Corrected Total   47     439.156667        .

                  Type 1 Analysis of Variance

          Source              Expected Mean Square

          blocks              Var(Error) + Q(blocks)
          variety             Var(Error) + 4 Var(time*variety) + Q(variety)
          time                Var(Error) + 4 Var(time*variety) + 16 Var(time)
          time*variety        Var(Error) + 4 Var(time*variety)
          Error               Var(Error)
          Corrected Total       .

                        Type 1 Estimates

              Variance Component      Estimate
              Var(time)               4.67396
              Var(time*VARIETY)       1.15384
              Var(Error)              2.10465
```

15.8.5   Random Effects Model IIc

Just as in the random effect model IIb, let us now suppose that in Example 15.5, only variety levels are random and that the covering times are fixed. This will again lead to a *random effects mixed model* of type IIc. The resulting ANOVA model for this mixed model would be

$$y_{ijk} = \mu + \alpha_i + B_j + C_{ij} + \varepsilon_{ijk}, \tag{15.19}$$

where

$$
\begin{aligned}
B_j &\sim \mathrm{N}(0, \sigma_B^2) & j &= 1, 2, \ldots, b \\
C_{ij} &\sim \mathrm{N}(0, \sigma_{AB}^2) & i &= 1, 2, \ldots, a \\
\varepsilon_{ijk} &\sim \mathrm{N}(0, \sigma^2) & k &= 1, 2, \ldots, r.
\end{aligned}
$$

Further, B_j, C_{ij} and ε_{ijk} are independently distributed with the above properties. The expected mean squares are given in Table 15.19.

As in the previous case, the fixed effects model hypotheses in (15.13a) to (15.13c) become respectively under model IIc,

$$H_0 : \alpha_1 = \alpha_2 = \cdots = \alpha_b = 0 \tag{15.20a}$$

$$H_0 : \sigma_B^2 = 0 \tag{15.20b}$$

$$H_0 : \sigma_{AB}^2 = 0. \tag{15.20c}$$

Source	d.f.	MS	E(MS)	F
A	$a-1$	MS_a	$\sigma^2 + r\sigma^2_{AB} + rb\sum_{i=1}^{a}\frac{\alpha_i^2}{a-1}$	$\mathrm{MS}_a/\mathrm{MS}_{ab}$
B	$b-1$	MS_b	$\sigma^2 + r\sigma^2_{AB} + ra\sigma^2_B$	$\mathrm{MS}_b/\mathrm{MS}_{ab}$
AB	$(a-1)(b-1)$	MS_{ab}	$\sigma^2 + r\sigma^2_{AB}$	$\mathrm{MS}_{ab}/\mathrm{MS}_e$
Error	$ab(r-1)$	MS_e	σ^2	
Total	$abr-1$			

Table 15.19: Table of expected mean squares under model IIc

Instead of using **proc glm**, we could also use **proc mixed** for the analysis of variance components models. We will illustrate the use of **proc mixed** for the three cases IIa, IIb and IIc that we have discussed in this section.

15.8.6   Use of PROC MIXED in SAS

The **proc mixed** is based on the Generalized Least Squares. Unlike **proc glm**, only fixed effects are listed in the model statement, and it does not produce sum of squares of effects and hence no expected mean squares are applicable. Random effects on the other hand are listed in the random statement that immediately follows the model statement. Estimation of variance components is either by maximum likelihood or restricted maximum likelihood estimation.

The **proc glm** is based on the general linear model of the form that has been discussed previously in our preceding chapters, namely, the model is written in the form

$$\mathbf{Y} = \mathbf{X}\boldsymbol{\beta} + \boldsymbol{\varepsilon}. \tag{15.21}$$

The **glm** model in (15.21) assumes that $\boldsymbol{\varepsilon}$ is distributed independently as normal. In most cases, this distributional assumption sometimes places restrictions on the random error terms $\boldsymbol{\varepsilon}$. The mixed model on the other hand allows a more flexible structural form of the distribution of the variance-covariance structure of $\boldsymbol{\varepsilon}$. The mixed model therefore extends the general linear model (GLM) and has the form

$$\mathbf{Y} = \mathbf{X}\boldsymbol{\beta} + \mathbf{Z}\boldsymbol{\gamma} + \boldsymbol{\varepsilon}, \tag{15.22}$$

where the additional term has $\mathbf{Z}$ as the design matrix and $\boldsymbol{\gamma}$ is the vector of random effects coefficients. Thus the model in (15.22) could contain both fixed and random effects and hence the name mixed model. Again the model assumes that both $\boldsymbol{\varepsilon}$ and $\boldsymbol{\gamma}$ are normally distributed with means zero and non zero variance structures (say A1 and A2 respectively).

We now employ **proc mixed** in SAS to implement our variance components models in IIa to IIc. Thus for model IIa, where both factors are considered random, the following SAS codes will provide the variance components in **proc mixed**.

```
set ex102;
proc mixed order=data ratio ic;
class blocks time variety;
model y=blocks;
random time variety time*variety;
run;
                        The Mixed Procedure

                    Model Information

        Data Set                   WORK.ex102
        Dependent Variable         y
        Covariance Structure       Variance Components
        Estimation Method          REML
        Residual Variance Method   Profile
        Fixed Effects SE Method    Model-Based
        Degrees of Freedom Method  Containment

                Class Level Information
```

```
Class      Levels   Values
blocks        4     1 2 3 4
time          3     FEB MAR APR
variety       4     V R F G
```

Dimensions

```
Covariance Parameters        4
Columns in X                 5
Columns in Z                19
Subjects                     1
Max Obs Per Subject         48
```

Number of Observations

```
Number of Observations Read        48
Number of Observations Used        48
Number of Observations Not Used     0
```

Iteration History

Iteration	Evaluations	-2 Res Log Like	Criterion
0	1	234.64191192	
1	1	191.38419594	0.00000000

Convergence criteria met.

Covariance Parameter Estimates

Cov Parm	Ratio	Estimate
time	2.2208	4.6740
variety	1.7491	3.6813
time*variety	0.5482	1.1538
Residual	1.0000	2.1046

Fit Statistics

```
-2 Res Log Likelihood        191.4
AIC (smaller is better)      199.4
AICC (smaller is better)     200.4
```

Fit Statistics

```
BIC (smaller is better)      195.8
```

Information Criteria

Neg2LogLike	Parms	AIC	AICC	HQIC	BIC	CAIC
191.4	4	199.4	200.4	192.1	195.8	199.8

Type 3 Tests of Fixed Effects

Effect	Num DF	Den DF	F Value	Pr > F
blocks	3	33	2.17	0.1104

In the above SAS codes, we have requested the following:

- In the model statement, we have specified the fixed effect (Blocks).

- In the random statement, we have specified that, both time, variety and their interactions should be treated as random.

- The option **ratio** requests that the ratio of the variance components with the residual variance components be computed

- The option **IC** requests various information criteria be displayed, these are usually Akaike, Schwarz, Hannan and Quinn and Bozdogan.

- The covariance parameter estimates give the variance components computed.

As we can see, the estimated variance components are exactly the same as those obtained with the **glm** procedure. Similarly, for models IIb and IIc, we have the following SAS codes and partial outputs.

```
set ex101;
/* Fit mixed model with Variety and blocks fixed */;
proc mixed order=data ratio ic;
class blocks time variety;
model y=blocks variety;
random time time*variety;
run;
```

Covariance Parameter Estimates

Cov Parm	Ratio	Estimate
time	2.2208	4.6740
time*variety	0.5482	1.1538
Residual	1.0000	2.1046

Fit Statistics

-2 Res Log Likelihood	177.1
AIC (smaller is better)	183.1
AICC (smaller is better)	183.8
BIC (smaller is better)	180.4

Information Criteria

Neg2LogLike	Parms	AIC	AICC	HQIC	BIC	CAIC
177.1	3	183.1	183.8	177.7	180.4	183.4

Type 3 Tests of Fixed Effects

Effect	Num DF	Den DF	F Value	Pr > F
blocks	3	33	2.17	0.1104
variety	3	6	7.57	0.0183

```
set ex101;
/* Fit mixed model with Time and blocks fixed */;
proc mixed order=data ratio ic;
class blocks time variety;
model y=blocks time;
random variety time*variety;
run;
```

Covariance Parameter Estimates

Cov Parm	Ratio	Estimate
variety	1.7491	3.6813
time*variety	0.5482	1.1538
Residual	1.0000	2.1046

Fit Statistics

-2 Res Log Likelihood	181.4
AIC (smaller is better)	187.4
AICC (smaller is better)	188.0
BIC (smaller is better)	185.5

Information Criteria

Neg2LogLike	Parms	AIC	AICC	HQIC	BIC	CAIC
181.4	3	187.4	188.0	183.3	185.5	188.5

Type 3 Tests of Fixed Effects

Effect	Num DF	Den DF	F Value	Pr > F
blocks	3	33	2.17	0.1104
time	2	6	12.13	0.0078

The estimated variance components for both mixed models agree with our earlier results when **proc varcomp** was used.

15.9 Exercises

15.1 The data in the following table is taken from the bacteria count data submitted by Binnie, N.S. to the Journal of Statistics Education Data Archive. The treatment means μ_{ij} (number of strain 3 bacteria counts in millions) in a two factor (time in hours and temperature in ^0C) study are presented in the table.

Time	Temperature 27	35	43
24	37.0	108.0	99.0
48	170.2	220.4	152.8

(a) Obtain the temperature main effects

(b) Obtain the main effects of time

(c) Prepare a treatment means plot and use this to determine whether temperature and time interact. Explain your findings.

(d) Make a logarithmic transformation of the μ_{ij} and plot the transformed values to check whether the transformation reduces the interaction. Explain your findings.

15.2 The data in the following table is taken from the bacteria count data submitted by Binnie, N.S. to the Journal of Statistics Education Data Archive. The treatment means μ_{ij} (number of strain 4 bacteria counts in millions) in a two factor (time in hours and temperature in ^0C) study are presented in the table.

Time	Temperature 27	35	43
24	19.0	124.0	97.2
48	131.8	175.4	149.0

(a) Obtain the temperature main effects

(b) Obtain the main effects of time

(c) Prepare a treatment means plot and use this to determine whether temperature and time interact. Explain your findings.

(d) Make a square root transformation of the μ_{ij} and plot the transformed values to check whether the transformation reduces the interaction. Explain your findings.

15.3 The data in the following table is taken from the bacteria count data submitted by Binnie, N.S. to the Journal of Statistics Education Data Archive. The treatment means μ_{ij} (number of strain 3 bacteria counts in millions) in a two factor (time in hours and concentration as a percentage) study are presented in the table.

Time	Concentration 0.6	0.8	1.0	1.2	1.4
24	48.33	77.33	78.33	110.67	93.67
48	131.00	153.67	188.67	253.67	178.67

(a) Obtain the concentration main effects

(b) Obtain the main effects of time

(c) Prepare a treatment means plot and use this to determine whether concentration and time interact. Explain your findings.

(d) Make a square root transformation of the μ_{ij} and plot the transformed values to check whether the transformation reduces the interaction. Explain your findings.

	Concentration				
Time	0.6	0.8	1.0	1.2	1.4
24	49.00	64.00	80.67	113.00	94.67
48	116.33	132.00	127.33	190.33	194.33

15.4 The data in the following table is taken from the bacteria count data submitted by Binnie, N.S. to the Journal of Statistics Education Data Archive. The treatment means μ_{ij} (number of strain 4 bacteria counts in millions) in a two factor (time in hours and concentration as a percentage) study are presented in the table.

   (a) Obtain the concentration main effects

   (b) Obtain the main effects of time

   (c) Prepare a treatment means plot and use this to determine whether concentration and time interact. Explain your findings.

   (d) Make a logarithmic transformation of the μ_{ij} and plot the transformed values to check whether the transformation reduces the interaction. Explain your findings.

15.5 The data in the following table is taken from the bacteria count data submitted by Binnie, N.S. to the Journal of Statistics Education Data Archive. The data contains measurement of bacteria counts of Staphylococcus aureus (strain 1), the time (in hours) of incubation and the temperature (in ^0C) of incubation. In order to test for the effects of time, temperature and their interactions, an analysis of variance model is proposed.

	Temperature		
Time	27	35	43
24	9, 16, 22, 30, 27	66, 93, 147, 199, 168	98, 82, 120, 148, 132
48	97, 123, 132, 263, 145	110, 149, 189, 263, 197	123, 146, 106, 232, 163

   (a) Obtain the ANOVA model for the data.

   (b) Obtain the predicted values for the number of bacteria counts

   (c) Obtain the residuals and check if they sum to zero for each of the factors time and temperature.

   (d) Obtain a normal probability plot of the residuals. Is the normality assumption reasonable for this data? Explain.

15.6 The data in the following table is taken from the bacteria count data submitted by Binnie, N.S. to the Journal of Statistics Education Data Archive. The data contains measurement of bacteria counts of Staphylococcus aureus (strain 2), the time (in hours) of incubation and the temperature (in ^0C) of incubation. In order to test for the effects of time, temperature and their interactions, an analysis of variance model is proposed.

	Temperature		
Time	27	35	43
24	3, 12, 37, 45, 32	71, 76, 63, 162, 155	67, 79, 113, 127, 118
48	84, 161, 153, 189, 191	123, 181, 202, 214, 233	107, 91, 189, 216, 141

   (a) Obtain the ANOVA model for the data.

(b) Obtain the predicted values for the number of bacteria counts

(c) Obtain the residuals and check if they sum to zero for each of the factors time and temperature.

(d) Obtain a normal probability plot of the residuals. Is the normality assumption reasonable for this data? Explain.

15.7 The data in the following table is taken from the bacteria count data submitted by Binnie, N.S. to the Journal of Statistics Education Data Archive. The data contains measurement of bacteria counts of Staphylococcus aureus (strain 1), the concentration of tryptone (a nutrient) and the temperature (in ^0C) of incubation. In order to test for the effects of concentration, temperature and their interactions, an analysis of variance model is proposed.

	Temperature		
Concentration	27	35	43
0.6	9, 97	66, 110	98, 123
0.8	16, 123	93, 149	82, 146
1.0	22, 132	147, 189	120, 106
1.2	30, 263	199, 263	148, 232
1.4	27, 145	168, 197	132, 163

(a) Obtain an estimated treatment means plot. Does it appear that concentration and temperature have any effects? Explain.

(b) Obtain the ANOVA model for the data.

(c) Set up the ANOVA table.

(d) Test for the interaction effects. Use $\alpha = 0.05$. State the null and the alternative hypotheses, your decision, and conclusion. State the p-value of the test.

(e) Test to see if concentration and temperature have significant effects on the number of counts at $\alpha = 0.05$. State the null and the alternative hypotheses, your decision, and conclusion. State the p-value of the test.

15.8 The data in the following table is taken from the bacteria count data submitted by Binnie, N.S. to the Journal of Statistics Education Data Archive. The data contains measurement of bacteria counts of Staphylococcus aureus (strain 2), the concentration of tryptone (a nutrient) and the temperature (in ^0C) of incubation. In order to test for the effects of concentration, temperature and their interactions, an analysis of variance model is proposed.

	Temperature		
Concentration	27	35	43
0.6	3, 84	71, 123	67, 107
0.8	12, 161	76, 181	79, 91
1.0	37, 153	63, 202	113, 189
1.2	45, 189	162, 214	127, 216
1.4	32, 191	155, 233	118, 141

(a) Obtain an estimated treatment means plot. Does it appear that concentration and temperature have any effects? Explain.

(b) Obtain the ANOVA model for the data.

(c) Set up the ANOVA table.

(d) Test for the interaction effects. Use $\alpha = 0.05$. State the null and the alternative hypotheses, your decision, and conclusion. State the p-value of the test.

(e) Test to see if concentration and temperature have significant effects on the number of counts at $\alpha = 0.05$. State the null and the alternative hypotheses, your decision, and conclusion. State the p-value of the test.

Chapter 16

Analysis of Covariance

16.1 Introduction

The analysis of covariance is a statistical technique which is a combination of **Regression** and **Analysis of Variance**. It is used in experiments where besides the observations of primary interests (variates), one or more other observations are taken on each experimental unit, called **concomitant** variables or **covariates**. Measurements on the covariates are made for the purpose of adjusting the measurements on the variate. These can be used to increase precision of the experimental comparisons or to throw further light on the treatment effects, or to remove environmental effects. It is assumed that the concomitant variable (x) cannot be controlled by the experimenter but can be observed along with the variable of interest (y). Thus, analysis of covariance is a method of adjusting for the effects of an uncontrollable nuisance variable. We present examples of the use of analysis of covariance.

Suppose in an experiment to study the effects of various diets (treatments) on the increase in body weight (y) of cows. Thus it would be necessary to have a group of cows at a fixed age and record initial weight (x) of each cow. Then x is the concomitant variable, and in the analysis, we shall try to adjust the experimental results y on the basis of their x values.

Another example concerns an experiment to consider drugs that are hypothesized to reduce blood pressure of adults. Since the blood pressure of adults before the administration of the drugs (treatments) varies considerably from one adult to another, a grouping of the adults according to their initial blood pressure is sometimes possible, albeit cumbersome in practice. Thus as in the earlier example, we may record the blood pressure x of each adult before the administration of the drugs and use this information to adjust the treatment effects y in our analysis.

It must be noted that not all concomitant observations are taken before treatments. On certain occasions, they are taken either during the experiment or at the end of the experiment. Consider the following example in which 40 plants (e.g. corn) had originally been planted in each plot. At harvest time, however, some of the plots have only 25 or even 20 plants left, the rest of them being eaten by wild animals. The yield from such plots is naturally lower than those from 40 plots. The number of plots may be recorded as x and later used to correct the yield y. We note here the assumption is that the number of plants left is not due to treatments but is due to an uncorrelated factor (animals) which introduces heterogeneity to the experimental plots.

A further use of analysis of covariance is in the missing plot technique where dummy variables are used as concomitants. Several examples will be given in this chapter to further illustrate the type of experimental situations in which the analysis of covariance can be profitably employed.

Example 16.1: Single Factor Case

The data in Table 16.1 were obtained in an experiment to compare three methods of applying a rust arrestor compound to steel coupons. These methods were brushing, spraying and dipping. Fifteen steel coupons were used, divided randomly into three groups. All of the steel coupons were in an initial state of rust (measured as x) and they were all exposed to a salt spray, the additional amount of rust due to this being measured as y.

A question of primary interest is whether the method of application affects the additional amount of rust. The observation x can be used to improve the precision of experimental comparisons here.

Brushing			Spraying			Dipping		
Coupon	y	x	Coupon	y	x	Coupon	y	x
1	63	16	6	81	48	11	72	40
2	77	45	7	73	40	12	54	31
3	81	50	8	59	24	13	57	40
4	60	19	9	74	33	14	59	33
5	63	18	10	77	41	15	52	20
Totals	344	148	Totals	364	186	Totals	294	164
Means	68.8	29.6	Means	72.8	37.2	Means	58.8	32.8

Table 16.1: Data for the experiment on steel coupons

The analysis of covariance is a method of adjusting the treatment means to what might have been obtained if a common value of x had been used throughout. If in an experiment, there exists a relationship between the observation y and a concomitant variable x, the observed (unadjusted) means of y could indicate completely wrong results.

16.2 One-Way ANOVA Review

For the one-way ANOVA, the factor effects model is given by

$$y_{ij} = \mu + \tau_i + \varepsilon_{ij},$$

with $i = 1, 2, 3$ and $j = 1, 2, 3, 4, 5$ (the number of replications). Interest centers on the hypotheses

$$H_0 : \tau_1 = \tau_2 = \tau_3$$
$$H_a : \text{at least one of the } \tau\text{'s is not zero,}$$

which is equivalent to

$$H_0 : \mu_1 = \mu_2 = \mu_3$$
$$H_a : \text{at least two of the } \mu\text{'s are not equal.}$$

A regression equivalent of the above ANOVA model can also be written and the appropriate hypothesis re-defined. In our analysis of the data in Table 16.1, we will first analyze the data as a one-way ANOVA (ignoring the effect of the covariate). In the second part, we would re-analyze the data using analysis of covariance.

16.3 Model and Assumptions

An analysis of covariance (ANCOVA) model takes the following form:

$$y_{ij} = \mu + \tau_i + \gamma(x_{ij} - \bar{x}_{..}) + \varepsilon_{ij},$$
$$i = 1, 2, 3 \quad \text{(treatments)},$$
$$j = 1, 2, 3, 4, 5 \quad \text{(observations on each treatment)}.$$

It is assumed that

(i) The x's are fixed, measured without error, and independent of treatments.

(ii) The ε_{ij} are independently and normally distributed with mean 0 and variance σ^2.

(iii) The regression coefficient is the same for all treatments, that is, equality of slope(s) of the different treatment regression lines.

(iv) The concomitant variable is unaffected by the particular assignment of treatments to units used. A further implicit assumption is that the concomitant variable does not itself contain errors.

(v) Equality of error variances for different treatments, that is, homogeneity of variances.

(vi) Linearity of regression relation with concomitant variable.

16.4   Hypothesis of Interest

Key statistical inference in analysis of covariance is whether the treatments have any effects, and if so, what these effects are. The test involves

$$H_0 : \tau_1 = \tau_2 = \cdots = 0$$
$$H_a : \text{not all the } \tau\text{'s are equal to zero.} \tag{16.1}$$

If the treatment effects differ, one wishes to do pairwise comparisons such as $\tau_i - \tau_j = 0$ with $i \neq j$. The full regression model with interaction is

$$y_{ij} = \mu + \tau_1 z_{ij1} + \tau_2 z_{ij2} + \gamma(x_{ij} - \bar{x}_{..}) + \beta_1 z_{ij1}(x_{ij} - \bar{x}_{..}) + \beta_2 z_{ij2}(x_{ij} - \bar{x}_{..}) + \varepsilon_{ij}, \tag{16.2}$$

where z_{ijk} can be from cell reference scheme or effect coding scheme. Here z_1 and z_2 are defined for the cell reference scheme as

$$z_1 = \begin{cases} 1, & \text{if brushing} \\ 0, & \text{elsewhere} \end{cases}, \qquad z_2 = \begin{cases} 1, & \text{if spraying} \\ 0, & \text{elsewhere.} \end{cases} \tag{16.3}$$

Similarly, if we wish to use the effect coding scheme, then z_1 and z_2 are again defined respectively as

$$z_1 = \begin{cases} 1, & \text{if brushing} \\ -1, & \text{if dipping} \\ 0, & \text{elsewhere} \end{cases}, \qquad z_2 = \begin{cases} 1, & \text{if spraying} \\ -1, & \text{if dipping} \\ 0, & \text{elsewhere.} \end{cases} \tag{16.4}$$

The interaction terms are the 5th and the 6th terms in the general model (16.2). A test of parallelism is given by the hypotheses

$$H_0 : \beta_1 = \beta_2 = 0$$
$$H_a : \text{at least one of the } \beta\text{'s is not zero.} \tag{16.5}$$

Analysis of covariance needs the assumption of parallelism to be true and once this is the case, our model of interest in the above example now becomes

$$y_{ij} = \mu + \tau_1 z_{ij1} + \tau_2 z_{ij2} + \gamma\, x_{ij} + \varepsilon_{ij}. \tag{16.6}$$

The model in (16.6) assumes there is no interaction between x and the treatment or factor effect. The hypothesis of interest in this case becomes

$$H_0 : \tau_1 = \tau_2 = 0$$
$$H_a : \text{at least one of the } \tau\text{'s is not zero.}$$

The comparisons, their estimators and population variances of the estimators are presented below for the cell reference case.

Cell Reference Coding Case

Comparison	Estimator	Variance
$\tau_1 - \tau_2 = 0$	$\hat{\tau}_1 - \hat{\tau}_2$	$\text{var}(\hat{\tau}_1) + \text{var}(\hat{\tau}_2) - 2\text{cov}(\hat{\tau}_1, \hat{\tau}_2)$
$\tau_1 - \tau_3 = \tau_1$	$\hat{\tau}_1$	$\text{var}(\hat{\tau}_1)$
$\tau_2 - \tau_3 = \tau_2$	$\hat{\tau}_2$	$\text{var}(\hat{\tau}_2)$

Similarly, for the comparison in the effect coding scheme, their estimators and the corresponding population variances of the estimators are presented below.

Effect Coding Case

Comparison	Estimator	Variance
$\tau_1 - \tau_2 = 0$	$\hat{\tau}_1 - \hat{\tau}_2$	$\text{var}(\hat{\tau}_1) + \text{var}(\hat{\tau}_2) - 2\text{cov}(\hat{\tau}_1, \hat{\tau}_2)$
$\tau_1 - \tau_3 = 2\tau_1 + \tau_2$	$2\hat{\tau}_1 + \hat{\tau}_2$	$4\text{var}(\hat{\tau}_1) + \text{var}(\hat{\tau}_2) + 4\text{cov}(\hat{\tau}_1, \hat{\tau}_2)$
$\tau_2 - \tau_3 = \tau_1 + 2\tau_2$	$\hat{\tau}_1 + 2\hat{\tau}_2$	$\text{var}(\hat{\tau}_1) + 4\text{var}(\hat{\tau}_2) + 4\text{cov}(\hat{\tau}_1, \hat{\tau}_2)$

Adjusted Treatment Means

Under the regression model, the estimators of the adjusted treatment means under both coding schemes, together with their accompanying variances are presented in the following tables.

Cell Reference Coding Case

Mean Response at $x = \bar{x}_{..}$	Estimator	Variance
$\mu + \tau_1$	$\hat{\mu} + \hat{\tau}_1$	$\text{var}(\hat{\mu}) + \text{var}(\hat{\tau}_1) + 2\text{cov}(\hat{\mu}, \hat{\tau}_1)$
$\mu + \tau_2$	$\hat{\mu} + \hat{\tau}_2$	$\text{var}(\hat{\mu}) + \text{var}(\hat{\tau}_2) + 2\text{cov}(\hat{\mu}, \hat{\tau}_2)$
$\mu + \tau_3 = \mu$	$\hat{\mu}$	$\text{var}(\hat{\mu})$

Effect Coding Case

Mean Response at $x = \bar{x}_{..}$	Estimator	Variance
$\mu + \tau_1$	$\hat{\mu} + \hat{\tau}_1$	$\text{var}(\hat{\mu}) + \text{var}(\hat{\tau}_1) + 2\text{cov}(\hat{\mu}, \hat{\tau}_1)$
$\mu + \tau_2$	$\hat{\mu} + \hat{\tau}_2$	$\text{var}(\hat{\mu}) + \text{var}(\hat{\tau}_2) + 2\text{cov}(\hat{\mu}, \hat{\tau}_2)$
$\mu + \tau_3$	$\hat{\mu} - \hat{\tau}_1 - \hat{\tau}_2$	$\text{var}(\hat{\mu}) + \text{var}(\hat{\tau}_1) + \text{var}(\hat{\tau}_2)$ $-2\text{cov}(\hat{\mu}, \hat{\tau}_1) - 2\text{cov}(\hat{\mu}, \hat{\tau}_2) + 2\text{cov}(\hat{\tau}_1, \hat{\tau}_2)$

Analysis of Data in Table 16.1-Cell Reference Coding Scheme

In this case we would create two dummy variables z_1 and z_2 for the three treatments as defined in (16.3). The SAS implementation is presented below. The first model statement in the SAS program fits the model in (16.2). That is, the full model

$$y_{ij} = \mu + \tau_1 z_{ij1} + \tau_2 z_{ij2} + \gamma(x_{ij} - \bar{x}_{..}) + \beta_1 z_{ij1}(x_{ij} - \bar{x}_{..}) + \beta_2 z_{ij2}(x_{ij} - \bar{x}_{..}) + \varepsilon_{ij}. \tag{16.7}$$

The hypothesis that $H_0 : \beta_1 = \beta_2 = 0$ which corresponds to that in (16.5) and which stipulates that the interaction is zero is tested by the first **test1** in the SAS program. This test gives a p-value of 0.6072, which indicates that we would fail to reject H_0. That is, we can reasonably assume that the interactions are zero in the above model-a necessary condition for the analysis of covariance. The second model statement in the SAS program therefore fits the covariance regression model

$$y_{ij} = \mu + \tau_1 z_{ij1} + \tau_2 z_{ij2} + \gamma(x_{ij} - \bar{x}_{..}) + \varepsilon_{ij}, \tag{16.8}$$

to the data in Table 16.1. This model has the following parameter estimates:

$$\hat{\mu} = 59.05483, \quad \hat{\tau}_1 = 12.03861, \quad \hat{\tau}_2 = 11.19692, \quad \text{and} \quad \hat{\gamma} = 0.63706.$$

```
data example;
input trt $ y x @@;
if trt='brus' then z1=1;
else z1=0;
if trt='spry' then z2=1;
else z2=0;
xx=x-33.2;
xz1=xx*z1;
xz2=xx*z2;
datalines;
brus 63 16 brus 77 45 brus 81 50 brus 60 19 brus 63 18
spry 81 48 spry 73 40 spry 59 24 spry 74 33 spry 77 41
dipp 72 40 dipp 54 31 dipp 57 40 dipp 59 33 dipp 52 20
;
run;
proc reg;
model y=z1 z2 xx xz1 xz2;
test1: test xz1, xz2;
model y=z1 z2 xx/covb;
test2: test z1, z2;
test3: test z1-z2;
test4: test z1;
test5: test z2;
run;
```

The REG Procedure
Model: MODEL1
Dependent Variable: y

Analysis of Variance

Source	DF	Sum of Squares	Mean Square	F Value	Pr > F
Model	5	1225.55584	245.11117	12.33	0.0008
Error	9	178.84416	19.87157		
Corrected Total	14	1404.40000			

Root MSE	4.45775	R-Square	0.8727	
Dependent Mean	66.80000	Adj R-Sq	0.8019	
Coeff Var	6.67328			

Parameter Estimates

Variable	DF	Parameter Estimate	Standard Error	t Value	Pr > |t|
Intercept	1	59.06263	1.99651	29.58	<.0001
z1	1	11.77622	2.86316	4.11	0.0026
z2	1	10.32504	2.98689	3.46	0.0072
xx	1	0.65657	0.27089	2.42	0.0384
xz1	1	-0.09023	0.30281	-0.30	0.7725
xz2	1	0.19651	0.36531	0.54	0.6037

The REG Procedure
Model: MODEL1

Test test1 Results for Dependent Variable y

Source	DF	Mean Square	F Value	Pr > F
Numerator	2	10.48329	0.53	0.6072
Denominator	9	19.87157		

The REG Procedure
Model: MODEL2
Dependent Variable: y

Analysis of Variance

Source	DF	Sum of Squares	Mean Square	F Value	Pr > F
Model	3	1204.58926	401.52975	22.11	<.0001
Error	11	199.81074	18.16461		
Corrected Total	14	1404.40000			

Root MSE	4.26200	R-Square	0.8577	
Dependent Mean	66.80000	Adj R-Sq	0.8189	
Coeff Var	6.38023			

```
                           Parameter Estimates

                        Parameter      Standard
        Variable    DF    Estimate       Error    t Value    Pr > |t|
        Intercept    1    59.05483      1.90647     30.98     <.0001
        z1           1    12.03861      2.71590      4.43      0.0010
        z2           1    11.19692      2.73392      4.10      0.0018
        xx           1     0.63706      0.10377      6.14     <.0001

                           Covariance of Estimates

  Variable        Intercept              z1               z2                 xx
  Intercept    3.6346455752    -3.619138673    -3.651875467      0.0043074728
  z1          -3.619138673      7.3761164766    3.4812995424      0.0344597827
  z2          -3.651875467      3.4812995424    7.4743268572     -0.047382201
  xx           0.0043074728     0.0344597827   -0.047382201       0.0107686821

                          Model: MODEL2
              Test test2 Results for Dependent Variable y

                                Mean
        Source         DF      Square    F Value    Pr > F
        Numerator       2    225.82520    12.43      0.0015
        Denominator    11     18.16461

              Test test3 Results for Dependent Variable y

                                Mean
        Source         DF      Square    F Value    Pr > F
        Numerator       1      1.63144     0.09      0.7700
        Denominator    11     18.16461

              Test test4 Results for Dependent Variable y

                                Mean
        Source         DF      Square    F Value    Pr > F
        Numerator       1    356.90346    19.65      0.0010
        Denominator    11     18.16461

              Test test5 Results for Dependent Variable y

                                Mean
        Source         DF      Square    F Value    Pr > F
        Numerator       1    304.68495    16.77      0.0018
        Denominator    11     18.16461
```

From the program above, the **test2** statement tests the null hypothesis $H_0 : \tau_1 = \tau_2 = 0$ against the alternative hypothesis H_a : at least one of the τ's is not zero. The null hypothesis states that there are no significant differences between the adjusted treatment means. The p-value for this test is 0.0015, which indicates that there are significant differences between the adjusted treatment means at $\alpha = 0.05$ since the p-value is less than 0.05. **Tests** 3, 4, and 5 conduct pairwise test on adjusted treatment means. For instance,

1. **Test 3** tests the hypotheses

$$H_0 : \tau_1 = \tau_2$$
$$H_a : \tau_1 \neq \tau_2.$$

The above hypotheses test whether there is a significant difference in the adjusted treatment means of brushing and spraying. The p-value for this test is 0.7700 indicating that there is no significant difference between the adjusted means of brushing and spraying.

2. **Test 4** similarly tests the hypotheses

$$H_0 : \tau_1 = \tau_3$$
$$H_1 : \tau_1 \neq \tau_3.$$

That is, it tests whether there is a significant difference in the adjusted treatment means of brushing and dipping. The p-value for this test is 0.0010 indicating that there is a significant difference between the adjusted means of brushing and dipping.

3. Finally, the hypotheses in **test 5** is

$$H_0 : \tau_2 = \tau_3$$
$$H_1 : \tau_2 \neq \tau_3.$$

Again, this hypothesis tests whether there is a significant difference in the adjusted treatment means of spraying and dipping. The p-value for this test is 0.0018 indicating that there is a significant difference between the adjusted means of spraying and dipping.

Calculating Adjusted Treatment Means

Based on the expressions in sub-section 16.4.1, the adjusted treatment means are computed as follows:

$$\text{Adj(brushing)} = \hat{\tau}_1 + \hat{\mu} = 12.03861 + 59.05483 = 71.0934$$
$$\text{Adj(spraying)} = \hat{\tau}_2 + \hat{\mu} = 11.19692 + 59.05483 = 70.2518$$
$$\text{Adj(dipping)} = \hat{\mu} \qquad\qquad = 59.0548.$$

We can also calculate the standard errors of the adjusted treatment means by using the expressions earlier established in this chapter. For instance, the variance of the adjusted treatment mean for brushing would be computed as

$$\text{Var (Adjusted brushing)} = \text{Var}(\hat{\tau}_1) + \text{Var}(\hat{\mu}) + 2\text{Cov}(\hat{\tau}_1, \hat{\mu})$$
$$= 7.3761164766 + 3.6346455752 - 2(3.619138673)$$
$$= 3.772484705.$$

Hence, the standard error for the adjusted brushing mean is $\sqrt{3.772484705} = 1.94229$. Similar calculations lead to the standard errors of adjusted treatment means for spraying and dipping being 1.95070 and 1.90647 respectively.

Analysis of Data in Table 16.1-Effect Coding Scheme

In this case we would again create two dummy variables z_1 and z_2 for the three treatments as defined in (16.4). Again, the full model of the form

$$y_{ij} = \mu + \tau_1 z_{ij1} + \tau_2 z_{ij2} + \gamma(x_{ij} - \bar{x}_{..}) + \beta_1 z_{ij1}(x_{ij} - \bar{x}_{..}) + \beta_2 z_{ij2}(x_{ij} - \bar{x}_{..}) + \varepsilon_{ij}, \qquad (16.9)$$

with interaction terms is first employed. **Test1b**, which is analogous to **test1** in the previous scheme tests for the interaction term. The p-value obtained here is 0.6072, same result as in the previous case, and this indicates that there are no significant interaction effects in the model. This subsequently leads to a reduced analysis of covariance model.

```
data new;
set example;
if trt='brus' then x1=1;
else if trt='dipp' then x1=-1;
else x1=0;
if trt='spry' then x2=1;
else if trt='dipp' then x2=-1;
else x2=0;
xx=x-33.2;
xx1=xx*x1;
xx2=xx*x2;

proc reg;
model y=x1 x2 xx xx1 xx2;
test1b: test xx1, xx2;
model y=x1 x2 xx/covb;
test2b: test x1, x2;
test3b: test x1-x2;
test4b: test 2*x1+x2;
test5b: test x1+2*x2;
run;
```

                    The REG Procedure
                      Model: MODEL1
                 Dependent Variable: y

```
                              Analysis of Variance

                                Sum of         Mean
Source                 DF       Squares        Square    F Value    Pr > F
Model                   5    1225.55584     245.11117     12.33     0.0008
Error                   9     178.84416      19.87157
Corrected Total        14    1404.40000

            Root MSE             4.45775    R-Square      0.8727
            Dependent Mean      66.80000    Adj R-Sq      0.8019
            Coeff Var            6.67328

                            Parameter Estimates

                       Parameter     Standard
       Variable    DF    Estimate       Error    t Value    Pr > |t|
       Intercept    1    66.42972     1.20799      54.99     <.0001
       z1           1     4.40913     1.69207       2.61     0.0285
       z2           1     2.95795     1.76193       1.68     0.1275
       xx           1     0.69200     0.12986       5.33     0.0005
       xz1          1    -0.12565     0.15155      -0.83     0.4285
       xz2          1     0.16108     0.19206       0.84     0.4233

          Test test1b Results for Dependent Variable y

                                Mean
       Source           DF     Square    F Value    Pr > F
       Numerator         2    10.48329      0.53     0.6072
       Denominator       9    19.87157


                      Dependent Variable: y

                      Analysis of Variance

                                Sum of         Mean
Source                 DF       Squares        Square    F Value    Pr > F
Model                   3    1204.58926     401.52975     22.11     <.0001
Error                  11     199.81074      18.16461
Corrected Total        14    1404.40000

            Root MSE             4.26200    R-Square      0.8577
            Dependent Mean      66.80000    Adj R-Sq      0.8189
            Coeff Var            6.38023

                          Parameter Estimates

                       Parameter     Standard
       Variable    DF    Estimate       Error    t Value    Pr > |t|
       Intercept    1    66.80000     1.10044      60.70     <.0001
       z1           1     4.29343     1.60047       2.68     0.0213
       z2           1     3.45174     1.61067       2.14     0.0553
       xx           1     0.63706     0.10377       6.14     <.0001

                        Covariance of Estimates

Variable        Intercept             z1              z2               xx
Intercept     1.2109741954    1.101832E-16    -1.22426E-16     3.060644E-17
z1            1.101832E-16    2.5615105105    -1.366043217     0.0387672555
z2           -1.22426E-16    -1.366043217     2.594247304     -0.043074728
xx            3.060644E-17    0.0387672555    -0.043074728     0.0107686821

                          Model: MODEL2
          Test test2b Results for Dependent Variable y

                                Mean
       Source           DF     Square     F Value    Pr > F
       Numerator         2   225.82520      12.43     0.0015
       Denominator      11    18.16461


          Test test3b Results for Dependent Variable y

                                Mean
       Source           DF     Square     F Value    Pr > F
       Numerator         1     1.63144       0.09     0.7700
       Denominator      11    18.16461


          Test test4b Results for Dependent Variable y

                                Mean
       Source           DF     Square     F Value    Pr > F
```

Numerator	1	356.90346	19.65	0.0010
Denominator	11	18.16461		

Test test5b Results for Dependent Variable y

Source	DF	Mean Square	F Value	Pr > F
Numerator	1	304.68495	16.77	0.0018
Denominator	11	18.16461		

Test2b tests for significant differences among the adjusted treatment means. **Test3b** tests for significance between adjusted treatment means 1 and 2 (brushing and spraying) and the null is formulated as $H_0 : \tau_1 = \tau_2$. The result from this test agrees with that obtained under the cell reference coding scheme. **Test4b** tests the hypothesis $H_0 : \tau_1 = \tau_3$ which becomes $\tau_1 - \tau_3 = \tau_1 + \tau_1 + \tau_2 = 2\tau_1 + \tau_2 = 0$. That is, the hypothesis becomes $H_0 : 2\tau_1 + \tau_2 = 0$. Again, the results from this test gives a p-value of 0.0010, which is the same as the result under the cell reference coding scheme. Finally, the test in **test5b** tests the hypothesis that $\tau_2 - \tau_3 = 0$, which can be written as $H_0 : \tau_1 + 2\tau_2 = 0$. The result from this test gives a p-value of 0.0018. This is the same result as under the cell reference coding scheme. The parameter estimates under the covariance model are

$$\hat{\mu} = 66.80000, \quad \hat{\tau}_1 = 4.29343, \quad \hat{\tau}_2 = 3.45174, \quad \text{and} \quad \hat{\gamma} = 0.63706.$$

Hence, the estimated adjusted treatment means are computed using the results in sub-section 16.4.1. as

$$\text{Adj(brushing)} = \hat{\tau}_1 + \hat{\mu} \qquad = 66.80 + 4.29343 \qquad = 71.0934$$
$$\text{Adj(spraying)} = \hat{\tau}_2 + \hat{\mu} \qquad = 66.80 + 3.45174 \qquad = 70.2518$$
$$\text{Adj(dipping)} = \hat{\mu} - \hat{\tau}_1 - \hat{\tau}_2 = 66.80 - 4.29343 - 3.45174 = 59.0548.$$

The estimated variance of the adjusted brushing mean is computed as

$$\text{Var (Adjusted brushing)} = \text{Var}(\hat{\tau}_1) + \text{Var}(\hat{\mu}) + 2\text{Cov}(\hat{\tau}_1, \hat{\mu})$$
$$= 2.5615105105 + 1.2109741954 + 2(1.101832E - 16)$$
$$= 3.772484705,$$

which leads to a standard error for the adjusted brushing mean of $\sqrt{3.772484705} - 1.94229$. This is the same result under the cell reference coding scheme. Similar calculations lead to the standard errors of adjusted treatment means for spraying and dipping.

16.5 Using ANOVA Model

Under the ANOVA model, the analysis of covariance model becomes

$$y_{ij} = \mu + \tau_i + \gamma(x_{ij} - \bar{x}_{..}) + \varepsilon_{ij}, \tag{16.10}$$

with $\sum \tau_i = 0$. This model can be implemented in SAS with **proc glm** as follows:

```
data example;
input trt $ y x @@;
xx=x-33.20;
datalines;
brus 63 16 brus 77 45 brus 81 50 brus 60 19 brus 63 18
spry 81 48 spry 73 40 spry 59 24 spry 74 33 spry 77 41
dipp 72 40 dipp 54 31 dipp 57 40 dipp 59 33 dipp 52 20
;
run;
proc glm order=data;
class trt;
model y=trt xx;
means trt/tukey;
lsmeans trt/stderr pdiff e adjust=tukey;
output out=aa r=resid;
run;
proc univariate normal data=aa;
var resid;
run;
```

```
                              The GLM Procedure

                           Class Level Information

                      Class     Levels   Values
                      trt            3   brus spry dipp

                     Number of observations    15
```

Dependent Variable: y

```
                                    Sum of
Source                 DF          Squares     Mean Square   F Value   Pr > F
Model                   3      1204.589258      401.529753     22.11   <.0001
Error                  11       199.810742       18.164613
Corrected Total        14      1404.400000

            R-Square     Coeff Var     Root MSE      y Mean
            0.857725      6.380234     4.261996     66.80000

Source                 DF       Type I SS     Mean Square   F Value   Pr > F
trt                     2     520.0000000     260.0000000     14.31   0.0009
xx                      1     684.5892578     684.5892578     37.69   <.0001


Source                 DF     Type III SS     Mean Square   F Value   Pr > F
trt                     2     451.6504016     225.8252008     12.43   0.0015
xx                      1     684.5892578     684.5892578     37.69   <.0001
```

             Tukey's Studentized Range (HSD) Test for y

NOTE: This test controls the Type I experimentwise error rate, but it generally has
            a higher Type II error rate than REGWQ.

```
          Alpha                                   0.05
          Error Degrees of Freedom                  11
          Error Mean Square                    18.16461
          Critical Value of Studentized Range   3.81952
          Minimum Significant Difference         7.2801
```

         Means with the same letter are not significantly different.

```
          Tukey Grouping        Mean     N    trt
                        A      72.800     5    spry
                        A
                        A      68.800     5    brus

                        B      58.800     5    dipp
```

                        Least Squares Means

                Coefficients for trt Least Square Means

```
                                       trt Level
Effect                              brus        spry        dipp
Intercept                              1           1           1
trt      brus                          1           0           0
trt      spry                          0           1           0
trt      dipp                          0           0           1
xx                            -2.842E-15   -2.842E-15   -2.842E-15
```

                        Least Squares Means
           Adjustment for Multiple Comparisons: Tukey-Kramer

```
                            Standard            LSMEAN
       trt      y LSMEAN       Error   Pr > |t|   Number
       brus   71.0934313   1.9422885    <.0001        1
       spry   70.2517429   1.9506977    <.0001        2
       dipp   59.0548257   1.9064746    <.0001        3
```

                Least Squares Means for effect trt
                Pr > |t| for H0: LSMean(i)=LSMean(j)

                        Dependent Variable: y

```
         i/j          1          2          3
           1                 0.9519     0.0027
           2     0.9519                 0.0046
           3     0.0027     0.0046
```

                        The UNIVARIATE Procedure

```
                    Variable:  resid

                  Tests for Normality

 Test                    --Statistic---    -----p Value------
 Shapiro-Wilk            W    0.969343     Pr < W      0.8482
 Kolmogorov-Smirnov      D    0.118841     Pr > D     >0.1500
 Cramer-von Mises        W-Sq 0.032479     Pr > W-Sq  >0.2500
 Anderson-Darling        A-Sq 0.23083      Pr > A-Sq  >0.2500
```

The results obtained with **proc glm** are exactly the same as those obtained in the previous subsections. Notice that the adjusted treatment means are obtained with the **lsmeans trt/stderr pdiff e** command. The **stderr** asked SAS to also output the standard errors of the adjusted means, while the **pdiff** asks that pairwise adjusted treatment means comparison be conducted and the **adjust=tukey** asks that multiple comparisons be conducted on the adjusted treatment means using the Tukey Procedure. We also note here that the Type I SS for treatment ('trt') is the unadjusted treatment SS and would be equivalent to the one presented in the one-way ANOVA model. The Type III SS however, gives us the adjusted treatment SS and enables us to test the treatment effects adjusted for all other factors in the model. Thus, Type III SS provides the appropriate SS for the analysis of covariance model. The adjusted treatment means under the ANOVA model are obtained from

$$\text{Adj}(\bar{y}_i) = \bar{y}_i - \hat{\gamma}(x_{i.} - \bar{x}_{..}).$$ (16.11)

Normality Assumption

The assumption of normality of errors are also satisfied from the Anderson-Darling test p-value of > 0.2500. We present the unadjusted and adjusted means in the following table.

Treatment	Unadjusted Mean	Adjusted Mean
Spraying	72.80	70.252
Brushing	68.80	71.093
Dipping	58.8	59.055

The result of Tukey's test on the unadjusted means are presented below. Notice the order of magnitude of the treatment means here.

$$\begin{array}{ccc} \bar{y}_{\text{spry}} & \bar{y}_{\text{brus}} & \bar{y}_{\text{dipp}} \\ 72.80 & 68.80 & 58.80 \end{array}$$

Again, we present the results of our pairwise tests on the adjusted means. We notice that the order of magnitude of the adjusted means for brushing and spraying are interchanged. The results still show that there are no significant differences between the adjusted treatment means for brushing and spraying but both are significantly different from the adjusted mean for dipping.

$$\begin{array}{ccc} \bar{y}_{\text{brus}} & \bar{y}_{\text{spry}} & \bar{y}_{\text{dipp}} \\ 71.093 & 70.252 & 59.055 \end{array}$$

16.6   Factorial Case: Two Factors

For a two-factor factorial case with only one covariate (x) with factors A and B having a and b levels respectively, the ANOVA model has the linear model formulation of the form

$$y_{ijk} = \mu + \alpha_i + \beta_j + (\alpha\beta)_{ij} + \varepsilon_{ijk},$$ (16.12)

where $i = 1, 2, 3, \ldots, a$; $j = 1, 2, 3, \ldots, b$; $k = 1, 2, 3, \ldots, n$. We would assume here that factors A and B have fixed effects.

Example 16.2

The data in Table 16.2 is from Winer (1971, p. 788) and relate to a 2×3 factorial experiment having $r = 5$ observations per cell (balanced). Here A represents methods of instruction in teaching map reading and B represents three instructors. The covariate measure is the score on an achievement test on map reading prior to the training, while the dependent variable is the score on a comparable form of the achievement test after training is completed. We further assume that $r = 5$ subjects are assigned at random to each cell of the factorial study.

Method	b_1 Instructor 1		b_2 Instructor 2		b_3 Instructor 3	
	x	y	x	y	x	y
a_1	40	95	30	85	50	90
	35	80	40	100	40	85
	40	95	45	85	40	90
	50	105	40	90	30	80
	45	100	40	90	40	85
a_2	50	100	50	100	45	95
	30	95	30	90	30	85
	35	95	40	95	25	75
	45	110	45	90	50	105
	30	88	40	95	35	85

Table 16.2: Data on methods of instruction

The following is the SAS program and the output.

```
data cov2;
do A=1 to 2;
do B=1 to 3;
do REP=1 to 5;
input x y @@;
output;
end; end; end;
datalines;
40 95 35 80 40 95 50 105 45 100 30 85 40 100 45 85 40 90 40 90
50 90 40 85 40 90 30 80 40 85 50 100 30 95 35 95 45 110 30 88
50 100 30 90 40 95 45 90 40 95 45 95 30 85 25 75 50 105 35 85
;
proc means;
var y x;
run;
proc glm;
class A B;
model y=A|B;
means A B /tukey;
run;
```

                          The GLM Procedure

Dependent Variable: y

```
                            Sum of
Source               DF     Squares      Mean Square   F Value   Pr > F
Model                 5   466.666667      93.333333      1.57    0.2060
Error                24  1425.200000      59.383333
Corrected Total      29  1891.866667


          R-Square     Coeff Var      Root MSE       y Mean
          0.246670     8.382224       7.706058      91.93333


Source               DF     Type I SS     Mean Square   F Value   Pr > F
A                     1    76.8000000     76.8000000      1.29    0.2667
B                     2   387.2666667    193.6333333      3.26    0.0559
A*B                   2     2.6000000      1.3000000      0.02    0.9784


Source               DF    Type III SS    Mean Square   F Value   Pr > F
A                     1    76.8000000     76.8000000      1.29    0.2667
B                     2   387.2666667    193.6333333      3.26    0.0559
A*B                   2     2.6000000      1.3000000      0.02    0.9784
```

```
                        The GLM Procedure

            Tukey's Studentized Range (HSD) Test for y

NOTE: This test controls the Type I experimentwise error rate, but it generally has
                a higher Type II error rate than REGWQ.

        Alpha                                    0.05
        Error Degrees of Freedom                   24
        Error Mean Square                    59.38333
        Critical Value of Studentized Range   2.91880
        Minimum Significant Difference         5.8075

    Means with the same letter are not significantly different.

        Tukey Grouping         Mean      N    A
                    A        93.533     15    2
                    A
                    A        90.333     15    1

                        The GLM Procedure

            Tukey's Studentized Range (HSD) Test for y

NOTE: This test controls the Type I experimentwise error rate, but it generally has
                a higher Type II error rate than REGWQ.

        Alpha                                    0.05
        Error Degrees of Freedom                   24
        Error Mean Square                    59.38333
        Critical Value of Studentized Range   3.53170
        Minimum Significant Difference         8.6063

    Means with the same letter are not significantly different.

        Tukey Grouping         Mean      N    B
                    A        96.300     10    1
                    A
                B   A        92.000     10    2
                B
                B        87.500     10    3
```

The ANOVA for the data indicates that the interaction effects are not significant, which leads us to consider the main effects of factors A and B. The results also indicate that there are no significant differences in the main effects of both factors A and B. Moreover, as expected both the Type I and Type III SS results are identical in this case.

Two-factor ANCOVA Model

The two-factor ANCOVA model has the linear form

$$y_{ijk} = \mu + \alpha_i + \beta_j + (\alpha\beta)_{ij} + \gamma(x_{ijk} - \bar{x}...) + \varepsilon_{ijk}, \tag{16.13}$$

where $i = 1, 2, 3, \ldots, a$; $j = 1, 2, 3, \ldots, b$; $k = 1, 2, 3, \ldots, n$. The analysis of covariance is accomplished with the following SAS program and partial output.

```
data new2;
set cov2;
xx=x-39.50;
run;
proc glm;
class A B;
model y=A|B xx/solution;
lsmeans A B A*B/stderr pdiff e adjust=tukey;
output out=aa r=resid;
run;
proc univariate normal data=aa;
var resid;
run;
                        The GLM Procedure

Dependent Variable: y
                                Sum of
Source              DF          Squares      Mean Square    F Value    Pr > F
Model                6      1291.419114       215.236519       8.24    <.0001
Error               23       600.447552        26.106415
```

```
Corrected Total          29    1891.866667

            R-Square     Coeff Var     Root MSE      y Mean
            0.682616     5.557771      5.109444      91.93333

Source                   DF    Type I SS     Mean Square   F Value   Pr > F
---------------------------------------------------------------------------
A                        1     76.8000000    76.8000000    2.94      0.0998
B                        2     387.2666667   193.6333333   7.42      0.0033
A*B                      2     2.6000000     1.3000000     0.05      0.9515
xx                       1     824.7524476   824.7524476   31.59     <.0001

Source                   DF    Type III SS   Mean Square   F Value   Pr > F
---------------------------------------------------------------------------
A                        1     147.4230966   147.4230966   5.65      0.0262
B                        2     292.8106367   146.4053184   5.61      0.0104
A*B                      2     14.4123463    7.2061732     0.28      0.7613
xx                       1     824.7524476   824.7524476   31.59     <.0001

                                      Standard
          Parameter        Estimate     Error    t Value   Pr > |t|
          Intercept     90.89860140 B  2.30984513   39.35   <.0001
          A      1       -5.27832168 B  3.25681934   -1.62   0.1187
          A      2        0.00000000 B     .          .        .
          B      1        7.84055944 B  3.23431946    2.42   0.0236
          B      2        1.96223776 B  3.27638000    0.60   0.5551
          B      3        0.00000000 B     .          .        .
          A*B    1 1     -0.35944056 B  4.57202236   -0.08   0.9380
          A*B    1 2      2.79720280 B  4.61969026    0.61   0.5508
          A*B    1 3      0.00000000 B     .          .        .
          A*B    2 1      0.00000000 B     .          .        .
          A*B    2 2      0.00000000 B     .          .        .
          A*B    2 3      0.00000000 B     .          .        .
          xx              0.75944056     0.13511563    5.62   <.0001
```

NOTE: The X'X matrix has been found to be singular, and a generalized inverse was
used to solve the normal equations. Terms whose estimates are followed by the
letter 'B' are not uniquely estimable.

Least Squares Means
Adjustment for Multiple Comparisons: Tukey-Kramer

```
                                               H0:LSMean1=
                            Standard  H0:LSMEAN=0   LSMean2
     A      y LSMEAN         Error      Pr > |t|    Pr > |t|
     1     89.7004662      1.3240490    <.0001      0.0262
     2     94.1662005      1.3240490    <.0001
```

Least Squares Means
Adjustment for Multiple Comparisons: Tukey-Kramer

```
                          Standard              LSMEAN
     B      y LSMEAN        Error    Pr > |t|   Number
     1     95.9202797     1.6171597   <.0001      1
     2     91.6202797     1.6171597   <.0001      2
     3     88.2594406     1.6213876   <.0001      3
```

Least Squares Means for effect B
Pr > |t| for H0: LSMean(i)=LSMean(j)

Dependent Variable: y

```
     i/j        1           2           3
      1                   0.1666      0.0077
      2       0.1666                  0.3257
      3       0.0077      0.3257
```

Least Squares Means
Adjustment for Multiple Comparisons: Tukey-Kramer

```
                              Standard              LSMEAN
     A    B    Y LSMEAN         Error    Pr > |t|   Number
     1    1    93.1013986     2.3098451   <.0001      1
     1    2    90.3797203     2.2860112   <.0001      2
     1    3    85.6202797     2.2860112   <.0001      3
     2    1    98.7391608     2.2939833   <.0001      4
     2    2    92.8608392     2.2939833   <.0001      5
     2    3    90.8986014     2.3098451   <.0001      6
```

Least Squares Means for effect A*B
Pr > |t| for H0: LSMean(i)=LSMean(j)

Dependent Variable: y

i/j	1	2	3	4	5	6
1		0.9576	0.2315	0.5327	1.0000	0.9839
2	0.9576		0.6847	0.1415	0.9707	1.0000
3	0.2315	0.6847		0.0059	0.2590	0.5940
4	0.5327	0.1415	0.0059		0.4821	0.1893
5	1.0000	0.9707	0.2590	0.4821		0.9901
6	0.9839	1.0000	0.5940	0.1893	0.9901	

The UNIVARIATE Procedure
Variable: resid

Tests for Normality

Test	--Statistic---		-----p Value------	
Shapiro-Wilk	W	0.968529	Pr < W	0.4998
Kolmogorov-Smirnov	D	0.14636	Pr > D	0.0974
Cramer-von Mises	W-Sq	0.073419	Pr > W-Sq	0.2462
Anderson-Darling	A-Sq	0.412772	Pr > A-Sq	>0.2500

Here, we focus on the Type III SS results. Again, the interaction term is not significant but the main effects of A and B are now significant after adjusting for the covariate. Thus, method a_2 produces a better result than method a_1. Similarly, for factor B, The Tukey-Kramer results indicate that there are no significant differences between levels 1 and 2, 2 and 3, but there is significant difference between adjusted means for 1 and 3-a contradiction. Thus our test is not powerful enough to detect these differences.

$\bar{y}_{B1}$	$\bar{y}_{B2}$	$\bar{y}_{B3}$
95.9203	91.6203	88.2594

Suppose instead, we conduct the following hypotheses for the levels of factor B indexed by levels 1, 2 and 3:

1.

$$\begin{aligned} H_0 &: \frac{\mu_1+\mu_2}{2} = \mu_3 \\ H_a &: \frac{\mu_1+\mu_2}{2} \neq \mu_3. \end{aligned} \tag{16.14}$$

2.

$$\begin{aligned} H_0 &: \mu_1 = \mu_2 \\ H_a &: \mu_1 \neq \mu_2. \end{aligned} \tag{16.15}$$

The two hypotheses above are implemented with the contrast statements in the ANCOVA model implementation as indicated in SAS program and partial output below.

```
proc glm data=new2;
class A B;
model y=A|B xx;
contrast '3 vs others' B 1 1 -2;
contrast 'Between 1 & 2' B 1 -1 0;
run;
```

Source	DF	Type III SS	Mean Square	F Value	Pr > F
A	1	147.4230966	147.4230966	5.65	0.0262
B	2	292.8106367	146.4053184	5.61	0.0104
A*B	2	14.4123463	7.2061732	0.28	0.7613
xx	1	824.7524476	824.7524476	31.59	<.0001

Contrast	DF	Contrast SS	Mean Square	F Value	Pr > F
3 vs others	1	200.3606367	200.3606367	7.67	0.0109
Between 1 & 2	1	92.4500000	92.4500000	3.54	0.0726

The contrasts analysis indicate that there is no significant difference between the means of instructors 1 and 2, however, the average results from instructors 1 and 2 are much better than the average result from instructor 3 (adjusting for all the other factors and the concomitant variable).

Unequal Cell Counts

In the preceding analysis, we analyzed a balanced two-way ANOVA. That is, each treatment combination is replicated the same number of times, in this case, five times. The next example deals with a situation where not all the treatment combinations are replicated the same number of times leading to unequal cell counts. The analysis of such data will fully be discussed in a later chapter. However, here, we illustrate the form ANCOVA takes in this case.

	b_1 Instructor 1		b_2 Instructor 2		b_3 Instructor 3	
Method	x	y	x	y	x	y
a_1	3	8	2	14	3	16
	5	16	1	11	2	10
	1	10	8	20	1	14
	9	24	7	15	2	14
			4	12	6	22
					2	16
a_2	7	18	0	8	0	10
	0	7	4	16	1	15
	4	10	8	20	9	26
	6	15	5	18	4	18
	9	23			4	18
					27	26
					8	24

Table 16.3: Example of unequal cell counts data

The data in Table 16.3 is analyzed by the following SAS program and the results are presented in the accompanying partial output.

```
data cov3;
input A B x y @@;
xx=x-4.2580645;
datalines;
1 1 3 8 1 1 5 16 1 1 1 10 1 1 9 24 1 2 2 14 1 2 1 11 1 2 8 20 1 2 7 15
1 2 4 12 1 3 3 16 1 3 2 10 1 3 1 14 1 3 2 14 1 3 6 22 1 3 2 16 2 1 7 18
2 1 0 7 2 1 4 10 2 1 6 15 2 1 9 23 2 2 0 8 2 2 4 16 2 2 8 20 2 2 5 18
2 3 0 10 2 3 1 15 2 3 9 26 2 3 4 18 2 3 4 18 2 3 7 26 2 3 8 24
;
proc means;
var y x;
run;
proc glm;
class A B;
model y=A|B xx/solution ss3;
lsmeans A B A*B/stderr pdiff e;
output out=aa r=resid;
run;
proc univariate normal data=aa;
var resid;
run;
```

                        The GLM Procedure

              Class Level Information

        Class   Levels   Values
        A   2   1 2
        B   3   1 2 3

        Number of observations   31

              The GLM Procedure

Dependent Variable: y

                          Sum of
Source   DF   Squares   Mean Square   F Value   Pr > F
Model   6   755.3092527   125.8848755   26.37   <.0001

```
Error                         24    114.5617150      4.7734048
Corrected Total               30    869.8709677
```

```
              R-Square    Coeff Var    Root MSE      y Mean
              0.868300    13.71036     2.184812     15.93548
```

Source	DF	Type III SS	Mean Square	F Value	Pr > F
A	1	1.1894585	1.1894585	0.25	0.6222
B	2	132.6948373	66.3474186	13.90	<.0001
A*B	2	7.3731595	3.6865798	0.77	0.4731
xx	1	630.8859040	630.8859040	132.17	<.0001

Parameter		Estimate	Standard Error	t Value	Pr > |t|
Intercept		18.82944232 B	0.82829976	22.73	<.0001
A	1	-0.90790117 B	1.24955711	-0.73	0.4745
A	2	0.00000000 B	.	.	.
B	1	-5.76138158 B	1.28113911	-4.50	0.0001
B	2	-3.31632643 B	1.37097785	-2.42	0.0235
B	3	0.00000000 B	.	.	.

```
.........................................................
```

xx		1.62637385	0.14146829	11.50	<.0001

A	y LSMEAN	Standard Error	H0:LSMEAN=0 Pr > |t|	H0:LSMean1=LSMean2 Pr > |t|
1	15.3990745	0.5747252	<.0001	0.6222
2	15.8035397	0.5645667	<.0001	

B	y LSMEAN	Standard Error	Pr > |t|	LSMEAN Number
1	13.5872916	0.7375774	<.0001	1
2	14.8411379	0.7328695	<.0001	2
3	18.3754918	0.6130400	<.0001	3

Least Squares Means

Least Squares Means for effect B
Pr > |t| for H0: LSMean(i)=LSMean(j)

Dependent Variable: y

i/j	1	2	3
1		0.2393	<.0001
2	0.2393		0.0011
3	<.0001	0.0011	

NOTE: To ensure overall protection level, only probabilities associated with
   pre-planned comparisons should be used.

Least Squares Means

A	B	y LSMEAN	Standard Error	Pr > |t|	LSMEAN Number
1	1	14.1065225	1.0929422	<.0001	1
1	2	14.1691598	0.9772841	<.0001	2
1	3	17.9215412	0.9199196	<.0001	3
2	1	13.0680608	0.9861225	<.0001	4
2	2	15.5131159	1.0924067	<.0001	5
2	3	18.8294423	0.8282998	<.0001	6

Least Squares Means for effect A*B
Pr > |t| for H0: LSMean(i)=LSMean(j)

Dependent Variable: y

i/j	1	2	3	4	5	6
1		0.9663	0.0137	0.4864	0.3717	0.0021
2	0.9663		0.0102	0.4349	0.3683	0.0013
3	0.0137	0.0102		0.0017	0.1046	0.4745
4	0.4864	0.4349	0.0017		0.1097	0.0001
5	0.3717	0.3683	0.1046	0.1097		0.0235
6	0.0021	0.0013	0.4745	0.0001	0.0235	

NOTE: To ensure overall protection level, only probabilities associated with
   pre-planned comparisons should be used.

The UNIVARIATE Procedure
Variable: resid

Tests for Normality

```
Test                 --Statistic---     -----p Value------
Shapiro-Wilk         W    0.962173      Pr < W      0.3327
Kolmogorov-Smirnov   D    0.093961      Pr > D     >0.1500
Cramer-von Mises     W-Sq 0.055833      Pr > W-Sq  >0.2500
Anderson-Darling     A-Sq 0.388931      Pr > A-Sq  >0.2500
```

In the results, only factor B and the covariate are significant. Multiple comparison tests show that instructors 1 and 2 are not different, but they both differ from instructor 3. The test for normality shows that the residuals appear to follow a normal distribution.

16.7   Other Covariance Models

Two or More Concomitant Variables

A single concomitant variable is often sufficient to reduce the error variability substantially in an analysis of covariance. However, the model can be extended in a straight forward fashion to include two or more concomitant variables. The single-factor covariance model for two concomitant variables x_1 and x_2, to the first order would be

$$y_{ij} = \mu + \tau_i + \gamma_1(x_{ij1} - \bar{x}_{..1}) + \gamma_2(x_{ij2} - \bar{x}_{..2}) + \varepsilon_{ij}. \tag{16.16}$$

The model in (16.16) can be extended to accommodate more than two concomitant variables. To use SAS to carry out the analysis, we only need to specify all the covariates in the model as we did for one covariate.

Multi-factor Covariance Analysis

A covariance model for two-factor (factor A has a levels and factor B has b levels) experiment with one concomitant variable is written as

$$y_{ijk} = \mu + \alpha_i + \beta_j + (\alpha\beta)_{ij} + \gamma(x_{ijk} - \bar{x}_{...}) + \varepsilon_{ijk}, \quad i = 1, 2, \ldots, a; \quad j = 1, 2, \ldots, b; \quad k = 1, 2, \ldots, n,$$

The above model can be extended to more than two factors in a straightforward manner. As in the other models considered in this chapter, SAS can be used to perform the data analysis.

16.8   Exercises

16.1 **Foot measurements.** The data in the file **kidsfeet.txt** is taken from the foot measurements data submitted by Meyer, M.C. to the Journal of Statistics Education Data Archive (see also Meyer, 2006). The variables in the data are the birth month, birth year, length of longer foot (in centimeters), width of longer foot (in centimeters), gender (B = male, G = female), foot measured (R = right foot, L = left foot), and handedness (R = right, L = left). In order to test if there is a difference between the average widths for male and female, a covariance model with one factor and length as a concomitant variable is proposed.

    (a) Obtain the residuals for the covariance model.

    (b) For each gender, plot the residuals against the predicted values.

    (c) State the regression model to be employed for testing whether or not the gender regression lines have the same slope. Use $\alpha = 0.10$. State the null and the alternative hypotheses, the decision rule, and your conclusion. What is the p-value of your test?

    (d) Could you conduct a test as to whether the regression functions are linear? If so, how many degrees of freedom are there for the denominator mean square in the test statistic?

    (e) Compare the gender main effects by means of a 99% confidence interval. Interpret your interval estimate.

16.2 **Foot measurements.** The **kidsfeet** data is described in Exercise 16.1. For this question, use a subset of the complete data for children whose left feet are measured.

(a) Draw a scatter diagram of the data. Does it appear that there are effects of gender on the mean of width? Explain.

(b) State the regression model equivalent to a covariance model with one factor, gender, and one concomitant variable, length. Use the effect size coding scheme. State the reduced regression model for testing for gender effects.

(c) Fit the full and reduced regression models and test for gender effects. Use $\alpha = 0.10$. State the null and the alternative hypotheses, the decision rule, and your conclusion. What is the p-value of the test?

(d) Estimate the mean width of longer foot for boys with length 24 centimeters. Use a 99% confidence interval.

16.3 **Foot measurements.** The **kidsfeet** data is described in Exercise 16.1. For this question, use a subset of the complete data for children who are right handed.

(a) Draw a scatter diagram of the data. Does it appear that there are effects of gender on the mean of longer foot width? Explain.

(b) State the regression model equivalent to the covariance model with one factor, gender, and one covariate, length. Use the reference cell coding scheme. State the reduced regression model for testing for gender effects.

(c) Fit the full and reduced regression models and test for gender effects. Use $\alpha = 0.10$. State the null and the alternative hypotheses, the decision rule, and your conclusion. What is the p-value of the test?

(d) Estimate the mean width of longer foot for girls with length 22 centimeters. Use a 99% confidence interval.

(e) Compare the gender main effects by means of a 99% confidence interval. Interpret your interval estimate.

16.4 **Public Schools Expenditures.** The data in the file **sat.txt** is taken from the SAT data submitted by Guber, D.L. to the Journal of Statistics Education Data Archive (see also Guber, D.L., 1999). The variables in the data are the name of state (two-letter state abbreviations), section (1 = Northeast, 2 = Midwest, 3 = South, 4 = West), 1994-95 expenditure per pupil (in $'0000), 1994 average pupil/teacher ratio, 1994-95 annual teachers' salary (in $'000), percentage of all eligible students taking SAT, average verbal SAT score, average math SAT score, and average total SAT score. In order to test if there is a difference between the average total SAT scores between the low-spending states and the high-spending states, a covariance model with one factor and percent of all eligible students taking SAT as a concomitant variable is proposed.

(a) Partition the 50 USA states into three groups of low-spending (if 1994-95 expenditure $<= 5.16$), medium-spending (if $5.16 < 1994 - 95$ expenditure $<= 6.16$), and high-spending states (if 1994-95 expenditure > 6.16).

(b) Conduct a one-way analysis of covariance to test the effect of group. Make all pairwise comparisons between the group effects; use either the Bonferroni or the Scheffé procedure with a 95% family confidence coefficient, whichever is more efficient. State your findings.

(c) Obtain the residuals for the covariance model.

(d) For each group, plot the residuals against the predicted values.

(e) State the regression model to be employed for testing whether or not the group regression lines have the same slope. Use $\alpha = 0.05$. State the null and the alternative hypotheses, the decision rule, and your conclusion. What is the p-value of your test?

(f) Could you conduct a test as to whether the regression functions are linear? If so, how many degrees of freedom are there for the denominator mean square in the test statistic?

(g) Make all pairwise comparisons between the group effects in one-way analysis of variance model; use either the Bonferroni or the Scheffé procedure with a 95% family confidence coefficient, whichever is more efficient. State your findings. Compare your findings with the results in (b).

(h) State the regression model equivalent to the analysis of variance model in this problem. Use the effect size coding scheme. State the reduced regression model for testing for group effects.

16.5 **Public Schools Expenditures.** The **sat** data is described in Exercise 16.4. In order to test if there is a difference between the average total SAT scores between the sections, a covariance model with one factor and percent of all eligible students taking SAT as a concomitant variable is proposed.

(a) Carry out a one-way analysis of covariance, using average total SAT as response and sections as factor. Make all pairwise comparisons between the section effects; use either the Bonferroni or the Scheffé procedure with a 95% family confidence coefficient, whichever is more efficient. State your findings.

(b) Obtain the residuals for the covariance model.

(c) For each section, plot the residuals against the predicted values.

(d) State the regression model to be employed for testing whether or not the section regression lines have the same slope. Use $\alpha = 0.05$. State the null and the alternative hypotheses, the decision rule, and your conclusion. What is the p-value of your test?

(e) Could you conduct a test as to whether the regression functions are linear? If so, how many degrees of freedom are there for the denominator mean square in the test statistic?

(f) Make all pairwise comparisons between the section effects in one-way analysis of variance model; use either the Bonferroni or the Scheffé procedure with a 95% family confidence coefficient, whichever is more efficient. State your findings. Compare your findings with the results in (a).

(g) State the regression model equivalent to the analysis of variance model in this problem. Use the effect size coding scheme. State the reduced regression model for testing for section effects.

16.6 **Public Schools Expenditures.** The **sat** data is described in Exercise 16.4. We will be using a group variable similar to the one defined in Exercise 16.4 and the section variable similar to the one defined in Exercise 16.5. In order to test if there is a difference between the average total SAT scores between the different groups and the different sections, a covariance model involving two factors (group and section) with percent of all eligible students taking SAT as a concomitant variable is proposed.

(a) Partition the 50 USA states into two groups of low-spending (if 1994-95 expenditure $<= 5.72$), and high-spending states (if 1994-95 expenditure > 5.72). Combine the Northeast section with the Midwest section to obtain three levels of section.

(b) State the covariance model involving the two factors group and section as defined in (a).

(c) Test for group main effects. Use $\alpha = 0.05$. State the null and the alternative hypotheses, the decision rule, and your conclusion. What is the p-value of the test?

(d) Test for section main effects. Use $\alpha = 0.05$. State the null and the alternative hypotheses, the decision rule, and your conclusion. What is the p-value of the test?

(e) Test for interaction effects. Use $\alpha = 0.05$. State the null and the alternative hypotheses, the decision rule, and your conclusion. What is the p-value of the test?

(f) For each factor, make all pairwise comparisons between factor level main effects. Use the Bonferroni procedure with a 90% family confidence coefficient. State your findings.

16.7 (a) State the analysis of covariance model for a single-factor study with four factor levels when there are two concomitant variables, each with linear term in the model.

(b) State the analysis of covariance model for a single-factor study with four factor levels when there are two concomitant variables, each with linear and quadratic terms in the model.

(c) State the covariance model that has a single factor with three levels and one concomitant variable with linear and quadratic terms in the model. Explain how you will test that only the linear term should be retained in the model.

Chapter 17

Randomized Complete Block Design

17.1 Introduction

In Chapter 14, we discussed the one-way ANOVA in the context of the completely randomized design (CRD), where treatments are applied completely at random within the experimental units. As we mentioned in that chapter, the CRD is best suited for situations where we believe that the experimental material is homogeneous or uniform. Consequently, we advised that the "CRD should therefore be used where extraneous factors can easily be controlled, such as in laboratories or green houses". Thus when the experimental units are not homogeneous, the CRD design usually have very high experimental error, and we would usually have to use the concept of *local control*, by forming homogeneous groups-called **blocks** from our experimental units. In this case, the experimental units within each block are homogeneous and have the same effect on the response variable of interest in the study. Further, for a complete randomized block design, each block contains as many units as are treatments. In the example in Chapter 14 relating to a CRD design of an experiment involving four diets each replicated 5 times, we have the following layout and treatment assignments to the labeled animals based on the SAS random generation of assignments of treatments to animals.

Rats	Diet	Rats	Diet	Rats	Diet	Rats	Dict
1	A	6	D	11	C	16	C
2	A	7	B	12	D	17	B
3	A	8	A	13	D	18	B
4	B	9	C	14	A	19	C
5	C	10	B	15	D	20	D

Table 17.1: A CRD layout with 4 treatments and 5 replications

The CRD design in Table 17.1 does not take into account such extraneous factors as gender, age, etc. of the rats in the above design. Since rats 1, 2, and 3 received diet A for instance, what if the three rats have all been of the same gender, while for instance all rats assigned to say, diet D have all been of the opposite gender. Then comparisons between the effects of diets A and D for instance will not only involve their response variable mean effects but also gender differences. We would therefore need to control for gender in other to have a more reliable basis for comparison of the diet effects. One way we can stratify the experimental material for instance is to set out to buy rats from the same litters (that is offspring from the same mother). We can then group the rats into five groups of four litters and the treatments can be assigned, again using SAS in the following random order.

```
/* Randomized Complete Block Design With 5 Blocks of Size 4 */
data rcb;
proc plan seed=1234;
factors block=5 cell=4;
treatments Diet=4 ordered;
output out=rcbd;
run; quit;
proc sort data=rcbd;
by block cell;
```

```
run;
proc print data=rcbd;
run;
```

The PLAN Procedure

Plot Factors

Factor	Select	Levels	Order
block	5	5	Random
cell	4	4	Random

Treatment Factors

Factor	Select	Levels	Order
Diet	4	4	Ordered

Obs	block	cell	Diet
1	1	1	1
2	1	2	2
3	1	3	3
4	1	4	4
5	2	1	1
6	2	2	3
7	2	3	2
8	2	4	4
9	3	1	2
10	3	2	1
11	3	3	4
12	3	4	3
13	4	1	2
14	4	2	1
15	4	3	3
16	4	4	4
17	5	1	3
18	5	2	1
19	5	3	2
20	5	4	4

A typical layout for the above randomized complete block design (RCBD) is displayed below. Notice that each block contains as many treatments as are in the study, and the treatments are randomly assigned within each block (litters). It is assumed that the animals within each block are homogeneous since they come from the same mothers.

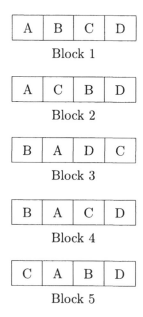

The ANOVA model for the RCBD with t treatments and b blocks of t units, such that the total number of observations in the design equals $n = bt$ can be written as

$$y_{ij} = \mu + \tau_i + \beta_j + \varepsilon_{ij}, \quad i = 1, 2, \ldots, t; \quad j = 1, 2, \ldots, b, \tag{17.1}$$

where

- y_{ij} is the random response for treatment i observed in block j,

- μ is the overall mean,

- τ_i is the effect of the i-th treatment,

- β_j is the effect of the j-th block, and

- ε_{ij} is the random error term for the i-th treatment in the j-th block. That is, the ε_{ij} are independent and identically distributed as $N(0, \sigma^2)$. We will assume that both treatments and blocks are viewed as being fixed factors and that $\sum_{i=1}^{t} \tau_i = 0$, and $\sum_{j=1}^{b} \beta_j = 0$.

A typical table of observations for a randomized block design experiment is given in Table 17.2,

Treatments	Blocks					Totals
	1	2	3	$\cdots$	b	
1	y_{11}	y_{12}	y_{13}	$\cdots$	y_{1b}	$y_{1.}$
2	y_{21}	y_{22}	y_{23}	$\cdots$	y_{2b}	$y_{2.}$
3	y_{31}	y_{32}	y_{33}	$\cdots$	y_{3b}	$y_{3.}$
$\vdots$	$\vdots$	$\vdots$	$\vdots$	$\vdots$	$\vdots$	$\vdots$
t	y_{t1}	y_{t2}	y_{t3}	$\cdots$	y_{tb}	$y_{t.}$
Total	$y_{.1}$	$y_{.2}$	$y_{.3}$	$\cdots$	$y_{.b}$	$y_{..} = G$

Table 17.2: Typical table of observations

where

$$y_{i.} = \sum_{j=1}^{b} y_{ij}, \quad y_{.j} = \sum_{i=1}^{t} y_{ij}, \quad \text{and} \quad y_{..} = G = \sum_{i=1}^{t} \sum_{j=1}^{b} y_{ij}.$$

Analysis

There are a total of bt experimental units (plots) in this experiment. Each treatment is said to be replicated b times. The total SS (SSTot), Blocks SS (SSBlk), Treatments SS (SSTrt) and Error SS (SSE) are obtained as follows:

$$\text{SSTot} = \sum_{i=1}^{t} \sum_{j=1}^{b} (y_{ij} - \bar{y}_{..})^2, \tag{17.2}$$

$$\text{SSTrt} = b \sum_{i=1}^{t} (\bar{y}_{i.} - \bar{y}_{..})^2, \tag{17.3}$$

$$\text{SSBlk} = t \sum_{j=1}^{b} (\bar{y}_{.j} - \bar{y}_{..})^2, \tag{17.4}$$

$$\text{SSE} = \sum_{i=1}^{t} \sum_{j=1}^{b} (y_{ij} - \bar{y}_{i.} - \bar{y}_{.j} + \bar{y}_{..})^2. \tag{17.5}$$

Alternatively, we can obtain the above SS as follows:

$$\text{Correction factor (CF)} = \frac{y_{..}^2}{bt} = \frac{G^2}{bt}$$

$$\text{SSTot} = y_{11}^2 + y_{12}^2 + \cdots + y_{bt}^2 - \text{CF} \quad \text{based on } (bt - 1) \text{ d.f.}$$

$$\text{SSTrt} = \frac{y_{1.}^2}{b} + \frac{y_{2.}^2}{b} + \cdots + \frac{y_{t.}^2}{b} - \text{CF} \quad \text{based on } (t - 1) \text{ d.f.}$$

$$\text{SSBlk} = \frac{y_{.1}^2}{t} + \frac{y_{.2}^2}{t} + \cdots + \frac{y_{.b}^2}{t} - \text{CF} \quad \text{based on } (b - 1) \text{ d.f.}$$

The Error SS is obtained by subtraction as

$$\text{Error SS (SSE)} = \text{Total SS} - \text{Treatments SS} - \text{Blocks SS}$$

The Error SS is based on $(bt - 1) - (b - 1) - (t - 1) = (b - 1)(t - 1)$ degrees of freedom. The structure of the analysis of variance table is presented in Table 17.3.

Source	d.f.	SS	MS	F
Blocks	$b - 1$	$\dfrac{\sum y_{.j}^2}{t} - \dfrac{y_{..}^2}{bt}$	$\dfrac{\text{SSBlk}}{b-1}$	
Treatments	$t - 1$	$\dfrac{\sum y_{i.}^2}{b} - \dfrac{y_{..}^2}{bt}$	$\dfrac{\text{SSTrt}}{t-1} = A$	A/S^2
Error	$(b - 1)(t - 1)$	SSE	$\dfrac{\text{SSE}}{(b-1)(t-1)} = S^2$	
Total	$bt - 1$	$\displaystyle\sum_i \sum_j y_{ij}^2 - \dfrac{y_{..}^2}{bt}$		

Table 17.3: Analysis of variance for a randomized block design

Hypothesis tests concerning treatment and block main effects

We are interested in the hypotheses relating to the equality of treatment means, namely,

$$\begin{aligned}
&H_0 : \mu_1 = \mu_2 = \cdots = \mu_t \\
&H_a : \mu_i \neq \mu_j \text{ for at least one pair } i, j.
\end{aligned} \quad (17.6)$$

In terms of the linear model formulation in (17.1), the hypotheses in (17.6) become

$$\begin{aligned}
&H_0 : \tau_1 = \tau_2 = \cdots = \tau_t = 0 \\
&H_a : \text{at least one } \tau_i \neq 0.
\end{aligned} \quad (17.7)$$

The test of the equality of treatment effects in (17.6) is tested with the variance ratio test, which is based on the F-statistic. That is,

$$F_{\text{Trt}} = \frac{\text{SSTrt}}{t - 1} = A \sim F_{(t-1),(t-1)(b-1)}. \quad (17.8)$$

That is, we compare the computed F-ratio for treatment with an F distribution with $(t-1)$ and $(t-1)(b-1)$ degrees of freedom at say, significance level α. Similarly, the block effects can also be tested from the following expression:

$$F_{\text{Blk}} = \frac{\text{SSBlk}}{b - 1} = B \sim F_{(b-1),(t-1)(b-1)}. \quad (17.9)$$

Again the computed F value will be compared with an F distribution with $(b - 1)$ and $(t - 1)(b - 1)$ degrees of freedom at say, significance level α.

Example 17.1

In an experiment to compare the performance of six newly introduced varieties of maize, it was thought fit to introduce four control varieties a, b, c and d. The experiment was laid out in a randomized block design with 10 plots per block and five blocks. Table 17.4 gives the data for this experiment.

Treatment	Blocks					Totals
	I	II	III	IV	V	
1a	1.63	1.48	1.43	1.76	1.17	7.47
1b	1.73	1.42	1.50	1.06	0.76	6.47
1c	1.49	1.70	1.52	1.48	0.85	7.04
1d	1.25	1.36	0.93	1.38	0.68	5.60
2	1.07	1.28	1.28	1.83	1.16	6.62
3	0.73	1.42	1.30	1.38	0.70	5.53
4	0.69	1.69	1.61	1.61	1.17	6.67
5	0.52	1.50	1.17	1.05	1.04	5.28
6	1.63	1.34	1.07	1.01	0.87	5.92
7	1.08	1.33	0.63	1.21	0.51	4.76
Block Totals	11.82	14.42	12.44	13.77	8.91	61.36

Table 17.4: The yields in (kg/plot) for the experiment in Example 17.1

The structure of the analysis of variance (degrees of freedom only) for this example is given in the following Table.

Source	d.f.
Blocks	4
Treatments	9
Error	36
Total	49

The 9 d.f. for treatments can be partitioned into the following components

(i) Control versus rest with 1 d.f.

(ii) Between controls with 3 d.f.

(iii) Between other varieties with 5 d.f.

Thus the total degrees of freedom is $1 + 3 + 5 = 9$.

Analysis

First we obtain the totals for variety 1 and varieties 2 to 7, viz:

$$\text{Total for variety 1 (Controls)} = 26.58$$
$$\text{Total for varieties 2-7} = 34.78$$

The average of each variety is displayed in the following table.

Variety	1	2	3	4	5	6	7
Means (kg/plot)	1.33	1.33	1.11	1.33	1.06	1.18	0.95

Next we compute the various SS as shown below:

$$\text{Correction Factor (CF)} = \frac{61.36^2}{50} = 75.3010$$

$$\text{Control versus Rest SS} = \frac{26.58^2}{20} + \frac{34.78^2}{30} - \text{CF} = 0.3454$$

$$\text{Between Control SS} = \frac{7.47^2}{5} + \frac{6.47^2}{5} + \frac{7.04^2}{5} + \frac{3.60^2}{5} - \frac{26.58^2}{20} = 0.3919$$

$$\text{Between other Varieties} = \frac{6.62^2}{5} + \cdots + \frac{4.76^2}{5} - \frac{34.78^2}{30} = 0.5737$$

$$\text{Total SS} = 1.63^2 + 1.48^2 + \cdots + 0.51^2 - \frac{61.36^2}{50} = 5.8068$$

$$\text{Blocks SS} = \frac{11.82^2}{10} + \frac{14.42^2}{10} + \cdots + \frac{8.91^2}{10} - \text{CF} = 1.8393$$

The analysis of variance of the data in this example is displayed in Table 17.5.

Source	d.f.	SS	MS	F
Blocks	4	1.8393		
Treatments	9	1.3110	0.1457	1.97
(a) Controls vs rest	1	0.3454	0.3454	4.68*
(b) Between Controls	3	0.3919	0.1306	1.77
(c) Between Others	5	0.5737	0.1147	1.55
Error	36	2.6565	0.0738	
Total	49	5.8068		

Table 17.5: Analysis of variance table
Significant at 5% level

From the ANOVA table, $S^2 = 0.0738$. Hence, the standard error (s.e.) for comparing control (Variety 1) mean with any other treatment mean equals

$$\sqrt{\left(\frac{S^2}{20} + \frac{S^2}{5}\right)} = \sqrt{\left(\frac{0.0738}{20} + \frac{0.0738}{5}\right)} = 0.136 \quad \text{(36 d.f.)}.$$

The s.e. for comparing any two means, not including control equals

$$\sqrt{\left(\frac{S^2}{5} + \frac{S^2}{5}\right)} = \sqrt{\left(\frac{0.0738}{5} + \frac{0.0738}{5}\right)} = 0.172 \quad \text{(36 d.f.)}.$$

The controls mean = 1.33 and the other varieties mean = 1.16, hence the standard error of difference equals

$$\sqrt{\left(\frac{S^2}{20} + \frac{S^2}{30}\right)} = \sqrt{\left(\frac{0.0738}{20} + \frac{0.0738}{30}\right)} = 0.078 \quad \text{(36 d.f.)}.$$

The results in Table 17.5 show that none of the six new varieties is better than the control. Variation between the mean yields for the six varieties is not significant and the control gives a significantly better yield than the average variety mean. The above analysis is implemented in SAS as follows:

```
data rcb1;
do trt=1 to 10;
 do block=1 to 5;
input y @@;
output;
```

```
end; end;
datalines;
1.63 1.48 1.43 1.76 1.17 1.73 1.42 1.50 1.06 0.76
1.49 1.70 1.52 1.48 0.85 1.25 1.36 0.93 1.38 0.68
1.07 1.28 1.28 1.83 1.16 0.73 1.42 1.30 1.38 0.70
0.69 1.69 1.61 1.61 1.17 0.52 1.50 1.17 1.05 1.04
1.63 1.34 1.07 1.01 0.87 1.08 1.33 0.63 1.21 0.51
run;
proc print;
run;
proc glm;
class block trt;
model y=block trt;
run;
```

```
Dependent Variable: y
```

Source	DF	Sum of Squares	Mean Square	F Value	Pr > F
Model	13	3.21727600	0.24748277	3.33	0.0021
Error	36	2.67189200	0.07421922		
Corrected Total	49	5.88916800			

R-Square	Coeff Var	Root MSE	y Mean
0.546304	22.16336	0.272432	1.229200

Source	DF	Type I SS	Mean Square	F Value	Pr > F
block	4	1.88310800	0.47077700	6.34	0.0006
trt	9	1.33416800	0.14824089	2.00	0.0686

Source	DF	Type III SS	Mean Square	F Value	Pr > F
block	4	1.88310800	0.47077700	6.34	0.0006
trt	9	1.33416800	0.14824089	2.00	0.0686

Again the computed F value will be compared with an F distribution with $(b-1)$ and $(t-1)(b-1)$ degrees of freedom at say, significance level α. Values of computed F greater than the tabulated F value indicate rejection of the null hypothesis.

Example 17.2

The following example is from Kutner et al. (2005, pp. 1082-1083) and reproduced with permission of The McGraw-Hill Companies. A physiologist studied the effects of three reagents on muscle tissue in dogs. Ten litters of three dogs each were randomly selected and the three reagents were randomly assigned to the three dogs in each litter. The response is the effects of reagents where the higher the value, the higher is the activity level of the muscle in dogs. Suppose reagents 2 and 3 were expected to be similar to each other, construct appropriate contrasts. The data are presented in Table 17.6.

Litter	Reagent			Litter	Reagent		
i	1	2	3	i	1	2	3
1	10	15	14	6	7	9	10
2	8	12	13	7	24	30	27
3	21	27	25	8	16	18	20
4	14	17	17	9	23	29	32
5	12	18	16	10	18	22	21

Table 17.6: Muscle tissue data

Although the ten litters were randomly selected, let for now assume that we will be fitting a fixed effect model to our data. The SAS implementation is presented below.

```
data rcb2;
do litter=1 to 10;
 do rgt=1 to 3;
input y @@;
output;
end; end;
datalines;
10 15 14 8 12 13 21 27 25 14 17 17 12 18 16 7 9 10 24 30 27 16 18 20 23 29 32 18 22 21
```

```
;
run;
proc glm;
class litter rgt;
model y=litter rgt;
means rgt/tukey;
means rgt/ dunnett ('1');
contrast 'one' rgt 0 1 -1;
contrast 'two' rgt 2 -1 -1;
output out=aa r=resid p=pred;
run;
proc univariate data=aa plot normal;
var resid;
run;
```

The GLM Procedure

Class Level Information

Class	Levels	Values
litter	10	1 2 3 4 5 6 7 8 9 10
rgt	3	1 2 3

Number of Observations Read	30
Number of Observations Used	30

The GLM Procedure

Dependent Variable: y

Source	DF	Sum of Squares	Mean Square	F Value	Pr > F
Model	11	1318.966667	119.906061	73.91	<.0001
Error	18	29.200000	1.622222		
Corrected Total	29	1348.166667			

R-Square	Coeff Var	Root MSE	y Mean
0.978341	7.010999	1.273665	18.16667

Source	DF	Type I SS	Mean Square	F Value	Pr > F
litter	9	1195.500000	132.833333	81.88	<.0001
rgt	2	123.466667	61.733333	38.05	<.0001

Source	DF	Type III SS	Mean Square	F Value	Pr > F
litter	9	1195.500000	132.833333	81.88	<.0001
rgt	2	123.466667	61.733333	38.05	<.0001

The GLM Procedure

Tukey's Studentized Range (HSD) Test for y

NOTE: This test controls the Type I experimentwise error rate, but it generally has a higher Type II error rate than REGWQ.

Alpha	0.05
Error Degrees of Freedom	18
Error Mean Square	1.622222
Critical Value of Studentized Range	3.60930
Minimum Significant Difference	1.4537

Means with the same letter are not significantly different.

Tukey Grouping	Mean	N	rgt
A	19.7000	10	2
A			
A	19.5000	10	3
B	15.3000	10	1

The GLM Procedure
Dunnett's t Tests for y

NOTE: This test controls the Type I experimentwise error for comparisons of all treatments against a control.

Alpha	0.05
Error Degrees of Freedom	18
Error Mean Square	1.622222
Critical Value of Dunnett's t	2.39864
Minimum Significant Difference	1.3663

Comparisons significant at the 0.05 level are indicated by ***.

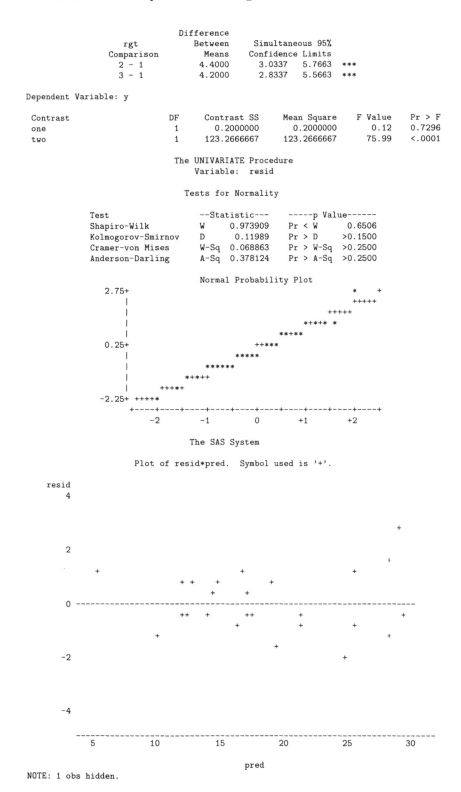

```
                         Difference
              rgt        Between      Simultaneous 95%
           Comparison     Means      Confidence Limits
              2 - 1       4.4000      3.0337   5.7663   ***
              3 - 1       4.2000      2.8337   5.5663   ***

Dependent Variable: y

Contrast              DF    Contrast SS    Mean Square    F Value    Pr > F
one                   1      0.2000000      0.2000000       0.12     0.7296
two                   1    123.2666667    123.2666667      75.99    <.0001

                     The UNIVARIATE Procedure
                        Variable:  resid

                     Tests for Normality

   Test                   --Statistic---       -----p Value------
   Shapiro-Wilk         W    0.973909      Pr < W        0.6506
   Kolmogorov-Smirnov   D    0.11989       Pr > D       >0.1500
   Cramer-von Mises     W-Sq 0.068863      Pr > W-Sq    >0.2500
   Anderson-Darling     A-Sq 0.378124      Pr > A-Sq    >0.2500

                     Normal Probability Plot
       2.75+                                          *    +
           |                                        ++++++
           |                                   +++++
           |                              *++** *
           |                        ++++*
       0.25+                     ++***
           |                  *****
           |              ******
           |           *+*++
           |       ++++*+
      -2.25+ +++++*
           +----+----+----+----+----+----+----+----+----+----+
              -2        -1         0        +1        +2

                        The SAS System

           Plot of resid*pred.  Symbol used is '+'.

     resid
        4


                                                          +


        2
                                                       '
              +                        +            +
                      +  +    +              +
                            +          +
        0 --------------------------------------------------------
                      ++    +        ++       +             +
                              +    +         +       +
                  +                                            +
                                   +
       -2                                +


       -4

          ---------------------------------------------------------
             5         10        15        20        25        30

                              pred
NOTE: 1 obs hidden.
```

The results suggest that there are significant differences among the activity levels of the reagents on the dogs. Tukey's test indicate that while there are no significant differences between the levels of reagents 2 and 3, however, reagents 2 and 3 means are significantly different from the mean of reagent 1. The reagent 1 has the smallest mean muscle activity. The contrast statements and results also confirm the above conclusions as well as the Dunnett multiple comparison procedure, which employs reagent 1 as the 'control' treatment and compares 2 and 3 means with this control mean.

The assumption of normality of error terms is tested by the Anderson-Darling test which indicates that we would fail to reject the null hypothesis that the error terms are normally distributed. Similarly, the plot of the residuals versus the predicted indicate that there is no concern with the assumption that the error terms have constant variance.

17.2   Relative Efficiency of the RCBD

The RCBD will be more preferable to the completely randomized design (CRD) if the variation among blocks in an RCBD is large, otherwise, the RCBD might not be much effective. This comparison between the two designs can be accomplished by obtaining the *relative efficiency* of our design. The relative efficiency for an RCBD can be computed from the following expression in (17.10) which relates the relative efficiency of design 1 to design 2.

$$RE = \frac{(f_1 + 1)(f_2 + 3)\text{MSE}_2}{(f_2 + 1)(f_1 + 3)\text{MSE}_1}, \tag{17.10}$$

where f_i is the error degrees of freedom associated with design i. If $RE > 1$, then design 1 provides more information than design 2 and is consequently more efficient. Since for a RCBD problem, we do not often know the MSE under the CRD, this can be estimated as

$$\text{MSE}_{CRD} = \frac{n_b\text{MSB}_b + (n_t + n_e)\text{MSE}_e}{n_b + n_t + n_e},$$

where, f_b, f_t and f_e are respectively the block, treatment and error degrees of freedom, while MSB_b and MSE_e are also respectively the blocks and error mean squares under the RCBD model. We can substitute the computed mean squares obtained in the ANOVA table for these mean squares. In Example 17.2, we have $b = 10, t = 3$, $n_t = 2, n_b = 9, n_e = 18$ and $\text{MSB}_b = 132.8333, \text{MSE}_e = 1.6222$. Hence,

$$\text{MSE}_{CRD} = \frac{(9 \times 132.8333) + (2 + 18) \times 1.6222}{2 + 9 + 18} = 42.3429.$$

Hence, the relative efficiency is computed as

$$\widehat{RE} = \frac{(f_{rcbd} + 1)(f_{crd} + 3)\text{MSE}_{crd}}{(f_{crd} + 1)(f_{rcbd} + 3)\text{MSE}_{rcbd}} = \frac{(18 + 1)(27 + 3)42.3429}{(27 + 1)(18 + 3)1.6222} = 25.30.$$

Alternatively, we could compute this as

$$\widehat{RE} = c + \frac{(1 - c) \times \text{MSE}_b}{\text{MSE}_e} \quad \text{where } c = \frac{b(t - 1)}{bt - 1}.$$

For our data, $c = 20/29 = 0.689655$ and hence,

$$\widehat{RE} = 0.689655 + \frac{0.310345 \times 132.833333}{1.622222} = 26.10.$$

In both cases the relative efficiency of the RCBD to the CRD is at least 25. Indicating that you will need about 25 replications of the CRD to produce the same kind of information as one replication of the RCBD.

Advantages and Disadvantages of the RCBD

We present here some of the advantages and disadvantages of the RCBD.

- The RCBD design is simple to implement both in terms of actual layout and analysis.

- It always lead to reduced error mean square when the blocking is justified.

- Some treatments can be replicated more times than others.

- It is far more efficient than the completely randomized design when blocking is justified.

- A missing observation or plot can easily be estimated.

- With missing values, the treatment means are biased and we can always use **glm** for the analysis. The RCBD may be less efficient than a completely randomized design.

- For large treatments or treatments combinations, the assumption of homogeneity of units within blocks may be suspect. However, we can overcome this by using incomplete block designs (not covered in this text), where the number of treatments of interest is greater than the number of experimental units per block.

- The error degrees of freedom is smaller than that of completely randomized design.

17.3 Tukey's Test for Additivity

Part of the assumption of the randomized complete block design is that there is no interaction effect between blocks and treatments. Hence, we are able to use the interaction variation as an error term in the model. To check for the validity of this assumption, Tukey's one degree of freedom *test for non-additivity* is often employed. While we do not want to go into the mathematics of it here, we have shown how this test can be readily implemented in SAS, with the data in Example 17.2. Readers can refer to Chapter 20 of Kutner et al. (2005) for details on Tukey's test for additivity.

```
data rcb2b;
set rcb1;
y=y-18.16667;
proc glm;
class litter rgt;
model y=litter rgt;
lsmeans litter rgt;
run;
data new;
set rcb2b;
if litter=1 then b=-5.16667;
else if litter=2 then b=-7.16667;
else if litter=3 then b=6.16663;
else if litter=4 then b=-2.16667;
else if litter=5 then b=-2.83333;
else if litter=6 then b=-9.50000;
else if litter=7 then b=8.83333;
else if litter=8 then b=-0.16667;
else if litter=9 then b=9.83333;
else b=2.16666;
if rgt=1 then t=-2.86667;
else if rgt=2 then t=1.53333;
else t=1.33333;
u=b*t;
proc print;
proc glm data=new;
class litter rgt;
model y=litter rgt u;
run;
```

```
                           The GLM Procedure
Dependent Variable: y


                             Sum of
Source               DF      Squares      Mean Square   F Value   Pr > F
Model                11    1318.966667     119.906061     73.91   <.0001
Error                18      29.200000       1.622222
Corrected Total      29    1348.166667


             R-Square    Coeff Var     Root MSE      y Mean
             0.978341   -38209946     1.273665     -0.000003


Source               DF     Type I SS     Mean Square   F Value   Pr > F
litter                9    1195.500000     132.833333     81.88   <.0001
rgt                   2     123.466667      61.733333     38.05   <.0001


                      The GLM Procedure
                    Least Squares Means

                    litter      y LSMEAN
                      1       -5.16667000
                      2       -7.16667000
                      3        6.16666333
```

```
            4         -2.16667000
            5         -2.83333667
            6         -9.50000333
            7          8.83333000
            8         -0.16667000
            9          9.83333000
           10          2.16666333


            rgt        y LSMEAN
             1       -2.86667000
             2        1.53333000
             3        1.33333000
```

Obs	litter	rgt	x	y	b	t	u
1	1	1	10	-8.1667	-5.16667	-2.86667	14.8111
2	1	2	15	-3.1667	-5.16667	1.53333	-7.9222
3	1	3	14	-4.1667	-5.16667	1.33333	-6.8889
4	2	1	8	-10.1667	-7.16667	-2.86667	20.5445
5	2	2	12	-6.1667	-7.16667	1.53333	-10.9889
6	2	3	13	-5.1667	-7.16667	1.33333	-9.5555
7	3	1	21	2.8333	6.16663	-2.86667	-17.6777
8	3	2	27	8.8333	6.16663	1.53333	9.4555
9	3	3	25	6.8333	6.16663	1.33333	8.2222
10	4	1	14	-4.1667	-2.16667	-2.86667	6.2111
11	4	2	17	-1.1667	-2.16667	1.53333	-3.3222
12	4	3	17	-1.1667	-2.16667	1.33333	-2.8889
13	5	1	12	-6.1667	-2.83333	-2.86667	8.1222
14	5	2	18	-0.1667	-2.83333	1.53333	-4.3444
15	5	3	16	-2.1667	-2.83333	1.33333	-3.7778
16	6	1	7	-11.1667	-9.50000	-2.86667	27.2334
17	6	2	9	-9.1667	-9.50000	1.53333	-14.5666
18	6	3	10	-8.1667	-9.50000	1.33333	-12.6666
19	7	1	24	5.8333	8.83333	-2.86667	-25.3222
20	7	2	30	11.8333	8.83333	1.53333	13.5444
21	7	3	27	8.8333	8.83333	1.33333	11.7777
22	8	1	16	-2.1667	-0.16667	-2.86667	0.4778
23	8	2	18	-0.1667	-0.16667	1.53333	-0.2556
24	8	3	20	1.8333	-0.16667	1.33333	-0.2222
25	9	1	23	4.8333	9.83333	-2.86667	-28.1889
26	9	2	29	10.8333	9.83333	1.53333	15.0777
27	9	3	32	13.8333	9.83333	1.33333	13.1111
28	10	1	18	-0.1667	2.16666	-2.86667	-6.2111
29	10	2	22	3.8333	2.16666	1.53333	3.3222
30	10	3	21	2.8333	2.16666	1.33333	2.8889

```
                    The GLM Procedure

                 Class Level Information

        Class        Levels    Values
        litter          10      1 2 3 4 5 6 7 8 9 10
        rgt              3      1 2 3

        Number of Observations Read        30
        Number of Observations Used        30
```

Dependent Variable: y

Source	DF	Sum of Squares	Mean Square	F Value	Pr > F
Model	12	1323.503192	110.291933	76.02	<.0001
Error	17	24.663475	1.450793		
Corrected Total	29	1348.166667			

R-Square	Coeff Var	Root MSE	y Mean
0.981706	-36134656	1.204489	-0.000003

Source	DF	Type I SS	Mean Square	F Value	Pr > F
litter	9	1195.500000	132.833333	91.56	<.0001
rgt	2	123.466667	61.733333	42.55	<.0001
u	1	4.536525	4.536525	3.13	0.0949

The Tukey's test for additivity gives a p-value of 0.0949 which is clearly not significant at $\alpha = 0.05$ level of significance. Hence we would fail to reject the null hypotheses of additivity, and conclude that in this example the assumption of additivity is correct and we can therefore use the interaction as an error term. The above test is only suitable for one observation per block/treatment combination design. Alternatively, we can conduct Tukey's one degree of freedom test as follows:

```
set rcgb;
```

```
proc glm;
class litter rgt;
model y=litter rgt;
output out=aa p=pred r=resid;
data new;
set aa;
u=pred*pred;
proc glm data=new;
class litter rgt;
model y=litter rgt u;
run;
```

                         The GLM Procedure

Dependent Variable: y

		Sum of			
Source	DF	Squares	Mean Square	F Value	Pr > F
Model	12	1323.503190	110.291932	76.02	<.0001
Error	17	24.663477	1.450793		
Corrected Total	29	1348.166667			

R-Square	Coeff Var	Root MSE	y Mean
0.981706	6.630212	1.204489	18.16667

Source	DF	Type I SS	Mean Square	F Value	Pr > F
litter	9	1195.500000	132.833333	91.56	<.0001
rgt	2	123.466667	61.733333	42.55	<.0001
u	1	4.536523	4.536523	3.13	0.0949

The result gives p-values that agree with the previous model. Here we use the product of the predicted values from the original model as a covariate in the new RCBD model.

17.4   Random Effects in RCBD

Blocks are Random

In Example 17.2, we have assumed that the blocks (litters) effects are fixed. Suppose the litters were randomly selected from a possible finite number of litters of three dogs. However, the blocks were not drawn with any random sampling method, then the blocks can still be viewed as fixed. In this case, since the litters are randomly selected, we would consider their effects to be random. Further, if the three reagents are all the treatment of interest, then the treatment effects will still be fixed and we would therefore have a *mixed* model with the following model formulation:

$$y_{ij} = \mu + \tau_i + B_j + \varepsilon_{ij}, \quad i = 1, 2, 3; \quad j = 1, 2, \ldots, 10, \tag{17.11}$$

where

- y_{ij} is the random response for treatment i observed in block j,

- μ is the overall mean,

- τ_i is the fixed effect of the i-th treatment such that $\sum \tau_i = 0$,

- B_j is the random effect of the j-th block, distributed normal with mean 0 and variance σ_b^2, that is, $B_j \sim N(0, \sigma_b^2)$, and

- ε_{ij} is the random error term for the i-th treatment in the j-th block. That is, the ε_{ij} are independent and identically distributed as $N(0, \sigma^2)$ and are independent of the B_j.

The hypotheses of interest now are

$$H_0 : \tau_1 = \tau_2 = \tau_3 = 0$$
$$H_a : \text{at least one } \tau_i \neq 0, \tag{17.12}$$

and

$$H_0 : \sigma_b^2 = 0$$
$$H_a : \sigma_b^2 \neq 0. \tag{17.13}$$

Source	df	MS	E(MS)	Estimate
Blocks	$b-1$	MS_b	$\sigma^2 + t\sigma_b^2$	$\hat{\sigma}_b^2 = (MS_b - MSE)/t$
Treatments	$t-1$	MS_t	$\sigma^2 + b\frac{\sum\tau_i^2}{t-1}$	MST/MSE
Error	$(b-1)(t-1)$	MSE	σ^2	MSE

Table 17.7: Expected mean squares under the mixed model (blocks random)

The expected mean squares (MS) for the mixed model (block random) for an RCBD with t treatments and b blocks are displayed in Table 17.7.

From the above expected mean squares, the test of the hypotheses in (17.12) will be conducted with a computed F value of MS_t/MSE and compared with an F distribution with $(t-1)$ and $(b-1)(t-1)$ degrees of freedom at a specified α level of significance.

Similarly, the hypotheses in (17.13) will also be tested by computing an F value of MS_b/MSE and compared with an F distribution with $(b-1)$ and $(b-1)(t-1)$ degrees of freedom at a specified α level of significance. This is just the same as the fixed effect model discussed earlier. On the surface, it does not seem to matter here whether the blocks are random or fixed since test of significance for the treatments effects is the same. This is probably true for the overall testing of treatment mean differences as in hypotheses (17.12), however, if interest centers on confidence intervals and correct estimates of treatment means, then the random model is very superior. We present below the SAS analysis of the mixed effects RCBD using **proc glm** and **proc mixed**.

```
set rcb2;
proc glm;
class litter rgt;
model y=litter rgt;
random litter/test;
lsmeans rgt/stderr pdiff;
run;
                      The GLM Procedure

                 Class Level Information

        Class      Levels    Values
        litter        10      1 2 3 4 5 6 7 8 9 10
        rgt            3      1 2 3

           Number of Observations Read        30
           Number of Observations Used        30
```

Dependent Variable: y

		Sum of			
Source	DF	Squares	Mean Square	F Value	Pr > F
Model	11	1318.966667	119.906061	73.91	<.0001
Error	18	29.200000	1.622222		
Corrected Total	29	1348.166667			

R-Square	Coeff Var	Root MSE	y Mean
0.978341	7.010999	1.273665	18.16667

Source	DF	Type I SS	Mean Square	F Value	Pr > F
litter	9	1195.500000	132.833333	81.88	<.0001
rgt	2	123.466667	61.733333	38.05	<.0001

Source	DF	Type III SS	Mean Square	F Value	Pr > F
litter	9	1195.500000	132.833333	81.88	<.0001
rgt	2	123.466667	61.733333	38.05	<.0001

```
                      The GLM Procedure
      Source       Type III Expected Mean Square
      litter       Var(Error) + 3 Var(litter)
      rgt          Var(Error) + Q(rgt)

                      The GLM Procedure
      Tests of Hypotheses for Mixed Model Analysis of Variance
```

Dependent Variable: y

Source	DF	Type III SS	Mean Square	F Value	Pr > F
litter	9	1195.500000	132.833333	81.88	<.0001

```
rgt                      2       123.466667      61.733333      38.05     <.0001
Error: MS(Error)        18        29.200000       1.622222
```

```
                          The SAS System

                        Least Squares Means

                                 Standard                      LSMEAN
          rgt      y LSMEAN         Error      Pr > |t|        Number
           1     15.3000000      0.4027682      <.0001            1
           2     19.7000000      0.4027682      <.0001            2
           3     19.5000000      0.4027682      <.0001            3

                   Least Squares Means for effect rgt
                  Pr > |t| for H0: LSMean(i)=LSMean(j)

                      Dependent Variable: y

          i/j           1             2             3
           1                       <.0001        <.0001
           2         <.0001                      0.7296
           3         <.0001        0.7296
```

NOTE: To ensure overall protection level, only probabilities associated with
      pre-planned comparisons should be used.

In the above program, the **random** statement specifies that litter is considered a random effect. The **test** option statements asks SAS to conduct the relevant significance tests. We also obtain the least squares means for the reagents as well as the standard errors of the difference between any two pairs of means indexed by i and j. Thus, estimates of $\sigma^2 = \text{MSE} = 1.62222$. Similarly, from the expected mean squares provided in SAS, the estimate of the variance components for blocks is

$$\hat{\sigma}_b^2 = \frac{\text{MS}_b - \text{MSE}}{t} = \frac{(132.83333 - 1.62222)}{3} = 43.7370.$$

On the other hand, the use of **proc mixed** is implemented again as follows:

```
set rcb2;
proc mixed;
class litter rgt;
model y=litter rgt;
random litter;
lsmeans rgt/pdiff;
run;
```

```
                       The Mixed Procedure

                       Model Information

          Data Set                  WORK.rcb2
          Dependent Variable        y
          Covariance Structure      Variance Components
          Estimation Method         REML
          Residual Variance Method  Profile
          Fixed Effects SE Method   Model-Based
          Degrees of Freedom Method Containment

                    Class Level Information

          Class    Levels   Values
          litter      10    1 2 3 4 5 6 7 8 9 10
          rgt          3    1 2 3

                          Dimensions

              Covariance Parameters        2
              Columns in X                14
              Columns in Z                10
              Subjects                     1
              Max Obs Per Subject         30

                    Number of Observations

          Number of Observations Read        30
          Number of Observations Used        30
          Number of Observations Not Used     0

                       Iteration History
```

```
     Iteration    Evaluations     -2 Res Log Like      Criterion
             0              1        74.28281310
             1              1        74.28281310      0.00000000
```

Convergence criteria met but final hessian is not positive
definite.

Covariance Parameter
Estimates

```
     Cov Parm        Estimate

     litter            253133
     Residual          1.6222
```

Fit Statistics

```
     -2 Res Log Likelihood            74.3
     AIC (smaller is better)          76.3
     AICC (smaller is better)         76.5
     BIC (smaller is better)          76.6
```

The Mixed Procedure

Type 3 Tests of Fixed Effects

```
                  Num      Den
     Effect        DF       DF    F Value    Pr > F
     litter         9        0       0.00       .
     rgt            2       18      38.05    <.0001
```

Least Squares Means

```
                          Standard
     Effect   rgt   Estimate    Error    DF    t Value    Pr > |t|
     rgt       1     15.3000    0.4028    18     37.99      <.0001
     rgt       2     19.7000    0.4028    18     48.91      <.0001
     rgt       3     19.5000    0.4028    18     48.41      <.0001
```

Differences of Least Squares Means

```
                              Standard
     Effect   rgt   _rgt   Estimate    Error    DF    t Value    Pr > |t|
     rgt       1     2     -4.4000     0.5696    18     -7.72      <.0001
     rgt       1     3     -4.2000     0.5696    18     -7.37      <.0001
     rgt       2     3      0.2000     0.5696    18      0.35      0.7296
```

The mixed model uses the **reml** (restricted maximum likelihood) procedure to estimate the parameters. The estimated variance components are provided by the covariance parameter estimates. Thus $\hat{\sigma}_b^2 = 253133$, while $\hat{\sigma}^2 = 1.6222$. The fixed effects test indicate that there are significant differences among the three reagent means (p-value is < 0.0001). The least squares estimates of the means are exactly the same as those obtained under the **glm**, and the pairwise comparison indicates that there are no significant difference between the means of reagents 2 and 3, but one is clearly different from both. While the variance components are not necessarily the same in both analysis, the standard error of each treatment mean are also different. In this case the standard error under the mixed model correctly calculates this to be

$$\text{s.e}(\hat{\mu}_i) = \sqrt{\frac{(\hat{\sigma}_b^2 + \hat{\sigma}^2)}{b}} = \sqrt{\frac{(253133 + 1.6222)}{10}} = 159.1020.$$

The mixed model correctly computes this standard error as it includes the variance components corresponding to litters. The **glm** on the other hand computes this as $\sqrt{\frac{\hat{\sigma}^2}{b}} = \sqrt{\frac{1.62222}{10}} = 0.4028.$

Blocks and Treatments are Random

In situations where both blocks and treatment effects can be assumed to be random, then the linear model formulation in this case becomes

$$y_{ij} = \mu + T_i + B_j + \varepsilon_{ij}, \quad i = 1, 2, \ldots, t; \ j = 1, 2, \ldots, b, \tag{17.14}$$

where now in addition to the case of the mixed model, $T_i \sim N(0, \sigma_T^2)$ and the corresponding hypotheses of interest now are

$$H_0 : \sigma_T^2 = 0$$
$$H_a : \sigma_T^2 \neq 0,$$

$$(17.15)$$

and

$$H_0 : \sigma_b^2 = 0$$
$$H_a : \sigma_b^2 \neq 0.$$

$$(17.16)$$

The expected mean squares (MS) for the random effects model for an RCBD with t treatments and b blocks are displayed in Table 17.8.

Source	df	MS	E(MS)	Estimate
Blocks	$b-1$	MS_b	$\sigma^2 + t\sigma_b^2$	$\hat{\sigma}_b^2 = (\mathrm{MS}_b - \mathrm{MSE})/t$
Treatments	$t-1$	MS_t	$\sigma^2 + b\sigma_T^2$	$\hat{\sigma}_T^2 = (\mathrm{MS}_t - \mathrm{MSE})/b$
Error	$(b-1)(t-1)$	MSE	σ^2	MSE

Table 17.8: Expected mean squares when blocks and treatments are random

17.5   Missing Values in RCBD

Sometimes an observation in one of the blocks of a randomized block design may be missing. This could be due to carelessness or error, or for reasons beyond our control such as unavoidable damage, no germination in say, a germination experiment for example, or an animal death in an experiment. A missing observation introduces a new problem into the analysis, since treatments are no longer orthogonal to blocks, that is, every treatment does not occur in every block. The general approach to missing value problem is an approximate analysis in which the missing observation is estimated and then the usual analysis of variance is performed proceeding just as if the missing observation were real data, with both the error and total degrees of freedom reduced by one each.

If the plot corresponding to the i-th treatment and the j-th block (y_{ij}) is missing, then it can be estimated approximately from the remaining data by the expression

$$y_0 = \frac{tT_i' + bB_j' - G'}{(b-1)(t-1)},$$

$$(17.17)$$

where T_i' is the total for all known yields for plots receiving treatment i, B_j' is the sum of all known yields from plots in Block j and G' is the sum of all known yields. The variance of the mean of a treatment estimate $\hat{t}_i$ (in which an observation is missing) is given by

$$\mathrm{Var}\,(\hat{t}_i) = \frac{\sigma^2}{b} \left[1 + \frac{t}{(b-1)(t-1)} \right].$$

$$(17.18)$$

The variance of any other treatment mean is still given by $\dfrac{\sigma^2}{b}$. Situations when more than one plot is missing is complicated and the standard procedure is to use (17.17) repeatedly. This will not be described in this text. Readers may consult Keppel and Wickens (2004, Chapter 17).

Example 17.3

Let us consider the data in Example 17.2. Suppose the value for reagent 1 in litter (block) 8 is 'missing' (as a result of one animal being dead). Then without this value, we have

$$\text{Reagent 1 total} = T_1' = 137.0$$
$$\text{Litter 8 total} = B_8' = 38.0$$
$$\text{Grand Total} = G' = 529.0.$$

Missing value estimate from (17.17) is therefore estimated as

$$y_0 = \frac{3 \times 137.0 + 10 \times 38.0 - 529.0}{2 \times 9} = 14.56.$$

This value of 14.56 compares favorably with the real value of 16.0. This estimated value is entered in the table with the observed values and the analysis of variance is performed as usual with reduced degrees of freedom in both the total and error lines. Including this missing value, the new

$$\text{Reagent 1 Total} = 151.56$$
$$\text{Litter 8 Total} = 52.56$$
$$\text{Grand Total} = 543.56.$$

In Table 17.9, we present the analysis of variance with the estimated missing value included.

Source	df	SS	MS	F
Litters	9	1196.644	132.960	
Reagents	2	118.719	59.359	36.11
Error	17	27.948	1.644	
Total	28	1343.310		

Table 17.9: ANOVA table for a missing value analysis

Note the new degrees of freedom for Error $= (18 - 1) = 17$ and for Total $= (29 - 1) = 28$ since one parameter y_0 was already estimated from the data. The estimated reagents (treatment) means are displayed as follows:

$$\text{Reagents means} \quad \begin{array}{cc} 1 & 15.156 \\ 2 & 19.700 \\ 3 & 19.500 \end{array}$$

Standard error (s.e.) for comparing any two means of reagents 2 and 3 is computed as

$$\sqrt{\frac{2S^2}{8}} = \sqrt{\frac{2 \times 1.644}{8}} = 0.6410.$$

Standard error (s.e.) for comparing reagent 1 with any other reagents from the expression in (17.18) is calculated as

$$\sqrt{\left\{ \frac{S^2}{8} \left(2 + \frac{3}{2 \times 9} \right) \right\}} = \sqrt{\left(\frac{13 \times S^2}{38} \right)} = 0.6673.$$

Both standard errors are based on 17 d.f. A SAS implementation of the missing value problem is given in what follows, but first the data are read in with the particular observation read in with a **period** indicating to SAS to treat that observation as missing. The analysis is then carried out as usual.

```
data rcb2;
do litter=1 to 10;
do rgt=1 to 3;
input y @@;
output;
end; end;
datalines;
10 15 14 8 12 13 21 27 25 14 17 17 12 18 16 7 9 10 24 30 27 . 18 20 23 29 32 18 22 21
;
proc print;
run;
proc means;
var litter rgt;
run;
proc glm;
class litter rgt;
model y=litter rgt;
```

```
means rgt/tukey;
lsmeans rgt/stderr;
run;
```

```
                            The GLM Procedure

                        Class Level Information

                Class      Levels   Values
                litter        10    1 2 3 4 5 6 7 8 9 10
                rgt            3     1 2 3

                Number of Observations Read        30
                Number of Observations Used        29
```

Dependent Variable: y

Source	DF	Sum of Squares	Mean Square	F Value	Pr > F
Model	11	1315.362197	119.578382	72.74	<.0001
Error	17	27.948148	1.644009		
Corrected Total	28	1343.310345			

R-Square	Coeff Var	Root MSE	y Mean
0.979195	7.029014	1.282189	18.24138

Source	DF	Type I SS	Mean Square	F Value	Pr > F
litter	9	1196.643678	132.960409	80.88	<.0001
rgt	2	118.718519	59.359259	36.11	<.0001

Source	DF	Type III SS	Mean Square	F Value	Pr > F
litter	9	1196.207407	132.911934	80.85	<.0001
rgt	2	118.718519	59.359259	36.11	<.0001

```
                        The GLM Procedure

               Tukey's Studentized Range (HSD) Test for y

        NOTE: This test controls the Type I experimentwise error rate.

                Alpha                                   0.05
                Error Degrees of Freedom                  17
                Error Mean Square                    1.644009
                Critical Value of Studentized Range  3.62796
```

Comparisons significant at the 0.05 level are indicated by ***.

rgt Comparison	Difference Between Means	Simultaneous 95% Confidence Limits		
2 - 3	0.2000	-1.2710	1.6710	
2 - 1	4.4778	2.9665	5.9891	***
3 - 2	-0.2000	-1.6710	1.2710	
3 - 1	4.2778	2.7665	5.7891	***
1 - 2	-4.4778	-5.9891	-2.9665	***
1 - 3	-4.2778	-5.7891	-2.7665	***

```
                        The GLM Procedure
                        Least Squares Means
```

rgt	y LSMEAN	Standard Error	Pr > |t|
1	15.1555556	0.4379509	<.0001
2	19.7000000	0.4054638	<.0001
3	19.5000000	0.4054638	<.0001

Summary of Results of Analysis

The summary of results indicates that reagent 1 is significantly different from each of reagents 2 and 3, while there are no significant differences in the means of reagents 2 and 3 at the 0.05 significance level. It could therefore be concluded based on the analysis above that the overall differences among the reagents means were significant at 5%. This result can be succinctly summarized as follows:

$$
\begin{array}{ccc}
19.70 & 19.50 & 15.16 \\
\underline{\quad 2 \quad} & \underline{\quad 3 \quad} & 1
\end{array}
$$

17.6   Exercises

17.1   (a) What are the advantages of using randomized complete block designs? Give examples from your area of studies.

   (b) Under what condition(s) would the Completely Randomized Designs have been better than Randomized Complete Block Designs?

17.2 Four different types of graphite coats on light box readings are to be studied. These readings might differ from day to day. Explain how to set up an experiment for the above study using a randomized complete block design. State a suitable model for the above experiment. Considering the following data, complete the analysis.

Day	Types			
	A	B	C	D
1	2	3	3	4
2	3	6	7	10
3	2	8	10	12
4	5	9	10	14

If one considers an experiment using one sample from each of 4 graphite coats and then taking a random order of readings of these coats in each day for a period of 4 days, discuss whether the experiment was conducted in a randomized complete block design.

17.3 A Supplementary Example using SAS

```
Title 'Randomized Complete Block Design';
Data rcb;
input block treat $ yield worth;
datalines;
1 A 32.6 112
1 B 36.4 130
1 C 29.5 106
2 A 42.7 120
2 B 33.1 143
2 C 30.1 152
3 A 29.2 122
3 B 21.2 132
3 C 20.1 129
;
proc anova;
class block treat;
model yield worth = block treat;
run;
```

Study how to modify the above program. How do you handle unbalanced data? Use SAS **proc glm**.

17.4   (a) At a local farm, USGS soil maps identified 5 different soil types. Within each irregularly shaped plot defined by a single soil type, 6 different fertilizer/watering regiments were applied. What is the block? Under what circumstances would it be considered a random block? A fixed block?

   (b) Suppose you had an experiment with 5 blocks and 4 treatments. Explain how to randomize treatments for a RCBD.

   (c) Construct and run SAS code to analyze the following RCBD data. Is the test on Block significant? In general, why might a test on Block be inappropriate?

Treatment	Blocks			
	1	2	3	4
A	8.7	8.9	9.2	8.3
B	8.8	8.4	9.6	8.3
C	7.4	8.0	8.9	7.9
D	7.9	7.3	9.0	7.6
E	10.2	9.3	11.3	9.5

17.5 An information-systems manager is testing four data-base management systems for possible use. A key variable is speed of execution of programs. The manager chooses six representative tasks and writes programs within each management system. The following times were recorded:

System	Task					
	1	2	3	4	5	6
A	58	324	206	94	39	418
B	47	331	163	75	30	397
C	73	355	224	106	59	449
D	38	297	188	72	25	366

(a) Compute means for each system and each task. Find the grand mean.

(b) Obtain SS(system), SS(task), SS(total), and SS(error).

(c) Is there a statistically detectable difference among system means? Use $\alpha = 0.05$.

(d) Should a post-hoc analysis be performed on system means? If yes, carry out the test.

17.6 A study was conducted to compare automobile gasoline mileage for k brands of gasoline. In the study b automobiles, all of the same make and model, were employed in the experiment. [Brands of gasoline are the treatments and automobiles are the blocks.] The following is a partially completed ANOVA table for the randomized block design.

Source	df	SS	MS	F
Brand	2	2.895	—	—
Automobile	—	2.520	—	—
Error	6	—	—	
Total	—	6.760		

(a) Copy and complete the ANOVA table.

(b) How many brands (k) of gasoline were used in the study?

(c) How many automobiles (b) were used in the study?

(d) How many observations were used in the analysis?

(e) Is there evidence of a difference in mean mileage for the k brands of gasoline?

(f) Is there evidence of a difference in mean mileage for the b automobiles?

17.7 The analysis of variance for a randomized block design produced the ANOVA table entries shown here.

Source	df	SS	MS	F
Treatments	2	27.2	—	—
Blocks	5	14.90	—	—
Error	—	33.4	—	
Total	—	—		

(a) Complete the ANOVA table.

(b) How many observations, treatments, and blocks are used in the analysis?

(c) Ley $\bar{y}_b = 12.1$ and $\bar{y}_c = 8.2$ for the Treatments, find a 95% confidence interval for $\mu_b - \mu_c$.

(d) Do the data provide sufficient evidence to indicate a difference among the treatment means? Test using $\alpha = 0.05$.

(e) Do the data provide sufficient evidence to indicate that blocking was useful design strategy to employ for this experiment? Test using $\alpha = 0.05$.

17.8 A machine-shop supervisor is interested in knowing whether there is significant difference among the production times of some machines running a certain number of different jobs. An experiment was set up from which data values were obtained. Part of the analysis of variance table is shown below:

Source	df	SS	MS	F
Machines	2	—	1.6239	—
Job Types	5	77.0444	—	—
Error	—	—	—	
Total	—	81.1778		

(a) What is the underlying design used in the experiment?

(b) How many machines and job types were used?

(c) Complete the analysis of variance table.

(d) At the $\alpha = 0.05$ level, test the hypothesis that there is no mean difference in the production time for each of the machines. Show all the steps in your test procedure.

Chapter 18

Non Orthogonal Classification

18.1 Introduction

In chapter 15, we discussed the two-factor analysis of variance for the case when the treatment combinations are equally replicated. In this chapter, we would consider two factor studies in which the treatment combinations are not equally replicated, leading to what is commonly referred to as *Unbalanced two-way ANOVA* or *Two-way non-orthogonal model*.

Let n_{ij} denote the number of observations corresponding to the (i, j) treatment combination, where $i = 1, 2, \ldots, a$, and $j = 1, 2, \ldots, b$. Thus we have a total of ab treatment combinations each having n replicates. Hence, a total of $N = n_{..}$ observations in the experiment. Here,

$$n_{i.} = \sum_{j=1}^{b} n_{ij}, \quad n_{.j} = \sum_{i=1}^{a} n_{ij}, \quad \text{and} \quad n_{..} = \sum_{i=1}^{a} \sum_{j=1}^{b} n_{ij}$$

The following table shows the number of observations in an experiment with two factors A and B.

Factor A Levels	Factor B Levels						Total
	1	2	$\cdots$	j	$\cdots$	b	
1	n_{11}	n_{12}	$\cdots$	n_{1j}	$\cdots$	n_{1b}	$n_{1.}$
2	n_{21}	n_{22}	$\cdots$	n_{2j}	$\cdots$	n_{2b}	$n_{2.}$
$\vdots$	$\vdots$	$\vdots$	$\cdots$	$\vdots$	$\cdots$	$\vdots$	$\vdots$
i	n_{i1}	n_{i2}	$\cdots$	n_{ij}	$\cdots$	n_{ib}	$n_{i.}$
$\vdots$	$\vdots$	$\vdots$	$\vdots$	$\vdots$	$\vdots$	$\vdots$	$\vdots$
a	n_{a1}	n_{a2}	$\cdots$	n_{aj}	$\cdots$	n_{ab}	$n_{a.}$
Total	$n_{.1}$	$n_{.2}$	$\cdots$	$n_{.j}$	$\cdots$	$n_{.b}$	$n_{..}$

Unbalanced situations often arise as either

1. The treatments or treatment combinations are not equally replicated. That is, the $n_{ij} \neq n$.

2. Some cells corresponding to a treatment or treatment combination is completely missing. That is, at least one of the $n_{ij} = 0$.

Let the kth observation on the (i, j) treatment combination be denoted by y_{ijk}, then the corresponding table of observations is as follows:

Factor A	Factor B					
	1	2	$\cdots$	j	$\cdots$	b
1	$y_{111},\ldots,y_{11n_{11}}$	$y_{121},\ldots,y_{12n_{12}}$	$\cdots$	$y_{1j1},\ldots,y_{1jn_{1j}}$	$\cdots$	$y_{1b1},\ldots,y_{1bn_{1b}}$
2	$y_{211},\ldots,y_{21n_{21}}$	$y_{221},\ldots,y_{22n_{22}}$	$\cdots$	$y_{2j1},\ldots,y_{2jn_{2j}}$	$\cdots$	$y_{2b1},\ldots,y_{2bn_{2b}}$
$\vdots$	$\vdots$	$\vdots$	$\cdots$	$\vdots$	$\cdots$	$\vdots$
i	$y_{i11},\ldots,y_{i1n_{i1}}$	$y_{i21},\ldots,y_{i2n_{i2}}$	$\cdots$	$y_{ij1},\ldots,y_{ijn_{ij}}$	$\cdots$	$y_{ib1},\ldots,y_{ibn_{ib}}$
$\vdots$	$\vdots$	$\vdots$	$\vdots$	$\vdots$	$\vdots$	$\vdots$
a	$y_{a11},\ldots,y_{a1n_{a1}}$	$y_{a21},\ldots,y_{a2n_{a2}}$	$\cdots$	$y_{aj1},\ldots,y_{ajn_{aj}}$	$\cdots$	$y_{ab1},\ldots,y_{abn_{ab}}$

By using

$$\bar{y}_{i..} = \frac{1}{n_{i.}} \sum_{j=1}^{b} \sum_{k=1}^{n_{ij}} y_{ijk}, \quad \bar{y}_{.j.} = \frac{1}{n_{.j}} \sum_{i=1}^{a} \sum_{k=1}^{n_{ij}} y_{ijk}, \quad \text{and} \quad \bar{y}_{...} = \frac{1}{n_{..}} \sum_{i=1}^{a} \sum_{j=1}^{b} \sum_{k=1}^{n_{ij}} y_{ijk},$$

the table of means are given by Table 18.1.

Factor A Levels	Factor B Levels						Total
	1	2	$\cdots$	j	$\cdots$	b	
1	$\bar{y}_{11.}$	$\bar{y}_{12.}$	$\cdots$	$\bar{y}_{1j.}$	$\cdots$	$\bar{y}_{1b.}$	$\bar{y}_{1..}$
2	$\bar{y}_{21.}$	$\bar{y}_{22.}$	$\cdots$	$\bar{y}_{2j.}$	$\cdots$	$\bar{y}_{2b.}$	$\bar{y}_{2..}$
$\vdots$	$\vdots$	$\vdots$	$\cdots$	$\vdots$	$\cdots$	$\vdots$	$\vdots$
i	$\bar{y}_{i1.}$	$\bar{y}_{i2.}$	$\cdots$	$\bar{y}_{ij.}$	$\cdots$	$\bar{y}_{ib.}$	$\bar{y}_{i..}$
$\vdots$	$\vdots$	$\vdots$	$\vdots$	$\vdots$	$\vdots$	$\vdots$	$\vdots$
a	$\bar{y}_{a1.}$	$\bar{y}_{a2.}$	$\cdots$	$\bar{y}_{aj.}$	$\cdots$	$\bar{y}_{ab.}$	$\bar{y}_{a..}$
Total	$\bar{y}_{.1.}$	$\bar{y}_{.2.}$	$\cdots$	$\bar{y}_{.j.}$	$\cdots$	$\bar{y}_{.b.}$	$\bar{y}_{...}$

Table 18.1: Table of means

18.2   Models

We shall consider both the analysis of variance model and regression analysis approaches for analyzing unbalanced data from two factor factorial experiments.

ANOVA Model

A fixed effect ANOVA model has the form

$$y_{ijk} = \mu + \alpha_i + \beta_j + (\alpha\beta)_{ij} + \varepsilon_{ijk}. \tag{18.1}$$

Example 18.1

A research laboratory was developing a new compound for the relief of severe cases of hay fever. The experiment was designed for the amounts of two ingredients (factors A and B) in the compound varied over three levels (Low, Medium and High). Four volunteers were randomly assigned to each of the nine treatment combinations. The time when the subjects began to suffer again from hay fever were recorded as the response variable. Unfortunately, three subjects did not immediately record these times, resulting in an unbalanced design with

$$n_{12} = n_{13} = n_{21} = n_{23} = n_{31} = n_{32} = n_{33} = 4, \quad n_{11} = 3 \text{ and } n_{22} = 2,$$

and the resulting data is presented in Table 18.2.

| Factor A | Factor B (ingredient 2) | | |
(ingredient 1)	$j = 1$ Low	$j = 2$ Medium	$j = 3$ High
$i = 1$ Low	2.4, *, 2.7, 2.5	4.6, 4.2, 4.9, 4.7	4.8, 4.5, 4.4, 4.6
$i = 2$ Medium	5.8, 5.2, 5.5, 5.3	*, 9.1, 8.7, *	9.1, 9.3, 8.7, 9.4
$i = 3$ High	6.1, 5.7, 5.9, 6.2	9.9, 10.5, 10.6, 10.1	13.5, 13.0, 13.3, 13.2

Table 18.2: Times when subjects began to suffer from hay fever (* = missing)

```
data fever;
input y A B REP;
datalines;
............
;
proc glm;
class A B;
model y=A|B/E E1 E2 E3 E4 SS1 SS2 SS3 SS4;
means A|B;
lsmeans A|B/stderr E;
run;
```

18.3 Types of Sum of Squares and Estimable Functions

There are four types of sum of squares (SS) for the general analysis of variance. In SAS, these are designated as SS1, SS2, SS3 and SS4 corresponding respectively to SS Types I, II, III and IV. The use of each depend on one or more of the following situations:

(a) The treatments or treatment combinations are equally replicated- that is, a balanced design. In this case, all the four types of SS give the same results.

(b) Each of the treatments are not equally replicated but each of the cells or treatment combinations has at least one observation. This would lead to an unbalanced or nonorthogonal case I. For this case, Types III and IV give the appropriate test of significance, although Type III is often preferred for its orthogonality properties.

(c) As in (b), but, interest is in testing for main effects and we assume that the weights are proportional to sample sizes, n_{ij}. In this case, the Type II SS will be most appropriate.

(d) As in (b), but one or more cells are completely empty, in this case Type IV SS will be most appropriate. Thus Types II and IV differ only when there are empty cells in the design.

The Type I SS is sequential and the relevant hypotheses depend on the order in which the effects are specified in the model. The SS for all effects are orthogonal in this case. That is, the SS for all the individual effects sum to the overall effects total SS. In SAS, the four types of SS and their accompanying estimable functions can be obtained as an option statement in the model statement viz: **model y = A|B/E1 E2 E3 E4 SS1 SS2 SS3 SS4.**

In this example we have the following with the option E (general estimable function). This option produces the general form of estimable functions as follows in SAS.

```
              The GLM Procedure
        General Form of Estimable Functions

     Effect              Coefficients

     Intercept           L1

     A        1          L2
     A        2          L3
     A        3          L1-L2-L3

     B        1          L5
     B        2          L6
     B        3          L1-L5-L6

     A*B      1 1        L8
```

```
A*B      1 2    L9
A*B      1 3    L2-L8-L9
A*B      2 1    L11
A*B      2 2    L12
A*B      2 3    L3-L11-L12
A*B      3 1    L5-L8-L11
A*B      3 2    L6-L9-L12
A*B      3 3    L1-L2-L3-L5-L6+L8+L9+L11+L12
```

		B			
		1	2	3	
	1	L8	L9	L10	L2
		$(\alpha\beta)_{11}$	$(\alpha\beta)_{12}$	$(\alpha\beta)_{13}$	α_1
A	2	L11	L12	L13	L3
		$(\alpha\beta)_{21}$	$(\alpha\beta)_{22}$	$(\alpha\beta)_{23}$	α_2
	3	L14	L15	L16	L4
		$(\alpha\beta)_{31}$	$(\alpha\beta)_{32}$	$(\alpha\beta)_{33}$	α_3
		L5	L6	L7	L1
		β_1	$\beta2$	β_3	μ

For example,

$$L2 = L8 + L9 + L10$$
$$L1 = L2 + L3 + L4.$$

Thus if L1 = 0, then we have L2 + L3 + L4 = 0 which implies that L4 = −L2 − L3. Similarly, L8 + L9 + L10 = 0 implies that L10 = −L8 − L9, and similarly for all the others. This result is succinctly presented in the table below (Note we will set L1 = 0).

Parameters	Coefficients	
μ	L1	
α_1	L2	
α_2	L3	
α_3	L4	L4 = L1−L2−L3
β_1	L5	
β_2	L6	
β_1	L7	L7 = L1−L5−L6
$(\alpha\beta)_{11}$	L8	
$(\alpha\beta)_{12}$	L9	
$(\alpha\beta)_{13}$	L10	L10 = L2−L8−L9
$(\alpha\beta)_{21}$	L11	
$(\alpha\beta)_{22}$	L12	
$(\alpha\beta)_{23}$	L13	L13 = L3−L11−L12
$(\alpha\beta)_{31}$	L14	L14 = L5−L8−L11
$(\alpha\beta)_{32}$	L15	L15 = L6−L9−L12
$(\alpha\beta)_{33}$	L16	L16 = L1−L2−L3−L5−L6+L8+L9+L11+L12

18.3.1   Type I Sum of Squares

For the Type I SS, for the model in (18.1), each source in the model is added sequentially to the model. The options in the GLM model statement **E1, SS1** will produce the following results for the data in Example 18.1.

```
                        The GLM Procedure
                    Type I Estimable Functions

                   ----------------------Coefficients----------------------
Effect             A                      B                    A*B

Intercept          0                      0                    0

A          1       L2                     0                    0
A          2       L3                     0                    0
A          3       -L2-L3                 0                    0

B          1       -0.0606*L2+0.0667*L3   L5                   0
B          2       0.0303*L2-0.1333*L3    L6                   0
B          3       0.0303*L2+0.0667*L3    -L5-L6               0

A*B        1 1     0.2727*L2              0.291*L5-0.0224*L6   L8
A*B        1 2     0.3636*L2              0.0199*L5+0.3831*L6  L9
A*B        1 3     0.3636*L2              -0.3109*L5-0.3607*L6 -L8-L9
A*B        2 1     0.4*L3                 0.3532*L5+0.0498*L6  L11
A*B        2 2     0.2*L3                 -0.0075*L5+0.2313*L6 L12
A*B        2 3     0.4*L3                 -0.3458*L5-0.2811*L6 -L11-L12
A*B        3 1     -0.3333*L2-0.3333*L3   0.3557*L5-0.0274*L6  -L8-L11
A*B        3 2     -0.3333*L2-0.3333*L3   -0.0124*L5+0.3856*L6 -L9-L12
A*B        3 3     -0.3333*L2-0.3333*L3   -0.3433*L5-0.3582*L6 L8+L9+L11+L12

Source             DF      Type I SS     Mean Square   F Value   Pr > F
A                   2     195.6031212    97.8015606    1488.73   <.0001
B                   2     119.3300968    59.6650484     908.22   <.0001
A*B                 4      28.0707214     7.0176803     106.82   <.0001
```

To conduct the hypotheses for the effects of A, we have two degrees of freedom, hence we need to state two null and corresponding alternative hypotheses. However, to formulate these hypotheses correctly, we examine the column labeled A under the coefficients and particularly, the A*B interaction coefficients. Since the effects of A are indexed by L2 and L3 respectively, hence we would have for the first test, the case when L2 = 1 and L3 = 0. The second set of hypotheses are also obtained when L2 = 0 and L3 = 1. We present the resulting coefficients for the AB interaction in the following tables.

0.2727	0.3636	0.3636	1
0	0	0	0
−0.3333	−0.3333	−0.3333	−1

L2 = 1, L3 = 0

0	0	0	0
0.40	0.20	0.40	1
−0.3333	−0.3333	−0.3333	−1

L2 = 0, L3 = 1

1. For the case when L2 = 1 and L3 = 0, the resulting hypotheses are:

$$H_0 : 0.2727\mu_{11} + 0.3636\mu_{12} + 0.3636\mu_{13} = 0.3333\mu_{31} + 0.3333\mu_{32} + 0.3333\mu_{33}$$
$$H_a : \text{equality does not hold.}$$

(18.2)

2. For the second case when L2 = 0 and L3 = 1, we also have,

$$H_0 : 0.4\mu_{21} + 0.2\mu_{22} + 0.4\mu_{23} = 0.3333\mu_{31} + 0.3333\mu_{32} + 0.3333\mu_{33}$$
$$H_a : \text{equality does not hold.}$$

(18.3)

18.3.2   Type II Sum of Squares

By using the SAS Type II estimable functions (output not included), the coefficients for AB interaction are presented in the following tables.

0.291	0.353	0.356	1
0.020	−0.008	−0.012	0
−0.311	−0.346	−0.343	−1

L2 = 1, L3 = 0

−0.022	0.050	−0.028	0
0.383	0.231	0.386	1
−0.361	−0.281	−0.358	−1

L2 = 0, L3 = 1

1. For the case when L2 = 1 and L3 = 0, the resulting hypotheses are:

$$H_0 : .291\mu_{11} + .353\mu_{12} + .356\mu_{13} + .02\mu_{21} = .008\mu_{22} + .012\mu_{23} + .311\mu_{31} + .345\mu_{32} + .344\mu_{33}$$
$$H_a : \text{equality does not hold.}$$

(18.4)

2. For the second case when L2 = 0 and L3 = 1, we also have,

$$H_0 : .05\mu_{12} + .383\mu_{21} + .231\mu_{22} + .386\mu_{23} = .022\mu_{11} + .028\mu_{13} + .361\mu_{31} + .281\mu_{32} + .358\mu_{33}$$
$$H_a : \text{equality does not hold.}$$

(18.5)

We notice that the above null hypotheses depend on the n_{ij}, the number of observations per cell. Thus Type II SS compares weighted averages of the means in the three rows. The hypotheses in (18.4) and (18.5) are implemented in SAS with the following statements and output.

```
proc glm data=fever;
class A B;
model y=A|B/E2 SS2;
contrast 'cont3a' A 1 0 -1
                  A*B .291 .353 .356 .02 -.008 -.012 -.311 -.345 -.344;
contrast 'cont3b' A 0 1 -1
                  A*B -0.022 0.05 -0.028 .383 0.231 0.386 -0.361 -0.281 -0.358;
run;
```

Source	DF	Type II SS	Mean Square	F Value	Pr > F
A	2	212.9975513	106.4987757	1621.12	<.0001
B	2	119.3300968	59.6650484	908.22	<.0001
A*B	4	28.0707214	7.0176803	106.82	<.0001

The GLM Procedure

Dependent Variable: y

Contrast	DF	Contrast SS	Mean Square	F Value	Pr > F
cont3a	1	208.9767777	208.9767777	3181.04	<.0001
cont3b	1	23.1794804	23.1794804	352.84	<.0001

Notice that the sum of squares for cont3a and cont3b do not add up to the Type II SS for effect A, which was 212.9976.

From the estimable functions for the effect of B, a similar set of hypotheses can be developed and tested accordingly in SAS. We leave this to the reader to do. For the interaction term that is based on four degrees of freedom, the relevant hypotheses can be obtained from the following coefficients in the Tables labeled (i) to (iv), where

- In Table (i) L8 = 1, L9 = L11 = L12 = 0
- In Table (ii), L9 = 1, L8 = L11 = L12 = 0
- In Table (iii), L11 = 1, L8 = L9 = L12 = 0
- In Table (iv), L12 = 1, L8 = L9 = L11 = 0.

1	0	−1
0	0	0
−1	0	1

(i)

0	1	−1
0	0	0
0	−1	1

(ii)

0	0	0
1	0	−1
−1	0	1

(iii)

0	0	0
0	1	−1
0	−1	1

(iv)

Hence you have the four components of the interaction presented as hypotheses

$$H_0 : \mu_{11} + \mu_{33} = \mu_{13} + \mu_{31} \tag{18.6a}$$
$$H_0 : \mu_{12} + \mu_{33} = \mu_{13} + \mu_{32} \tag{18.6b}$$
$$H_0 : \mu_{21} + \mu_{33} = \mu_{23} + \mu_{31} \tag{18.6c}$$
$$H_0 : \mu_{22} + \mu_{33} = \mu_{23} + \mu_{32} \tag{18.6d}$$

The alternative hypotheses for each of the hypotheses in (18.6a) to (18.6d) are that LHS is not equal to RHS. The hypotheses are implemented in SAS with the contrasts statements below with the corresponding output.

```
proc glm data=fever;
class A B;
model y=A|B/E2 SS2;
means A|B;
contrast 'INTA' A*B 1 0 -1 0 0 0 -1 0 1;
contrast 'INTB' A*B 0 1 -1 0 0 0 0 -1 1;
contrast 'INTC' A*B 0 0 0 1 0 -1 -1 0 1;
contrast 'INTD' A*B 0 0 0 0 1 -1 0 -1 1;
run;
```
                              The GLM Procedure
Dependent Variable: y

Contrast	DF	Contrast SS	Mean Square	F Value	Pr > F
INTA	1	25.28102564	25.28102564	384.83	<.0001
INTB	1	9.00000000	9.00000000	137.00	<.0001
INTC	1	12.96000000	12.96000000	197.28	<.0001
INTD	1	6.05000000	6.05000000	92.09	<.0001

If there is no interaction between the two factor variables, then our model becomes

$$y_{ijk} = \mu + \alpha_i + \beta_j + \varepsilon_{ijk}. \tag{18.7}$$

In this case, the Type II SS is equivalent to the balanced case with the appropriate contrasts formulated as desired.

18.3.3 Type III Sum of Squares

The Type III SS is invoked with the option **Model y=A|B /E3 SS3;** The resulting output from our analysis is presented below.

                         The GLM Procedure
                    Type III Estimable Functions

                    -----------------------Coefficients-----------------------
Effect		A	B	A*B
Intercept		0	0	0
A	1	L2	0	0
A	2	L3	0	0
A	3	-L2-L3	0	0
B	1	0	L5	0
B	2	0	L6	0
B	3	0	-L5-L6	0
A*B	1 1	0.3333*L2	0.3333*L5	L8
A*B	1 2	0.3333*L2	0.3333*L6	L9
A*B	1 3	0.3333*L2	-0.3333*L5-0.3333*L6	-L8-L9
A*B	2 1	0.3333*L3	0.3333*L5	L11
A*B	2 2	0.3333*L3	0.3333*L6	L12
A*B	2 3	0.3333*L3	-0.3333*L5-0.3333*L6	-L11-L12
A*B	3 1	-0.3333*L2-0.3333*L3	0.3333*L5	-L8-L11
A*B	3 2	-0.3333*L2-0.3333*L3	0.3333*L6	-L9-L12
A*B	3 3	-0.3333*L2-0.3333*L3	-0.3333*L5-0.3333*L6	L8+L9+L11+L12

                         The GLM Procedure
Dependent Variable: y

Source	DF	Sum of Squares	Mean Square	F Value	Pr > F
Model	8	343.0039394	42.8754924	652.65	<.0001
Error	24	1.5766667	0.0656944		
Corrected Total	32	344.5806061			

R-Square	Coeff Var	Root MSE	y Mean
0.995424	3.547905	0.256309	7.224242

Dependent Variable: y

Source	DF	Type III SS	Mean Square	F Value	Pr > F
A	2	204.0610220	102.0305110	1553.11	<.0001
B	2	112.9626258	56.4813129	859.76	<.0001
A*B	4	28.0707214	7.0176803	106.82	<.0001

With the Type III SS, the two degrees of freedom of main effect A (characterized by L2 and L3), can be written down from the estimable functions, viz: For instance, when L2 = 1 and L3 = 0, we would have, for main effect of A,

0.333	0.333	0.334	1
0	0	0	0
−0.333	−0.333	−0.334	−1

L2 = 1, L3 = 0

0	0	0	0
0.333	0.333	0.334	1
−0.333	−0.333	−0.334	−1

L2 = 0, L3 = 1

1. For the case when L2 = 1 and L3 = 0, the resulting hypotheses are

$$H_0 : (\mu_{11} + \mu_{12} + \mu_{13})/3 = (\mu_{31} + \mu_{32} + \mu_{33})/3$$
$$H_a : \text{equality does not hold.}$$

(18.8)

2. Similarly, for the case when L2 = 0 and L3 = 1, the resulting hypotheses are

$$H_0 : (\mu_{21} + \mu_{22} + \mu_{23})/3 = (\mu_{31} + \mu_{32} + \mu_{33})/3$$
$$H_a : \text{equality does not hold.}$$

(18.9)

We see that in both cases, the hypotheses test the equality of sample average of row means which is exactly what we have in a balanced case. Hence the coefficients of the estimable functions are independent of the cell counts, that is, n_{ij}'s.

The main effects of B

These are indexed by L5 and L6, and again we have the following tables of estimable coefficients:

0.333	0	−0.333
0.333	0	−0.333
0.334	0	−0.334

L5 = 1, L6 = 0

0	0.333	−0.333
0	0.333	−0.333
0	0.334	−0.334

L5 = 0, L6 = 1

The corresponding hypotheses are again given as follows:

1. For the case when L5 = 1 and L6 = 0, the resulting hypotheses are

$$H_0 : (\mu_{11} + \mu_{21} + \mu_{31})/3 = (\mu_{13} + \mu_{23} + \mu_{33})/3$$
$$H_a : \text{equality does not hold.}$$

(18.10)

2. Similarly, for the case when L5 = 0 and L6 = 1, the resulting hypotheses are

$$H_0 : (\mu_{12} + \mu_{22} + \mu_{32})/3 = (\mu_{13} + \mu_{23} + \mu_{33})/3$$
$$H_a : \text{equality does not hold.}$$

(18.11)

These hypotheses are implemented in SAS with the contrast statements and the corresponding output. Note that we had to make some of the cells to have weights ±0.334 so that the contrast can be estimable.

```
proc glm;
class A B;
model y=A|B/E3 SS3;
means A|B;
contrast 'A1' A 1 0 -1
         A*B .333 .333 .334 0 0 0 -.333 -.333 -.334;
contrast 'A2' A 0 1 -1
         A*B 0 0 0 .333 .333 .334 -.333 -.333 -.334;
contrast 'B1' B 1 0 -1
         A*B .333  0 -.333 .333 0 -.333 .334 0 -.334;
contrast 'B2' B 0 1 -1
         A*B 0 .333 -.333 0 .333 -.333 0 .334 -.334;
run;


Dependent Variable: y
```

Contrast	DF	Contrast SS	Mean Square	F Value	Pr > F
A1	1	200.1279016	200.1279016	3046.34	<.0001
A2	1	20.7928638	20.7928638	316.51	<.0001
B1	1	106.7560824	106.7560824	1625.04	<.0001
B2	1	5.7828818	5.7828818	88.03	<.0001

18.4   Least Squares Estimates

The least squares estimates of the main effects of A and their corresponding standard errors are computed as follows: For level A1, we have the least square mean estimate as

$$\text{A1 LSMEAN} = (\bar{y}_{11} + \bar{y}_{12} + \bar{y}_{13})/3$$
$$= (2.5333 + 4.6000 + 4.5750)/3$$
$$= 3.9028.$$

The corresponding standard error of this estimate is given by

$$\text{s.e.(A1)} = \frac{1}{3}\sqrt{\left(\frac{1}{n_{11}} + \frac{1}{n_{12}} + \frac{1}{n_{13}}\right) \times S}$$
$$= \frac{1}{3}\sqrt{\left(\frac{1}{3} + \frac{1}{4} + \frac{1}{4}\right) \times S}$$
$$= 0.30429 \times 0.25631$$
$$= 0.07799,$$

where S is the root mean square $= 0.25631$, obtained from the analysis of variance table. Similarly, the A2 and A3 LS Means are computed respectively from the observed means as

$$\text{A2 LSMEAN} = (\bar{y}_{21} + \bar{y}_{22} + \bar{y}_{23})/3 = 7.8250,$$
$$\text{A3 LSMEAN} = (\bar{y}_{31} + \bar{y}_{32} + \bar{y}_{33})/3 = 9.8333,$$

and the standard errors for the two levels are computed as

$$\text{s.e.(A2)} = \frac{1}{3}\sqrt{\left(\frac{1}{n_{21}} + \frac{1}{n_{22}} + \frac{1}{n_{23}}\right) \times S} = \frac{1}{3}\sqrt{\left(\frac{1}{4} + \frac{1}{2} + \frac{1}{4}\right) \times S} = 0.08544$$

$$\text{s.e.(A3)} = \frac{1}{3}\sqrt{\left(\frac{1}{n_{31}} + \frac{1}{n_{32}} + \frac{1}{n_{33}}\right) \times S} = \frac{1}{3}\sqrt{\left(\frac{1}{4} + \frac{1}{4} + \frac{1}{4}\right) \times S} = 0.07399.$$

The implementation of the interaction hypotheses are exactly as those given in (18.6a) to (18.6d).

```
                    The GLM Procedure

        Level of             --------------y--------------
        A            N           Mean          Std Dev
        1            11       4.02727273      0.97988868
        2            10       7.61000000      1.87820837
        3            12       9.83333333      3.12797737

        Level of             --------------y--------------
        B            N           Mean          Std Dev
        1            11       4.84545455      1.51681484
        2            10       7.73000000      2.75884436
        3            12       8.98333333      3.70670972

Level of    Level of         --------------y--------------
A           B          N         Mean          Std Dev
1           1          3       2.5333333      0.15275252
1           2          4       4.6000000      0.29439203
1           3          4       4.5750000      0.17078251
2           1          4       5.4500000      0.26457513
2           2          2       8.9000000      0.28284271
2           3          4       9.1250000      0.30956959
3           1          4       5.9750000      0.22173558
3           2          4      10.2750000      0.33040379
3           3          4      13.2500000      0.20816660

                    The GLM Procedure
                    Least Squares Means
```

```
                Coefficients for A Least Square Means

                                  A Level
    Effect                            1            2            3

    Intercept                         1            1            1
    A       1                         1            0            0
    A       2                         0            1            0
    A       3                         0            0            1
    B       1                  0.33333333   0.33333333   0.33333333
    B       2                  0.33333333   0.33333333   0.33333333
    B       3                  0.33333333   0.33333333   0.33333333
    A*B     1 1                0.33333333            0            0
    A*B     1 2                0.33333333            0            0
    A*B     1 3                0.33333333            0            0
    A*B     2 1                         0   0.33333333            0
    A*B     2 2                         0   0.33333333            0
    A*B     2 3                         0   0.33333333            0
    A*B     3 1                         0            0   0.33333333
    A*B     3 2                         0            0   0.33333333
    A*B     3 3                         0            0   0.33333333

                                         Standard
                A       y LSMEAN            Error    Pr > |t|
                1      3.90277778       0.07799243     <.0001
                2      7.82500000       0.08543642     <.0001
                3      9.83333333       0.07399011     <.0001
```

The least squares estimates of effect means for factor B and their corresponding standard errors can similarly be obtained by employing the same procedure as above. These results are given in the following SAS output.

```
                Coefficients for B Least Square Means

                                  B Level
    Effect                            1            2            3

    Intercept                         1            1            1
    A       1                  0.33333333   0.33333333   0.33333333
    A       2                  0.33333333   0.33333333   0.33333333
    A       3                  0.33333333   0.33333333   0.33333333
    B       1                         1            0            0
    B       2                         0            1            0
    B       3                         0            0            1
    A*B     1 1                0.33333333            0            0
    A*B     1 2                         0   0.33333333            0
    A*B     1 3                         0            0   0.33333333
    A*B     2 1                0.33333333            0            0
    A*B     2 2                         0   0.33333333            0
    A*B     2 3                         0            0   0.33333333
    A*B     3 1                0.33333333            0            0
    A*B     3 2                         0   0.33333333            0
    A*B     3 3                         0            0   0.33333333

                        Least Squares Means

                                         Standard
                B       y LSMEAN            Error    Pr > |t|
                1      4.65277778       0.07799243     <.0001
                2      7.92500000       0.08543642     <.0001
                3      8.98333333       0.07399011     <.0001
```

LS Estimates for the interaction effects

For the interactions, the least square estimates for each of the nine means are computed as

$$\text{LSMEAN } AB_{ij} = \bar{y}_{ij}$$

$$\text{s.e.}(AB_{ij}) = \sqrt{\frac{1}{n_{ij}}} \times S.$$

For example, the LS estimate of the mean for A1B1 is $\bar{y}_{11} = 2.5333$ with a corresponding standard error of $\sqrt{\frac{1}{3}} \times S = 0.25631 \times 0.57735 = 0.14798$. The following SAS output display these results.

```
              Coefficients for A*B Least Square Means

                              A*B Level
                              1   1   1   2   2   2   3   3   3
     Effect                   1   2   3   1   2   3   1   2   3

     Intercept                1   1   1   1   1   1   1   1   1
     A       1                1   1   1   0   0   0   0   0   0
     A       2                0   0   0   1   1   1   0   0   0
     A       3                0   0   0   0   0   0   1   1   1
     B       1                1   0   0   1   0   0   1   0   0
     B       2                0   1   0   0   1   0   0   1   0
     B       3                0   0   1   0   0   1   0   0   1
     A*B     1 1              1   0   0   0   0   0   0   0   0
     A*B     1 2              0   1   0   0   0   0   0   0   0
     A*B     1 3              0   0   1   0   0   0   0   0   0
     A*B     2 1              0   0   0   1   0   0   0   0   0
     A*B     2 2              0   0   0   0   1   0   0   0   0
     A*B     2 3              0   0   0   0   0   1   0   0   0
     A*B     3 1              0   0   0   0   0   0   1   0   0
     A*B     3 2              0   0   0   0   0   0   0   1   0
     A*B     3 3              0   0   0   0   0   0   0   0   1
```

| | | | Standard | |
A	B	y LSMEAN	Error	Pr > |t|
1	1	2.5333333	0.1479802	<.0001
1	2	4.6000000	0.1281546	<.0001
1	3	4.5750000	0.1281546	<.0001
2	1	5.4500000	0.1281546	<.0001
2	2	8.9000000	0.1812380	<.0001
2	3	9.1250000	0.1281546	<.0001
3	1	5.9750000	0.1281546	<.0001
3	2	10.2750000	0.1281546	<.0001
3	3	13.2500000	0.1281546	<.0001

The following SAS program will produce a plot of the interaction effects based on the least squares estimates of means. The plot is presented in Figure 18.1.

```
data example;
do A=1 to 3;
do B=1 to 3;
input means @@;
output;
end; end;
datalines;
2.533 4.60 4.575 5.45 8.90 9.125 5.975 10.275 13.250
;
proc print;
goptions vsize=6 hsize=6;
symbol1 c=black v=diamond i=join;
symbol2 c=black v=star i=join;
symbol3 c=black v=plus i=join;
proc gplot;
plot means*B=A/noframe;
run;
```

18.5 Regression Implementation

The regression approach is carried out by employing the effect coding scheme, where, for factor A, we have

$$x_1 = \begin{cases} 1, & \text{if A1} \\ -1, & \text{if A3} \\ 0, & \text{elsewhere,} \end{cases} \qquad x_2 = \begin{cases} 1, & \text{if A2} \\ -1, & \text{if A3} \\ 0, & \text{elsewhere.} \end{cases} \qquad (18.12)$$

Similarly for factor B,

$$z_1 = \begin{cases} 1, & \text{if B1} \\ -1, & \text{if B3} \\ 0, & \text{elsewhere,} \end{cases} \qquad z_2 = \begin{cases} 1, & \text{if B2} \\ -1, & \text{if B3} \\ 0, & \text{elsewhere.} \end{cases} \qquad (18.13)$$

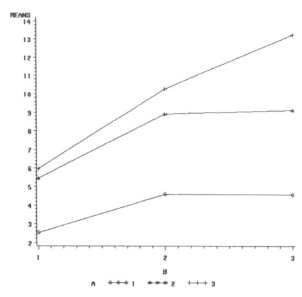

Figure 18.1: Plot of the interaction effects

Here, x_1 and x_2 will account for the two degrees of freedom for the factor A main effects, while z_1 and z_2 will similarly represent the two degrees of freedom for factor B main effects. The four degrees of freedom for interaction terms are obtained as the product of the main effects as discussed in our previous chapters. That is,

$$xz_{11} = x_1 * z_1$$
$$xz_{12} = x_1 * z_2$$
$$xz_{21} = x_2 * z_1$$
$$xz_{22} = x_2 * z_2$$

and the model of interest is

$$y_i = \beta_0 + \underbrace{\beta_1 x_1 + \beta_2 x_2}_{\text{A main effects}} + \underbrace{\beta_3 z_1 + \beta_4 z_2}_{\text{B main effects}} + \underbrace{\beta_5 xz_{11} + \beta_6 xz_{12} + \beta_7 xz_{21} + \beta_8 xz_{22}}_{\text{AB interaction effects}} + \varepsilon_i. \tag{18.14}$$

The interaction effects are first tested with what is often called the '*chunk*' test. This is accomplished by testing the hypotheses

$$H_0 : \beta_5 = \beta_6 = \beta_7 = \beta_8 = 0 \quad \text{versus} \quad H_a : \text{at least one of } \beta_5, \beta_6, \beta_7, \beta_8 \neq 0. \tag{18.15}$$

If the null hypothesis is rejected, then this implies that the interaction is significant and efforts should be concentrated in interpreting the significant interaction terms and main effects become secondary. The main effects of factor B are tested with the following hypotheses in (18.16).

$$H_0 : \beta_3 = \beta_4 = 0 \quad \text{versus} \quad H_a : \text{at least one of } \beta_3, \beta_4 \neq 0. \tag{18.16}$$

Similarly, the main effects of A are tested with the hypotheses in (18.17).

$$H_0 : \beta_1 = \beta_2 = 0 \quad \text{versus} \quad H_a : \text{at least one of } \beta_1, \beta_2 \neq 0. \tag{18.17}$$

The SAS program for implementing the regression approach together with the implementation of the hypotheses in equations (18.15) to (18.17) are displayed in the following along with a partial output.

```
data more;
set fever;
array xll(2) x1-x2;
array zll(2) z1-z2;
do i=1 to 2;
if A=i then xll(i)=1;
else if A=3 then xll(i)=-1;
```

```
else xll(i)=0;
end;
do j=1 to 2;
if B=j then zll(j)=1;
else if B=3 then zll(j)=-1;
else zll(j)=0;
end;
xz11=x1*z1;
xz12=x1*z2;
xz21=x2*z1;
xz22=x2*z2;
output;
drop i j;
proc print data=more; run;
proc reg data=more;
model y=x1-x2 z1-z2 xz11 xz12 xz21 xz22/ss1 ss2;
INT: test xz11, xz12, xz21, xz22;
BB: test z1, z2;
AA: test x1, x2;
run;
```

The REG Procedure
Model: MODEL1
Dependent Variable: y

Number of Observations Read   33
Number of Observations Used   33

Analysis of Variance

Source	DF	Sum of Squares	Mean Square	F Value	Pr > F
Model	8	343.00394	42.87549	652.65	<.0001
Error	24	1.57667	0.06569		
Corrected Total	32	344.58061			

Root MSE	0.25631	R-Square	0.9954
Dependent Mean	7.22424	Adj R-Sq	0.9939
Coeff Var	3.54791		

Dependent Variable: y

Parameter Estimates

Variable	DF	Parameter Estimate	Standard Error	t Value	Pr > |t|	Type I SS	Type II SS
Intercept	1	7.18704	0.04577	157.01	<.0001	1722.25939	1619.58720
x1	1	-3.28426	0.06421	-51.15	<.0001	192.38505	171.87444
x2	1	0.63796	0.06729	9.48	<.0001	3.21807	5.90449
z1	1	2.53426	0.06421	-39.47	<.0001	111.20023	102.33837
z2	1	0.73796	0.06729	10.97	<.0001	8.12987	7.90061
xz11	1	1.16481	0.09265	12.57	<.0001	25.91381	10.38428
xz12	1	-0.04074	0.09155	-0.45	0.6603	0.17424	0.01301
xz21	1	0.15926	0.09155	1.74	0.0947	1.27940	0.19882
xz22	1	0.33704	0.10301	3.27	0.0032	0.70327	0.70327

Test INT Results for Dependent Variable y

Source	DF	Mean Square	F Value	Pr > F
Numerator	4	7.01768	106.82	<.0001
Denominator	24	0.06569		

Test BB Results for Dependent Variable y

Source	DF	Mean Square	F Value	Pr > F
Numerator	2	56.48131	859.76	<.0001
Denominator	24	0.06569		

Test AA Results for Dependent Variable y

Source	DF	Mean Square	F Value	Pr > F
Numerator	2	102.03051	1553.11	<.0001
Denominator	24	0.06569		

The partial F test for the interaction is obtained as

$$F^* = \frac{\text{SS}(xz_{11}, xz_{12}, xz_{21}, xz_{22} | x_1, x_2, z_1, z_2)/4}{\text{MSE}}$$
$$= \frac{(25.9138 + 0.1742 + 1.2794 + 0.7033)/4}{0.06569}$$
$$= 106.82.$$

The extra SS are obtained from the Type I SS column.

18.6 Case of No Significant Interaction Effects

Consider the following data from Kutner et al. (2005, p. 866) and reproduced with permission of The McGraw-Hill Companies. Six male and six female volunteers were used in each age group, and the observations in hundreds of dollars, classified by age (factor A) and gender of owner (factor B) are displayed in Table 18.3. One of the observations in cells (middle, Male) and (Elderly, Female) are missing because the offer received in each case was a trade-in offer, not a cash offer. Thus leading to a non-balanced situation.

Factor A	Factor B (gender of owner)	
	$j = 1$	$j = 2$
(age)	Male	Female
$i = 1$ Young	21, 23, 19, 22, 22, 23	21, 22, 20, 21, 19, 25
$i = 2$ Middle	30, 29, 26, *, 27, 27	26, 29, 27, 28, 27, 29
$i = 3$ Elderly	25, 22, 23, 21, 22 21	23, 19, *, 21, 20, 20

Table 18.3: Male and female volunteers (* = missing)

A fixed effect ANOVA model for the analysis of the data in Table 18.3 is given by

$$y_{ijk} = \mu + \alpha_i + \beta_j + (\alpha\beta)_{ij} + \varepsilon_{ijk}, \tag{18.18}$$

such that for all $n_{ij} > 0$,

$$\sum_i \alpha_i = \sum_j \beta_j = \sum_i (\alpha\beta)_{ij} = \sum_j (\alpha\beta)_{ij} = 0.$$

The data in Table 18.3 can be analyzed by using the Types II and III sum of squares.

18.6.1 Type II Sum of Squares Analysis

The following SAS program is employed to analyze the data in Table 18.3 together with a partial output. Notice that we have requested the Type II sum of squares along with their corresponding estimable functions.

```
data volunteer;
input y a b rep;
datalines;
  ...............
;
proc print noobs;
run;
proc glm;
class a b;
model y=a|b/ss2 e2;
run;
                    Class Level Information

          Class          Levels      Values
          a                   3      1 2 3
          b                   2      1 2
```

```
            Number of Observations Read        34
            Number of Observations Used        34

            Type II Estimable Functions

                    ----------------Coefficients----------------
       Effect        a                      b            a*b

       Intercept     0                      0            0

       a        1    L2                     0            0
       a        2    L3                     0            0
       a        3    -L2-L3                 0            0

       b        1    0                      L5           0
       b        2    0                      -L5          0

       a*b     1 1   0.5161*L2+0.0323*L3    0.3548*L5    L7
       a*b     1 2   0.4839*L2-0.0323*L3    -0.3548*L5   -L7
       a*b     2 1   0.0147*L2+0.4839*L3    0.3226*L5    L9
       a*b     2 2   -0.0147*L2+0.5161*L3   -0.3226*L5   -L9
       a*b     3 1   -0.5308*L2-0.5161*L3   0.3226*L5    -L7-L9
       a*b     3 2   -0.4692*L2-0.4839*L3   -0.3226*L5   L7+L9
```

Dependent Variable: y

Source	DF	Sum of Squares	Mean Square	F Value	Pr > F
Model	5	295.1372549	59.0274510	23.17	<.0001
Error	28	71.3333333	2.5476190		
Corrected Total	33	366.4705882			

R-Square	Coeff Var	Root MSE	y Mean
0.805350	6.783537	1.596126	23.52941

Source	DF	Type II SS	Mean Square	F Value	Pr > F
a	2	289.8907481	144.9453740	56.89	<.0001
b	1	4.3880743	4.3880743	1.72	0.2000
a*b	2	4.1876833	2.0938416	0.82	0.4499

In this example, the interaction effects is not significant (p-value $= 0.4499$). Since this is the case, we can now focus our attention on the main effects. Again the gender main effect is not significant indicating that there are no significant differences between the cash obtained by male and female after adjusting for age. However, the main effect of age is highly significant. The main effect of age is best tested with Type II SS since the interaction is not significant. The two degrees of freedom hypotheses can be tested again as follows: The factor A main effects are indexed in the estimable functions by L2 and L3. Thus we have case (i) L2 = 1, L3 = 0 and case (ii), L2 = 0 and L3 = 1. The corresponding contrasts coefficients are presented as

0.5161	0.4839
0.0147	−0.0147
−0.5308	−0.4692

L2 = 1, L3 = 0

0.0322	−0.0322
0.4839	0.5161
−0.5161	−0.4839

L2 = 0, L3 = 1

The corresponding hypotheses are again given as follows:

1. For the case when L2 = 1 and L3 = 0, the resulting hypotheses are

$$H_0 : \mu_{1.} = \mu_{3.}$$
$$H_a : \mu_{1.} \neq \mu_{3..}$$

(18.19)

The null hypotheses in (18.19) can alternatively be written as

$$H_0 : 0.5161\mu_{11} + 0.4839\mu_{12} + 0.0147\mu_{21} = 0.0147\mu_{22} + 0.5308\mu_{31} + 0.4692\mu_{32},$$

and the above can be tested in SAS with the following statement

```
contrast 'A1 versus A2' a 1 0 -1
              a*b .5161 .4839 .0147 -.0147 -0.5308 -0.4692;
```

2. Similarly, for case (ii), that is, when L2 = 0 and L3 = 1, the resulting hypotheses are

$$H_0 : \mu_{2.} = \mu_{3.}$$
$$H_a : \mu_{2.} \neq \mu_{3.}.$$

(18.20)

The null hypotheses in (18.20) can also alternatively be written as

$$H_0 : 0.0322\mu_{11} + 0.4839\mu_{21} + 0.5161\mu_{22} = 0.032\mu_{12} + 0.5161\mu_{31} + 0.4839\mu_{32},$$

and the above can be tested in SAS with the following statement:

```
contrast 'A1 versus A2' a 0 1 -1
              a*b .0322 -.0322 .4839 .5161 -.5161 -.4839;
```

In both hypotheses (18.19) and (18.20), $\mu_{1.}, \mu_{2.}$ and $\mu_{3.}$ are the respective weighted means for the first, second and third levels of factor A (that is, weighted over those of factor B).

These contrasts as well as Tukey's test is implemented in the following program with a partial output.

```
proc glm data=volunteer;
class a b;
model y=a|b/ss2;
contrast 'a1 versus a3' a 1 0 -1
          a*b .5161 .4839 0.0147 -.0147 -.5308 -.4692;
contrast 'a2 versus a3' a 0 1 -1
          a*b .0322 -.0322 .4839 .5161 -.5161 -.4839;
means a/tukey;
lsmeans A/stderr E;
run;
```

Source	DF	Type II SS	Mean Square	F Value	Pr > F
a	2	289.8907481	144.9453740	56.89	<.0001
b	1	4.3880743	4.3880743	1.72	0.2000
a*b	2	4.1876833	2.0938416	0.82	0.4499

Contrast	DF	Contrast SS	Mean Square	F Value	Pr > F
a1 versus a3	1	0.0009283	0.0009283	0.00	0.9849
a2 versus a3	1	213.5134482	213.5134482	83.81	<.0001

The GLM Procedure

Tukey's Studentized Range (HSD) Test for y

NOTE: This test controls the Type I experimentwise error rate.

Alpha	0.05
Error Degrees of Freedom	28
Error Mean Square	2.547619
Critical Value of Studentized Range	3.49918

Comparisons significant at the 0.05 level are indicated by ***.

a Comparison	Difference Between Means	Simultaneous 95% Confidence Limits		
2 - 3	6.1818	4.4978	7.8658	***
2 - 1	6.2273	4.5788	7.8758	***
3 - 2	-6.1818	-7.8658	-4.4978	***
3 - 1	0.0455	-1.6031	1.6940	
1 - 2	-6.2273	-7.8758	-4.5788	***
1 - 3	-0.0455	-1.6940	1.6031	

a	y LSMEAN	Standard Error	Pr > |t|
1	21.5000000	0.4607620	<.0001
2	27.7333333	0.4832512	<.0001
3	21.4666667	0.4832512	<.0001

Clearly, there is no significant difference between the weighted means of level 1 and level 3 from the first contrast result. However, there is very strong significant difference between levels 2 and 3 (second contrast), what remains is between level 2 and level 1. We can, because of multiple comparisons problem accomplish this with Tukey's test. Our results indicate that while there are no significant differences in cash offers between the young and the elderly age volunteers, there is a significant difference in cash offers between the middle age and the two other levels.

18.6.2 Type III Sum of Squares Analysis

With Type III SS, the most appropriate for this situation, again the partial output below indicate that only the main effect of A is highly significant. Again the estimable functions and their coefficients are presented and the main effect of A are indexed by L2 and L3 respectively, leading to the following estimable contrasts of coefficients.

0.5	0.5
0.0	0.0
−0.5	−0.5

L2 = 1, L3 = 0

0.0	0.0
0.5	0.5
−0.5	−0.5

L2 = 0, L3 = 1

```
proc glm data=volunteer;
class a b;
model y=a|b/e3 ss3;
run;
```

```
                     Type III Estimable Functions

                      -------------Coefficients-------------
          Effect          a                b               a*b

          Intercept       0                0               0

          a      1        L2               0               0
          a      2        L3               0               0
          a      3        -L2-L3           0               0

          b      1        0                L5              0
          b      2        0                -L5             0

          a*b    1 1      0.5*L2           0.3333*L5       L7
          a*b    1 2      0.5*L2           -0.3333*L5      -L7
          a*b    2 1      0.5*L3           0.3333*L5       L9
          a*b    2 2      0.5*L3           -0.3333*L5      -L9
          a*b    3 1      -0.5*L2-0.5*L3   0.3333*L5       -L7-L9
          a*b    3 2      -0.5*L2-0.5*L3   -0.3333*L5      L7+L9
```

Source	DF	Type III SS	Mean Square	F Value	Pr > F
a	2	288.6060606	144.3030303	56.64	<.0001
b	1	4.5375000	4.5375000	1.78	0.1928
a*b	2	4.1876833	2.0938416	0.82	0.4499

The corresponding hypotheses are given as follows:

1. For the case when L2 = 1 and L3 − 0, the resulting hypotheses are

$$H_0 : \mu_{1.} = \mu_{3.}$$
$$H_a : \mu_{1.} \neq \mu_{3.}.$$

(18.21)

The null hypotheses in (18.21) can alternatively be written as

$$H_0 : \frac{\mu_{11} + \mu_{12}}{2} = \frac{\mu_{31} + \mu_{32}}{2}.$$

2. Similarly, for case (ii), that is, when L2 = 0 and L3 = 1, the resulting hypotheses are

$$H_0 : \mu_{2.} = \mu_{3.}$$
$$H_a : \mu_{2.} \neq \mu_{3.}.$$

(18.22)

The null hypotheses in (18.22) can also alternatively be written as

$$H_0 : \frac{\mu_{21} + \mu_{22}}{2} = \frac{\mu_{31} + \mu_{32}}{2}.$$

The hypotheses in (18.21) and (18.22) can be tested in SAS with the following statements and partial output.

```
proc glm data=volunteer;
class a b;
model y=a|b/ss3;
contrast 'a1 versus a3' a 1 0 -1
                        a*b .5 .5 0 0 -.5 -.5;
contrast 'three1'       a 1 0 -1;
contrast 'a2 versus a3' a 0 1 -1
                        a*b 0 0 .5 .5 -.5 -.5;
contrast 'three2'       a 0 1 -1;
means a/tukey;
lsmeans A/stderr;
run;
```

Source	DF	Type III SS	Mean Square	F Value	Pr > F
a	2	288.6060606	144.3030303	56.64	<.0001
b	1	4.5375000	4.5375000	1.78	0.1928
a*b	2	4.1876833	2.0938416	0.82	0.4499

Contrast	DF	Contrast SS	Mean Square	F Value	Pr > F
a1 versus a3	1	0.0063492	0.0063492	0.00	0.9605
three1	1	0.0063492	0.0063492	0.00	0.9605
a2 versus a3	1	214.2060606	214.2060606	84.08	<.0001
three2	1	214.2060606	214.2060606	84.08	<.0001

Tukey's Studentized Range (HSD) Test for y

NOTE: This test controls the Type I experimentwise error rate.

Alpha	0.05
Error Degrees of Freedom	28
Error Mean Square	2.547619
Critical Value of Studentized Range	3.49918

Comparisons significant at the 0.05 level are indicated by ***.

a Comparison	Difference Between Means	Simultaneous 95% Confidence Limits		
2 - 3	6.1818	4.4978	7.8658	***
2 - 1	6.2273	4.5788	7.8758	***
3 - 2	-6.1818	-7.8658	-4.4978	***
3 - 1	0.0455	-1.6031	1.6940	
1 - 2	-6.2273	-7.8758	-4.5788	***
1 - 3	-0.0455	-1.6940	1.6031	

a	y LSMEAN	Standard Error	Pr > |t|
1	21.5000000	0.4607620	<.0001
2	27.7333333	0.4832512	<.0001
3	21.4666667	0.4832512	<.0001

Note that the contrasts designated as 'three1' and 'three2' are equivalent respectively to the two hypotheses in their implementation in SAS. That is, the contrasts are implemented as in the balanced two-way ANOVA model.

18.6.3 Regression Implementation

To implement the regression approach, we code the factor A levels as

$$x_1 = \begin{cases} 1, & \text{if Young} \\ -1, & \text{if Elderly} \\ 0, & \text{elsewhere,} \end{cases} \qquad x_2 = \begin{cases} 1, & \text{if Middle} \\ -1, & \text{if Elderly} \\ 0, & \text{elsewhere.} \end{cases} \tag{18.23}$$

Similarly for factor B,

$$z = \begin{cases} 1, & \text{if Male} \\ -1, & \text{if Female.} \end{cases} \tag{18.24}$$

Here, x_1 and x_2 will account for the two degrees of freedom for the factor A main effects, while z will similarly represent the one degree of freedom for gender effect. The two degrees of freedom for the interaction effects are again obtained as the product of the main effects. That is,

$$x_1 z = x_1 * z; \qquad x_2 z = x_2 * z,$$

and the model of interest is

$$y_i = \beta_0 + \underbrace{\alpha_1 x_1 + \alpha_2 x_2}_{\text{A main effects}} + \underbrace{\beta_1 z}_{\text{B main effects}} + \underbrace{\gamma_1 x_1 z + \gamma_2 x_2 z}_{\text{AB interaction effects}} + \varepsilon_i. \tag{18.25}$$

The interaction effects are first tested with the following hypotheses

$$H_0 : \gamma_1 = \gamma_2 = 0 \quad \text{versus} \quad H_a : \text{at least one of } \gamma_1, \gamma_2 \neq 0. \tag{18.26}$$

```
data more;
set volunteer;
array x11(2) x1-x2;
do i=1 to 2;
if A=i then x11(i)=1;
else if A=3 then x11(i)=-1;
else x11(i)=0;
end;
if B=1 then z=1;
else z=-1;
x1z=x1*z;
x2z=x2*z;
output;
drop i;
proc print data=more; run;
proc reg data=more;
model y=x1-x2 z x1z x2z;
INT: test x1z, x2z;
AA: test x1, x2;
BB: test z;
run;
```

The REG Procedure
Model: MODEL1
Dependent Variable: y

Number of Observations Read   34
Number of Observations Used   34

Analysis of Variance

Source	DF	Sum of Squares	Mean Square	F Value	Pr > F
Model	5	295.13725	59.02745	23.17	<.0001
Error	28	71.33333	2.54762		
Corrected Total	33	366.47059			

Root MSE	1.59613	R-Square	0.8054	
Dependent Mean	23.52941	Adj R-Sq	0.7706	
Coeff Var	6.78354			

Model: MODEL1
Dependent Variable: y

Parameter Estimates

Variable	DF	Parameter Estimate	Standard Error	t Value	Pr > |t|
Intercept	1	23.56667	0.27475	85.78	<.0001
x1	1	-2.06667	0.38243	-5.40	<.0001
x2	1	4.16667	0.39157	10.64	<.0001
z	1	0.36667	0.27475	1.33	0.1928
x1z	1	-0.20000	0.38243	-0.52	0.6051
x2z	1	-0.30000	0.39157	-0.77	0.4500

Model: MODEL1
Test INT Results for Dependent Variable y

Source	DF	Mean Square	F Value	Pr > F
Numerator	2	2.09384	0.82	0.4499
Denominator	28	2.54762		

Model: MODEL1
Test AA Results for Dependent Variable y

Source	DF	Mean Square	F Value	Pr > F
Numerator	2	144.30303	56.64	<.0001
Denominator	28	2.54762		

```
                        Model: MODEL1
           Test BB Results for Dependent Variable y

                              Mean
      Source          DF      Square    F Value    Pr > F
      Numerator       1       4.53750     1.78     0.1928
      Denominator     28      2.54762
```

The results from the regression analysis agree with those obtained earlier with **proc glm** using the Type III SS. The hypotheses of no interaction effects, no factor B and factor A main effects (denoted above respectively as tests INT, AA and BB) give results that are consistent with those obtained earlier. Since only the main effect of factor A is significant, the hypotheses in (18.19) becomes in the regression case,

$$
\begin{aligned}
H_0 &: \alpha_1 = \alpha_3 \\
H_a &: \alpha_1 \neq \alpha_3,
\end{aligned}
\tag{18.27}
$$

and these become since $\alpha_3 = -(\alpha_1 + \alpha_2)$,

$$
\begin{aligned}
H_0 &: 2\alpha_1 + \alpha_2 = 0 \\
H_a &: 2\alpha_1 + \alpha_2 \neq 0.
\end{aligned}
\tag{18.28}
$$

Similarly, the hypotheses in (18.20) can also be written following a similar argument as

$$
\begin{aligned}
H_0 &: \alpha_1 + 2\alpha_2 = 0 \\
H_a &: \alpha_1 + 2\alpha_2 \neq 0.
\end{aligned}
\tag{18.29}
$$

Both of these are implemented in SAS with the following commands and partial output.

```
proc reg data=more;
model y=x1-x2 z x1z x2z;
A1: test 2*x1+x2=0;
A2: test x1+2*x2=0;
run;
                        The REG Procedure
                         Model: MODEL1

           Test A1 Results for Dependent Variable y

                              Mean
      Source          DF      Square    F Value    Pr > F
      -----------------------------------------------------
      Numerator       1       0.00635     0.00     0.9605
      Denominator     28      2.54762

           Test A2 Results for Dependent Variable y

                              Mean
      Source          DF      Square    F Value    Pr > F
      -----------------------------------------------------
      Numerator       1       214.20606   84.08    <.0001
      Denominator     28      2.54762
```

Again our results agree with those obtained under the contrasts implementation in **proc glm** using the Type III SS. Because the interaction is not significant, it has often been suggested that we can fit a model without the interaction terms, that is fit ANOVA and regression models respectively as

$$
y_{ijk} = \mu + \alpha_i + \beta_j + \varepsilon_{ijk}
\tag{18.30a}
$$

$$
y_i = \beta_0 + \underbrace{\alpha_1 x_{1i} + \alpha_2 x_{2i}}_{\text{A main effects}} + \underbrace{\beta_1 z_i}_{\text{B main effects}} + \varepsilon_i,
\tag{18.30b}
$$

for all $n_{ij} > 0$. In (18.30a), we have $\sum_i \alpha_i = \sum_j \beta_j = 0$. The implementation of the model in (18.30a) in SAS is presented below and in this case both the Type II and Type III SS give identical results. In this case, the Type II SS is often much preferred.

```
proc glm data=volunteer;
class a b;
model y=a b/ss2 ss3 e2 e3;
contrast 'a1 versus a3' a 1 0 -1;
contrast 'a2 versus a3' a 0 1 -1;
means a/tukey;
lsmeans A/stderr E;
run;
```

The GLM Procedure

Class Level Information

Class	Levels	Values
a	3	1 2 3
b	2	1 2

Number of Observations Read	34
Number of Observations Used	34

Type II Estimable Functions

Effect		-Coefficients- a	b
Intercept		0	0
a	1	L2	0
a	2	L3	0
a	3	-L2-L3	0
b	1	0	L5
b	2	0	-L5

Type III Estimable Functions

Effect		-Coefficients- a	b
Intercept		0	0
a	1	L2	0
a	2	L3	0
a	3	-L2-L3	0
b	1	0	L5
b	2	0	-L5

Dependent Variable: y

Source	DF	Sum of Squares	Mean Square	F Value	Pr > F
Model	3	290.9495716	96.9831905	38.53	<.0001
Error	30	75.5210166	2.5173672		
Corrected Total	33	366.4705882			

R-Square	Coeff Var	Root MSE	y Mean
0.793923	6.743141	1.586621	23.52941

Source	DF	Type II SS	Mean Square	F Value	Pr > F
a	2	289.8907481	144.9453740	57.58	<.0001
b	1	4.3880743	4.3880743	1.74	0.1967

Source	DF	Type III SS	Mean Square	F Value	Pr > F
a	2	289.8907481	144.9453740	57.58	<.0001
b	1	4.3880743	4.3880743	1.74	0.1967

Dependent Variable: y

Contrast	DF	Contrast SS	Mean Square	F Value	Pr > F
a1 versus a3	1	0.0009255	0.0009255	0.00	0.9848
a2 versus a3	1	213.5110689	213.5110689	84.82	<.0001

Tukey's Studentized Range (HSD) Test for y

NOTE: This test controls the Type I experimentwise error rate.

Alpha	0.05
Error Degrees of Freedom	30
Error Mean Square	2.517367
Critical Value of Studentized Range	3.48651

Comparisons significant at the 0.05 level are indicated by ***.

```
                        Difference
            a           Between       Simultaneous 95%
         Comparison      Means       Confidence Limits
           2 - 3         6.1818      4.5139   7.8497   ***
           2 - 1         6.2273      4.5945   7.8600   ***
           3 - 2        -6.1818     -7.8497  -4.5139   ***
           3 - 1         0.0455     -1.5873   1.6782
           1 - 2        -6.2273     -7.8600  -4.5945   ***
           1 - 3        -0.0455     -1.6782   1.5873
```

Least Squares Means

Coefficients for a Least Square Means

```
                                     a Level
       Effect                      1      2      3

       Intercept                   1      1      1
       a          1                1      0      0
       a          2                0      1      0
       a          3                0      0      1
       b          1               0.5    0.5    0.5
       b          2               0.5    0.5    0.5
```

```
                            Standard
         a     y LSMEAN       Error     Pr > |t|
         1    21.5000000    0.4580181    <.0001
         2    27.7600196    0.4790269    <.0001
         3    21.5127077    0.4790269    <.0001
```

Similarly the regression implementation of the model in (18.30b) in SAS is accomplished with the following SAS statements and partial output.

```
proc reg data=more;
model y=x1-x2 z;
AA: test x1, x2;
BB: test z;
A1: test 2*x1+x2=0;
A2: test x1+2*x2=0;
run;
```

```
                    The REG Procedure
                      Model: MODEL1
                  Dependent Variable: y
```

```
         Number of Observations Read        34
         Number of Observations Used        34
```

Analysis of Variance

```
                              Sum of        Mean
Source              DF       Squares       Square    F Value   Pr > F
Model                3     290.94957     96.98319     38.53    <.0001
Error               30      75.52102      2.51737
Corrected Total     33     366.47059
```

```
         Root MSE            1.58662    R-Square    0.7939
         Dependent Mean     23.52941    Adj R-Sq    0.7733
         Coeff Var           6.74314
```

```
                   Model: MODEL1
              Dependent Variable: y
```

Parameter Estimates

```
                   Parameter    Standard
Variable    DF     Estimate       Error     t Value    Pr > |t|

Intercept    1     23.59091      0.27233     86.63      <.0001
x1           1     -2.09091      0.37959     -5.51      <.0001
x2           1      4.16911      0.38867     10.73      <.0001
z            1      0.36022      0.27283      1.32      0.1967
```

```
                   Model: MODEL1
      Test AA Results for Dependent Variable y
```

```
                              Mean
Source            DF        Square     F Value   Pr > F
Numerator          2      144.94537     57.58    <.0001
```

```
Denominator          30        2.51737

                    Model: MODEL1
        Test BB Results for Dependent Variable y

                          Mean
Source              DF    Square    F Value    Pr > F
Numerator            1   4.38807       1.74    0.1967
Denominator         30   2.51737

                    Model: MODEL1
        Test A1 Results for Dependent Variable y

                          Mean
Source              DF    Square    F Value    Pr > F
Numerator            1 0.00092549       0.00    0.9848
Denominator         30   2.51737

                    Model: MODEL1
        Test A2 Results for Dependent Variable y

                          Mean
Source              DF    Square    F Value    Pr > F
Numerator            1 213.51107      84.82    <.0001
Denominator         30   2.51737
```

Again, the results from the regression analysis agree with those obtained under the ANOVA model using **proc glm**.

18.7   Case of Empty Cell Entries

Suppose in Table 18.3, there were no elderly female getting a cash back, resulting in the following Table 18.4, that is $n_{32} = 0$. In this case, the Type IV SS is the most appropriate.

Factor A (age)	Factor B (gender of owner)	
	$j = 1$ Male	$j = 2$ Female
$i = 1$ Young	21, 23, 19, 22, 22, 23	21, 22, 20, 21, 19, 25
$i = 2$ Middle	30, 29, 26, 28, 27, 27	26, 29, 27, 28, 27, 29
$i = 3$ Elderly	25, 22, 23, 21, 22 21	

Table 18.4: The otherwise balanced data with missing cell

Since there are no sample information on elderly females, the resulting table of population means are as presented in Table 18.5. Thus there are no sample information for mean μ_{32}.

Factor A (age)	Factor B (gender of owner)	
	$j = 1$ Male	$j = 2$ Female
$i = 1$ Young	μ_{11}	μ_{12}
$i = 2$ Middle	μ_{21}	μ_{22}
$i = 3$ Elderly	μ_{31}	

Table 18.5: Table of population means

Partial information can be obtained on the interaction if we exclude sample information on elderly subjects. That is, if we restrict ourselves to young and middle aged subjects. The necessary contrast for this can be obtained from Table 18.5 by considering, $\mu_{11} - \mu_{12}$ and $\mu_{21} - \mu_{22}$, which leads to the contrast

$$L = \mu_{11} - \mu_{12} - \mu_{21} + \mu_{22}.$$

```
data miss;
input y a b rep;
```

```
datalines;
   21.0       1       1       1
   23.0       1       1       2
   ........................
   22.0       3       1       5
   21.0       3       1       6
;
proc print noobs;
run;
proc glm;
class a b;
model y=a|b/ss4 e4;
contrast 'int' a*b 1 -1 -1 1 0;
contrast 'a1 vs a3' a 1 0 -1
                 a*b 1 0 0 0 -1;
contrast 'a2 vs a3' a 0 1 -1
                 a*b 0 0 1 0 -1;
contrast 'b' b 1 -1
         a*b .5 -.5 .5 -.5 0;
run;
```

The GLM Procedure

Type IV Estimable Functions

Effect		------Coefficients------		
		a	b	a*b
Intercept		0	0	0
a	1	L2	0	0
a	2	L3	0	0
a	3	-L2-L3	0	0
b	1	0	L5	0
b	2	0	-L5	0
a*b	1 1	L2	0.5*L5	L7
a*b	1 2	0	-0.5*L5	-L7
a*b	2 1	L3	0.5*L5	-L7
a*b	2 2	0	-0.5*L5	L7
a*b	3 1	-L2-L3	0	0

NOTE: Other Type IV estimable functions exist.

Dependent Variable: y

Source	DF	Sum of Squares	Mean Square	F Value	Pr > F
Model	4	260.0000000	65.0000000	26.14	<.0001
Error	25	62.1666667	2.4866667		
Corrected Total	29	322.1666667			

R-Square	Coeff Var	Root MSE	y Mean
0.807036	6.525173	1.576917	24.16667

Source	DF	Type IV SS	Mean Square	F Value	Pr > F
a	2*	137.4444444	68.7222222	27.64	<.0001
b	1*	0.3750000	0.3750000	0.15	0.7011
a*b	1	0.0416667	0.0416667	0.02	0.8980

* NOTE: Other Type IV Testable Hypotheses exist which may yield different SS.

Dependent Variable: y

Contrast	DF	Contrast SS	Mean Square	F Value	Pr > F
int	1	0.04166667	0.04166667	0.02	0.8980
a1 vs a3	1	1.33333333	1.33333333	0.54	0.4708
a2 vs a3	1	90.75000000	90.75000000	36.49	<.0001
b	1	0.37500000	0.37500000	0.15	0.7011

The analysis of the data in Table 18.4 is carried out in SAS with the use of Type IV SS only (complete cell missing). The contrast for testing the interaction designated 'int' is as specified earlier and implemented in SAS as a contrast statement. Notice that the ANOVA Type IV SS is exactly the same as those obtained with the contrast statement. The corresponding p-value is 0.8980 which clearly indicates that the interaction in the partial table is not significant at $\alpha = 0.05$ level of significance.

Again, the results indicate that only the main effects of factor A are significantly different. From the estimable functions and their coefficients, the two degrees of freedom for these tests are obtained in (a) when L2 = 1 and L3

= 0 and (b) when L2 = 0 and L3 = 1 respectively.

1	0	1
0	0	0
−1		−1

(a)

0	0	0
1	0	1
−1		−1

(b)

The contrasts result indicate that significant differences exist between the means of the middle and the elderly participants while ignoring the effect of gender. Of course if there are no interactions present as in this example, we can fit a model without the interaction terms as in the next output with its implementing program.

```
proc glm data=miss;
class a b;
model y=a b/ss4 e4;
lsmeans a;
contrast 'a1 vs a3' a 1 0 -1;
contrast 'a2 vs a3' a 0 1 -1;
run;
```

```
                      Type IV Estimable Functions

                                  -Coefficients-
              Effect            a          b

              Intercept         0          0

              a        1        L2         0
              a        2        L3         0
              a        3        -L2-L3     0

              b        1        0          L5
              b        2        0          -L5
```

Dependent Variable: y

Source	DF	Sum of Squares	Mean Square	F Value	Pr > F
Model	3	259.9583333	86.6527778	36.22	<.0001
Error	26	62.2083333	2.3926282		
Corrected Total	29	322.1666667			

R-Square	Coeff Var	Root MSE	y Mean
0.806906	6.400603	1.546812	24.16667

Source	DF	Type IV SS	Mean Square	F Value	Pr > F
a	2	257.7361111	128.8680556	53.86	<.0001
b	1	0.3750000	0.3750000	0.16	0.6954

```
              Least Squares Means

              a        y LSMEAN
              1        21.5000000
              2        27.7500000
              3        22.2083333
```

Dependent Variable: y

Contrast	DF	Contrast SS	Mean Square	F Value	Pr > F
a1 vs a3	1	1.7202381	1.7202381	0.72	0.4042
a2 vs a3	1	105.2916667	105.2916667	44.01	<.0001

The partial inference drawn from this analysis again agrees with those obtained for the main effect of A (age) given that the interaction can be assumed to be zero.

Additional Comments

1. If the design is balanced, all the four types of SS give identical results. SAS **proc glm** will automatically give results of Types I and III.

2. For the general case of unbalanced design, use Type III or IV, although Type III is often preferred.

3. However, if you have an empty cell(s), use Type IV. However, care must be taken in the interpretations of main effects. It is sometimes difficult to compare main effects because some marginal totals may not exist.

4. Types I and II as we have seen are functions of the cell counts n_{ij}, and we further note here that Type I SS is sequential and therefore tests carried out with Type I SS depends on the order in which the variables are entered.

5. Type IV depends on the label of the factor levels and we would expect the SS to change if we were to change the labels of the factor. Thus, tests with Type IV SS depend on the order and labels of the factor levels. On the other hand, Type III tests do not depend on the order or labels of factor levels

18.8   Exercises

18.1 **Bacteria counts.** The data in the following table is taken from the bacteria count data submitted by Binnie, N.S. to the Journal of Statistics Education Data Archive. The data contains measurement of bacteria counts of Staphylococcus aureus (strain 3), the concentration of tryptone (a nutrient) and the temperature (in ^0C) of incubation. In order to test for the effects of concentration, temperature and their interactions, an analysis of variance model is proposed.

| | Temperature | | |
Concentration	27	35	43
0.6	10, 129	93, 146	42, 118
0.8	26, 145	98, 217	108, 99
1.0	50, 156	89, 269	96, 141
1.2	52, 243	149, 284	131, 234
1.4	47, 178	113, 186	121, 172

(a) Set up the analysis of variance table. Which source account for most of the total variability in the response variable? Explain.

(b) Test the interaction effects at $\alpha = 0.05$. State the null and the alternative hypotheses, your decision, and conclusion. State the p-value of the test.

(c) Test for whether or not the main effects of temperature and concentration are present. For each test, state the null and the alternative hypotheses, your decision, and conclusion. State the p-value.

(d) Estimate μ_{22} with a 90% confidence interval. Interpret the interval.

(e) Use a 95% confidence interval to estimate the difference $\mu_{1.} - \mu_{2.}$. Based on your interval, can we say that there is a difference between $\mu_{1.}$ and $\mu_{2.}$?

(f) Use a 95% confidence interval to estimate the difference $\mu_{11} - \mu_{12}$. Interpret the interval estimate.

18.2 **Bacteria counts.** The data in the following table is taken from the bacteria count data submitted by Binnie, N.S. to the Journal of Statistics Education Data Archive. The data contains measurement of bacteria counts of Staphylococcus aureus (strain 3) and the temperature (in ^0C) of incubation. Suppose the temperature is randomly chosen from all possible temperatures. In order to test for the effects of temperature, an analysis of variance model is proposed.

Temperature		
27	35	43
10, 26, 50, 52, 47	93, 98, 89, 149, 113	42, 108, 96, 131, 121
129, 145, 156, 243, 178	146, 217, 269, 284, 186	118, 99, 141, 234, 172

(a) Test whether the mean bacteria counts is the same for all temperatures. Use $\alpha = 0.05$. For the test, state the null and the alternative hypotheses, your decision, and conclusion. State the p-value of the test.

(b) Estimate the mean bacteria counts for all temperatures by setting a 95% confidence interval. Interpret the interval.

(c) Use a 90% confidence interval to estimate the ratio $\sigma_\mu^2/(\sigma_\mu^2 + \sigma^2)$. Interpret the interval.

(d) What are the point estimate for σ_μ^2 and σ^2?

18.3 Bacteria counts. The data in the following table is taken from the bacteria count data submitted by Binnie, N.S. to the Journal of Statistics Education Data Archive. The data contains measurement of bacteria counts of Staphylococcus aureus (strain 3), the time (in hours) of incubation and the temperature (in ^0C) of incubation. Suppose the temperature is randomly chosen from all possible temperatures. In order to test for the effects of time, temperature and their interactions, an analysis of variance model is proposed.

Time	Temperature		
	27	35	43
24	10, 26, 50, 52, 47	93, 98, 89, 149, 113	42, 108, 96, 131, 121
48	129, 145, 156, 243, 178	146, 217, 269, 284, 186	118, 99, 141, 234, 172

(a) Test whether the mean bacteria counts is the same for all times. Use $\alpha = 0.05$.

(b) Estimate the mean bacteria counts for all temperatures by setting a 95% confidence interval. Interpret the interval.

(c) Use a 90% confidence interval to estimate the ratio $\sigma_\mu^2/(\sigma_\mu^2 + \sigma^2)$. Interpret the interval.

(d) What are the point estimate for σ_μ^2 and σ^2?

(e) Test whether temperature and time interact at $\alpha = 0.05$. State the null and the alternative hypotheses, your decision, and conclusion. State the p-value of the test.

(f) Test for the significance of temperature main effect. For the test, state the null and the alternative hypotheses, your decision, and conclusion. State the p-value of the test.

18.4 Bacteria counts. The data in the following table is taken from the bacteria count data submitted by Binnie, N.S. to the Journal of Statistics Education Data Archive. The data contains measurement of bacteria counts of Staphylococcus aureus (strain 4), the time (in hours) of incubation and the temperature (in ^0C) of incubation. Suppose both time and temperature are randomly chosen. In order to test for the effects of time, temperature and their interactions, an analysis of variance model is proposed.

Time	Temperature		
	27	35	43
24	14, 20, 17, 29, 18	102, 110, 106, 178, 124	31, 62, 119, 132, 142
48	102, 109, 129, 161, 158	136, 139, 156, 233, 213	111, 148, 97, 177, 212

(a) Test whether the mean bacteria counts is the same for all times. Use $\alpha = 0.05$.

(b) Estimate the mean bacteria counts for all times by setting a 95% confidence interval. Interpret the interval.

(c) Use a 90% confidence interval to estimate the ratio $\sigma_\mu^2/(\sigma_\mu^2 + \sigma^2)$. Interpret the interval.

(d) What are the point estimate for σ_μ^2 and σ^2?

(e) Test whether temperature and time interact at $\alpha = 0.05$. State the null and the alternative hypotheses, your decision, and conclusion. State the p-value of the test.

(f) Test for the significance of temperature main effect. For the test, state the null and the alternative hypotheses, your decision, and conclusion. State the p-value of the test.

Appendix

Table 1: Standard normal probabilities (Area between 0 and z)

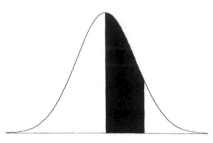

z	0.00	0.01	0.02	0.03	0.04	0.05	0.06	0.07	0.08	0.09
0.0	0.0000	0.0040	0.0080	0.0120	0.0160	0.0199	0.0239	0.0279	0.0319	0.0359
0.1	0.0398	0.0438	0.0478	0.0517	0.0557	0.0596	0.0636	0.0675	0.0714	0.0753
0.2	0.0793	0.0832	0.0871	0.0910	0.0948	0.0987	0.1026	0.1064	0.1103	0.1141
0.3	0.1179	0.1217	0.1255	0.1293	0.1331	0.1368	0.1406	0.1443	0.1480	0.1517
0.4	0.1554	0.1591	0.1628	0.1664	0.1700	0.1736	0.1772	0.1808	0.1844	0.1879
0.5	0.1915	0.1950	0.1985	0.2019	0.2054	0.2088	0.2123	0.2157	0.2190	0.2224
0.6	0.2257	0.2291	0.2324	0.2357	0.2389	0.2422	0.2454	0.2486	0.2517	0.2549
0.7	0.2580	0.2611	0.2642	0.2673	0.2704	0.2734	0.2764	0.2794	0.2823	0.2852
0.8	0.2881	0.2910	0.2939	0.2967	0.2995	0.3023	0.3051	0.3078	0.3106	0.3133
0.9	0.3159	0.3186	0.3212	0.3238	0.3264	0.3289	0.3315	0.3340	0.3365	0.3389
1.0	0.3413	0.3438	0.3461	0.3485	0.3508	0.3531	0.3554	0.3577	0.3599	0.3621
1.1	0.3643	0.3665	0.3686	0.3708	0.3729	0.3749	0.3770	0.3790	0.3810	0.3830
1.2	0.3849	0.3869	0.3888	0.3907	0.3925	0.3944	0.3962	0.3980	0.3997	0.4015
1.3	0.4032	0.4049	0.4066	0.4082	0.4099	0.4115	0.4131	0.4147	0.4162	0.4177
1.4	0.4192	0.4207	0.4222	0.4236	0.4251	0.4265	0.4279	0.4292	0.4306	0.4319
1.5	0.4332	0.4345	0.4357	0.4370	0.4382	0.4394	0.4406	0.4418	0.4429	0.4441
1.6	0.4452	0.4463	0.4474	0.4484	0.4495	0.4505	0.4515	0.4525	0.4535	0.4545
1.7	0.4554	0.4564	0.4573	0.4582	0.4591	0.4599	0.4608	0.4616	0.4625	0.4633
1.8	0.4641	0.4649	0.4656	0.4664	0.4671	0.4678	0.4686	0.4693	0.4699	0.4706
1.9	0.4713	0.4719	0.4726	0.4732	0.4738	0.4744	0.4750	0.4756	0.4761	0.4767
2.0	0.4772	0.4778	0.4783	0.4788	0.4793	0.4798	0.4803	0.4808	0.4812	0.4817
2.1	0.4821	0.4826	0.4830	0.4834	0.4838	0.4842	0.4846	0.4850	0.4854	0.4857
2.2	0.4861	0.4864	0.4868	0.4871	0.4875	0.4878	0.4881	0.4884	0.4887	0.4890
2.3	0.4893	0.4896	0.4898	0.4901	0.4904	0.4906	0.4909	0.4911	0.4913	0.4916
2.4	0.4918	0.4920	0.4922	0.4925	0.4927	0.4929	0.4931	0.4932	0.4934	0.4936
2.5	0.4938	0.4940	0.4941	0.4943	0.4945	0.4946	0.4948	0.4949	0.4951	0.4952
2.6	0.4953	0.4955	0.4956	0.4957	0.4959	0.4960	0.4961	0.4962	0.4963	0.4964
2.7	0.4965	0.4966	0.4967	0.4968	0.4969	0.4970	0.4971	0.4972	0.4973	0.4974
2.8	0.4974	0.4975	0.4976	0.4977	0.4977	0.4978	0.4979	0.4979	0.4980	0.4981
2.9	0.4981	0.4982	0.4982	0.4983	0.4984	0.4984	0.4985	0.4985	0.4986	0.4986
3.0	0.4987	0.4987	0.4987	0.4988	0.4988	0.4989	0.4989	0.4989	0.4990	0.4990
3.1	0.4990	0.4991	0.4991	0.4991	0.4992	0.4992	0.4992	0.4992	0.4993	0.4993
3.2	0.4993	0.4993	0.4994	0.4994	0.4994	0.4994	0.4994	0.4995	0.4995	0.4995
3.3	0.4995	0.4995	0.4995	0.4996	0.4996	0.4996	0.4996	0.4996	0.4996	0.4997
3.4	0.4997	0.4997	0.4997	0.4997	0.4997	0.4997	0.4997	0.4997	0.4997	0.4998

Table 2: Values of t_α in a t distribution with 'df' degrees of freedom [Shaded area: $P(t > t_\alpha) = \alpha$]

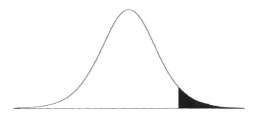

df	$t_{.100}$	$t_{.050}$	$t_{.025}$	$t_{.010}$	$t_{.005}$	df
1	3.078	6.314	12.706	31.821	63.657	1
2	1.886	2.920	4.303	6.965	9.925	2
3	1.638	2.353	3.182	4.541	5.841	3
4	1.533	2.132	2.776	3.747	4.604	4
5	1.476	2.015	2.571	3.365	4.032	5
6	1.440	1.943	2.447	3.143	3.707	6
7	1.415	1.895	2.365	2.998	3.499	7
8	1.397	1.860	2.306	2.896	3.355	8
9	1.383	1.833	2.262	2.821	3.250	9
10	1.372	1.812	2.228	2.764	3.169	10
11	1.363	1.796	2.201	2.718	3.106	11
12	1.356	1.782	2.179	2.681	3.055	12
13	1.350	1.771	2.160	2.650	3.012	13
14	1.345	1.761	2.145	2.624	2.977	14
15	1.341	1.753	2.131	2.602	2.947	15
16	1.337	1.746	2.120	2.583	2.921	16
17	1.333	1.740	2.110	2.567	2.898	17
18	1.330	1.734	2.101	2.552	2.878	18
19	1.328	1.729	2.093	2.539	2.861	19
20	1.325	1.725	2.086	2.528	2.845	20
21	1.323	1.721	2.080	2.518	2.831	21
22	1.321	1.717	2.074	2.508	2.819	22
23	1.319	1.714	2.069	2.500	2.807	23
24	1.318	1.711	2.064	2.492	2.797	24
25	1.316	1.708	2.060	2.485	2.787	25
26	1.315	1.706	2.056	2.479	2.779	26
27	1.314	1.703	2.052	2.473	2.771	27
28	1.313	1.701	2.048	2.467	2.763	28
29	1.311	1.699	2.045	2.462	2.756	29
30	1.310	1.697	2.042	2.457	2.750	30
z	1.282	1.645	1.960	2.326	2.576	z

Table 3: Values of $\chi^2_{\alpha,\mathrm{df}}$ in a chi-square distribution with 'df' degrees of freedom [Shaded area: $P(\chi^2 > \chi^2_{\alpha,\mathrm{df}}) = \alpha$]

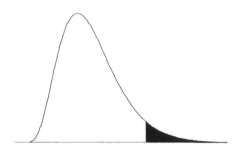

df	$\alpha = .995$	$\alpha = .990$	$\alpha = .975$	$\alpha = .950$	$\alpha = .050$	$\alpha = .025$	$\alpha = .010$	$\alpha = .005$	df
1	0.0000393	0.000157	0.000982	0.00393	3.841	5.024	6.635	7.879	1
2	0.0100	0.0201	0.0506	0.103	5.991	7.378	9.210	10.597	2
3	0.0717	0.115	0.216	0.352	7.815	9.348	11.345	12.838	3
4	0.207	0.297	0.484	0.711	9.488	11.143	13.277	14.860	4
5	0.412	0.554	0.831	1.145	11.070	12.833	15.086	16.750	5
6	0.676	0.872	1.237	1.635	12.592	14.449	16.812	18.548	6
7	0.989	1.239	1.690	2.167	14.067	16.013	18.475	20.278	7
8	1.344	1.646	2.180	2.733	15.507	17.535	20.090	21.955	8
9	1.735	2.088	2.700	3.325	16.919	19.023	21.666	23.589	9
10	2.156	2.558	3.247	3.940	18.307	20.483	23.209	25.188	10
11	2.603	3.053	3.816	4.575	19.675	21.920	24.725	26.757	11
12	3.074	3.571	4.404	5.226	21.026	23.337	26.217	28.300	12
13	3.565	4.107	5.009	5.892	22.362	24.736	27.688	29.819	13
14	4.075	4.660	5.629	6.571	23.685	26.119	29.141	31.319	14
15	4.601	5.229	6.262	7.261	24.996	27.488	30.578	32.801	15
16	5.142	5.812	6.908	7.962	26.296	28.845	32.000	34.267	16
17	5.697	6.408	7.564	8.672	27.587	30.191	33.409	35.718	17
18	6.265	7.015	8.231	9.390	28.869	31.526	34.805	37.156	18
19	6.844	7.633	8.907	10.117	30.144	32.852	36.191	38.582	19
20	7.434	8.260	9.591	10.851	31.410	34.170	37.566	39.997	20
21	8.034	8.897	10.283	11.591	32.671	35.479	38.932	41.401	21
22	8.643	9.542	10.982	12.338	33.924	36.781	40.289	42.796	22
23	9.260	10.196	11.689	13.091	35.172	38.076	41.638	44.181	23
24	9.886	10.856	12.401	13.848	36.415	39.364	42.980	45.559	24
25	10.520	11.524	13.120	14.611	37.652	40.646	44.314	46.928	25
26	11.160	12.198	13.844	15.379	38.885	41.923	45.642	48.290	26
27	11.808	12.879	14.573	16.151	40.113	43.195	46.963	49.645	27
28	12.461	13.565	15.308	16.928	41.337	44.461	48.278	50.993	28
29	13.121	14.256	16.047	17.708	42.557	45.722	49.588	52.336	29
30	13.787	14.953	16.791	18.493	43.773	46.979	50.892	53.672	30

Table 4: Values of f_{α,ν_1,ν_2} in an F distribution [Shaded area: $P(F > f_{\alpha,\nu_1,\nu_2}) = \alpha$]
Numerator degrees of freedom is ν_1 and denominator degrees of freedom is ν_2.

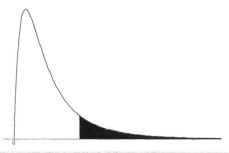

ν_2	α	1	2	3	4	5	6	7	8	9	10	11	12	15	20	25	30	40	1000
1	0.100	39.86	49.50	53.59	55.83	57.24	58.20	58.91	59.44	59.86	60.19	60.47	60.71	61.22	61.74	62.05	62.26	62.53	63.30
	0.050	161.45	199.50	215.71	224.58	230.16	233.99	236.77	238.88	240.54	241.88	242.98	243.91	245.95	248.01	249.26	250.10	251.14	254.19
	0.025	647.79	799.50	864.16	899.58	921.85	937.11	948.22	956.66	963.28	968.63	973.03	976.71	984.87	993.10	998.08	1001.41	1005.60	1017.75
	0.010	4052.18	4999.50	5403.35	5624.58	5763.65	5858.99	5928.36	5981.07	6022.47	6055.85	6083.32	6106.32	6157.28	6208.73	6239.83	6260.65	6286.78	6362.68
2	0.100	8.53	9.00	9.16	9.24	9.29	9.33	9.35	9.37	9.38	9.39	9.40	9.41	9.42	9.44	9.45	9.46	9.47	9.49
	0.050	18.51	19.00	19.16	19.25	19.3	19.33	19.35	19.37	19.38	19.40	19.40	19.41	19.43	19.45	19.46	19.46	19.47	19.49
	0.025	38.51	39.00	39.17	39.25	39.3	39.33	39.36	39.37	39.39	39.40	39.41	39.41	39.43	39.45	39.46	39.46	39.47	39.50
	0.010	98.5	99.00	99.17	99.25	99.3	99.33	99.36	99.37	99.39	99.40	99.41	99.42	99.43	99.45	99.46	99.47	99.47	99.50
3	0.100	5.54	5.46	5.39	5.34	5.31	5.28	5.27	5.25	5.24	5.23	5.22	5.22	5.20	5.18	5.17	5.17	5.16	5.13
	0.050	10.13	9.55	9.28	9.12	9.01	8.94	8.89	8.85	8.81	8.79	8.76	8.74	8.70	8.66	8.63	8.62	8.59	8.53
	0.025	17.44	16.04	15.44	15.1	14.88	14.73	14.62	14.54	14.47	14.42	14.37	14.34	14.25	14.17	14.12	14.08	14.04	13.91
	0.010	34.12	30.82	29.46	28.71	28.24	27.91	27.67	27.49	27.35	27.23	27.13	27.05	26.87	26.69	26.58	26.50	26.41	26.14
4	0.100	4.54	4.32	4.19	4.11	4.05	4.01	3.98	3.95	3.94	3.92	3.91	3.90	3.87	3.84	3.83	3.82	3.80	3.76
	0.050	7.71	6.94	6.59	6.39	6.26	6.16	6.09	6.04	6.00	5.96	5.94	5.91	5.86	5.80	5.77	5.75	5.72	5.63
	0.025	12.22	10.65	9.98	9.6	9.36	9.20	9.07	8.98	8.90	8.84	8.79	8.75	8.66	8.56	8.50	8.46	8.41	8.26
	0.010	21.2	18.00	16.69	15.98	15.52	15.21	14.98	14.80	14.66	14.55	14.45	14.37	14.20	14.02	13.91	13.84	13.75	13.47
5	0.100	4.06	3.78	3.62	3.52	3.45	3.40	3.37	3.34	3.32	3.30	3.28	3.27	3.24	3.21	3.19	3.17	3.16	3.11
	0.050	6.61	5.79	5.41	5.19	5.05	4.95	4.88	4.82	4.77	4.74	4.70	4.68	4.62	4.56	4.52	4.50	4.46	4.37
	0.025	10.01	8.43	7.76	7.39	7.15	6.98	6.85	6.76	6.68	6.62	6.57	6.52	6.43	6.33	6.27	6.23	6.18	6.02
	0.010	16.26	13.27	12.06	11.39	10.97	10.67	10.46	10.29	10.16	10.05	9.96	9.89	9.72	9.55	9.45	9.38	9.29	9.03
6	0.100	3.78	3.46	3.29	3.18	3.11	3.05	3.01	2.98	2.96	2.94	2.92	2.90	2.87	2.84	2.81	2.80	2.78	2.72
	0.050	5.99	5.14	4.76	4.53	4.39	4.28	4.21	4.15	4.10	4.06	4.03	4.00	3.94	3.87	3.83	3.81	3.77	3.67
	0.025	8.81	7.26	6.6	6.23	5.99	5.82	5.70	5.60	5.52	5.46	5.41	5.37	5.27	5.17	5.11	5.07	5.01	4.86
	0.010	13.75	10.92	9.78	9.15	8.75	8.47	8.26	8.10	7.98	7.87	7.79	7.72	7.56	7.40	7.30	7.23	7.14	6.89
7	0.100	3.59	3.26	3.07	2.96	2.88	2.83	2.78	2.75	2.72	2.70	2.68	2.67	2.63	2.59	2.57	2.56	2.54	2.47
	0.050	5.59	4.74	4.35	4.12	3.97	3.87	3.79	3.73	3.68	3.64	3.60	3.57	3.51	3.44	3.40	3.38	3.34	3.23
	0.025	8.07	6.54	5.89	5.52	5.29	5.12	4.99	4.90	4.82	4.76	4.71	4.67	4.57	4.47	4.40	4.36	4.31	4.15
	0.010	12.25	9.55	8.45	7.85	7.46	7.19	6.99	6.84	6.72	6.62	6.54	6.47	6.31	6.16	6.06	5.99	5.91	5.66
8	0.100	3.46	3.11	2.92	2.81	2.73	2.67	2.62	2.59	2.56	2.54	2.52	2.50	2.46	2.42	2.40	2.38	2.36	2.30
	0.050	5.32	4.46	4.07	3.84	3.69	3.58	3.50	3.44	3.39	3.35	3.31	3.28	3.22	3.15	3.11	3.08	3.04	2.93
	0.025	7.57	6.06	5.42	5.05	4.82	4.65	4.53	4.43	4.36	4.30	4.24	4.20	4.10	4.00	3.94	3.89	3.84	3.68
	0.010	11.26	8.65	7.59	7.01	6.63	6.37	6.18	6.03	5.91	5.81	5.73	5.67	5.52	5.36	5.26	5.20	5.12	4.87
9	0.100	3.36	3.01	2.81	2.69	2.61	2.55	2.51	2.47	2.44	2.42	2.40	2.38	2.34	2.30	2.27	2.25	2.23	2.16
	0.050	5.12	4.26	3.86	3.63	3.48	3.37	3.29	3.23	3.18	3.14	3.10	3.07	3.01	2.94	2.89	2.86	2.83	2.71
	0.025	7.21	5.71	5.08	4.72	4.48	4.32	4.20	4.10	4.03	3.96	3.91	3.87	3.77	3.67	3.60	3.56	3.51	3.34
	0.010	10.56	8.02	6.99	6.42	6.06	5.80	5.61	5.47	5.35	5.26	5.18	5.11	4.96	4.81	4.71	4.65	4.57	4.32
10	0.100	3.29	2.92	2.73	2.61	2.52	2.46	2.41	2.38	2.35	2.32	2.30	2.28	2.24	2.20	2.17	2.16	2.13	2.06
	0.050	4.96	4.10	3.71	3.48	3.33	3.22	3.14	3.07	3.02	2.98	2.94	2.91	2.85	2.77	2.73	2.70	2.66	2.54
	0.025	6.94	5.46	4.83	4.47	4.24	4.07	3.95	3.85	3.78	3.72	3.66	3.62	3.52	3.42	3.35	3.31	3.26	3.09
	0.010	10.04	7.56	6.55	5.99	5.64	5.39	5.20	5.06	4.94	4.85	4.77	4.71	4.56	4.41	4.31	4.25	4.17	3.92
11	0.100	3.23	2.86	2.66	2.54	2.45	2.39	2.34	2.30	2.27	2.25	2.23	2.21	2.17	2.12	2.10	2.08	2.05	1.98
	0.050	4.84	3.98	3.59	3.36	3.2	3.09	3.01	2.95	2.90	2.85	2.82	2.79	2.72	2.65	2.60	2.57	2.53	2.41
	0.025	6.72	5.26	4.63	4.28	4.04	3.88	3.76	3.66	3.59	3.53	3.47	3.43	3.33	3.23	3.16	3.12	3.06	2.89
	0.010	9.65	7.21	6.22	5.67	5.32	5.07	4.89	4.74	4.63	4.54	4.46	4.40	4.25	4.10	4.01	3.94	3.86	3.61
12	0.100	3.18	2.81	2.61	2.48	2.39	2.33	2.28	2.24	2.21	2.19	2.17	2.15	2.10	2.06	2.03	2.01	1.99	1.91
	0.050	4.75	3.89	3.49	3.26	3.11	3.00	2.91	2.85	2.80	2.75	2.72	2.69	2.62	2.54	2.50	2.47	2.43	2.30
	0.025	6.55	5.10	4.47	4.12	3.89	3.73	3.61	3.51	3.44	3.37	3.32	3.28	3.18	3.07	3.01	2.96	2.91	2.73
	0.010	9.33	6.93	5.95	5.41	5.06	4.82	4.64	4.50	4.39	4.30	4.22	4.16	4.01	3.86	3.76	3.70	3.62	3.37

Table 4: Values of f_{α, ν_1, ν_2} in an F distribution (continued)

ν_2	α	1	2	3	4	5	6	7	8	9	10	11	12	15	20	25	30	40	1000
13	0.100	3.14	2.76	2.56	2.43	2.35	2.28	2.23	2.20	2.16	2.14	2.12	2.10	2.05	2.01	1.98	1.96	1.93	1.85
	0.050	4.67	3.81	3.41	3.18	3.03	2.92	2.83	2.77	2.71	2.67	2.63	2.60	2.53	2.46	2.41	2.38	2.34	2.21
	0.025	6.41	4.97	4.35	4.00	3.77	3.60	3.48	3.39	3.31	3.25	3.20	3.15	3.05	2.95	2.88	2.84	2.78	2.60
	0.010	9.07	6.70	5.74	5.21	4.86	4.62	4.44	4.30	4.19	4.10	4.02	3.96	3.82	3.66	3.57	3.51	3.43	3.18
14	0.100	3.10	2.73	2.52	2.39	2.31	2.24	2.19	2.15	2.12	2.10	2.07	2.05	2.01	1.96	1.93	1.91	1.89	1.80
	0.050	4.60	3.74	3.34	3.11	2.96	2.85	2.76	2.70	2.65	2.60	2.57	2.53	2.46	2.39	2.34	2.31	2.27	2.14
	0.025	6.30	4.86	4.24	3.89	3.66	3.50	3.38	3.29	3.21	3.15	3.09	3.05	2.95	2.84	2.78	2.73	2.67	2.50
	0.010	8.86	6.51	5.56	5.04	4.69	4.46	4.28	4.14	4.03	3.94	3.86	3.80	3.66	3.51	3.41	3.35	3.27	3.02
16	0.100	3.05	2.67	2.46	2.33	2.24	2.18	2.13	2.09	2.06	2.03	2.01	1.99	1.94	1.89	1.86	1.84	1.81	1.72
	0.050	4.49	3.63	3.24	3.01	2.85	2.74	2.66	2.59	2.54	2.49	2.46	2.42	2.35	2.28	2.23	2.19	2.15	2.02
	0.025	6.12	4.69	4.08	3.73	3.50	3.34	3.22	3.12	3.05	2.99	2.93	2.89	2.79	2.68	2.61	2.57	2.51	2.32
	0.010	8.53	6.23	5.29	4.77	4.44	4.20	4.03	3.89	3.78	3.69	3.62	3.55	3.41	3.26	3.16	3.10	3.02	2.76
18	0.100	3.01	2.62	2.42	2.29	2.20	2.13	2.08	2.04	2.00	1.98	1.95	1.93	1.89	1.84	1.80	1.78	1.75	1.66
	0.050	4.41	3.55	3.16	2.93	2.77	2.66	2.58	2.51	2.46	2.41	2.37	2.34	2.27	2.19	2.14	2.11	2.06	1.92
	0.025	5.98	4.56	3.95	3.61	3.38	3.22	3.10	3.01	2.93	2.87	2.81	2.77	2.67	2.56	2.49	2.44	2.38	2.20
	0.010	8.29	6.01	5.09	4.58	4.25	4.01	3.84	3.71	3.60	3.51	3.43	3.37	3.23	3.08	2.98	2.92	2.84	2.58
20	0.100	2.97	2.59	2.38	2.25	2.16	2.09	2.04	2.00	1.96	1.94	1.91	1.89	1.84	1.79	1.76	1.74	1.71	1.61
	0.050	4.35	3.49	3.10	2.87	2.71	2.60	2.51	2.45	2.39	2.35	2.31	2.28	2.20	2.12	2.07	2.04	1.99	1.85
	0.025	5.87	4.46	3.86	3.51	3.29	3.13	3.01	2.91	2.84	2.77	2.72	2.68	2.57	2.46	2.40	2.35	2.29	2.09
	0.010	8.10	5.85	4.94	4.43	4.10	3.87	3.70	3.56	3.46	3.37	3.29	3.23	3.09	2.94	2.84	2.78	2.69	2.43
22	0.100	2.95	2.56	2.35	2.22	2.13	2.06	2.01	1.97	1.93	1.90	1.88	1.86	1.81	1.76	1.73	1.70	1.67	1.57
	0.050	4.30	3.44	3.05	2.82	2.66	2.55	2.46	2.40	2.34	2.30	2.26	2.23	2.15	2.07	2.02	1.98	1.94	1.79
	0.025	5.79	4.38	3.78	3.44	3.22	3.05	2.93	2.84	2.76	2.70	2.65	2.60	2.50	2.39	2.32	2.27	2.21	2.01
	0.010	7.95	5.72	4.82	4.31	3.99	3.76	3.59	3.45	3.35	3.26	3.18	3.12	2.98	2.83	2.73	2.67	2.58	2.32
24	0.100	2.93	2.54	2.33	2.19	2.10	2.04	1.98	1.94	1.91	1.88	1.85	1.83	1.78	1.73	1.70	1.67	1.64	1.54
	0.050	4.26	3.40	3.01	2.78	2.62	2.51	2.42	2.36	2.30	2.25	2.22	2.18	2.11	2.03	1.97	1.94	1.89	1.74
	0.025	5.72	4.32	3.72	3.38	3.15	2.99	2.87	2.78	2.70	2.64	2.59	2.54	2.44	2.33	2.26	2.21	2.15	1.94
	0.010	7.82	5.61	4.72	4.22	3.90	3.67	3.50	3.36	3.26	3.17	3.09	3.03	2.89	2.74	2.64	2.58	2.49	2.22
26	0.100	2.91	2.52	2.31	2.17	2.08	2.01	1.96	1.92	1.88	1.86	1.83	1.81	1.76	1.71	1.67	1.65	1.61	1.51
	0.050	4.23	3.37	2.98	2.74	2.59	2.47	2.39	2.32	2.27	2.22	2.18	2.15	2.07	1.99	1.94	1.90	1.85	1.70
	0.025	5.66	4.27	3.67	3.33	3.10	2.94	2.82	2.73	2.65	2.59	2.54	2.49	2.39	2.28	2.21	2.16	2.09	1.89
	0.010	7.72	5.53	4.64	4.14	3.82	3.59	3.42	3.29	3.18	3.09	3.02	2.96	2.81	2.66	2.57	2.50	2.42	2.14
28	0.100	2.89	2.50	2.29	2.16	2.06	2.00	1.94	1.90	1.87	1.84	1.81	1.79	1.74	1.69	1.65	1.63	1.59	1.48
	0.050	4.20	3.34	2.95	2.71	2.56	2.45	2.36	2.29	2.24	2.19	2.15	2.12	2.04	1.96	1.91	1.87	1.82	1.66
	0.025	5.61	4.22	3.63	3.29	3.06	2.90	2.78	2.69	2.61	2.55	2.49	2.45	2.34	2.23	2.16	2.11	2.05	1.84
	0.010	7.64	5.45	4.57	4.07	3.75	3.53	3.36	3.23	3.12	3.03	2.96	2.90	2.75	2.60	2.51	2.44	2.35	2.08
30	0.100	2.88	2.49	2.28	2.14	2.05	1.98	1.93	1.88	1.85	1.82	1.79	1.77	1.72	1.67	1.63	1.61	1.57	1.46
	0.050	4.17	3.32	2.92	2.69	2.53	2.42	2.33	2.27	2.21	2.16	2.13	2.09	2.01	1.93	1.88	1.84	1.79	1.63
	0.025	5.57	4.18	3.59	3.25	3.03	2.87	2.75	2.65	2.57	2.51	2.46	2.41	2.31	2.20	2.12	2.07	2.01	1.80
	0.010	7.56	5.39	4.51	4.02	3.70	3.47	3.30	3.17	3.07	2.98	2.91	2.84	2.70	2.55	2.45	2.39	2.30	2.02
40	0.100	2.84	2.44	2.23	2.09	2.00	1.93	1.87	1.83	1.79	1.76	1.74	1.71	1.66	1.61	1.57	1.54	1.51	1.38
	0.050	4.08	3.23	2.84	2.61	2.45	2.34	2.25	2.18	2.12	2.08	2.04	2.00	1.92	1.84	1.78	1.74	1.69	1.52
	0.025	5.42	4.05	3.46	3.13	2.90	2.74	2.62	2.53	2.45	2.39	2.33	2.29	2.18	2.07	1.99	1.94	1.88	1.65
	0.010	7.31	5.18	4.31	3.83	3.51	3.29	3.12	2.99	2.89	2.80	2.73	2.66	2.52	2.37	2.27	2.20	2.11	1.82
1000	0.100	2.71	2.31	2.09	1.95	1.85	1.78	1.72	1.68	1.64	1.61	1.58	1.55	1.49	1.43	1.38	1.35	1.30	1.08
	0.050	3.85	3.00	2.61	2.38	2.22	2.11	2.02	1.95	1.89	1.84	1.80	1.76	1.68	1.58	1.52	1.47	1.41	1.11
	0.025	5.04	3.70	3.13	2.80	2.58	2.42	2.30	2.20	2.13	2.06	2.01	1.96	1.85	1.72	1.64	1.58	1.50	1.13
	0.010	6.66	4.63	3.80	3.34	3.04	2.82	2.66	2.53	2.43	2.34	2.27	2.20	2.06	1.90	1.79	1.72	1.61	1.16

Table 5: Durbin-Watson test bounds, where $p - 1$ = number of predictors

	Level of Significance $\alpha = 0.05$									
	$p-1=1$		$p-1=2$		$p-1=3$		$p-1=4$		$p-1=5$	
n	d_L	d_U	d_L	d_U	d_L	d_U	d_L	d_U	d_L	d_U
15	1.08	1.36	0.95	1.54	0.82	1.75	0.69	1.97	0.56	2.21
16	1.10	1.37	0.98	1.54	0.86	1.73	0.74	1.93	0.62	2.15
17	1.13	1.38	1.02	1.54	0.90	1.71	0.78	1.90	0.67	2.10
18	1.16	1.39	1.05	1.53	0.93	1.69	0.82	1.87	0.71	2.06
19	1.18	1.40	1.08	1.53	0.97	1.68	0.86	1.85	0.75	2.02
20	1.20	1.41	1.10	1.54	1.00	1.68	0.90	1.83	0.79	1.99
21	1.22	1.42	1.13	1.54	1.03	1.67	0.93	1.81	0.83	1.96
22	1.24	1.43	1.15	1.54	1.05	1.66	0.96	1.80	0.86	1.94
23	1.26	1.44	1.17	1.54	1.08	1.66	0.99	1.79	0.90	1.92
24	1.27	1.45	1.19	1.55	1.10	1.66	1.01	1.78	0.93	1.90
25	1.29	1.45	1.21	1.55	1.12	1.66	1.04	1.77	0.95	1.89
26	1.30	1.46	1.22	1.55	1.14	1.65	1.06	1.76	0.98	1.88
27	1.32	1.47	1.24	1.56	1.16	1.65	1.08	1.76	1.01	1.86
28	1.33	1.48	1.26	1.56	1.18	1.65	1.10	1.75	1.03	1.85
29	1.34	1.48	1.27	1.56	1.20	1.65	1.12	1.74	1.05	1.84
30	1.35	1.49	1.28	1.57	1.21	1.65	1.14	1.74	1.07	1.83
31	1.36	1.50	1.30	1.57	1.23	1.65	1.16	1.74	1.09	1.83
32	1.37	1.50	1.31	1.57	1.24	1.65	1.18	1.73	1.11	1.82
33	1.38	1.51	1.32	1.58	1.26	1.65	1.19	1.73	1.13	1.81
34	1.39	1.51	1.33	1.58	1.27	1.65	1.21	1.73	1.15	1.81
35	1.40	1.52	1.34	1.58	1.28	1.65	1.22	1.73	1.16	1.80
36	1.41	1.52	1.35	1.59	1.29	1.65	1.24	1.73	1.18	1.80
37	1.42	1.53	1.36	1.59	1.31	1.66	1.25	1.72	1.19	1.80
38	1.43	1.54	1.37	1.59	1.32	1.66	1.26	1.72	1.21	1.79
39	1.43	1.54	1.38	1.60	1.33	1.66	1.27	1.72	1.22	1.79
40	1.44	1.54	1.39	1.60	1.34	1.66	1.29	1.72	1.23	1.79
45	1.48	1.57	1.43	1.62	1.38	1.67	1.34	1.72	1.29	1.78
50	1.50	1.59	1.46	1.63	1.42	1.67	1.38	1.72	1.34	1.77
55	1.53	1.60	1.49	1.64	1.45	1.68	1.41	1.72	1.38	1.77
60	1.55	1.62	1.51	1.65	1.48	1.69	1.44	1.73	1.41	1.77
65	1.57	1.63	1.54	1.66	1.50	1.70	1.47	1.73	1.44	1.77
70	1.58	1.64	1.55	1.67	1.52	1.70	1.49	1.74	1.46	1.77
75	1.60	1.65	1.57	1.68	1.54	1.71	1.51	1.74	1.49	1.77
80	1.61	1.66	1.59	1.69	1.56	1.72	1.53	1.74	1.51	1.77
85	1.62	1.67	1.60	1.70	1.57	1.72	1.55	1.75	1.52	1.77
90	1.63	1.68	1.61	1.70	1.59	1.73	1.57	1.75	1.54	1.78
95	1.64	1.69	1.62	1.71	1.60	1.73	1.58	1.75	1.56	1.78
100	1.65	1.69	1.63	1.72	1.61	1.74	1.59	1.76	1.57	1.78

Reprinted, with permission, from J. Durbin and G.S. Watson, and Oxford University Press "Testing for serial correlation in least squares regression. II", Biometrika, 38 (1951), pp. 159-178.

Table 5: Durbin-Watson test bounds (continued)

	Level of Significance $\alpha = 0.01$									
	$p-1=1$		$p-1=2$		$p-1=3$		$p-1=4$		$p-1=5$	
n	d_L	d_U	d_L	d_U	d_L	d_U	d_L	d_U	d_L	d_U
15	0.81	1.07	0.70	1.25	0.59	1.46	0.49	1.70	0.39	1.96
16	0.84	1.09	0.74	1.25	0.63	1.44	0.53	1.66	0.44	1.90
17	0.87	1.10	0.77	1.25	0.67	1.43	0.57	1.63	0.48	1.85
18	0.90	1.12	0.80	1.26	0.71	1.42	0.61	1.60	0.52	1.80
19	0.93	1.13	0.83	1.26	0.74	1.41	0.65	1.58	0.56	1.77
20	0.95	1.15	0.86	1.27	0.77	1.41	0.68	1.57	0.60	1.74
21	0.97	1.16	0.89	1.27	0.80	1.41	0.72	1.55	0.63	1.71
22	1.00	1.17	0.91	1.28	0.83	1.40	0.75	1.54	0.66	1.69
23	1.02	1.19	0.94	1.29	0.86	1.40	0.77	1.53	0.70	1.67
24	1.04	1.20	0.96	1.30	0.88	1.41	0.80	1.53	0.72	1.66
25	1.05	1.21	0.98	1.30	0.90	1.41	0.83	1.52	0.75	1.65
26	1.07	1.22	1.00	1.31	0.93	1.41	0.85	1.52	0.78	1.64
27	1.09	1.23	1.02	1.32	0.95	1.41	0.88	1.51	0.81	1.63
28	1.10	1.24	1.04	1.32	0.97	1.41	0.90	1.51	0.83	1.62
29	1.12	1.25	1.05	1.33	0.99	1.42	0.92	1.51	0.85	1.61
30	1.13	1.26	1.07	1.34	1.01	1.42	0.94	1.51	0.88	1.61
31	1.15	1.27	1.08	1.34	1.02	1.42	0.96	1.51	0.90	1.60
32	1.16	1.28	1.10	1.35	1.04	1.43	0.98	1.51	0.92	1.60
33	1.17	1.29	1.11	1.36	1.05	1.43	1.00	1.51	0.94	1.59
34	1.18	1.30	1.13	1.36	1.07	1.43	1.01	1.51	0.95	1.59
35	1.19	1.31	1.14	1.37	1.08	1.44	1.03	1.51	0.97	1.59
36	1.21	1.32	1.15	1.38	1.10	1.44	1.04	1.51	0.99	1.59
37	1.22	1.32	1.16	1.38	1.11	1.45	1.06	1.51	1.00	1.59
38	1.23	1.33	1.18	1.39	1.12	1.45	1.07	1.52	1.02	1.58
39	1.24	1.34	1.19	1.39	1.14	1.45	1.09	1.52	1.03	1.58
40	1.25	1.34	1.20	1.40	1.15	1.46	1.10	1.52	1.05	1.58
45	1.29	1.38	1.24	1.42	1.20	1.48	1.16	1.53	1.11	1.58
50	1.32	1.40	1.28	1.45	1.24	1.49	1.20	1.54	1.16	1.59
55	1.36	1.43	1.32	1.47	1.28	1.51	1.25	1.55	1.21	1.59
60	1.38	1.45	1.35	1.48	1.32	1.52	1.28	1.56	1.25	1.60
65	1.41	1.47	1.38	1.50	1.35	1.53	1.31	1.57	1.28	1.61
70	1.43	1.49	1.40	1.52	1.37	1.55	1.34	1.58	1.31	1.61
75	1.45	1.50	1.42	1.53	1.39	1.56	1.37	1.59	1.34	1.62
80	1.47	1.52	1.44	1.54	1.42	1.57	1.39	1.60	1.36	1.62
85	1.48	1.53	1.46	1.55	1.43	1.58	1.41	1.60	1.39	1.63
90	1.50	1.54	1.47	1.56	1.45	1.59	1.43	1.61	1.41	1.64
95	1.51	1.55	1.49	1.57	1.47	1.60	1.45	1.62	1.42	1.64
100	1.52	1.56	1.50	1.58	1.48	1.60	1.46	1.63	1.44	1.65

Table 6: Orthogonal polynomial coefficients

						X						
k	POLYNOMIAL	1	2	3	4	5	6	7	8	9	10	(Σp_j^2)
3	Linear	−1	0	1								2
	Quadratic	1	−2	1								6
4	Linear	−3	−1	1	3							20
	Quadratic	1	−1	−1	1							4
	Cubic	−1	3	−3	1							20
5	Linear	−2	−1	0	1	2						10
	Quadratic	2	−1	−2	−1	2						14
	Cubic	−1	2	0	−2	1						10
	Quartic	1	−4	6	−4	1						70
6	Linear	−5	−3	−1	1	3	5					70
	Quadratic	5	−1	−4	−4	−1	5					84
	Cubic	−5	7	4	−4	−7	5					180
	Quartic	1	−3	2	2	−3	1					28
	Quintic	−1	5	−10	10	−5	1					252
7	Linear	−3	−2	−1	0	1	2	3				28
	Quadratic	5	0	−3	−4	−3	0	5				84
	Cubic	−1	1	1	0	−1	−1	1				6
	Quartic	3	−7	1	6	1	−7	3				154
	Quintic	−1	4	−5	0	5	−4	1				84
	Sextic	1	−6	15	−20	15	−6	1				924
8	Linear	−7	−5	−3	−1	1	3	5	7			168
	Quadratic	7	1	−3	−5	−5	−3	1	7			168
	Cubic	−7	5	7	3	−3	−7	−5	7			264
	Quartic	7	−13	−3	9	9	−3	−13	7			616
	Quintic	−7	23	−17	−15	15	17	−23	7			2184
	Sextic	1	−5	9	−5	−5	9	−5	1			264
	Septic	−1	7	−21	35	−35	21	−7	1			3,432
9	Linear	−4	−3	−2	−1	0	1	2	3	4		60
	Quadratic	28	7	−8	−17	−20	−17	−8	7	28		2,772
	Cubic	−14	7	13	9	0	−9	−13	−7	14		990
	Quartic	14	−21	−11	9	18	9	−11	−21	14		2,002
	Quintic	−4	11	−4	−9	0	9	4	−11	4		468
	Sextic	4	−17	22	1	−20	1	22	−17	4		1,980
	Septic	−1	6	−14	14	0	−14	14	−6	1		858
	Octic	1	−8	28	−56	70	−56	28	−8	1		12,870
10	Linear	−9	−7	−5	−3	−1	1	3	5	7	9	330
	Quadratic	6	2	−1	−3	−4	−4	−3	−1	2	6	132
	Cubic	−42	14	35	31	12	−12	−31	−35	−14	42	8,580
	Quartic	18	−22	−17	3	18	18	3	−17	−22	18	2,860
	Quintic	−6	14	−1	−11	−6	6	11	1	−14	6	780
	Sextic	3	−11	10	6	−8	−8	6	10	11	3	660
	Septic	−9	47	−86	92	56	−56	−42	86	−47	9	29,172
	Octic	1	−7	20	−28	14	14	−28	20	−7	1	2,860
	Novic	−1	9	−36	84	−126	126	−84	36	−9	1	48,620

Table 7: Upper α point of Studentized range, $q_\alpha(k, \nu)$,
where $k = r =$ number of treatments to be compared and $\nu =$ the number of degrees of freedom

$$1 - \alpha = .95$$

ν	2	3	4	5	6	7	8	9	10	11	12	13	14	15	16	17	18	19	20
1	18.0	27.0	32.8	37.1	40.4	43.1	45.4	47.4	49.1	50.6	52.0	53.2	54.3	55.4	56.3	57.2	58.0	58.8	59.6
2	6.08	8.33	9.80	10.9	11.7	12.4	13.0	13.5	14.0	14.4	14.7	15.1	15.4	15.7	15.9	16.1	16.4	16.6	16.8
3	4.50	5.91	6.82	7.50	8.04	8.48	8.85	9.18	9.46	9.72	9.95	10.2	10.3	10.5	10.7	10.8	11.0	11.1	11.2
4	3.93	5.04	5.76	6.29	6.71	7.05	7.35	7.60	7.83	8.03	8.21	8.37	8.52	8.66	8.79	8.91	9.03	9.13	9.23
5	3.64	4.60	5.22	5.67	6.03	6.33	6.58	6.80	6.99	7.17	7.32	7.47	7.60	7.72	7.83	7.93	8.03	8.12	8.21
6	3.46	4.34	4.90	5.30	5.63	5.90	6.12	6.32	6.49	6.65	6.79	6.92	7.03	7.14	7.24	7.34	7.43	7.51	7.59
7	3.34	4.16	4.68	5.06	5.36	5.61	5.82	6.00	6.16	6.30	6.43	6.55	6.66	6.76	6.85	6.94	7.02	7.10	7.17
8	3.26	4.04	4.53	4.89	5.17	5.40	5.60	5.77	5.92	6.05	6.18	6.29	6.39	6.48	6.57	6.65	6.73	6.80	6.87
9	3.20	3.95	4.41	4.76	5.02	5.24	5.43	5.59	5.74	5.87	5.98	6.09	6.19	6.28	6.36	6.44	6.51	6.58	6.64
10	3.15	3.88	4.33	4.65	4.91	5.12	5.30	5.46	5.60	5.72	5.83	5.93	6.03	6.11	6.19	6.27	6.34	6.40	6.47
11	3.11	3.82	4.26	4.57	4.82	5.03	5.20	5.35	5.49	5.61	5.71	5.81	5.90	5.98	6.06	6.13	6.20	6.27	6.33
12	3.08	3.77	4.20	4.51	4.75	4.95	5.12	5.27	5.39	5.51	5.61	5.71	5.80	5.88	5.95	6.02	6.09	6.15	6.21
13	3.06	3.73	4.15	4.45	4.69	4.88	5.05	5.19	5.32	5.43	5.53	5.63	5.71	5.79	5.86	5.93	5.99	6.05	6.11
14	3.03	3.70	4.11	4.41	4.64	4.83	4.99	5.13	5.25	5.36	5.46	5.55	5.64	5.71	5.79	5.85	5.91	5.97	6.03
15	3.01	3.67	4.08	4.37	4.59	4.78	4.94	5.08	5.20	5.31	5.40	5.49	5.57	5.65	5.72	5.78	5.85	5.90	5.96
16	3.00	3.65	4.05	4.33	4.56	4.74	4.90	5.03	5.15	5.26	5.35	5.44	5.52	5.59	5.66	5.73	5.79	5.84	5.90
17	2.98	3.63	4.02	4.30	4.52	4.70	4.86	4.99	5.11	5.21	5.31	5.39	5.47	5.54	5.61	5.67	5.73	5.79	5.84
18	2.97	3.61	4.00	4.28	4.49	4.67	4.82	4.96	5.07	5.17	5.27	5.35	5.43	5.50	5.57	5.63	5.69	5.74	5.79
19	2.96	3.59	3.98	4.25	4.47	4.65	4.79	4.92	5.04	5.14	5.23	5.31	5.39	5.46	5.53	5.59	5.65	5.70	5.75
20	2.95	3.58	3.96	4.23	4.45	4.62	4.77	4.90	5.01	5.11	5.20	5.28	5.36	5.43	5.49	5.55	5.61	5.66	5.71
24	2.92	3.53	3.90	4.17	4.37	4.54	4.68	4.81	4.92	5.01	5.10	5.18	5.25	5.32	5.38	5.44	5.49	5.55	5.59
30	2.89	3.49	3.85	4.10	4.30	4.46	4.60	4.72	4.82	4.92	5.00	5.08	5.15	5.21	5.27	5.33	5.38	5.43	5.47
40	2.86	3.44	3.79	4.04	4.23	4.39	4.52	4.63	4.73	4.82	4.90	4.98	5.04	5.11	5.16	5.22	5.27	5.31	5.36
60	2.83	3.40	3.74	3.98	4.16	4.31	4.44	4.55	4.65	4.73	4.81	4.88	4.94	5.00	5.06	5.11	5.15	5.20	5.24
120	2.80	3.36	3.68	3.92	4.10	4.24	4.36	4.47	4.56	4.64	4.71	4.78	4.84	4.90	4.95	5.00	5.04	5.09	5.13
∞	2.77	3.31	3.63	3.86	4.03	4.17	4.29	4.39	4.47	4.55	4.62	4.68	4.74	4.80	4.85	4.89	4.93	4.97	5.01

The Analysis of Variance by Scheffé, H. ©1959.
Reproduced with permission of John Wiley & Sons, Inc.

Table 7: Upper α point of Studentized range (continued)

$1 - \alpha = .99$

r

ν	2	3	4	5	6	7	8	9	10	11	12	13	14	15	16	17	18	19	20
1	90.0	135	164	186	202	216	227	237	246	253	260	266	272	277	282	286	290	294	298
2	14.0	19.0	22.3	24.7	26.6	28.2	29.5	30.7	31.7	32.6	33.4	34.1	34.8	35.4	36.0	36.5	37.0	37.5	37.9
3	8.26	10.6	12.2	13.3	14.2	15.0	15.6	16.2	16.7	17.1	17.5	17.9	18.2	18.5	18.8	19.1	19.3	19.5	19.8
4	6.51	8.12	9.17	9.96	10.6	11.1	11.5	11.9	12.3	12.6	12.8	13.1	13.3	13.5	13.7	13.9	14.1	14.2	14.4
5	5.70	6.97	7.80	8.42	8.91	9.32	9.67	9.97	10.2	10.5	10.7	10.9	11.1	11.2	11.4	11.6	11.7	11.8	11.9
6	5.24	6.33	7.03	7.56	7.97	8.32	8.61	8.87	9.10	9.30	9.49	9.65	9.81	9.95	10.1	10.2	10.3	10.4	10.5
7	4.95	5.92	6.54	7.01	7.37	7.68	7.94	8.17	8.37	8.55	8.71	8.86	9.00	9.12	9.24	9.35	9.46	9.55	9.65
8	4.74	5.63	6.20	6.63	6.96	7.24	7.47	7.68	7.87	8.03	8.18	8.31	8.44	8.55	8.66	8.76	8.85	8.94	9.03
9	4.60	5.43	5.96	6.35	6.66	6.91	7.13	7.32	7.49	7.65	7.78	7.91	8.03	8.13	8.23	8.32	8.41	8.49	8.57
10	4.48	5.27	5.77	6.14	6.43	6.67	6.87	7.05	7.21	7.36	7.48	7.60	7.71	7.81	7.91	7.99	8.07	8.15	8.22
11	4.39	5.14	5.62	5.97	6.25	6.48	6.67	6.84	6.99	7.13	7.25	7.36	7.46	7.56	7.65	7.73	7.81	7.88	7.95
12	4.32	5.04	5.50	5.84	6.10	6.32	6.51	6.67	6.81	6.94	7.06	7.17	7.26	7.36	7.44	7.52	7.59	7.66	7.73
13	4.26	4.96	5.40	5.73	5.98	6.19	6.37	6.53	6.67	6.79	6.90	7.01	7.10	7.19	7.27	7.34	7.42	7.48	7.55
14	4.21	4.89	5.32	5.63	5.88	6.08	6.26	6.41	6.54	6.66	6.77	6.87	6.96	7.05	7.12	7.20	7.27	7.33	7.39
15	4.17	4.83	5.25	5.56	5.80	5.99	6.16	6.31	6.44	6.55	6.66	6.76	6.84	6.93	7.00	7.07	7.14	7.20	7.26
16	4.13	4.78	5.19	5.49	5.72	5.92	6.08	6.22	6.35	6.46	6.56	6.66	6.74	6.82	6.90	6.97	7.03	7.09	7.15
17	4.10	4.74	5.14	5.43	5.66	5.85	6.01	6.15	6.27	6.38	6.48	6.57	6.66	6.73	6.80	6.87	6.94	7.00	7.05
18	4.07	4.70	5.09	5.38	5.60	5.79	5.94	6.08	6.20	6.31	6.41	6.50	6.58	6.65	6.72	6.79	6.85	6.91	6.96
19	4.05	4.67	5.05	5.33	5.55	5.73	5.89	6.02	6.14	6.25	6.34	6.43	6.51	6.58	6.65	6.72	6.78	6.84	6.89
20	4.02	4.64	5.02	5.29	5.51	5.69	5.84	5.97	6.09	6.19	6.29	6.37	6.45	6.52	6.59	6.65	6.71	6.76	6.82
24	3.96	4.54	4.91	5.17	5.37	5.54	5.69	5.81	5.92	6.02	6.11	6.19	6.26	6.33	6.39	6.45	6.51	6.56	6.61
30	3.89	4.45	4.80	5.05	5.24	5.40	5.54	5.65	5.76	5.85	5.93	6.01	6.08	6.14	6.20	6.26	6.31	6.36	6.41
40	3.82	4.37	4.70	4.93	5.11	5.27	5.39	5.50	5.60	5.69	5.77	5.84	5.90	5.96	6.02	6.07	6.12	6.17	6.21
60	3.76	4.28	4.60	4.82	4.99	5.13	5.25	5.36	5.45	5.53	5.60	5.67	5.73	5.79	5.84	5.89	5.93	5.98	6.02
120	3.70	4.20	4.50	4.71	4.87	5.01	5.12	5.21	5.30	5.38	5.44	5.51	5.56	5.61	5.66	5.71	5.75	5.79	5.83
∞	3.64	4.12	4.40	4.60	4.76	4.88	4.99	5.08	5.16	5.23	5.29	5.35	5.40	5.45	5.49	5.54	5.57	5.61	5.65

Bibliography

Agresti, A. (1990). *Categorical Data Analysis*. John Wiley & Sons, Inc., New York, NY.

Basu, B. and Famoye, F. (2004). Domestic violence against women and their economic dependence: A count data analysis. *Review of Political Economy*, 16(4):457–472.

Bates, D. and Watts, D. (1981). A relative offset orthogonality convergence criterion for nonlinear least squares. *Technometrics*, 123:179–183.

Bates, D. and Watts, D. (1988). *Nonlinear Regression Analysis and Its Applications*. John Wiley & Sons, Inc., New York, NY.

Binnie, N. (2004). Using eda, anova and regression to optimize some microbiology data. *Journal of Statistics Education*, 12(2).

Bliss, C. (1935). The calculation of the dosage-mortality curve. *Annals of Applied Biology*, 22:134–167.

Böhning, D. (1994). A note on a test for Poisson overdispersion. *Biometrika*, 81:418–419.

Böhning, D. (1998). Zero-inflated Poisson models and C.A.MAN: A tutorial collection of evidence. *Biometrical Journal*, 40:833–843.

Box, G. and Cox, D. (1964). An analysis of transformation. *Journal of the Royal Statistical Society B*, 26:211–243.

Breslow, N. and Day, N. (1980). *Statistical Methods in Cancer Research. Vol. 1: The Analysis of Case-control Studies*. International Agency on Cancer, Lyon, France.

Breusch, T. and Pagan, A. (1979). A simple test for heteroscedasticity and random coefficient variation. *Econometrica*, 47:1287–1294.

Cameron, A. and Trivedi, P. (1998). *Regression Analysis of Count Data*. Cambridge University Press, Cambridge, UK.

Cameron, A., Trivedi, P., Milne, F., and Piggot, J. (1988). A microeconometric model of the demand for health care and health insurance in Australia. *Review of Economic Studies*, 55:85–106.

Canbolat, A. (1997). *Population Biology of the Loggerhead Sea Turtle Caretta caretta (LINNAEUS. 1785) in Dalyan and Patara*. PhD thesis, University of Hacettepe, Turkey.

Chu, S. (1996). Diamond ring pricing using linear regression. *Journal of Statistics Education*, 4(3).

Consul, P. and Famoye, F. (1992). Generalized Poisson regression model. *Communications in Statistics: Theory & Methods*, 21(1):89–109.

Cox, D. (1972). Regression models and life tables. *Journal of the Royal Statistical Society, Series B*, 34(2):187–220.

Daniel, C. and Wood, F. (1980). *Fitting Equations to Data*. John Wiley & Sons, Inc., New York, NY, revised edition.

Daniel, W. (1999). *Biostatistics: A Foundation for Analysis in the Health Sciences*. John Wiley & Sons, Inc., New York, NY, seventh edition.

Dickey, D. and Arnold, J. (1995). Teaching statistics with data of historic significance: Galileo's gravity and motion experiments. *Journal of Statistics Education*, 3(1).

Dieckmann, A. (1981). Ein einfaches stochastisches modell zur analyse von häufigkeitsverteilungen abweichenden verhaltens. *Zeitschrift für Soziologie*, 10:319–325.

Dobson, A. (1990). *An Introduction to generalized Linear Models*. Chapman and Hall, London.

Dunn, O. and Clark, V. (2009). *Basic Statistics: A Primer for the Biomedical Sciences*. John Wiley & Sons, Inc., New York, NY, fourth edition.

Durbin, J. and Watson, G. (1951). Testing for serial correlation in least squares regression. *Biometrika*, 37:409–428.

Famoye, F. (1993). Restricted generalized Poisson regression model. *Communications in Statistics: Theory & Methods*, 22(5):1335–1354.

Famoye, F. (1995). Generalized binomial regression model. *Biometrical Journal*, 37:581–594.

Famoye, F. and Singh, K. (2006). Zero-inflated generalized Poisson regression model with an application to domestic violence data. *Journal of Data Science*, 4:117–130.

Finney, D. (1947). The estimation from individual records of the relationship between dose and quantal response. *Biometrika*, 34:320–334.

Fox, J. (1991). *Regression Diagnostics: An Introduction*. SAGE Publishers, Thousand Oaks, CA.

Friendly, M. (2000). *Visualizing Categorical Data*. SAS Institute Inc., Cary, NC.

Frome, E. (1983). The analysis of rates using Poisson regression models. *Biometrics*, 39:665–674.

Frome, E., Kutner, M., and Beauchamp, J. (1973). Regression analysis of Poisson-distributed data. *Journal of the American Statistical Association*, 68:935–940.

Goldfeld, S. and Quandt, R. (1965). Some tests for homoskedasticity. *Journal of the American Statistical Association*, 60:539–547.

Graybill, F. and Iyer, H. (1994). *Regression Analysis: Concepts and Applications*. Duxbury Press, Belmont, CA.

Grizzle, J., Starmer, C., and Koch, G. (1969). Analysis of categorical data by linear models. *Biometrics*, 25:489–504.

Guber, D. (1999). Getting what you pay for: The debate over equity in public school expenditures. *Journal of Statistics Education*, 7(2).

Guerrero, V. and Johnson, R. (1982). Use of the box-cox transformation with binary response data. *Biometrika*, 69:309–314.

Harrison, M. and McCabe, B. (1979). A test for heteroscedasticity based on ordinary least squares residuals. *Journal of the American Statistical Association*, 74:494–499.

Harvey, A. and Collier, P. (1977). Testing for functional misspecification in regression analysis. *Journal of Econometrics*, 6:103–119.

Hedlund, R. (1978). Cross-over voting in a 1976 presidential primary. *Public Opinion Quarterly*, 41:498–514.

Heinz, G., Peterson, L., Johnson, R., and Kerk, C. (2003). Exploring relationships in body dimensions. *Journal of Statistics Education*, 11(2).

Hettmansperger, T. (1984). *Statistical Inference Based on Ranks*. John Wiley & Sons, Inc., New York, NY.

Hilbe, J. (2007). *Negative Binomial Regression*. Cambridge University Press, Cambridge, UK.

Hoerl, A. and Kennard, R. (1970). Ridge regression: Biased estimation for nonorthogonal problems. *Technometrics*, 12:55–67.

Hoerl, A., Kennard, R., and Baldwin, K. (1975). Ridge regression: Some simulations. *Communications in Statistics*, 4:105–123.

Holford, T. (1983). The estimation of age, period and cohort effects of vital rates. *Biometrics*, 39:311–324.

Hosmer, D. and Lemeshow, S. (1999). *Applied Survival Analysis: Regression Modeling of Time to Event Data*. John Wiley & Sons, Inc., New York, NY.

Huber, P. (1973). Robust regression: Asymptotics, conjectures and Monte Carlo. *Annals of Statistics*, 1:799–821.

Hubert, J. (1992). *Bioassay*. Kendall/Hunt Publishing Company, Dubuque, Iowa.

Keppel, G. and Wickens, T. (2004). *Design and Analysis: A Researcher's Handbook*. Prentice Hall, Upper Saddle River, NJ, fourth edition.

Kleinbaum, D., Kupper, L., Nizam, A., and Muller, K. (2008). *Applied Regression Analysis and Other Multivariable Methods*. Duxbury Press, Belmont, CA, fourth edition.

Koch, G. and Edwards, S. (1988). Clinical efficiency trials with categorical data. In Peace, K., editor, *Biopharmaceutical Statistics for Drug Development*, pages 403–451. Marcel Dekker, New York, NY.

Koenker, R. and Bassett, G. (1978). Regression quantiles. *Econometrika*, 46:33–50.

Kutner, M., Nachtsheim, C., Neter, J., and Li, W. (2005). *Applied Linear Statistical Models*. McGraw-Hill & Irwin, New York, NY, fifth edition.

Lambert, D. (1992). Zero-inflated Poisson regression, with an application to defects in manufacturing. *Technometrics*, 34:1–14.

Lawal, B. (2003). *Categorical Data Analysis with SAS and SPSS Applications*. Lawrence Erlbaum Associates, Mahwah, NJ.

Lawless, J. (1987). Negative binomial and mixed Poisson regression. *The Canadian Journal of Statistics*, 15(3):209–225.

Lindsey, J. and Merch, G. (1992). Fitting and comparing probability distributions with log linear models. *Computational Statistics & Data Analysis*, 13:373–384.

Long, J. (1997a). *Regression Models for Categorical and Limited Dependent Variables*. SAGE Publishers, Thousand Oaks, CA.

Long, J. (1997b). *Regression Models for Categorical and limited Dependent Variables*. Sage Publications, Thousand Oaks, CA, third edition.

Lunneborg, C. (2000). *Modeling Experimental and Observational Data*. Authors Choice Press, New York, NY.

Meyer, M. (2006). Wider shoes for wider feet? *Journal of Statistics Education*, 14(1).

Michaelis, L. and Menten, M. (1913). Die kinetik der invertinwirkung. *Biochemische Zeitschrift*, 49:333–369.

Montgomery, D. (2009). *Design and Analysis of Experiments*. John Wiley & Sons, Inc., Hoboken, NJ, seventh edition.

Myers, R. (1990). *Classical and Modern Regression with Applications*. PWS-Kent, Boston, MA, second edition.

Myers, R., Montgomery, D., Vining, G., and Robinson, T. (2010). *Generalized Linear Models with Applications in Engineering and the Sciences*. John Wiley & Sons, Inc., New York, NY, second edition.

Özmen, I. and Famoye, F. (2007). Count regression models with an application to zoological data containing structural zeros. *Journal of Data Science*, 5:491–502.

Pregibon, D. (1981). Logistic regression diagnostics. *Annals of Statistics*, 9:705–724.

Ridout, M., Demetrio, C., and Hinde, J. (1998). Models for count data with many zeros. In *Proceedings of the Nineteenth International Biometric Conference*, pages 179–190, Cape Town, South Africa.

Schroeder, R. and Vangilder, L. (1997). Tests of wildlife habitat models to evaluate oak mast production. *Wildlife Society Bulletin*, 25:639–646.

Sedmak, D., Meineke, T., Knechtges, D., and Anderson, J. (1989). Prognostic significance of cytokeratin-positive breast cancer metastases. *Modern Pathology*, 2:516–520.

Simonoff, J. (2003). *Analyzing Categorical Data*. Springer-Verlag, New York, NY.

U.S. Department of Health and Human Services (2003). *Behavior in School-Aged Children, 2001-2002 [United States]*, Ann Arbor, MI. ICPSR 4372-v1, Calverton, MD: ORC Macro (Macro International Inc.) [producer].

Utts, J. (1982). The Rainbow test for lack of fit in regression. *Communications in Statistics - Theory and Methods*, 11:1801–1815.

Wang, W. and Famoye, F. (1997). Modeling household fertility decisions with generalized Poisson regression. *Journal of Population Economics*, 10:273–283.

Watnik, M. (1998). Pay for play: Are baseball salaries based on performance? *Journal of Statistics Education*, 6(2).

White, H. (1980). A heteroskedasticity-consistent covariance matrix estimator and a direct test for heteroskedasticity. *Econometrics*, 48:817–838.

Winer, B. (1971). *Statistical Principles in Experimental Design*. McGraw-Hill Book Company, New York, NY, second edition.

Woodward, G., Lange, S., Nelson, K., and Calvert, H. (1941). The acute and oral toxicity of acetic, chloracetic, dichloracetic and trichloracetic acids. *Journal of Industrial Hygiene and Toxicology*, 23:78–81.

Index

Accelerated failure time, 333
Additive factor effects, 402
Adjusted treatment means, 440, 443
Autocorrelated Error Terms, 113
Autocorrelation coefficient, 114

Binary, 5, 8

Categorical response variable, 229
Cell means model, 376
Censored data, 319, 321
Centering, 124, 220
Centering and Scaling, 127
Coding scheme
    cell reference, 184, 370, 393
    effect, 187, 372, 394
Coefficient
    of correlation, 31, 141
    of determination, 30
    of multiple determination, 41
    of multiple partial correlation, 147
    of partial correlation, 142
    of semi-partial correlation, 148
Completely randomized design, 369
Condition index, 82
Condition number, 83
Conditional failure rate, 332
Confidence interval, 26, 29, 33
Contrast, 383
    for interaction, 420
    orthogonal, 384
Covariance ratios, 73
Cumulative distribution function, 2, 321
Cumulative hazard function, 332

DFBETAS, 72
DFFITS, 73
Dispersion parameter, 287
Distribution
    F, 4
    binomial, 5
    chi-squared, 4
    exponential, 332
    generalized gamma, 333
    log-normal, 334
    negative binomial, 287
    normal, 2
    Poisson, 6, 273
    standard normal, 333
    Student's t, 3
    Weibull, 334
Dummy variables in piece-wise regression, 199

Effect
    interaction, 404, 405
    main, 404
    simple, 404
End time, 319
Error
    Type I, 8
    Type II, 8
Estimable functions, 481
Estimation
    constrained least squares, 52
    Kaplan-Meier method, 322
    life-table method, 325
    M method, 97
    maximum likelihood, 15, 232
    MM method, 99
    ordinary least squares, 14
Experiment-wise error rate, 382
Exponential growth curves, 350

Factor effects model, 376, 438
Fisher's Z transformation, 32
Fixed effects model, 423

Goodness-of-fit, 274
Graphical residual analysis, 63–66

Hazard function, 332
Heteroscedasticity, 63, 95
Homogeneity of variances, 13, 104
Homoscedasticity, 13
Hypothesis
    alternative, 7
    null, 7
Hypothesis testing on residuals, 66–67

Influence function, 95
Influential Observations
    detecting, 70
Interaction effects, 276

Leverage, 73

Matrix
    hat, 52, 62
Minimum variance unbiased estimation, 129
Missing values in RCBD, 473
Mixed effects model, 423
Mixed model, 430
Model
    k-th order polynomial, 209
    ANCOVA, 438
    ANOVA, 375, 438, 480
    ANOVA model for RCBD, 458
    ARIMA, 121
    ARMA, 121
    assumptions, 13–14
    autoregressive error, 116
    baseline category, 229
    binary response, 230
    cumulative logit, 263
    deterministic, 8
    drug responsiveness, 351
    first-order autoregressive, 115, 116
    first-order moving average, 121
    full, 50
    generalized Poisson regression, 291
    linear, 9
    linear probability, 230
    logistic, 277
    Michaelis-Menten, 352
    negative binomial regression, 287
    negative exponential, 350
    nonlinear, 9
    Poisson regression, 273
    proportional hazard, 333, 337
    proportional odds, 264
    reduced, 50
    regression, 9
    standardized regression, 128
    truncated regression, 339
    two-factor ANCOVA, 449
    validation, 168–177
    with two or more categorical independent variables,
       197
    zero-inflated, 292
    zero-truncated, 300
Monotone decreasing function, 333
Multicollinearity
    detecting, 79

in polynomial regression, 210, 220
    methods for solving, 124
    other remedial methods, 134
    selection method for solving, 163
Multiple comparison
    Bonferroni test, 385
    Fisher's LSD, 381
    Scheffé's test, 384
    Tukey test, 382
Multiple comparisons procedures, 381–386

Normal approximation to binomial, 6
Normal probability plot, 68

Offset, 277
Omnibus test, 43
Orthogonal polynomials, 222
Outliers
    Detecting, 69
Over-dispersion, 273
    causes, 284
    effect of, 283
    in Poisson regression, 284

P-value, 8
Partial F test, 47
Prediction interval, 29
Probability density function, 2, 321
Probability mass function, 2, 273

Qualitative factor, 388
Quantile, 2
Quantitative factor, 388, 411

Random effects in RCBD, 469
Random effects model, 423–432
Randomized block design, 416
Randomized complete block design, 458
Regression
    breakpoint, 352
    count data, 273
    generalized Poisson, 291
    linear logistic, 231
    locally weighted (LOESS), 224
    logistic, 229
    matrix form, 52
    multiple logistic, 242
    negative binomial, 287
    non-parametric, 224
    nonlinear, 349
    piece-wise, 199, 352
    piece-wise polynomial, 224
    Poisson, 273
    polynomial, 209
    quantile, 104

ridge, 124, 129–134
robust, 78, 95
Tobit, 343
weighted least squares, 84
Regression equation, 15, 18, 26, 42
Relative efficiency of RCBD, 466
Residual plots, 63
Residuals, 27
    deleted, 69
    diagnostics, 63–67
    PRESS, 69
    raw, 62
    standardized, 62
    Studentized, 62
    Studentized deleted, 71

Sampling distributions, 3
Sampling zeros, 292
Selection strategy
    all possible regression, 156
    backward, 161
    forward, 158
    stepwise, 124, 162
Serially correlated, 113
Shrinkage parameter, 131
Smoothing parameter, 225
Standardized regression coefficients, 128
Start time, 319
Statistic
    Cook's D, 72
    deviance, 233
    Durbin-Watson d, 115
    Pearson chi-square, 233
    PRESS, 74
Structural zeros, 292
Sum of squares
    lack of fit, 213
    pure error, 213
Survival function, 321
Survival probability, 326

Test
    for homogeneity of variance, 378
    for normality, 67, 377
    for over-dispersion, 284
    lack of fit, 212–216
    multiple comparison, 381
    of coincidence, 185, 187, 193, 196
    of common intercept, 185, 188, 193, 196
    of no interaction, 184
    of parallelism, 184, 187, 192, 196
Tolerance, 82
Transformation
    Box-Cox, 88

reciprocal, 89
Tukey's test for additivity, 467
Two-factor ANOVA, 447
Type I sum of squares, 482
Type II sum of squares, 483
Type III sum of squares, 485
Types of sum of squares, 481

Unbalanced two-way ANOVA, 479
Unbiasedness, 19
Under-dispersion, 273
Uses of regression, 9

Variable
    binary, 229
    categorical, 229
    concomitant, 1, 437
    dependent, 1, 8
    explanatory, 1, 8, 229
    independent, 1, 8
    Likert type, 229
    multi category response, 261
    qualitative, 183
    response, 1, 8
Variables-added-in-order, 47–49
Variables-added-last, 49–50
Variance components, 426
Variance inflation factor, 81, 221

Zero-inflated
    generalized Poisson regression, 295
    negative binomial regression, 294
    Poisson regression, 293
Zero-inflated models, 292–299
Zero-order correlation, 141
Zero-truncated
    generalized Poisson, 301
    negative binomial, 301
    Poisson, 300
Zero-truncated models, 300–305

About the Authors

Bayo Lawal is Professor of Statistics at Kwara State University, Malete, Kwara State Nigeria. He received his B.Sc. (Hons) degree in Mathematics from the Ahmadu Bello University, Nigeria and his Master's degree in Biometry from the University of Reading, UK. His Ph.D. in Statistics is from the University of Essex, UK. Professor Lawal has taught for several years at the University of Ilorin, Nigeria, St Cloud State University, St Cloud, MN and Temple University in Philadelphia, USA. He was a visiting professor in the department of Biometry and Epidemiology at the Medical University of South Carolina, Charleston, SC between 2009 and 2000. He also served as chair of the Departments of Statistics at St Cloud State University, MN, USA and at the University of Ilorin in Nigeria. He also served as Dean of the School of Sciences, Auburn University at Montgomery between 2004 and 2008 as well as the dean of the School of Arts and Sciences at the America University of Nigeria, Yola, Nigeria between 2008 and 2011. He is currently serving as head of the Department of Statistics and Mathematical Sciences at Kwara State University, Malete, Nigeria.

Felix Famoye is a Professor and a Consulting Statistician in the Department of Mathematics at Central Michigan University, in Mount Pleasant, Michigan. He received his B.Sc. (Hons) degree in Statistics from the University of Ibadan, Nigeria and his Ph.D. degree in Statistics from the University of Calgary under the Canadian Commonwealth Scholarship. He received the College of Science and Technology Outstanding Teaching awards and the University Excellence in Teaching Award. He received the College of Science and Technology Outstanding Research Award and the University President's Award for Outstanding Research and Creative Activity. Under the United States Fulbright scholarship award, he visited the University of Lagos, Lagos, Nigeria.